Syntropy

In Defense of Quantum Mechanics

The Answer to Life, the Universe, and Everything

Book Three of the Quantum Mechanics Series

By: Mark My Words

Get some Syntropy today. You have nothing better going on in your life.

Syntropy or the Perpetual Motion Cycle is Hard Science that the materialists, naturalists, nihilists, and atheists will never discover, because these people have already rejected it and concluded that it does not exist. That's the power of blind faith and pre-chosen beliefs – they can prevent you from making new and interesting discoveries in science and religion. You have never heard of Syntropy nor the Perpetual Motion Cycle because the materialists and naturalists have already concluded that it doesn't exist.

Remember, you can't make scientific discoveries by denying their existence.

Syntropy is the answer to life, the universe, and everything; and, our scientists don't even know it because they have already rejected it and concluded that it doesn't exist. Nevertheless, it has been observed that psyche, or intelligence, or knowledge can reverse entropy or eliminate entropy at will. I discuss this Science within this book and many other books that I have written.

Many modern-day scientists and philosophers of science are starting to say that it's time for a paradigm shift in Science. I'm here to tell people what that new paradigm should look like and be like. The new paradigm in Science should identify and eliminate the things within "science" that are false, while at the same time pointing us to the things that are obviously true.

The second law of thermodynamics is obviously false. The second law predicts that we shouldn't exist. The fact that you exist and are reading this right now is scientific proof that the second law of thermodynamics is false. It's time for scientists like me to search for and find a better and more truthful replacement for the falsified second law of thermodynamics. I attempt to do so within this book.

In this book, I state and summarize my case. I discuss the changes that Science and Philosophy need to make in order to match with reality and the observational evidence that the human race has experienced as a whole.

We each have a purpose for being on this planet. My mission or calling is to fix Science and to complete Philosophy by bringing Quantum Mechanics, Psyche, and Syntropy into the mix. I have done so within my books. Quantum Mechanics is Syntropy. Psyche is Syntropy. Invisible non-physical forces and fields are Syntropy. The Perpetual Motion Cycle is syntropic and conserved. Syntropy is eternal and everlasting, without a beginning of days or an end of years. Physical matter is based upon mass or entropy and accumulates entropy or heat. Physical matter had a beginning. Everything else is Syntropy. Quantum Mechanics is Supernatural Mechanics. Quantum Mechanics is Non-Local Spiritual Mechanics. Quantum Mechanics is Transdimensional Physics. Quantum Mechanics is Action at a Distance. Quantum Mechanics is Syntropy.

Some type of Syntropy must exist or all of that subsequent entropy wouldn't have been possible. Syntropy is the opposite of entropy; and, a massive initial infusion of Order or Syntropy was necessary to make entropy and physical matter possible. Simple. Logical. Parsimonious. True. There's the truth, and then there's everything else.

I introduce the Ultimate Law of Thermodynamics and the Quantum Law of Thermodynamics within this book. I identify Psyche and introduce the Law of Psyche. The goal here is to produce a Model of Reality that matches with and explains the observations and experiences of the human race as a whole. We are going to have to rethink Science if we are going to make it match with what has been experienced and observed by Out-of-Body Travelers.

My ultimate goal is to expose and eliminate the false and the falsified, thereby setting people free to find and know the truth.

My Amazon Author Page: https://amazon.com/author/science

The Associated Facebook Page: https://www.facebook.com/MarkMyScience/

The Associated Twitter Page: https://twitter.com/Mark_Me_Words

The Associated Website: https://syntropy.website/

The Sister Website: https://syntropy.site/

A Similar Forum: http://ldssoul.com/forum/viewforum.php?f=102

First Edition

"This is the science book that I wish I would have had in high school and college because it explains how everything works." D. M. B.

"The coolest book ever written by a man." H. S.

Table of Contents

Table of Contents

Dedication

Dedicated to you!

May it set you free.

Reason for Being

It is interesting to note that my doctorate or PhD is in Quantum Neuroscience.

(See: https://www.amazon.com/gp/product/B079Z6QQQB/ and https://www.amazon.com/gp/product/B01J023TGU/ and https://quantum-neuroscience.com/).

The problem is that it's a self-awarded degree, because there is no college or university on the planet that would actually grant a doctorate in Quantum Neuroscience, because the materialists, naturalists, and atheists who control and run our public schools have already concluded in advance that the non-physical, or the transdimensional, or the quantum does not exist. They have already concluded that there is no such thing as Quantum Neuroscience, and they come unglued at the very suggestion of the idea.

My other doctorate or PhD in hard science is in Syntropy and the Perpetual Motion Cycle.

(See: https://www.amazon.com/gp/product/B07BPT3W8R/ and https://syntropy.website/ and https://syntropy.site/).

Again, there is no college or university on the planet that would actually grant a doctorate in Syntropy or the Perpetual Motion Cycle because the materialists, naturalists, nihilists, and atheists who control and run and peer review our public schools have already concluded that Syntropy and the Perpetual Motion Cycle do not exist. These people have already concluded that you can't have a PhD in a science that they don't believe exists.

The Materialists, Naturalists, Darwinists, Nihilists, Behaviorists, and Atheists deny the existence of the transdimensional or the non-physical. Consequently, these denialistic philosophies of science have no explanatory power when it comes to the transdimensional or the non-physical. Therefore, these people refuse to study and do science in this topic. These people deliberately delete, censor, ban, and destroy any scientific evidence that falsifies their beliefs. They pretend that it doesn't exist. This is Bad Faith and Bad Science; and, they don't even know it, because they refuse to think about it or study it.

Denying the existence of something you don't like or don't want to believe in is not the proper way to do science; and, it's definitely not the best nor the most efficient way to make new and interesting scientific discoveries. Denying the existence of Quantum Mechanics or Non-Physical Mechanisms won't help you make new and interesting discoveries in Non-Physical Physics, Transdimensional Physics, Quantum Field Theory, or Quantum Mechanics. Denying the existence of God won't help you to find and know God. Every philosophy of science that denies the existence of one thing or another, as one of its primary axioms, is destined to fail in the end. You can't make scientific discoveries by denying their existence. This is logical common sense.

I'm a scientist, so when it comes to psyche or mind, I want to explain it scientifically, not deny its existence. I want to examine what has actually been experienced and observed. Consequently, I also have a self-awarded doctorate in Psyche Ontology that I have called the Ultimate Cause or the Ultimate Model of Reality. I discovered that Psyche or Intelligence is the fundamental unit of reality and existence. That's a Psyche Ontology. As far as I know, I'm the first person on the planet to propose a Psyche Ontology, just as I am the first person on the planet to propose Quantum Neuroscience.

(See: https://ultimate-model-of-reality.com/ and https://mypsyche.us/ and https://psyche-ontology.com/ and https://bio-psycho-social.com/).

I've written four different books on this Ultimate Cause or Psyche Ontology topic. Obviously, the materialists, naturalists, and atheists who control and run our public schools would never grant a PhD in Psyche Ontology because they have already concluded in advance that Psyche, Non-Local Consciousness, Quantum Wave Processors, Quantum Information Processors, and Quantum Information Storage Devices do not exist. They are not going to grant a PhD in something that they have already concluded does not exist.

I originally wanted to work on a doctorate degree in Origin Science. All of my books have been in pursuit of that goal. I have written twenty different tomes that explore Origin Science from many different angles. I have discovered things that they don't want me to discover. I have discovered the Perpetual Motion Cycle, Syntropy, a Psyche Ontology, the Law of Psyche, Ultimate Cause and Ultimate Causality, the Ultimate Model of Reality and Existence, Science 2.0, Transdimensional Physics or Non-Physical Physics, Scientific Proofs of God's Existence, the Ultimate Law of Thermodynamics, and the Quantum Law of Thermodynamics. The materialists and naturalists erroneously teach and believe that entropy and physical matter are conserved. I have used the Ultimate Law of Thermodynamics and the Quantum Law of Thermodynamics to correct that erroneous belief.

(See: https://origin-science.org/ and https://quantum-law-of-thermodynamics.com/ and https://ultimate-law-of-thermodynamics.com/).

I have been busy trying to find all of the Truths in Science that were just sitting there waiting to be found. I wanted to find and fix everything that is wrong with science. I have observed that the false is falsified by the truth, and that the truth is repeatedly and constantly experienced and observed by Someone Psyche or Someone Intelligent somewhere sometime somehow. I'm not afraid of stirring up controversy because nobody is reading what I'm writing anyway. They aren't looking for this stuff and don't want to find it. During the past five years, I have written these books and made these websites for my benefit only. There's no way in the universe that I can remember everything that I have learned, so I have to write it down, or it will be lost forever. I'm not afraid to let the truth or the evidence take me wherever it wants to lead me. I'm an open-minded scientist.

(See: https://amazon.com/author/science and https://markme.us/ and https://markme.website/ and https://tripping.website/ and https://god-is-light.com/ and https://stealth-ascendancy.com/).

I also have a self-awarded PhD in the Philosophy of Science.

(See: https://philosophy-of-science.com/ and https://science-2-0.com/ and https://scientific-proof-of-god.com/ and https://evolution-is-entropy.com/ and).

I have taken that topic further than anyone else on the planet has ever taken it. I have written half a dozen different books on the topic, because it's an important one to understand. I continue to research them and enhance them. Now, I might be able to find a university that would grant a PhD in the Philosophy of Science because these people don't consider philosophy to be a science; however, I have discovered that a true philosophy of science, or a philosophy of science based upon the truth, actually falsifies materialism, naturalism, nihilism, atheism, the theory of evolution, the second law of thermodynamics, creation ex nihilo, and creation by chance. That's not going to fly on this planet.

The scientists on this planet are not ready for what I'm trying to present them. They don't get it. They don't want it. They aren't looking for it yet. Therefore, they haven't found it, and they probably never will, because they have already rejected it.

Mark My Words

The Key to Entropy

Our scientists have the wrong definition for entropy. They erroneously define entropy as disorder. Heat or thermodynamics has zero correlation with disorder. Disorder doesn't cause heat, and disorder doesn't eliminate heat. There's no observed correlation between the two. According to the equation for heat, heat actually has a great deal of order and organization within it, and our scientists don't even know it. Likewise, the absolute zero temperature space all around us and all through us also has a great deal of order and organization within it, and our scientists don't even know it.

Thanks to the quantum fields that were designed by the Gods and made by the Controlling Psyches within Nature, there is perfect order and organization at the quantum level in the transdimensional realm or the conserved realm at absolute zero heat or temperature. Disorder doesn't cause thermodynamics, and it doesn't eliminate thermodynamics either. There's no correlation between the two.

Conserved means entropyless or no-entropy. Conserved means syntropic or exergic. Conserved means perpetual motion. Conserved means eternal and everlasting. Therefore, entropy means "not conserved". In other words, entropy is not conserved. Entropy is an aging process that is built exclusively into physical matter and nothing else. This is the Ultimate Law of Thermodynamics. It defines entropy in terms of what is and is not conserved.

https://ultimate-law-of-thermodynamics.com/

This is science that our scientists haven't discovered yet, because they have already rejected it without even giving it a hearing or knowing that it exists.

This is the key to entropy:

Study the following equation for heat (Q). There is something important here that they completely overlooked, because they aren't searching for it and don't want it to be true because it falsifies their second law of thermodynamics.

$Q = mc\Delta T$

Thermodynamics is the transfer of heat from a hot system to a cold system. According to thermodynamics and the equation for heat (Q), anything that has NO mass ($m = 0$) has NO heat storage capacity ($mc = 0$), which means that the massless has NO heat ($Q = 0$) and is natively at absolute zero instead ($T = 0$). Everything goes to zero when mass goes to zero!

Remember that, it's important!

Now, let's look at the official heat-based equation for entropy, and see what we can learn.

$S = Q/\Delta T$

The heat-based equation for entropy (S) is based upon heat (Q). When heat goes to zero ($Q \rightarrow 0$), entropy goes to zero ($S \rightarrow 0$)! It's that simple! This is the KEY to entropy and thermodynamics, and it has been completely overlooked by the scientists on our earth. They have never looked at this nor thought about it even once during their entire lives. It's there in plain sight and obvious, but they have already rejected it, choosing to go with materialism and the second law of thermodynamics instead. Their materialism and naturalism prevent them from discovering this simple truth.

Now, let's combine the two equations and solve the equation for entropy (**S**).

S = mcΔT/ΔT

So, what do we get?

We get:

S = mc

This is science that our scientists haven't discovered yet, and will never discover, because they have already rejected it and chosen to go with materialism and their erroneous second law of thermodynamics instead.

Entropy is mass's heat storage capacity (**S = mc**). Entropy (**S**) is mass (**m**) times a mass's specific heat capacity (**c**). No mass (**m = 0**), then no entropy (**S = 0**). It really is that simple. No mass (**m = 0**), then no heat (**Q = 0**), according to the equation for heat. No heat storage capacity (**mc = 0**), then no heat (**Q = 0**) and no entropy (**S = 0**), according to the equations for heat and entropy.

The equations for heat and entropy falsify the second law of thermodynamics which states that entropy can never go to zero and cease to exist. The second law erroneously states that the total amount of entropy or disorder in the universe is constantly increasing and can never decrease and go to zero. This is false. The second law of thermodynamics erroneously states that entropy, disorder, and physical matter are conserved. It's wrong. Fntropy is not conserved. Disorder is not conserved. Mass is not conserved! Entropy and mass come and go in a Perpetual Motion Cycle (**E = mc²**) as Nature's Psyche sees fit. Entropy means "not conserved". The truth is internally consistent and self-supportive, while at the same time completely falsifying the second law of thermodynamics as well as materialism and naturalism.

The Ultimate Law of Thermodynamics states that entropy is not conserved, mass is not conserved, disorder is not conserved, heat is not conserved, resistance to acceleration is not conserved, and heat storage capacity is not conserved. It also states that energy's form is not conserved. ONLY psyche and energy are truly conserved. Psyche is a quantum computer that can form and transform information and energy making them into anything that this psyche or intelligence wants the energy and information under its control to become.

https://ultimate-law-of-thermodynamics.com/

Mass and entropy are purely physical phenomena and only exist at the physical level. They completely cease to exist at the quantum level. This is what we actually experience and observe. It's there to be found within the equations for heat and entropy. This is the Quantum Law of Thermodynamics, which states that there is no heat or thermodynamics at the massless level, entropyless level, heatless level, quantum level, or conserved level. Heat is a function of mass. Entropy is also a function of mass. No mass or resistance to acceleration, then no heat and no entropy. It really is that simple.

According to the combined equations for heat and entropy (**S = mc**), entropy, mass, heat storage capacity, and resistance to acceleration do not exist in the quantum realm, transdimensional realm, syntropic realm, absolute zero temperature realm, or conserved realm. In other words, the perpetual motion machines that we call the "quantum fields" are completely massless, heatless, entropyless, and conserved. This is the Quantum Law of Thermodynamics, which uses the equations for heat and entropy to tell us that heat and entropy don't exist at the quantum level in the massless realm, or the quantum realm, or the heatless realm, or the syntropic realm, or the exergic realm, or the conserved realm where there is no resistance to acceleration.

https://quantum-law-of-thermodynamics.com/

Our scientists have already rejected all of this and have chosen to go with their second law of thermodynamics instead, which erroneously states that the total amount of entropy or disorder in our universe is always increasing and can never go to zero. This is not what we experience and observe, therefore, it is obviously false. WE KNOW it is false because their own equations for heat and entropy falsify their second law of thermodynamics, replacing it with the truth instead.

Quantum Field Theory and the Perpetual Motion Cycle also falsify the second law of thermodynamics, and our scientists don't even know it.

Nature's Psyche is the thing that made and is now running the quantum fields. The quantum fields are the medium through which quantum waves travel and psyches interact.

Nature's Psyche makes quantum waves or photons at will from mass, heat, resistance to acceleration, or entropy; and then later, when that same psyche chooses to stop the quantum wave or photon it has made, that psyche transforms the infinite acceleration omnipresent quantum wave it has made INTO some type of localized mass, heat, resistance to acceleration, or entropy (mass's heat storage capacity) instead. This is the Perpetual Motion Cycle ($E = mc^2$). It has always existed, and it will always exist, because the underlying energy and psyche are always conserved. The quantum fields are syntropic and conserved by Nature's Psyche; and, the Perpetual Motion Cycle is Syntropy, or the conservation of energy and psyche. Our scientists will never discover Syntropy or the Perpetual Motion Cycle because they have already rejected these things.

A physical atom is one specific form of energy, and it is made from a choreographed combination of different energies or information, by a bunch of different psyches or intelligences. There's a lot of information or energy within a physical atom, and it's controlled by a lot of different psyches or intelligences.

Nature's Psyche makes the quantum waves or photons from mass, heat, entropy, or resistance to acceleration; and then later, the same psyche or intelligence within the quantum wave chooses to stop that wave or chooses to collapse the wave, thereby transforming that quantum wave or photon INTO some type of mass, heat, entropy, or resistance to acceleration – whatever it chooses for that energy or information to become. Then later, Nature's Psyche transforms some of that mass, heat, or entropy back into massless, heatless, chargeless, and entropyless infinite acceleration quantum waves or photons instead within our stars. This is the Perpetual Motion Cycle and Syntropy; and, our scientists deny its existence.

According to the Law of Psyche, each psyche or intelligence has a certain amount of energy that's under its control, and that controlling psyche can form or transform the energy under its control into anything that it wants that energy to be, anytime and anywhere that it chooses to do so.

Quantum information can be formed and transformed, while the underlying energy and psyche are always conserved. This is what we have actually experienced and observed. Energy's form, or quantum information, or memory is never conserved unless some psyche or intelligence chooses to conserve it, chooses to remember it, or chooses to store it somewhere within its memory banks at the quantum level. Quantum information is processed and stored within Psyche or Intelligence. Energy and information are synonymous with each other. Information is transmitted within energy. Energy or information can be transformed anytime that its controlling psyche chooses to transform it. All the while, the psyche, quantum information processor, or quantum information storage device is conserved and preserved. This is the way things really work at the quantum level.

Nature's Psyche makes the quantum waves or quantum information in the first place from mass or energy, and then Nature's Psyche later collapses or processes that quantum wave in some fashion transforming that energy into something else instead, including information, resistance to acceleration, mass, mass's heat storage capacity (entropy), or possibly entropic physical matter or a physical atom. Nature's Psyche can then take any atom of physical matter and transform it instantly into massless and entropyless and heatless quantum waves or photons as is currently happening within our suns. Then when that photon reaches your skin, the psyche within it can choose to stop the photon and transform that infinite acceleration quantum wave INTO heat, mass, or resistance to acceleration instead. Nature's Psyche could even choose to transform quantum waves or photons back into physical atoms if it wants to do so. This is the Perpetual Motion Cycle.

Our scientists haven't discovered any of this, and they never will, because they have already rejected it and decided that it does not exist. Denying the existence of something will never help you to discover it. You can't make scientific discoveries by denying their existence. You can't find and know the Truths in Science by denying their existence. This is logical common sense.

Our scientists deny the existence of the Perpetual Motion Cycle. Our scientists deny the existence of Syntropy. Our scientists deny the existence of consciousness, intelligence, life force, or psyche. Most of our scientists deny the existence of the quantum fields, because they are invisible, massless, heatless, intangible, entropyless, and non-physical. Our scientists deny the existence of Quantum Waves, Quantum Wave Processors, Quantum Information Processors, and Quantum Information Storage Devices. Yet something has to exist at the quantum level that is capable of making, transmitting, receiving, processing, analyzing, choosing among, and storing quantum waves and quantum information. Psyche is that Quantum Computer, Quantum Wave Processor, Quantum Information Storage Device, Infinite Singularity, Ultimate Point Particle, and Quantum Machine; and, our scientists don't even know it because they have already rejected it and denied its existence.

Thanks to their erroneous and falsified second law of thermodynamics, our scientists haven't yet discovered Syntropy, Psyche, or the Perpetual Motion Cycle. They don't know what any of this is, how it works, why it is important, or that it even exists. Instead, our scientists have erroneously concluded in advance that these things do not exist. Therefore, our scientists aren't looking for these things and will never find them. They don't want to find them. They don't want these things to be true.

Our scientists can't have the Truths in Science by constantly denying their existence or by constantly rejecting them. Scientific discoveries cannot be made by denying their existence. Our scientists will never discover these things because they don't want to.

This is the greatest science ever discovered, but our scientists have already rejected it, choosing to go with materialism, naturalism, creation by chance, and their second law of thermodynamics instead.

We each receive precisely what we are willing to accept. Our scientists don't want the truth. They prefer their deceptions and lies instead.

Mark My Words

My Books

I have always loved science; and, ever since I got rid of My Materialism, My Naturalism, My Nihilism, My Scientism, and My Atheism, I have been a very open-minded scientist more than willing to think outside the box. I have developed a great love for Syntropy or Quantum Mechanics.

The best way to connect with these concepts related to Syntropy is through my many different books.

In all of my books, I make detailed comparisons between Syntropy and Entropy.

How does this work out in practice?

Well, it essentially means that I end up making detailed comparisons between Quantum Mechanics and Naturalism. Quantum Mechanics, Spirit Matter, and Psyche are Syntropy. In contrast, Materialism, Naturalism, Darwinism, Nihilism, Atheism, and the Theory of Evolution are based upon Entropy, Chaos, Random Diffusion, Random Mutations, Physical Matter, and Classical Physics. There's a huge difference between the two. Syntropy is order and organization and life. Entropy is chaos, death, disease, cancer, and extinction. Evolution is entropy. Random mutations are entropy. Natural selection results in death, extinction, and entropy. Entropy cannot do order, organization, and life. It's physically impossible.

See my Author Page on Amazon for books wherein I discuss these concepts and ideas in much greater detail.

https://amazon.com/author/science

As an open-minded scientist, I hope that you will welcome any comparison between Syntropy and Entropy.

Mark My Words

Sandy Says "Go Buy Some Books and Keep Me Fed"

Go Buy Some Books

And

Keep Me Fed!

I'll Thank You if

You Do.

This is Sandy. She wanted me to tell you to go buy some books in order to keep the food and doggy treats a coming.

Her Amazon Page:

https://amazon.com/author/science

Sandy kindly requests that you click the link and make a donation to her cause.

Conserved Means Entropyless

You have got to figure out how everything works in the Transdimensional Realm or the Non-Physical Realm, if you truly want to find and know the Truths in Science. You have to be willing to accept the fact that the Non-Physical is entropyless or conserved – to one extent or another. You have to realize and accept the fact that everything in the Transdimensional Realm or Quantum Realm is entropyless – that it is conserved or preserved or ageless, which means that it cannot get old, it cannot wear out, and it cannot die.

Once you understand and accept this Truth in Science, then instantly you know why our Quantum Information Processor or Psyche has always existed and will always exist, because it is quantum or transdimensional and therefore always conserved. This Truth in Science explains the LAW of Quantum Information Conservation. There has to be some type of Quantum Information Processor, Quantum Wave Processor, and Quantum Memory Storage Device at the quantum level capable of making, transmitting, receiving, analyzing, processing, and storing quantum waves or quantum information; otherwise, the LAW of Quantum Information Conservation could never possibly be true. This is logical common sense, which is why everyone seems to overlook it and reject it. Most human beings aren't in the habit of thinking rationally and logically.

Have you ever wondered what "conserved" truly means when it comes to physics?

Conserved means entropyless or syntropic. Conserved means that all of the entropy in the system has gone to zero and ceased to exist.

So, what is entropy?

Entropy is a catch-all term that is used to describe a variety of different physical phenomena.

Entropy is the aging process, which means that entropy consists of all the different clocks or timing mechanisms that are built into physical matter or a physical atom. Entropy is exclusively a physical phenomenon. Entropy doesn't exist at the quantum level in the Ageless Realm, or the Conserved Realm, or the Entropyless Realm.

Entropy is often defined as a transfer of heat from a hot system to a cold system. NO heat or NO thermodynamics, then NO entropy. There is NO entropy at absolute zero, both at the physical level and at the quantum level. Entropy goes to zero and ceases to exist, when the temperature or heat goes to absolute zero and ceases to exist.

Entropy is also mass or resistance to acceleration. A transdimensional or non-physical system, that has NO mass or NO resistance to acceleration, has NO entropy. That means that a transdimensional system, or a quantum system, or a massless system, or a non-physical system is conserved because it is entropyless. When mass goes to zero, entropy goes to zero.

Entropy is purely a physical phenomenon. Entropy doesn't exist at the quantum level in the Quantum Realm or the Massless Realm. Whenever mass or resistance to acceleration goes to zero, entropy goes to zero and ceases to exist. This truth is called the Quantum Law of Thermodynamics, which states that there is no mass, no resistance to acceleration, no thermodynamics, and therefore NO entropy at the quantum level in the Transdimensional Non-Physical Realm. Entropy doesn't exist at the quantum level, which means that everything is conserved at the quantum level.

Powerful. Logical. Rational. Obvious. Simple. Real. True.

These ideas are demonstrably true. By studying the equations for heat and entropy, one can see with his own eyes that this is true.

The equation for heat (**Q**):

Q = mcΔT

The typical equation for entropy:

S = Q/ΔT

Put the equation for heat into the equation for entropy, and then solve for entropy, in order to find another powerful and true definition for entropy:

S = mcΔT/ΔT

Solving this equation, we get the following: **S = mc**.

Entropy (**S**) is the heat storage capacity of mass (**mc**). Simple. Parsimonious. Logical. Real. True.

Entropy is mass's heat storage capacity. No mass, no heat, or no heat storage capacity, then NO entropy. Entropy (**S**) goes to zero whenever mass (**m**), heat (**Q**), thermodynamics (**Q** or **T**), temperature (**T**), or mass's heat storage capacity (**mc**) goes to zero.

Scientists lie, but the equations don't lie.

S = mc.

Entropy is mass's heat storage capacity. Entropy goes to zero when mass goes to zero. This is one of the greatest scientific discoveries in the whole universe, due to its explanatory power.

Conserved means entropyless.

—

True Definitions for Entropy

Here we have presented five true and demonstrable definitions for entropy.

1. Conserved or Syntropic or Exergic is the antonym for Entropy. Conserved means entropyless. Entropy means non-conserved. Entropy, mass, thermodynamics, and heat are not conserved. This is the Ultimate Law of Thermodynamics, which tells us what is not being conserved. Conserved means entropyless or syntropic. Conserved means that all of the entropy in the system has gone to zero and ceased to exist. This scientific truth explains why we exist. Energy is conserved. The Quantum Realm or the Transdimensional Non-Physical Realm is conserved.

2. Entropy is the aging process or the different clocks that are built into a physical atom. Entropy doesn't exist at the quantum level in the Transdimensional Realm, or the Quantum Realm, or the Ageless Realm, or the Conserved Realm. Conserved or Syntropic means eternal and everlasting. Entropy is not conserved. Entropy is temporary.

3. Entropy is the transfer of heat from a hot system to a cold system. NO heat, then NO entropy. When heat or temperature or thermodynamics goes to zero, then entropy goes to zero. When the temperature is absolute zero, then the amount of entropy in the system is absolute zero, whether we are talking about a transdimensional system or a physical system.

4. Entropy is mass or resistance to acceleration. NO mass, then NO entropy. Whenever mass goes to zero and ceases to exist, entropy goes to zero and ceases to exist. This is obviously true because it has been experienced and observed. Photons or quantum waves are obviously massless, heatless, chargeless, entropyless, conserved, and have NO resistance to acceleration. Photons, quantum waves, virtual particles, thoughts, and superconductors function best and most efficiently at absolute zero, where entropy has gone to zero and ceased to exist. This is what has been experienced and observed.

5. Entropy is mass's heat storage capacity. **S = mc**. No mass, no heat, or no heat storage capacity, then no entropy. This is obviously true. The equations don't lie. Scientists lie, but the equations don't lie. This is one of the greatest scientific discoveries in the universe due to its explanatory power. It explains everything that has ever been experienced and observed. It explains why we are here and why we exist. Everything at the quantum level or the massless level or the entropyless level is being conserved, in one form or another. Entropy is purely a physical phenomenon. Entropy ceases to exist at the quantum level, or the massless level, or the conserved level. Entropy is not conserved.

—

The Mass-Based Definition for Entropy

Here are the details for the mass-based definition for entropy:

Q = mcΔT

Heat (**Q**) is equal to mass (**m**) times that mass's heat storage capacity (**c**) times the change in temperature (**ΔT**).

S = Q/ΔT

Entropy (**S**) is equal to heat (**Q**) divided by the change in temperature (**ΔT**). Combining these two equations into one, we get the following equation. Notice that the temperature (**T**) is irrelevant, because the temperature cancels out.

S = mcΔT/ΔT

Solving this equation, we get the following: **S = mc**.

Entropy (**S**) is the heat storage capacity of mass (**mc**).

Simple. Parsimonious. Demonstrable. Replicable. Verifiable. Logical. Conservative. True.

This is the mass-based definition for entropy. It's there in plain sight just waiting for someone to find it and present it to the universe.

Notice that the temperature (**T**) is irrelevant, because the temperature cancels out. This process works and this equation holds true no matter what the temperature might happen to be. It even works at absolute zero Kelvin (**T or K = 0**). Everything continues to function perfectly fine at the quantum level, even at absolute zero. In fact, by studying photons and superconductors, we observe that absolute zero maximizes the efficiency and functionality of quantum objects such as photons, quantum waves, superconductors, virtual particles (thoughts), and quantum fields. Absolute zero is maximum efficiency and maximum order at the quantum level thanks to the Quantum Fields and the Conservation of Energy.

S = mc

This simple equation for entropy is possibly the greatest scientific discovery of all time. (It falsifies your earth's second law of thermodynamics.) It's demonstrably true.

c = the specific heat capacity of the mass under consideration

m = mass

Entropy (**S**) is mass's heat storage capacity (**mc**). No heat storage capacity, then no heat. No mass and no heat, then NO entropy. NO thermodynamics, then NO entropy. When temperature goes to absolute zero, entropy goes to absolute zero. Simple. Logical. Parsimonious. True. Science doesn't get any better than this.

Entropy (**S**) equals mass (**m**), particularly the heat storage capacity of mass (**mc**). Entropy is mass's heat storage capacity (**S = mc**). Entropy is also resistance to acceleration or mass. This is the mass-based definition for entropy. No mass, then no entropy. It's that simple.

Conserved means entropyless.

According to the Ultimate Law of Thermodynamics, mass, heat, entropy, and thermodynamics are NOT conserved, whereas, everything in the Ageless Realm, the Massless Realm, the Transdimensional Realm, or the Quantum Realm is conserved in one form or another. According to the Quantum Law of Thermodynamics, there is no entropy and no thermodynamics at the quantum level in the Transdimensional Realm or the Massless Realm. This is obviously true, or we wouldn't exist in the first place.

Mark My Words

Obviously Organized

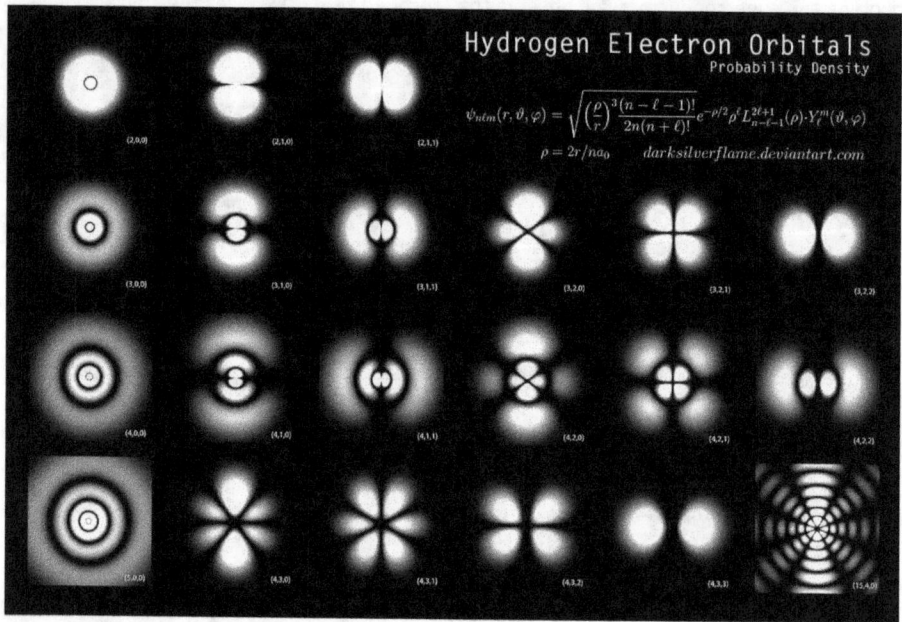

Atomic Orbitals or Electron Orbital Shells

https://en.wikipedia.org/wiki/Atomic_orbital

Do you see any order, organization, structure, planning, intention, or purpose in any of this?

Anything that was obviously organized obviously has an Organizer who organized it.
Anything that was obviously made obviously has a Maker who designed it and made it.

Who is making the electrons act this way? Who is communicating this information to the electrons? How are the electrons able to understand and obey these Laws or Commands?

Order and organization such as this do not just spontaneously generate out of thin air from nothing as the Materialists and Naturalists claim.

Electrons are sentient, perceptive, conscious, alive, and aware; and, they act accordingly. They KNOW what they are, where they are, what they are supposed to be doing, and how they are supposed to act. There's intelligence and consciousness within them.

Psyche is the innate intelligence within all the different forms of energy which gives that energy the inherent ability to understand, follow, and obey God's Laws and God's Commands. Each massless elementary particle is some type of Psyche or Intelligence.

The Second Law of Thermodynamics Is Obviously False

The second law of thermodynamics states that the total amount of entropy or disorder in our universe is constantly increasing and that it can never decrease and go to zero. This claim is obviously false.

The very fact that we exist is scientific proof that this claim is false. We haven't observed anything that the second law of thermodynamics predicts. We don't observe proton decay. We don't observe an ever-encroaching gray goo coming in at us from all sides. We don't observe ever-increasing disorder or chaos. We don't even observe heat death. Heat death doesn't exist. Heat death or absolute zero represents Maximum Efficiency when it comes to energy, photons, quantum waves, psyches, quantum consciousness, quantum information, quantum fields, and superconductors. Does it not? These quantum objects function perfectly and function best at absolute zero. We don't observe anything that the second law of thermodynamics predicts, which is why it is false.

Instead, we observe constantly conserved order and organization thanks to the massless and entropyless Quantum Fields, the Conservation of Energy and Psyche, the Conservation of Quantum Information within Psyche, and the Perpetual Motion Cycle $E = mc^2$. The observation of these things falsifies the second law of thermodynamics as well as creation by chance, or creation by entropy, or creation by random disorder.

The ONLY part of the second law of thermodynamics that is demonstrably true is the obvious fact that physical matter cannot be used to design and run perpetual motion machines. Physical matter wasn't designed to work that way. Physical matter has physical limitations which prevent it from functioning as a usable and dependable perpetual motion machine in the real world. In other words, the ONLY part of the second law that is demonstrably true is the "thermodynamics"; and, thermodynamics or heat has absolutely nothing to do with "disorder" or "chaos".

Isn't that ironic? Disorder and the "disorder definition for entropy" have absolutely nothing to do with thermodynamics or heat. Disorder doesn't produce heat, and disorder doesn't eliminate heat. There's NO correlation between the two, which means that "disorder" is the wrong definition for thermodynamics or entropy.

The sun is often portrayed as maximum chaos or maximum disorder, and there's tons of mass and heat and therefore "entropy" that's associated with our sun. Heat death doesn't maximize disorder as the second law of thermodynamics claims. There's NO correlation between heat and disorder!

For contrast, notice that there is NO mass, heat, or entropy associated with photons, quantum waves, and the quantum fields, which is why these massless and entropyless quantum objects represent perfectly conserved syntropic Order and Organization at the quantum level in the Spirit World or the Quantum Realm. The Quantum Fields are massless, entropyless, syntropic, exergic, and conserved perpetual motion machines. The Perpetual Motion Cycle $E = mc^2$ runs on energy, which is why it is constantly and forever conserved. The Conservation of Energy and Psyche, as well as the Conservation of Quantum Information within Psyche or Intelligence, is what makes these quantum perpetual motion machines possible at the quantum level in the Transdimensional Non-Physical Realm. Quantum objects don't need heat as a lubricant to make them run – in fact, these perpetual motion machines run best and most efficiently at "heat death" or absolute zero. This is what we have experienced and observed, is it not?

The second law of thermodynamics will never be true as long as the Quantum Fields exist. You would have to destroy the Quantum Fields in order to return our universe to the Chaos Realm from whence it was originally organized. You would have to destroy the Quantum Fields in order to make the second law of thermodynamics even remotely true. Even then, the second law of thermodynamics would still be false thanks to the Conservation of Energy and Psyche. Photons, Quantum Waves, Quantum Fields, Conservation of Energy and Psyche, and Syntropy ARE scientific proof that the second law of thermodynamics is false.

When our scientists finally figure out how things really work in the real world, they will vote to replace the falsified second law of thermodynamics with Quantum Field Theory, Conservation of Energy and Psyche, and the Conservation of Quantum Information within Psyche. They will replace the second law of thermodynamics with the things that have actually been experienced and observed.

According to the Philosophy of Science, there are many different Scientific Methods or scientific methodologies that we can use in our pursuit of the truth. I used a few of them in this essay. I use many more of them in my books.

ALL of the different Scientific Methods can be used to FALSIFY the Theory of Evolution, the Second Law of Thermodynamics, Materialism, Naturalism, Darwinism, Nihilism, Atheism, Creation by Chance, Creation Ex Nihilo, or anything else that is by definition produced by chance alone because chance by definition cannot do causation. When we scientists FALSIFY a theory or an idea, we do indeed PROVE that it is false. Negating the Consequent or Falsifying a Theory is a philosophically valid and logically sound Scientific Method than can be used to PROVE that our ideas and theories are false.

Therefore, it is NO exaggeration and it is NO joke whenever I state the fact that the Scientific Methods PROVE that the Theory of Evolution and the Second Law of Thermodynamics are false, because that's precisely what they were designed to do.

OBSERVATION is one of the Scientific Methods; and, what do we observe?

We observe that we exist.

The very fact that we exist is scientific proof that the second law of thermodynamics is false, because the second law of thermodynamics predicts or says that we shouldn't exist. It's obvious that the second law is false.

If disorder or entropy were truly a conserved substance as the second law of thermodynamics claims, then we wouldn't exist. There has to be some type of Syntropy within the system in order for us to exist in the first place. That massless entropyless Syntropy or Conservation exists and is stored within the Energy, the Quantum Waves, the Quantum Information, the Psyches or Intelligences, the Photons, and the Quantum Fields. These things are different types of non-physical immaterial perpetual motion machines. The Quantum Realm or Psyche Realm is the Perpetual Motion Realm or the Energy Realm. It will never get old, never wear out, and never die because it is constantly conserved. It is eternal and everlasting. It has always existed, and it will always exist. Energy cannot be made, and it cannot be destroyed. It's that simple.

You would have to destroy the Energy, the Psyches or Quantum Consciousness, the Quantum Information within Psyches or Intelligences, the Quantum Realm, the Perpetual Motion Cycle $E = mc^2$, and the Quantum Fields in order to make the second law of thermodynamics true.

Mark My Words

What Do Genes Code For?

What is science trying to tell us? Science is observation and experience. What is observation and experience trying to tell us?

Let's start this adventure by asking, "What do the genes code for?" A lot can be learned by asking and answering this simple question, especially when we apply knowledge from computer science to gene expression. Let's look at the science.

Genes code for proteins. Genes contain the coded formula needed by the cell to produce proteins. Proteins are the most common of the complex molecules. Proteins and genes are molecules. Gene expression is protein synthesis – the production of proteins from genes.

Research in the 1970s showed that numerous **non-coding** sequences — **introns** — are also found within genes, interrupting the protein-coding regions, or exons. It is estimated that only about five percent of human **DNA** encodes protein.

The genome of an organism is inscribed in DNA, or in some viruses within RNA. The portion of the genome that **codes for a protein** or an RNA is referred to as a **gene**. Those **genes** that **code for proteins** are composed of tri-nucleotide units called codons, each coding for a single amino acid. The **genetic code** allows DNA and RNA sequences to be "decoded" into the amino acids of a protein.

There are an estimated **19,000-20,000** human protein-coding genes. The estimate of the number of human genes has been repeatedly revised down from initial predictions of **100,000** or more as genome sequence quality and gene finding methods have improved and could continue to drop further.

The **2.9 billion** base pairs of the haploid human genome correspond to a maximum of about **725 megabytes** of data since every base pair can be coded by 2 bits.

The haploid human genome (**23 chromosomes**) is estimated to be about 3.2 billion bases long and to contain 20,000–25,000 distinct protein-coding genes. That means there is roughly **800 megabytes** of data storage capacity within the human genome.

Transcription of messenger RNA and translation of messenger RNA into proteins both occur on the time scale of **1 minute** for a protein of typical length. However, longer transcripts and bigger proteins take proportionally longer to make. The largest protein in the human body is titin. It would take approximately an hour to translate its ~30,000 amino acids.

Transfer RNA **activation** refers to the attachment of an amino acid to its Transfer RNA codon and Transfer RNA molecule. In the genetic code, codons made of three bases specify an amino acid. With three bases, there are **64** possible permutations. With three codons corresponding to STOP codons, this leaves **61** combinations that can code for an amino acid.

In a mammalian cell, there can be as many as **10 million** ribosomes. Ribosomes are a cell structures that make proteins. There are **37.2 trillion** cells in your body.

In summary, for a typical human of **70** kg, there are almost $7*10^{27}$ atoms (that's a **7** followed by **27** zeros!) Another way of saying this is "**seven billion billion billion**." Of this, almost **2/3** is hydrogen, **1/4** is oxygen, and about **1/10** is carbon. These **three** atoms add up to **99**% of the total!

1 Petabyte = 1000000000000000B = 10^{15} bytes = 1000 terabytes.

1 Yottabyte = 1000^8 bytes = 10^{24} bytes = 1000000000000000000000000 bytes = 1000 zettabytes = 1 trillion terabytes.

A yottabyte is one septillion bytes. To save all those bytes you need a data center as big as the states of Delaware and Rhode Island combined. It doesn't seem like much, until they tell you the price tag: $100 trillion.

It would take thousands of yottabytes of usable information storage to handle the Command and Control for all the different atoms in your physical body, telling them where to go and what to do when they get there. Who is storing that information? How is that information transmitted to the atoms and the molecules? How are the atoms and molecules able to receive and understand that information?

Finally, code of any kind requires an intelligent Programmer to design and produce that code. Code doesn't just spring into existence from thin air. Code also requires an intelligent Engineer to design and make the hardware that the code runs on.

— Adapted from Uncle Google

For command and control and the transfer of usable information, we require Bytes at a minimum because a Bit doesn't mean anything except ON or OFF. That's why information storage estimates are done in bytes and not bits. A byte is where physical information storage and program codes begin to be useful.

I have been a computer scientist all of my adult life for decades of my life; and, memory storage capacity has been of supreme concern for me in my work throughout the years. I understand how physical information storage works. Data storage at the physical level is highly inefficient because it is based upon atoms; and, atoms take up a lot of physical space. There's an infinitely better and more efficient way to store data and information – at the quantum level or the psyche level.

Scientific Observations: Any type of programming code requires a Programmer to code it or make it. Any type of hardware requires an Engineer to design it and make it.

Scientific Observations: Genes and a genome are obviously programming code. Genes code for proteins; and, a genome is both software and hardware. A genome is hardware made from DNA molecules.

Scientific Conclusion: Our genes and genomes obviously have a Programmer who programmed them; and, our genes and genomes obviously have an Engineer who designed them and made them.

Look at the numbers above from Uncle Google. What are they trying to tell us?

In order to control all of the atoms within a physical body individually at the physical level would require at least thousands of yottabytes of physical data storage and some way to transmit that information between the atoms and molecules. Such a thing is physically impossible. It can't be done, which means that it isn't being done – not at the physical level. However, the psyche or intelligence who is controlling the atoms within your physical body is indeed controlling, and transmitting, and storing yottabytes of information. How's that being done? Where is that information being stored? How is this information being organized, stored, accessed, and transmitted? Who is doing all of this information storage and organization? Who knows what it all means, or who understands it all? Who knows how to access it all and use it all? It can't be our genes, because there are NO genes within our atoms! It can't be evolution because natural selection doesn't touch our genes nor our proteins.

The Materialists, Naturalists, Darwinists, Nihilists, Behaviorists, Determinists, Physical Reductionists, Evolutionists, Classical Physicists, and Atheists assure us that it's all being handled by our genes at the physical level. Are they right, or is there something important that they are missing?

Scientific Observation: Anything that was obviously made obviously had a Maker who made it.

Scientific Observation: Proteins and their matching genes were obviously made. A genome was obviously made.

Scientific Conclusion: Therefore, it is logical and rational to conclude that proteins, their matching genes, and the genomes had a Maker who made them.

Who coded the genomes? I'm a computer programmer and computer scientist; and, I know for a fact that elegant programming code doesn't write itself. This reality and truth has been experienced and observed. Code is designed and made by Someone Intelligent.

Scientific Observation: Programming code obviously requires an Intelligent Programmer to design it, write it, program it, and make it.

Scientific Observation: A genome is the ultimate software or programming code that has ever been made or written. In fact, a genome is both hardware and software simultaneously.

Scientific Conclusion: A genome obviously has an Intelligent Programmer who designed it, wrote it, programmed it, and made it.

Now, with this information securely in place, let's study protein synthesis or gene expression. We are particularly interested in trying to figure out how all of that information and data being used in protein synthesis is being transmitted and received by the individual atoms, molecules, enzymes, and proteins within your physical cells. The atoms and molecules are communicating with each other wirelessly at a distance. How do they do that? Well, they aren't doing it at the physical level because action at a distance is physically impossible.

Protein Synthesis

These online articles provide some of the details for protein synthesis or gene expression:

https://gizmodo.com/5557676/how-much-money-would-a-yottabyte-hard-drive-cost

https://en.wikipedia.org/wiki/Protein_biosynthesis

https://en.wikipedia.org/wiki/Gene_expression

https://en.wikipedia.org/wiki/Amino_acid_activation

https://en.wikipedia.org/wiki/Transcription_(biology)

https://en.wikipedia.org/wiki/Messenger_RNA

https://en.wikipedia.org/wiki/Transfer_RNA

Discussing Protein Synthesis or Gene Expression

Our genes code for proteins. How do our genes get translated into proteins? It is a complex process that's been called Gene Expression or Protein Synthesis.

Transcription enzymes are molecules. Transcription factors or transcription enzymes read genes and produce messenger RNA molecules from those genes. How do the transcription enzymes know that the cell needs a new protein? Who tells them? How do the transcription enzymes know which type of protein is needed and where it is needed? How is that information transmitted and understood? How do the transcription enzymes know which one of the 20,000 genes needs to be read in order to produce that newly requisitioned protein, and where that specific gene is located within the genome? How do the transcription enzymes go directly to the specific gene that needs to be read? Where does that wireless command-and-control information come from? How did the transcription enzymes learn how to read, translate, and transcribe the DNA and the genes? What makes the transcription enzymes move? Targeting, target acquisition, and docking require some kind of intelligent command and control, do they not? What are the transcription molecules using to see, read, and perceive the genome and their environment? They go to the specific gene that they need to read, know where it is, know how to get there, and recognize it when they see it. How are these transcription molecules doing that? It's as if they have eyes and a brain; yet, they are only molecules. These individual molecules are smarter and more intelligent than we are. How's that possible? These are scientific questions.

If I told you to make a specific protein, how long would it take for you to find the correct gene and then transcribe that gene into an mRNA molecule? The mapping of our human genome took decades. The transcription enzymes typically do transcription in a matter of minutes. They KNOW precisely which protein has been requisitioned by the cell, and they KNOW precisely where its associated gene is located within the genome. How do they KNOW? These are molecules. How would you transmit information and knowledge to atoms and molecules? It's NOT going to be done at the physical level. It's not being done by our genes or evolution or natural selection! It has to be done at the quantum level or the psyche level. It has to be WiFi at the quantum level! When it comes to Science, this is where the tires really hit the pavement. It does indeed generate a lot of friction among the science community at-large. The Materialists and Naturalists assure us that the quantum level or the psyche level does not exist; therefore, these people have no idea how all this information is being transmitted between atoms and molecules, or by whom.

Why are the introns there? Who knows what they mean and how to use them? Do they serve any purpose? God knows! I don't. Random chance or entropy clearly couldn't have designed and produced a genome, so who did? It couldn't be evolution because evolution is entropy or death. There's no intelligence or life in the various different types of evolution.

The introns are skipped during the encoding or transcription of messenger RNA, which is used by ribosomes to produce proteins. How does this invisible psyche or force know the difference between the introns and the exons? Who or what is telling the transcription enzymes to skip the introns? How does each transcription enzyme KNOW which gene to read and where that gene is located in the genome? It goes straight to what it needs and reads only what it needs. How does it do that? How does it know to skip the introns and only read the exons? How is that information being transmitted, received, and understood? Who is handling the logistics for all of this within a living cell at the quantum level?

How is all this 3D addressing for all the different genes being stored? It's not being stored within the genes – not at the physical level at least. By definition, in principle, the genes cannot communicate with each other and other molecules at the physical level. This communication has to be taking place at the quantum level or the psyche level because it isn't taking place at the physical level. There are NO wires in our brains! Likewise, there are NO wires connecting the individual atoms together into a communication network. There's NO WiFi for genes and molecules and atoms at the physical level because action at a distance is physically impossible according to classical physics; therefore, if wireless communication between atoms exists, it exists at the quantum level or the sub-atomic level.

Telepathy is WiFi at the quantum level or the psyche level. Thoughts and memories are quantum waves – they survive the death of our physical brain.

How does a cell decide that it needs a new protein? Who issues that instruction or command to produce a new protein? How is that information or command-set transmitted from molecule to molecule? How do the transcription factors or transcription enzymes know in advance, at a distance, telepathically which gene to read and where that gene is located in the genome? How does Command and Control KNOW where each gene is located and what each gene does? Who is Command and Control at the quantum level? How would you send messages and commands to atoms and molecules knowing that you can't do so at the physical level because it's physically impossible at the physical level?

How do these molecules communicate with each other and coordinate with each other? How do the transcription enzymes do targeting, docking, and manipulation? From whom are these molecules getting their instructions and location addressing? The transcription enzymes KNOW precisely what part of the chromosome to unzip. How do they know? How did they get this information? Who told them? How is that information transmitted to the chromosome making it unzip? Where is all this information being stored, since our 750-megabyte genome clearly doesn't have enough memory storage capacity to store and transmit petabytes of data and information? The Materialists, Naturalists, and Darwinists assure us that it's all being controlled by our genes, natural selection, and evolution; but, command and control can't be taking place within our genes because there are no genes within atoms! How are genes able to control atoms at a distance at the physical level? They can't! It's physically impossible! There's NO physical mechanism for doing so!

In a eukaryotic cell, messenger RNA is transcribed in the nucleus; and then, it is transported or moved through the nuclear membrane to the cytoplasm and then moved to a specific ribosome where it is translated into a protein. Who tells the mRNA molecule how to get out of the nucleus and into the cytoplasm where the ribosomes reside? Who is telling each mRNA molecule which one of the 10 million ribosomes to move towards? What is making the mRNA molecule move? Most of the ribosomes are conveniently located near where the constructed protein is needed and will be used. How is all of this invisible non-physical information being transmitted and received by the different atoms and molecules?

Furthermore, someone psyche or someone intelligent has to requisition and make the necessary amino acids and the tRNA codons BEFORE they can be combined or activated and turned into transfer RNA. The transfer RNA is made in the nucleus of the cell from DNA. The protein itself is constructed from dozens to thousands of tRNA molecules outside of the nucleus at some distant ribosome. How do the activation enzymes KNOW in advance which amino acid and codon combination will be needed in the future at some distant ribosome? Who tells the activation enzymes which type of tRNA molecule to make? There are theoretically 64 different types of tRNA molecules that can be made. Who decides which type to make and when to make it? Who tells the newly made tRNA molecules where to go? Each one has to go to a specific ribosome outside the nucleus in the cytoplasm. There are over 10 million different ribosomes in the cell. How are the tRNA molecules told which ribosome to move towards? How is this information transmitted to the tRNA molecule? Who or what makes the tRNA molecules move? Who is handling the logistics or the Command and Control for the activation and transportation of tRNA molecules?

The transfer RNA and messenger RNA are made within the nucleus; and, the ribosomes reside outside the nucleus. Who is doing all of this telepathic communication and action at a distance making sure that the tRNA molecules, mRNA molecules, and ribosomes coordinate with each other and work together like a finely tuned machine? How do each of the manufactured transfer RNA molecules KNOW in advance which ribosome to move towards? Who makes this assignment? A tRNA molecule moves towards ONE of 10 million different ribosomes. How does it know which ribosome to move towards? Who is telling each tRNA molecule where to go after it is made? Who or what makes them move?

In eukaryotic cells, tRNAs are transcribed by RNA polymerase III as pre-tRNAs in the nucleus. The ribosomes reside outside of the nucleus. Who is handling the logistics for all of this? It can't be entropy or random chance! By definition, in principle, entropy or random chance can't do logistics or Command and Control. It can't be evolution or our genes because the genes have NO physical means for commanding and controlling tRNA and mRNA molecules at a distance – not at the physical level at least. This information exchange has to be taking place wirelessly or telepathically at the quantum level or the psyche level because it is physically impossible at the physical level.

Who is requisitioning each and every one of the billions of different tRNA molecules that are being made within a cell in real time as they are needed? Who is getting the correct amino acids there on time for the activation process? Transfer RNA is made in the nucleus of a cell and has to make its way outside the nucleus to ONE of 10 million different ribosomes for protein synthesis. Who is handling the logistics for all of this and getting each tRNA molecule to the correct place at the correct time? Who is requisitioning each tRNA molecule and amino acid in the first place and getting all the atoms and amino acids and codons into place for that physical tRNA activation process? Who designed and created this whole process? There's NO such thing as spontaneous generation or creation ex nihilo.

Remember, each transfer RNA molecule that is made is sent to only ONE of the 10 million different ribosomes within the cell. Who chooses its destination? Who tells each tRNA molecule where its target ribosome is located? How does a tRNA molecule do this targeting and docking procedure? How does the tRNA know how to do so? Who or what makes the messenger molecules move? Just like in this essay, this protein synthesis and tRNA manufacturing process is done over and over and over again trillions of times within each living cell every day of your life. It is deliberate and intelligent. Who is controlling it all?

The tRNA and mRNA molecules are manufactured within the nucleus of the cell. They can literally go anywhere within the cell after they are manufactured and released from the nucleus; so, why do they pick that one specific ribosome to migrate towards after they have been manufactured or made? Who or what makes these tRNA and mRNA molecules move? How do they know where to go? Someone intelligent or someone psyche has to be coordinating the long-distance interaction between the mRNA molecule, the ribosome, and all the dozens or hundreds of tRNA molecules that are incoming towards the same ribosome. Someone has to handle the logistics and keep all of this straight! How are all of these molecules and ribosomes communicating with each other and coordinating with each other at a distance? Action at a distance and telepathy are quantum mechanical processes, NOT physical processes. It's physically impossible for our genes to control all of this.

Scientific Observation: Anything that was obviously made obviously had a Maker or Someone Intelligent who made it. This scientific truth is obviously true because

it has been experienced and observed trillions of different times by billions of different people. You could say that this particular premise is axiomatic or LAW.

Scientific Observation: Genes, proteins, transcription enzymes, activation enzymes, transfer RNA, messenger RNA, and ribosomes were obviously made. They didn't just spontaneously generate out of thin air ex nihilo. Spontaneous generation or creation ex nihilo is physically impossible. Spontaneous generation, abiogenesis, chemical evolution, or macro-evolution is prevented from happening by entropy or the second law of thermodynamics. This is what has been experienced and observed. Entropy or death cannot design and create.

Scientific Conclusion: Therefore, it is logical and rational to conclude that genes, proteins, transcription enzymes, activation enzymes, transfer RNA, messenger RNA, and ribosomes have Someone Intelligent or Someone Psyche who designed them and made them in the first place. Anything that was obviously made obviously had a Maker who made it. This is Observational Science 101.

Who is storing the 3D addressing information for each of the 10 million different ribosomes within a cell? Who is making these assignments or handling the logistics for all of this? Who is making sure that each of the trillions of different tRNA molecules are making it to the correct ribosome on time? The tRNA molecules and mRNA molecules could go anywhere after leaving the nucleus of the cell; so, why do they go to that one specific ribosome? What's making these messenger molecules move? They are "transporting". They are moving. Why? How? Molecules normally just sit there and vibrate under the effects of Brownian motion; so, what's making the atoms and molecules within a living cell move towards a specific targeted destination? It's as if these things are alive while they are within a living cell. How's that possible? It's not possible according to the Materialists and Naturalists! It's definitely NOT possible at the physical level!

How is all that information being transmitted in advance from the mRNA molecule to the individual tRNA molecules that will be docking with the different ribosomes in the future? How is this communication at a distance between atoms and molecules taking place? The tRNA molecules are being told where to go. How do they receive, translate, and understand this information? Who taught them this language? The activation enzymes are also being told in advance which amino acid to bind with which codon during the production of transfer RNA molecules. The amino acids and codons are being told where to go too. Who is coordinating all of this so that it works flawlessly rather than gumming up and screeching to a halt? Who enlivens and activates these formerly inert and inactive atoms and molecules once they are drawn into a living cell? Can you sense the potential for discovery?

Where did the first ribosomes come from? Who tells the different proteins to self-assemble into ribosomes? Ribosomes are made from proteins; and, proteins are made by ribosomes? So, where did the ribosomes come from in the first place? Who made the proteins that were used to construct the first ribosomes? Who made the first ribosomes? Who taught the ribosomes how to manufacture proteins from mRNA molecules and tRNA molecules? Who taught the mRNA and tRNA to go to the ribosomes rather than someplace else? Who designed and produced this gene expression or protein synthesis process in the first place? Or did it all just spontaneously generate out of thin air as the Materialists and Naturalists claim? This is science. These questions are science.

Who is making all of these assignments and timing all of these appointments? Who is overseeing all of this? Remember, this is atoms and molecules that we are talking about here, NOT genes! There aren't any genes within an atom! Who is doing all of this communication between atoms and molecules? How is this information transmitted? How do the atoms and molecules translate and understand this information? How do the atoms and molecules know where to go? They can go anywhere; yet within a living cell, they go precisely where they are told to go! How is all this Command and Control being done? Who is doing it? How are these requests being transmitted to the atoms and the molecules?

Command and Control obviously needs an intelligent Commander of some sort; or, nothing would ever happen!

Amino acids and codons are made from atoms. Who or what is telling each atom that it has an appointment at a specific location at a specific time for amino acid synthesis and tRNA codon synthesis? Who designed and made all of this fine tuning? Atoms can be assembled into anything, theoretically; so, who or what is assembling these particular atoms into amino acids and codons; and, how does it know how to do so? Outside of a living cell, atoms NEVER self-assemble into functional proteins and their matching genes. They can't because it's physically impossible. That's why the protein synthesis process was designed and created in the first place.

Fine-tuning requires a Fine Tuner; or, there would be NO fine-tuning! Random chance can't do fine-tuning. Random chance produces disorder, entropy, chaos, and death. Fine-tuning is Syntropy. Entropy or evolution can't do fine-tuning.

Design and Creation require a Designer and a Creator. The protein synthesis process was obviously designed and created. Therefore, it is obvious that the whole gene expression process has a Designer and Creator who made it.

How do the activation enzymes KNOW in advance at a distance which amino acid to combine with which codon during the activation process and the production of transfer RNA? How are those activation enzymes reading that future mRNA molecule from a distance so that they KNOW in advance which type of transfer RNA molecule to make? Who or what is doing all of the Command and Control for the production of transfer RNA; and, how are all of these petabytes of information being transmitted to the different individual atoms? Who is keeping track of it all? How are the atoms able to communicate with each other and coordinate with each other? What makes these atoms and molecules move towards each other and bind with each other? In a beaker, the atoms and molecules do nothing but Brownian motion; but, within a living cell the very same atoms and molecules literally come alive with purpose and activity! They combine together on command.

How? Why? Who is transmitting all of these different commands and instructions to the individual atoms and molecules? Who is keeping it straight and making it work? Why do the atoms and molecules choose to obey while they are within a living cell, while the very same atoms and molecules don't do anything but Brownian motion and random diffusion while they are outside a living cell? How do the atoms and molecules KNOW that they are within a living cell? How are the atoms able to understand these instructions and how do the atoms see and know where to go? Where is all of this information being stored in the first place? There aren't any genes within an atom!

The Materialists, Naturalists, Darwinists, and Atheists assure us that only physical matter or physical atoms exist; but, clearly this is NOT the case. Clearly these people are wrong! Materialism, Naturalism, and their derivatives are based exclusively on entropy. Entropy is death. Entropy or death cannot design and create.

How would you communicate between atoms and make atoms do things and construct things if given the assignment to do so?

How do the activation enzymes know at a distance in advance which amino to combine with which codon so that the transfer RNA molecules arrive in the correct order and the correct number at each ribosome for RNA translation or protein synthesis? Protein synthesis happens rapidly, which means that the transfer RNA molecules arrive pre-sorted in the correct sequence BEFORE they get to the ribosome for RNA translation into a functional protein. There's NO hunting and pecking and randomness going on throughout protein synthesis – there's ONLY purposeful activity. How's that possible? It's NOT possible at the physical level. It's only possible through Psyche or Intelligence at the quantum level.

How is this information transmitted from molecule to molecule and by whom? Who or what makes these individual molecules move or swim through the cytoplasm to the specific ribosome where they are needed in a nice and orderly manner? How does each tRNA molecule know which of the 10 million ribosomes it is supposed to move towards? Who is storing and organizing all of this information? Who set up this information network and logistics highway in the first place? Who is controlling all of this, and how is all of this information being transmitted to the molecules and being received and understood by the molecules? Who taught the atoms and the molecules how to communicate with each other?

Who is handling all of the telepathic communication and coordination taking place between the targeted ribosome, the incoming mRNA molecule, and the hundreds of different incoming tRNA molecules? The Materialists and Naturalists and Darwinists will tell you that it is evolution and/or the genes who are handling all of this. Are they right? Do they know what they are talking about? Or are they making things up out of thin air?

Some people have read the first couple of pages of some of my books and then asked, "Where's the Science?" Do you see any science in any of this? The Materialists and Naturalists will tell you that none of what I have written here is real science. These people will tell you that I don't understand science. They have told me that I don't understand the power and the beauty and the ability of evolution to be able to design and create. These people have told me that what I have written here is NOT science. Instead, they will assure you that Materialism, Naturalism, Darwinism, and Atheism ARE science. Are they right, or are they employing *circular reasoning* in order to make their case? *Circular reasoning* or *begging the question* is a logic fallacy the last time I checked.

Entropy is death. Materialism, Naturalism, Darwinism, Nihilism, Atheism, Classical Physics, and the Theory of Evolution are based exclusively on entropy. These people teach that ONLY entropic physical matter exists. Are they right? Or did they get their science wrong? Evolution is entropy. Evolution is death. The theory of evolution is Creation by Entropy or Creation by Death. These people will assure you that evolution, entropy, or death can design and create anything it wants to at will. These people are wrong! Entropy or death cannot design and create. It's physically impossible. Furthermore, entropy or death cannot do protein synthesis or gene expression. Design and creation require some type of Intelligence, Psyche, or Syntropy, and so does protein synthesis.

Could You Do the Logistics for Protein Synthesis?

You are intelligent. Could you handle and do the logistics for protein synthesis within a single living cell in your body?

If I were to give you the assignment to handle the logistics for 10 million ribosomes, millions of messenger RNA molecules, billions of transfer RNA molecules, billions of amino acids and codons, and trillions of atoms within a single living cell, would you be able to handle the logistics for all of that and keep all of it straight in your head? Would the individual atoms and molecules and proteins make it to their assignments on time and arrive at the correct location in time?

Of course not! A single physical brain doesn't have anywhere near enough memory storage capacity and processing power for such a task; yet, someone psyche or someone intelligent is doing all of this with ease at the quantum level or the psyche level for the trillions of different cells within your physical body; otherwise, a cell would be nothing but entropy, disorder, and random chaos. A cell would be dead or entropic without all of this Psychic Command and Control. Whoever is handling the logistics for all of this is infinitely more intelligent and infinitely more powerful than we are. Yet, the Materialists, Naturalists, Darwinists, and Atheists want you to believe that all of this happens by random chance! Do you believe them? I don't. Not anymore. I used to be a Materialist, Naturalist, Nihilist, and Atheist until I actually started looking at the science and the evidence.

It's physically impossible for a single human being to handle all of the logistics for even one single living cell; yet, these little critters (the atoms and molecules) know precisely where they are supposed to go and show up on time like clockwork as if they are infinitely more intelligent and skillful than our smartest human being or most powerful computer. Outside of a living cell, the atoms and molecules are dumb as rocks; but, inside a living cell, these very same atoms and molecules are infinitely more intelligent and competent than even our best and brightest human beings. How do the atoms and molecules get so smart once they find themselves within a living cell? What's going on here?

The Materialists, Naturalists, Darwinists, and Atheists assure us that our genes and evolution are handling all of this. Are they right? Or, are they trying to trick you and deceive you? Who is handling the timing and the logistics for protein synthesis or gene expression and keeping it purring like a finely tuned machine?

Remember, anything that is obviously fine-tuned obviously had a Fine Tuner who fine-tuned it! The logistics for protein synthesis has obviously been fine-tuned. Therefore, protein synthesis or gene expression obviously has a Fine Tuner.

Remember, any type of Command and Control obviously needs a Commander or Controller. The logistics for protein synthesis is obviously being commanded and controlled, or nothing would happen in the first place except Brownian motion and random diffusion. The logistics for protein synthesis obviously has a Commander or a Psyche who is in control of it all at the quantum level.

The Materialists, Naturalists, Darwinists, Nihilists, Behaviorists, Determinists, Physical Reductionists, Evolutionists, Classical Physicists, and Atheists assure us that protein synthesis is happening by random chance and was produced by random chance. Are they right? Or did they get their science wrong?

According to these people, only entropic physical matter exists. Entropy is death. Materialism, Naturalism, and the Theory of Evolution are Creation by Entropy or Creation by Death. According to these people, entropy or death can design and create at will. Are they right? Or did they get their science wrong?

Do you see any signs of intelligence or psyche behind any of this, or do you still believe that it all happens by random chance? Entropy is death. Do you really believe that entropy or death can do protein synthesis? Do you really believe that entropy or death is running the show? The Theory of Evolution is Creation by Entropy or Creation by Death. The Physicalists, Darwinists, and Naturalists assure us that only entropic physical atoms exist. Their science is based exclusively on entropy or death. The Materialists and Naturalists tell us that there is no such thing as Psyche or Syntropy. Do you think they are right? Do you really think that Creation by Death or the Theory of Evolution works as advertised? Or do you think that someone intelligent is running the show?

It's as if there is some type of Overlord Psyche or Command and Control Psyche; and, there also seems to be some type of intelligence or psyche within each and every atom too! Someone Intelligent or Someone Psyche is overseeing all of this and controlling all of this cellular activity at the quantum level; or, it would be nothing but Brownian motion, random diffusion, entropy, chaos, and death. Death can't design and create. These very same atoms and molecules in a beaker don't do anything but Brownian motion and random diffusion; but, within a living cell they literally come alive with purpose and activity. How's that possible? The answer isn't found within our genes. The atoms don't have genes – at least not at the physical level! Who knows what the atoms have at the quantum level or the psyche level, though? God knows! I certainly don't! For all we know, there is a quantum computer within each and every physical atom handling all of the data processing and memory storage at the quantum level which isn't possible at the physical level.

When the Human Psyche and the Human Spirit leave the physical body, then all of these different physical processes shut down and stop working; and, entropy, random diffusion, chaos, and death take over instead. Entropy is death. Syntropy is the opposite of entropy. Syntropy is life. Syntropy is eternal life. Psyche is Syntropy – Psyche is eternal and everlasting, without a beginning of days or an end of years. Syntropy is Conservation of Psyche. It requires some type of Syntropy in order to produce life. Some type of Syntropy MUST exist, or all of that subsequent entropy and life wouldn't have been possible in the first place. While doing science, that is what I have experienced and observed.

Syntropy really is the answer to life, the universe, and everything – but only if you choose to believe that it is real and truly exists. You can also go with Materialism, Naturalism, Darwinism, Nihilism, and Atheism; and remain in ignorance if that's what you want to do. The choice is yours to make. Nobody else can do that for you. Science is choice. Science is observation and experience. This is science. These questions are science.

Do you believe the science – the observations and the experiences; or, do you believe the philosophical speculation and wishful thinking of the Materialists, Naturalists, Darwinists, Nihilists, and Atheists? Beliefs are chosen into existence by the Human Psyche. Your beliefs and actions are chosen into existence by your Psyche. Choice is purely a function of Psyche or Intelligence. The rocks and entropy haven't been caught in the act of doing science and choice. Raw physical matter has no choice. According to the Materialists, Naturalists, and Behaviorists, the rocks and entropy cannot do choice. They are right. There is no choice at the physical level. Physical matter simply does what it is told to do. Told by whom? It is physically impossible for the rocks and entropy to do science and make choices. This is what has been experienced and observed. Science IS observation and experience. It requires some type of Syntropy or Psyche in order to be able to do science and make choices. It takes some type of Syntropy, or Psyche, or Intelligence in order to make life. It takes Psyche or Intelligence to do science at every level of existence.

Materialism and Naturalism cannot explain how protein synthesis is made to work within a living cell; but, Quantum Mechanics, Syntropy, and Psyche certainly can.

Remember, gene expression and protein synthesis NEVER happen outside of a living cell. They can't. It's physically impossible. Outside of a living cell, protein synthesis is prevented from happening by random diffusion, entropy, or the second law of thermodynamics. This is what has been experienced and observed. Protein synthesis requires some kind of Intelligent, Purposeful, and Syntropic Command and Control at the quantum level in order to work right at the physical level; otherwise, it would be nothing but random chaos, random diffusion, and Brownian motion within a cell; and, nothing would ever get done. This is what has been experienced and observed.

Conclusions Regarding Command and Control

Proteins, enzymes, amino acids, tRNA, and mRNA are molecules that swim freely throughout the cytoplasm of a living cell. Who requisitions them? Who tells them where to go and what to do after they are made? How do you communicate with molecules and control molecules at the quantum level? It requires Someone Psyche and Someone Intelligent; otherwise, it can't be done. The Materialists, Naturalists, Darwinists, Nihilists, Behaviorists, Determinists, Physical Reductionists, and Atheists teach, preach, and believe that psyche and the non-physical DO NOT EXIST. They are wrong. These people don't know what they are talking about. Within a living cell, atoms and molecules are extremely intelligent, psychic, sentient, perceptive, purposeful, conscious, and alive. Intelligence is necessary. Psyche or Intelligence has been experienced and observed.

Psyche or Non-Local Consciousness is experienced first-hand by Out-of-Body Explorers as an immaterial viewpoint in space. Whenever someone goes looking for their own Human Psyche while out-of-body on the astral plane, there's nothing there to be found because they ARE their Psyche. Your spirit body and physical body don't have experiences and don't form memories of events while your Human Psyche is separated from them. It's your Human Psyche who experiences events and then forms memories of those events – both at the quantum level and the physical level. This is what has been experienced.

Psyche is seen or observed by Out-of-Body Travelers as a pinpoint of light, a point particle of light, a spark of light, or a Photon. A Photon is one type of Psyche or Intelligence. This is what has been observed. Science is observation and experience. Remember, Photons are Psyches or Intelligences; and, Photons have already been experienced and observed on BOTH sides of the veil – both on the spiritual side and the physical side.

Furthermore, each massless quantum particle or massless elementary particle that is discovered looks like and ACTS like a Photon and is therefore also a type of Psyche or Intelligence. Remember, ONLY Psyche or Intelligence can do command, control, requisition, logistics, targeting, perception, target acquisition, docking, and choice at the quantum level.

Mark My Words

—

Source

The Scientific Method Proves That the Theory of Evolution Is False

https://www.amazon.com/dp/B01IAAIRT2

I Am Not a Creationist: So What Am I?

https://www.amazon.com/dp/B071XTM8XY

Summary Of: The Theory of Evolution Proves that God Exists

https://www.amazon.com/dp/B01GQCWED6

The Second Comforter: Supping with Our Resurrected Lord Jesus Christ

https://www.amazon.com/dp/B01IAKHTY6

I'm Not a Creationist

I'm not a Creationist.

What does this mean in practice?

It means that I don't believe in creation ex nihilo, which the Christian Catholics and Protestants choose to believe in and promote. Creation ex nihilo is impossible and patently absurd.

I'm not a Creationist – I don't believe in creation ex nihilo.

Furthermore, it means that I don't believe in spontaneous generation or abiogenesis. Whether they realize it or not, the Evolutionists and Darwinists are also creationists. They too have chosen to believe in Chemical Evolution, or Macro-Evolution, or Creation by Spontaneous Generation. Spontaneous generation or chemical evolution is the same exact thing as creation ex nihilo – same principle but different "nothing". Spontaneous generation, or chemical evolution, or macro-evolution is physically impossible and patently absurd. It's prevented from happening by entropy.

I'm not a Creationist – I don't believe in spontaneous generation or the theory of evolution.

As I see it, creation ex nihilo is a type of Atheism – creation of something from nothing by nothing. Nothing can create something from nothing. Not even God can create something from nothing. Creation ex nihilo is patently absurd, just like Atheism. You will never catch "nothing" in the act of designing and creating something. It can't be done, which means that it wasn't done.

Likewise, creation by nothing, or spontaneous generation, or chemical evolution, or macro-evolution is patently absurd, just like Atheism. Design and creation by "Nothing" or Atheism is illogical, irrational, and absurd. It has never been experienced nor observed because it's impossible. Nothing can design and create something.

Creation ex nihilo is spontaneous generation or abiogenesis, which means that creation ex nihilo is a type of chemical evolution, magical evolution, or macro-evolution. Technically, creation ex nihilo is creation from nothing; and, spontaneous generation is creation by nothing; but, where both are concerned "Nothing" is the operative word. They are nothing. They don't exist. Creation from nothing and creation by nothing are patently absurd. They are impossible! That means that they never happened. That means that they have been falsified.

Genes and proteins don't just spontaneously generate out of thin air, and neither does physical matter. These things are organized or made by Someone Intelligent or Someone Psyche.

I don't believe that God is a creationist or magician. I believe that God is a scientist, the Ultimate Scientist.

God organized this physical universe from pre-existing Dark Matter or pre-existing Spirit Matter – not ex nihilo. There's no such thing as creation ex nihilo. It's impossible, which means that it didn't happen because it can't happen.

Like God, I'm a scientist; and, as a scientist, creation ex nihilo doesn't make any logical sense to me. Creation ex nihilo demands way too much blind-faith from me. Nobody can create something from nothing, not even God.

Likewise, since I'm a scientist, I KNOW for a fact that spontaneous generation, chemical evolution, or macro-evolution was falsified in 1859 by Louis Pasteur – the very same year that Charles Darwin published "On the Origin of Species". We've KNOWN since the very beginning that the Theory of Evolution is false; but, most of the scientists chose to ignore that evidence preferring the convenient fiction instead. Charles Darwin was telling them what they wanted to hear; and, Louis Pasteur was not. Self-deception works, and it works every time. I KNOW because I used to be a Physicalist, Naturalist, Nihilist, and Atheist.

God, or Psyche, or Syntropy had to exist BEFORE God did "creation ex nihilo"; and, the very existence of God or the pre-existence of God FALSIFIES creation ex nihilo, because God ain't nothing.

—

Observation or Premise: Anything that was obviously made obviously had a Maker or Creator who made it.

Observation or Premise: A genome was obviously made.

Logical Conclusion: Therefore, a genome obviously has a Maker or a Creator who designed it, programmed it, engineered it, field-tested it, fine-tuned it, made it, manufactured it, and deployed it.

—

Your genome is God's Signature, and it is written on every cell in your body. Your genome didn't just spontaneously generate out of thin air as the Materialists, Naturalists, and Darwinists claim because that's physically impossible. Spontaneous generation or chemical evolution is prevented from happening by random diffusion and the second law of thermodynamics. Entropy cannot design and create proteins, genes, eyes, brains, and life forms. It has never been caught in the act of doing so; and, it never will be because it's impossible.

Remember, Materialism, Naturalism, Darwinism, Nihilism, and Atheism are based exclusively on entropy; and, it always requires some type of Syntropy or Intelligence in order to be able to design and create something new and useful from atoms or from scratch.

The false is falsified by the truth; and, the truth is repeatedly experienced and observed. Remember, creation ex nihilo, spontaneous generation, macro-evolution, and chemical evolution have NEVER been experienced nor observed. They are impossible which means that they are false and have been falsified.

What Has Been Experienced and Observed

The Materialists, Naturalists, Darwinists, Nihilists, Behaviorists, Determinists, Physical Reductionists, and Atheists teach, preach, and believe that psyche and the non-physical DO NOT EXIST. They are wrong. These people don't know what they are talking about. Psyche or Intelligence has been experienced and observed.

Psyche or Non-Local Consciousness is experienced first-hand by Out-of-Body Explorers as an immaterial viewpoint in space. Whenever someone goes looking for their own Human Psyche while out-of-body on the astral plane, there's nothing there to be found

42

because they ARE their Psyche. Your spirit body and physical body don't have experiences and don't form memories of events while your Human Psyche is separated from them. It's your Human Psyche who experiences events and then forms memories of those events – both at the quantum level and the physical level. This is what has been experienced.

Psyche is seen or observed by Out-of-Body Travelers as a pinpoint of light, a point particle of light, a spark of light, or a Photon. A Photon is one type of Psyche or Intelligence. This is what has been observed. Science is observation and experience. Remember, Photons are Psyches or Intelligences; and, Photons have already been experienced and observed on BOTH sides of the veil – both on the spiritual side and the physical side.

Furthermore, each massless quantum particle or massless elementary particle that is discovered ACTS like a Photon and is therefore also a type of Psyche or Intelligence. Remember, ONLY Psyche or Intelligence can do command, control, and choice.

Mark My Words

—

Source

I Am Not a Creationist: So What Am I?

https://www.amazon.com/dp/B071XTM8XY

References

God Is in the Light: God is light, and in Him is no darkness at all.

https://www.amazon.com/dp/B07168S37N

Quantum Mechanics from a Non-Physical Spiritual Perspective.

https://www.amazon.com/dp/B01J023TGU

https://www.amazon.com/dp/1521132380

Quantum Neuroscience: The Answer to Life, the Universe, and Everything.

https://www.amazon.com/dp/B079Z6QQQB

Science 2.0: I Upgraded My Science.

https://www.amazon.com/dp/B0771K6WTX

Scientific Proof of God's Existence: Finding God Where the Atheists Refuse to Look for Him

https://www.amazon.com/dp/B07B26CRHX

Proof of God's Existence

Observation or Premise: Anything that was obviously made obviously had a Maker or Creator who made it.

Observation or Premise: A genome was obviously made.

Logical Conclusion: Therefore, a genome obviously has a Maker or a Creator who designed it, programmed it, engineered it, field-tested it, fine-tuned it, made it, manufactured it, and deployed it.

—

This syllogism is both a Philosophical Proof of God's Existence and a Scientific Proof of God's Existence. It works. It's true. Every aspect of it is true, because both the premises and the conclusion have been experienced and observed in real life by real people. This Scientific Proof of God's Existence is both philosophically *valid* and observationally *sound*. It, and similar ones like it, convinced me that God does indeed exist.

Variations on this syllogism were my first Scientific Proof of God's Existence; and, I first built one based upon the non-existence of abiogenesis, or the non-existence of macro-evolution. After science itself proved to me that God must of necessity exist, then I simply KNEW that God exists, even though I still have yet to get to know God.

In 2016, when I first realized that NO type of evolution – chemical evolution, micro-evolution, macro-evolution, spontaneous generation, abiogenesis, random mutations, or natural selection – is capable of designing, creating, and implementing genomes from atoms or from scratch, I just KNEW that God exists because I KNEW why He must of necessity exist. God must exist in order to have done ALL of the science, organization, proteins, and genomes which the different types of evolution could NEVER have done.

In a very real sense, the Theory of Evolution proved to me that God exists. In other words, the inability of the different types of evolution to design, create, manufacture, and deploy proteins and the matching genes to go along with those proteins BECAME convincing scientific proof of God's necessity, which effectively became Scientific Proof of God's Existence.

Based upon this same exact theme, I developed many other Scientific Proofs of God's Existence and eventually used the Scientific Method itself to falsify Materialism, Naturalism, Darwinism, Atheism, and the Theory of Evolution.

I kept a log of my ongoing discoveries in the following books:

1. *Summary Of: The Theory of Evolution Proves that God Exists*

     https://www.amazon.com/dp/B01GQCWED6

     https://www.amazon.com/dp/1521130485

2. *The Theory of Evolution Proved to Me that God Exists: Why I Am No Longer an Atheist and Why I No Longer Believe in the Theory of Evolution*

     https://www.amazon.com/dp/B01HZYBZ7K

     https://www.amazon.com/dp/1521131228

3. *The Scientific Method Proves That the Theory of Evolution Is False*

4. *Using the Scientific Method to Eliminate the Usual Suspects and to Prove the Truth*

This was an era of transition for me. I used to be a Materialist, Naturalist, Nihilist, and Atheist. To develop a Scientific Proof of God's Existence that I actually KNEW to be true, as well as using the Scientific Methods to falsify Materialism, Naturalism, Darwinism, Atheism, and the Theory of Evolution represented a MAJOR paradigm shift in my way of thinking and doing science.

Once I finally KNEW the Truth, there was no going back to the deceptions and the lies; and, I was able to progress forward from there and make additional adjustments and improvements to my science, my worldview, and my philosophical underpinnings.

Although it is technically true – due to the *affirming the consequent* logic fallacy – that it's impossible to use the Scientific Methods to prove anything true, there is a legitimate work-around that actually produces the same result. This work-around is a philosophically *valid* and an observationally *sound* way to find the truth through a process of elimination.

The way this is done in practice is that we use the Scientific Methods and *negating the consequent* to falsify and eliminate everything that is false such as Materialism, Naturalism, Darwinism, Atheism, and their derivatives so that ONLY the Truth remains.

Think about it logically.

If you successfully eliminate everything that is false – everything that has NEVER been experienced nor observed by anyone, not even God – then ONLY the Truth will remain, because ONLY the observed and the verified and the proven will remain. The Truth is whatever has been experienced, lived, and observed by Someone Psyche sometime somewhere. Science is observation, or it should be.

Sherlock Holmes taught me how this process of elimination works in principle or practice.

How often have I said to you that when you have eliminated the impossible [and the false]**, whatever remains, however improbable, must be the truth? — Sherlock Holmes.**

Sherlock Holmes and the Philosophy of Science taught me how to use the Scientific Method to falsify the Theory of Evolution, Materialism, Naturalism, and Atheism; and, observations from real life people like you and me falsified Nihilism for me.

If you successfully eliminate everything that is false, then ONLY the Truth will remain; and, that's precisely how we use the Scientific Methods to "prove" the truth, by using the Scientific Methods to eliminate everything that is false or everything that has NEVER been experienced nor observed. It works!

God Is Light

The Materialists, Naturalists, Darwinists, Nihilists, Behaviorists, Determinists, Physical Reductionists, and Atheists teach, preach, and believe that psyche and the non-physical DO NOT EXIST. They are wrong. These people don't know what they are talking about. Psyche or Intelligence has been experienced and observed.

Psyche or Non-Local Consciousness is experienced first-hand by Out-of-Body Explorers as an immaterial viewpoint in space. Whenever someone goes looking for their own Human Psyche while out-of-body on the astral plane, there's nothing there to be found because they ARE their Psyche. Your spirit body and physical body don't have experiences and don't form memories of events while your Human Psyche is separated from them. It's your Human Psyche who experiences events and then forms memories of those events – both at the quantum level and the physical level. This is what has been experienced.

Psyche is seen or observed by Out-of-Body Travelers as a pinpoint of light, a point particle of light, a spark of light, or a Photon. A Photon is one type of Psyche or Intelligence. This is what has been observed. Science is observation and experience. Remember, Photons are Psyches or Intelligences; and, Photons have already been experienced and observed on BOTH sides of the veil – both on the spiritual side and the physical side.

Furthermore, each massless quantum particle or massless elementary particle that is discovered ACTS like a Photon and is therefore also a type of Psyche or Intelligence. Remember, ONLY Psyche or Intelligence can do command, control, and choice.

The Materialists, Naturalists, Darwinists, Nihilists, Behaviorists, Determinists, Physical Reductionists, and Atheists will tell you the Psyche DOES NOT EXIST. They don't know what they are talking about. Massless particles are psyches. A Psyche is a photon, a massless elementary particle. Psyches look like and ACT like photons, both at the physical level and the quantum level. Psyches are intelligent, conscious, sentient, prescient, aware, and alive. Psyches are also quantum mechanical or supernatural which means that when they are not being held back by mass or physical restrictions, Psyches have God-like supernatural powers such as omniscience, omnipresence, telepathy, telekinesis, teleportation, quantum tunneling, action at a distance, instantaneous communication, levitation, phase-shifting, teleology, syntropy, indestructibility, immortality, and eternal life.

Mark My Words

—

Source

Science 2.0: I Upgraded My Science

https://www.amazon.com/dp/B0771K6WTX

Communication between Atoms and Molecules

You can see the signature of God, the intelligence of God, and the hand of God within the genome, protein synthesis, ATP synthesis, tRNA activation, transcription factors, cellular target acquisition, and the biochemistry of a living cell. God's influence and interaction is obvious and clear to anyone who is willing to look and see. It was science that convinced me that God exists.

If I were to give you the task to make atoms communicate with each other and interact with each other, how would you accomplish that task wirelessly at a distance? If I were to give you the task to make molecules communicate with each other within a living cell, move throughout that cell to specific targets, and interact with each other when they get there, how would you accomplish that task wirelessly at a distance?

How do you communicate with atoms and molecules? How do you make them do targeting, selection, movement, docking, teleology, transactions, logistics, and acquisition? How would you make a protein tell you that it is broken or will soon be broken, and that your cell will soon need a new one? How would you handle the communications and logistics for protein synthesis wirelessly at a distance within a living cell? How would you requisition a new protein for a living cell wirelessly at a distance? How would you tell transcription enzymes (RNA polymerases) wirelessly at a distance which type of protein the cell currently needs, which one of the 20,000 genes it needs to read in order to produce that protein, and where that specific gene is located within the genome so that the transcription enzyme can go directly to that specific gene, read it, and transcribe it? How would you make the transcription enzymes move within a living cell? How would you do targeting and acquisition within a living cell wirelessly at a distance?

Targeting ATP Synthase Wirelessly at a Distance

https://en.wikipedia.org/wiki/ATP_synthase

https://science-2-0.com/wp-content/uploads/2018/07/ATP-Synthase.pdf

ATP molecules were designed to provide energy to all the different cellular machinery within a living cell. After the ATP molecule is charged by the ATP Synthase enzyme, the ATP molecule is dumped into the cytoplasm of the living cell. It can literally go anywhere. Who or what makes it swim or move through the cytoplasm instead of piling up at the site of ATP synthesis? More importantly, who tells the ATP molecules where each one of them is needed? Who handles the logistics for all of this? How do they learn wirelessly at a distance that they are going to be needed at a specific location sometime in the future? Who does target acquisition for ATP molecules within a living cell? Targeting and acquisition require intelligent logistics; otherwise, it would be nothing but random chaos within a living cell.

Target acquisition is the detection, identification, and location of a **target** in sufficient detail to permit the effective employment of lethal and non-lethal means. The term is used for a broad area of applications.

Getting cruise missiles to do target acquisition requires a massive infusion of intelligence from Someone Psyche. The same thing remains true when it comes to atoms and molecules within a living cell. Target acquisition between atoms and molecules within a living cell requires a massive infusion of intelligence from Someone Psyche.

In principle, target acquisition between atoms and molecules is physically impossible, which means that we need a quantum mechanical or psychic explanation for how it's being done within a living cell.

Who does target acquisition for transcription enzymes and ATP molecules? How do these atoms, enzymes, and molecules detect, perceive, identify, and locate their targets wirelessly at a distance? Who or what makes atoms, enzymes, proteins, and other molecules move to a specific location and then dock with other molecules at that location? Brain cells migrate to a different location after they are made. Who does target acquisition for brain cells? Synaptogenesis also requires target acquisition between an axon and a dendrite. Who handles the logistics and target acquisition for synaptogenesis? There's custom-made machinery that was designed for synaptogenesis; but, who is running or operating that machinery? Who is handling target acquisition for that machinery? How is all that information communicated between the pre-synaptic neuron and the post-synaptic neuron while they are choosing a mutual site to build a synapse or a gap-in-space? How would you communicate through a gap-in-space wirelessly at a distance?

Target acquisition gets to the very heart of what makes activity within a living cell completely different than the non-activity, quantum fluctuations, entropy, random diffusion, and Brownian motion taking place external to a living cell. Does it not?

Targeting Ribosomes Wirelessly at a Distance

https://en.wikipedia.org/wiki/Ribosome

https://science-2-0.com/wp-content/uploads/2018/07/Ribosome.pdf

How would you tell a messenger RNA molecule (mRNA) wirelessly at a distance which one of the ten million ribosomes to move towards? Ten million different possibilities! Could you handle the logistics for ten million ribosomes? How would you do target acquisition for the ten million associated mRNA molecules telling them which ribosome to move towards and dock with? Furthermore, how would you tell the billions of associated transfer RNA molecules (tRNA) wirelessly at a distance, which one of the ten million ribosomes to move towards? How would you communicate with messenger RNA molecules and transfer RNA molecules wirelessly at a distance and make them move or swim through the cytoplasm of a living cell towards a specific target at a specific location within the cell? How would you communicate wirelessly at a distance with a messenger RNA molecule (mRNA) and transfer RNA molecule (tRNA) telling them where their target ribosome is located within a living cell? How would you coordinate between mRNA, tRNA, and ribosomes wirelessly at a distance? It's being done, so how is it being done? These are molecules and cellular machinery that we are talking about here.

Remember, messenger RNA and transfer RNA are made within the nucleus of the cell, but then they have to migrate through the nuclear membrane and move to the outside of the nucleus into the cytoplasm where their target ribosome is located. They can go anywhere after they are made, so how do you make mRNA and tRNA molecules move towards a specific ribosome? How would you communicate wirelessly at distance with mRNA and tRNA molecules? How would you handle target acquisition for mRNA and tRNA molecules? Someone Psyche or Someone Intelligent has to be issuing these orders; otherwise, it would be nothing but random chaos within a living cell.

Some of the live-motion animations of protein synthesis listed below in the Reference Section show the two subunits of a ribosome selecting, targeting, and acquiring one specific mRNA molecule for the translation process wirelessly at a distance. In other animations and biology textbooks, the mRNA is shown targeting a specific fixed ribosome. What gives? Well, there are in fact two different locations for ribosomes. Ribosomes are classified as being either "free" or "membrane-bound". Some ribosomes are free-floating in the cytoplasm and actively target and acquire specific mRNA molecules. Other ribosomes fix themselves into a membrane, and the mRNA molecules may in fact actively target them instead. Either way, ribosomes and mRNA molecules are targeting each other wirelessly at a distance within a living cell.

Any way you choose to look at it, trillions of physical molecules floating free in the cytoplasm of living cells are doing target acquisition wirelessly at a distance every second of the day. These molecules know precisely what they are and what they are supposed to do; and, they proceed in doing their mission or assignment wirelessly at a distance as if they were intelligent and alive. It's as if each one of these molecules has a Psyche or a Soul.

Transfer RNA molecules also actively target and acquire specific ribosomes and specific codons on mRNA molecules wirelessly at a distance as if the tRNA molecules are conscious, sentient, alive, intelligent, insightful, and self-aware. When it comes to communication between molecules, target acquisition taking place wirelessly at a distance is a telepathic phenomenon which means that it involves Psyche and Quantum Mechanics.

Transfer RNA Molecules Target Activation Enzymes and Ribosomes Wirelessly at a Distance

https://en.wikipedia.org/wiki/Transfer_RNA

https://en.wikipedia.org/wiki/Aminoacyl_tRNA_synthetase

https://science-2-0.com/wp-content/uploads/2018/07/Transfer-RNA.pdf

https://science-2-0.com/wp-content/uploads/2018/07/Aminoacyl-tRNA-synthetase.pdf

In eukaryotic cells, tRNA are made by a special protein that reads the DNA code and makes an RNA copy, or pre-tRNA. This process is called transcription and it's used for making tRNA. It's done by RNA polymerase III. Pre-tRNA are processed once they leave the nucleus. The anti-codons on the tRNA molecules are made in advance from DNA within the nucleus of the cell and eventually sent to attach to or bind with the codons on a specific mRNA molecule at some specific ribosome outside of the nucleus of the cell.

Activation enzymes in the cytoplasm make tRNA molecules bind with specific amino acids. The tRNA molecules carry amino acids to ribosomes within the cell. How would you handle target acquisition and logistics for anti-codons, amino acids, tRNA production, and activation enzymes wirelessly at a distance in order to successfully produce a fully charged tRNA molecule?

The logistics and target acquisition for all of these physical processes is handled wirelessly at a distance by Nature's Psyche or God's Psyche. There's NO physical explanation for how molecules are communicating with each other and targeting each other wirelessly at a distance within a living cell; therefore, we must look for some kind of psychic or quantum mechanical explanation if we want a scientific explanation for how this is being done.

Each tRNA molecule is recognized by a tRNA-activating enzyme that binds a specific amino acid to the tRNA, using ATP for energy. These different molecules are recognizing each other and targeting each other wirelessly at a distance. That's physically impossible which means that it has to be taking place at the quantum level or the psyche level.

There are 64 different anti-codons possible for tRNA. One book I was reading stated that 64 different activation enzymes are used to make or charge the 64 different types of tRNA molecules that are possible within a living cell. Again, we have the problem of logistics and target acquisition that we must explain. Someone intelligent has to decide which activation enzyme to use and when to use it so that the correct number of charged tRNA molecules are made and deployed to each of the ten million ribosomes within the cell. Who is handling the logistics for all of this? How are all these different molecules sensing each other and targeting each other wirelessly at a distance? It's being done, so how is it being done? To complicate matters, when it comes to tRNA production and tRNA charging, the live motion videos don't always match with the descriptions in the textbooks. When it comes to tRNA molecules, there's more than meets the eye.

Each tRNA molecule, after it has been made from DNA in the nucleus of the cell, will then target a specific amino acid to transport to a specific ribosome; and, it will find that amino acid docked within an activation enzyme in the cytoplasm of the cell, which means that the tRNA molecule will also have to target and dock with a specific type of activation enzyme. We have molecules targeting and docking with other molecules within a living cell wirelessly at a distance.

Each type of tRNA molecule carries a specific amino acid to a specific ribosome and then docks with a specific mRNA codon being processed by that specific ribosome. Could you handle the logistics and target acquisition for ten million mRNA molecules and billions of tRNA molecules? How would you communicate with them wirelessly at a distance? How would you explain to them where to go and how to get there?

How do the activation enzymes, who are charging the tRNA molecules, as well as the tRNA molecules themselves "read" the mRNA molecule (that the tRNA molecules will be docking with in the future) wirelessly at a distance so that the right number of charged tRNA molecules and the right type of tRNA molecules are produced in advance in the right sequence before they arrive at the ribosome and dock with the mRNA molecule for translation into a protein? Who is handling the logistics for all of this? How would you make tRNA activation enzymes communicate with an mRNA molecule wirelessly at a distance? How would you communicate with an mRNA molecule and dozens of tRNA molecules wirelessly at a distance telling them which specific ribosome to travel towards after they have been produced? It can't be done at the physical level, which means that it's being done at the quantum level or the psyche level. There is no other logical explanation for what we are observing. Target acquisition requires Intelligence or Psyche. It's not going to work any other way. Target acquisition between atoms and molecules is a sure sign of Psyche, or Intelligence, or Conscious Intent.

The tRNA molecules arrive at the correct ribosome in the right sequence, deposit their amino acid at the correct ribosome, and then the discharged tRNA molecules set off in search of another amino acid in order to start the process all over again. The tRNA molecules transport amino acids to a specific ribosome and then dock with a specific mRNA codon being processed by that ribosome. The wireless logistics and target acquisition taking place wirelessly at a distance between tRNA molecules and specific mRNA molecules is off the charts. A living cell is alive with such activity; and, it's all being handled wirelessly at a distance. It's all being handled by Someone Psyche or Someone Intelligent. It's purposeful, deliberate, and intentional. There's nothing random about it. Imagine the inefficiency if the ribosome had to cycle through 64 different anti-codons every time hoping to get lucky and find the correct anti-codon sooner rather than later. There's nothing random about it. The correct tRNA molecule with the correct anti-codon and the correct amino acid arrives at the correct time at the correct ribosome and docks with the correct mRNA codon depositing its correct amino acid into the growing polypeptide chain. It's molecular clockwork. It's logistics at its finest.

The Materialists, Physicalists, Naturalists, Darwinists, Nihilists, and Atheists deliberately ignore and reject the wireless communication taking place at a distance between the tRNA molecule and its targeted mRNA molecule. These people have chosen to believe that Psyche, Telepathy, Telekinesis, and Action at a Distance do not exist. They are wrong. These quantum mechanical processes have been experienced and observed at the molecular level within all the living cells on this planet.

Billions of tRNA molecules within a living cell are actively doing target acquisition and logistics wirelessly at a distance as if they were alive, sentient, perceptive, and self-aware.

The tRNA molecule sings, "Ain't got no strings on me!"

The tRNA molecules are alive with intelligence, purpose, intention, deliberation, motion, choice, target acquisition, logistics, and activity. They are self-aware. They know what they are and what they are supposed to do; and, they proceed to do it.

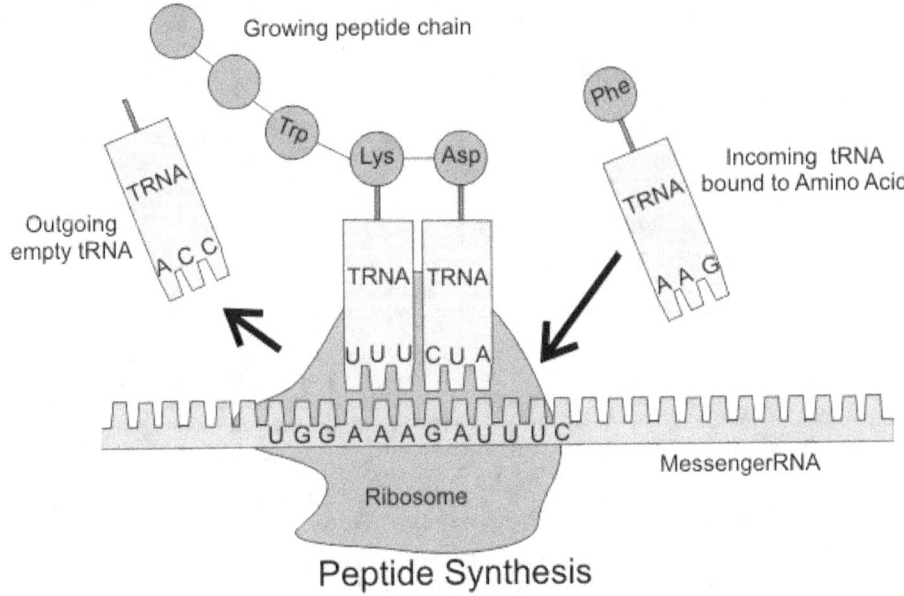

Peptide Synthesis

https://upload.wikimedia.org/wikipedia/commons/0/0f/Peptide_syn.png

https://science-2-0.com/wp-content/uploads/2018/07/Peptide_syn.png

https://en.wikipedia.org/wiki/Transfer_RNA

https://en.wikipedia.org/wiki/Aminoacyl_tRNA_synthetase

The tRNA molecules KNOW in advance precisely where they are needed, and they head there immediately after they are made or recharged. Then, the discharged tRNA molecules go in search of another activation enzyme and amino acid after the discharged tRNA molecule has left its amino acid behind at the ribosome. All of this target acquisition, intelligence, and logistics is taking place wirelessly at a distance within a living cell.

What's making these molecules move or swim through the cytoplasm of the cell? It has to be some kind of invisible, non-physical, immaterial tractor beam or repulsor beam making these molecules move because there is no physical explanation for what we are observing. Invisible non-physical forces and fields are quantum mechanical in nature and origin, not physical.

https://www.youtube.com/watch?v=kmrUzDYAmEI

https://science-2-0.com/wp-content/uploads/2018/07/Protein-synthesis.zip

According to information from this video, the tRNA molecule has to be manufactured within the nucleus; whereas, the activation enzyme that binds an amino acid to a specific type of tRNA molecule resides in the cytoplasm of the cell. So, once again, we observe tRNA molecules and amino acids purposefully targeting and acquiring activation enzymes within the cytoplasm of a living cell wirelessly at a distance. In other words, we observe molecules actively targeting and docking with other molecules as if they can see each other and recognize each other wirelessly or telepathically at a distance. There is no physical explanation for how this is being done, so we must look for a quantum explanation or a psychic explanation.

We observe Intelligence, Psyche, Target Acquisition, Logistics, Wireless Communication at a Distance, and the Hand of God in action whenever we observe tRNA molecules targeting and interacting with other molecules wirelessly at a distance.

Molecules Targeting Molecules Wirelessly at a Distance

Wireless communication at a distance between molecules within a living cell is physically impossible according to the Materialists, Naturalists, Darwinists, Nihilists, and Atheists; yet, it's happening, nonetheless. Someone has gotten their science wrong.

The target acquisition between molecules taking place wirelessly at a distance within a living cell is off the scale. It appears to be infinite and eternal. Kill the cell, though, and it all stops. There's no purpose or reason for it to continue. This is amazing science that has been observed and caught in the act. Molecules are communicating with molecules wirelessly at a distance within all the living cells on this planet. That's physically impossible, but completely possible at the quantum level or the psyche level. This is proof of concept where Psyche or Intelligence is concerned. It's also proof-of-concept where Quantum Mechanics or Supernatural Mechanics is concerned.

Wireless communication between atoms and molecules as well as target acquisition for atoms and molecules has to be handled at the quantum level because it's impossible at the physical level. This is the smoking gun where Psyche, Intelligence, Quantum Consciousness, or Non-Local Consciousness is concerned. There is no physical explanation for how atoms and molecules recognize each other wirelessly at a distance and target each other wirelessly at a distance. Action at a Distance between atoms and molecules is by definition, in principle, a quantum mechanical process and not a physical process. There's no physical explanation for how this is taking place; but, it does take place, nonetheless.

The Materialists, Naturalists, Darwinists, and Atheists want you to believe that only physical matter exists; but, we have all of that non-physical, intangible, immaterial Energy or Psyche to have to deal with. Psyche is Energy. Psyche is the intelligence within energy. Psyche is the fundamental unit of reality. Thoughts and memories are quantum waves. There's a whole non-physical universe out there that the Physicalists and Naturalists choose to ignore and reject. It's all based upon Energy, and Psyche is the innate inherent intelligence within that Energy. Energy is psychic, intelligent, syntropic, conscious, teleological, perceptive, deliberate, intentional, purposeful, and aware. Energy or Psyche is conserved, which means that it can neither be created nor destroyed, which means that it is eternal and everlasting without a beginning of days or an end of years. These realities and truths explain everything that we have ever experienced and observed. Remember, Psyche or Energy is the fundamental unit of reality. This is what has been experienced and observed.

Some of the scientists have claimed that there's nothing fundamental about energy. They may be right. In fact, the Psyche or Intelligence or Consciousness who commands and controls energy might in fact be the fundamental unit of reality, resulting in a Psyche Ontology. It has been observed that the smaller can reside within and control the larger. If energy is represented by vibrating strings; then, Psyche or Intelligence would be the vibration in the string or the thing making the string vibrate at a specific frequency. Thought waves might in fact be the fundamental unit of reality. As I see it though, it all comes down to Psyche, Thought, Consciousness, Awareness, and Intelligence. Energy doesn't spontaneously form into higher levels of order. Instead, Someone Psyche makes that energy change its form or makes that energy conform. In a very real sense, this would make Choice the fundamental unit of reality. Choice would then determine everything else, including the different forms of energy.

Transcription Factors Targeting Specific Genes Wirelessly at a Distance

In molecular biology, a transcription factor (TF or sequence-specific DNA-binding factor) is a protein that controls the rate of transcription of genetic information from DNA to messenger RNA, by binding to a specific DNA sequence. The function of TFs is to regulate – turn on and off – genes in order to make sure that they are expressed in the right cell at the right time and in the right amount throughout the life of the cell and the organism. Groups of TFs function in a coordinated fashion to direct cell division, cell growth, and cell death throughout life; cell migration and organization (body plan) during embryonic development; and intermittently in response to signals from outside the cell, such as a hormone. There are up to 2600 TFs in the human genome. TFs work alone or with other proteins in a complex, by promoting (as an activator), or blocking (as a repressor) the recruitment of RNA polymerase (the enzyme that performs the transcription of genetic information from DNA to RNA) to specific genes.

Source: https://en.wikipedia.org/wiki/Transcription_factor

https://science-2-0.com/wp-content/uploads/2018/07/Transcription-Factor.pdf

Transcription factors bind to a specific DNA sequence. In other words, transcription factors are doing target acquisition wirelessly at a distance within the nucleus of a cell. Transcription factors are doing recruitment of other molecules wirelessly at a distance! Transcription factors also coordinate with each other and other molecules within the cell wirelessly at a distance. Action at a Distance is a quantum mechanical process or phenomenon – not a physical process.

According to this description from the Wikipedia, transcription factors contain or have access to an immaterial, non-physical, non-local, transdimensional, spiritual, "quantum body plan" or "quantum map" separate from the genome itself. Transcription factors have their own source of information other than the genome. Transcription factors know when to turn genes on and when to turn genes off. The transcription factors are running and controlling the genes, not the other way around. This is heavy duty science that falsifies the claims of the Naturalists and Darwinists. These people claim that our genes, and therefore evolution and natural selection, are running the show. These people are wrong. Transcription factors are running the show; but, who is running the transcription factors?

Who is telling the transcription factors when to act and when to refrain? These are molecules. Someone Psyche is communicating with them wirelessly at a distance telling them where to go and when to act. Could you handle the logistics for 2,600 different types of transcription factors? Would you know when they should repress a specific gene or activate a specific gene? Would you know where that gene is located and what it does? During cell differentiation, cell replication, synaptogenesis, manufacture of a physical brain, and the development of a physical body, would you know which genes need to be turned on, which genes need to be turned off, and when to do so? Transcription factors know wirelessly at a distance. Someone is commanding them and controlling them wirelessly at a distance. Transcription factors and transcription enzymes (RNA Polymerases) have access to logistics, information, quantum maps, quantum scheduling, and 3D quantum blueprints that are not contained within the genome itself.

Clearly, Someone Psyche or Someone Intelligent is controlling and running the transcription factors because random activation and random deactivation of genes simply isn't going to work. The Physicalists, Naturalists, Darwinists, and Atheists tell us that all of this is happening randomly through blind luck or random chance under the command and control of evolution and natural selection. They wrong. There's no way physically possible for evolution or natural selection to reach into a transcription factor telepathically at a distance and tell that transcription factor what to do and when to do it.

These transcription factors are running off a non-physical non-local template, blueprint, or schedule that isn't stored within the genome. The transcription factors are turning genes on and off, which means that someone else is running them besides the genes. Transcription factors are also doing target acquisition wirelessly at a distance within a living cell.

Transcription factors are another smoking gun where Psyche or Intelligence is concerned. Think about it. Transcription factors are obviously following a timing schedule as well as a blueprint or a map that isn't contained within the genome. Transcription factors are deciding when to turn specific genes off, and when to turn them back on. That's powerful command and control. Command and control of any kind requires Intelligence as well as timing, scheduling, and logistics. Transcription factors are molecules; but clearly, they are intelligent, sentient, aware, perceptive, conscious, and alive. That's physically impossible but completely possible at the quantum level or the psyche level.

Some people look at all of this evidence and choose to see the Hand of God in all of it. Others look at the same evidence and choose to see nothing but random processes and blind luck. It's a choice. The Human Psyche makes choices; and, that's why we are here on this earth – to make choices and decide for ourselves what we want to believe and what we want to ridicule and reject.

Transcription Enzymes Targeting Specific Genes Wirelessly at a Distance

There are at least three different types of transcription enzymes. Typically, they are called RNA Polymerases. Transcription requires wireless target acquisition at a distance within the nucleus of the cell. Someone Psyche or Someone Intelligent has to tell a Transcription Factor and an RNA Polymerase which specific protein is needed by the cell, which specific gene will produce that needed protein, and where that specific gene is located within the genome. The intelligence, logistics, wireless communication, mapping, and scheduling for all of this is off the charts. It has been estimated that there are between 2 million and 4 million proteins within a living cell. Would you know which gene produces them and where that gene is currently located in the genome? Somehow the Transcription Factors and RNA Polymerases know, wirelessly at a distance. Would you know which protein the cell currently needs and where the matching gene for that protein is located in the genome? Somehow the Transcription Factors and RNA Polymerases know, wirelessly at a distance.

Wireless communication within a living cell, Action at a Distance within a living cell, telekinesis or the movement of molecules within a living cell, super intelligence within a living cell, target acquisition and logistics within a living cell, quantum mapping or quantum blueprints of a living cell, scheduling and timing within a living cell, as well as foresight and planning and teleology within a living cell are all quantum mechanical or supernatural processes – not physical processes. There's no physical explanation for how atoms and molecules are communicating with each other wirelessly at a distance within a living cell through invisible, intangible, non-physical quantum waves; therefore, we must look to Quantum Mechanics and Psyche if we want a scientific explanation for these phenomena.

https://science-2-0.com/wp-content/uploads/2018/07/Transcription.pdf

https://science-2-0.com/wp-content/uploads/2018/07/Eukaryotic-Transcription.pdf

https://science-2-0.com/wp-content/uploads/2018/07/Post-Transcriptional-Modification.pdf

https://science-2-0.com/wp-content/uploads/2018/07/RNA-Polymerase-I.pdf

https://science-2-0.com/wp-content/uploads/2018/07/RNA-Polymerase-II.pdf

https://science-2-0.com/wp-content/uploads/2018/07/RNA-Polymerase-III.pdf

The Materialists, Naturalists, Darwinists, Nihilists, and Atheists will tell you that I got my science wrong and that all of this is being handled randomly by evolution and natural selection. However, all of the target acquisition and logistics and scheduling taking place wirelessly at a distance within every living cell on this planet convinced me that the Materialists, Naturalists, and Atheists are wrong. I used to be a Materialist, Naturalist, Nihilist, and Atheist until the scientific evidence convinced me that I was wrong.

I'm not going to fight it any longer. I'm going with the scientific evidence on this one, not the wishful thinking and philosophical speculation of the Materialists, Naturalists, Nihilists, and Atheists. I finally realized that we atheists don't know what we're talking about after all.

Wireless Communication at a Distance

How does a wife communicate wirelessly at a distance with her husband telling him where to go and what to do when he gets there? Obviously, it requires some kind of common language. It requires sound waves, visible light, or some kind of wavy scribbles on a piece of paper for a wife to communicate her demands to her husband. It also requires some kind of purpose, reason, desire, or need on the part of the wife. "Command and Control" requires intelligence, consciousness, awareness, perception of a need, language, meaning, purpose, teleology, thoughts, desires, and some kind of immaterial non-physical quantum mechanism for transmitting the message from one physical entity to another; otherwise, we end up with random chaos.

How are atoms and molecules communicating with each other wirelessly at a distance within a living cell? Obviously, it requires some kind of common language, psyche or intelligence, quantum waves or sub-atomic waves, and a reason or a purpose for communicating.

Transfer RNA (tRNA) with its unique anti-codon is charged or attached to the correct amino acid by a specific type of activation enzyme. How would you communicate wirelessly at a distance with specific amino acids, tRNA molecules, and the associated anti-codons telling them where to go and what activation enzyme to interact with when they get there, where that activation enzyme is currently located, and then telling the charged tRNA molecules with foresight in advance which mRNA molecule and ribosome to target and dock with? How would you handle the communications and logistics for all of this wirelessly at a distance? How would you make the amino acids, tRNA molecules with their anti-codons, and activation enzymes swim or move through the cytoplasm of a living cell to a specific location at a specific time? How would you get specific anti-codons on tRNA and specific amino acids to actively target and acquire specific activation enzymes, or vice versa? How would you handle target acquisition between molecules within a living cell?

Genes are used to produce proteins. Used by whom? Who is using them, and how is this person communicating to the genes, transcription factors, and transcription enzymes (RNA polymerases) that a specific protein needs to be made? Transcription factors and transcription enzymes target specific genes. How would you handle target acquisition for each and every transcription enzyme and transcription factor within a living cell? Would you be able to tell the transcription factors and transcription enzymes what gene produces which specific protein, and where that gene is located within the genome? How would you communicate with genes, proteins, enzymes, molecules, and atoms wirelessly at a distance if I were to give you the task of doing so?

How do atoms communicate with each other wirelessly at a distance? The chemical bonding of atoms into molecules happens wirelessly at a distance; and, the logistics for all of this only gets more complex from there onward.

How would you tell a newly made protein wirelessly at a distance where to go and what to become when it gets there? A newly made protein can go anywhere within a cell, so how do you tell it wirelessly at a distance where to go and what to do or become when it gets there? How do you tell a protein where it's currently needed and how to interact with other proteins when it gets there? How would you handle targeting, movement, and docking when it comes to atoms and molecules? How would you communicate with atoms and molecules? How do you make a protein or enzyme swim or move through the cytoplasm of a living cell to a specific destination? How would you communicate with a protein or enzyme wirelessly at a distance telling it where its target location is currently located and how to get there? There are 100 trillion atoms within a living cell. Could you handle the logistics and the target acquisition for the trillions of atoms, amino acids, enzymes, and proteins within a single living cell? Someone can – someone much more intelligent, omniscient, and omnipresent than we currently are.

Physical atoms and molecules within a living cell do target acquisition wirelessly at a distance. That's physically impossible – they are atoms and molecules after all. Targeting and acquisition require intelligent command and control; otherwise, random chaos is the result. Some kind of invisible, immaterial, non-physical force or field is required to make atoms and molecules move telekinetically through a living cell towards a specific destination in 3D space. Some kind of invisible, intangible, non-physical force or field is needed to communicate with atoms and molecules telepathically or psychically at a distance through space telling them where to go, how to get there, and what to do when they get there. Intangible, invisible, non-physical chemical bonds are needed to get atoms to bind with atoms to form molecules, to get amino acids to bind with amino acids to form proteins, and to get proteins to bind with other proteins in order to produce cell membranes and cellular machinery.

It has been observed that the very same atoms and molecules that do absolutely nothing within a beaker will literally come alive with purpose and activity and motion within a living cell. How's that possible? Who is telling them where to go and what to do when they get there? Who is handling the logistics or the "command and control" for all of this? How is all of this information being communicated between the molecules and the atoms within a living cell wirelessly at a distance? How would you make atoms and molecules do target acquisition wirelessly from a distance? How would you communicate with atoms and molecules and make them do what you want them to do?

Communication between atoms and molecules within a living cell wirelessly at a distance is where we see the hand of God in action within a living cell. It's where we catch Nature's Psyche in action. Meaning, purpose, deliberation, choice, wireless telepathic information exchange, stored information within a genome, use of information within proteins and enzymes, logistics, targeting, docking, deliberate interaction between atoms and molecules, motion, manufacturing, and construction are where we see the signature of God or the handiwork of God within a living cell.

Communication between atoms and molecules within a living cell requires Superior Intelligence – an intelligence or a psyche which the Materialists, Naturalists, Darwinists, Nihilists, Behaviorists, Determinists, and Atheists assure us does not exist. Wireless communication at a distance between atoms and molecules within a living cell requires deliberation, planning, purpose, meaning, intention, teleology, foresight, and intelligence.

Clearly, someone has gotten their science wrong. If the atoms and molecules within a living cell are NOT communicating with each other wirelessly at a distance, then it should be nothing but random chaos within a living cell just as it is outside of a living cell. There should be absolutely no difference between the insides of a living cell and the outsides of a living cell if the Physicalists, Naturalists, Darwinists, Nihilists, and Atheists are right and have gotten their science correct.

Obviously, though, the Materialists, Naturalists, Darwinists, Nihilists, and Atheists are wrong. These people are always wrong when it comes to science and the wireless communication between atoms and molecules that is taking place at a distance within a living cell. The Materialists, Naturalists, Darwinists, Nihilists, and Atheists assure as that wireless action at a distance and psyche do not exist. They are wrong. The observational evidence tells us that they are wrong.

By definition in principle, there is no physical mechanism in place for making atoms and molecules communicate with each other wirelessly at a distance at the physical level. It requires some kind of non-physical, intangible, quantum force or wave to get atoms and molecules to communicate with each other and interact with each other within a living cell wirelessly at a distance. Action at a Distance or Communication at a Distance is a quantum mechanical phenomenon. Quantum waves are WiFi at the spiritual level, the psychic level, the transdimensional level, or the quantum level. Action at a Distance, Telepathy, and Transpersonal Communication between atoms and molecules are psychic processes or quantum mechanical processes – not physical processes.

Clearly, communication between atoms and molecules is being done within a living cell; so, our task as scientists is to figure out how it is being done and by whom. The task of a scientist is to find the truth – not to make up convenient fictions in an attempt to explain away the facts, experiences, and observations of the human race. Science is observation and experience – not philosophical speculation and wishful thinking.

Defining Psyche

Law of Psyche: The Law of Psyche states that every massless quantum particle is some type of Psyche or Intelligence. Psyche is the innate intelligence within all the different forms of energy which gives that energy the inherent ability to understand, follow, and obey God's Laws and God's Commands. Every massless elementary particle is some type of Psyche because it is psychic, quantum, telepathic, makes choices, can self-generate or self-propagate, can think, can communicate, can transmute, can move, can phase-shift, can collapse its own wave function, is perceptive, is sentient, is omniscient, is omnipresent, is intelligent, is conscious, is alive, must travel at the speed-of-light from our perspective at the physical level, is prescient, and without any physical limitations imposed upon it can quantum tunnel at will in its native original environment at the quantum level. Remember, every massless quantum particle is some type of Psyche, or Intelligence, or Life Force. Psyche is seen by Out-of-Body Travelers as a pinpoint of light or a Photon. Each Photon is a type of Psyche or Intelligence. A Psyche is a Photon, a massless elementary particle. If it looks like a Photon and acts like a Photon, then it is some type of Psyche. Every massless quantum is a Psyche. Psyche USES mass, energy, or matter to get things done. This is what has been experienced and observed.

Psyche has been hiding in plain sight all the time where nobody can see it nor find it because they aren't looking for it. But, there's no great mystery there. Psyche is seen as a Photon BOTH at the quantum level and the physical level. Psyche or Intelligence IS what it appears to be, a Photon of Light or a Pinpoint of Light. No surprise there. Psyche is what it has been experienced and observed to be – a massless quantum of energy that looks like and acts like a Photon. Psyche, whenever it has been seen both at the quantum level and the physical level, looks like a Photon or a Pinpoint of Light.

Atoms and molecules are different forms of energy. Physical matter is comprised of different forms of energy. Communication between atoms and molecules wirelessly at a distance at the quantum level within a living cell is precisely where we catch Psyche, Syntropy, Intelligence, Conscious Intention, Awareness, Choice, Teleology, or some type of deliberate purposeful Life Force in the act because there is no physical explanation for how atoms and molecules are communicating with each other wirelessly at a distance at the physical level.

Target acquisition between atoms and molecules requires intelligence. Psyche is intelligence. Psyche is also Energy. In fact, Psyche is the intelligence within all the different forms of energy. Some of that energy has been set aside to ACT, and some of that energy has been set aside to be acted upon. Psyche or Intelligence is the type of energy that ACTS, and matter of any kind is the type of energy that is acted upon or the type of energy that was designed to react to psyche's commands.

Psychic telepathic communication between atoms and molecules wirelessly at a distance at the quantum level within a living cell is precisely where we catch science, observation, and experience in the act; and, it's precisely what the Materialists, Naturalists, Darwinists, Nihilists, and Atheists assure does not exist. Are these people right, or have they gotten their science wrong?

Science is observation and experience. We know what the observations and experiences are telling us must be true. I'm going to go with that. I'm going to go with the science on this one, and not the philosophical speculation and wishful thinking of the Physicalists, Naturalists, Darwinists, and Atheists because these people really don't know what they are talking about at the most fundamental level.

Target acquisition between atoms and molecules within a living cell is where we catch Psyche or Intelligence in the act. Somehow all of this logistical information is being communicated between the atoms and the molecules within a living cell wirelessly at a distance. By definition, in principle, target acquisition between atoms and molecules is physically impossible, which means that we need a quantum mechanical or psychic explanation for the target acquisition between molecules and atoms that's taking place wirelessly at a distance within a living cell. Action at a Distance, or transpersonal communication, or telepathy, or telekinesis is a spiritual (quantum mechanical) phenomenon, not a physical phenomenon.

The Gods created entropy and random diffusion to prevent target acquisition from happening naturally outside of a living cell. In other words, the Gods created entropy and random diffusion in order to make spontaneous generation, target acquisition, chemical evolution, or macro-evolution physically impossible outside of a living cell. Atoms and molecules function differently outside of a living cell than they do within a living cell. This is what has been experienced and observed. Our task as scientists is to explain all of this scientifically which involves and requires the use of Quantum Mechanics, Action at a Distance, Supernatural Mechanisms, Intelligence, Intention, Teleology, and Psyche if we want to succeed.

By allowing Quantum Mechanics and Psyche in to play, we can literally explain everything that comes our way. With Quantum Mechanics and Psyche, explanatory power approaches infinity. In contrast, we observe that Materialism, Naturalism, Darwinism, Nihilism, Behaviorism, Determinism, and Atheism were designed to prevent us from discovering Quantum Mechanics, Supernatural Mechanisms, Action at a Distance, Transpersonal Communication, Telepathy, Telekinesis, Quantum Tunneling, Psyche, Choice, Syntropy, and God. Physicalism, Materialism, Naturalism, and Atheism were designed to prevent us from discovering science, knowledge, and truth. This is what we have experienced and observed.

The truth is whatever has been repeatedly experienced and observed. Knowledge consists of what has been experienced and observed. Science is observation and experience. Materialism, Naturalism, Darwinism, and Atheism were designed to prevent us from discovering the truth. These falsified philosophies are anti-science because they are against observations and personal experiences that are of a non-physical nature and origin. The Materialists, Naturalists, Darwinists, Nihilists, and Atheists actively and deliberately hide, ridicule, reject, and destroy any observational evidence that falsifies their chosen point-of-view. This is what has been experienced and observed.

Only Psyche or Intelligence Can Do Target Acquisition

ONLY Psyche or Intelligence can do target acquisition at the quantum level between atoms and molecules. Psyche or Intelligence has been experienced and observed. Psyche or Non-Local Consciousness is experienced first-hand by Out-of-Body Explorers as an immaterial viewpoint in space. Whenever someone goes looking for their own Human Psyche while out-of-body on the astral plane, there's nothing there to be found because they ARE their Psyche. Your spirit body and physical body don't have experiences and don't form memories of events while your Human Psyche is separated from them. It's your Human Psyche who experiences events and then forms memories of those events – both at the quantum level and the physical level. This is what has been experienced.

Psyche is seen or observed by Out-of-Body Travelers as a pinpoint of light, a point particle of light, a spark of light, or a Photon. A Photon is one type of Psyche or Intelligence. This is what has been observed. Science is observation and experience. Remember, Photons are Psyches or Intelligences; and, Photons have already been experienced and observed on BOTH sides of the veil – both on the spiritual side and the physical side. Furthermore, each massless quantum particle or massless elementary particle that is discovered ACTS like a Photon and is therefore also a type of Psyche or Intelligence. Remember, ONLY Psyche or Intelligence can do target acquisition.

Mark My Words

—

Source

BioPsychoSocial: Including Psyche or Light into our Theoretical Models

https://www.amazon.com/dp/B0713NDHVW

Reference Material

Protein Synthesis

https://www.youtube.com/watch?v=2BwWavExcFI

https://science-2-0.com/wp-content/uploads/2018/07/DNA-Transcription-and-Translation.zip

https://www.youtube.com/watch?v=kmrUzDYAmEI

https://science-2-0.com/wp-content/uploads/2018/07/Protein-synthesis.zip

https://www.youtube.com/watch?v=gG7uCskUOrA

https://www.youtube.com/watch?v=bKIpDtJdK8Q

https://www.youtube.com/watch?v=D3fOXt4MrOM

Molecular Machines

https://www.youtube.com/watch?v=WsofH466lqk

https://www.youtube.com/watch?v=WFCvkkDSfIU

https://www.youtube.com/watch?v=zm-3kovWpNQ

https://www.youtube.com/watch?v=X_tYrnv_o6A

https://www.youtube.com/watch?v=39UKSfsc9Z0

https://www.youtube.com/watch?v=rdF3mnyS1p0

https://www.youtube.com/watch?v=A2my52zQA6k

https://www.youtube.com/watch?v=b_cp8MsnZFA

https://www.youtube.com/watch?v=GM9buhWJjlA

Rana, F. (2008). *The Cell's Design: How Chemistry Reveals the Creator's Artistry*. Grand Rapids, MI: Baker Books.

Rana, F. (2011). *Creating Life in the Lab*. Grand Rapids, MI: Baker Books.

In the Beginning There Was Syntropy

Do you believe that the Big Bang happened?

The majority of our scientists have chosen to believe that the Big Bang happened and that the Big Bang Model is true. In fact, it has been claimed at times that there is a 100% consensus among scientists that the Big Bang is real and that the Big Bang happened. Are you one of them? Your chosen beliefs matter.

https://www.space.com/8066-big-bang-solid-theory-mysteries-remain.html

By the way, that claim is posturing, but it has been made, nonetheless. It has been stated that 99% or 100% of our scientists are Big Bangers. In reality, it's impossible for every scientist to believe in the Big Bang Model because human beings don't work that way. When have you ever known human beings to be in complete consensus or complete unity? It doesn't happen where physical fallen mortal human beings are concerned. Like everything else in science, the Big Bang is in dispute. Nevertheless, a lot of scientists believe that the Big Bang happened, and that Big Bang Theory is real and true.

Your beliefs are chosen into existence by your Psyche, on both sides of the veil. Our chosen beliefs have consequences and ramifications, or they should. What are some of the ramifications of a belief in the Big Bang Theory?

Scientific Observation: Events have causes. Events are caused. Anything that obviously began obviously has Someone Intelligent who made it or caused it to begin. This premise is obviously true. This is Science 101. It has been experienced and observed.

Scientific Observation: Our physical universe and physical matter obviously began. The associated entropy obviously began. The associated physical laws, physical limitations, and physical constants obviously began. Planets, stars, and galaxies began.

Scientific Conclusion: Therefore, it is logical and rational to conclude that physical matter, this physical universe, the physical laws, the physical constants, and entropy had Someone Intelligent or Someone Psyche who caused them to begin.

Scientific Corollary: If you have chosen to believe that our physical universe and physical matter began with a Big Bang, then it should be obvious to you that Someone Intelligent or Someone Psyche pushed the button that made it go bang. If the Big Bang truly happened, we KNOW that someone triggered it and made it go bang. There's NO such thing as spontaneous generation or creation ex nihilo. Creation ex nihilo or spontaneous generation has NEVER been experienced nor observed. Spontaneous generation has been falsified by science; and, creation ex nihilo violates the first law of thermodynamics or the conservation of energy and mass. Every scientist should know this to be true when they are thinking logically and rationally.

Discussion of the Big Bang Model – Science or Fiction:

https://www.youtube.com/watch?v=PwodpHGfI4k

https://www.youtube.com/watch?v=gufYmnj0Gjw

https://www.youtube.com/watch?v=dPTcnZXN8WY

The Big Bang Singularity was Pure Syntropy – the complete absence of entropy. It would take a God to make such a thing. If you believe in the Big Bang Model for origins, then you simply KNOW that Someone Psyche or Someone Intelligent triggered it and made it go bang. This is what logical common sense tells us must be true. Of course, there will always be people who will choose to go contrary to logic and common-sense because their chosen agenda demands that they do so. Nevertheless, if you believe that the Big Bang happened, then both logic and common-sense demand that you MUST also believe in God – the Person who prepared it, triggered it, and made it go bang. The Big Bang Singularity wasn't produced by a random fluctuation in entropy. That's physically impossible.

Defining and Discussing Syntropy

Syntropy means without beginning of days or end of years. Syntropy means eternal, immortal, and everlasting. Syntropy is the core construct. Syntropy is Eternal Life.

There are different types of Syntropy. Psyche is Syntropy. Intelligence is Syntropy. Quantum Mechanics is Syntropy. Energy or light is Syntropy. Quantum waves or thoughts are Syntropy. Spirit matter is a type of Syntropy. Eternal Life is Syntropy. These things are eternal and everlasting. They have always existed and will always exist. They are indestructible. They are always conserved.

In contrast, physical matter and the associated entropy had a beginning, which means that they can theoretically have an end. Entropy is death. Entropy is disorder, chaos, random diffusion, and death. The theory of evolution is Creation by Entropy or Creation by Death. Death or entropy cannot design and create. Entropy is restricted exclusively to physical matter. The entropy or the space-time is what distinguishes a particle of spirit matter from a particle of physical matter. The physical limitations or the physical laws and physical constants are what are used to convert a particle of spirit matter into a particle of physical matter.

It's also possible to have a type of has-always-existed syntropy that is represented by a complete absence of entropy but is nonetheless disorganized or chaotic in nature. These realms are spirit matter unorganized, non-consensus realities, or what some people call the Primal Construct or the Chaos Construct. It's eternal and it's everlasting or Syntropic meaning that it has NO entropy and experiences NO passage of time or NO aging process; but, it is chaotic and unorganized and disordered, nonetheless.

Entropy, the passage of time, or the aging process is tied exclusively to physical matter. Everything else is a type of Syntropy – eternal and everlasting. It's the physical matter or the entropy that began, NOT everything else. The classical physics began; whereas, the Quantum Mechanics is eternal and everlasting.

Syntropy is typically defined as being the opposite of entropy. Entropy brings physical things to an end. Entropy is death. Syntropy is the opposite of entropy. Syntropy is life. Syntropy is immortality and eternal life.

In the beginning was syntropy, and the syntropy was good.

In the beginning of what?

Think about it logically.

They are talking about the beginning of physical matter and entropy – the beginning of physical limitations, physical laws, and the physical constants – the beginning of planets, stars, and galaxies.

Think about it logically. Some type of Syntropy MUST exist or all of that subsequent entropy and life wouldn't have been possible in the first place.

Entropy is often defined as a disordering process or a decay process. Entropy or disorder is increasing within this physical universe. This is the second law of thermodynamics. This means that Someone Intelligent had to infuse order or syntropy into the whole thing in the beginning, or all of that subsequent entropy wouldn't have been possible.

At the beginning of our physical universe, Someone Psyche filled the whole thing FULL of syntropy or energy or order; otherwise, there would be NO entropy right now, and our whole physical universe would have been "born" in heat death FULL of entropy to begin with. The very existence of physical matter, the very existence of this physical universe, and the very existence of entropy ARE proof positive that Someone somewhere KNOWS how to do syntropy.

The existence of syntropy or order or potential within physical matter, the existence of our physical universe FULL of syntropy at its beginning, the very existence of life within this physical universe, and the ongoing existence of entropy are Scientific Proof of God's Existence whether we realize it or not.

Syntropy is the KEY to understanding everything; and, it all begins with a realistic understanding of physical matter, entropy, random diffusion, and classical physics. We have to know and understand syntropy's opposite before we can gain some kind of sense as to what syntropy really is. Entropy is obvious and staring us in the face; so, syntropy should be obvious and staring us in the face as well. So, where is it and what is it? Well, start by asking yourself what the opposite of physical matter, entropy, random diffusion, chaos, and classical physics might be.

In the beginning of our physical universe, some 13.8 billion years ago by recent scientific estimates, our physical universe was FULL of syntropy. That's physically impossible, because physical matter does NOT produce syntropy. Instead, physical matter collects or accumulates entropy, disorder, chaos, and random diffusion, according to the second law of thermodynamics and classical physics.

So, how did our physical universe get filled FULL of syntropy, in the beginning? That syntropy had to come from somewhere or someone, or our physical universe would not exist, and you and I would not exist as physical beings. In the beginning, our physical universe was new and FULL of syntropy or potential, so what is syntropy? Syntropy is the opposite of entropy. Syntropy is life. Entropy is death.

Without an infusion of syntropy, order, or potential, our universe would have been "born" at heat death FULL of entropy and would never have done anything useful during its endless existence at absolute zero. That's the way classical physics works. Physical matter moves towards FULL entropy. Everything physical moves towards disorder and death.

So, the real question is, **"Who filled our physical universe FULL of syntropy and life in the beginning?"**

Well, it couldn't have been evolution. Evolution or random mutation is entropy! Furthermore, evolution, random mutations, and natural selection didn't exist yet, at the beginning of our physical universe. In fact, evolution, random mutations, and natural selection didn't begin or start until AFTER God had designed, created, and produced the physical matter, proteins, genes, genomes, eyes, brains, and life forms in the first place. Something that doesn't exist yet, like evolution or random mutation, can't create something that does exist like physical matter, proteins, genes, genomes, and life forms.

"Who filled our physical universe FULL of syntropy in the beginning?"

Obviously, it couldn't have been evolution. The chemical evolution of genes and proteins from atoms is physically impossible because entropy, random diffusion, or classical physics prevents chemical evolution from happening in the physical world. Abiogenesis, spontaneous generation, and all the other forms of macro-evolution are prevented from happening by entropy, random diffusion, and classical physics. Chemical evolution, abiogenesis, spontaneous generation, and macro-evolution are physically impossible. They can't happen, which means that they didn't happen.

Physical matter, once it has been created or produced by the Gods, can only collect entropy – it can't release syntropy or do syntropy.

Since evolution (genetic change) or random mutation is entropy, and since any other type of evolution (chemical evolution or spontaneous generation or macro-evolution) is prevented from happening by entropy, where did ALL the proteins, genes, genomes, and life forms come from?

"Where did the syntropy needed to create proteins, genes, genomes, eyes, brains, and life forms come from, since it couldn't have come from the physical matter itself?"

I can only think of one logical answer to these questions.

The syntropy exists separate from our physical universe; and, the syntropy pre-dates or preceded our physical universe. The syntropy is endless, eternal, and uncreated, meaning that the syntropy has always existed and will always exist. Syntropy is the core construct. Syntropy is consciousness or life.

The Atonement of Christ Is Syntropy

As I continued to study these obvious truths, lots of interesting things were revealed to me.

Dialogue: Do you think it's the atonement itself that causes syntropy?

Nibley: Well, "atonement" means syntropy — bringing back to its former state, restoring to its former state. You see, when something breaks down, it becomes disorganized and fractured. "At one" means unified again — returned to its unity, returned to its former integrity and structure.

https://dialoguejournal.com/wp-content/uploads/sbi/articles/Dialogue_V12N04_12.pdf

The Atonement of Christ is syntropy.

Atonement means to be restored to our former state of existence. We are made at-one with God or restored to our former life. We are made immortal again. We return to the presence of God once again. The Atonement of Christ brings us back to Syntropy or God if we choose to let it do so. We can fight it and resist it and choose to pursue entropy and death and hell instead if we want to; and, many people want to fight against Syntropy and God. There are those who prefer entropy, death, and hell. They prefer the evil and the darkness over the goodness and light. That's just the way it is.

Christ's Resurrection, and our eventual resurrection from the dead, are a function of syntropy.

The Atonement of Christ and Resurrection from the Dead are the opposite of entropy or physical death. That means that the Atonement of Christ and the Resurrection of the Dead are pure syntropy because they are life or an end to death and entropy.

Now we are starting to get somewhere.

So, what else is syntropy?

What's the opposite of physical matter? Since physical matter was designed to collect entropy or to fill full of entropy, then what's the opposite of physical matter? The opposite of physical matter should be syntropy.

The polar opposite of physical matter is spirit matter. Spirit matter is pure syntropy. Spirit matter has NO entropy, meaning that it doesn't age and deteriorate, meaning that it doesn't collect entropy or move towards heat death.

As physical beings, all we know about with our physical senses and physical bodies is entropy or death.

As spirit beings, during our pre-mortal existences, all we knew about was syntropy or life.

Quantum Superposition is the idea that two or more quantum waves or quantum objects can exist in the same space at the same time, as long as they are out of phase with each other.

Psyche, Spirit Matter, and Physical Matter can exist in the same location or the same space at the same time because they are out of phase with each other – each one existing on a different frequency or wave. Combined together, though, in the same space or added together in the same space at the same time, they make up a human being. Someone Psyche links them together or ties them together into a single functional whole, using the silver cords that out-of-body travelers have seen.

The Silver Cord

https://www.youtube.com/watch?v=yzVfPQoWJas

https://science-2-0.com/wp-content/uploads/2018/04/The-Silver-Cord.zip

The spirit body is something completely different than the physical body. And, I have encountered a lot of evidence suggesting that the Psyche or the Intelligence can function separately from the spirit body. It's the Human Psyche who has the experiences and makes the memories, NOT the spirit body, and definitely NOT the physical body. The spirit body and physical body have NO experiences and make NO memories while the Psyche is separated from them. Our thoughts and memories are contained within the Human Psyche, and not the spirit body and not the physical brain; therefore, our thoughts, memories, and personalities survive the death of our physical brain. They also survive separation from our spirit body.

The Psyche, Spirit Body, and Physical Body can exist in the same location or the same space at the same time because they are out of phase with each other – each one existing on a different frequency, wave, or dimension. This is possible due to Quantum Superposition – the ability to be in more than one place at the same time, and the ability for different parts of you to exist in the same space at the same time. The Psyche and Spirit Body can also separate from the Physical Body as needed.

A Spirit or a Ghost is a spirit body and psyche combination. Psyche or Intelligence is syntropic because it is conserved. Psyche cannot die. It is eternal and everlasting. Information about Psyche is hidden in plain sight where the Materialists, Naturalists, Darwinists, Nihilists, and Atheists will NEVER go looking for it because these people refuse to look at the scriptures and revelations where God talks about this topic.

Doctrine and Covenants 93: 29-30, 36:

Man was also in the beginning with God. Intelligence, or the light of truth, was not created or made, neither indeed can be. All truth is independent in that sphere in which God has placed it, to act for itself, as all intelligence also; otherwise there is no existence. The glory of God is intelligence, or, in other words, light and truth.

Abraham 3: 17-25:

There is nothing that the Lord thy God shall take in his heart to do but what he will do it. Howbeit that he made the greater star; as, also, if there be two spirits, and one shall be more intelligent than the other, yet these two spirits [psyches], notwithstanding one is more intelligent than the other, have no beginning; they existed before, they shall have no end, they shall exist after, for they are gnolaum, or eternal. And the Lord said unto me: These two facts do exist, that there are two spirits, one being more intelligent than the other; there shall be another more intelligent than they; I am the Lord thy God, I am more intelligent than they all. I dwell in the midst of them all; I now, therefore, have come down unto thee to declare unto thee the works which my hands have made, wherein my wisdom excelleth them all, for I rule in the heavens above, and in the earth beneath, in all wisdom and prudence, over all the intelligences thine eyes have seen from the beginning; I came down in the beginning in the midst of all the intelligences thou hast seen. Now the Lord had shown unto me, Abraham, the intelligences that were organized before the world was; and among all these there were many of the noble and great ones; and God saw these souls that they were good, and he stood in the midst of them, and he said: These I will make my rulers; for he stood among those that were spirits, and he saw that they were good; and he said unto me: Abraham, thou art one of them; thou wast chosen before thou wast born. And there stood one among them that was like unto God, and he said unto those who were with him: We will go down, for there is space there, and we will take of these materials, and we will make an earth whereon these may dwell; and we will prove them herewith, to see if they will do all things whatsoever the Lord their God shall command them.

How do you organize an Intelligence or a Psyche since it is already eternal and everlasting? You provide a spirit body for it. A spirit body is comprised of organized spirit matter. You can also provide a physical body for it. A physical body is comprised of organized physical matter, which is made from different forms of energy.

Psyche or Intelligence has always existed; and, it will always exist. There is NO existence without Psyche. Out-of-Body Travelers and Near-Death Experiencers have observed that it is their Human Psyche who has experiences and remembers events – not their spirit body, and NOT their physical body. Your Psyche, Intelligence, Memories, and Personality continue to exist long after your physical brain is dead and gone. This is what has been experienced and observed. Science is observation and experience, or it should be.

Doctrine and Covenants 131: 6-8:

It is impossible for a man to be saved in ignorance. There is no such thing as immaterial matter. All spirit is matter, but it is more fine or pure, and can only be discerned by purer eyes; we cannot see it; but when our bodies are purified we shall see that it is all matter.

The Biblical God Jesus Christ has told us that spirit matter and physical matter are the same thing. Spirit matter and physical matter are quantum objects; but, they exist in a different phase, frequency, or dimension. I got the impression that God knows what He's talking about. It's as if He were there.

Spirit matter is syntropic. It's not weighed down by entropy or death. In contrast, entropic physical matter has been phase-shifted, filled with space-time, slowed down to sub-light speeds, and made subject to the time-passage feature or aging feature that we call entropy. Spirit matter and physical matter are the same thing, just in a different phase of existence. They both are different forms of energy. Spirit matter is eternal and everlasting – it never ages, never dies, and never ceases to exist. Spirit matter has always existed and will always exist. Entropic physical matter is spirit matter that has been filled with space-time and the ability to experience the passage of time, aging, entropy, and death. Remove the entropy, or the death, or the physical limitations from physical matter, and it will become spirit matter once again.

So, what else is syntropy?

Well, it's obvious that classical physics and random diffusion are entropy, so what's the opposite of classical physics and random diffusion?

The opposite of random diffusion or chaos is order and life. Quantum Mechanics is the opposite of classical physics and entropy, which means that Quantum Mechanics is pure syntropy. Quantum Mechanics is eternal, endless, without beginning or end. Quantum Mechanics is the Priesthood Power of God. Quantum Mechanics is Pure Syntropy.

What do we know about the Priesthood Power of God or Quantum Mechanics?

Only what the Biblical God has revealed to us. If we truly want to know what the Priesthood Power of God is, then we have to go where the Biblical God talks about it in order to find out what it is.

Hebrews 7: 3: [The Priesthood:] **Without father, without mother, without descent, having neither beginning of days, nor end of life; but made like unto the Son of God; abideth a priest continually.**

The Priesthood of God or Quantum Mechanics is without father or mother, having neither the beginning of days nor the end of life, but is made like unto the Son of God, the Biblical God Jesus Christ or Christ Jehovah.

JST, Hebrews 7: 3: **For this Melchizedek was ordained a priest after the order of the Son of God, which order was without father, without mother, without descent, having neither beginning of days, nor end of life. And all those who are**

ordained unto this priesthood are made like unto the Son of God, abiding a priest continually.

It's this Priesthood Power or Quantum Mechanics that makes you like a God, like unto the Son of God, the Biblical God Jesus Christ. The Gods confer a fullness of this Priesthood Power on those that the Gods have learned to trust.

The Priesthood is an everlasting principle, and existed with God from eternity, and will to eternity, without beginning of days or end of years. The keys have to be brought from heaven whenever the Gospel is sent. When they are revealed from heaven, it is by Adam's authority.

https://www.lds.org/manual/teachings-joseph-smith/chapter-8?lang=eng

The Priesthood or Quantum Mechanics is without beginning of days or end of years.

Moses 6: 64-68:

And it came to pass, when the Lord had spoken with Adam, our father, that Adam cried unto the Lord, and he was caught away by the Spirit of the Lord, and was carried down into the water, and was laid under the water, and was brought forth out of the water. And thus he was baptized, and the Spirit of God descended upon him, and thus he was born of the Spirit, and became quickened in the inner man. And he heard a voice out of heaven, saying: Thou art baptized with fire, and with the Holy Ghost. This is the record of the Father, and the Son, from henceforth and forever; and thou art after the order of him who was without beginning of days or end of years, from all eternity to all eternity. Behold, thou art one in me, a son of God; and thus, may all become my sons. Amen.

The Priesthood Power of God or Quantum Mechanics is without beginning of days or end of years. It has always existed and will always exist, because it is pure syntropy. The Biblical God has been teaching us about the Priesthood Power of God or Quantum Mechanics for millennia; but, few of us have been listening.

Alma 13: 7-9:

This high priesthood being after the order of his Son, which order was from the foundation of the world; or in other words, being without beginning of days or end of years, being prepared from eternity to all eternity, according to his foreknowledge of all things — now they were ordained after this manner — being called with a holy calling, and ordained with a holy ordinance, and taking upon them the high priesthood of the holy order, which calling, and ordinance, and high priesthood, is without beginning or end — thus they become high priests forever, after the order of the Son, the Only Begotten of the Father, who is without beginning of days or end of years, who is full of grace, equity, and truth. And thus it is. Amen.

The Priesthood or Quantum Mechanics is without beginning of days or end of years, being prepared from eternity to all eternity. It's this Priesthood Power that makes you without beginning of days or an end of years. It's this Priesthood Power, a fullness of this Priesthood Power, which makes you into a God.

Doctrine and Covenants 84: 17-21:

Which priesthood continueth in the church of God in all generations and is without beginning of days or end of years. And the Lord confirmed a priesthood also upon Aaron and his seed, throughout all their generations, which priesthood

also continueth and abideth forever with the priesthood which is after the holiest order of God. And this greater priesthood administereth the gospel and holdeth the key of the mysteries of the kingdom, even the key of the knowledge of God. Therefore, in the ordinances thereof, the power of godliness is manifest. And without the ordinances thereof, and the authority of the priesthood, the power of godliness is not manifest unto men in the flesh; for without this no man can see the face of God, even the Father, and live.

The Priesthood or the Fullness of Quantum Mechanics continues in the Church of God from one generation to the next. No matter which generation you find yourself living in, all you have got to do is to find the True Church of God, the True Church of Jesus Christ, in order to find the Priesthood of God. The Greater Priesthood or the Fullness of Quantum Mechanics is called the Key of the Knowledge of God, the Ordinances of God, the Power of Godliness, and the Key of the Mysteries of the Kingdom. The Priesthood or Quantum Mechanics abides forever, and it is without beginning of days or end of years. This Priesthood or Quantum Mechanics administers the Gospel of Christ and the Church of Jesus Christ.

Moses 1: 1-5:

The words of God, which he spake unto Moses at a time when Moses was caught up into an exceedingly high mountain, and he saw God face to face, and he talked with him, and the glory of God was upon Moses; therefore, Moses could endure his presence. And God spake unto Moses, saying: Behold, I am the Lord God Almighty, and Endless is my name; for I am without beginning of days or end of years; and is not this endless? And, behold, thou art my son; wherefore look, and I will show thee the workmanship of mine hands; but not all, for my works are without end, and also my words, for they never cease. Wherefore, no man can behold all my works, except he behold all my glory; and no man can behold all my glory, and afterwards remain in the flesh on the earth.

Are you starting to see a pattern here?

I started to notice the trend while doing research into Quantum Neuroscience. He's God because He has a fullness of this Priesthood Power or holds all the KEYS to Quantum Mechanics. His power and glory are a fullness of Quantum Mechanics. Quantum Mechanics has NO physical limitations, because Quantum Mechanics or the Priesthood of God is pure syntropy.

I talk about Quantum Mechanics is much greater detail in the following book.

Quantum Neuroscience: The Answer to Life, the Universe, and Everything:

https://www.amazon.com/dp/B079Z6QQQB

Quantum Mechanics is syntropy, or the Priesthood Power of God. The Biblical God Jesus Christ and His Father hold all the KEYS to this Priesthood Power; and, they only delegate that power and authority to those of their choosing. That is as it should be. You don't want to give omnipotence to someone like a Militant Atheist, the Christian Inquisition, a Muslim Terrorist, the Mossad, the CIA, the Mad Scientists, Hitler, Stalin, or Satan, because they would only use that power to destroy everything around them. God and Christ use this power to create and give life. The primary reason we are here on this planet being subjected to physical limitations is to give each one of the chance or opportunity to decide for himself or herself whether he or she is going to be a Hitler or a Christ. The whole purpose of this mortal life is to decide for yourself whether you want to be a Devil or a Saint. The choice is yours.

Do you find any of this as fascinating as I do?

It's ALL there. It's consistent. It fits. It makes logical sense. Its explanatory power is through the roof! Are you starting to see what's possible by choosing to allow ALL of the evidence into evidence? Since entropy is obvious and staring us in the face, syntropy should also be obvious and should also be staring us in the face just as much. It is! Syntropy is Quantum Mechanics or the Priesthood Power of God. It's been staring us in the face all the way along, and most of us have been trying to ignore it and pretending that it doesn't exist. The very existence of Quantum Mechanics or the Priesthood Power of God FALSIFIES Materialism, Naturalism, Darwinism, Nihilism, Behaviorism, Determinism, and Atheism.

Who Wound Up the Clock in the First Place?

When our physical universe was created and brought into existence, it was FULL of syntropy or full of potential. Thereafter, the entropy or the passage of time started. Things started to age, run-down, and die. Who wound up the clock in the first place? That's the question we should be asking, isn't it? That's the question that they are avoiding, isn't it?

As scientists, we are obligated to try to find an explanation for the reversal of entropy that took place at the beginning of our physical universe. We are obligated to try to explain where that initial syntropy came from and how it might have gotten there. It's a legitimate scientific question. Who or what wound up the clock? Scientists are supposed to ask difficult questions and then try to pursue logical answers to those questions. Who reversed the entropy? Who wound up the clock? How was it done?

These questions can only be answered by choosing to allow ALL of the evidence into evidence, and then by pursuing a preponderance of that evidence. That's precisely what I'm doing here. I'm using ALL of the evidence that I have on hand to answer the question.

This is the way that science should be done, but typically isn't done.

The way that the Materialists, Naturalists, Darwinists, Nihilists, and Atheists do science is by ridiculing, mocking, banning, censoring, and blocking evidence. Observe for yourself if it isn't so. What will they do with the evidence that I presented here? You guessed it. Some of them even go as far as to destroy the evidence. Materialism, Naturalism, Darwinism, and Atheism are based upon a refusal to look at evidence, because ALL the evidence we have on hand as a race FALSIFIES Materialism, Naturalism, and their derivatives.

In contrast, Science 2.0 allows ALL of the evidence into evidence and then pursues a preponderance of the evidence.

Science 2.0: I Upgraded My Science:

https://www.amazon.com/dp/B0771K6WTX

https://www.amazon.com/author/science

I have observed that Quantum Mechanics and the Priesthood Power of God are the same thing; and, it has become clear to me that Quantum Mechanics or the Priesthood is syntropy – the opposite of entropy. The Atonement of Christ and our eventual Resurrection from the Dead are based upon syntropy, or quantum mechanisms, or the Priesthood Power of God.

Do you see what becomes possible by choosing to allow ALL of the evidence into evidence? The explanatory power of your science goes through the roof! There are no limits to what you can explain because you have deliberately removed ALL of the physical limitations or materialistic limitations from the equation. In fact, under Science 2.0, we actually allow what's wrong with Materialism, Naturalism, Darwinism, Nihilism, and Atheism to point us directly at the Truth. Powerful, is it not?

I've become a big fan of Science 2.0 and choosing to allow ALL of the evidence into evidence. I'm now answering questions that I couldn't answer before. In fact, I'm now asking questions that I refused to ask before, because I KNOW that I can answer them now.

I used to be a Materialist, Naturalist, Nihilist, and Atheist. I used to be an active proponent and practitioner of Scientism. All I can say now is that Science 2.0 is infinitely better, because the explanatory power of Science 2.0, Quantum Mechanics, Syntropy, and Quantum Neuroscience is through the roof. There are NO limits anymore to what I can accomplish and learn through science, thanks to Science 2.0 and my decision to allow ALL of the evidence into evidence.

Quantum Mechanics is syntropy. Quantum Mechanics is eternal and everlasting, without beginning of days or an end of years. Quantum Mechanics is the opposite of Entropy, Classical Physics, Materialism, Naturalism, Darwinism, Nihilism, and Atheism. After we have successfully eliminated everything that is false, then only the truth remains. After we have successfully used the Scientific Methods to falsify and eliminate Classical Physics, Materialism, Naturalism, Darwinism, Scientism, and Atheism, then only Quantum Mechanics remains. Quantum Mechanics is the true science that remains after ALL the pseudo-sciences have been falsified and eliminated. Fascinating, is it not?

So, What Else Is Pure Syntropy?

Well, what else was there in the beginning? What else is without the beginning of days or the end of years? What else is immortal, eternal, endless, and everlasting?

WE KNOW that Spirit Matter, Quantum Mechanics, and the Priesthood of God were without beginning of days or end of years; so, what else fits in that category and would qualify as syntropy?

You and I would! Our Human Psyche would qualify as Pure Syntropy.

You and I existed before the first God designed and created the first particle of physical matter. You and I – our intelligence, psyche, or non-local consciousness – have always existed and will always exist. Psyche is without beginning of days or end of years. Psyche or Intelligence or Life is PURE SYNTROPY.

Syntropy, Psyche, or Intelligence is some kind of living organizational force. Without syntropy, there would be no order, no organization, no genomes, no proteins, and no physical life forms in this physical universe because entropy or random diffusion prevents these things from happening spontaneously at the physical level. It requires syntropy at the quantum level or the psyche level in order to bring order, organization, and life to this physical universe at the physical level. Without syntropy, entropy would rule supreme; and, everything would be random chaos.

Psyche or Intelligence or Life is something completely different than spirit matter or a quantum object. Matter of any kind, whether spirit matter or physical matter, responds or

REACTS to the demands and commands of Psyche or Quantum Non-Local Consciousness. Intelligence or Psyche ACTS, chooses, and decides what to do with the matter that's under its control – both the spirit matter and the physical matter. Psyche ACTS. Matter of any kind REACTS.

Intelligent design and intelligent creation are syntropy in action because they involve Someone Psyche in order to get the job done. There's NO Psyche involved in chemical evolution, random mutations, natural selection, random diffusion, or entropy, which means that evolution of any kind cannot design, create, and produce genes, proteins, genomes, eyes, brains, and life forms. Evolution is entropy. Entropy of any kind cannot design and create, because there is NO Psyche there to do the work. Psyche provides the syntropy, or the design and creation. Matter of any kind simply reacts. Matter cannot design and create anything, because matter has been designed and programmed to react, and nothing more.

This is the way things really work, in every realm of existence.

The Human Psyche, Nature's Psyche, and God's Psyche were there in existence BEFORE the first particle of physical matter was designed, created, and brought into existence. Physical matter has a beginning, and can therefore theoretically have an end; but, Psyche or Intelligence has NO beginning and therefore can have no end. Psyche of any kind is PURE SYNTROPY. Psyche is consciousness, sentience, and life. Psyche is the Ultimate Causal Agent. Psyche is the source of design and creation.

With the Ultimate Model of Reality series, I repair and correct philosophy so as to bring it into line with the observed realities of our existence by implementing a Psyche Ontology and a Psyche Epistemology, wherein Psyche is the fundamental unit of reality and Psyche is the best way to find and know the truth.

See the following for a detailed explanation of Ultimate Causality.

The Ultimate Model of Reality: Psyche Is the Ultimate Cause.

https://www.amazon.com/dp/B071NC9JK6

The BEST way to find and know the truth about Psyche or Intelligence is to live it and experience it first-hand for yourself, or to choose to trust someone who has.

As far as I'm concerned, the BEST source of public evidence concerning Psyche or Intelligence comes from the Biblical God Jesus Christ, because He has lived it and experienced it first-hand, and I have learned to trust that He will tell me how things really are in the Quantum Non-Local Realm or the Spirit Worlds. I have observed that we must turn to the True and Living God, if we want to find out how things really are in the unseen non-physical worlds that surround us. In contrast, we know that Satan and the evil spirits are trying to trick us and deceive us. They are trying to convince us that there's no such thing as God or the spirit world.

In the books that the Biblical God Jesus Christ had a hand in writing and producing, He consistently uses the word "Intelligence" to define and describe Psyche or Quantum Non-Local Consciousness.

In contrast, a Quantum Object is some type of matter – spirit matter or physical matter. According to the Quantum Law of Complementarity, a Quantum Object can either be in a spiritual state or a physical state, but it can't be in both states simultaneously. This means that we human beings have both a spirit body and a physical body. The Quantum Law of Superposition teaches us that a spirit body and a physical body can coexist in the same space and same location at the same time, because they are out of phase with each other. They each exist on a different wavelength or frequency, meaning that they are out of phase with each other, meaning that they can superpose each other or coincide with each other in the same space at the same time.

So, What About Psyche?

We are going to have to rethink this thing if we are going to make it match with what has been experienced and observed by Out-of-Body Travelers.

Psyche is something completely different than a Quantum Object or matter. Psyche ACTS. Matter of any kind REACTS to the commands and demands of Psyche. Psyche or Intelligence is conscious and alive. Psyche or Consciousness is the thing that collapses the quantum wave function. Psyche ACTS. Matter REACTS. Psyche makes decisions and choices, as to what it wants to do with its assigned spirit body and physical body.

God told some of us about Actors and Reactors over 2,500 years ago.

2 Nephi 2: 11-14, 25-27:

For it must needs be, that there is an opposition in all things. If not so, my firstborn in the wilderness, righteousness could not be brought to pass, neither wickedness, neither holiness nor misery, neither good nor bad. Wherefore, all things must needs be a compound in one; wherefore, if it should be one body it must needs remain as dead, having no life neither death, nor corruption nor incorruption, happiness nor misery, neither sense nor insensibility. Wherefore, it must needs have been created for a thing of naught; wherefore there would have been no purpose in the end of its creation. Wherefore, this thing must needs destroy the wisdom of God and his eternal purposes, and also the power, and the mercy, and the justice of God. And if ye shall say there is no law, ye shall also say there is no sin. If ye shall say there is no sin, ye shall also say there is no righteousness. And if there be no righteousness there be no happiness. And if there be no righteousness nor happiness there be no punishment nor misery. And if these things are not there is no God. And if there is no God we are not, neither the earth; for there could have been no creation of things, neither to act nor to be acted upon; wherefore, all things must have vanished away. And now, my sons, I speak unto you these things for your profit and learning; for there is a God, and he hath created all things, both the heavens and the earth, and all things that in them are, both things to act and things to be acted upon. Adam fell that men might be; and men are, that they might have joy. And the Messiah cometh in the fulness of time, that he may redeem the children of men from the fall. And because that they are redeemed from the fall they have become free forever, knowing good from evil; to act for themselves and not to be acted upon, save it be by the punishment of the law at the great and last day, according to the commandments which God hath given. Wherefore, men are free according to the flesh; and all things are given them which are expedient unto man. And they are free to choose liberty and eternal life, through the great Mediator of all men, or to choose captivity and death, according to the captivity and power of the devil; for he seeketh that all men might be miserable like unto himself. And now, my sons, I would that ye should look to the great Mediator, and hearken unto his great commandments; and be faithful unto his words, and choose eternal life, according to the will of his Holy Spirit; and not choose eternal death, according to the will of the flesh and the evil which is therein, which giveth the spirit of the devil power to captivate, to bring you down to hell, that he may reign over you in his own kingdom. I have

spoken these few words unto you all, my sons, in the last days of my probation; and I have chosen the good part, according to the words of the prophet. And I have none other object save it be the everlasting welfare of your souls. Amen.

Corruption is entropy. Incorruption is syntropy.

This individual, this prophet of God, knew about entropy and syntropy over 2,600 years ago. Think about it. Most of our scientists on this planet still know nothing about syntropy. They haven't even thought about syntropy. They don't want to know anything about syntropy, except to "know" and believe that it does not exist.

There has to be an opposition in all things. If there is entropy, then there has to be syntropy! If there is death, then there has to be some kind of life. If there is hell, then there has to be a heaven. If there's misery, then there has to be some type of happiness available. If there's sin, then there has to be some type of righteousness. If there's punishment, then there has to be some type of reward. If there is destruction, then there has to be some kind of order or organization. If there is evil, then there has to be good. If there is Satan, then there has to be a God. If there are things that react, then there have to be Actors.

Not only do his statements feel like scripture, they are also the pinnacle of truth and science. He talks about what things would be like if there were NO syntropy. It would be a "compound in one". It would be chaos, random diffusion, entropy! Our physical universe would be nothing but one homogenous ball of gas if it even existed at all.

A "compound in one" is the Gray Goo Scenario.

Grey goo (also spelled gray goo) is a hypothetical end-of-the-world scenario involving molecular nanotechnology in which out-of-control self-replicating robots consume all biomass on Earth while building more of themselves, a scenario that has been called ecophagy ("eating the environment", more literally "eating the habitation"). The original idea assumed machines were designed to have this capability, while popularizations have assumed that machines might somehow gain this capability by accident.

https://en.wikipedia.org/wiki/Grey_goo

How did this prophet know about the Gray Goo over 2,600 years ago? Well, that's exactly what we would have if there were no God, no syntropy, no psyche, and no life. Without God's Psyche and Syntropy, the whole universe would be nothing more than one homogenous ball of gas – or a gray goo, if you like.

The very existence of physical matter, especially finely-tuned physical matter or organized physical matter, is Scientific Proof of God's Existence.

Without God's Psyche and Syntropy, there would be NO purpose for our existence because we wouldn't have a physical existence. In fact, without of Psyche of our own, we wouldn't exist.

Psyche can choose to act! Matter of any kind simply reacts. He even says as much.

We – our Psyche or Intelligence – can choose eternal life or eternal death. The choice is ours. We can have, and we will have, anything that we want. Our very future is chosen into existence, thanks to Syntropy or the Atonement of Christ.

Remember, every massless elementary particle or every massless quantum is some type of Psyche. Psyches have unique assignments, functions, or positions to fulfill.

Another Way to Define Psyche

Psyche has been observed by Out-of-Body Travelers as a pinpoint of light, a pinprick of light, a point particle of light, or a spark of light. That's precisely the way that the physicists describe the fermions – quarks, leptons, neutrinos, and electrons. These "particles" of light OBEY the Pauli exclusion principle. Obedience to Laws requires intelligence! The components of the quarks and the electrons are individual pinpricks of Psyche or Intelligence; and, the psyches within fermions USE the bosons, gluons, photons, quantum waves, and light to communicate with each other, interact with each other, and bind themselves to each other. The elementary parts of Fermions are Psyche or Intelligence. That would make Psyche or Intelligence the fundamental unit of reality and existence. Photons, gluons, and W and Z bosons are the four force-carrying gauge bosons of the Standard Model. There's NO mass in the elementary components of the fermions, the psyches, or the pinpricks of light. Over 99% of the mass within an atom is associated with the gluons. Psyche or Intelligence is driving the mass or driving the gluons. It's all made from Energy which is always conserved.

God Is Psyche, Light, or Syntropy

The existence of entropy, when properly understood, is scientific proof of syntropy; and, Syntropy, Order, Organization, Life, Psyche, Physical Matter, Proteins, Genes, Genomes, Eyes, Brains, Physical Life Forms, and Quantum Mechanics are Scientific Proof of God's Existence. These things wouldn't exist without God or Syntropy. Without God and Syntropy, everything would be Gray Goo or one homogenous ball of gas. It would ALL be random diffusion, disorder, chaos, and entropy. There would be NO physical life anywhere in this universe. That's just the way it is. Science tells us that it is so.

In the books that the Biblical God Jesus Christ had a hand in writing and producing, He consistently uses the word "Intelligence" to define and describe Psyche or Quantum Non-Local Consciousness. There are many different types of Psyche or Intelligence. A Photon is a Psyche. The other massless elementary particles that have been discovered are Psyches. It's interesting to observe that a Psyche or Intelligence looks like a Photon and acts like a Photon BOTH at the quantum level and at the physical level. Photons or Psyches have been experienced and observed. Science is observation and experience, or it should be.

Doctrine and Covenants 93: 29-30, 36:

Man was also in the beginning with God. Intelligence, or the light of truth, was not created or made, neither indeed can be. All truth is independent in that sphere in which God has placed it, to act for itself, as all intelligence also; otherwise there is no existence. The glory of God is intelligence, or, in other words, light and truth.

What were we in the beginning? We were Intelligence, or Psyche, or Quantum Non-Local Consciousness. We were massless Photons of Light.

We were some kind of light, photon, or quantum wave. Thoughts, memories, and consciousness are quantum waves. WE KNOW, because thoughts and memories survive the death of our physical body and physical brain, according to the empirical evidence from

Near-Death Experiences (NDEs), Out-of-Body Experiences (OBEs), Shared-Death Experiences (SDEs), and our after-death Life Reviews.

Knowledge of all these things becomes possible by choosing to allow ALL of the evidence into evidence, and then choosing to pursue a preponderance of the evidence.

Intelligence, or Psyche, or the Light of Truth, or Consciousness was not created or made, nor indeed can be. Psyche or Non-Local Consciousness is what we truly are. We have always existed, and we will always exist. There is NO existence and NO life without Psyche or Consciousness. That's just the way it is.

Abraham 3: 18-19, 21-23:

There is nothing that the Lord thy God shall take in his heart to do but what he will do it. Howbeit that he made the greater star; as, also, if there be two spirits, and one shall be more intelligent than the other, yet these two spirits, notwithstanding one is more intelligent than the other, have no beginning; they existed before, they shall have no end, they shall exist after, for they are gnolaum, or eternal. And the Lord said unto me: These two facts do exist, that there are two spirits, one being more intelligent than the other; there shall be another more intelligent than they; I am the Lord thy God, I am more intelligent than they all. I dwell in the midst of them all; I now, therefore, have come down unto thee to declare unto thee the works which my hands have made, wherein my wisdom excelleth them all, for I rule in the heavens above, and in the earth beneath, in all wisdom and prudence, over all the intelligences thine eyes have seen from the beginning; I came down in the beginning in the midst of all the intelligences thou hast seen. Now the Lord had shown unto me, Abraham, the intelligences that were organized before the world was; and among all these there were many of the noble and great ones; and God saw these souls that they were good, and he stood in the midst of them, and he said: These I will make my rulers; for he stood among those that were spirits, and he saw that they were good; and he said unto me: Abraham, thou art one of them; thou wast chosen before thou wast born.

Intelligence is Psyche. Intelligences are Psyches. Psyche is consciousness and life.

This also feels like scripture to me because of all the different scientific truths that it contains. Notice how God compares Psyches or Intelligences to stars – pinpoints of light.

Let's say that you are a God, and you are standing among a group of Intelligences, or Psyches, or Living Lights, or Conscious Lights. How would you organize them? There is NO such thing as creation ex nihilo; so, from the Gods' perspective, creation is simply organization or reorganization of something that already exists. How would you organize Psyches or Intelligences?

God tells us in the last verse how He organized these Intelligences. God organized them into spirits by giving them spirit bodies. God gave them greater order or organization. Spirits or souls are a dynamic living holistic combination of psyche and spirit body. These Intelligences or Psyches were organized or given greater order by placing them into spirit bodies made from spirit matter.

Intelligence has several meanings, three of which are:

(1) It is the light of truth that gives life and light to all things in the universe. It has always existed.

(2) The word intelligences may also refer to spirit children of God.

(3) The scriptures also may speak of intelligence as referring to the spirit element [and psyche or non-local consciousness] **that existed before we were begotten as spirit children.**

https://www.lds.org/scriptures/gs/intelligence-intelligences?lang=eng&clang=eng#p1

Psyche or Intelligence or Consciousness is PURE SYNTROPY. It is without beginning of days or end of years. Your Psyche has always existed and will always exist. We, our Psyche, is co-eternal with God. God simply achieved order or organization faster than we did. Our God, our Heavenly Father and Heavenly Mother, simply started the whole process a lot sooner than we did.

Spirits are syntropy, because they are comprised of psyche and spirit matter, both of which are eternal, everlasting, and without beginning of days or an end of years. Spirit bodies have a beginning or a birth or a date of organization; but, spirit matter and psyche do not. God is a fullness of Syntropy – a fullness of glory, or a fullness of Quantum Mechanics, or a fullness of Priesthood Power. God is Syntropy.

In contrast, random diffusion, disorder, chaos, and death are entropy. WE KNOW that Syntropy must exist, because there is entropy. There would be nothing, not even entropy or physical matter, if Syntropy did not exist.

In the beginning, there was Syntropy, and that Syntropy was a good thing. Without Syntropy, we wouldn't exist; and, this reality once again becomes Scientific Proof of God's Existence, when we finally understand what Syntropy is and how it really works.

Cool, huh?

Well, I think it is. Your mileage may vary, of course, especially if you are at war with God. I used to hate God, so I know how it goes. I used to be a Materialist, Naturalist, Nihilist, and Atheist. Obviously, I have changed. I didn't like where that road was taking me, so one day I decided to turn around and go the other way. My life has been getting better and better ever since.

Syntropy is the KEY to understanding everything else, in my humble opinion. Syntropy is the answer to life, the universe, and everything. It explains why there is something rather than nothing. The purpose of this essay was to reveal where I got the idea of Syntropy or Eternal Life. I found it interesting and essential for my understanding of Science because Quantum Mechanics, Energy, and Psyche ARE Syntropy.

Mark My Words

—

Source

God Is in the Light: God is light, and in Him is no darkness at all.

https://www.amazon.com/dp/B07168S37N

Science 2.0: I Upgraded My Science

https://www.amazon.com/dp/B0771K6WTX

Defining Syntropy

So, what do we observe?

Science is all about observation and experience, or it should be! Philosophical speculation is worthless if it isn't backed up by observations and experience.

So, what do we observe?

We observe entropy.

We observe things running down, wearing out, and ceasing to exist. We observe disease, death, cancer, corruption, and extinction – all the result of entropy or random mutations. We see disorder, chaos, and death ever on the increase.

We observe the stars burning out and blowing up. We observe our physical universe moving slowly towards heat death.

We observe that Materialism, Naturalism, Darwinism, Nihilism, Atheism, Classical Physics, and the Theory of Evolution are based exclusively on entropy. Materialism and Naturalism are entropy. Nihilism is death or entropy. Physical matter is subject to entropy. Evolution is entropy. Natural selection is Creation by Entropy. The Materialists, Darwinists, and Naturalists teach that only entropic physical matter exists. The whole thing is moving towards entropy and heat death. Entropy is death. Death is what we observe, is it not?

So, where did all the Order, Organization, Syntropy, and Life come from in the first place? There had to be an initial infusion of Syntropy, or all of this subsequent entropy and death wouldn't have been possible. This is Logic 101.

Where's the Syntropy? It should be observable too. Syntropy has to exist, or there wouldn't be all this entropy; and, our physical universe would have burned out and died an eternity or two ago. So, what is Syntropy and where is it?

Syntropy has a number of different definitions.

> **The negentropy has different meanings in information theory and theoretical biology. In a biological context, the negentropy (also negative entropy, syntropy, extropy, ectropy, or entaxy) of a living system is the entropy that it exports to keep its own entropy low; it lies at the intersection of entropy and life. In other words, negentropy is reverse entropy. It means things becoming more orderly. By 'order' is meant organization, structure, and function: the opposite of randomness or chaos.**
>
> https://en.wikipedia.org/wiki/Negentropy

Syntropy is typically defined as being the opposite of entropy. So, what's entropy, and what's the opposite of entropy? Well, according to this quote, randomness, disorder, or chaos is entropy, which means that Syntropy is organization, order, structure, functionality, purpose, and life.

Entropy is a tricky and confusing bastard. Entropy has no parentage, causal force, existence, or being. When it comes to entropy, there's nothing there to detect or measure. Entropy is an emergent property, which basically means that it doesn't really exist as a person, place, or thing. Like evolution or natural selection, entropy really doesn't exist as a thing, a force, a field, a quantum particle, or even a law. Entropy is simply a measure of the unavailable energy in a system – the energy that we humans can no longer get at. The energy is still there because ALL energy is conserved, we just can't get at it because it has reached thermal equilibrium and consequently has been made unavailable to us. There's no force, field, quantum, or energy particle keeping track of entropy that we know of. So, how are we supposed to measure it? Well, should we ever reach the point of ZERO available energy or maximum entropy, we won't be there to measure it, so why worry about it in the first place?

I chose the word "bastard" carefully to describe and define entropy because entropy is death. To add insult to injury, I've come across dozens of different definitions for entropy – many of them contradictory, inconsistent, and self-defeating. Since entropy doesn't really exist as an entity, quantum, field, force, object, or thing, we have a devil of a time explaining what it is and how it works. The best we can do is to turn to Uncle Google and hope for the best.

Entropy was extremely low at the instant of the Big Bang, and it has been increasing ever since. That means that the Big Bang Singularity was Pure Syntropy. Entropy is an emergent property or a statistical property of physical matter. Entropy is a measure of how far from equilibrium a system is. Entropy is a measure of the proportion of energy in a system that can be used for useful work. Entropy is a thermodynamic quantity representing the unavailability of a system's thermal energy for conversion into mechanical work, often interpreted as the degree of disorder or randomness in the system. Entropy is a lack of order or predictability, a gradual decline into disorder. Entropy is a lack of information. Entropy is a measurement of the unavailable energy in a system. Maximum entropy is maximum thermal equilibrium which means that NO work or ZERO work can be done in that system.

I eventually discovered that "anergy", or "thermal equilibrium", or "no available energy" is an excellent and descriptive definition for entropy. The energy is still there, it's just not available for our use to do work.

Exergy is the energy that is available to be used. After the system and surroundings reach thermal equilibrium, the exergy is zero. Determining exergy was the first goal of thermodynamics. Energy is neither created nor destroyed during a process. Energy changes from one form to another. In contrast, exergy is always destroyed when a process is irreversible, for example loss of heat to the environment. This destruction is proportional to the entropy increase of the system together with its surroundings. The destroyed exergy has been called anergy. Anergy means no available energy. For an isothermal process, exergy and energy are interchangeable terms, and there is no anergy. In an isothermal system, all of the energy is available for use all of the time – all of the energy is exergy.

Adapted from: https://en.wikipedia.org/wiki/Exergy

"Unavailable Energy" means "NO available energy" or "anergy". In a very real sense, entropy is a measurement of nothingness or a measurement of nothing. Well, that's going to take us far, isn't it? How do you measure nothing or the amount of nothing? We are talking about unavailable energy here, which means that it isn't going to be available for measurement either, if you get right down to it. "Zero available energy" or "zero available work" is entropy's maximum measurement or what they call Maximum Entropy.

In contrast, zero entropy is Syntropy. An Isothermal System is Pure Syntropy or Pure Exergy because in an Isothermal System ALL of the energy is available for use all of the time. For an Isothermal System, exergy and energy are interchangeable terms, and there is NO anergy or entropy. In other words, in an Isothermal System ALL of the energy is exergy which means that ALL of the energy is available all the time. This is extremely powerful Science. The Quantum Realm, Psyche Realm, or Syntropy Realm is an eternally Isothermal System or a perfectly Isothermal System. An Isothermal System is always conserved. It is eternal and everlasting. Psyche, Intelligence, Life Force, or Energy is an Isothermal System. It cannot be made, and it cannot be destroyed. It has always existed, and it will always exist. It is always conserved.

They describe entropy as an "emergent law" or "statistical law" which means that it doesn't really exist or doesn't stand on its own. Maximum entropy implies ZERO Syntropy, or ZERO available energy, or ZERO exergy. I wasn't exaggerating when I said that entropy is a tricky and confusing illegitimate bastard. Since we are talking about ZERO or NOTHING, it's hard to nail it down. It's kind of hard to measure NOTHING or the amount of NOTHING. Maximum entropy is ZERO available energy, which means that there's nothing to measure and nobody to measure it.

So, how are we to make sense of entropy since it doesn't really exist and isn't measurable with our physical instruments? Well, death is my favorite definition for entropy, because I understand death. Death is all or nothing. If the Gods don't intervene and make some of that energy available to us for use once again through Syntropy, then the physicists assure us that our universe is going to end in Heat Death. We can all understand how that will work.

I have watched a lot of Space Time videos hoping that he would have some brilliant insight into entropy, and he basically describes entropy as "heat death", the "end of life", and the "end of our universe". He makes it clear that entropy is death. He's even had T-shirts made and produced dozens of videos to celebrate Heat Death, the end of our earth, and the end of our universe. Clearly, he teaches and believes that entropy or death is conserved. According to the Nihilists and Atheists, Entropy is the Conservation of Death. That's what ALL of the Materialists, Naturalists, and Atheists teach and believe. So, how do you measure or quantify death? Death really isn't a quantum field, quantum wave, quantum particle, or quantum object; so, there's really nothing there to measure or detect.

Technically, entropy isn't constructed nor made because it really doesn't exist as a quantum, field, force, or entity. Only the Gods know which energy or quanta they have made unavailable to us and which energy or quanta are still available for our use. We just assume that we are running out of energy, but we have no way to prove that we really are. I mean, where's that massive infusion of Dark Energy coming from? Is what we are losing through Entropy being fed back to us in the form of Dark Energy? Will there be a new massive wave of galactic formation sometime in the distant future from all of that new Dark Energy that the Gods are currently infusing into our expanding universe?

Since energy is always conserved, the Gods can reach into the system anytime they want and make that unavailable energy available to us once again whenever they choose to do so. When it comes to entropy, the Gods are the ones who are accounting for it, keeping track of it, and causing it. The Gods are the ones who are making that energy unavailable to us. The energy is still there and available for use anytime the Gods want to give us access to it again. The Gods can reach into our system and warm things up anytime they choose to do so.

Entropy isn't God contrary to what the Materialists, Naturalists, and Atheists claim, teach, and believe. Heat death is coming, and then comes our afterlife in the spirit world, resurrection from death, immortality, and eternal life – all of which will FALSIFY entropy or death and signify Syntropy instead. Based upon what we know of God the Father, Jesus

Christ, the Atonement of Christ, Psyche or Non-Local Consciousness, Conservation of Energy, our After-Life in the Spirit World, Resurrection from Death, and the forthcoming New Heaven and New Earth, it's hard to make a case that entropy is going to have any lasting effect on any of us. That's why Jesus called it the "Gospel" or the "Good News" – entropy isn't going to have the last word after all.

Nevertheless, we're still in the mess for now. In order for us fallen mortal beings to overcome entropy here and now, we would have to persuade the Gods to remove that entropy from our system and make ALL of the energy available to us once again. Psyche or Energy is the ultimate perpetual motion machine; and, the Gods designed entropy or death to prevent us from gaining access to perpetual motion machines, immortality, and eternal life while we are fallen, physical, entropic mortal beings.

Fascinating, is it not?

So, what is entropy? Entropy is the opposite of Syntropy. From our limited mortal perspective, entropy is corruption, random mutation, evolution, disease, disorder, death, and extinction. Entropy is the arrow of time. The different types of evolution are entropy, and the different types of evolution are prevented from happening by entropy. Entropy is death. The Theory of Evolution is based exclusively on entropy or death. It doesn't consider anything else to be possible. The Theory of Evolution is Creation by Entropy or Creation by Death – that's never going to work.

Syntropy is the opposite of entropy. Syntropy is what the theory of evolution pretends to be but isn't. This means that Syntropy must be some type of incorruption, healing, order, organization, life, light, psyche, intelligence, atonement, resurrection, immortality, eternal life, timelessness, mercy, grace, and love. It sounds like heaven to me. It sounds like God.

Syntropy is order, organization, structure, life, light, mercy, grace, and love – the opposite of randomness, chaos, entropy, death, extinction, classical physics, materialism, naturalism, evolution, darkness, and creation ex nihilo or atheism. Syntropy is NOT magic. Syntropy has always existed and will always exist. Syntropy is eternal and everlasting, without a beginning of days or an end of years. Psyche is Syntropy. Quantum Mechanics is a type of Syntropy. The Priesthood Power of God is Syntropy. Relationships or friendships are a type of Syntropy. The Conservation of Energy is Syntropy. Syntropy is an eternally isothermal system. There is NO heat flow and therefore no entropy in a perfectly isothermal system. Consequently, the second law of thermodynamics doesn't exist in an eternally isothermal system. The Syntropic Realm or Quantum Realm is such a system.

Love is Syntropy – the more you give away, the more you have. According to the Near-Death Experiencers, all of the different universes are based upon Love or Syntropy. The whole shebang is based upon Love, God's Love.

In contrast, entropy was designed by the Gods to limit us and restrict us. The physicists call entropy a law, the Second Law of Thermodynamics; but, entropy is nevertheless a measurement of unavailability or a measurement of nothingness, and what's unavailable to us is fully available to the Gods because the Gods made the rules or the laws that are currently making that energy unavailable to us but not to them.

From our mortal perspective, entropy reigns supreme. Consequently, we make up science fiction in order to get around it – things like the theory of evolution, natural selection, spontaneous generation, abiogenesis, and creation ex nihilo or atheism, which are "the creation of something from nothing by Nothing". Creation ex nihilo is impossible. Creation ex nihilo is magic; and, there's no such thing as magic. Creation ex nihilo is spontaneous generation; and, spontaneous generation, abiogenesis, or chemical evolution was falsified in 1859 by Louis Pasteur, the very same year that Charles Darwin published "On the Origin of Species by Means of Natural Selection".

Syntropy is what chemical evolution and macro-evolution pretend to be but are not. Instead, spontaneous generation, or chemical evolution, or macro-evolution is prevented from happening by entropy. The Gods created entropy and our genes to prevent us from evolving into something else or devolving into nothing, respectively. Evolution is NOT a good thing – it results in disease, cancer, death, and extinction. Entropy or death prevents the different types of evolution from happening to our species; and, our genes were designed to prevent us from devolving into extinction. Evolution is not a good thing. If the theory of evolution were actually true, there would be NO life on this planet. Evolution is death and extinction. The Theory of Evolution is Creation by Entropy or Creation by Death.

Remember, creation ex nihilo is based upon nothing; whereas, Syntropy is based upon Quantum Mechanics, Psyche, Relationships, Energy, Life, and Love. Syntropy is Eternal Life. In contrast, entropy is death. There is a difference between the two.

Remember, Syntropy MUST of necessity exist, or all of that subsequent entropy would not have been possible. In the beginning, our physical universe had to receive a huge infusion of Syntropy, or none of that subsequent entropy could have happened or been possible. Some type of Syntropy or God MUST exist, or we would not exist as physical beings. Entropy or physical matter cannot design and create. Entropy can only corrupt and destroy. Entropy moves everything towards disorder and death. In contrast, Syntropy moves everything towards order, life, and love. Syntropy is LIFE and LOVE. Psyche or Intelligence is Syntropy. Energy is Syntropy. Psyche, Energy, or Life Force is conserved. Syntropy is the Conservation of Energy or the Conservation of Psyche.

John 10: 10:

The thief cometh not, but for to steal, and to kill, and to destroy: I am come that they might have life, and that they might have it more abundantly.

To have life more abundantly!

How?

SYNTROPY!

How?

Atonement, Resurrection, Priesthood Power, and Eternal Life.

Entropy cometh to steal, kill, and destroy. Entropy is the thief.

Jesus Christ or Syntropy came that we might have life and have it more abundantly. Syntropy is like mercy, grace, and love – the more you give away, the more you have. Syntropy is expansive and eternal. Entropy deflates and destroys. Entropy is death. Syntropy is Eternal Life.

Materialism, Naturalism, Darwinism, Nihilism, Behaviorism, Determinism, Physical Reductionism, Behaviorism, Atheism, Classical Physics, Psychiatry, and Pharmacology are based exclusively on entropy. Entropy is death and extinction. In contrast, everything good is based upon Syntropy, Relationships, Quantum Mechanics, Psyche, Energy, and Love. The drugs can't do relationships or psyche therapy; but, the various different types of Syntropy certainly can. Syntropy is psyche therapy. Syntropy is friendship and love.

Remember, everything we observe is based upon quantum laws and physical laws. The quantum laws and physical laws were ordered or organized by God, or Syntropy, or Psyche, or Intelligence, or Life. They work because God or Syntropy makes them work.

Syntropy is Divine Light, Life Energy, Order, Organization, Atonement, Life, the Spark of Life, Life Force, Psyche, Intelligence, Quantum Mechanics, the Priesthood Power of God, and Reverse Entropy. The Atonement of Christ is Syntropy. It reverses the effects of death and entropy.

Syntropy means "without beginning of days or an end of years". Syntropy means eternal and everlasting. Syntropy means immortal. Psyche is Syntropy. Quantum Mechanics is Syntropy. Spirit Matter or Dark Matter seems to be syntropic or based exclusively upon energy. God's Priesthood Power is Syntropy. The Light of Christ, or Dark Energy, or the Zero-Point Field is Syntropy. God is Syntropy. Energy is Syntropy. Syntropy is the Conservation of Energy or the First Law of Thermodynamics. They are without a beginning of days or an end of years. They are eternal and everlasting. They are perpetual and self-renewing. They are conserved.

Death, extinction, or natural selection is entropy.

Random mutations are entropy.

Classical physics is entropy. Materialism and Naturalism are based upon entropy. Darwinism is based exclusively on entropy. Entropy cannot design and create. It can only destroy. Entropy is death.

The chemical evolution of genes and proteins from atoms is prevented from happening by entropy and random diffusion. Macro-evolution, chemical evolution, abiogenesis, and spontaneous generation are physically impossible because they are prevented from happening by random diffusion, chaos, genetics, and entropy. Only Psyche or Intelligence can design and create. Psyche is Energy. Energy or Psyche can remove entropy from a system at will.

Evolution of any kind is entropy, randomness, chance, or chaos. The opposite of evolution, random mutations, and natural selection are design, planning, teleology, purpose, programming, creation, order, organization, life, law, quantum mechanics, psyche, intelligence, and Syntropy. The Theory of Evolution is based upon entropy, and entropy prevents evolution of any kind from happening in the first place. Entropy or death was designed to prevent our physical bodies from evolving into something else. The opposite of the Theory of Evolution is Syntropy. Materialism, Naturalism, Darwinism, Nihilism, Atheism, and the Theory of Evolution are Creation by Entropy or Creation by Death. Please explain to me how that's supposed to work. It can't. It's physically impossible!

Remember, Syntropy or Quantum Mechanics is incomplete, ineffective, illogical, inefficient, meaningless, purposeless, and worthless if it is defined by, restrained by, limited by, restricted by, and polluted by Entropy, Chaos, Randomness, Physicalism, Materialism, Naturalism, Darwinism, Nihilism, and Atheism. Syntropy is the opposite of entropy. Syntropy is the opposite of Materialism and Naturalism. Syntropy is effectively the opposite of Classical Physics and Determinism.

To those of us who are restricted by, limited by, and restrained by entropy, physical bodies, and this physical universe, Quantum Mechanics or Transdimensional Physics seems to be magic; but, Quantum Mechanisms or Syntropic Mechanisms are the LAW and therefore are the NORM where the spirit world, or transdimensional world, or quantum world, or the prime construct is concerned. Syntropy, or Psyche, or Power, or Light brings order out of randomness and chaos. In other words, Syntropy reverses entropy or Syntropy overcomes entropy at will. Syntropy is the NORM, and entropy is the exception to the norm.

Syntropy is the Prime Construct. Psyche or Syntropy is the organizing force of the multiverse. Syntropy is God's Love. Syntropy is Quantum Mechanics or Transdimensional Physics. Syntropy is the power by which everything was made. Action at a Distance is based upon Syntropy. Action at a Distance implies NO entropy or NO physical restrictions. Syntropy is eternal and everlasting. The Priesthood Power of God is described as being "without beginning of days or an end of years". The Priesthood Power of God is Syntropy or Quantum Mechanics. Syntropy renews itself and enhances itself as it renews us and enhances us. The more God gives away, the more He has when He is done.

Action at a Distance involves perception at a distance – some kind of instantaneous transpersonal communication between psyches at a distance across space-time. Action at a Distance between atoms and molecules involves perception at a distance – the psyche or intelligence within atoms and molecules sensing and perceiving each other instantaneously at a distance across space-time. These are quantum mechanical or spiritual phenomena, not physical phenomena. Action at a Distance is syntropic. Action at a Distance is the ultimate perpetual motion machine. If we could tap into that directly, there would be NO entropy in our lives.

I used to be a Physicalist, Naturalist, Nihilist, and Atheist; but, I'm no longer satisfied being dumb and blind without a soul or a mind. Nowadays, I have to know what things are and how they work.

Psyche is the innate intelligence within all the different forms of energy or quanta. Quanta are organized packets of energy. Quanta or "particles" are waves of energy moving through a quantum field. Someone Psyche has to start the quanta moving in the first place or nothing would ever happen. Syntropy is the conservation of energy. Syntropy is the First Law of Thermodynamics. Psyche or Energy is syntropic, which means that Psyche or Energy is conserved.

In contrast, entropic physical matter is made from different forms of energy. The form is NEVER conserved. Physical matter, entropy, and time are made or caused to begin, which means that they are NOT conserved. The Gods can transform entropic physical matter into a different form of energy anytime they choose to do so. This is the Ultimate Law of Thermodynamics. It states that Psyche, Energy, Life Force, and Syntropy are conserved; whereas, entropic physical matter is NOT conserved. The Ultimate Law of Thermodynamics simply differentiates between what is being conserved and what is NOT conserved.

The Ultimate Law of Thermodynamics is a necessary adjustment or refinement to the classical laws of thermodynamics because the Materialists, Naturalists, Darwinists, Nihilists, and Atheists erroneously teach and believe that physical matter, death, and entropy are conserved because these people erroneously teach and believe that entropic physical matter is the ONLY thing that exists. The Ultimate Law of Thermodynamics corrects that flaw by stating that entropy, death, and physical matter are NOT conserved. Therefore, if there is any conflict between the first law of thermodynamics and the second law of thermodynamics, the Ultimate Law of Thermodynamics declares the first law of thermodynamics the winner every time. The Ultimate Law of Thermodynamics states that Psyche, Energy, Life Force, and Syntropy are always conserved; whereas, entropy, death, and physical matter are NOT conserved. This is what has been experienced and observed by Out-of-Body Travelers and Near-Death Experiencers. Science is observation and experience, not wishful thinking and philosophical speculation. The Ultimate Law of Thermodynamics gives precedence to whatever has been experienced and observed.

Syntropy or Quantum Mechanics ends up being Scientific Proof of God's Existence. Entropy or physical matter also ends up being Scientific Proof of God's Existence. Physical matter and entropy wouldn't exist if it weren't for the causal intervention from Psyche, Syntropy, and God thereby bringing entropy, time, and physical matter into existence in the first place.

Scientific Observation: Anything that was obviously made obviously has a Maker who made it. Anything that obviously began obviously has Someone Psyche or Someone Intelligent who caused it to begin.

Scientific Observation: Entropic physical matter was obviously made from raw energy or unorganized energy. Entropic physical matter is highly organized energy. According to physicist David Bohm, physical matter is condensed light or frozen light. Condensed or frozen by whom? Physical matter and the associated entropy were obviously made. They obviously had a beginning of some type.

Scientific Conclusion: Therefore, it is logical and rational to conclude that physical matter and the associated entropy have a Maker who made the matter and caused the entropy or the flow of time to begin. Entropic physical matter is organized energy. Order and organization don't just spontaneously generate out of thin air from nothing. They are obviously made and caused to begin by some kind of Creator or Maker.

Entropy or the Second Law of Thermodynamics proves that God exists.

Does the Second Law of Thermodynamics Prove the Existence of God?

- By John M. Cimbala

Professor of Mechanical Engineering

The Pennsylvania State University

In this short article, I summarize my ideas about the second law of thermodynamics, and why I believe it points to a creator God.

This article also appears in the book *In Six Days - Why Fifty Scientists Choose to Believe in Creation*, edited by John F. Ashton, and published by Master Books, Green Forest, AR. Copyright 2000 by John F. Ashton. The book is available on-line at Creation Ministries International and at Answers in Genesis.

A formal definition of the second law of thermodynamics is "In any closed system, a process proceeds in a direction such that the unavailable energy (the entropy) increases." In other words, in any closed system, the amount of disorder always increases with time.

Things progress naturally from order to disorder, or from an available energy state to one where energy is more unavailable. A good example: a hot cup of coffee cools off in an insulated room. The total amount energy in the room remains the same (which satisfies the first law of thermodynamics). Energy is not lost, it is simply transferred (in the form of heat) from the hot coffee to the cool air, warming up the air slightly. When the coffee is hot, there is available energy because of the temperature difference between the coffee and the air. As the coffee cools down, the available energy is slowly turned to unavailable energy. At last, when the coffee is room temperature, there is no temperature difference between the coffee and the air, i.e. the energy is all in an unavailable state. The closed system (consisting of the room and the coffee) has suffered what is technically called a "heat death." The system is "dead" because no further work can be done since there is no more available energy. The second law says that the reverse cannot happen! Room temperature coffee will not get hot all by itself, because this would require turning unavailable energy into available energy.

Now consider the entire universe as one giant closed system. Stars are hot, just like the cup of coffee, and are cooling down, losing energy into space. The hot stars in cooler space represent a state of available energy, just like the hot coffee in the room. However, the second law of thermodynamics requires that this available energy is constantly changing to unavailable energy. In another analogy, the entire universe is winding down like a giant wind-up clock, ticking down and losing available energy. Since energy is continually changing from available to unavailable energy, someone had to give it available energy in the beginning! (I.e. someone had to wind up the clock of the universe at the beginning.)

Who or what could have produced energy in an available state in the first place? Only someone or something not bound by the second law of thermodynamics. Only the creator of the second law of thermodynamics could violate the second law of thermodynamics and create energy in a state of availability in the first place.

As time goes forward (assuming things continue as they are), the available energy in the universe will eventually turn into unavailable energy. At this point, the universe will be said to have suffered a heat death, just like the coffee in the room. The present universe, as we know it, cannot last forever. Furthermore, imagine going backwards in time. Since the energy of the universe is constantly changing from a state of availability to one of less availability, the further back in time one goes, the more available the energy of the universe. Using the clock analogy again, the further back in time, the more wound up the clock. Far enough back in time, the clock was completely wound up.

The universe therefore cannot be infinitely old. One can only conclude that the universe had a beginning, and that beginning had to have been caused by someone or something operating outside of the known laws of thermodynamics.

Is this scientific proof for the existence of a Creator God? I think so. Evolutionary theories of the universe cannot counteract the above arguments for the existence of God. Evidence such as this helped to convince me to believe in God, and to accept His plan of salvation through His son Jesus Christ. For further details about my conversion to Christianity, I have written a short testimony.

http://www.personal.psu.edu/faculty/j/m/jmc6/second_law.html

I believe 100% that this is the way things truly are. Some type of Syntropy, Psyche, Intelligence, or God MUST EXIST in order to have provided that initial infusion of Order, Organization, Available Energy, or Syntropy into this physical universe so that all of the subsequent entropy would have been possible in the first place. Entropy or physical matter couldn't have provided that initial infusion of Syntropy. WE KNOW for a fact that that Available Energy or Syntropy had to come from someplace else besides this physical universe, because this physical universe is based upon entropy, and restricted to entropy, and can only produce entropy.

Somebody somewhere knows how to wind up a universe, or we wouldn't be here. If the original construct were nothing but physical matter or entropy, then we wouldn't be here. Heat death would prevail. Someone Psyche has to know how to reverse entropy or remove entropy, or we wouldn't be here. Entropy doesn't exist in any useful form. There has to be Syntropy somewhere in some kind of Syntropy Realm or Quantum Realm, or we wouldn't exist. It's time that we, as scientists, finally choose to face reality and accept the fact that the ONLY reason we exist as physical beings is due to some type of Syntropy, which was infused into this physical universe at the beginning of this physical universe. The Big Bang, if it happened, was Pure Syntropy. It was made and caused to begin.

Syntropy is the opposite of entropy. Entropy is corruption, disorder, and death, which means that Syntropy has to be some kind of incorruption, order, organization, and life. Syntropy is the Conservation of Energy. Syntropy is infinitely more tangible and real than entropy.

Mark My Words

Energy Is Syntropic and Entropy Is Not

Syntropy is the opposite of entropy.

Can you hear the Materialists, Naturalists, Atheists, and Darwinists complaining in the background?

These people will tell you and assure you that I don't understand Science. They have told me many times that I don't understand Science. They will tell you that Syntropy is a fictional concept that I have made up out of thin air in order to convince people that God exists. They will complain that: "If Syntropy actually existed then it would have been discovered by now, and experienced and observed by now. Why haven't the scientists discovered Syntropy? Syntropy should have been seen, experienced, and found by now if it actually existed. The fact that nobody has ever heard of Syntropy is proof positive that Syntropy does not exist."

Are they right?

Do they have a point?

By stating that Syntropy is the opposite of entropy, it really doesn't explain what Syntropy is because entropy is intangible. We really can't get our hands on entropy, now can we?

However, over time, my research revealed to me that Energy is Syntropy – or energy is syntropic. We all understand Power or Energy, or we should if we are scientists. Energy is some type of invisible, immaterial, intangible, non-physical force, or field. Energy or Quantum Waves are the basis for Quantum Mechanics. Quanta are organized packets of energy. It's all invisible, intangible, and non-physical – except for the physical matter. Physical matter is highly organized energy, and extremely limited and restricted energy. Physical matter is energy, too. Physical matter is made up of the invisible, intangible, non-physical stuff that we call energy. Energy is the power by which things are made, including physical matter and the emergent property of entropy. Syntropy has been found and observed. Syntropy is Energy, and Energy is conserved. Psyche is the innate intelligence within all the different forms of energy, which gives that energy the inherent ability to follow, understand, and obey God's Laws and God's Commands.

The Materialists, Naturalists, Darwinists, Nihilists, and Atheists teach and believe that the invisible, the intangible, the non-local, the transdimensional, the spiritual, the psychic, and the non-physical do not exist. These people teach that Psyche, Syntropy, Intelligence, Spirit Matter, and Action at a Distance do not exist. The Scientific Naturalists don't spend any time studying Psyche or Syntropy because they have convinced themselves that Psyche does not exist. The Materialists, Naturalists, and Atheists have belabored their point-of-view for thousands of years in our public schools; so now, I'm going to spend a few minutes in this essay and belabor my observations or power-points. It will get repetitive; but, for some of you this will be the first time in your life that you have heard of this science, so it will all be new for you.

Syntropy or syntropic means eternal and everlasting, without a beginning of days or an end of years. The syntropic cannot be made nor destroyed. Energy is syntropic. Psyche is syntropic. Psyche is Energy. Psyche is the innate intelligence within all the different forms of energy. Entropy is death. Syntropy is Eternal Life. Psyche, Energy, Intelligence, or the Life Force within Energy is conserved. Energy is the ultimate perpetual motion machine. Energy is always conserved. In contrast, Out-of-Body Travelers and Near-Death Experiencers have observed that entropy or death is NOT conserved. Syntropy really is the opposite of entropy. Syntropy is the First Law of Thermodynamics. Syntropy is the conservation of Psyche or the conservation of Energy.

According to the First Law of Thermodynamics, energy is syntropic because energy or syntropy is conserved. The amount of physical matter and the amount of entropy changes over time; but, the amount of Energy, Psyche, Life Force, or Syntropy is always conserved. Energy, Psyche, or Syntropy is eternal and everlasting, without a beginning of days or an end of years. It can neither be created nor destroyed. Energy is Syntropy. Energy is syntropic. Psyche or Energy is conserved.

Remember, Syntropy is the conservation of energy. Syntropy is the First Law of Thermodynamics. Syntropy is the conservation of psyche. Syntropy is the foundation of physics. Psyche is Energy. Psyche is the intelligence within all the different forms of energy. Psyche or Energy is syntropic meaning that Psyche, or Intelligence, or Life Force, or Energy is conserved. Choice is the surest sign of Psyche at every level of existence. Whenever you witness a choice being made between competing alternatives, you have in fact witnessed Psyche in action. Our Psyche is identified by the choices that it makes.

So, why hasn't Syntropy been discovered before now? It's because Materialism, Naturalism, Darwinism, Nihilism, Atheism, and their derivatives were designed to prevent us from discovering Syntropy and Psyche. These people teach and believe that only entropic physical matter exists; therefore, these people erroneously teach and believe that physical matter and entropy are conserved. Entropy is death. These people literally teach and believe that entropy or death is conserved. Do they not? That's why we have never heard of Syntropy before. You will NEVER discover the **Law of the Conservation of Psyche** by denying Psyche's existence. You will NEVER discover what Psyche is and how it works by denying Psyche's existence.

Syntropy is the conservation of Psyche or the conservation of your Life Force – not the conservation of entropy or death. The Physicalists and Naturalists teach that Psyche and Syntropy do not exist. Syntropy is Immortality and Eternal Life. Syntropy, or the Conservation of Energy, is the ultimate perpetual motion machine. The Materialists, Naturalists, and Darwinists teach that there is no such thing as Eternal Life; consequently, these people teach that there is no such thing as Syntropy. Instead, these people teach and believe that death or entropy is conserved. These people teach that physical matter and entropy are conserved.

They are wrong!

Entropic physical matter is made from energy. Entropy is a measure of unavailable energy. Their quantity changes over time. Does it not? Physical matter and entropy come and go as God sees fit. Anything that is made is not conserved. The quantity or the amount of physical matter and entropy is NOT conserved. It changes over time. Physical matter and entropy were made by the Gods, which means that physical matter and entropy can be disassembled by the Gods and transformed into something else instead. Entropic physical matter is comprised of different forms of energy. The form is never conserved. Only the underlying energy is conserved. This is the Ultimate Law of Thermodynamics. It explains everything that has ever been experienced and observed.

The Ultimate Law of Thermodynamics emphasizes that Energy is Conserved, and that entropy is NOT conserved. Psyche or Life Force is conserved; whereas, physical matter, entropy, and death are not conserved. That explains a lot doesn't it? The dreaded and feared entropy is not conserved. Entropy is death. Death is not conserved. Syntropy or the Ultimate Law of Thermodynamics is the answer to life, the universe, and everything. Is it not? You are NEVER going to discover the Ultimate Law of Thermodynamics by denying its existence as the Materialists and Naturalists do. You will NEVER figure out what it is and how it works by denying its existence. That's the reality of our existence.

Entropic physical matter is a construct, which means that entropic physical matter can be deconstructed and turned back into Raw Energy or Syntropy any time the Gods choose to do so. Entropy was made and implemented by the Gods to prevent immortality, superpowers, and perpetual motion machines from being possible in a fallen physical mortal world such as ours. Entropy and death were made to prevent us from evolving into a different species and going extinct as a race. Evolution is extinction. Evolution is entropy and death. Entropy is an expiration date. Remove the entropy and the other physical restrictions, and instantly you are looking at perpetual motion machines, quantum mechanical superpowers, action at a distance, syntropy, and immorality. Remove the entropy, and you are instantly looking at Syntropy or the Conservation of Energy and Psyche. Physical matter, entropy, the passage of time, time itself, locality, and even certain types of space are constructs, which means that they were made or constructed; and, it also means that they can be deconstructed and turned back into Raw Energy, Usable Energy, or Syntropy instead.

Materialism, Naturalism, Darwinism, Nihilism, and Atheism were designed to prevent us from discovering Psyche, Syntropy, Action at a Distance, Supernatural Mechanics, God, the Quantum Law of Thermodynamics, and the Ultimate Law of Thermodynamics; and, they work as advertised. You are NEVER going to figure out what these things are and how they work by denying their existence.

Remember, Psyche is the intelligence within energy. Psyche or Energy is conserved. Psyche or Energy is syntropic which means that it is eternal and everlasting. It cannot be made, and it cannot be destroyed. Intelligence or the Life Force within energy is syntropic which means that it is conserved.

In contrast, the Materialists, Naturalists, Darwinists, Nihilists, Atheists, and Classical Physicists erroneously teach and believe that physical matter and entropy are conserved. These people teach that ONLY entropic physical matter exists. These people teach that Psyche, Action at a Distance, the Non-Physical, and Syntropy do not exist.

They are wrong!

Entropic physical matter is NOT conserved! Physical matter isn't the only thing that exists. It is Energy, Psyche, Life Force, Intelligence, or Syntropy that's being conserved according to the First Law of Thermodynamics, NOT physical matter and NOT entropy. We human beings have made and destroyed physical matter. We make physical matter and the associated entropy in our particle accelerators or atom smashers. We destroy physical matter and the associated entropy in our atomic bombs. Physical matter or entropy is NOT conserved. Its amount changes over time!

It's the Syntropy, Intelligence, Psyche, Life Force, or Energy that's being conserved; whereas, the amount of entropy changes over time. Does it not? If its amount changes over time, then it's not being conserved. If it is made, then it is not conserved. Entropy is made or implemented, and its amount changes over time; therefore, entropy is not conserved.

Remember, Syntropy is the Conservation of Energy and Psyche. Syntropy is the First Law of Thermodynamics. Energy, Psyche, Life Force, or Syntropy is conserved, meaning that it is eternal and everlasting without a beginning of days or an end of years. Energy, Psyche, Life Force, Intelligence, or Syntropy cannot be created nor destroyed. Physical matter can be made and destroyed. Entropy can be made and eliminated. Locality comes and goes. Time is a construct. In contrast, Energy, or Psyche, or Syntropy is conserved. It can neither be made nor destroyed. Energy, Psyche, Life Force, or Syntropy has always existed and it will always exist. It is eternal and everlasting.

Syntropy is raw energy, available energy, or syntropic energy. Physical matter is entropic energy or limited and restricted energy. Physical matter is organized energy – it's organized by limiting it and restricting it with physical laws, space-time, locality, entropy, and sub-light speed limits. Remove the physical restraints from physical matter or entropic matter or localized matter, and it goes back to be syntropic matter, spirit matter, quantum matter, or Energy. Syntropy is Energy. In fact, Syntropy is the conservation of energy. Psyche is Energy. Psyche is also Intelligence. In fact, Psyche is the intelligence within energy.

It's one eternal round. This is cool science. It explains everything.

Physical matter is made by converting syntropic matter into entropic matter. Physical matter is made by taking Syntropy or Energy and then imposing space-time, locality, entropy, and sub-light velocity limitations onto it. Remove the space-time, locality, entropy, and other physical limitations, and that physical matter goes back to being Energy, Spirit Matter, or Syntropy. The Syntropy or Energy is always conserved; whereas, the amount of physical matter and entropy come and go as God sees fit.

Physical matter or entropy comes and goes; but, the Energy or Syntropy is always conserved. Take Syntropy or syntropic spirit matter, and then impose space-time, entropy, locality, and sub-light speed limits onto that Syntropy or spirit matter and it becomes entropic physical matter instead. Remove the locality and entropy from physical matter and it goes back to being intangible Syntropy, Spirit Matter, or Energy. Physical matter is made by imposing physical limitations onto Syntropy, Spirit Matter, or Energy. Physical matter is tangible Energy. Physical matter is tangible Syntropy. Entropic physical matter is entropic Energy. Energy or Syntropy can take on many different FORMS. The FORM is never conserved. It's the Energy or the Syntropy that's being conserved. Physical matter is made by imposing physical limitations on Syntropy or Energy. That is the answer to life, the universe, and everything. Is it not? It explains everything, does it not?

Whenever the Gods interact directly with Raw Energy, the Gods are in fact interacting directly with the Psyche or the Intelligence within that Energy. Psyche and Energy are essentially synonymous, or reside at the same level of existence, namely the quantum level or the psyche level. Psyche and Energy are syntropic, which means that they are conserved. Syntropy is conservation of energy. Psyche and Energy are eternal and everlasting, without a beginning of days or an end of years. Psyche and Energy are syntropic, not entropic. Energy is intelligent, conscious, psychic, perceptive, and aware.

There's a spark of psyche or intelligence within every physical atom and within every part of a physical atom because physical atoms and their component parts are made by the Gods from energy. Psyche is the innate intelligence within all the different forms of energy.

Energy can neither be created nor destroyed. Psyche can neither be created nor destroyed. Psyche is the intelligence or the life force within energy that gives energy the ability to obey God's Laws and God's Commands. Psyche or Energy is conserved. Psyche and Energy are basically synonymous. The First Law of Thermodynamics tells us that the energy is constant, and that the energy is conserved – not the FORM of that energy. Energy can be transformed from one form to another, which means that the FORM is never conserved.

Energy or Syntropy is some type of invisible, immaterial, intangible, non-physical force, or field. Psyche is some type of invisible, immaterial, non-physical, intangible force, or field. Psyche is a type of Energy. Psyche is a type of Syntropy. Psyche, Syntropy, Life Force, or Energy is conserved. It's eternal and everlasting, without a beginning of days or an end of years.

Syntropy really is the opposite of entropy. Syntropy is conserved; and, entropy is NOT conserved. Entropy is ONLY associated with physical matter or entropic matter. Everything else is Syntropy. Entropy is a function of time. Entropy is the result of the passage of time. Entropy is an aging process. Entropy is an expiration date. Entropy is a clock that winds down. Entropic physical matter is limited to space-time. Entropic physical matter is limited to and limited by locality and entropy. Locality is a function space; and, entropy is a function of time. Locality and entropy are a function of space-time, respectively. Entropy or the aging process doesn't exist in the timeless and ageless Quantum Realm or Syntropic Realm.

Locality and entropy are NOT conserved – their quantity, number, and position change over time. Entropic physical matter is NOT conserved! Entropy and locality are NOT conserved. Physical matter is a function of space-time. Space-time is NOT conserved. Its dimensions are changing all the time. Its quantity seems to be changing as well. Everything else is Syntropy. Syntropy or Energy is conserved. Syntropy is Energy. Remove the entropy from physical matter and it goes back to being Syntropy. Entropic physical matter is Syntropy or Energy that has had physical limitations imposed upon it. Remove the physical limitations, and physical matter goes back to being Syntropy or Energy. Parsimonious. Logical. Simple. True.

Ironically, the proven and verified existence of Syntropy, the First Law of Thermodynamics, the Conservation of Psyche, or the Conservation of Energy falsifies the claims of Materialism, Naturalism, Darwinism, Nihilism, Atheism, and Classical Physics which state that Psyche or Syntropy does not exist. The Materialists and Naturalists teach that entropic physical matter is the only thing that exists, which means that the Materialists and Naturalists erroneously teach and believe that physical matter and entropy are conserved. They are wrong. Entropy and physical matter are NOT conserved! Their quantity changes over time. It's Psyche, Syntropy, Life Force, or Energy that's actually being conserved according to the First Law of Thermodynamics. In fact, physical matter is organized energy, localized energy, entropic energy, limited energy, or restricted energy. Physical matter is energy that has had physical limitations placed upon it. Energy can take on many different FORMS, including spirit matter, psyche, and physical matter. The Energy or the Syntropy is conserved; but, the physical matter, locality, and entropy come and go as God sees fit.

Can you see how severely Physicalism, Materialism, Naturalism, Darwinism, Nihilism, and Atheism can mess you up when it comes time to do science, study science, learn science, and understand science? Because these people have chosen to believe that ONLY entropic physical matter exists, these people erroneously claim and teach that physical matter and entropy are conserved; and, it destroys everything else that they encounter in science because they can't explain anything else that they encounter in science. These people have NO scientific explanation for Psyche, Syntropy, Non-Locality, Non-Physical Forces and Fields, Intangibles, Quantum Mechanics, Action at a Distance, Transdimensional Physics, Eternity, Immortality, Life Force, Consciousness, Intelligence, Thoughts, Quantum Waves, After-Death Memories, Perpetual Motion Machinery, and the many different Supernatural Mechanisms that have been experienced and observed. You are NEVER going to figure out what these things are and how they work by denying their existence. Materialism, Naturalism, Darwinism, and Atheism are Bad Science because they deny the existence of science at its most fundamental level.

The very existence of something intangible, like entropy, falsifies Materialism, Naturalism, Darwinism, Nihilism, and Atheism which claim that intangibles DO NOT EXIST. Naturalists and Atheists claim that intangibles do not exist. These people claim that only entropic physical matter exists; and, they are wrong. Physicalism, Naturalism, and their derivatives are self-defeating because intangibles such as entropy, gravity, space, non-locality, dark energy, action at a distance, radio waves, and time can be demonstrated and proven to exist. Likewise, Syntropy or Energy can also be demonstrated and proven to exist despite the fact that it is intangible most of the time. In contrast, the Naturalists and Atheists assure us that the intangible or the non-physical does not exist. They are wrong.

Remember, Syntropy is the Conservation of Energy. Syntropy is the First Law of Thermodynamics. Energy is Syntropy. Psyche is Energy. Syntropy is Energy. Psyche is the intelligence within all energy. Syntropy, Psyche, or Energy is syntropic which means that Syntropy or Energy is conserved. Syntropy or the Conservation of Energy has indeed been experienced and observed. It's infinitely more tangible and real than entropy. Syntropy or the First Law of Thermodynamics has been verified and proven to be true. The Materialists and Naturalists are wrong to say that it does not exist. The Materialists and Naturalists misunderstand the First Law of Thermodynamics, what it really is, and how it really works. The Materialists and Naturalists claim that physical matter and entropy are conserved, and that entropic physical matter is the only thing that exists. These people are demonstrably wrong. Materialism, Naturalism, Darwinism, Nihilism, and Atheism have been falsified by Syntropy, Psyche, the First Law of Thermodynamics, the Conservation of Psyche, and the Conservation of Energy.

I used to be a Materialist, Naturalist, Nihilist, and Atheist; but, science convinced me that I was wrong. Science convinced me that God exists. I'm too far gone now for the Naturalists and Atheists to reclaim me because I have seen through to the truth that is being hid behind all the lies. The Creator or the Maker has to be syntropic; otherwise, chaos, death, and entropy would reign supreme. The Order, Organization, and Syntropy have to come from someplace. They can't just spontaneously generate out of thin air from nothing as the Materialists and Naturalists claim.

Entropic physical matter is NOT the only thing that exists, contrary to what the Naturalists and Atheists claim. The double slit experiment of Quantum Mechanics tells us that the fundamental nature of reality is NOT physical at all. The Copenhagen interpretation of Quantum Mechanics says that the wave function doesn't have a physical nature. Quantum Mechanics, Syntropy, the Conservation of Psyche, and Quantum Non-Local Consciousness FALSIFY Physicalism, Materialism, Naturalism, Darwinism, Nihilism, Behaviorism, Determinism, Physical Reductionism, Classical Physics, Atheism, Entropy, and the Second Law of Thermodynamics.

At the quantum level, the First Law of Thermodynamics FALSIFIES and OVERRIDES the Second Law of Thermodynamics. In other words, Syntropy, the Conservation of Energy, or the Conservation of Psyche FALSIFIES or TRUMPS entropy or the second law of thermodynamics on both sides of the veil. Furthermore, it has been observed by Out-of-Body Travelers that there is NO thermodynamics or heat flow at the quantum level or the psyche level. Thermodynamics, or heat flow, or entropy is purely a classical mechanical physical phenomenon taking place at the physical level. Entropy or the Second Law of Thermodynamics has nothing to do with Quantum Mechanics and doesn't exist at the quantum level. Entropy is NOT a quantum particle, quantum force, or quantum field. Entropy doesn't exist as a substance or a thing. Entropy doesn't reign supreme after all. This just might be one of the most significant scientific discoveries of all time.

In fact, Thermodynamics doesn't exist at the quantum level, which means that both the First Law and the Second Law are technically meaningless and don't exist at the quantum level or the psyche level. The Conservation of Energy or the Conservation of Psyche ends up being the only thing that exists at the quantum level and in quantum mechanics. Physical Laws end up being nothing more than emergent properties of the underlying and REAL LAWS of Psyche, Energy, and Quantum Mechanics. In other words, Syntropy, Intelligence, or the Conservation of Psyche FALSIFIES or NULLIFIES entropy and the second law of thermodynamics.

This reality and truth is demonstrated to be true in the following video; but, you have to read between the lines in order to understand why it is so.

Reversing Entropy with Maxwell's Demon

https://www.youtube.com/watch?v=KR23aMjIHIY

https://evolution-is-entropy.com/wp-content/uploads/2018/08/Reversing-Entropy-with-Maxwells-Demon.zip

Psyche and Energy are being conserved, not physical matter and entropy. This is the Ultimate Law of Thermodynamics. The Ultimate Law of Thermodynamics differentiates between what is being conserved and what is not being conserved. Entropic physical matter is comprised of different forms of energy. The form is made, which means that the form is never conserved. Only the underlying energy is conserved. Psyche is the innate intelligence within all the different types of energy. Energy is psychic, telepathic, supernatural, intelligent, conscious, perceptive, sentient, responsive, alive, and aware. Energy is the ultimate perpetual motion machine because energy is conserved. Psyche or Energy is syntropic. It is conserved.

Syntropy, the Conservation of Psyche, or the Conservation of Energy really is the answer to life, the universe, and everything. Is it not? It explains everything. Does it not?

The Ultimate Law of Thermodynamics differentiates between what is conserved and what is not conserved. When it comes to science, it is absolutely essential to get this right from the very beginning; otherwise, all of your subsequent science and interpretations of science will be wrong. Energy and Psyche are conserved. They are syntropic. Syntropy is the First Law of Thermodynamics. Syntropy is the conservation of energy. Entropic physical matter is made from raw energy, which means that it can be disassembled and formed into something else instead. Consequently, entropic physical matter is NOT conserved. Physical matter and entropy are NOT syntropic. Entropic physical matter is comprised of different forms of energy. The form is never conserved because the form can be transformed into something else as the Gods see fit to do so. Only the underlying energy is conserved.

Technically, the Ultimate Law of Thermodynamics is not a new scientific discovery – it is carefully hidden within the First Law of Thermodynamics where the Atheists and

Naturalists cannot see it nor find it. I make a big deal of the Ultimate Law of Thermodynamics because it repudiates and falsifies Materialism, Naturalism, Darwinism, Nihilism, and even Atheism. The Ultimate Law of Thermodynamics teaches that Psyche or Energy is conserved; whereas, entropic physical matter is NOT conserved. Think about it logically. Entropy is some type of invisible, intangible, non-physical, emergent property of physical matter; and, the declared existence of something intangible like entropy falsifies Materialism, Naturalism, Darwinism, Nihilism, and Atheism which teach that ONLY physical matter exists and that intangibles do not exist. Scientific Naturalism is shot through with holes. Syntropy is the First Law of Thermodynamics – the conservation of energy; and, the verified and proven existence of Syntropy falsifies Materialism, Naturalism, Darwinism, Nihilism, Atheism, and their derivatives. Energy and Psyche are syntropic which means they are conserved. The false is falsified by the truth; and, the truth is repeatedly experienced and observed.

Remember, physical matter and entropy are made or caused to begin which means that they are NOT conserved. This explains everything that we have ever experienced and observed. Psyche or Syntropy is infinitely more tangible and real than entropy because Psyche or Energy is conserved; whereas, entropy is not.

The verified and proven existence of Syntropy, the Conservation of Psyche, or the Conservation of Energy falsifies Materialism, Naturalism, Darwinism, Nihilism, Atheism, Classical Physics, and even the Theory of Evolution which state that Psyche or Syntropy does not exist. The false is falsified by the truth; and, the truth is repeatedly experienced and observed. Psyche, Syntropy, Life Force, or Energy has been experienced and observed; whereas, physical matter and entropy are never conserved.

There you have the answer to life, the universe, and everything.

Mark My Words

A Philosophical Justification for Psyche

Kelly, E. F., Kelly, E. W., Crabtree, A., Grosso, M., & Gauld, A. (2007). *Irreducible Mind: Toward a Psychology for the 21st Century*. Plymouth, United Kingdom: Rowman and Littlefield.

The book, *Irreducible Mind*, introduced an interesting phenomenon that I had never encountered before. Some of the Materialists, Naturalists, Darwinists, Nihilists, and Atheists use the First Law of Thermodynamics or the Conservation of Energy to FALSIFY Cartesian Dualism and the existence of Psyche, Soul, or Non-Local Consciousness. These people equate energy with physical matter through $E = mc^2$; and then, they state that they can't possibly imagine how some invisible intangible non-physical substance can influence and control physical matter. In a fascinating twist, these people state that Cartesian Dualism, Non-Physical Substances, and Psyche violate the First Law of Thermodynamics which they define as the Conservation of Physical Matter. These people equate physical matter with energy, and then claim that the First Law of Thermodynamics or the Conservation of Physical Matter falsifies Psyche, Syntropy, Non-Physical Substances, and God. Their sophistry is simply amazing.

Your chosen worldview determines the definitions that you give to scientific concepts irrespective of what has actually been experienced and observed. The Materialists, Naturalists, Darwinists, Nihilists, and Atheists teach and preach that ONLY entropic physical matter exists; therefore, these people consistently define BOTH the First Law of

Thermodynamics and the Second Law of Thermodynamics as the "Conservation of Physical Matter" and the "Conservation of Entropy". One error leads to another, which then leads to another.

I own and have read from many different books and articles that use philosophical sophistry to prove that psyche, mind, consciousness, choice, or syntropy does not exist. The best they can do is to convince you that psyche, spirit, or soul does not exist. They really can't prove that it doesn't exist. It's impossible to prove that something doesn't exist; but, the Materialists, Naturalists, and Atheists try to do so anyway.

I also own and have read from many different books and philosophers who use philosophical sophistry to provide philosophical justification or a philosophical foundation for psyche, mind, consciousness, awareness, choice, and syntropy. Although interesting to study and think about at times, it's really not convincing either if you choose not to believe in it. Our beliefs are chosen into existence by our Human Psyche; and, the Human Psyche can choose to believe that the Human Psyche doesn't exist.

I prefer the scientific approach. I define science as observation and experience. Psyche has been experienced and observed; therefore, we KNOW that it exists. Quod erat demonstrandum! Simple. Logical. Parsimonious. Real. True.

There are times when I don't like philosophy and philosophers. They can say a lot without actually saying anything useful or interesting. It ends up being lots of obtuse, unintelligible, circular, tautological wordiness that gets us nowhere when they are done. The Materialists, Naturalists, Nihilists, and Atheists are philosophers, not scientists. These people deny the observations and experiences of the human race; and then, they replace those observations and experiences with philosophical sophistry instead.

I have learned to despise philosophical sophistry that never actually gets to the point and never gets to the point of making sense. An essay and book called "Reconsidering Psychology" by Faulconer and Williams is a case in point – there are sixty pages of words within the first essay of the book that never seem to come to a point and never seem to make a point. They say a lot without actually saying anything useful or interesting; and, when you are done, you still have no idea what point they were trying to make. Science needs a philosophical and metaphysical foundation; but, it also needs to make sense when they are done with it.

Instead of philosophical sophistry, I like getting to the point, revisiting the point over and over again from many different angles, and hammering on the point to the point of exhaustion until it becomes solid, obvious, and clear to me how things really work. I learn best by trying to explain my scientific theories, hypotheses, and ideas to someone else. I want everything to be plain and simple to understand. I want everything to be obvious, clear, and to-the-point when I'm done with it. I like power points. I like scoring points. It's what I choose to do.

Okay, half-way into their essay "Reconsidering Psychology" Faulconer and Williams make the point that psyche, or consciousness, was defined differently by the ancients than it currently is by the moderns. The ancients defined it as Divine-centered; and, the moderns define it as self-centered. I prefer to define it as what it really is.

Psyche is the innate intelligence within all the different forms of energy which gives that energy the inherent ability to understand, follow, and obey God's Laws and God's Commands.

My definition is axiomatic and LAW. It's the Law of Psyche. The Gods have observed that all the different forms of energy have some level of intelligence or some degree of intelligence within them, which the Gods call "Intelligence" and we humans call "Psyche" or "Soul". Energy can be formed into many different things, including entropic physical matter; and, energy's intelligence or psyche comes with it no matter what that energy gets formed into. Psyche is the intelligence within energy; and, all the different forms of energy are intelligent and psychic.

The first law of thermodynamics is a version of the law of conservation of energy, adapted for thermodynamic systems. The law of conservation of energy states that the total energy of an isolated system is constant; energy can be transformed from one FORM to another but can be neither created nor destroyed.

https://en.wikipedia.org/wiki/First_law_of_thermodynamics

I choose to give pre-eminence to the Conservation of Energy or the Conservation of Psyche. The First Law of Thermodynamics is the Conservation of Energy or the Conservation of Psyche. Syntropy is the First Law of Thermodynamics. Syntropy is Eternal Life. The Life Force or Psyche is always conserved. Technically, my definition for the First Law of Thermodynamics also falsifies or contradicts Cartesian Dualism. I treat energy or psyche as the primal construct. Therefore, physical matter and spirit matter are simply different FORMS of Organized Energy. Physical matter is Organized Energy or Concentrated Energy. It's ALL energy. Entropic physical matter, syntropic physical matter, spirit matter, and dark matter are ALL comprised of Energy; and, Psyche is the innate intelligence within all the different FORMS of energy, including physical matter and spirit matter or dark matter.

Gravity, magnetism, the nuclear forces, quantum waves, dark energy, invisible intangible forces or fields, quantum fields, and psyche are different FORMS of Energy. Entropy is an emergent property of entropic physical matter, which is made from different FORMS of Energy. The spiritual or the quantum or the non-physical is NOT a different substance. It's ALL energy, including both the physical matter and the spirit matter or dark matter. Physical matter is comprised of different FORMS of energy. Physical matter is comprised of the same substance as spirit matter and psyche. It's ALL energy. Energy can influence energy. Psyche is Energy. Psyche or Non-Local Consciousness can influence and control the different FORMS of energy, including spirit matter and physical matter.

Energy or Psyche is conserved meaning that it is syntropic. It cannot be made, and it cannot be destroyed. It is eternal and everlasting. Entropic physical matter is made from raw energy or available energy, which means that entropic physical matter is NOT conserved. The quantity of physical matter and entropy changes over time, and they can be transformed into something else besides entropic physical matter anytime the Gods decide to do so. These truths are the Ultimate Law of Thermodynamics.

The Ultimate Law of Thermodynamics differentiates between what is being conserved and what is not conserved. Anything that is made is NOT conserved. Energy and the innate intelligence within that energy are being conserved. Psyche or Energy is conserved. In contrast, entropic physical matter is made from raw energy. Anything that is made is not being conserved. Entropic physical matter is made from different FORMS of energy. The FORM is never conserved. The FORM can always be transformed into something else. Physical matter and entropy are NOT conserved. This explains everything that has been observed. This is the Ultimate Law of Thermodynamics.

When it comes to the answer to life, the universe, and everything – it ALL comes down to how you choose to define "Conservation of Energy". The Physicalists, Naturalists, Darwinists, Nihilists, and Atheists define it as the "Conservation of Physical Matter" or the

"Conservation of Entropy" because these people erroneously believe that ONLY entropic physical matter exists. In contrast, I define "Conservation of Energy" as Syntropy, Eternal Life, Life Force, or the Conservation of Psyche. We both can't be right. These ideas are mutually exclusive. The verification of the one falsifies the other. Science is observation and experience. The Conservation of Psyche or the Life Force has been experienced and observed; whereas, physical matter and entropy are never conserved – their amount or quantity is changing all the time.

So, how is Psyche or Intelligence identified or recognized? It all comes down to choice. Choice is a product of Psyche, Intelligence, Consciousness, or Awareness. Psyche is synonymous with choice, and a specific Psyche or Intelligence is cornered, pin-pointed, and identified by the choices that it makes.

The Gods can form energy into anything and can control everything, except for our choices or our psyche. Psyche, Choice, or Intelligence is the fundamental unit of reality. This is a Psyche Ontology. The Gods cannot create Psyche or Intelligence, and the Gods can't destroy it either, which means that the Gods cannot control our choices. The Gods can determine consequences for our choices; but, they cannot make our choices for us because then they would no longer be choices. It all comes down to choice. Choice is a function of psyche or intelligence. Choices are produced by psyche or intelligence. Psyche or Intelligence is identified by the choices that it makes. Choice is the surest sign of Psyche's existence.

As I see it, the only philosophical justification that psyche or mind really needs is the empirical fact that it has been experienced and observed. Science is observation and experience; and, Psyche has been experienced and observed, therefore we KNOW that it is real and truly exists. That's all the philosophical justification that it needs. Observation and experience trump philosophical sophistry every time. Once it has been experienced and observed, the philosophical debate should be over. Of course, that doesn't stop the Materialists, Naturalists, Darwinists, Nihilists, and Atheists from trying to convince you otherwise; but, these people are wrong to do so.

Psyche or Intelligence is experienced first-hand by out-of-body explorers as an immaterial viewpoint in space. Whenever they go looking for their own psyche, there's nothing there to be seen or found. Psyche or Intelligence is seen as a pinprick of light, a point particle of light, or a spark of light.

If there is such a thing as a point particle or an infinite singularity, Psyche is it. Psyche, Intelligence, or Thought is the fundamental unit of reality. Psyche is the "elementary particle" that the physicists are looking for. A psyche ontology is the ultimate model of reality and existence. Psyche is the ultimate cause, which means that psyche is the ultimate causal agent or the ultimate causal force. Psyche gets straight to the point.

Out-of-body explorers have noticed that it is their immaterial intangible Psyche or Consciousness who has experiences and forms memories, not their spirit body and not their physical body. Your spirit body and physical body don't have experiences and don't form memories while your psyche is separated from them. It's the Psyche or the Intelligence who has experiences and forms memories of those experiences. That is what has been experienced and observed.

Psyche or Intelligence has been experienced and observed by out-of-body explorers and during near-death experiences; therefore, we KNOW that it is real and truly exists. That's all the philosophical justification and philosophical foundation that it needs.

Mark My Words

—

Source

The Ultimate Model of Reality: Psyche Is the Ultimate Cause

https://www.amazon.com/dp/B071NC9JK6

Reference

Buhlman, William L. (1996). *Adventures Beyond the Body: How to Experience Out-of-Body Travel*. New York: HarperCollins.

Psyche and Energy Are Essentially Synonymous

I'm a scientist. I choose to believe in what has been experienced and observed.

I define Science as observation and experience. Science is the study of everything that has ever been experienced and observed. Nothing should be off-limits. As scientists, our ultimate task is to figure out what things are and how they work. Our ultimate goal is to try to find a truthful, logical, and realistic explanation for everything that has ever been experienced and observed on both sides of the veil. The truth, or reality, consists of everything that has ever been experienced or observed. The ultimate goal of the scientist is to pursue and find the truth – namely, to pursue and find everything that has ever been experienced and observed. The BEST and FASTEST way to find and know the truth is to live it, experience it, observe it, and witness it first-hand for yourself – or to choose to trust someone who has.

Many of our modern-day "scientists" take it upon themselves to ridicule evidence, mock evidence, censor evidence, ban evidence, block evidence, hide evidence, suppress evidence, and destroy evidence. I used to be a Materialist, Naturalist, Nihilist, and Atheist. These falsified philosophies, religions, or worldviews are based exclusively upon a refusal to look at any evidence that falsifies them. The hallmark of Atheism is a refusal to look at evidence. I KNOW because I have been there and done that.

I have changed. I realize now that the job of a scientist is to figure out what everything is and how it really works rather than censoring, banning, suppressing, hiding, and destroying the evidence. I want to discover and explore everything that has ever been experienced and observed, and then try to figure out what things are and how they work. I have now done so. I abandoned my former path and have started down this path instead.

Scientific Definition: Psyche is the innate intelligence within all the different forms of energy which gives that energy the inherent ability to understand, follow, and obey God's Laws and God's Commands.

Scientific Observation: Events must have causes.

Scientific Observation: Psyche has been experienced and observed. Out-of-Body Travelers have experienced first-hand their own individual psyche as an immaterial viewpoint in space. Whenever they are separated from their spirit body looking at their spirit body, there's nothing to be seen or found whenever they go looking for their psyche. Psyche is experienced as an immaterial viewpoint in space. While out-of-body, psyche is seen or observed as a pinprick of light, a spark of light, or a point particle of light. That is what has been experienced and observed.

Scientific Observation: Out-of-Body Travelers have observed that while separated from their spirit body looking at their spirit body, it is their immaterial psyche who is having the experience and forming memories of the event – NOT their spirit body and NOT their physical body. A physical body and a spirit body don't have experiences and don't form memories while the Human Psyche is separated from them. It is the Psyche or Intelligence who experiences events and then forms memories of those experiences. That is why our thoughts, memories, and personality survive the death of our physical body and physical brain.

Scientific Conclusions: Based upon these Scientific Observations, it is logical and rational to conclude that Psyche is the Ultimate Cause. In other words, events are caused by Psyche, and events are experienced and remembered by Psyche. Events are chosen into existence by Psyche. The events associated with Quantum Mechanics are caused by Nature's Psyche, the Human Psyche, or God's Psyche. Psyche is the ultimate causal agent. Furthermore, there is NO event that caused Psyche or Energy to begin. Energy, and the psyche or intelligence within that energy, has always existed and will always exist. Psyche or Energy is syntropic meaning that it is conserved. Psyche or Energy cannot be created, and it cannot be destroyed. It is eternal and everlasting. Psyche is indeed the Ultimate Cause and the ultimate causal agent. Events must have causes. Some sort of Psyche is ultimately the cause of every event. In contrast, there was NO event that produced Psyche and Energy, or that caused Psyche and Energy to begin. These scientific truths explain everything that has ever been experienced or observed. These scientific truths are the answer to life, the universe, and everything.

Whenever the Gods interact directly with Raw Energy, the Gods are in fact interacting directly with the Psyche or the Intelligence within that Energy. Psyche and Energy are essentially synonymous, or reside at the same level of existence, namely the quantum level or the psyche level. Psyche and Energy are syntropic, which means that they are always conserved. Syntropy is conservation of energy. Psyche and Energy are eternal and everlasting, without a beginning of days or an end of years. Psyche and Energy are syntropic, not entropic. Energy is intelligent, conscious, psychic, perceptive, and aware. There's a spark of psyche or intelligence within every physical atom and within every part of a physical atom because physical atoms and their component parts are made by the Gods from energy.

The Gods can FORM raw energy into anything, which means that the Gods can control everything, except for a Psyche's choices. Psyche, Energy, or Intelligence cannot be made which means that it cannot be destroyed. It has always existed, and it will always exist. Psyche or Energy is conserved, which means that Psyche's choices are also conserved or allowed to stand. The Gods cannot control our choices nor make our choices for us because Psyche or Energy is conserved. Psyche is synonymous with choice. Each Psyche or Intelligence is identified by the choices that it makes. Psyche is also synonymous with energy. Psyche is the intelligence or consciousness within energy that gives energy the ability to obey God's Laws. Any massless elementary particle is some type of Psyche.

Quantum Mechanics is ultimately disappointing for the natural man or the physical man because physical matter was designed by the Gods to prevent us from gaining direct access to and control over Quantum Mechanics or Raw Energy. It's hard to make money from spiritual mechanisms or quantum mechanisms that we have no reliable control over. The physical laws, physical limitations, and physical restrictions within physical matter typically prevent us from controlling quantum mechanisms directly with our psyche or our mind. Physical matter was designed by the Gods to do just that – restrict our access to

Quantum Mechanics or Supernatural Mechanisms. The explanatory power of Psyche and Quantum Mechanics is through the roof to infinity and beyond; but, the practical applications are rather limited unless the Gods choose to give you direct access to Quantum Mechanics. Quantum Mechanics is God's Priesthood Power. God is selective as to whom He chooses to give that power or a portion of that power.

The Human Psyche and the choices that it makes are also ultimately disappointing for the natural man or the physical man as well because there is no way to reliably predict, control, and determine what an individual Psyche might choose to do next. Psyche can change its mind at any time and choose to do something completely different next time. Failure is always in the past, and there is no guarantee that the Human Psyche will continue to fail. There's no way to reliably predict what the Human Psyche might choose to do next time around. He or she might choose to do the same thing or might choose to do something completely different instead. There's no way to know for sure.

The physical matter within our physical body serves as a dampener or a veil over our Psyche preventing us from gaining direct access to and control over Quantum Mechanics or God's Priesthood Power without God's permission. Physical matter is the veil of forgetfulness that the Gods occasionally talk about whenever they try to explain to us why we no longer have direct access to and control over the spiritual or the quantum mechanical. The physical laws, physical restrictions, and physical limitations were deliberately designed by the Gods to limit and restrict our access to Quantum Mechanics; and, they work as designed. It works so well that some of us have chosen to believe that Psyche, Syntropy, Supernatural Mechanisms, Action at a Distance, Non-Locality, Spirituality, God, and Spirit Matter do not exist.

I know how it goes. I used to be a Materialist, Naturalist, Nihilist, and Atheist until the observations and experiences of the human race (the scientific evidence or the observational evidence) convinced me that I was wrong.

According to the Ultimate Law of Thermodynamics, Psyche, Intelligence, Energy, Syntropy, or our Life Force is conserved. In contrast, the different FORMS of energy are not conserved. The Gods can change energy's FORM anytime they choose to do so. Remember, the FORM can be changed, which means that it isn't conserved. Entropic physical matter is comprised of many different FORMS of energy. Entropic physical matter was designed and made by the Gods, which means that entropic physical matter can be disassembled or unmade and turned back into raw energy or available energy instead. When it comes to Energy, the FORM can be changed by the psyche who controls the Energy, which means that the FORM is never conserved. Only the underlying Energy or Psyche is being conserved.

In summary, the FORM is never conserved according to the Ultimate Law of Thermodynamics. Entropic physical matter is comprised of different FORMS of energy. This means that physical matter and entropy are not conserved. Their amount or quantity is constantly changing as the Gods see fit. Human beings, the children of the Gods, create new physical matter and its associated entropy in their particle accelerators or atom smashers. Human beings also destroy physical matter in their atomic bombs converting parts of it back into raw energy or unorganized energy. Anything that was formed, organized, or made is never conserved. It can be unmade and formed into something else instead. The FORM is never conserved. Physical matter, physical laws, physical constants, physical restrictions, and entropy were formed by the Gods or made by the Gods, which means that these things can be disassembled and formed into something else instead. In contrast, Psyche and Energy cannot be made which means that they cannot be destroyed, which means that they are always conserved. Energy and Psyche are syntropic which means that they are eternal and everlasting without a beginning of days or an end of years.

Do you see how that works?

This is the Ultimate Law of Thermodynamics. It's the answer to life, the universe, and everything.

The Ultimate Law of Thermodynamics states that physical matter and entropy are NOT conserved. Anything that was obviously formed or made obviously can be disassembled and formed into something else, which means that the form is not being conserved. Entropic physical matter is comprised of different forms of energy. Physical matter and the emergent entropy were obviously formed or made from raw energy, which means that physical matter and entropy can be disassembled, and the resulting raw energy or available energy then transformed into something else besides entropic physical matter. The First Law of Thermodynamics states that energy or psyche is conserved, not the form of that energy. The form is never conserved according to the First Law of Thermodynamics and the Ultimate Law of Thermodynamics. Entropy is death. Entropy is an expiration date. Death is NOT conserved according to the Ultimate Law of Thermodynamics. In contrast, your Psyche or Energy or Life Force is eternal and everlasting. Your Life Force, Psyche, or Energy is syntropic which means that it's always conserved. Energy or Psyche cannot be made, and it cannot be destroyed which means that it is conserved.

Life Force, Psyche, or Energy is the ultimate perpetual motion machine because it's always conserved. In fact, entropic physical matter, the aging process, the passage of time, and entropy were designed and made by the Gods in order to put a temporary end to perpetual motion machines where physical matter and a fallen physical world are concerned. Do you see how that works? It's the answer to life, the universe, and everything. It explains everything that has ever been experienced or observed.

Entropy is death. In the Garden of Eden, Adam and Eve had immortal syntropic physical bodies. The fruit from the tree of knowledge introduced entropy or death into their physical bodies. That fruit converted their immortal syntropic physical bodies into mortal entropic physical bodies instead. Their syntropic physical bodies were perpetual motion machines – eternal and everlasting. Entropy puts a temporary end to immortal physical bodies and perpetual motion machines. In order to produce a perpetual motion machine, remove the entropy from it. In order to put a temporary end to immortal physical bodies and perpetual motion machines, introduce entropy into it. That's how things really work. It has been experienced and observed.

With the internal psyche's permission, the Gods can FORM energy into anything that they want it to be, including physical matter, entropy or unavailable energy, space, time, locality, spirit matter (tachyons and dark matter), gravity, magnetism, visible and invisible light, radio waves, microwaves, x-rays, gamma rays, the strong and weak nuclear forces, chemical bonds, gluons, bosons, quarks, vibrating strings, physical laws, physical constants, invisible intangible forces and fields, quantum waves, quantum mechanics, information, quantum memory storage, intangible non-physical quantum computers, quantum fields, and the zero-point field of light or the vacuum energy (dark energy). It's ALL energy. These are all different FORMS of Energy, and the FORM is never conserved. The FORM can be changed or transformed.

Remember, physical matter, physical laws, and entropy were formed by the Gods or made by the Gods, which means that the Gods can change their FORM anytime they choose to do so. Physical matter and entropy are never conserved. Entropic physical matter is formed or made by the Gods. The Gods can FORM energy into anything that they want it to be. That reality means that the Gods can also make or form syntropic physical matter if they choose to do so – physical matter that doesn't have any entropy or an aging process built into it – immoral physical matter or syntropic physical matter. In 1 Corinthians 15, the

Apostle Paul calls syntropic physical matter "spiritual" matter – a resurrected physical body is a physical body that has no entropy within it, which means that a resurrected physical body or a "spiritual" physical body is immortal and cannot age or die. ALL of the energy within a syntropic physical body is available for use.

Physical matter doesn't have to be made with entropy within it or associated with it. The entropy can be left out of the physical matter if the Gods so desire. In other words, the entropy is never conserved. Entropy is an add-on that the Gods can remove anytime they see fit to do so. Entropy is death. There's nothing eternal or everlasting about death or entropy contrary to what the Materialists, Naturalists, and Atheists are trying to get us to believe. These people are trying to convince us that entropy or death is conserved; but, they are wrong.

Entropy is the result of the fall of Adam and Eve, where our physical bodies are concerned. Adam and Eve were made with syntropic physical bodies or immortal physical bodies, and when they fell God introduced entropy into their physical bodies. Entropy is death. Entropy is an expiration date. There's nothing special, sacred, or sacrosanct about entropy because entropy is not conserved. Death is not conserved. Entropy was made by the Gods which means that entropy can be unmade or eliminated whenever the Gods decide to do so. Death is impossible at the psyche level because Psyche or Energy is always conserved or preserved.

The Gods created entropic physical matter to prevent us from developing superpowers at the physical level. It was designed to keep us humble. The Gods created entropy to prevent spontaneous generation, chemical evolution, and macro-evolution where entropic physical matter is concerned. Our genes will never turn us into X-Men. Instead, our genes and the entropic physical matter within our physical bodies prevent us from becoming X-Men and gaining direct access to Quantum Mechanics and Superpowers. The whole purpose of entropy and entropic physical matter is to put an expiration date on our physical bodies and to prevent us from becoming immortal and gaining direct access to Quantum Mechanics, Superpowers, or God's Priesthood Power.

These are some of the lessons that we learn from the Ultimate Law of Thermodynamics. The Ultimate Law of Thermodynamics explains everything that we have ever encountered, experienced, or observed as a race. The Ultimate Law of Thermodynamics differentiates between what is being conserved and what is not conserved.

Physical atoms are obviously made or organized at the quantum level by Someone Psyche and thereby brought into existence or born at the physical level. The Gods have to organize and form raw energy or unorganized energy into physical atoms and physical laws, or physical matter would never exist.

Dark energy or the light of Christ is the power by which the planets and stars were made. It takes a massive infusion of dark energy, the light of Christ, or vacuum energy to give birth to a physical planet, or a physical star, or a physical galaxy. The raw energy is there waiting to be used. We look through our telescopes and we observe that physical galaxies are born all at once, and then they start to wind up from there. It takes a massive infusion of dark energy, the light of Christ, or vacuum energy to stretch and expand a physical universe as is currently happening to our physical universe. It takes a constant infusion of dark energy, the light of Christ, or vacuum energy to make our physical earth expand and grown. The explanatory power of dark energy or the light of Christ is impressive. It proves that our physical universe is not a closed system but is currently being made and formed by the Gods.

Anything that is obviously organized obviously has an Organizer who organized it. Any Law that's being obeyed obviously has a Law Maker and a Law Enforcer who brought it into existence. It also has Someone Psyche who is choosing to obey that Law. Anything that was obviously formed or made obviously has a Maker and a Fine-Tuner who formed it and made it. Physical matter, entropy, time, space, the physical constants, physical restrictions, physical limitations, and physical laws were obviously formed and made which means that they obviously have some kind of Maker or Fine-Tuner who made them. Despite the claims of the Materialists and Naturalists, physical matter and physical organization don't just spontaneously generate out of thin air from nothing. Spontaneous generation and creation ex nihilo are physically impossible. They are prevented from happening by entropy and the first law of thermodynamics respectively.

Physical machinery such as proteins, genes, nano-machines, and other molecules are made, organized, and constructed from the quantum level as well by Someone Psyche. Entropy prevents molecules from spontaneously generating into complex FORMS such as genes, functional genomes, nano-machines, proteins, enzymes, and living cells. Entropy was designed by the Gods to prevent spontaneous generation, chemical evolution, or macro-evolution. Because the Scientific Naturalists teach that only entropic physical matter exists, Materialism, Naturalism, Darwinism, Natural Selection, and the Theory of Evolution are defined as Creation by Entropy or Creation by Death which means that evolution of any kind is physically impossible. There's no way in the universe that entropy or death will ever be caught in the act of design and creation. A random fluctuation in entropy is NEVER going to produce life. It's physically impossible.

The atoms within molecules are linked together through a reaction known as chemical bonding. Chemical bonding is an invisible, non-physical, intangible force or field that binds atoms together into molecules. Chemical bonding is Action at a Distance, or some type of telekinesis. Technically, Action at a Distance is a quantum mechanical process, not a physical process. Action at a Distance is one of the surest signs that you have caught Quantum Mechanics and Someone Psyche in the act. Some atoms will spontaneously self-assemble into molecules due to the natural effects of chemical bonding; but, the complex assembly of molecules into functional genomes, proteins, enzymes, nano-machines, physical cells, and physical bodies is handled by Someone Psyche at the quantum level because entropy prevents it from happening naturally at the physical level. However, chemical bonding is just another FORM of energy; and, the FORM is always formed or made by the Gods which means that the FORM can be disassembled back into raw energy or available energy and then formed into something else instead. The FORM is never conserved.

According to the Ultimate Law of Thermodynamics, the Psyche or Energy is always conserved; but, the FORM of that energy is never conserved. The God's can change energy's FORM anytime they choose to do so. The God's can FORM energy into anything they want it to be, including entropic physical matter, which means that physical matter and entropy are NEVER conserved. They can be unmade and formed into something else by the Gods whenever the Gods see fit to do so.

Entropic physical matter is not the only thing that exists. There are many different FORMS of energy in addition to entropic physical matter. Entropic physical matter is just one form of energy, and one form of matter. There are many others. The FORM is never conserved. Entropy is death. It's nice to know that entropy or death is not conserved. It's also nice to know that Psyche, or Energy, or the Life Force is always conserved. Syntropy means conservation of energy, or conservation of the Life Force. Psyche or Intelligence is always conserved; whereas, physical matter and entropy are never conserved. This is the Ultimate Law of Thermodynamics.

The Ultimate Law of Thermodynamics is hidden within the First Law of Thermodynamics which states that energy is conserved.

The first law of thermodynamics is a version of the law of conservation of energy, adapted for thermodynamic systems. The law of conservation of energy states that the total energy of an isolated system is constant; energy can be transformed from one form to another but can be neither created nor destroyed.

https://en.wikipedia.org/wiki/First_law_of_thermodynamics

Energy can neither be created nor destroyed. Psyche can neither be created nor destroyed. Psyche is the intelligence or the life force within energy that gives energy the ability to obey God's Laws and God's Commands. Psyche or Energy is conserved. The First Law of Thermodynamics tells us that the energy is constant, and that the energy is conserved – not the FORM of that energy. Energy can be transformed from one form to another, which means that the FORM is never conserved.

Transformed by whom?

Transformation of any kind requires Someone Psyche or Someone Intelligent to instigate, originate, initiate, and start that transformation. Any type of beginning requires Someone Psyche to cause it to begin; otherwise, random chaos would continue to reign supreme. Events must have causes. Ultimately, Someone Psyche ends up being the cause of every event. Every causal chain can eventually be traced back to Someone Psyche.

The Ultimate Law of Thermodynamics emphasizes that the FORM is never conserved, which means that the different FORMS or emergent properties of energy such as physical matter, entropy, space, locality, and time are never conserved. Whenever something is formed or made or organized, that means that its FORM is not being conserved. Only Psyche, Life Force, or Energy is syntropic or conserved. Energy, Intelligence, or Psyche cannot be created, and it cannot be destroyed. However, energy can be transformed into many different things by the Gods or by Someone Psyche. The FORM is never conserved. Only the underlying energy is conserved.

In summary, entropic physical matter is created by the Gods from raw energy or available energy which means that physical matter and entropy can be destroyed and turned back into raw energy or available energy once again. The Energy, or the Psyche, or the Life Force cannot be made, and it cannot be destroyed, which means that it's always conserved. However, the form of that energy is never conserved. The Gods can transform energy into anything they desire including physical matter and entropy. The Ultimate Law of Thermodynamics differentiates between what is being conserved and what is not being conserved, as a result it is good science.

The Materialists, Naturalists, Darwinists, and Atheists erroneously teach and believe that physical matter and entropy are being conserved because these people erroneously teach that only entropic physical matter exists. One error or falsehood leads to another.

Materialism, Naturalism, Physicalism, Darwinism, Nihilism, Behaviorism, Determinism, Physical Reductionism, and Atheism were designed to prevent us from discovering the Ultimate Law of Thermodynamics; and, they have done just that. Have they not? Have you ever heard of the Ultimate Law of Thermodynamics? Of course not. The Materialists, Naturalists, Darwinists, and Atheists have successfully hidden it from us; and, these are the people who currently control our public schools.

The Materialists, Naturalists, Darwinists, and Atheists erroneously teach that entropy or death is conserved. Do they not? The observed and proven fact that physical matter, entropy, and death are not conserved falsifies Materialism, Naturalism, Darwinism, and their derivatives as does the observed and proven fact that entropic physical matter is not the only thing that exists. The Ultimate Law of Thermodynamics and Scientific Naturalism are mutually exclusive. They both can't be true simultaneously. If one of them is true, then the other one is automatically false. They falsify each other. Your task in all of this, as with the whole of science, is to decide which model is true and which model is false.

I have observed that the false is falsified by the truth; and, the truth is repeatedly experienced and observed. It has been repeatedly observed by near-death experiencers and out-of-body travelers that entropy or death is not conserved. How else is that fact supposed to be experienced and observed?

Entropy is death. Entropy is an integral part of entropic physical matter; and, entropic physical matter is not conserved which means that the entropy associated with it isn't conserved either. There's nothing sacrosanct, fundamental, or primal about entropy or death contrary to what the Materialists and Naturalists claim, teach, and choose to believe. These people are wrong whenever they assert and teach that entropy or death is conserved. That's good news, isn't it?

The Ultimate Law of Thermodynamics states that physical matter and entropy are NOT conserved. Only Psyche, Life Force, or Energy is conserved. The Ultimate Law of Thermodynamics, or Syntropy, is the pinnacle of science. It explains everything else and overrides everything else. It explains everything that has ever been experienced or observed. The Ultimate Law of Thermodynamics is the answer to life, the universe, and everything.

Ultimately, death is impossible because entropy is NOT conserved. Entropy or death doesn't exist at the psyche level. Psyche, Energy, or Intelligence was not made, and it cannot be destroyed. Psyche or Energy is always conserved in every dimension of reality and existence. Psyche is existence. Your Psyche or Intelligence cannot die. This is the Quantum Law of Thermodynamics which states that heat death is impossible at the quantum level or the psyche level because the second law of thermodynamics doesn't exist in the Quantum Realm, Psyche Realm, Syntropy Realm, or Spirit World.

Psyche is Energy. All you need is energy to remove entropy from any system. Psyche is Intelligence or Knowledge. All you need is knowledge or information to remove entropy or disorder from any system. Psyche is Eternal Life or Life Force. All you need is Eternal Life or Life Force to remove entropy or death from any system. Psyche is Syntropy. All you need is Syntropy to remove entropy from any system.

Psyche is the innate intelligence within all the different forms of energy. Energy, Psyche, or Life Force is conserved according to the First Law of Thermodynamics. Entropy is death. The Ultimate Law of Thermodynamics states that Psyche, Syntropy, Energy, Intelligence, and Life Force are conserved; whereas, entropy, death, and physical matter are NOT conserved.

Based upon the First Law of Thermodynamics and the Ultimate Law of Thermodynamics, the Quantum Law of Thermodynamics states that heat death is impossible at the quantum level and that death is impossible at the psyche level because the second law of thermodynamics or entropy doesn't exist in the Quantum Realm or the Spirit World. This is what has been experienced and observed. The Quantum Law of Thermodynamics or Syntropy states that Energy, and the Psyche or Intelligence within that Energy, is eternal and everlasting and always conserved. The Quantum Law of

Thermodynamics also states that the Quantum Realm is an Isothermal System which means that there is NO anergy or NO entropy within the Quantum Realm and also means that ALL of the energy within the Quantum Realm is always available for use all the time. This has also been experienced and observed. Much more on this later!

The Quantum Law of Thermodynamics or Syntropy ends up providing us with the answer to life, the universe, and everything. It explains everything that has ever been experienced and observed.

Through the Quantum Law of Thermodynamics and the Ultimate Law of Thermodynamics, my goal and desire is to present the world with the greatest, most comprehensive, most useful, most realistic, and most truthful Model of Reality that has ever been given to the world to date. In order to succeed, I had to falsify and eliminate everything that is false in the hope that only the truth will remain when I am done. It's been fascinating to experience and observe what remains after we have eliminated everything that is false and everything that has been falsified such as Materialism, Physicalism, Naturalism, Darwinism, Nihilism, Behaviorism, Determinism, Physical Reductionism, Atheism, and Classical Physics. The truth remains because the truth has been experienced and observed by Someone Psyche.

The Ultimate Law of Thermodynamics and the Quantum Law of Thermodynamics are based exclusively on the observations and experiences of the human race as a whole. The observations and experiences of the human race led me to discover the Ultimate Law of Thermodynamics and the Quantum Law of Thermodynamics and convinced me that they are real and true. It's all been experienced and observed. That's what happens when you finally allow all of the evidence into evidence. You start to make scientific discoveries that nobody has ever thought of before.

Mark My Words

Another Way to Define Psyche

Psyche has been observed by Out-of-Body Travelers as a pinpoint of light, a pinprick of light, a point particle of light, or a spark of light. That's precisely the way that the physicists describe the fermions – quarks, leptons, neutrinos, and electrons. These "particles" of light OBEY the Pauli exclusion principle. Obedience to Laws requires intelligence! The components of the quarks and the electrons are individual pinpricks of Psyche or Intelligence; and, the psyches within fermions USE the bosons, gluons, photons, quantum waves, and light to communicate with each other, interact with each other, and bind themselves to each other. The elementary parts of Fermions are Psyche or Intelligence. That would make Psyche or Intelligence the fundamental unit of reality and existence. Photons, gluons, and W and Z bosons are the four force-carrying gauge bosons of the Standard Model. There's NO mass in the elementary components of the fermions, the psyches, or the pinpricks of light. Over 99% of the mass within an atom is associated with the gluons. Psyche or Intelligence is driving the mass or driving the gluons. It's all made from Energy which is always conserved.

Every massless elementary particle is some type of Psyche or Intelligence. Psyche controls mass or energy at the quantum level to get things done for us at the physical level. This truth is hiding in plain sight where nobody can see it.

"The property of the strong nuclear force is known as asymptotic freedom, and the particles that mediate this force are known as gluons. Somehow, the energy binding the proton together, the other 99.0% of the proton's mass, comes from these gluons."

https://www.forbes.com/sites/startswithabang/2016/08/03/where-does-the-mass-of-a-proton-come-from/

http://origin-science.org/wp-content/uploads/2018/09/Proton.pdf

https://medium.com/starts-with-a-bang/where-does-the-mass-of-a-proton-come-from-57129a06fdfc

http://origin-science.org/wp-content/uploads/2018/09/MASS.pdf

Technically, the elementary particles or psyches that make up a gluon are massless like photons; but, the force mediated or controlled by the gluon has or produces a lot mass or resistance to acceleration. Likewise, the elementary particles or psyches that make up the quarks and the electrons are massless like photons; but, the Higgs Field that interacts with the quarks and the electrons gives them their small amount of mass. Massless particles are psyches.

"Quantum fields contain excited states that we observe as particles. These quantum fields can be divided into matter fields (whose particles are electrons, quarks, etc.) and force fields (whose particles are photons, gluons, bosons, etc.). The Higgs mechanism gives mass not only to weak particles, but also to electrons, quarks, and other fundamental particles. The more strongly a particle interacts with the Higgs Field, the more massive it is. It's important to note, however, that most of the mass in composite particles, like protons, nuclei, and atoms, does not come from the Higgs mechanism, but from the binding energy that holds these particles together."

https://physics.aps.org/articles/v6/111

http://origin-science.org/wp-content/uploads/2018/09/Why-Particles-Have-Mass.pdf

This explains how everything works at the quantum level and the psyche level.

Mark My Words

The Law of Psyche

Every Psyche, Intelligence, Life Force, or Consciousness has a certain amount of energy that's under its control. The controlling psyche can form the energy under its control into anything that it wants that energy to be. Energy is infinitely malleable or infinitely transformable, because energy or psyche is always conserved.

Psyche has been experienced and observed by Out-of-Body Explorers and Near-Death Experiencers. Psyche or Consciousness is experienced as an immaterial viewpoint in space. Psyche is seen as a pinpoint of light or a spark of light. The First Law of Thermodynamics tells us that Psyche or Energy is always conserved, which means that it is eternal and everlasting. It cannot be made, and it cannot be destroyed. Psyche or Energy is syntropic which means that it is always conserved both at the quantum level and the physical level.

I prefer the scientific approach. I define science as observation and experience. Psyche has been experienced and observed; therefore, we KNOW that it exists. Quod erat demonstrandum!

Law of Psyche: The Law of Psyche states that every massless quantum particle is (or has) some type of Psyche or Intelligence. Psyche is the innate intelligence within all the different forms of energy which gives that energy the inherent ability to understand, follow, and obey God's Laws and God's Commands. Every massless elementary particle is some type of Psyche because it is psychic, quantum, telepathic, makes choices, can self-generate or self-propagate, can think, can communicate, can transmute, can move, can phase-shift, can collapse its own wave function, is perceptive, is sentient, is omniscient, is omnipresent, is intelligent, is conscious, is alive, must travel at the speed-of-light from our perspective at the physical level, and is prescient. Without any physical limitations imposed upon it, Psyche can quantum tunnel at will in its native original environment at the quantum level.

Remember, every massless quantum particle is (or has) some type of Psyche, or Intelligence, or Life Force. Psyche is seen by Out-of-Body Travelers as a pinpoint of light or a Photon. Each Photon is a type of Psyche or Intelligence. A Psyche is a Photon, a massless elementary particle. Psyche is energy or light. If it looks like a Photon and acts like a Photon, then it is some type of Psyche. Every massless quantum particle is a Psyche or has a Psyche. Psyche USES mass, energy, or matter to get things done. This is what has been experienced and observed.

Psyche has been hiding in plain sight all the time where nobody can see it nor find it because they aren't looking for it; but, there's no great mystery here. Psyche is seen as a Photon BOTH at the quantum level and the physical level because it is a Photon. Psyche or Intelligence IS what it appears to be, a Photon of Light or a Pinpoint of Light. No surprise there. Psyche is what it has been experienced and observed to be – a massless quantum of energy that looks like and acts like a Photon. Psyche, whenever it has been seen both at the quantum level and the physical level, looks like a Photon or a Pinpoint of Light.

I know that it's hard to believe because we have been trained, brainwashed, and conditioned all of our lives not to believe it; but, it is true, nonetheless. Psyche has been experienced and observed. Every massless quantum or every massless elementary particle looks like a photon and acts like a photon because it is effectively a Controlling Psyche who USES energy, mass, or matter to get things done both at the quantum level and the physical level.

Mark My Words

Explaining and Defining Entropy

I was erroneously led to believe that Entropy is one of the elementary forces, fields, or quanta. I'm sure that I'm not the only scientist who has been given that impression. I truly believed that Entropy is made from energy, and that Entropy is one of those detectable and measurable "particles" or quanta, or one of those detectable and usable forces or fields within the universe. I visualized entropy as something that has been organized and made from raw energy. I thought Entropy was a real thing. I thought that

Entropy is one of the elementary forces in our universe. I was wrong. Entropy has no useful form. Entropy is an emergent property, and not an actual thing. Entropy doesn't exist as a person, place, or thing.

I used to be a Materialist, Naturalist, Nihilist, and Atheist; and, it's still very much with me to this very day. As a result, I have had to unlearn many of the things that I was taught and used to believe. We Materialists and Naturalists teach and believe that ONLY entropic physical matter exists. As a result, we truly believe that entropy or death is conserved. We believe in the Supremacy of Entropy or the Supremacy of Death. We believe in the Conservation of Entropy or the Conservation of Death. We believe that the universe will end in Heat Death. We teach, preach, and believe that Heat Death is permanent, that Heat Death is conserved, and that Heat Death is the Fundamental Law of our Universe. Within our minds, the Second Law of Thermodynamics overrides everything else and dominates everything else. For us the Second Law is sacred and inviolate. You could even say that it is our god. We worship it with reverence, honor, and respect.

We come to see Entropy as a thing – an omnipresent and omnipotent overlord, field, force, or god – much in the same way that the Darwinists view Evolution and Natural Selection. We come to see Entropy as one of the fundamental forces of nature – some type of quantum field. We even call it a LAW, the Second Law of Thermodynamics. In jest, we also call it Murphy's Law. For us, Entropy and Heat Death are sacrosanct. The Second Law is SACRED. That's no exaggeration. That's precisely the word that we use to describe the Second Law of Thermodynamics whenever we talk about it. Across the internet, we declare Entropy to be the Ultimate Law in the Universe. We really, truly hope and believe that entropic physical matter is the ONLY thing that exists, and that death or entropy is conserved. But, are we right? No, we are not. We are demonstrably wrong. Everything that has ever been experienced and observed at the quantum level proves that we are wrong. Science is observation and experience – not wishful thinking.

Breaking the groupthink, brainwashing, and conditioning hasn't been easy for me. My error in regard to Entropy was first pointed out to me in the following PBS Space Time YouTube Videos, which I will now try to analyze in order to expose my thought processes:

The Misunderstood Nature of Entropy

https://www.youtube.com/watch?v=kfffy12uQ7g

https://evolution-is-entropy.com/wp-content/uploads/2018/08/The-Misunderstood-Nature-of-Entropy.zip

I had a ton of cognitive dissonance and was severely disturbed the first couple of times that I watched this video.

Why?

Well, it's subtle because this video does indeed promote Materialism, Atheism, Naturalism, and the Party Line. I used to be a Materialist, Naturalist, Nihilist, and Atheist; and like the person in this video, I truly believed that entropy is one of the fundamental laws of nature and one of the fundamental forces in the universe. Unlike this individual, though, I thought that entropy was some type of energy, force, or field. I thought that entropy was a tangible and real thing.

So, what's the problem?

That's what everyone believes, isn't it? We believe in the Supremacy of Entropy. Our lives are based upon entropy. Death surrounds us on all sides; and, entropy is death. The Materialists and Naturalists teach and believe that entropy or death is conserved. Do they not? The existence of death is obvious. Is it not? Death is real, so entropy is real, and entropy or heat death is unavoidable. Everyone knows that!

So, what's the problem?

Well, the problem was that the evidence doesn't support the truthfulness of our chosen beliefs. Instead, the actual evidence falsifies those beliefs; and thus, ALL of the cognitive dissonance that I experienced while watching this video. I could sense the inconsistencies and contradictions. I had convinced myself that entropy is made from energy, and that entropy is some type of fundamental force or field within nature and throughout the universe. I was surprised to find out that I was wrong; and, I didn't like the fact that I was forced to change the way that I think about entropy and talk about entropy.

It's annoying when your favorite ideas and theories are proven false. It takes a while to get used to it and accept it.

There's nothing fundamental, permanent, or lawful about entropy. Just introduce Psyche, or Energy, or Intelligence into the system, and that's the end of entropy! Just transform entropic physical matter into something else, and that's the end of entropy! Entropy is NOT conserved.

Ouch!

Entropy isn't anything like what I thought it was. I didn't like that at first. I thought entropy was something. Instead, it ended up being nothing. Entropy is much ado about nothing.

Entropy is NOT one of the elementary particles. Entropy is NOT a quantum particle. Entropy is NOT a quantum field. Entropy is NOT one of the fundamental forces of nature and the universe. In fact, entropy doesn't really exist except only in our mind's eye or imagination as a thought experiment. Entropy is an emergent property of statistical mechanics. Entropy shouldn't be elevated to the status of a LAW because entropy is nothing but a figment of our imagination. Entropy is a thought experiment. There's NOTHING fundamental or lawful about entropy at the quantum level because entropy or thermodynamics doesn't exist at the quantum level and the psyche level. Entropy cannot be a fundamental law because entropy has NO representation or correlate at the quantum level.

Double ouch!

Can you imagine the cognitive dissonance that I was experiencing? Entropy didn't end up being what I thought it was. I had to deal with it and change a lot of this book as a result. I thought I was done; but now, I'm going to have to proofread and modify everything all over again. I didn't like that. I had some thinking to do and some choices to make. The ramifications and repercussions began to multiply.

In another video about the Unruh Effect, Matt mentions that as a physical object reaches the speed of light, **it uses up ALL the energy in the universe** and suddenly everything in the universe catches up with it. He talks about energy as if energy is a usable commodity and as if energy can be used up and cease to exist; but, this is physically impossible if the First Law of Thermodynamics is true. In other words, his ongoing claim from one video to the next is that it's the entropic physical matter that is being conserved and therefore ends up being the last thing standing; and, NOT the energy. All throughout his various different videos, he actually uses the Second Law of Thermodynamics to FALSIFY and ELIMINATE the First Law of Thermodynamics or the Conservation of Energy. It doesn't have to be that way. It's simply a preferential choice that he makes. In fact, it isn't that way; but, he doesn't know that. He has convinced himself that our universe will end in Heat Death or Entropy. That's what he wants to believe; but, his belief doesn't make it true.

ALL of the Materialists, Naturalists, Darwinists, Nihilists and Atheists turn the whole of physics upside-down making entropic physical matter the substance that is conserved and making energy the commodity that can be used up and cease to exist. They are wrong. Physical matter and entropy are NOT conserved. It's the Energy, or the Psyche, or the Life Force that ends up being conserved. Even in the complete and total Heat Death of a physical universe, ALL of the Energy is still there waiting for Someone Psyche to come along at the quantum level, reorganize it, and make it useful once again at the physical level. It's the Energy or the Psyche who is REAL, and NOT the entropy or heat death. Entropy is death. Eternal Life is REAL, not eternal death. They can't see it because they don't want to see it and don't want it to be true; but, it is true, nonetheless. We KNOW that the Conservation of Psyche or the Conservation of Personality is true because it has been experienced and observed.

Your Psyche or Intelligence continues to exist long after your physical brain is dead and gone; and, your Psyche or Intelligence will continue to exist long after any part of this physical universe has suffered heat death. You can be born again; and, you will still be you, even though your previous physical body, physical planet, or physical universe is currently dead and gone.

The truth is hidden within this video, and he can't see it because he isn't looking for it and doesn't want it. Within this video and his other videos, he emphasizes that **entropy is an emergent property of statistical mechanics. Entropy is a statistical phenomenon. Entropy is an emergent property of the things that are REAL.** That was a bit of a shock to say the least. I understand "emergent properties". We Materialists, Naturalists, and Atheists teach that psyche or mind is an emergent property of the physical brain. We use the word "emergent" to talk about things that don't really exist. "Emergent" is the code-word for imaginary. Whether he realizes it or not, he's telling me that entropy is imaginary; and, I just KNEW that he is right. Entropy is emergent or imaginary.

By using the word "emergent" to describe entropy, he is in fact saying that entropy doesn't really exist as a force, field, object, or thing. That completely contradicted my view of entropy and required some adjustment on my part because I instinctively KNEW that he is right. He convinced me that he is right, even though he simultaneously continued to erroneously state that entropy is one of the fundamental laws of the universe. By truthfully and accurately calling entropy an "emergent property" he literally FALISFIED his claim that entropy is one of the fundamental forces of nature – hence, all of the cognitive dissonance that I was experiencing. I was forced to take the time and think it through.

Here are some of the ideas expressed in this video, along with my reply. When it comes to the quotes from the video, I adapted them and edited them rather than quoting them directly. It looks better and makes more sense because the visuals, gestures, and diagrams from the video aren't easily included into this document.

"Entropy is a law that seems emergent from deeper laws – it's statistical in nature – and yet may ultimately be more fundamental and unavoidable than any other law of physics."

The Materialists, Naturalists, Nihilists, and Atheists have elevated entropy to the level of a Fundamental Unavoidable LAW, even though entropy seems to be emergent from deeper more fundamental laws and is statistical, theoretical, or intangible in nature. Entropy is a philosophical concept rather than a physical reality. There's NOTHING fundamental and unavoidable about an emergent property. Entropy is a thought experiment rather than a physical thing.

When he used the word "emergent" to describe entropy, that's when I fully realized that entropy is a philosophical concept or a figment of the imagination rather than something actual, tangible, and real. I just KNEW that he is right. This was the straw that broke the camel's back. His claim spurred me to dig in deeper and try to figure out what entropy truly is. This essay was the result.

Eventually I realized that entropy doesn't really exist. That idea is not going to go over well with the Materialists, Naturalists, Darwinists, Nihilists, and Atheists who have based their religion on entropy. Nevertheless, Out-of-Body Travelers have observed that every time their Human Psyche enters into a non-consensus reality or a disorganized reality, that particular spiritual reality automatically reorganizes itself to match the demands and expectations of the Human Psyche. In other words, whenever a Human Psyche or Human Life Force enters into a non-consensus reality, ALL of the entropy or disorder within the system completely disappears. That's when I realized that entropy doesn't really exist as a fundamental force, field, or elementary particle. Entropy or disorder completely ceases to exist the very moment a Human Psyche enters into the system. This is what has been experienced and observed. There's nothing fundamental about entropy or disorder!

Psyche is Energy, and Energy is all that's needed to eliminate entropy from a system. Psyche automatically brings order to disorder or entropy. In other words, Psyche or Syntropy immediately eliminates, falsifies, and trumps disorder or entropy. Furthermore, heat death is impossible at the quantum level or the psyche level. The second law of thermodynamics doesn't exist in the Quantum Realm or the Syntropic Realm.

What a fascinating idea? Don't you think?

He's right. Entropy is an emergent property of statistical mechanics, which means that entropy doesn't really exist. Entropy isn't a force or a field; therefore, there is nothing fundamental, permanent, unavoidable, or lawful about entropy. Entropy is simply a statistical measurement of the unavailable energy within a given system. Unavailable to whom? Unavailable to physical fallen mortal beings such as ourselves. ALL of that energy is still available to the Human Psyche and God's Psyche, even though some of it is not currently available to our physical body or physical environment.

Do you see how that works? It's the answer to life, the universe, and everything. It explains everything that has ever been experienced and observed.

The entropy of the physical universe must increase – so says the Second Law of Thermodynamics. It's a law that seems emergent from deeper laws. It's statistical in nature. Entropy will lead to the end of our universe. An increase in entropy means that the heat reservoirs are approaching the same temperature, reducing their capacity to do useful work. Entropy is a measure of how evenly spread out a system's energy is. The more evenly spread the energy is, the less useful the energy is. Entropy will always increase unless energy comes in from the outside to reestablish the temperature differential. This understanding of entropy is in terms of flowing heat. Entropy is an emergent property of statistical mechanics.

Entropy isn't going to lead to the end of our universe. After Heat Death, our universe will still be there waiting for Someone Psyche to enter into it and bring it back to life once again. All the stuff beyond the bounds or the horizon of our observable universe is still there and still in play. We have no idea what God has chosen to do with it. For all we know, God has chosen to resurrect it and successfully raised it all from the dead. Your Psyche and God's Psyche can eliminate entropy or disorder at will. Psyche or Energy is all that's needed to remove entropy from any system that it encounters; and, Psyche or Energy is always conserved.

Dark Energy or the Expanding Force of the Universe represents an ongoing infusion of Energy from someplace else besides our physical universe by Someone Else besides our physical universe. The Gods can take that Dark Energy any time they choose to do so and form portions of it into physical stars, physical planets, and physical galaxies. God's Psyche can transform energy into anything He wants it to be, including entropic physical matter, planets, stars, and galaxies. It would take some kind of God or Psyche to infuse Dark Energy into our physical universe throughout the whole of our physical universe at a universal scale. Would it not? That's what we currently observe.

Entropy is restricted exclusively to physical matter and is an emergent property of physical matter, mass, time, and thermal equilibrium. ALL of the Energy is still there; and while separated from its physical body, the Human Psyche can access that energy, use it, and organize it anytime the Human Psyche chooses to do so. The same truth would apply to God's Psyche as well. ALL of the Energy is still there just waiting for Someone Psyche to access it, use it, and reorganize it. Energy is conserved, NOT entropy.

Out-of-Body Travelers have observed that in the Spirit World, heat or temperature no longer applies and no longer exists. Thermodynamics does not exist in the Quantum Realm. It's all about Psyche or Energy. Entropy or thermodynamics ONLY applies to the physical realm. It doesn't touch nor impact the spiritual realm or the quantum realm. Psyche, Intelligence, or Life Force is not affected by entropy. In fact, Psyche or Intelligence eliminates entropy at will. This is what has been experienced and observed. Psyche is Energy. Entropy, disorder, and death are completely eliminated whenever Energy or the Human Psyche enters into a system. Energy, Information, or Intelligence is all that's needed to eliminate disorder or entropy. Psyche is the innate intelligence within all the different forms of energy. The Conservation of Psyche or the Conservation of Energy completely overrides and eliminates Entropy and the Second Law of Thermodynamics at the quantum level. This is what has been experienced and observed.

His words are atheistic and materialistic and support the party-line; but, the examples and the evidence and other parts of his words don't support the party line and contradict the atheistic and materialistic belief which states that Entropy is Supreme, and that Entropy is Conserved. Materialism, Naturalism, and Atheism are internally inconsistent and self-contradictory much of the time. It's a bad platform upon which to build a scientific theory or a scientific argument. Entropy doesn't work; whereas, Psyche, Intelligence, or Energy certainly does. Remember, there is NO useful form of entropy.

Statistical Mechanics inevitably leads to entropy and the second law. The inevitability of the rise of entropy is as fundamental as counting. Macrostates are entirely defined by thermodynamic properties – temperature, pressure, volume, and number of particles. A system left alone will eventually try out all the different microstates that are possible given the laws of physics. All particle arrangements will eventually happen. The chance of getting all the molecules on one side of the room is so small that it never happens. But, if you leave a system alone long enough, its particles and its energy will find its way into all the different forms that are possible. The vast majority of possible distributions of energy or microstates leave the system very close to a single macrostate, that's the state of thermal equilibrium or maximum entropy, in which energy is

maximally spread out and temperature, pressure, density, and volume have the values we expect from classical thermodynamics.

Statistical mechanics is used to propose and support the idea that a random fluctuation of entropy produced the Big Bang – the idea that everything randomly lined up into one corner of the room and then went bang. Statistical mechanics and entropy are self-contradictory and self-defeating. I don't know if they can sense it or not, but the idea that all particle arrangements will eventually happen falsifies the idea of entropy and the second law of thermodynamics which state that order or organization or work within the system is NO longer possible now that the system has reached thermal equilibrium.

I was taught to believe and used to believe that Entropy or the Second Law of Thermodynamics is one of the fundamental laws of nature and reality. I used to erroneously believe that Entropy or the Second Law is a fundamental force in the universe, a physical law, an essential and unavoidable ingredient of nature, a quantum field, an energy wave, a quantum particle, and a universal reality of space-time. I used to believe that Entropy is a real thing. However, Entropy or the Second Law of Thermodynamics doesn't have a quantum mechanical foundation and doesn't exist at the quantum level which means that it doesn't really exist at the physical level either. The physical is an extension of the quantum, and the Second Law of Thermodynamics doesn't exist at the quantum level. There is NO entropy and there are NO thermodynamics at the Quantum Level or the Psyche Level. Heat Death is impossible in the Quantum Realm. ALL of the energy is always available at the quantum level because Psyche or Energy is always conserved, especially at the quantum level. At the Quantum Level, the First Law of Thermodynamics completely falsifies, eliminates, and trumps the Second Law of Thermodynamics.

Because Entropy doesn't have a quantum mechanical foundation nor a quantum mechanical existence, Entropy is not and cannot be one of the fundamental forces or one of the fundamental laws in nature and the universe. Entropy cannot be a quantum, a quantum field, or a quantum force comprised from energy or made from energy. Entropy doesn't really exist. Entropy is purely an abstract theoretical concept – a thought experiment – an emergent property of the real and fundamental laws of reality and existence.

It has been observed by Out-of-Body Travelers that once the Human Psyche comes in to play, entropy or disorder goes away. If it happens at the quantum level, then it's also possible at the physical level. The physical level is based exclusively on the quantum level or the psyche level; and, death is impossible at the psyche level. Entropy is a convention and an invention that really doesn't exist as a quantum law or a physical law. Entropy doesn't exist as a quantum particle, a quantum force, a quantum field, or a quantum law, which means that entropy doesn't really exist at the physical level either.

Entropy or the Second Law of Thermodynamics is simply an illusion – an emergent property of more fundamental laws. Entropy was designed into entropic physical matter. Entropy was made and caused to happen by Someone Psyche, which means that it can be eliminated by Psyche. It has been observed that Syntropy, Intelligence, or Psyche overrides and eliminates entropy at will. Energy or Psyche or Information is all that's needed to eliminate entropy and bring order and life to a system.

Do you see how that works? It's the answer to life, the universe, and everything.

Psyche or Energy falsifies and eliminates entropy. The First Law of Thermodynamics falsifies the Second Law of Thermodynamics at all levels of existence; and, the Second Law of Thermodynamics or Entropy is used by the Naturalists and Atheists to FALSIFY the existence of God, Psyche, our upcoming After-Life, and the First Law of Thermodynamics.

We Materialists, Naturalists, and Atheists use the Second Law of Thermodynamics to prove that God and Psyche do not exist and cannot exist. That's how we use it, is it not? The net result of all of this is that the First Law and the Second Law are mutually exclusive, especially at the quantum level.

Think about it, even at the physical level, entropy doesn't really exist in any usable form. Entropy has no physical component. Entropy is NOT a quantum particle, quantum force, quantum field, quantum wave, or quantum law. Entropy really shouldn't be elevated to the status of law. I have observed that entropy is a bastard – an illegitimate child or an emergent property of the REAL, fundamental, and demonstrable aspects of reality, truth, and existence. Entropy has NO fundamental existence at the quantum level or the psyche level, which means that it doesn't really exist at the physical level either. It only seems to exist. That's the true nature of an emergent property such as entropy. Entropy comes in no useful form.

Entropy or the Second Law of Thermodynamics has NO quantum mechanical basis. The Second Law doesn't exist and doesn't apply at the quantum level or the psyche level. Heat death is impossible at the quantum level and the psyche level. Entropy is exclusively a classical physics phenomenon, and an emergent one at that. Therefore, with NO quantum mechanical component, entropy cannot be one of the fundamental forces in nature and our universe. In fact, it can be argued that entropy doesn't really exist at all. It only seems to exist. Entropy is accounting or statistics – NOT a real thing.

Order is not the same thing as low entropy, and the second law isn't always the tendency towards disorder. So, the macrostate that defines thermodynamic equilibrium is the one with the most microstates, which means the one with the maximum entropy. Any system not in equilibrium must increase in entropy. This is assuming you don't force the system from the outside. To reduce entropy, you must introduce an external source of energy. To reduce entropy, you must introduce information, knowledge, or intelligence.

Here he emphasizes that entropy isn't about disorder. This leads me to believe that the BEST definition for entropy is NOT "disorder" but rather "death". Entropy is death. Entropy is Heat Death. The Theory of Evolution is Creation by Entropy or Creation by Death. In other words, the Theory of Evolution is Abiogenesis or Creation Ex Nihilo, which are physically impossible. Creation Ex Nihilo is magic – it doesn't really exist. Entropy doesn't exist in any usable form either. Contrary to what the Materialists, Naturalists, and Atheists claim, entropy cannot design and create. Furthermore, Heat Death is impossible at the quantum level and the psyche level. The second law of thermodynamics doesn't exist in the Quantum Realm or Spirit World because thermodynamics don't exist in the Transdimensional Realms.

Entropy is an emergent physical property that only applies when you don't force the system from the outside. Entropy only works when you prevent Psyche or Energy from entering into the system. Psyche is Life Force. Psyche is Energy. Introduce Psyche, Intelligence, Life Force, Energy, or Syntropy into a system, and entropy or disorder ceases to exist. This is what has been experienced and observed.

Entropy only applies to the physical realm. Heat flow doesn't apply at the quantum level or the psyche level. Entropy or thermal equilibrium is only a physical phenomenon. Anytime the Human Psyche enters into a non-consensus reality in the Quantum Realm that Human Psyche forcefully organizes that system and automatically removes all of the entropy or disorder from that system. In other words, entropy doesn't apply and doesn't exist at the quantum level or the spiritual level; and since the quantum level is the basis or the foundation of the physical level, that also means that entropy technically doesn't apply and doesn't really exist at the physical level either. Entropy is an illusion – a trick of the mind. Entropy is an "emergent property" which means that it only seems to be real when in fact it is not. The energy is still there, and ALL of it is still available to any Psyche, Ruling Intelligence, or Active Agent who enters into the system.

To reduce entropy and increase order, you must introduce an external source of energy such as Psyche or Non-Local Consciousness. Psyche is Energy. Psyche is the innate intelligence within all the different forms of energy which gives that energy the ability to understand, follow, and obey the Human Psyche's desires and commands. There's the Energy or the Psyche who ACTS, and then there is the energy or the psyche that simply REACTS. The Human Psyche and God's Psyche CHOOSE and ACT; whereas, the energy or psyche associated with "matter" REACTS. This explains everything that has ever been experienced and observed.

In contrast, the Materialists, Naturalists, and Atheists erroneously teach that entropic physical matter is the ONLY thing that exists, and that psyche or non-local consciousness does not exist. The observational evidence and the experiential evidence falsify their claims. The false is falsified by the truth; and, the truth is repeatedly experienced and observed.

Entropy is statistical and emerges from the behavior of particles under the laws of motion. This is where the second law appears to add something new to the universe not seen in the more fundamental laws. It seems to add the arrow of time. The laws of motion don't care about the direction of time; and yet, the second law of thermodynamics clearly distinguishes between the past and the future. It's almost as if the concept of time is emergent and statistical just like entropy.

Entropy is an emergent property of the passage of time. Time is an emergent property of mass. Mass is an emergent property of the Higgs Field. The Higgs Field is God's Field; and, the Higgs Field was made by God and is maintained by God, which means that the Higgs Field and its effects can be eliminated anytime God chooses to do so on a selective basis.

Time and entropy are emergent concepts and NOT physical realities. Entropy and time are NOT elementary particles, quantum fields, NOR any of the fundamental forces of the universe. Entropy and the aging process don't even exist in the Quantum Realm or the Spirit World where everything is Syntropic, Eternal, Everlasting, and Conserved. Entropy, mortality, the aging process, and time ONLY apply to the physical realm and entropic physical matter. The Heat Death of our physical universe won't even touch nor affect your Psyche, Spirit Body, Soul, and Mind. You will go on long after your physical body is dead and gone. Entropy, the arrow of time, the aging process, and the second law do not exist at the quantum level or the psyche level. Heat Death is impossible at the quantum level. At that level, the Second Law of Thermodynamics is completely overridden by the First Law of Thermodynamics – the Conservation of Energy or the Conservation of Psyche. This is what has been experienced and observed.

Furthermore, it has also been experienced and observed that the Human Psyche or Human Intelligence immediately removes the "entropy" or "disorder" from the system, whenever it enters into a non-consensus reality. Likewise, when it comes to a physical world or a physical universe that has suffered Heat Death, ALL of the energy is still there and all of that "entropy" or "disorder" is automatically removed from the system whenever one of the Gods enters into the system and brings order, organization, and life to that system. Entropy and the Second Law of Thermodynamics are non-existent in the Spirit World or the Transdimensional Realm because thermodynamics and the aging process are non-existent at the Quantum Level or the Psyche Level. In the Spirit World, temperature and heat don't matter. It's all about energy, psyche, intelligence, life force, and consciousness. Some of the people who enter into the presence of God initially think that they are going to be burned to a crisp by all of the energy, light, and glory; but, they continue to exist because physical matter is NOT the fundamental unit of reality – psyche or energy is. Psyche cannot be destroyed, and it cannot die.

The idea that our physical universe is going to end in Heat Death is based upon the idea that our physical universe is a closed system. However, the ongoing infusion of Dark Energy into our physical universe and the accelerating expansion of our physical universe are scientific proof that our physical universe is NOT a closed system, and they actually falsify Entropy and the Second Law of Thermodynamics whether we realize it or not. The energy supposedly lost to entropy seems to be getting reintroduced to the system as dark energy instead.

There's nothing sacrosanct about Entropy. It's the Psyche or the Energy that's actually being conserved, NOT the physical matter and NOT the entropy. The amounts or quantities of physical matter and entropy are constantly changing which means that they are NOT conserved.

Furthermore, entropy cannot be a fundamental law because entropy does not have a quantum mechanical counterpart, quantum force, quantum field, or quantum particle that is associated with it. Thermodynamics or heat flow doesn't exist at the quantum level or the psyche level, so the concept of entropy or thermal equilibrium is meaningless when it comes to the Quantum Realm or the Psyche Realm. There is NO Second Law of Thermodynamics at the quantum level; therefore, entropy cannot be one of the fundamental laws or fundamental forces in our universe. Entropy is a measure of unavailable energy within any given physical system. Even after Heat Death at the physical level, the Energy or Psyche or Intelligence is still there and is still available and usable at the quantum level. The energy is just no longer available to physical fallen mortal beings at the physical level unless Someone Psyche or Someone Intelligent makes it available to them. Heat death is impossible at the quantum level; and, death is impossible at the psyche level. Entropy and death only exist at the physical level; and, it's temporary. Entropy or death is never conserved. It's Psyche, Intelligence, Energy, or Life Force that's actually being conserved.

Do you see how that works? It's the answer to life, the universe, and everything. It explains everything that has ever been experienced or observed.

Syntropy is the First Law of Thermodynamics on both sides of the veil. Psyche or Energy is syntropic which means that it is conserved. It has been observed that the Gods can override and eliminate entropy at will by choosing to bring order, organization, light, and life to an entropic or disorganized region of our physical universe. Information, Intelligence, Knowledge, or Psyche is all that's needed to eliminate disorder or entropy from any system. Contrary to what the Materialists and Atheists claim, Psyche or Energy is the fundamental unit of reality, and NOT entropy. Syntropy, Psyche, or Intelligence overrides and eliminates entropy at will.

Remember, Heat Death is impossible at the quantum level. Entropy or the Second Law of Thermodynamics doesn't exist as a quantum phenomenon and doesn't exist at the quantum level; therefore, there is NOTHING foundational, fundamental, permanent, or lawful about entropy. Syntropy is the thing that reigns supreme, NOT entropy.

Are You a Boltzmann Brain?

Are You a Random Fluctuation in Entropy?

https://www.youtube.com/watch?v=nhy4Z_32kQo

https://science-2-0.com/wp-content/uploads/2018/08/Are-You-a-Fluctuation-in-Entropy.zip

Was an incredible drop in entropy responsible for the Big Bang? If that's the case, this would lead us to conclude that a great many other things are possible, including the likelihood that you are a Boltzmann Brain. The Second Law of Thermodynamics tells us that everything moves towards a state of greater entropy, but it is low entropy that enables galaxies, stars, planets, and human beings to begin and to exist. The lowest entropy state that we're aware of in our universe was right before the Big Bang. The entirety of our own universe was condensed to a microscopic size. What caused this extreme low entropy? If it was caused by a random fluctuation in particles that implies that a great many other things are possible due to the random fluctuations of particles. One of the strangest of those possibilities is that of the Boltzmann Brain. If the Big Bang was caused by random entropic fluctuations, then it is much more likely you are a Boltzmann Brain than that you are a human being.

In this video, they give Boltzmann the credit for developing the mathematics behind Entropy and the Second Law of Thermodynamics – statistical mechanics. Entropy and the Second Law were developed to convince us that God does not exist, and that God cannot exist. Likewise, the purpose behind Boltzmann Brains is to convince you that you do not exist. Entropy and Boltzmann Brains are a couple of the cornerstones for Nihilism, which is the philosophical belief that your Psyche or Non-Local Consciousness DOES NOT EXIST, and that you cease to exist when your physical body and physical brain are dead. It's all part of the "You Are Your Brain" mind-set of the Materialists, Naturalists, Darwinists, Nihilists, Behaviorists, Determinists, and Atheists. These people teach and believe that random fluctuations in entropy can produce people, brains, and big bang singularities. These people teach and believe that entropy or death can design and create. It's magic! These people believe in magic. They believe in Creation Ex Nihilo which is magic.

The fact that the Naturalists and Atheists have to resort to Boltzmann Brains, Blind Luck, Creation Ex Nihilo, Creation by Entropy, Creation by Death, Creation by Quantum Fluctuations, and Creation by Random Fluctuations in Entropy in order to explain the origin and the cause of our physical universe tells me that these people really don't have a case. Boltzmann Brains, Creation Ex Nihilo, and the creative powers of Random Fluctuations in Entropy are such stupid ideas that they actually make belief in God seem logical and rational in comparison.

Entropy is death. The Second Law of Thermodynamics was designed to FALSIFY the First Law of Thermodynamics, or the Conservation of Psyche, or the Conservation of Your Life Force. Entropy and the Second Law were designed to FALSIFY the existence of Psyche and God.

Based upon this theory of Boltzmann Brains, they suggest that the Big Bang was a random fluctuation in entropy. In other words, the Big Bang was caused by entropy – a random fluctuation in entropy. "Random fluctuations in entropy" is a hack that they use to make the Big Bang possible without invoking God as the causal source of the Big Bang Singularity. Likewise, Boltzmann Brains (random fluctuations in entropy) are invoked as a causal explanation for the origin of your existence without any direct intervention from God. Entropy and Boltzmann Brains are prime examples of the "Anything But God" mentality of the Materialists, Naturalists, Darwinists, Nihilists, and Atheists. These people are willing to accept anything as the origin for you and the universe, so long as it isn't God.

Again, throughout the whole process, they imply that Entropy is a person or a thing who can cause other things to begin or can cause other things to happen if given enough time to do so. They arrive at this kind of desperate and illogical conclusion because they have convinced themselves that Entropy Is Supreme, that Psyche or God does not exist, and that entropy or physical matter (entropic physical matter) is the only thing that truly exists. Fascinating, is it not? Self-deception works, and it works every time. Our beliefs are chosen into existence by our Human Psyche, and we don't even know it.

Of course, the logical conclusion of Entropy and the Second Law is Heat Death, Nihilism, and the End of the Universe; and, since Entropy or the Second Law or Entropic Physical Matter is the ONLY thing that exists, a random fluctuation in entropy has to be the cause of the Big Bang and the origin of the Big Bang Singularity. Quod erat demonstrandum! It makes perfect logical sense and gets rid of all that nasty God business.

The only problem is that it is ridiculous and ludicrous. It's nothing but a thought experiment that has nothing to do with reality as it has been experienced and observed. It begs the question. A random fluctuation in entropy is NEVER going to make a Big Bang Singularity NOR a Boltzmann Brain. It's physically impossible!

It shows their desperate need to eliminate God from the equation.

Figuring out what they got wrong helps me to find and understand what is real and truly exists. Contrary to what they imply, Entropy is NOT a universal field, and Entropy is NOT one of the fundamental forces in nature. Entropy is a measurement of nothing. In other words, Entropy is a measurement of the unavailable energy within a given system.

Unavailable to whom?

Unavailable to physical, fallen, mortal, entropic beings such as us. The energy is still there waiting to be accessed and used again; but, as physical fallen mortal beings, we can't get at it because it is no longer available for our use. The Gods made that energy unavailable for our use, and it would take a God to make that energy available for our use once again. The energy is still there because ALL of the energy is always conserved. At the Quantum Level ALL of the energy is available for our Psyche's use. Psyche, Energy, Life Force, or Intelligence cannot be made, and it cannot be destroyed. It is always conserved. But, energy isn't always available for use by physical, fallen, mortal beings such as us. This is what has been experienced and observed.

I used to visualize the whole creation process or big bang process as "removing the entropy from a system", or "pulling the entropy out of a system", or "reversing entropy"; but, entropy really isn't a thing. Instead, all the Gods have to do is give us access to that energy, or make that energy available for our use, or organize that energy for us; and instantly, entropy has been reversed or entropy has ceased to exist. The presence of the Human Psyche or God's Psyche automatically removes and eliminates entropy or disorder. This is what has been experienced and observed.

This is fascinating stuff to think about.

Out-of-Body Travelers have observed that whenever they enter into a non-consensus reality, the spirit matter and energy within the system organizes itself according to the demands of their Psyche, Mind, or Consciousness. That non-consensus reality is "entropic" and in maximum disorder until their Human Psyche or Human Mind enters into that region of space and organizes it according to their wishes. Psyche literally reverses entropy and eliminates disorder. Psyche is the organizing force in the multiverse on both sides of the veil. The Human Psyche makes the demands and Nature's Psyche complies.

The Human Psyche or Mind clearly and obviously organizes chaotic, entropic, and disorganized spirit matter and energy whenever it enters into a non-consensus reality. This is what has been experienced and observed. In the Spirit World or Quantum Realm, the Human Psyche automatically removes the entropy from a system and organizes that system whenever that Psyche or Mind enters into a disorganized chaotic non-consensus reality. The Human Psyche literally reverses and removes the entropy. That is what it does! The Human Psyche brings order, organization, and life to non-consensus realities. This is how things really work. WE KNOW that it is so because it has been experienced and observed.

In one of the Space Time videos, he states that belief should wait for evidence. Well, I upgraded my science to Science 2.0 which allows all of the evidence into evidence and then pursues a preponderance of that evidence. Science 2.0 defines Science as observation and experience. I now treat Out-of-Body Experiences (OBEs), Near-Death Experiences (NDEs), Shared-Death Experiences (SDEs), and After-Death Life Reviews as scientific evidence. I now use that evidence to inform my beliefs. Out-of-Body Experiences (OBEs) FALSIFY Nihilism, Entropy, and the Second Law of Thermodynamics at the quantum level or the psyche level where it matters most.

It's essential to allow ALL of the evidence into evidence if we want to know what things are and how they truly work, which was something that I refused to do when I was a Materialist, Naturalist, Nihilist, and Atheist. These naturalistic and atheistic philosophies or religions are based exclusively on a refusal to look at and accept any evidence or information that falsifies them. They are a head-in-the-sand approach to science. When it comes to Psyche, Syntropy, Life Force, and Eternal Life, these atheistic people don't want these things cluttering up their theories, ideas, and beliefs; consequently, these people are ever-learning but never able to come to a knowledge of the truth. I know how it goes. I've been there and done that.

Refusal of evidence prevents us from being able to understand and explain that evidence. Explaining away the evidence isn't an explanation of the evidence. Atheism, Naturalism, Materialism, Darwinism, Nihilism, Behaviorism, Determinism, Physical Reductionism, and Classical Physics tell us that Psyche, Mind, or Non-Local Consciousness DOES NOT EXIST; consequently, these philosophies or religions reject and deny ALL of the observations and experiences of Psyche's or Mind's existing and functioning separately from its spirit body and physical body – events which have been experienced and observed by the human race as a whole.

In truth, human beings are descendants of the Gods. Both our physical body and our spirit body are descended from the Gods. We are the children of God. Our surname is God. We are Gods. Consequently, whenever our Human Psyche or Human Mind enters into a non-consensus reality in the quantum realm, the spirit matter and energy in that region of space reads our mind and organizes itself according to our desires and demands. The Human Psyche or Human Mind literally reverses the entropy or removes the entropy from that system bringing order, organization, and life to that system instead. In a non-consensus reality in the spirit world, the Human Psyche is God. Psyche or Intelligence reverses entropy and brings order, organization, and life to a system. This is what has been experienced and observed. It's a lot more interesting and real than Boltzmann Brains and random fluctuations in entropy.

It's kind of hard for me to believe that the Big Bang Singularity was a Boltzmann Brain or a random fluctuation in entropy. Entropy hasn't been observed to work that way.

In fact, if we get right down to it, entropy doesn't work. Entropy is the absence of work. There is no useful form of entropy; and, Heat Death is impossible in the Quantum Realm.

Nevertheless, because these people have unilaterally decided that God, or God's Psyche, or Intelligence cannot be used as a causal explanation for origins, these people are left with things like Entropy Fluctuations and Boltzmann Brains for their causal explanation of the Big Bang and Your Existence respectively. It demonstrates their desperation and lack of satisfactory causal explanations, now that they have chosen to eliminate God or Psyche as one of the candidates. These people are willing to accept and believe in anything but God. This is what has been experienced and observed.

Reversing Entropy with Maxwell's Demon

https://www.youtube.com/watch?v=KR23aMjIHIY

https://evolution-is-entropy.com/wp-content/uploads/2018/08/Reversing-Entropy-with-Maxwells-Demon.zip

Can a demon defeat the 2nd Law of Thermodynamics? Entropy is sometimes described as a measure of disorder or randomness. The second law of thermodynamics – the law that entropy must, on average, increase – has been interpreted as the inevitability of the decay of structure. This is misleading. As we saw in our episode on the physics of life, structure can develop in one region even as the entropy of the universe rises. Ultimately, entropy is a measure of the availability of free energy – of energy that isn't hopelessly mixed into thermal equilibrium. Pump energy into a small system and complexity can thrive.

From what I can tell, his message in this video is that Entropy Reigns Supreme at the universal scale and that the only way to overcome entropy is either through a random fluctuation in entropy or by pumping energy into the system.

His conclusion: the process of the Demon using energy while using its physical brain makes the overall entropy of the universe increase while at the same time only making it seem that the entropy or disorder in the experimental system has decreased. In other words, entropy in any physical system will always increase as long as any type of work is being done. His falsifying Maxwell's Demon supports and verifies the Supremacy of Entropy Doctrine and the Second Law of Thermodynamics. His takeaway message is that entropy always increases, and the Heat Death of our universe is inevitable and unavoidable. He begs the question and jumps to the conclusion that the Demon has to be a physical entity; and, it is obvious that physical entities produce entropy. In other words, he has to turn Maxwell's Demon into a physical entity in order to falsify the entropy reversing effects of Maxwell's Demon. This is necessary, because Entropy or Heat Death doesn't exist at the quantum level and the psyche level.

Of course, since I allow all of the evidence into evidence and pursue a preponderance of that evidence, the takeaway message that I get from this video is that Psyche or Intelligence can reverse entropy at will. Intelligent Demons and Intelligent Gods can devise ways to reverse entropy and bring order and organization to a particular system at will. Psyche or Intelligence reverses entropy even at the physical level whether we realize it or not. He actually uses a Go Board to demonstrate how Human Intelligence can maximize or minimize entropy at will. Information is all that's needed to increase or decrease entropy at will. Despite his attempts to falsify Maxwell's Demon, the truth of the situation is hidden within his thought experiments, nonetheless. Psyche is Energy after all; and, adding energy to a system always eliminates entropy or disorder within that system. Likewise, adding

Psyche, Information, or Intelligence to a system always eliminates entropy or disorder from that system.

Furthermore, even at the point of Heat Death, the energy or potential is still there waiting for Someone Psyche or Someone Intelligent to enter into the system and bring order, organization, and life to that system once again. The Quantum Fields are still there! This is what has been experienced and observed. Observations and experiences override philosophical speculation and wishful thinking every time. Heat Death is impossible at the quantum level. The Second Law of Thermodynamics doesn't exist in the Quantum Realm or Syntropy Realm.

Technically, a demon is an evil spirit not a physical entity; and, it has been observed by Out-of-Body Explorers that Psyche, Spirit, or Energy can override and eliminate entropy or disorder at will. It's impossible to FALSIFY Maxwell's Demon at the quantum level or the psyche level because Thermodynamics or Heat Flow of any kind doesn't exist at the quantum level. This is what has been experienced and observed. In other words, the Second Law of Thermodynamics doesn't exist at the quantum level or the psyche level; so, it is meaningless and pointless to claim that entropy or heat death is going to bring an end to the Quantum Universe, the Spirit World, or the Psyche Realm. It's physically impossible. Heat Death is impossible at the quantum level or the psyche level.

That's one of the reasons why the Christians and Jesus called the Plan of Salvation "The Gospel" or "The Good News" because it overcomes and eliminates entropy or the second law of thermodynamics at a fundamental level. The Atonement of Christ is Pure Syntropy; and, Syntropy always eliminates disorder, death, and entropy. Psyche or Energy is always conserved.

He ends this video by mentioning a couple of different types of entropy. There are probably dozens of different definitions for entropy, which is one of the reasons why it is so hard to figure out what entropy really is. Sometimes, it is easier to figure out what entropy is not. Entropy is NOT psyche or syntropy. Entropy is NOT a quantum particle, quantum force, quantum wave, quantum field, or quantum law. The Second Law of Thermodynamics doesn't exist at the quantum level or the psyche level. Heat Death is impossible in the Quantum Realm or Syntropy Realm.

This video contains ancillary concepts about entropy that I find interesting.

First of all, since entropy is NOT a quantum wave, quantum particle, quantum force, or quantum field, there is NO basis for the Second Law of Thermodynamics within Quantum Mechanics. Consequently, any attempt on the part of the Materialists and Naturalists to create a Quantum Theory of Entropy can only succeed in explaining the classical physics and emergent properties of particles at the physical level because Thermodynamics and Heat Flow do not exist at the quantum level or the psyche level. Entropy is simply an illusion. Even in Heat Death at the physical level, ALL of the energy is still there and available for use at the quantum level; and, both the Human Psyche and God's Psyche have full access to that energy, at will, at the quantum level even though that energy is temporarily unavailable at the physical level.

Entropy is a statistical phenomenon. Entropy defines how far a system is from thermal equilibrium; and, it also defines the availability of free energy that can be extracted to do work. Entropy is connected to disorder and randomness in a very real way. See, entropy is a measure of ignorance. Entropy very directly measures how much information we don't have about a system. But with knowledge comes power, literally. Know a system perfectly and you can return it to order, or you can extract its energy. Entropy is a direct measure of hidden

information. High entropy configurations can be transformed to low entropy as long as we have information about the system. Information about a system allows us to extract useful work. If you have perfect knowledge of the state of all the particles in the box, you could open and close the door in exactly the right sequence to sort the particles any way you liked. You could turn information into work or into structure. In other words, you could turn entropy into syntropy at will.

The "you" that he keeps talking about is your Human Psyche, whether he realizes it or not. Psyche doesn't burn energy or use energy while it works. Psyche is energy – conserved energy. Psyche is Maxwell's Demon; and, Psyche works precisely as his Demon works completely reversing or eliminating entropy as it sees fit.

The observation that "information is conserved at the quantum level" FALSIFIES or NULLIFIES the idea that entropy represents a loss of information at the physical level. The information is still there at the quantum level, even though it no longer exists or is no longer available at the physical level. Quantum Mechanics or Syntropy defeats, falsifies, and eliminates entropy or the Second Law of Thermodynamics. That's good news, is it not?

Psyche or Intelligence is Energy or Power. Knowledge is a function of Psyche or Intelligence; and, Psyche is Energy or Syntropy. KNOW and control a system perfectly, and you (your psyche) can return it to order and extract its energy at will. Better said, Psyche or Intelligence can form and organize energy or spirit matter into anything it wants it to be at will. In other words, the Second Law of Thermodynamics doesn't apply and doesn't exist at the quantum level or the psyche level. Even at the physical level, entropy submits to the influence – or the ordering and transforming effects – of Psyche or Intelligence. Entropy isn't the last word in reality after all.

God's Psyche or Intelligence can move ALL of the physical particles to one side of the room at will. Because the Human Psyche is bound by and limited by entropic physical matter, we currently don't have such a capability. However, Out-of-Body Travelers have observed that while their spirit body and psyche are separated from their physical body, their Psyche has the ability to organize spirit matter and energy at will simply by entering into a disorganized or randomized non-consensus reality in the quantum realm or the spirit world. In other words, even the Human Psyche can move ALL of the particles of spirit matter to one side of the room while that Human Psyche is separated from its physical body.

Do you see how that works? It's the answer to life, the universe, and everything. It explains everything that has ever been experienced and observed.

God gave the physical laws, under the control of Nature's Psyche, greater power and greater control over physical matter than what He gave the Human Psyche. God did this in order to create the ultimate consensus reality – a physical reality – where we can actually depend upon our house, our car, our spouse, and our dog actually being there tomorrow when we go looking for them. If human beings or Human Psyches could move around physical matter with their minds at will, then it would break the reliability, predictability, stability, and dependability of our physical consensus reality; and therefore, it would break the reason for having a physical reality in the first place. God limits us; and most of the time, He prevents us from moving physical matter and controlling physical matter with our minds; and, this limiting phenomenon or effect is known as Entropy or the Second Law of Thermodynamics.

In the thought experiment within this video, Maxwell's Demon has the ability to observe the particles in the system, can open and close the door to the system, and has the ability to determine where it wants each particle to be. In other words, the Demon is an intelligent psyche and therefore has the inherent ability to turn a high entropy situation into a low entropy situation. Said another way, the Demon has the inherent ability to design and create a Big Bang Singularity or Pure Syntropy. Psyche or Intelligence can eliminate entropy or thermal equilibrium at will. This is what has been experienced and observed.

In this thought experiment, in order to save the second law of thermodynamics and prevent it from being falsified, Matt had to turn the demon into a physical entity that burns energy while doing its work. He had to turn the demon into a physical entity, because thermodynamics, thermal equilibrium, and heat flow do NOT exist at the quantum level or the psyche level.

As this video demonstrates and proves, Psyche, or Pure Syntropy, Information, or Knowledgeable Intelligence really does eliminate disorder, randomness, and entropy at will. Information, Knowledge, and Intelligence are a function of Psyche. If your psyche has perfect knowledge of a system and perfect control over that system, it can do anything it wants to do with that system including completely reversing the entropy of that system. This reality and truth has actually been experienced and observed. The Human Psyche automatically organizes a non-consensus reality whenever it enters into such a system.

Psyche trumps entropy. Psyche is Energy after all, and an infusion of energy is all that's needed to eliminate entropy from a system. Psyche is Intelligence, and an infusion of information or intelligence is all that's needed to eliminate entropy or disorder from a system. Psyche is Life Force, and Psyche or Syntropy is all that's needed to eliminate death or entropy from a system.

Do you see how that works? It's the answer to life, the universe, and everything. It explains everything that has ever been experienced and observed. By allowing Psyche and Quantum Mechanics in to play, you can literally explain everything that comes your way.

The Physics of Life

https://www.youtube.com/watch?v=GcfLZSL7YGw

https://evolution-is-entropy.com/wp-content/uploads/2018/08/The-Physics-of-Life.zip

Our universe is prone to increasing disorder and chaos. So how did it generate the extreme complexity we see in life? Actually, the laws of physics themselves may demand it.

In this video, he uses a fundamental force of nature (abiogenesis) to override the fundamental force of the universe (entropy). He defines life as one of the emergent properties of the laws of physics. He pays homage to Darwinism, Abiogenesis, and the Theory of Evolution.

This video is nothing but science fiction.

Abiogenesis is NOT a fundamental force of nature or the universe. In fact, abiogenesis, spontaneous generation, creation ex nihilo, or chemical evolution was FALSIFIED in 1859 by Louis Pasteur – the same year that Charles Darwin published "On the Origin of Species by Means of Natural Selection". We have KNOWN since the very beginning of the Theory of Evolution that Materialism, Naturalism, Darwinism, Abiogenesis, Chemical Evolution, and Spontaneous Generation are FALSE; but, most scientists have chosen to ignore that evidence because it isn't telling them what they want to hear.

The Theory of Evolution is Creation by Entropy or Creation by Death. It's magic!

In this video, they teach that life is one of the emergent properties of the Laws of Physics. They reify or deify the Physical Laws. They have to do so because they have denied the existence of Psyche and God. The advantage to Psyche, Intelligence, and God is that they provide a realistic, possible, plausible, and credible Scientific Explanation for the causal origin of life which creation ex nihilo, spontaneous generation, and abiogenesis will NEVER be able to provide. Creation by Intelligence has been experienced and observed; and, Creation Ex Nihilo or Abiogenesis has not.

In this video they also imply that Entropy or the Second Law of Thermodynamics is one of the fundamental forces of the universe, which it is not. In fact, Entropy or the Second Law of Thermodynamics as well as Thermodynamics are NON-EXISTENT at the quantum level or the psyche level where the Spirit Body and Psyche reside. Energy or Psyche is the fundamental unit of reality, NOT entropy. Out-of-Body Travelers have observed that whenever their Human Psyche enters into a non-consensus reality in the Spirit Realm, the spirit matter and energy in that reality reorganizes itself according to the demands and expectations of their Human Psyche. In other words, Psyche, Energy, Intelligence, or Syntropy completely eliminates disorder and entropy. Psyche or Life Force designs and creates. Entropy or death does not.

Entropy really doesn't exist as a fundamental force, quantum field, or elementary particle. Entropy is simply a measure of the amount of unavailable energy within any given system. Within quantum systems at the quantum level, it has been experienced and observed that ALL of the energy in the system is available for use. In other words, entropy or thermodynamics doesn't exist at the quantum level; and since the quantum level is at the foundation of the physical level, technically entropy doesn't really exist at the physical level either. Entropy is the Grand Illusion. Entropy is a thought experiment – a figment of the imagination. Fascinating, is it not?

I'm a philosopher, theoretician, logician, and scientist. I like thinking about this stuff – what it is and how it works. My goal in life is to figure out what everything is and how it works; and, I believe that I'm finally beginning to do so.

The Quantum Law of Thermodynamics states that heat death is impossible at the quantum level or the psyche level, and that the second law of thermodynamics doesn't exist in the Quantum Realm or the Spirit World. This is what has been experienced and observed by Out-of-Body Explorers and Near-Death Experiencers. The Ultimate Law of Thermodynamics states that physical matter and entropy are NOT conserved.

Why E = mc² Is Wrong

https://www.youtube.com/watch?v=eOCKNH0zaho

https://science-2-0.com/wp-content/uploads/2018/08/Why-E-Equals-mc2-Is-Wrong.zip

It is wrong to equate energy with mass. Technically, they are not the same thing at all. Likewise, it is wrong to equate energy with matter. They definitely are not the same thing. The Materialists, Naturalists, Darwinists, Nihilists, and Atheists erroneously equate energy with matter, and then teach and believe that Physical Matter and Entropy are conserved. They are wrong to do so. Entropy and Physical Matter are NOT conserved. Entropic Physical Matter is NOT conserved. Only the underlying Energy, Psyche, Intelligence, or Life Force is conserved. The Materialists, Naturalists, and Atheists erroneously teach and believe that Entropy, Death, Heat Death, and Physical Matter are conserved because these people have chosen to believe that Entropic Physical Matter is the ONLY thing that exists. They are wrong; and, the Science falsifies their ideas and their beliefs whether they realize it or not. The Ultimate Law of Thermodynamics states that entropy and physical matter are NOT conserved.

White Holes

https://www.youtube.com/watch?v=S4aqGI1mSqo

https://scientific-proof-of-god.com/wp-content/uploads/2018/08/White-Holes.zip

Some people think that a white hole could be the origin of our universe. To make a white hole, you would need to reverse entropy. Entropy defines the direction of the flow of time. To reverse time, you need to break the second law of thermodynamics; and, you need to decrease entropy or eliminate entropy. The Big Bang looks mathematically like a white hole.

The Big Bang Singularity, if it existed, had to be Pure Syntropy or a complete elimination of entropy. It would have to have been a white hole. Based upon his Supremacy of Entropy Doctrine, he assures us that such a thing is physically impossible because a white hole breaks the second law of thermodynamics and the flow of time. A white hole reverses the direction of entropy and time, essentially by eliminating entropy or time. That's physically impossible, which means that the Big Bang Singularity if it existed would have been physically impossible as well. That means that a Big Bang Singularity or White Hole would have had to have been a quantum phenomenon completely outside the laws of classical physics. It would have had to have been created by Someone Psyche with God-like powers who lives and functions in the Quantum Realm or the Transdimensional Realm because a White Hole or Big Bang Singularity is physically impossible at the physical level in the entropic physical realm.

The Big Bang Theory, if true, is Scientific Proof of God's Existence. Someone Psyche or Someone Intelligent was needed to make the White Hole or the Big Bang Singularity; and then, Someone Psyche or Someone Intelligent was needed to push the button and make it go bang. Such a thing will never be produced by a quantum fluctuation or a random fluctuation in entropy. That's physically impossible. God is needed to do the physically impossible.

This video along with the one on Maxwell's Demon mentioned above describe what's needed to break entropy or the second law of thermodynamics. An infusion of Energy into a system can break the second law of thermodynamics or eliminate entropy. Psyche is Energy, which means that Psyche or Intelligence can break entropy or the second law of thermodynamics at will. Information or knowledge can break entropy and defeat the second law. Psyche is Intelligence, Knowledge, and Information. Psyche can break entropy at will. Order or organization breaks entropy and defeats the second law. Psyche is the Organizing Force of the Universe. Once again, Psyche can break entropy at will. Psyche is also Syntropy. By definition, in principle or practice, Syntropy breaks or eliminates entropy. All of this is internally consistent which tells me that I'm onto something important. Psyche, Intelligence, Energy, Information, or Organization is all that's needed to defeat entropy and break the second law of thermodynamics. Entropy or Heat Death is NOT conserved!

In this and other videos, he also suggests that removing the flow of time or reversing the flow of time breaks entropy. Time and entropy are an emergent property of mass, the Higgs Field, and physical matter. Removing the Higgs Field stops the flow of time, makes the elementary particles travel at the speed-of-light, and therefore puts an end to entropy. Photons moving at the speed-of-light experience no passage of time which means that they have NO entropy. Entropy is a function of time. Stop the passage of time, and you stop or eliminate entropy. The flow of time produces entropy, so eliminate the flow of time for an object and you eliminate entropy and defeat the second law of thermodynamics for that object.

All of these different ways for defeating entropy are scientific proof that entropy is NOT conserved which is one of the main points that I have been trying to make all the way along. The Ultimate Law of Thermodynamics states that entropy and physical matter are not conserved. This is precisely what has been experienced and observed. Furthermore, the Quantum Law of Thermodynamics states that entropy, heat death, and the second law of thermodynamics do not exist at the quantum level or the psyche level in the Quantum Realm. That means there is no aging process in the Quantum Realm which also means that death or annihilation of Psyche or Energy is impossible in the Quantum Realm. Psyche or Energy is always conserved. All of this is internally consistent which tells me that I'm onto something important once again. It works. It has all been experienced and observed. This tells me that it is GOOD SCIENCE and actually explains what is happening at the psyche level in the Quantum Realm.

What's Wrong with the Big Bang?

https://www.youtube.com/watch?v=JDmKLXVFJzk

https://science-2-0.com/wp-content/uploads/2018/08/What-Is-Wrong-With-the-Big-Bang-Theory.zip

There's a problem with the Big Bang Theory. The clues tell us that we have missed something huge!

What's wrong with the Big Bang?

The Big Bang requires a Syntropic No-Entropy Singularity, or a White Hole, which is physically impossible. A White Hole is Pure Syntropy. A White Hole is a complete reversal of entropy. Such a thing is physically impossible according to the Second Law of Thermodynamics and the Supremacy of Entropy Doctrine. The existence of a White Hole or a Big Bang Singularity would falsify the Second Law of Thermodynamics, Heat Death, and the Supremacy of Entropy Doctrine.

Scientists have observed that our physical universe is expanding. So, they rationally take that expansion and run it backwards in time. Then, they take a leap of faith and keep reversing that expansion and keep running it backwards until they have our whole universe residing in a single point that they call the Big Bang Singularity. Now, think about this process logically and rationally. While rewinding everything and moving backwards in time towards that hypothetical Big Bang Singularity, YOU CAN STOP anywhere along the way and still produce the same results that we see today when you start running the clock forward again. In other words, you don't have to run the clock all the way back to the Big Bang Singularity and still end up with the expanding universe that we see around us today. You could just as easily and more logically run the clock back to the time where the Gods started making physical matter, planets, stars, and galaxies from Raw Energy or Available Energy; and, it would still produce the same observable universe that we have today as long as the Gods start infusing huge amounts of dark energy on a universal scale somewhere along the way.

The Big Bang and the Big Bang Singularity are not necessary to produce what we see today. In fact, they would not produce what we see today. In truth, if the Big Bang really happened, then our universe would be nothing more than one huge homogeneous expanding cloud of hydrogen and helium gas because hydrogen molecules repel each other, and helium repels everything. It's physically impossible for expanding clouds of hydrogen and helium gas to coalesce into planets, stars, and galaxies. It can't be done which means that it wasn't done.

Remember, the Big Bang Singularity is unnecessary to produce what we see today. You can stop your rewind anywhere along the way and still produce the results that we see around us today. Remember, you don't have to rewind it all the way to the Big Bang Singularity in order to produce the results that we see today. In fact, it's better if you don't.

The existence of the Big Bang Singularity is in dispute. First of all, there's no physical explanation for how to make such a thing. It would have to be spiritual, quantum mechanical, and syntropic in nature and origin. Second, the Big Bang Singularity isn't necessary to produce what we see today. What is absolutely necessary is the production of entropic physical matter approximately 13.4 billion years ago. The physical matter and entropy had to be made relatively recently, or there would be no visible planets, stars, and galaxies because we wouldn't exist either. The construction of physical matter, planets, stars, and galaxies is the essential component for the existence of our physical universe – not the Big Bang Singularity. The second law of thermodynamics or entropy tells us that genomes, planets, stars, and galaxies would not exist right now unless entropic physical matter was made by the Gods relatively recently.

Based upon $E = mc^2$, the Gods can turn Dark Matter and Dark Energy into entropic physical matter anytime they choose to do so, and the Gods can also turn entropic physical matter into Dark Matter and Dark Energy anytime they choose to do so; consequently, we are not done yet with what the Gods can still choose to do. When it comes to $E = mc^2$, ALL of the stuff on the right side of the equation is MADE by the Gods. It's the energy that can't be made and can't be destroyed; but, that energy can be FORMED and TRANSFORMED as the Gods see fit. ALL of the energy is available to the Gods for use ALL of the time. The Gods just can't create the energy nor destroy the energy; but, the Gods can FORM the energy into anything they want any time they want.

What else is wrong with the Big Bang Theory?

Well, it requires a God to make it possible and to control the rate of expansion. It requires a God to make the White Hole or the Big Bang Singularity. It requires a God to trigger it and make it go bang. It requires a God to control the rate of expansion. It requires a God to start and stop Inflation at a universal scale. If you believe in the Big Bang Theory, then you have to believe in God whether you realize it or not.

The Big Bang requires a God to start and stop Inflation as well as to change the Cosmological Constant or the amount of Dark Energy which is being infused into the system to control the rate of expansion. The Big Bang Singularity, the Big Bang, the subsequent Inflation Period, and the Accelerating Expansion of our physical universe require a massive infusion of Syntropy, Energy, Dark Energy, or Organization from the Gods in order to make them happen. It would take a God to convert Raw Energy or Available Energy into the different forms of physical matter or organized energy.

It would take a God to produce the Higgs Field. Order and organization don't just spring into existence from nothing. They are MADE or caused to begin. Anything that was obviously made obviously has a Maker who made it. Order and organization of any kind are obviously made, which means that they obviously have a Maker who made them. Physical matter was obviously organized from different forms of energy. Physical matter was made. Physical matter didn't just spontaneously generate out of thin air from nothing. It required a God to design and make the first particles of physical matter.

Furthermore, expanding clouds of hydrogen and helium gas are NEVER going to coalesce into planets, stars, and galaxies. NEVER! It would take a God to make expanding clouds of hydrogen and helium gas coalesce into planets, stars, and galaxies. The very existence of planets, stars, and galaxies is Scientific Proof of God's Existence. They shouldn't exist if the Big Bang Theory is true and the Big Bang truly happened.

Finally, the Horizon Problem and Flatness Problem successfully and convincingly falsify the Big Bang Theory and require intervention from God in order to overcome them. The non-existence of Population III stars, comprised exclusively of hydrogen and helium, falsifies the Big Bang Theory. Entropy or the Second Law of Thermodynamics falsifies the Big Bang Theory; and, the Big Bang if it happened FALSIFIES entropy and the second law of thermodynamics. They both can't be true simultaneously because they are mutually exclusive and falsify each other. If it existed, the Big Bang Singularity was Pure Syntropy. Syntropy falsifies and eliminates entropy and the Second Law of Thermodynamics.

ALL of the scientific evidence that I encounter FALSIFIES the Big Bang Theory and makes it physically impossible. Only a God could make a Big Bang Singularity or a White Hole and contain it before letting it go. The Big Bang would require a God to make it happen, to start and stop inflation, and then to drive the rate of expansion. Only a God could change the universal constant at a universal scale. The Big Bang Theory also requires a God to make expanding clouds of hydrogen and helium gas coalesce into planets, stars, and galaxies.

I used to be a Materialist, Naturalist, Nihilist, and Atheist. I was a firm believer in Scientism, which is the philosophical belief that Science or the Scientific Method is the only way to find and know the truth. BEFORE I was going to be willing to believe in God, I was holding out for Scientific Proof of God's Existence, firm in the belief that I would never find any scientific proof of God's existence. God called my bluff.

If the Big Bang Theory is true (and most scientists believe that it is) and if the Big Bang truly happened (and most scientists believe that it did), then it FALSIFIES entropy or the Second Law of Thermodynamics, and it FALSIFIES the Theory of Relativity and the Speed-of-Light Limitation. In order to successfully produce the Big Bang Singularity as well as the Big Bang and Inflation, then God, Syntropy, the Conservation of Energy, or the Conservation of Psyche, or the First Law of Thermodynamics would have to completely override and negate Entropy or the Second Law of Thermodynamics. This reality, if true, would suggest that entropy doesn't really exist and that the Second Law of Thermodynamics is fundamentally flawed and false. These are some very painful scientific discoveries for the Materialist, Naturalist, Nihilist, and Atheist that I used to be. It means that with this knowledge, I can no longer be a Materialist, Naturalist, Nihilist, and Atheist in good conscience and remain true to Science as a whole.

That's what happens to you when you choose to allow all the evidence into evidence and then start to pursue a preponderance of that evidence – you start to make scientific discoveries that nobody has ever thought about before – and it falsifies your favorite beliefs and ideas. I'm not going to apologize for figuring this stuff out. I'm a scientist. We are supposed to examine the evidence, and then figure out what things are and how they work. We are supposed to figure out what's wrong with our theories and ideas. That's what scientists do, or that's what scientists are supposed to do.

There's nothing wrong with the Big Bang Theory so long as we realize that each part of it required a God to make it happen, if it did in fact happen. So long as we are fine with God's causing and producing each part of the Big Bang thereby making it happen, then the Big Bang Theory becomes a viable option for how God produced our physical universe; but, it's physically impossible for all of that to have happened on its own spontaneously through random fluctuations in entropy or quantum fluctuations. Anything that is obviously caused obviously has Someone Psyche or Someone Intelligent who caused it. The Big Bang, if it happened, was obviously caused; therefore, it MUST have some kind of Transdimensional God who caused it.

It has been observed by Out-of-Body Travelers that the Human Psyche removes entropy from any system that it encounters and introduces order and organization instead. That's what Psyche does; and, that's what God's Psyche did. The Human Psyche can do the physically impossible; and, the same can be said for God's Psyche. Within a non-consensus reality in the spirit world or the quantum realm, the Human Psyche is God. This is what has been experienced and observed. Psyche is Energy. Psyche or Intelligence is Syntropy; and, Syntropy is the opposite of entropy. Syntropy FALSIFIES and eliminates entropy. In the non-consensus realms in the Quantum Realm or Spirit World, the Human Psyche or Human Intelligence automatically removes entropy or disorder from a system thereby bringing order and organization instead. This is what has been experienced and observed.

Psyche is Energy. Energy is required to reverse and eliminate entropy. Psyche or Life Force is the innate intelligence within all the different forms of energy which gives that energy the inherent ability to understand, follow, and obey God's Laws and God's Commands. Energy is conscious, sentient, intelligent, perceptive, psychic, aware, and alive. When entering a non-consensus reality, the Human Psyche automatically removes the entropy or disorder and replaces it with order and organization instead. This is what has been experienced and observed. The Human Psyche is Pure Syntropy. The observed, verified, and proven existence of Psyche or Pure Syntropy ultimately falsifies entropy and the Second Law of Thermodynamics.

In contrast to all of this, the Materialists, Naturalists, Darwinists, Nihilists, and Atheists tell us and assure us that psyche or non-local consciousness DOES NOT EXIST. They are wrong to do so because God's Psyche and their psyche is the only reason that they exist.

Do you see how that works? It's the answer to life, the universe, and everything. It explains everything that has ever been experienced or observed. Science is observation and experience after all – or it should be.

Science 2.0 allows all of the evidence into evidence, and then pursues a preponderance of that evidence. If you successfully eliminate everything that is false and everything that has been falsified, then ONLY the truth will remain. The BEST and FASTEST way to find and know the truth is to live it, experience it, and observe it for yourself, or to choose to trust someone who has. Truth is based upon a preponderance of the evidence rather than philosophical speculation and wishful thinking. The false is falsified by the truth; and, the truth is constantly experienced and observed. Science is observation and experience. If you successfully eliminate everything that is false such as Materialism, Naturalism, Darwinism, Nihilism, and Atheism, then ONLY the truth will remain.

I used to be a Materialist, Naturalist, Nihilist, and Atheist; but, I no longer bow the knee in worship of Heat Death and Entropy. It's no longer my God. Instead, I have successfully used Science and Observation to falsify these things. My greatest desire is to set people free from the chains of Nihilism that have them bound. Due to a complete lack of hope, suicide is on the rise; but, it doesn't have to be that way. It isn't that way.

Christ's message is a message of Hope, Syntropy, Resurrection, Atonement, Redemption, Eternal Life, and an End to Death.

BEFORE we had Science, Quantum Mechanics, and Quantum Field Theory to tell us what's going on at the quantum level, the Biblical God Christ Jehovah revealed the truth of the situation to us – namely that entropy, anergy, heat death, or the second law of thermodynamics DOES NOT EXIST at the psyche level in the Quantum Realm or the Spirit World. In other words, Energy and the Psyche within that Energy is always conserved, which means that Psyche, Life Force, or Intelligence is eternal and everlasting without a beginning of days or an end of years. Psyche or your Immortal Soul cannot be made, and it cannot be destroyed. It has always existed and will always exist. The Energy in quantum fields and the Energy behind science and quantum mechanics cannot be made nor destroyed either. Heat death is impossible at the quantum level. This is the Quantum Law of Thermodynamics. It is Syntropy.

Psyche or Intelligence is eternal. Your spirit body was made, which means that your spirit body can theoretically be disassembled or destroyed. In contrast, your Psyche or Intelligence has always existed and will always exist. Psyche cannot be made nor destroyed. Psyche is syntropic or eternal. Jesus taught the Doctrine of Eternal Life. If Jesus is God like He claimed to be, then He would KNOW all about Eternal Life. Eternal Life goes in both directions, forwards and backwards, especially where Psyche or Intelligence is concerned. Syntropy is Eternal Life. God revealed the truth of it to us long before we had scientists to tell us that it isn't true, and that psyche or spirit does not exist.

Instead of these truths, we now have the modern-day Materialists, Naturalists, Darwinists, Nihilists, and Atheists who assure us that psyche does not exist, that the Heat Death of the Universe is eternal and everlasting, that death is permanent, and that the associated entropy will always be conserved. If they are right, then God is wrong. If God is right, then the Naturalists and Atheists are wrong. Your task as a scientist is to figure out who is right and who is wrong. All of eternity hangs in the balance.

Mark My Words

The Quantum Law of Thermodynamics

Psyche is the innate intelligence within all the different forms of energy which gives that energy the inherent ability to understand, follow, and obey God's Laws and God's Commands. Energy is intelligent, psychic, conscious, sentient, perceptive, alive, aware, and conserved. This is what has been experienced and observed.

The Quantum Law of Thermodynamics states that Heat Death is impossible at the psyche level or the quantum level because Psyche, Life Force, or Energy is always conserved in the Psyche Realm or the Quantum Realm. The Quantum Law of Thermodynamics states that there is NO Thermodynamics and NO Heat Flow at the quantum level or the psyche level. Thermal equilibrium, entropy, and heat death DO NOT EXIST and do not apply at the quantum level or the psyche level. This means that entropy or death is NOT conserved. This is what has been experienced and observed by Out-of-Body Explorers and those who have had a Near-Death Experiences.

The Quantum Law of Thermodynamics FALSIFIES Materialism, Naturalism, Darwinism, Nihilism, Behaviorism, Determinism, Physical Reductionism, Classical Physics, the Second Law of Thermodynamics, and Atheism which claim that entropic physical matter is the ONLY thing that exists, and that entropy or death is conserved. The Quantum Law of

Thermodynamics ultimately falsifies or trumps the Second Law of Thermodynamics. At the quantum level, Maxwell's Demon wins the battle after all and entropy is defeated. The Quantum Law of Thermodynamics is based upon the First Law of Thermodynamics or the idea that Psyche or Energy is always conserved. The Conservation of Psyche or the Conservation of Energy holds true on both sides of the veil – both in the Quantum Realm and in the Physical Realm.

When it comes to the Quantum Law of Thermodynamics, there is nothing to prove because it has already been experienced and observed. Nevertheless, ALL of the zeroes and infinities that we encounter in math is proof that the Quantum Law of Thermodynamics or Syntropy is true. Think about it. Syntropy is ZERO entropy. Syntropy is also Eternal Life or Infinite Life. Psyche or the Life Force is syntropic – it is eternal and everlasting. It cannot be made, and it cannot be destroyed. Psyche or Energy is always conserved on both sides of the veil – both on the spiritual side and the physical side. Syntropy means the Conservation of Psyche or the Conservation of Energy. This is what has been experienced and observed.

Remember, the Quantum Law of Thermodynamics states that Heat Death is impossible in the Quantum Realm or the Psyche Realm. Within the Quantum Realm or the Psyche Realm, entropy or the second law of thermodynamics does not exist; and, Syntropy, Psyche, or Eternal Life reigns supreme. This is what has been experienced and observed. Syntropy or the Quantum Law of Thermodynamics is the answer to life, the universe, and everything. It explains everything that has ever been experienced and observed.

Mark My Words

Syntropy Must Exist

I start with what has been experienced and observed; and then, I try to find a scientific explanation for it.

The Quantum Law of Thermodynamics states that heat death is impossible and non-existent at the quantum level and the psyche level. Psyche is the innate intelligence within all the different forms of energy which gives that energy the inherent ability to understand, follow, and obey God's Laws and God's Commands. Energy is psychic and syntropic. This is what has been experienced and observed.

Syntropy is ZERO entropy. Syntropy is also Eternal Life or INFINITE LIFE. The Quantum Law of Thermodynamics is Syntropy. The First Law of Thermodynamics is Syntropy. The Ultimate Law of Thermodynamics points us to Syntropy. Psyche or Life Force is Syntropy. Energy is Syntropy. Syntropy is the Conservation of Psyche or the Conservation of Energy. Syntropy is Exergy, or energy that is available for use. Organization is Syntropy. God is Syntropy. These things have been experienced and observed.

The Quantum Law of Thermodynamics has to be true. The First Law of Thermodynamics has to be true. Syntropy has to exist. The scientists don't realize it yet; but, some type of Syntropy MUST EXIST; otherwise, we wouldn't exist. So, where is it; and, how does it work? Our job as scientists is to figure out what things are and how they work.

The Big Bang Singularity, if it existed, was Pure Syntropy or Pure Exergy. ALL of the energy within that thing was available for use. It would take the Quantum Law of

Thermodynamics and the impossibility of heat death at the quantum level in order to produce a Big Bang Singularity. It would take Someone Psyche, Someone Intelligent, Someone Powerful, and Someone Syntropic to produce a Big Bang Singularity. You can't imbue or gift capabilities that you yourself do not have.

The moment when the Gods started making entropic physical matter, planets, stars, and galaxies in this corner of the universe 13 billion years ago was a moment of High Syntropy or High Exergy, when a lot of available energy was converted into the different particles of physical matter and then organized into planets, stars, and galaxies. The Gods had to make that energy available for use; otherwise, there would be NO physical matter, NO usable physical matter, and NO planets and stars in the first place. The Ultimate Law of Thermodynamics states that physical matter and entropy are NOT conserved which means that they are made or caused to begin. It takes huge amounts of Syntropy or Exergy in order make physical matter, planets, stars, and galaxies. This is what has been experienced and observed.

The Syntropy has to exist, or we wouldn't exist. The truth of it is obvious.

Most of the scientists don't want this to be true, however, because it FALSIFIES Materialism, Naturalism, Darwinism, Nihilism, Behaviorism, Determinism, Atheism, Classical Physics, the Theory of Relativity, Heat Death, and Entropy which are the pillars of their religion. Nevertheless, the existence of Syntropy has to be true; otherwise, we would not exist. There's no way around it if we want to exist.

The Atonement of Christ and the Resurrection are Pure Syntropy; and, their observed and experienced existence FALSIFIES heat death, entropy, the different types of atheism, and the second law of thermodynamics. The false is falsified by the truth; and, the truth is repeatedly experienced and observed. As scientists, it's time for us to start looking for Signs of Psyche or Signs of Syntropy because they are there to be found.

If you believe in the Big Bang Theory, believe that the Big Bang Singularity existed, and truly believe that the Big Bang happened, then you already believe in Syntropy whether you realize it or not. It would take a God to make such a thing, would it not? Someone Psyche had to make the thing, and then make it go bang. The Big Bang, if it happened, was a quantum event of Pure Syntropy because classical physics, physical limitations, physical laws, physical constants, physical matter, time, mass, and entropy didn't exist yet.

Remember, the Big Bang Singularity, if it existed, was Pure Exergy or Pure Syntropy. The Big Bang, if it happened, was a quantum event of Pure Syntropy. The construction of entropic physical matter, planets, stars, and galaxies was also an event of High Organization or High Syntropy. It had to be; otherwise, we wouldn't exist.

The design and production of genomes was an event of High Organization or High Syntropy that ran counter to the effects of entropy. Once again, it had to be syntropic, or we wouldn't exist. Your genome is God's Signature; and, it's written on every cell within your physical body.

Mark My Words

Being Watched by the Universe

The Quantum Law of Thermodynamics states that heat death and death are impossible at the psyche level in the Quantum Realm because Psyche or Life Force is always conserved. Syntropy is the Quantum Law of Thermodynamics. The Ultimate Law of

Thermodynamics states that physical matter and entropy are NOT conserved – only the underlying Energy or Psyche or Life Force is conserved. The Ultimate Law of Thermodynamics also points us directly to Syntropy. Syntropy has been experienced and observed.

The Quantum Zeno Effect tells us that the atoms KNOW when we are watching them. The atoms can read our mind. This is Action at a Distance. The atoms don't have any classical physics mechanism for being able to read our mind. All of this telepathic mind-reading or psyche-reading, that's taking place at a distance, has to be happening at the quantum level because it is physically impossible at the physical level. Telepathy is WiFi at the quantum level or the psyche level. The Universe or Nature's Psyche is watching us, reading our minds, and then responding accordingly.

The Quantum Law of Thermodynamics, the Ultimate Law of Thermodynamics, Syntropy, Action at a Distance, and the Quantum Zeno Effect have been experienced and observed which means that they have already been verified and proven to be real and true. Our task as scientists is to take them and run with them.

Syntropy is ZERO entropy. Syntropy is also Eternal Life or INFINITE LIFE. The math behind ZERO and INFINITY has already been discovered and has already been proven to exist. It's just sitting there waiting for someone to take it and run with it.

Atoms Are Sentient, Conscious, and Alive

https://phys.org/news/2015-10-zeno-effect-verifiedatoms-wont.html

https://quantum-neuroscience.com/wp-content/uploads/2018/07/Zeno-Effect-Verified-—-Atoms-Wont-Move-While-You-Watch-Them.pdf

The Quantum Zeno Effect tells us that Nature's Psyche within Energy and atoms is conscious, sentient, perceptive, intelligent, telepathic, psychic, alive, and aware. Psyche is the innate intelligence within all the different forms of energy which gives that energy the inherent ability to read your thoughts and KNOW your mind and intentions.

The Quantum Experiment that Broke Reality

https://www.youtube.com/watch?v=p-MNSLsjjdo

https://psyche-ontology.com/wp-content/uploads/2018/08/The-Quantum-Experiment-that-Broke-Reality.zip

As usual, I edited the contents of these videos and enhanced parts of the presentation so that it hopefully makes sense without the diagrams and the visuals.

In this video, he repeatedly mentions that photons, electrons, and atoms MAKE CHOICES where to land. He talks about these quantum objects KNOWING the interference pattern of a wave function and how it works. He also talks about "The Universe" or what I call "Nature's Psyche" making choices. "Knowing", "Intelligence", and "Making Choices" are functions of Psyche or Quantum Non-Local Consciousness. The Materialists, Naturalists, Darwinists, Nihilists, Behaviorists, Determinists, and Atheists assure us that choice or free will does not exist at the physical level. This means that "choice" or "free will" has to exist at the quantum level or the psyche level, just like Quantum Mechanics and the Copenhagen Interpretation predict. Where Quantum Mechanics, Supernatural Mechanics, or Spiritual Mechanics is concerned, Psyche is the BEST EXPLANATION for what we are actually experiencing and observing. Psyche is the ONLY thing we know of that is capable of making choices both at the quantum level and at the physical level.

The single particle double-slit experiment hints that the fundamental nature of reality may not be physical at all. Light is a wave in the electromagnetic field. Light also comes in indivisible little bundles of electromagnetic energy called "photons" – a photon wave packet. Each photon is a little bundle of waves, waves in the electromagnetic field. Photons are indivisible packets of energy. This means that each photon has to DECIDE whether it is going to go through one slit or the other. It can't split in half and then recombine on the other side, or can it? Each photon dumps all of its energy at a single point. Each photon reaches the screen KNOWING which regions are the most likely landing spots and which are the least likely. It KNOWS the interference pattern of a pure wave that passed through both slits equally; and, it CHOOSES its landing point based on that information or knowledge.

Individual electrons, atoms, and molecules also produce interference patterns when launched at a detector screen individually. They KNOW how they are supposed to act. Describing the behavior of the wave function is the heart of quantum mechanics. That wave holds the INFORMATION about all the possible final positions of the particle as well as the possible positions at every stage in the journey. For some reason, when the wave reaches the screen it CHOOSES a final location, and that implies CHOOSING from the various different possible paths towards its final location.

The Copenhagen Interpretation says that the wave function doesn't have a physical nature. Instead, the wave function is comprised of pure possibility or pure intelligence. It's almost like the universe is allowing all possibilities to exist simultaneously but holds off CHOOSING which actually happened until the last instant. In the Copenhagen Interpretation, that final CHOICE of the experiment or that final CHOICE of location is fundamentally random within the constraints of the wave function and the final collapse of the wave function. We know that light is a wave in the electromagnetic field; and, quantum field theory tells us all fundamental particles or quanta are waves in their own respective fields.

Here he talks about The Universe making CHOICES. This is synonymous with my ongoing claim that Nature's Psyche makes choices. Choice is a function of Psyche. Chosen behaviors are produced by Psyche or Intelligence. Studying information, understanding information, and using information are a function of Psyche or Intelligence. Energy is psychic and intelligent. It makes choices among competing alternatives. Each wave packet of energy is intelligent, knowledgeable, telepathic, psychic, prescient, sentient, perceptive, conscious, alive, aware, and makes choices. There's a little bit of Psyche or Life Force in each wave packet of energy. The Psyche or Intelligence within energy is the fundamental nature of reality and the fundamental unit of reality – NOT entropic physical matter.

The Universe or Nature's Psyche is studying us, observing us, reading our minds, and then making CHOICES and DECISIONS based upon that information. Psyche is the innate intelligence within all the different forms of energy. Energy is intelligent, psychic, telepathic, sentient, perceptive, conscious, alive, and aware. Only Psyche or Intelligence can make choices. Choice is a function of Psyche and a product of Psyche. Heat Death is impossible at the quantum level or the psyche level because Energy or Psyche is always conserved. Energy or Psyche is the only thing that's needed to completely eliminate entropy or disorder from any system, even a physical system. This is what has been experienced and observed.

Remember, Psyche is the BEST EXPLANATION for choice, work, and causality at the quantum level. When it comes time to make choices and to do work at the quantum level or sub-atomic level, genes, evolution, natural selection, and entropic physical matter just don't cut it. They can't deliver the goods. They can't do the work. They don't work.

What does all this have to do with entropy?

Well, a careful study of entropy reveals that information or knowledge is all that's needed to eliminate entropy from any system. In other words, Psyche or Intelligence can eliminate entropy from a system at will. There is NO entropy and NO heat death at the quantum level or the psyche level. Heat Death or Entropy is impossible in the Quantum Realm where psyche and energy are always conserved. This is the Quantum Law of Thermodynamics which states that entropy, heat death, and the second law of thermodynamics do not exist in the Spirit World or the Quantum Realm because Psyche or Intelligence eliminates entropy at will. This is what has been experienced and observed.

How the Quantum Eraser Rewrites the Past

https://www.youtube.com/watch?v=8ORLN_KwAgs

https://psyche-ontology.com/wp-content/uploads/2018/08/How-the-Quantum-Eraser-Rewrites-the-Past.zip

Technically, the collapse of the wave function doesn't require a conscious observer to resolve the wave function. An impenetrable barrier works just fine. The Psyche within the energy KNOWS that it has to resolve the situation whenever it encounters a barrier. According to the Orthodox Interpretation of Quantum Mechanics by Henry P. Stapp, it is Nature's Psyche or The Universe who collapses the wave function, and NOT the observer or the Human Psyche. Maybe "collapse of the wave function" is better defined as Resolution of the Wave Function? Either way, it's Nature's Psyche who collapses the wave function or resolves the wave function and NOT the Human Psyche. The Human Psyche simply decides what it wants to do with its physical body; and then, Nature's Psyche collapses the necessary wave functions and fires the necessary neurons to make the Human Psyche's choices a physical reality.

Physicists have tried hard to see which slit each particle actually travels through before it produces the famous interference pattern. Turns out the universe is onto us! Any experiment that determines unambiguously which slit the particle traverses destroys the interference pattern. This is even true if you place detectors on the far side of the slits after the wave particle should have already interfered with itself. It's like the wave function is collapsing retroactively, as if the universe is saying, "OK, guys. The human figured out which way you went. No more funny stuff. Better pretend like you were particles the whole time." Physicists hate being outsmarted by the universe.

The Universe or Nature's Psyche is watching us, reading our minds, and then responding accordingly. In this video and other videos, he talks about the CHOICES that the quantum particles or the wave functions make.

In a more complex experiment where the photon is split into an entangled pair of photons, it seems as if the entangled photons are communicating with each other and sharing information with each other. It's also as if the photons are reading your mind presciently while at the same time collapsing the wave function retroactively. If the photons KNOW that you haven't detected which slit they went through, they produce the interference pattern. If the photons KNOW that you have detected which slit they went through, then the interference pattern isn't produced. It's as if The Universe or Nature's Psyche is onto us, observing what we are doing, reading our minds presciently, and even sending information into the past retroactively based upon what it KNOWS will happen in the future. This is way out there far beyond anything that Materialism, Naturalism, and Atheism can explain to us scientifically or in scientific terms. It requires Psyche, Consciousness, Intelligence, and Quantum Mechanics in order to produce a scientific explanation for these kinds of quantum mechanical phenomena. Classical physics isn't going to cut it.

The Copenhagen Interpretation tells us that the wave function isn't physical. Quantum properties act and change instantly at any distance. This is called Non-Locality. The delayed choice quantum eraser double-slit experiment doesn't tell us whether the wave function is physical or not, but it tells us that the Copenhagen Interpretation or any other metaphysical interpretation of the wave function is no less crazy-sounding than a physical interpretation. In fact, the solution may lie in this fascinating phenomenon of quantum entanglement. Entangled particles really are able to influence each other instantaneously at a distance. Perhaps this thing we call observation is just entanglement between the observer and the experiment.

In the second of these two Space Time videos about the double-slit experiment and the Copenhagen Interpretation of Quantum Mechanics, this is the closest that you will see Matt get to the spiritual or the metaphysical side of Quantum Mechanics. Though reserved, he, too, seems to have the same sense of wonder, respect, and awe that most of the rest of us have when he realizes that The Universe or Nature's Psyche is watching us and responding accordingly. He says that it's amazing.

Where these double-slit experiments are concerned, observation is the telepathic interaction between the Human Psyche and Nature's Psyche. The Human Psyche does the observation, and Nature's Psyche KNOWS that it is being observed or messed with. It's Nature's Psyche who collapses the wave function. The Human Psyche isn't consciously aware of any of the quantum mechanical stuff. Instead, the Human Psyche just designs the experiments and makes the observations. The Universe or Nature's Psyche KNOWS whether we are watching or not, and it responds accordingly. Think about it! The Universe or Nature's Psyche can read our minds and responds accordingly. This is what has been experienced and observed.

The Universe's Psyche, Nature's Psyche, or the Psyche within the individual atoms and wave packets KNOWS what we are doing, reads our mind, and chooses the appropriate response. Choice is a function of Psyche, or Intelligence, or Consciousness. A Psyche is identified by the choices that it makes.

Energy is intelligent, psychic, telepathic, sentient, conscious, perceptive, alive, and aware of us. Psyche is the innate intelligence within all the different forms of energy which gives that energy the inherent ability to understand, follow, and obey God's Laws and God's Commands. The Universe or Nature's Psyche studies us, observes us, reads our mind, knows what we are doing, and responds accordingly. It's intelligent and alive.

He talks about photons sending information back into the past to their entangled counterparts. However, there is another possible explanation. From the photon's perspective, it quantum tunnels to its destination and experiences NO passage of time. It has been suggested that photons KNOW where they are going to land before they launch; otherwise, they don't launch if they have nowhere to land. Under this interpretation of the experiment, the photons aren't sending information into the past. Instead, the photons

KNOW the future before they launch; otherwise, they don't launch. Either way, photons are psychic and intelligent. Photons have a Psyche who can perceive the environment and process information from that environment.

Physical atoms are comprised of different forms of energy working together to accomplish a larger whole. The energy within atoms is intelligent, psychic, telepathic, can read our minds, and can then respond accordingly.

There is NO entropy and NO classical physics in Non-Locality or Quantum Entanglement. There is NO entropy and NO classical physics in the Quantum Zeno Effect. This stuff is pure Telepathy, Syntropy, and Action at a Distance. Action at a Distance functions at infinite velocities meaning that it is instantaneous. It's also prescient, retroactive, and effectively timeless. These are quantum phenomena or spiritual phenomena, NOT physical phenomena. Things work differently at the quantum level or the spiritual level in the Psyche Realm than they do at the physical level.

There is NO entropy and NO classical physics in the Quantum Realm or the Spirit World. The Quantum Law of Thermodynamics tells us that there is NO entropy, NO classical physics, and NO thermodynamics at the quantum level or the psyche level within the Transdimensional Realm. The Quantum Law of Thermodynamics tells us that Heat Death is impossible in the Quantum Realm. Everything in the Quantum Realm happens instantaneously and is conserved. There is NO passage of time, no aging process, and therefore NO entropy in the Quantum Realm or the Spirit World. Things work differently at the quantum level or the psyche level than they do at the physical level.

Nature's Psyche within Energy can read your mind, know what you are thinking, observe what you are doing, and then respond accordingly.

How is this possible?

It's possible because Psyche or Non-Local Consciousness in its native unrestricted state (no spirit body and no physical body limiting it or restricting it) is omniscient and omnipresent. From a photon's perspective, it experiences NO passage of time while it travels to its destination. In other words, from a photon's perspective, it quantum tunnels to its destination even though from our perspective it took that photon 13 billion years to reach us. Even more impressive, a photon KNOWS where it is going to land before it launches; otherwise, it doesn't form, and it doesn't launch. These realities explain the results that we are getting from the double-slit experiments. The photon KNOWS if you are going to be messing with its wave-path before it launches. Nature's Psyche, within the energy that comprises photons and electrons, is observing you and reading your mind. It KNOWS what you are doing and understands your intentions. It KNOWS whether you are messing with it or not, observing it or not.

In these and other videos, they emphasize that The Observer doesn't have to be a human being in order to collapse the wave function. The Observer can be a barrier or a destination such as a photo-electric detector. When a photon reaches a barrier or reaches its destination, it has to resolve the situation. This resolution is the collapse of the wave function. When the photon reaches a barrier or reaches its destination, the wave function achieves Locality, and the wave function collapses, or the wave function is resolved into a single point on the photon detector. Nature's Psyche within Energy obeys God's Laws; consequently, the chosen result is orderly, structured, and lawful.

ALL of the energy is sentient, alive, conscious, intelligent, psychic, and aware. Energy or Nature's Psyche responds to God's Laws and God's Commands because it has covenanted with God to do so. That's how God was able to bring order and organization to chaos. Laws, restrictions, limitations, and obedience to God's Commands bring order, organization, structure, purpose, and consensus to Energy and Nature's Psyche. Energy and Nature's Psyche get as much out of the deal as God does. Deliberate obedience to God's Laws brings order, organization, structure, purpose, meaning, and consensus to our Physical Reality; otherwise, we wouldn't exist, physical matter wouldn't exist, and the whole thing would still be random chaos.

Psyche is the innate intelligence within all the different forms of energy. The Ultimate Law of Thermodynamics states that Psyche, Intelligence, Life Force, Energy, the Quantum Realm, and Syntropy are conserved; whereas, physical matter, death, and entropy are NOT conserved. The Ultimate Law of Thermodynamics simply differentiates between what is being conserved and what is NOT conserved. The Quantum Law of Thermodynamics states that heat death is impossible at the quantum level because the second law of thermodynamics or entropy doesn't exist at the quantum level. The Quantum Realm is an Isothermal Realm which means that in the Quantum Realm or Spirit World all of the energy is available for use all of the time, and that energy is always conserved. There is NO anergy or entropy in the Quantum Realm. Heat death is impossible at the quantum level. Consequently, the Quantum Law of Thermodynamics states that death is impossible at the psyche level. Your psyche, life force, or intelligence was not made, and it cannot be destroyed. Your Psyche cannot die. It is always conserved. This is what has been experienced and observed.

By allowing Psyche and Quantum Mechanics in to play, we can literally explain everything that comes our way. This is what I have experienced and observed.

In contrast, Physicalism, Materialism, Naturalism, Darwinism, Nihilism, Behaviorism, Determinism, Physical Reductionism, and Atheism were designed to prevent us from discovering the Ultimate Law of Thermodynamics and the Quantum Law of Thermodynamics; and, that's precisely what they do.

Conclusions about Entropy

The authors of these Space Time videos are Atheists, Materialists, Naturalists, and Physicalists. They really, truly believe that Entropy is one of the Fundamental Laws of Nature. They even have t-shirts and videos celebrating Heat Death and the End of Our Universe. In every video, they remind us that Heat Death is unavoidable and inevitable. They preach and teach Heat Death and the Theory of Evolution. The Theory of Evolution is Creation by Entropy or Creation by Death; and, these people expect us to believe that this is real and true. These people teach and believe that death, heat death, physical matter, and entropy are conserved. They teach that only entropic physical matter exists; and therefore, entropic physical matter is the only thing that exists to be conserved.

Through these videos, I believe that I have gotten into everything that has anything to do with entropy. This section of my book gives a pretty solid introduction to entropy – defining what it is and what it isn't. The fact that entropy doesn't exist in any usable form, that entropy is an emergent property of statistical mechanics, that entropy is not a quantum field nor a quantum force, that entropy is not a person nor a place nor a thing, and that entropy doesn't exist at the quantum level tells us that entropy is a philosophical concept or figment of the imagination much like natural selection or evolution. Entropy doesn't really exist. Entropy is a thought experiment rather than an actual thing that can be measured, detected, made, and observed. Entropy is a theoretical measurement of the non-existence of something, like the non-existence of order, or the non-existence of information, or the non-existence of thermal disequilibrium, or the non-existence of usable energy, or the non-existence of available energy. It's kind of difficult to detect and measure the non-existence of something. It's fascinating to observe that entropy doesn't exist in any usable form.

This is ironic because the Materialists, Naturalists, Darwinists, Nihilists, and Atheists base ALL of their beliefs on the non-existence of psyche, spirit matter, non-locality, non-physicality, and God. They tell us that all of these things, which have been experienced and observed, are simply figments of our imagination. However, in truth, Entropy, Natural Selection, Creation by Entropy, Creation by Death, Creation by Blind Chance, Chemical Evolution, Abiogenesis, Spontaneous Generation, Macro-Evolution, and Creation Ex Nihilo are in fact the things that are really truly nothing more than figments of the imagination and wishful thinking. These things don't exist in any tangible, usable form. These people have chosen to believe in things that don't really exist and have chosen to make these things their God. It really is ironic that these people tell us that all of our spiritual experiences and supernatural experiences and quantum experiences are figments of our imagination, while at the same time all of their chosen beliefs are nothing more than figments of their imagination. I find it fascinating from a psychological perspective how these people repeatedly tell us that everything that we have experienced and observed is nothing but a figment of our imagination and doesn't exist; whereas in truth, everything they have chosen to believe in is nothing but a figment of their imagination and doesn't really exist.

Remember, entropy or the second law of thermodynamics doesn't exist at the quantum level and the psyche level according to the Quantum Law of Thermodynamics. This is what has been experienced and observed. Science is observation and experience, or it should be.

Remember, Psyche or Intelligence can remove entropy from a system at will. This is what has actually been experienced and observed. In his Go Board thought experiments, Matt used his intelligence or psyche to remove the entropy from the board at will. Out-of-Body Travelers have also observed that whenever their Human Psyche enters into a non-consensus reality in the Spirit World or the Quantum Realm, that reality or region organizes itself according to the demands and expectations of their Human Psyche. In the Quantum Realm, when it comes to non-consensus realities, the Human Psyche is God. The Human Psyche literally removes the entropy or the disorder from non-consensus realities at will. Thanks to Psyche, there is NO entropy and NO heat death at the quantum level and the psyche level. Psyche or Energy is always conserved. Psyche cannot die. This is what has actually been experienced and observed. Science is observation and experience after all, or it should be. This is GOOD SCIENCE because it has actually been experienced and observed. If you truly want to know how things really work at the psyche level and the quantum level, then go with what has already been experienced and observed.

Anytime scientists talk about "emergent properties" or "statistical measurements", they are in fact talking about things that don't really exist as real tangible forces, fields, and entities. They are talking about things that are a figment of the imagination. They are talking about religion and philosophy. They are talking about thought experiments. Within these videos, they define entropy as an emergent property of statistical mechanics. That tells us a lot about what entropy is, and about what entropy is not. Syntropy, Life Force, Intelligence, Information, or Psyche is infinitely more substantial and real than entropy.

The Darwinists erroneously portray Natural Selection (or Abiogenesis) as a living, breathing, thinking, purposeful, deliberate, and cunning entity or being who designed, planned, made, and created our DNA, our genomes, our proteins, our cells, our eyes, our brain, our temperament, our predispositions, our physical bodies, and our lives. The

Darwinists portray Natural Selection as one of the fundamental forces in nature. They are wrong. Natural Selection results in our death and extinction. The only time Natural Selection comes into play is when we die. Natural Selection doesn't touch our genes! Furthermore, Abiogenesis or Chemical Evolution is physically impossible. It can't happen which means that it didn't happen.

Likewise, we Naturalists, Atheists, and Physicalists erroneously portray Entropy as a living, breathing, thinking, purposeful, deliberate, and cunning entity or being who inexorably moves physical matter, planets, stars, galaxies, energy, intelligence, and life towards thermal equilibrium and Heat Death. We choose to believe and convince ourselves that Entropy is one of the fundamental forces in the universe. We choose to believe that entropy or death is conserved. We call entropy a LAW, and we teach that Heat Death is inevitable and unavoidable and permanent. We are wrong. Entropy is death and extinction. Death is the opposite of Eternal Life. Entropy and physical matter are NOT conserved. Syntropy or Eternal Life falsifies and eliminates death and entropy.

Maximum entropy or anergy is the absence of available energy. The Physicalists, Materialists, Naturalists, Darwinists, Nihilists, Behaviorists, Determinists, Atheists, and Classical Physicists teach and believe that once that energy has been made unavailable, then that energy is forever unavailable. They call it Heat Death. These people literally teach and believe that Entropy is Conserved. These people literally teach and believe that death is conserved. Valar morghulis.

Disorder is a good definition for entropy. Eventually, though, I realized that the BEST and MOST descriptive definition for Entropy is "Death" – the end of potential or the end of life. Entropy is death. Entropy is Heat Death. The Theory of Evolution is Creation by Entropy or Creation by Death. The Theory of Evolution is a random fluctuation in entropy – Pure Chance or Blind Luck. It's physically impossible for a random fluctuation in entropy to produce order, organization, structure, and life. Entropy doesn't exist as a person, place, or thing. Entropy is NOT one of the fundamental forces of the universe. The different types of evolution are entropy; and, evolution of any kind is prevented from happening by entropy or the random diffusion of particles.

So, what is entropy?

Entropy is no thing! Technically, Entropy doesn't do anything. Entropy is the complete absence of work. Entropy knows nothing and does nothing. Entropy doesn't exist in any useful form. Entropy is an emergent property. Entropy is a philosophical and statistical concept. By definition and in principle, Entropy is useless and worthless. Like Natural Selection, Entropy doesn't do anything. Entropy doesn't work. I could go on; but, Entropy can't. Entropy is the end. Entropy is death. Entropy is thermal equilibrium or heat death.

Entropy is the statistical measurement of the amount of unusable energy in any given system. Entropy is an emergent property of thermal equilibrium. Maximum entropy means that there is NO usable energy or NO available energy in the system that is currently under consideration. Yet, here we are outside of that system, looking at that system, studying that system, and realizing that that system has suffered Heat Death and no longer has any usable energy within that system. Entropy is a measurement of the amount of "nothingness" within a system. Entropy is NOT one of the fundamental forces or fundamental laws of our universe. The person or psyche outside of the system – who is looking at the system and seeing and realizing that that system no longer has any usable energy within it – is the fundamental unit or fundamental force within the multiverse.

Do you see how that works? It explains everything that has ever been experienced and observed.

Entropy is Death. Natural Selection results in Entropy or Death. Natural Selection doesn't do anything! Natural Selection can't touch and doesn't touch our genes! Natural Selection simply "sits around" and waits for us to die. Like Entropy, Natural Selection is NOT one of the fundamental forces in our universe. Natural Selection is one of the emergent properties of various different physical laws. Natural Selection doesn't really exist. They built whole sciences on something that doesn't exist – entropy or natural selection. Natural Selection and Entropy are the same thing in that they are no thing. They are simply statistical measurements of REAL THINGS, or emergent properties of REAL THINGS.

The Theory of Evolution is Creation by Death or Creation by Entropy; and, Entropy is no thing. Entropy is a statistical measurement of the amount of unusable energy within any given system. Technically, Entropy isn't even an emergent law or a fundamental law of physical matter. Entropy is simply a measurement of the amount of energy that has been made unavailable to fallen, physical, mortal beings such as us. The energy is still there waiting to be accessed and used to make new and different things; but, that energy is currently unavailable for use by physical, fallen, mortal, entropic beings.

Do you see how that works?

It was news to me. I always thought that Entropy was a universal force or field, and that Entropy was made from energy like all the other fundamental forces and fields in this universe. I was wrong. Entropy is NO thing. Entropy is NOT a person, place, or thing. Entropy is a measurement of nothing, or a measurement of nothingness. Contrary to what I used to believe, entropy is not a thing. Entropy wasn't made from energy nor caused to exist. Entropy is not a physical constant, physical law, or physical force. Entropy is NOT a quantum field. Entropy is NOT one of the "elementary particles" or quanta. Maximum entropy means ZERO available energy. ZERO is nothing. Entropy doesn't do any work. Entropy doesn't do anything. Like Natural Selection and the Theory of Evolution, Entropy is a philosophy or a religion, and nothing more.

The revelation that I received is that Heat Death is impossible at the quantum level in the Quantum Realm. Your Psyche, Intelligence, Energy, or Life Force cannot die. This is the Quantum Law of Thermodynamics.

Exergy is the energy that is available to be used. After the system and surroundings reach thermal equilibrium, the exergy is zero. Determining exergy was the first goal of thermodynamics. Energy is neither created nor destroyed during a process. Energy changes from one form to another. In contrast, exergy is always destroyed when a process is irreversible, for example loss of heat to the environment. This destruction is proportional to the entropy increase of the system together with its surroundings. The destroyed exergy has been called anergy. Anergy means no available energy. For an isothermal process, exergy and energy are interchangeable terms, and there is no anergy. In an isothermal system, all of the energy is available for use all of the time – all of the energy is exergy.

Heat Death, Entropy, Anergy, Unavailable Energy, and the Second Law of Thermodynamics do not exist at the quantum level. Thermodynamics as we know it doesn't exist at the quantum level or the psyche level. Everything at the quantum level is syntropic, meaning that it's all being conserved. Psyche, Intelligence, or Energy is eternal and everlasting without a beginning of days or

an end of years. **Syntropy is the Conservation of Psyche or the Conservation of Energy. The Quantum Realm or Spirit World is an Isothermal System which means that ALL of the energy is available for use all of the time at the quantum level and the psyche level. There is NO entropy. The Quantum Realm or Spirit World is Pure Syntropy or Pure Exergy. If the Big Bang Singularity existed, then it was Pure Exergy or Pure Syntropy. Syntropy or the Quantum Law of Thermodynamics is the thing that would have made it possible. Syntropy or the Quantum Law of Thermodynamics is the answer to life, the universe, and everything. It explains everything that has ever been experienced and observed on both sides of the veil.**

Heat Death or Entropy is impossible at the quantum level. We KNOW that this is true because we exist. If Psyche or Energy could die at the psyche level, then we wouldn't exist. If Heat Death or Entropy were happening at the quantum level, then we wouldn't exist. The Syntropy, or Conservation of Energy, or Reuse and Recycling of Energy has to be happening someplace, or we wouldn't exist. In an Isothermal System, ALL of the energy is available for use all of the time. There is NO anergy or NO entropy in an Isothermal System. The Quantum Realm or Spirit World is an Isothermal System. There is NO entropy, NO anergy, and NO unavailable energy in the Quantum Realm. Entropy and Heat Death do not exist at the quantum level. Psyche or Life Force cannot die. In the Quantum Realm or Syntropic Realm, ALL of the energy is available for use all of the time because it is conserved. This is what has been experienced and observed.

Even at the physical level, entropy doesn't exist in any usable form. You can't use entropy to get anything done. Entropy isn't a quantum particle, quantum wave, quantum force, or quantum field. Technically, entropy doesn't exist as a person, place, or thing at the physical level. Entropy is an emergent property of things that do exist; but, entropy itself doesn't exist in any real sense or usable form. Entropy is a measure of unavailable energy. Entropy is a measure of nothingness. The energy is still there, it's just not available for our use at the physical level because it has reached thermal equilibrium at the physical level; however, ALL of that energy is still available for use at the quantum level. These truths explain what everything is and how everything works on both sides of the veil – both in the spirit world and our physical world.

Entropy is death. Syntropy is Eternal Life. Syntropy is infinitely more constructive, tangible, and real than entropy.

Entropy is a measurement of the unavailable energy within a given system. In contrast, Syntropy is the First Law of Thermodynamics – the Conservation of Psyche, or the Conservation of Energy, or the Conservation of the System. Even if our whole physical universe were to suffer Heat Death, it's still there. ALL of its energy and mass is still there waiting to be reinvigorated by Someone Psyche. The Quantum Fields are still there!

Psyche is the innate intelligence within ALL the different forces, fields, and energy which gives that energy the inherent ability to understand, follow, and obey God's Laws and God's Commands. Life Force, Psyche, Intelligence, or Energy is syntropic, which means that Psyche or Energy is conserved. Psyche or Energy is eternal and everlasting. It cannot be made, and it cannot be destroyed. Psyche or Energy is conserved. Syntropy, Energy, or Psyche is infinitely more tangible and real than entropy. The Human Psyche organizes, structures, and enlivens every non-consensus reality that it enters into. This is what has been experienced and observed.

Psyche processes, produces, and stores information. Information, knowledge, and intelligence are always conserved at the psyche level within the Human Psyche. The scientists have proven that information is conserved at the quantum level or energy level within Quantum Mechanics. However, I get the impression that NOTHING is conserved at the physical level where entropic physical matter is concerned.

Materialism, Naturalism, Darwinism, Nihilism, and Atheism have been FALSIFIED by observation and experience which means that their corollaries the "Conservation of Physical Matter", the "Conservation of Death", and the "Conservation of Entropy" are fruit from the poisoned tree. Death is the best and most explanatory definition for entropy. It tells it as it is. We all understand death.

Death is impossible because entropy is not conserved.

You have to let your brain work on that one for a while in order to see the truth of it and the truths behind it. In order for death to be eternal and everlasting, entropy has to be conserved. Since entropy is NOT conserved, it's impossible for death to be eternal and everlasting.

Ask yourself what breaks the Conservation of Entropy? Ask yourself what breaks death? There you will find the answer to life, the universe, and everything for real.

Psyche, Life Force, Intelligence, Knowledge, Information, Order, Organization, Syntropy, God, the Atonement of Christ, and Energy **break** death and the Conservation of Entropy. Syntropy is infinitely more substantial and real than entropy. That is what has been experienced and observed; and, it's all because entropy or death is not conserved.

Death is impossible because entropy is not conserved. Entropy or death doesn't exist at the psyche level. Your Psyche cannot die; and, Your Psyche cannot be killed nor destroyed. Your Psyche or Intelligence was not made; and, it cannot be destroyed because it is the fundamental unit of existence and reality. Without Psyche, there would be NO existence or reality. Death is impossible at the quantum level because entropy or heat death doesn't exist at the quantum level. At the quantum level ALL of the energy is available for use all of the time. This is what has been experienced and observed.

Syntropy is ZERO entropy. Syntropy is also Eternal Life or INFINITE LIFE. The math behind Syntropy has already been discovered. I understand the mathematics behind ZERO entropy and INFINITE LIFE; and, so do you. It's really simple to understand. You just have to choose to do so. You – your psyche – has to make that choice. Nobody else can do it for you. Your beliefs are chosen into existence by your Human Psyche whether you realize it or not. Your beliefs, memories, and personality survive the death of your physical brain. It's your Human Psyche or Intelligence who has experiences and forms memories of those events – not your spirit body, and NOT your physical body. This is what has been experienced and observed.

If you want your theories, ideas, and laws to be true, then they absolutely must match with what has been experienced and observed. There's no other way to guarantee that you have found the truth. Science is observation and experience after all.

As scientists, we have got to find a way to explain what has been experienced and observed. The FACT that Psyche has been experienced and observed is telling me that it's not getting the attention that it's deserved. Information, knowledge, data, instructions, commands, targeting information, logistics, and intelligence are being transmitted between atoms and molecules at the sub-atomic level or the quantum level. This is what is being experienced and observed. The order, organization, laws, restrictions, structure, and thoughtful planning have to come from Someone Psyche at the quantum level because communication between atoms and molecules is physically impossible at the physical level according to the Naturalists and Atheists.

A functional genome isn't going to spontaneously generate out of thin air from nothing as the Materialists and Darwinists claim is being done. The Big Bang Singularity, if it existed, would have had to have been a white hole; and, a white hole represents a complete reversal of entropy or Pure Syntropy. It would take a God to make such a thing. The ONLY reason we exist is because Someone Psyche can break and destroy death and entropy at will. Syntropy, the Conservation of Psyche, the First Law of Thermodynamics, and the Quantum Law of Thermodynamics explain why there is something rather than nothing. Our job as scientists is to explain what things are and how they work. I believe I have now done so.

Out-of-Body Travelers, Seers, and Near-Death Experiences have observed that the omnipresent nature of Psyche in its native environment or native format makes instantaneous psyche-to-psyche communication across the vast distances of our physical universe possible so long as each Psyche or Intelligence is not being limited by and restricted by physical matter and its associated locality, entropy, time, mass, and other physical limitations. This reality has also been experienced and observed. Psyche-to-psyche telepathy or transpersonal communication is real and truly happens. The same can be said for telekinesis and quantum tunneling. However, physical matter and entropy were designed to limit and restrict these superpowers or our innate quantum mechanical capabilities while we reside within an entropic physical body in a mortal fallen physical entropic realm.

Nevertheless, Quantum Mechanics or Superpowers are the NORM; and, our entropic physical body is the exception to the norm. Heat death and entropy are impossible at the quantum level because Psyche or Energy is always conserved. Heat Death is impossible at the quantum level because entropy is not conserved at the quantum level. In fact, entropy or the second law of thermodynamics doesn't exist at the quantum level. Ultimately, death is impossible because entropy and death do not exist at the psyche level. This is what has been experienced and observed. This is good news for anyone who loves life.

My message to the world is that Psyche and Quantum Mechanics are the fundamental reality of our universe and NOT physical matter or entropy. In fact, I use the First Law of Thermodynamics, the Conservation of Energy, and the Conservation of Psyche to FALSIFY entropy and the second law of thermodynamics, especially at the quantum level or the psyche level where it matters most. The human race as a whole has experienced and observed that thermodynamics, heat, anergy, and entropy DO NOT EXIST at the quantum level or the psyche level. Heat Death is impossible at the quantum level. Entropy is NOT one of the fundamental laws or fundamental forces within our universe. Entropy or heat death doesn't even exist in the Quantum Realm or the Spirit World. At the quantum level, all of the energy is available for use all of the time. It's an Isothermal System. This is what has been experienced and observed; and, Science is observation and experience after all. Go with what has been experienced and observed; and then, try to find a scientific explanation for those observations and experiences. That's what I try to do.

I like this stuff. I'm a theoretician, philosopher, scientist, and logician. I want to figure out what these things are and how they work. I believe that I'm finally beginning to do so. It's interesting; and, it's a lot of fun.

Mark My Words

Laws Are Limitations and Restrictions

In its primal original state, Syntropy would represent chaos – the complete absence of laws, limits, and restrictions. Syntropy means the conservation of energy. Syntropy means eternal and everlasting, without a beginning of days or an end of years. Energy or psyche or intelligence cannot be created, and it cannot be destroyed, which means that it is always conserved. Energy or Psyche has always existed, and it will always exist. Energy and Psyche are syntropic. They are conserved. Syntropy is the First Law of Thermodynamics, and it applies on both sides of the veil. This is the cornerstone of physics both in the quantum realm and the physical realm. Energy, Psyche, or Intelligence is always conserved. It is syntropic.

Laws are abnormal. Laws are made. Laws are restrictions and limitations. Laws therefore provide order and organization, the exact opposite of chaos. Laws or commandments bring order and organization out of chaos, but only if they are followed and obeyed.

Laws, to be real and effective, require some kind of universal Law Giver and Law Enforcer. Furthermore, Laws require someone intelligent enough and powerful enough to be able to understand the Laws and then choose to follow and obey those Laws. Laws are completely worthless without a Law Giver or Law Standardizer, a Law Enforcement System or a Law Teaching System, and someone smart enough and powerful enough to be able to choose to follow and obey those Laws.

Psyche is the innate intelligence or life force within all the different forms of energy which gives that energy the ability to understand, follow, and obey God's Laws and God's Commands.

In the beginning – BEFORE there was order, organization, laws, limits, and restrictions – it was pure chaos. It was syntropic; but, it was chaos. Then God brought the offer of order, organization, laws, limits, restrictions, light, and love into chaos in exchange for a promise – from the intelligences within all the different forms of energy within our area of space – to follow and obey God's Laws, God's Commandments, and God's Restrictions. Thereby, order and organization were born or made in this part of the multiverse or this part of chaos. Order and organization are the direct result of Laws, Limits, and Restrictions carefully understood and carefully obeyed.

The Quantum Realm has been organized and is being organized by the Gods, but it doesn't have any physical restrictions. The Physical Realm has a lot more restrictions, limitations, rules, laws, order, constants, predictability, control, structure, and organization than the Quantum Realm or the Spirit World. The Physical Realm is the ultimate consensus reality because it has a lot more restrictions and limitations. Entropy is one of those restrictions. In contrast, there is no entropy in the Spirit World. Everything there is syntropic. Spirit matter can be formed into different things; but, the energy and intelligence within it cannot be destroyed and is always conserved.

A willingness on the part of Psyche to submit to limitations and restrictions is what provides order and organization within the universe, on both sides of the veil. Without God's Laws and the subsequent Physical Constants, our corner of the universe would still be nothing but random chaos. Yes, it would still be syntropic, but it would be chaos.

Energy and Psyche cannot be made, and they cannot be destroyed. However, they do have the innate intelligence necessary to be able to understand, follow, and obey God's Laws and God's Commands. The innate Psyche within Energy can choose to submit to order, organization, restrictions, and limitations. Consequently, through obedience to God's Laws and God's Commands, the Gods can form raw energy, unorganized matter, or available energy into anything they want it to be, including physical matter and entropy. Entropic physical matter is comprised of different forms of energy. Physical matter is organized energy, which means that physical matter is limited and restricted energy. Physical matter was made by limiting and restricting the energy within it thereby bringing order and organization to it.

The science convinced me that God exists. Syntropy, or some kind of God, MUST EXIST, or all of that subsequent entropy and death wouldn't have been possible in the first place. This is logical common sense. Entropy is one of the physical laws. Entropy is made. Entropy has a beginning which means it can have an end. Syntropy means "unmade" – eternal and everlasting, without a beginning or an end. Entropy is NOT syntropic, which means that entropy was made. Entropic physical matter is made, which means it can be disassembled and formed into something else. Entropy is one of the physical laws. However, Entropy or Anergy doesn't exist at the quantum level. Heat death is impossible at the quantum level in the Quantum Realm or Spirit World. Your Psyche or Life Force cannot die.

Laws of any kind are made; therefore, they require some kind of Law Giver or Law Maker. The fact that Laws are made and then enforced is what brings order and organization to chaos. Laws that aren't enforced or universalized are worthless. Laws also require Someone Psyche who is intelligent enough and powerful enough to choose to understand, follow, and obey those laws. Laws and Physical Constants are worthless if they aren't followed and obeyed. Psyche is the innate intelligence within all the different forms of energy which gives that energy its intrinsic ability to understand, follow, and obey God's Laws and God's Commands.

Laws, limits, commandments, and restrictions bring forth order, organization, structure, physical matter, and the possibility of physical life; but, these laws and restrictions would be absolutely worthless and ineffective if the psyche or intelligence within energy didn't have the inherent ability to understand, to follow, and to choose to obey those Laws. Laws don't do anything without someone intelligent enough to follow and obey those Laws. These realities and truths explain everything that has ever been experienced or observed.

Mark My Words

Entropy of Any Kind Cannot Design and Create

Entropy is death. The Materialists, Naturalists, Darwinists, Nihilists, Behaviorists, Determinists, Physical Reductionists, Atheists, and Classical Physicists erroneously teach that ONLY physical matter or entropy exists – which means that these people erroneously teach that physical matter or entropy is being conserved. These people erroneously teach that physical matter or death is conserved. These people literally preach and teach that death is conserved. That's what they teach, do they not? But, they are wrong.

What's really being conserved?

The first law of thermodynamics is a version of the law of conservation of energy, adapted for thermodynamic systems. The law of conservation of energy states that the total energy of an isolated system is

constant; energy can be transformed from one form to another but can be neither created nor destroyed.

https://en.wikipedia.org/wiki/First_law_of_thermodynamics

According to the First Law of Thermodynamics, it's the Energy or the Syntropy that's actually being conserved and NOT the physical matter or the entropy. Entropic physical matter is comprised of different FORMS of Energy. The FORM is never conserved. It's the Energy or the Syntropy that's actually being conserved. The FORM changes. Entropic physical matter is never conserved – their amount or their quantity is constantly changing. We humans create new physical matter in our particle accelerators; and, we destroy physical matter in our atomic bombs. The amount of physical matter and entropy are constantly changing, which means that physical matter and entropy are NOT conserved. Death is not conserved. It's the Energy, Life Force, Psyche, Intelligence, or Syntropy that's actually being conserved according to the First Law of Thermodynamics – not the physical matter, death, and entropy. Psyche is Energy. Psyche is the intelligence within energy. Psyche or Energy is conserved.

The Ultimate Law of Thermodynamics goes out of its way to differentiate between what is being conserved and what is not conserved. Psyche is the intelligence within energy. Syntropy is the First Law of Thermodynamics – the conservation of energy or the conservation of psyche. Psyche or Energy is conserved; whereas, physical matter and entropy are not conserved. Entropic physical matter is comprised of different FORMS of energy. The FORM is never conserved.

The FORM of the Energy is constantly changing. The FORM of the Energy is never conserved. It's the Energy, Psyche, or Syntropy that's being conserved. Physical matter, spirit matter, dark matter, gravity, dark energy, forces, fields, the strong and weak nuclear forces, electricity, magnetism, the zero-point field, visible and invisible light, locality, space, time, and entropy are different FORMS of Energy. The FORM is never conserved. The FORM is constantly changing. Its amount is constantly changing. It's the underlying Energy, Syntropy, Psyche, Intelligence, or Life Force that's being conserved and remaining constant and eternal – not the FORM of that energy. Energy or Syntropy can take on many different FORMS; but, the Energy or the Syntropy is always conserved.

Space-time – locality and entropy – are just another FORM of Energy; and, the FORM is never conserved. Physical matter is just another FORM of Energy; and, the FORM is never conserved. Physical matter, space-time, locality, and entropy are MADE, which means that they are NEVER conserved. They are just different FORMS of energy. It's the underlying Energy or the Syntropy that's being conserved, and NOT the physical matter or the entropy. Realizing that it's the Energy or the Syntropy that's being conserved, and not the physical matter or the entropy, ends up being the answer to life, the universe, and everything. It explains everything. Does it not?

Entropy is death. Entropy or death cannot design, create, and originate anything. It's physically impossible. There's no such thing as the spontaneous generation of Order and Syntropy from entropy and physical matter. That's physically impossible. Entropy is a dead-end. Entropy is randomness, disorder, heat death, extinction, evolution, death, and chaos. Death cannot design and create. Entropy or death can only destroy. Entropy or death is not conserved. Physical matter is not conserved. Their quantity changes over time.

Evolution is entropy. The Theory of Evolution, Materialism, Naturalism, Darwinism, Atheism, Nihilism, and Classical Physics are based exclusively upon entropy and restricted to entropy. Evolution or entropy couldn't have made us. It's physically impossible. The different types of evolution or spontaneous generation are prevented from happening by entropy. The theory of evolution is Creation by Entropy or Creation by Death. That's physically impossible. It's never going to happen. Entropy and death prevent the different types of evolution from happening. Natural selection is death – survival of the fittest. Natural selection doesn't touch our genes. Random mutations are entropy.

The chemical evolution or the spontaneous generation of functional genes and proteins from atoms, RNA, DNA, or amino acids is prevented from happening by entropy or random diffusion. Macro-evolution – chimp-like ancestors giving birth to chimpanzees and humans – is prevented from happening by genetics and entropy. It's physically impossible for genetically compatible Mr. and Mrs. Mutants to be born at the same place and at the same time for millions of years on end. It can't be done, which means that it wasn't done. Macro-evolution is physically impossible. It's prevented from happening by genetics and entropy.

Evolution is entropy or death. All the different types of evolution are based exclusively on entropy or death. So, where did all the Life come from? Where did the Order and the Organization come from? Where did the genomes, proteins, genes, eyes, brains, and life forms come from? Who made them? How were they made? Where did the planets, stars, and galaxies come from? Who made them? How were they made? Where did the physical matter come from? Who made it; and, how was it made? Evolution isn't the answer. Evolution is entropy and death. Entropy and death cannot design and create. In fact, evolution (genetic change), natural selection, and random mutations didn't even exist until after God made them in the first place. Evolution, natural selection, and random mutations didn't even exist until after God made the genomes, proteins, genes, and life forms to begin with.

There has to be some counterforce opposing entropy, or we wouldn't exist. Someone Psyche had to load this physical universe with Syntropy, or the subsequent entropy would not have been possible.

Syntropy is available energy. The available energy had to come from someplace, or we wouldn't exist. It had to come from someplace besides this physical universe, because this physical universe is based exclusively on entropy. The Syntropy came first, and the Syntropy has always existed. If it didn't, then we wouldn't exist. This is Logic 101.

Remember, entropic physical matter is comprised of different FORMS of Energy. Physical matter, entropy, and death are never conserved. It's the underlying Energy, Psyche, Life Force, or Syntropy that's being conserved, and not the entropy or the physical matter.

Syntropy or Quantum Mechanics is the BEST, most-realistic, most-explanatory, and most all-inclusive way to support, validate, and understand science. Syntropy or Psyche is the BEST way to explain the order and organization within this physical universe. According to the Big Bang Theory, this physical universe should be nothing more than one big homogenous ball of expanding gas with every atom moving away from every other atom.

It is not!

Why?

The answer is Syntropy. Someone Psyche has infused Syntropy, Order, Organization, Restrictions, and LAW into certain parts of this physical universe so as to bring order, organization, genomes, life, stars, planets, and galaxies into existence. This Syntropy could only have come from the quantum realm or the psyche realm, because Syntropy does NOT exist natively or naturally here in this physical realm from what we have observed. The physical is ruled by entropy. The very existence – the observed and proven existence – of Quantum Mechanics, or Transdimensional Physics, or Syntropy FALSIFIES Materialism, Naturalism, Darwinism, Nihilism, Behaviorism, Determinism, Scientism, Atheism, and even Classical Physics. The very existence of Syntropy negates or counteracts entropy. Thanks to Syntropy, our physical universe is NOT a closed system.

Syntropy is like love – the more of it you give away, the more of it that you have when you are done. Syntropy or Energy is the ultimate perpetual motion machine. Syntropy, order, or love is completely different than entropy, corruption, disorder, and death. Syntropy, Energy, Intelligence, or Psyche is perpetual, eternal, and everlasting because it's being conserved. God's Psyche is being conserved. It's the quantity of physical matter or entropy that's constantly changing. Is it not?

It's interesting to observe that the Materialists, Naturalists, Darwinists, Nihilists, and Atheists have succeeded in eliminating Syntropy from our vocabulary. Our spell-checkers have never seen the word "Syntropy". According to the Materialists and Naturalists, Syntropy or Psyche does not exist. These people don't believe in Syntropy or the Atonement of Christ. These people don't believe in Syntropy, or the Non-Physical, or Action at a Distance, or Quantum Mechanics. It's impossible to believe in, understand, and accept something that you have never heard of and refuse to learn about. You will never find Syntropy if you never go looking for it. I KNOW, because I used to be a Materialist, Naturalist, Nihilist, and Atheist; and, I had never heard of Syntropy, and didn't believe that such things exist. Syntropy is a new discovery for me. I'd never really thought of it or thought about it, until the past year or so; and, I'm 57 years old.

When it comes to Syntropy, I practically had to invent the term, just so that I would have a word to use to describe the phenomenon. The whole thing started when I asked myself what the opposite of entropy is. Everything has its opposite. I realized that the different types of evolution are entropy and are prevented from happening by entropy. There's no such thing as spontaneous generation or macro-evolution thanks to entropy. The Theory of Evolution is impossible thanks to entropy. The Materialists, Naturalists, Darwinists, and Atheists are prevented from understanding and accepting Syntropy, thanks to entropy. Entropy is so dominant in our society that practically no one has ever heard of Syntropy. Your typical scientist won't have a clue when it comes to Syntropy; and, they're supposed to be the best and brightest among us.

Remember, the theory of evolution is based exclusively and deliberately on physical matter and entropy. These people teach that only physical matter or entropy exists. Remember, entropy is death. The theory of evolution is Creation by Entropy or Creation by Death because these people have chosen to believe that only entropy or death exists. By definition, in principle, death or entropy cannot design and create anything. That means that the different types of evolution cannot design and create anything whatsoever because the theory of evolution is based exclusively on death or entropy. The theory of evolution is self-defeating because entropy or death cannot design and create.

Syntropy is the opposite of entropy. Entropy is death; therefore, Syntropy is Eternal Life. Syntropy is the Conservation of Energy, the Conservation of Psyche, the Conservation of Personality, the Conservation of Intelligence, or the Conservation of the Life Force.

Materialism, Naturalism, Darwinism, Nihilism, Behaviorism, Determinism, Physical Reductionism, and Atheism create a whole host of unanswered and unanswerable questions; and, Physicalism or Scientific Naturalism can't be used to answer those questions either. All of those unanswerable and unanswered questions produce dissonance, discord, and disharmony. It produces tension and contention. It lacks explanatory power.

In the science of Psychology, for example, Materialism, Naturalism, and Atheism produce a large number of unanswerable questions and unsolvable problems such as the Binding Problem, the Free-will vs. Determinism Problem, the Mind-Brain Problem, the Problem of Entropy or Death versus an After-Life, the Problem of Mental Causation, the Problem of Other Minds, the Problem of Consciousness or Thought, the Morality and Ethics Problem, the Problem of Personality or Self, the Problem of Unconsciousness, the Problem of Resiliency, and the Nature versus Nurture Problem. ALL of these unsolvable problems are immediately solved and immediately answered by allowing Psyche, Syntropy, and Quantum Mechanics into play. There are NO unanswerable problems once we eliminate Materialism, Naturalism, Atheism, and their derivatives from science; and then embrace Psyche, Quantum Mechanics, Action at a Distance, and Syntropy instead.

By allowing Psyche, Syntropy, Action at a Distance, and Quantum Mechanics in to play, we can literally explain everything that comes our way. Being able to answer every question that you encounter produces consonance, harmony, and peace of mind. The explanatory power of Quantum Mechanics, Psyche, Action at a Distance, and Syntropy is through the roof to infinity and beyond.

In contrast, the Materialists, Naturalists, Darwinists, Nihilists, and Atheists teach that Psyche, Syntropy, Supernatural Mechanisms, Superpowers, Miracles, Intangible Forces and Fields, Telepathy, Telekinesis, Teleportation, Transmutation, God's Psyche, God, and Action at a Distance DO NOT EXIST. How does claiming that these things do not exist explain what they are and how they work? It doesn't, now does it? It just creates a whole host of unanswerable questions. That's bad science. It completely lacks explanatory power. Scientists are supposed to explain what things are and how things work; and, the Materialists and Naturalists prevent themselves from doing so by claiming that the interesting half of science does not exist.

The Materialists, Naturalists, Darwinists, and Atheists cripple and destroy everything they get their hands on, including Science. It's what they do. They deny the existence of scientific evidence. They deny the existence of observations and experiences.

Materialism, Naturalism, Darwinism, Nihilism, and Atheism were designed to prevent us from discovering Psyche, Syntropy, the Ultimate Law of Thermodynamics, Action at a Distance, Supernatural Mechanisms, Transdimensional Physics, Non-Locality, Non-Physical Intangible Forces and Fields, Spirit Matter or Dark Matter, After-Death Life Reviews, Life after Death, the Vacuum Energy or Dark Energy, Quantum Waves, Quantum Mechanics, Consciousness, Intelligence, Choice, Eternal Life, and God. These falsified philosophies were designed to produce a wide variety of unanswered and unanswerable questions. They were designed to produce dissonance, disharmony, discord, contention, and confusion; and, that's precisely what they do.

If you successfully eliminate everything that is false, then only the truth will remain. You start by eliminating Materialism, Naturalism, Darwinism, Nihilism, and Atheism because they have been falsified or proven false by the observations and the experiences of the human race as a whole. Science is observation and experience. The false is falsified by the truth; and, the truth is repeatedly experienced and observed.

Whenever it comes time to integrate a wide variety of different concepts and subjects into one realistic whole, the Human Psyche ends up being the BEST Integrator. Science is supposed to be the pursuit of the Best Explanation. Claiming that something "does not exist" is the worst possible explanation that you can possibly give when it comes to a scientific object, a scientific concept, or a scientific subject. Explaining what it is and how it works is the BEST possible explanation that you can possibly give to a scientific object, scientific concept, or a scientific subject. By allowing Psyche and Quantum

Mechanics in to play, you can literally explain everything that comes your way. Try it! You will like it!

Mark My Words

Taking Quantum Mechanics Seriously

The people behind Quantum Field Theory were geniuses; but, geniuses say a lot of stuff, and not all of it is right. Many of these people deliberately ignore and reject the necessary influence of Psyche or Intelligence. Quantum Field Theory is worthless without the Psyche or Intelligence who designed and made the different quantum fields and the different types of quantum particles that move through those quantum fields. Quantum Field Theory is also worthless without the Psyche or Intelligence who made and enforces the laws and rules of engagement upon which the various quantum fields and quantum particles depend for their existence and interaction. Quantum Field Theory wouldn't work without the Psyche or Intelligence who chooses what type of particles an annihilating pair of particles can become and will become. ALL of the information and knowledge is conserved or preserved within Psyche, even if it ends up being lost someplace else within the system.

In the most explanatory, parsimonious, conservative, and logical interpretations of Quantum Mechanics, Someone Psyche is needed to make the choices necessary to collapse the wave function. Choice is a function of Psyche. According to the Behaviorists and Determinists, entropic physical matter can't make any choices. They are right. Choice is done by Psyche, and NOT physical matter. However, these people are wrong whenever they choose to believe that Psyche doesn't exist. Geniuses can say a lot of stuff, but they aren't always right. We KNOW that Psyche exists because it has been experienced and observed by Out-of-Body Travelers and Near-Death Experiencers. Our job as scientists is to find a scientific explanation for all this observational evidence rather than trying to explain it away.

Remember, a scientist is a pursuer of evidence, not a destroyer of evidence. Give me evidence, lots and lots of evidence. Evidence is based upon the personal experiences of the human race as a whole. In order to believe in Naturalism and Atheism, you have to dismiss and destroy a lot of evidence. That's what these people do. I KNOW because I used to be a Materialist, Naturalist, Nihilist, and Atheist until science, action at a distance, psyche, quantum field theory, and quantum mechanics convinced me that I was wrong. Science is observation and experience. Observation and experience convinced me that I was wrong. As a result, Science (observation and experience) eventually convinced me that God exists – the Biblical God Jesus Christ. We KNOW that He is real and truly exists because He has been experienced and observed.

Unlike the Materialists, Naturalists, Darwinists, Nihilists, Behaviorists, Determinists, Physical Reductionists, Classical Physicists, and Atheists, nowadays I choose to take Quantum Mechanics and Psyche seriously. Others have observed that Quantum Mechanics is our most used, best proven, most real, and truest science that we have.

I found the answer to life, the universe, and everything within Psyche, Quantum Mechanics, Supernatural Mechanisms, Syntropy, and Action at a Distance. It explains everything that we have ever experienced and observed as a race. Nowadays when it comes to Psyche, Quantum Mechanics, and Supernatural Mechanisms, I take it seriously because its explanatory power is through the roof to infinity and beyond. It's good science!

A quantum or "particle" is a packet of energy, particularly a wave packet of energy or a packet of energy waves. Quantum Mechanics is Energy Mechanics, or how energy behaves at the quantum level and the psyche level. Quantum Mechanics and Quantum Field Theory make Energy the fundamental unit of reality. Entropic physical matter is made from different forms of energy. The form is NEVER conserved which means that the form can NEVER be the fundamental unit of reality. Entropic physical matter cannot be the fundamental unit of reality because it is comprised of many different forms of energy. The First Law of Thermodynamics or the Conservation of Energy suggests that energy is the fundamental unit of reality. Quantum Mechanics and Quantum Field Theory tell us clearly and conclusively that "quantum" or "energy" is the fundamental unit of reality and existence. It's ALL made from energy.

Is there anything more fundamental than Energy? Well, Psyche is the innate intelligence within all the different forms of energy which gives that energy the inherent ability to understand, follow, comprehend, choose, and obey God's Laws and God's Commands. Energy is intelligent, psychic, conscious, sentient, perceptive, aware, and alive. In a very real sense, that would make the Psyche or the Intelligence within Energy the fundamental unit of reality, nature, and existence. Psyche or Intelligence is energy's Command and Control. That sounds pretty fundamental to me.

Physicists have suggested that Energy itself is a suggestion of or a hint of something more fundamental than Energy. Clearly, the Energy is following Laws. The Laws would be more fundamental than the Energy. Would they not? So, who made the Laws, and who is giving that Energy the ability to understand and follow those Laws? The Psyche or the Intelligence, who chooses what it wants the energy under its command and control to form into or to become, would in fact be more fundamental than the energy. Would it not? Choosing what you want Raw Energy or Available Energy to form into sounds pretty fundamental to me. Only Psyche or Intelligence can do choice. The Chooser would be more fundamental and essential than the things that have been chosen into existence.

Some of the philosophers have suggested that consciousness, thought, intelligence, symmetry, or law is more fundamental than energy. Of course, the Ultimate Law of Thermodynamics reminds us that entropic physical matter is MADE from many different forms of energy, which means that the energy is more fundamental than entropic physical matter. The fact that Energy or Psyche is conserved, while entropy and physical matter are NOT conserved, tells us clearly and conclusively that Energy or Psyche is more fundamental than entropic physical matter. However, when we are talking about Raw Energy, Unorganized Matter, or Chaotic Mass by itself, the Psyche or Person who forms that energy into different things like entropic physical matter would end up being more fundamental and essential than the energy itself.

Physical matter is organized energy. Physical matter is made from many different forms of energy. Anything that is obviously organized obviously has an Organizer who organized it. The Psyche or Intelligence who tells that energy to take that form or organizes energy into that form would be more fundamental and essential than the physical matter that is formed; and, the Psyche or Intelligence within those different forms of energy, who understands the instructions and chooses to form into entropic physical matter, would also be more fundamental and essential than the energy itself or the thing that the energy gets formed into. Psyche or Intelligence explains why one packet of energy forms into a quark and a different packet of energy forms into a gluon, and how those different packets of energy know how to find each other, know how to recognize each other, and know how to interact with each other.

Furthermore, Restrictions, Limitations, Order, Organization, Symmetry, and Laws require some kind of Law Giver as well as Someone Psyche who is intelligent enough and powerful enough to be able to understand and obey those Laws. Energy is smart enough, powerful enough, and sentient enough to understand and to choose to obey God's Laws. That explains why we observe order and organization rather than chaos. That explains why we exist. When all is said and done, the Law Giver as well as the Law Follower who is choosing to obey those Laws end up being more fundamental and essential than the Energy itself or the form that the Energy takes. Do you see how that works? It's the answer to life, the universe, and everything. It explains everything that has ever been experienced and observed.

Remember, Quantum Mechanics is Energy Mechanics. Quantum Mechanics, Psyche, and Quantum Field Theory explain how energy chooses to behave. They explain why there is order and organization rather than random chaos. They explain why we exist.

According to the Copenhagen Interpretation of Quantum Mechanics, some kind of Psyche or Consciousness is needed to make a choice and collapse the wave function thereby converting a spiritual intention or a psychic intention into a physical reality.

According to the Orthodox Interpretation of Quantum Mechanics by Henry P. Stapp, the Human Psyche (process 1) chooses what it wants to do with its physical body; and then, Nature's Psyche (processes 2 and 3) collapses the necessary wave functions and turns on or fires the necessary neurons in order to make the Human Psyche's choices a physical reality.

Psyche makes choices. Process 1 is represented by the choice of the observer or the choice made by the Human Psyche. The Human Psyche chooses what it wants to do with its physical body. Process 2 is the collapse of the wave function or the physical manifestation of the Human Psyche's choices. Nature's Psyche collapses the wave functions and turns on the appropriate neurons according to Orthodox Quantum Mechanics. Nature's Psyche also makes the physical brain, does neurogenesis, does neural and glial migration, maps the physical brain, does synaptogenesis, maps meaning and purpose onto synapses and neurons, and does axon and dendrite growth **at the quantum level**; and then, Nature's Psyche uses those quantum maps at the quantum level to get things done for us at the physical level. The Human Psyche isn't consciously aware of any of these quantum mechanical processes. Process 3 is the choice of the feedback (or the choice of the outcome) – what Dirac called: "a choice on the part of nature". Nature's Psyche chooses whether to comply with or to reject the Human Psyche's requests.

If I choose to put my hand and arm through a brick wall, Nature's Psyche responds with a "No" letting me know that such a thing is not permitted whenever I try to put my hand or fist through a brick wall. However, if God were to intervene and command Nature's Psyche to let my hand and arm pass through that brick wall, then that's precisely what would happen when I tried to put my hand and arm through that brick wall. They would slip through with ease. When it comes to the physically impossible, God's Psyche can persuade Nature's Psyche to say "Yes" if God has a reason for doing so. Nature's Psyche has covenanted to obey God's commands; and thereby, God brings order and organization and law to Nature's Psyche and our physical reality.

The Human Psyche has no idea which specific neuron it needs to fire in order to lift its left index finger off the table; but, Nature's Psyche KNOWS. Nature's Psyche made each physical brain on this planet, and then mapped that specific physical function onto each physical brain that it made. Due to this Quantum Map of Physical Functionality that Nature's Psyche has made, Nature's Psyche KNOWS precisely which neuron it needs to fire in order to lift your left index finger off the table into the air. You (your Human Psyche) issue the command; and then, Nature's Psyche does the actual work of collapsing the necessary wave functions and firing the necessary neurons. That's how things really work.

I'm not going to apologize for figuring out how things work. I'm a scientist; and, that's what scientists are supposed to do.

http://www.newdualism.org/papers/H.Stapp/Stapp-kout2009.pdf

https://quantum-neuroscience.com/wp-content/uploads/2018/07/The-Effect-of-Mind-upon-Brain.pdf

http://www-physics.lbl.gov/~stapp/PTRS.pdf

http://mypsyche.us/wp-content/uploads/2017/10/PTRS.pdf

All the memories that survive the death of your physical brain clearly are NOT stored with your physical brain. Some of them are stored by the Human Psyche; and, a more comprehensive set of your memories is stored by Nature's Psyche and the Angels in Heaven in preparation for your upcoming Life Review after you are dead. The brain registers physical sensations; and then, some of that information is passed on to the Human Psyche by Nature's Psyche. Over time, the brain is mapped by Nature's Psyche to facilitate the formation of long-term memories through input from the physical environment. It takes a while for Nature's Psyche to map memory formation onto our brains, which means that it takes a while before we start to form long-term memories of our physical experiences. It usually takes three or four years for Nature's Psyche to map long-term memory formation and retrieval functionality onto our physical brains. The Human Psyche learns and grows in the meantime; but, the memories of specific events are not retained until that functionality has been mapped onto the human brain by Nature's Psyche.

Over time, the brain gets mapped and made by Nature's Psyche. Additional functionality comes online as Nature's Psyche successfully maps that functionality into the physical brain at the quantum level and the physical level. Nature's Psyche makes our brain at the physical level and maps our brain at the quantum level. This is cool science. It explains everything that has ever been experienced and observed.

Destroy the part of the brain that Nature's Psyche has mapped and dedicated to the formation of short-term and long-term memories; and, that part of the brain can no longer be used to facilitate memory formation; and then, we go back to being like an infant unable to form long-term memories. The whole brain is used for short-term memory or working memory; so, that functionality is kind of hard to destroy. But, certain parts of the brain are mapped by Nature's Psyche specifically to handle and facilitate long-term memory storage; and, if that part of the brain is destroyed, then a person becomes an infant again unable to create or form long-term memories. The Human Psyche seems to retain access to what it has learned, linked, and stored in the past; but, the damaged physical brain can no longer be used to form new long-term memories. That's the way things really work. It has been experienced and observed.

Remember, while bound, married, linked, or tied to the physical body, the Human Psyche receives its sensory input and memories of those sensations through its physical brain, even though it doesn't store its memories within its brain. All of our after-death Life Review memories are clearly stored someplace else besides our physical brain. This science comes from taking Quantum Mechanics and Psyche seriously and treating these things as Science. This Science has been experienced and observed. It's real, and true. I've explained how things really work. By allowing Psyche and Quantum Mechanics in to play, we can literally explain everything that comes our way.

This is what happens to you when you start to take Psyche and Quantum Mechanics seriously – you start to make scientific discoveries that nobody has ever thought of before.

In Science, we employ Psyche and Quantum Mechanics in order to get the Best Explanation – the most parsimonious and most logical explanation – an explanation that actually works and makes sense when we are done. We employ Psyche, Syntropy, Action at a Distance, Intangible Supernatural Mechanisms, Invisible Forces and Fields, and Quantum Mechanics in order to do better science. Long-term memory is best understood by employing Psyche and Quantum Mechanics rather than limiting ourselves exclusively to the physical brain and physical mechanisms, especially when it comes to the memories that survive the death of our brain. This is logical common sense.

Thoughts and memories are quantum waves. They survive the death of our physical brain. Our brain simply registers physical input or physical sensations, and then immediately passes that information on to Nature's Psyche and the Human Psyche for additional processing and decision-making. There's nothing within the physical brain that does choice, has experiences, or forms memories while the Human Psyche is separated from the physical body and physical brain. This reality has been experienced and observed.

We can't seem to remember much before the age of three or four years because Nature's Psyche hasn't mapped long-term memory functionality onto our physical brain yet. Thereby, God shows us that the Human Psyche can learn and grow without any long-term access to memories or past events. Learning takes place in these infants, including the learning of language, even though they are unable to form long-term memories for specific events. It shows us that brain functionality is completely separate from the Human Psyche and what it can do. Even without long-term episodic memory functionality within the physical brain, the Human Psyche continues to learn and grow anyway. Cognitive systems, memory, attention, planning, intention, learning, knowledge, personality, preferences, and intelligence are an innate part of our Human Psyche – not our genes, and not our brain. For most of us, our memory continues to get better as Nature's Psyche maps more quantum memory functionality onto our physical brain. We remember more of that physical input and the associated feelings as that physical functionality gets mapped onto our brain by Nature's Psyche at the quantum level.

The knowledge that we gain in this life goes with us into the next life through the Human Psyche. A spirit or a ghost is a spirit body and psyche combination. Out-of-Body Travelers have observed that it is their Human Psyche who has experiences and forms memories while their Psyche is separated from their spirit body and their physical body. The spirit body and physical body don't have experiences and don't form memories while the Human Psyche is separated from them. That is what has been experienced and observed.

Sensory memory is the auditory memory, visual memory, and other types of sensation that register on the physical brain. It's only there for a second or two at most. The brain is NOT a video tape machine. We don't have instant replay. The brain does NOT store any of this sensory memory – it simply registers that sensory memory and passes it immediately on to Nature's Psyche. Nature's Psyche then decides what to pass on to the Human Psyche. There are many physical and quantum processes that we have NO awareness of whatsoever. They are completely unconscious. The unconscious part of us is in fact Nature's Psyche. Nature's Psyche handles all the stuff that we aren't consciously aware of.

When it comes to sensory memory and working memory, the Human Psyche chooses what parts of it that it wants to attend to and what parts of it that it wants to ignore. Working memory is what we are attending to during that brief moment of sensory input. Working memory consists of what you are currently processing and attending to. The whole brain is used for working memory because the whole brain is used to register physical sensations. The brain is NOT a video tape machine, however. Only what you are currently sensing registers on the brain, and then it is gone! That's why our long-term memory is so spotty and unreliable. The Human Psyche actually has to choose what to pay attention to and what it wants to remember in order to remember it for itself. That's also why Nature's Psyche has to map specific portions of the brain for long-term memory storage and recall capabilities at the quantum level because that functionality does not exist at the physical level.

A physical brain has no RAM. There's no long-term memory storage taking place within a physical brain. That explains a lot, doesn't it? That explains why our long-term memory is so weak when it comes to the details. When it comes to long-term memory, it's all being stored at the quantum level by Nature's Psyche; and, most of us don't gain complete access to it while here in mortality within a physical body. If our brain were really a hard drive or digital video recorder, then we would remember everything that we have ever experienced and observed; and, our brain would also need petabytes of RAM to store all of that information.

By definition, long-term memory is your unlimited store or repository of information that you can actively think about, process, remember, and retain. It's unlimited because it is stored for us in the quantum realm by Nature's Psyche and by choice of the Human Psyche. Apparently, Nature's Psyche actually has to map a part of the physical brain to gather or organize these long-term memories for quantum storage; and, Nature's Psyche might also have to map parts of the physical brain for retrieval of these long-term memories back into the Human Psyche for further processing. A few of us actually have superior autobiographical memory, eidetic memory, photographic memory, or savant capabilities. In other words, we have more access to the quantum realm than others do, meaning that we have better recall than others do.

Those with superior autobiographical memory have the ability to actually relive and re-experience the past events of their lives as if they were there. It's as if these people have direct access to their after-death Life Review while they are still alive in the flesh. That's a very fascinating spiritual gift that has nothing to do with their genes and their physical brain. Your physical brain doesn't have RAM, which means that your long-term memories or consolidated memories are NOT stored within your physical brain. There are NO memory engrams within a physical brain. All long-term memory storage is done in the quantum realm. Only sensory memory registers upon the physical brain, but it is immediately gone only to be replaced by the next sensations. Sensory memory doesn't last. Working memory doesn't last either, because working memory consists of the sensory memories that we (the Human Psyche) are currently choosing to pay attention to.

I'm no good at remembering or reliving past events. My feelings are suppressed most of the time. However, I seem to be able to paraphrase and put into words any concept, theory, or idea that I have ever learned during my life. I'm constantly thinking about and comparing theories, concepts, and ideas. I'm a theoretician, philosopher, logician, and scientist. I notice patterns and trends; therefore, I notice the things that others are completely oblivious to. I'm also highly critical and skeptical. It's as if God set me up to discover Action at a Distance, Quantum Tunneling, the Quantum Zeno Effect, the Conservation of Psyche, the Ultimate Law of Thermodynamics, the Quantum Law of Thermodynamics, Syntropy, Supernatural Mechanisms, and Quantum Mechanics when I finally chose to go looking for them.

The Materialists, Naturalists, Darwinists, Nihilists, Behaviorists, Determinists, and Atheists teach that Psyche, Syntropy, Choice, Supernatural Mechanisms, the Non-Physical, God, Quantum Waves, and Action at a Distance DO NOT EXIST. That's bad science. How does claiming that something does not exist explain what it is and how it works? It doesn't, now does it? It's sloppy, and it's weak. It's boring. It's lazy and worthless to claim on blind faith alone that something does not exist, especially after it has been experienced and observed by thousands if not billions of different people.

Quantum waves are thoughts and memories. Quanta are organized packets of energy. Psyche is the innate intelligence within all the different forms of energy. Quantum waves and quanta are formed into physical laws and physical particles by Someone Psyche or Someone Intelligent. In other words, Energy is formed into physical laws and physical particles by Someone Psyche or Someone Intelligent. Psyche or Energy is conserved; but, the form of that energy is not conserved. The form can change and be changed meaning that the form is not conserved. When it comes to energy, the form can be transformed into a different form by Someone Intelligent or Someone Psyche. This is the Ultimate Law of Thermodynamics.

The Ultimate Law of Thermodynamics states that physical matter and entropy are NOT conserved. Anything that was obviously formed or made obviously can be disassembled and formed into something else, which means that the form is not being conserved. Entropic physical matter is comprised of different forms of energy. Physical matter and entropy were obviously formed or made from raw energy, which means that physical matter and entropy can be disassembled. and the resulting raw energy or available energy then transformed into something else besides entropic physical matter. The First Law of Thermodynamics states that energy or psyche is conserved, not the form of that energy. The form is never conserved according to the Ultimate Law of Thermodynamics. Entropy is death. Entropy is an expiration date. Death is NOT conserved according to the Ultimate Law of Thermodynamics. In contrast, your Psyche or Energy is eternal and everlasting. Your Life Force, Psyche, Intelligence, or Energy is syntropic which means that it's being conserved. Energy or Psyche cannot be made, and it cannot be destroyed which means that it is conserved.

Technically, the Ultimate Law of Thermodynamics is not a new scientific discovery – it is carefully hidden within the First Law of Thermodynamics where the Atheists and Naturalists cannot see it nor find it. I make a big deal of the Ultimate Law of Thermodynamics because it repudiates and falsifies Materialism, Naturalism, Darwinism, Nihilism, and even Atheism.

The Ultimate Law of Thermodynamics teaches that Psyche or Energy is conserved; whereas, physical matter and entropy are NOT conserved. Think about it logically. Entropy is some type of invisible, intangible, non-physical force or field; and, the verified and proven existence of entropy falsifies Materialism, Naturalism, Darwinism, Nihilism, and Atheism which teach that ONLY physical matter exists. Scientific Naturalism is shot through with holes. Syntropy is the First Law of Thermodynamics – the conservation of energy or the conservation of psyche; and, the verified and proven existence of Syntropy falsifies Materialism, Naturalism, Darwinism, Nihilism, Atheism, and their derivatives. Energy and Psyche are syntropic which means they are conserved. The false is falsified by the truth; and, the truth is repeatedly experienced and observed.

Life Force, Psyche, Intelligence, or Energy is the ultimate perpetual motion machine because it's always conserved. It's the ultimate truth. In fact, entropic physical matter, the aging process, the passage of time, locality, and entropy were designed and made by the Gods in order to put a temporary end to perpetual motion machines where physical matter and a fallen physical world are concerned. Do you see how that works? It's the answer to life, the universe, and everything. It explains everything that has ever been experienced or observed. This is the Ultimate Law of Thermodynamics.

The Orthodox Interpretation of Quantum Mechanics also explains everything that has ever been experienced and observed when it comes to Quantum Mechanics and our physical reality. "Choices made by Psyche or Quantum Non-Local Consciousness" is the best explanation for how the wave function gets collapsed both in the Transdimensional Realm and here in the Physical Realm. Psyche or personality is identified by the choices that it makes. The Human Psyche makes choices deciding what it wants to do with its physical body. Nature's Psyche makes choices as to whether the Human Psyche's choices will be permitted or not. Psyche of any kind makes choices, thereby collapsing the associated wave functions. This experienced and observed reality explains everything that we have ever encountered as a race.

Psyche or Intelligence is syntropic, which means that it is eternal and everlasting without a beginning of days or an end of years. Psyche or Intelligence cannot be made, and it cannot be destroyed. Psyche or Intelligence is syntropic, which means that it is conserved. Syntropy is the conservation of energy or the conservation of psyche. Psyche and Energy are co-eternal in that they are syntropic and conserved. Quantum Mechanisms are the energy, forces, and fields that Psyche controls in order to get things done both in the Transdimensional Spiritual Quantum Realm and here in the Physical Realm.

Psyche is Energy. Psyche is the intelligence within energy. Psyche or Energy is syntropic which means that Psyche or Energy is conserved. Psyche or Energy is eternal and everlasting without a beginning of days or an end of years. Psyche is Quantum Waves. Thoughts and memories are different types of Quantum Waves, or different types of Energy. Syntropy is therefore the conservation of energy, or the conservation of psyche, or the conservation of quantum waves. Energy or Quantum Waves have been experienced and observed. Have they not?

Thoughts and memories are quantum waves or a type of organized energy. Thoughts and memories are transmitted quantum waves and stored quantum waves respectively. In all of the science fiction shows that we enjoy, telepathy is portrayed incorrectly. The Materialists, Naturalists, and Atheists who write our science fiction erroneously portray telepathy as brain-to-brain communication with lots of pain, headaches, broken eardrums, strokes, and bloody noses being the typical result of telepathy. Transpersonal communication, or telepathy, is defined as person-to-person communication or psyche-to-psyche communication. It doesn't hurt because it has nothing to do with the physical body and the physical brain. Even if telepathy did have something to do with our brains, there are no pain receptors within our brains to hurt. We can do telepathy without a physical brain, in fact, we can do it better without a physical brain.

Out-of-Body Travelers and Near-Death Experiencers have observed that we don't need a physical brain in order to do telepathy or transpersonal communication. In fact, a brain only gets in the way. The Gods designed and created physical matter and physical brains in part to decrease the innate telepathic, psychic, and quantum mechanical capabilities of our psyche. Entropic physical matter was designed to dampen and restrict our supernatural capabilities and thereby limit the amount of damage that we can do to each other through telepathic, psychic, or quantum mechanical means. Remember, the psyche is eternal – the psyche is conserved. The psyche cannot be damaged nor hurt by telepathy. Psyche is Energy; therefore, transpersonal communication or telepathy is in fact Energy communicating with Energy. It has nothing to do with our physical brain. The brain only gets in the way. If you take Quantum Mechanics and Psyche seriously, you end up with completely different results than the ones produced by the Materialists, Naturalists, Darwinists, Nihilists, Behaviorists, Determinists, Classical Physicists, and Atheists.

God's Psyche can form Energy or Quantum Waves into anything He wants, including physical matter, spirit matter, entropic matter, syntropic matter, the passage of time, space, locality, physical laws, quantum laws, and entropy. As physical beings, we cannot form Energy or Quantum Waves into anything we want because we are subject to physical limitations and physical restrictions.

The Energy is conserved but the form of that energy is never conserved because that energy can always be formed into something else by Someone Intelligent or Someone Psyche. Psyche, or Energy, or Quantum Waves are quantum mechanical phenomena – not physical phenomena. Psyche and Energy are conserved; whereas, physical matter and entropy are not conserved. Entropic physical matter is comprised of different forms of energy, and the form is never conserved. The form can be changed by Someone Psyche even though the underlying Energy is always conserved. This is the Ultimate Law of Thermodynamics.

Psyche is experienced first-hand as an immaterial viewpoint in space. Whenever near-death experiencers and out-of-body travelers are outside of their spirit body looking at their spirit body, there's nothing there or nothing to be found whenever they go looking for their psyche, intelligence, or consciousness. These people have observed that it is their Psyche who makes choices, has experiences, and forms memories – not their spirit body and not their physical body. While out-of-body, other Psyches are viewed, seen, or observed as being a pinprick of light, a point particle of light, or a spark of light. Psyche has been experienced and observed. Science is observation and experience; and, I want my science to match with what has been experienced and observed.

The Biblical God Jesus Christ has also been experienced and observed after He rose from the dead – experienced and observed both in the flesh and during our near-death experiences. Science is observation and experience – or it should be.

Anything that is formed, organized, or made from raw energy is not conserved. The form is never conserved. Anything that is formed, organized, or made obviously has Someone Psyche or Someone Intelligent who formed it, organized it, and made it – which means that that object can be disassembled and formed into something else instead. Only the underlying Energy or Psyche is conserved. The form of that energy is never conserved because energy can be transformed into something else whenever the Gods decide to do so. The transformation of energy takes place at the quantum level or the psyche level – not the physical level. Energy or Psyche is a quantum mechanical phenomenon – not a physical one. The explanatory power of Quantum Mechanics is infinitely superior to Materialism, Naturalism, Darwinism, Nihilism, Atheism, and Classical Physics.

In summary, the Energy or Psyche is conserved; whereas, the form of that energy is never conserved. Entropic physical matter is comprised of different forms of energy. The form is never conserved. Entropic physical matter was made from raw energy by Someone Psyche which means that it can be disassembled and formed into something else by Someone Psyche. Physical matter and entropy are never conserved. This is the Ultimate Law of Thermodynamics. The Ultimate Law of Thermodynamics differentiates between what is conserved and what is not conserved.

The Materialists, Naturalists, Darwinists, Nihilists, and Atheists erroneously teach that only entropic physical matter exists, which also means that these people erroneously teach that physical matter and entropy are conserved. One error leads to another, which then leads to another.

Anything that was obviously made obviously has a Maker who made it. Anything that has a beginning obviously has Someone Psyche who caused it to begin. Physical matter and entropy were obviously formed and made from raw energy, which means that physical matter and entropy obviously have a Maker who designed them, made them, and caused them to begin.

160

The Gods can form energy into anything they want it to be, including physical matter and entropy, which means that the form is not conserved. The form can change and be changed. Both physical matter and entropy can be formed into something else. Only the underlying Energy or Psyche is being conserved. This is the Ultimate Law of Thermodynamics. This is what has been experienced and observed.

Remember, Quantum Mechanics, Psyche, Syntropy, and Action at a Distance are supernatural in nature and origin. The proven and verified existence of quantum mechanisms or supernatural mechanisms falsifies the claims of Materialism, Naturalism, Darwinism, Nihilism, and Atheism which claim that such invisible, intangible, supernatural, non-physical forces and fields DO NOT EXIST. How does claiming that something does not exist explain what it is and how it works? It doesn't! It's worthless! Claiming that Psyche, or God, or Supernatural Mechanisms do not exist will NEVER explain what they are and how they work.

My ultimate goal in life is to figure out what things are and how they work. I'm a scientist – a pursuer of evidence not a destroyer of evidence. Destroying evidence and claiming that it doesn't exist DOES NOT EXPLAIN what these things are and how they work.

Materialism, Naturalism, Darwinism, Nihilism, Atheism, Classical Physics, and their derivatives completely lack explanatory power when it comes to invisible, non-physical, intangible forces and fields such as psyche, action at a distance, telepathy, telekinesis, mind-over-matter, quantum waves, thoughts, memories, after-death life reviews, spirit matter or dark matter, dark energy or the vacuum energy, the strong and weak nuclear forces, as well as gravity, magnetism, radio waves, microwaves, X-rays, gamma rays, and invisible light. These people don't take Psyche, Quantum Mechanics, and Action at a Distance seriously. I do.

If you successfully eliminate everything that is false, then only the truth will remain. Physicalism, Scientific Naturalism, Darwinism, Nihilism and Atheism have been falsified. Science has proven them false. Observation and experience have proven them false. There's nothing within them to verify or prove to be true. There's nothing to experience and observe within Materialism, Naturalism, Darwinism, Nihilism, and Atheism. They should be eliminated from science because they have been falsified by science. They have been falsified by observations and experience.

It's fascinating to study what remains after you have removed Materialism, Naturalism, Darwinism, Nihilism, and Atheism from science. The truth remains. The experienced and the observed remain. Quantum Mechanics, Psyche, Syntropy, Eternal Life, Action at a Distance, Energy, Supernatural Mechanisms, the Biblical God Jesus Christ, and invisible intangible non-physical Forces and Fields remain. The truth remains after you have successfully eliminated Materialism, Naturalism, Darwinism, Nihilism, and Atheism from science.

The Truth, Quantum Mechanics, Psyche, Intelligence, Supernatural Mechanisms, Priesthood Power, Syntropy, Conservation of Energy, Eternal Life, the Life Force within Energy, Non-Locality, Action at a Distance, and Jesus Christ end up being the answer to life, the universe, and everything.

Mark My Words

—

Reference Material

Quantum Mechanics from a Non-Physical Spiritual Perspective

Quantum Neuroscience: The Answer to Life, the Universe, and Everything

Finding the Truth about Physical Matter and Entropy

The Materialists, Naturalists, Darwinists, Nihilists, and Atheists teach that ONLY entropic physical matter exists. Based upon $E = mc^2$, these people define the First Law of Thermodynamics as the Conservation of Physical Matter; and, based upon their belief in the Supremacy of Entropy, these people define the Second Law of Thermodynamics as the Conservation of Entropy. Based upon their belief that only entropic physical matter exists, these people literally but erroneously teach and believe that physical matter and entropy are conserved. Their chosen beliefs are demonstrably false.

Mass is typically treated as being synonymous with physical matter. So technically, the word "massless" means non-physical. The verified and proven existence of Massless Particles such as Photons FALSIFIES the claims of Materialism, Naturalism, Darwinism, Nihilism, Behaviorism, Physical Reductionism, and Atheism which assert that the non-physical DOES NOT EXIST. Scientific Naturalism falls in the light of evidence.

Science is observation and experience; and, the observations and experiences of the human race falsify the claims of Materialism, Naturalism, Darwinism, Nihilism, and Atheism. In turn, these people deny the existence of these observations and experiences because the observations and experiences of the human race do indeed falsify Physicalism, Naturalism, Atheism, and their derivatives.

I'm interested in finding out what was really done to us and how things really work at all levels of existence. Others create "Bibles" for fictional worlds – for the comic books filled with superheroes and superpowers. I want to figure out how this world, the real world, works.

As I see it, superpowers do indeed exist in the form of quantum mechanics. Action at a distance and quantum mechanics are superpowers; and, these supernatural superpowers are an inherent part of our Psyche or Non-Local Consciousness. Our Psyche, Intelligence, Consciousness, or Life Force has NO physical limitations whatsoever. It's pure energy. It's a superpower. It's a superhero. It's syntropic. It's eternal and everlasting. It cannot be made, and it cannot be destroyed. In its native original format, Psyche or Intelligence has full access to quantum mechanics or energy mechanisms. Psyche is eternal, immortal, and conserved. Psyche is superpowered.

Psyche is the inherent intelligence in all the different forms of energy, which gives that energy the ability to understand, follow, and obey God's Commands and God's Laws. Energy or Psyche is conserved. It cannot be made, and it cannot be destroyed. It has always existed, and it will always exist. Syntropy is the First Law of Thermodynamics – the conservation of energy or the conservation of psyche. The Ultimate Law of Thermodynamics states that energy or psyche is conserved; whereas, physical matter and entropy are NOT conserved. Entropic physical matter is comprised of different forms of energy; and, the form is NEVER conserved. The conserved is syntropic, which means that the conserved is immortal and superpowered. Physical matter and entropy are not conserved. There are NO superpowers within entropic physical matter!

As I see it, entropic physical matter, entropy, space-time, locality, the sub-light speed limits of relativity, physical constants, physical restrictions, and physical laws were designed and made by the Gods to prevent us from gaining direct access to superpowers or quantum mechanics. As a result, the Gods have to perform our miracles for us.

That is the net effect of entropy and physical matter, is it not?

Entropy, physical matter, locality, space-time, classical physics, the theory of relativity, and physical restrictions prevent us from gaining direct access to teleportation, quantum tunneling, telekinesis, action at a distance, telepathy, immortality, eternal life, mind-over-matter, levitation, phase-shifting, superposition (being two or more places at once), and faster-than-light travel. Physical matter, genes, space-time, physical constants, physical laws, locality, and entropy were designed and made by the Gods to prevent us from getting superpowers. We'll NEVER become X-Men through our genes! It's physically impossible.

In order to successfully quantum tunnel physical matter, the Gods have to temporarily remove time or causality from physical matter. The speed-of-light is the speed-of-causality. The passage of time has to go to zero and time has to stop while at the same time the speed-of-light or the speed-of-causality has to go to infinity in order to quantum tunnel physical matter anywhere in the universe instantaneously. Giving physical matter an infinite De Broglie Wavelength would also allow it to quantum tunnel. The Gods designed and created these different physical laws or physical limitations in order to prevent the atoms within your physical body from quantum tunneling away from you at will – or randomly for that matter. Since the Gods created and imposed these physical limitations in the first place in order to make physical matter possible, the Gods can also temporarily remove these limitations from physical matter thereby allowing the physical matter to quantum tunnel.

I experienced quantum tunneling when God stopped the passage of time and teleported me and the car I was driving to safety in order to prevent a lot of people from dying or being hurt; therefore, I know it's possible, and I have a good sense for how it works. I felt time stop; and, I also momentarily felt the omnipresence or felt like I was omnipresent. It felt like the whole of eternity folded into me. I could sense everything around me 360 degrees, yet I was consciously aware and still able to think. It felt like I had an eternity or was surrounded by an eternity of peace and safety. I could sense where I wanted to be. Then instantly, I was someplace else with the car completely stopped and the car turned off. All I had time to do was to press the brakes. God took care of everything else for me.

That's the advantage of KNOWING that physical matter, the physical laws, the physical constants, and physical restrictions were MADE by the Gods. Instantly, you KNOW that the Gods can temporarily remove those physical restrictions as they see fit. Instantly, you can explain everything that has ever been experienced or observed. You KNOW that it's ALL made from raw energy; and, that organized energy can be changed, reorganized, rearranged, or even removed from this dimension as the Gods see fit. The FORM is never conserved. It can be changed whenever the Gods choose to do so.

Instantly, you KNOW why there is something rather than nothing. Instantly, you KNOW how everything works and what everything is comprised of. Instantly, you KNOW that entropic physical matter is comprised of different FORMS of energy. Instantly, you KNOW that the De Broglie Wavelength, Entropy, Time, the Higgs Field, Physical Matter, Dark Energy, Gravity, the Sub-Light-Speed Limit of Physical Matter, Physical Laws, Physical Constants, and Physical Restrictions were MADE and can be changed, manipulated, rearranged, or eliminated as the Gods see fit. Instantly, you have a scientific explanation for Action at a Distance, Quantum Mechanics, the Origin of Physical Matter and Physical Restrictions, Entropy, Time, Mass, Phase-Shifting, Quantum Tunneling, Telepathy, Superpowers, Supernatural Mechanisms, and what we sometimes call "Miracles". It's ALL energy; and, energy can be made to take on any FORM that the Gods see fit, including physical matter, physical laws, physical restrictions, physical constants, entropy, time, space, spirit matter or dark matter, dark energy or the vacuum energy, quantum fields, quantum waves, quantum particles, gravity, magnetism, light, nuclear forces, physical restrictions, or anything else that you can possibly imagine. It's ALL energy; and, ALL of that energy can be transformed into anything that the Gods see fit to transform it into. That's why there is something rather than nothing.

Space Time Videos

https://www.youtube.com/channel/UC7_gcs09iThXybpVgjHZ_7g

You can watch the Space Time videos on YouTube and see with your own eyes that it's ALL MADE from energy – many different FORMS of energy, each with its own specific specialized purpose, function, design, time, and place in the universe. It's highly organized – deliberately organized. Even "empty space" consists of a web or a mesh of organized energy that they call the "Quantum Field". It's ALL MADE from energy. It's ALL comprised of organized forms of energy. The reality and truth of it is obvious.

The Quantum Field Theory playlist on Space Time is particularly convincing scientific evidence that EVERYTHING is made from energy – different FORMS of energy that they call Quantum Fields. It's ALL energy – deliberately and purposefully organized energy. It has been ordered and organized. It was ALL MADE from energy. Anything that is obviously organized obviously has Someone Psyche or Someone Intelligent who organized it.

Quantum Field Theory Playlist

https://www.youtube.com/watch?v=hYkaahzFWfo&list=PLsPUh22kYmNBpDZPejCHG zxyfgitj26w9

The Higgs Mechanism Explained

https://www.youtube.com/watch?v=kixAljyfdqU

https://syntropy.site/wp-content/uploads/2018/08/The-Higgs-Mechanism-Explained.zip

Massless Speed-of-Light Particles or Massless Psyches USE mass to resist acceleration and to slow down to sub-light velocities. Mass produces the experience of time or the passage of time. Mass makes entropy possible. Mass is just a different FORM of energy. It's all energy, and energy is always conserved.

The most interesting science in this video is their discovery that, without the mass or the resistance to acceleration which the electron (and quark) gets from the Higgs Field and weak hyper-charge, the electron would be a massless particle and travel at the speed-of-light. In other words, the electron would have NO mass without the Higgs Field. The Higgs Field provides the mass, or resistance to acceleration, and slows the electron down.

Mass is resistance to acceleration. Mass slows particles down. Massless particles like photons travel at the speed-of-light, and they experience NO passage of time and therefore NO entropy. Photons and other Massless Particles are capable of being immortal and conserved. The Higgs Field produces mass, which slows quantum particles down to below the speed-of-light, which then produces the passage of time as a result. The passage of time makes entropy possible. The fundamental elementary components of the electrons and the quarks are Massless Speed-of-Light Particles; and, the Higgs Field gives the quarks and the electrons their small amount of mass and thereby slows them down to sub-light speeds.

Remember, mass produces resistance to acceleration and thereby slows Massless Particles down to sub-light speeds.

It's ALL made from Energy or Quantum Fields.

There technically is no such thing as physical matter. Physical matter as we call it is comprised of many different FORMS of energy, each with its own unique purpose, assignment, structure, alignment, quantum wave, quantum field, place, lattice, and functionality. Physical matter was obviously organized from many different FORMS of energy which were in turn organized from raw energy that has also been given a distinct purpose, form, or an assignment from the Gods. Anything that was obviously organized obviously has an Organizer who organized it. Physical matter is obviously organized. Quantum fields were obviously organized. So is space, time, and even entropy.

The organization within it ALL is obvious which means that it has Someone Psyche or Someone Intelligent who gave ALL of it purpose, order, structure, organization, and meaning. The order or organization within everything around us is obvious and undeniable, if you are willing to be honest with yourself. Everything has been organized from raw energy or chaotic energy, including physical matter and entropy. Everything was made by the Gods – by God's Psyche – from raw energy, unorganized matter, or available energy. Order and organization don't just spring into existence whole from nothing. They are deliberately and consciously MADE. Are they not?

I like having a scientific explanation for everything that I encounter, experience, and observe. KNOWING that the Gods MADE everything from raw energy gives me a scientific explanation for everything that I encounter, experience, and observe. I like KNOWING what everything is and how it works. It's ALL made from energy; and, the Gods can transform or organize energy into anything, including physical matter, physical restrictions, physical laws, and entropy. The underlying energy or psyche is always conserved; but, the FORM of that energy is NEVER conserved. Entropic physical matter is comprised of different FORMS of energy; therefore, physical matter and entropy are NOT conserved. The Gods can transform entropic physical matter into anything they want it to be anytime they choose to do so. Particles of matter are coming into existence and being annihilated all the time; and, energy is the currency of that transaction. Psyche, consciousness, or intelligence is identified by the choices that it makes. God's Psyche is identified by the choices that it makes. God is identified by the things that He made.

The very existence of physical laws, physical restrictions, physical constants, and physical matter is Scientific Proof of God's Existence. The order and organization within our physical universe are Scientific Proof of God's Existence. God is Energy. God is Light. God is in the Light.

Anything that was obviously organized obviously has an Organizer who organized it. All of these different things in the lists above were organized from raw energy. Physical matter is organized energy. Entropy is a "form" of energy in that entropy is a measure of energy's transition from a useful form to an unavailable form. God's Psyche can make unavailable energy available again because God designed and made the rules which made that energy unavailable in the first place. The Materialists and Naturalists teach that the Unavailable Forms of energy are forever conserved or forever unavailable. These people teach that entropy or death is conserved. They are wrong. The FORM is never conserved.

Energy is at the foundation of it all. It's ALL energy; and, psyche is the innate intelligence within all the different forms of energy which gives that energy the inherent ability to understand, follow, and obey God's Laws and God's Commands. The Gods bring order and organization to raw energy or chaotic energy; and in return, the raw energy chooses to follow and obey God's Laws and God's Commands. Order and organization are the result. Physical matter, dark matter, entropy, time, mass, the Higgs Field, gravity, dark energy, vacuum energy, quantum fields, and physical restrictions are some of the results of this ordering or organization of raw energy or chaotic energy. This is what has been experienced and observed. It all makes logical sense; and, it can ALL be explained in a scientific manner once you realize that the Gods made everything from energy. It has all been experienced and observed. Science is observation and experience after all.

The proven and verified existence of Quantum Tunneling and Action at a Distance falsifies Materialism, Naturalism, Darwinism, Nihilism, Atheism, Classical Physics, and their derivatives. Think about it logically. These people teach that ONLY physical matter exists. Therefore, the proven and verified existence of invisible, intangible, supernatural, non-physical forces and fields repudiates and falsifies Materialism, Naturalism, Darwinism, Nihilism, and even Atheism. These people are proven wrong by Science, particularly Quantum Mechanics or Transdimensional Physics.

The Physicalists, Naturalists, Darwinists, and Atheists deliberately try to remove the Psyche or the Intelligence behind it ALL; but, the order, organization, and Organizer remain, nonetheless. It's undeniable and unavoidable. Order and organization don't just spring forth from chaos or from nothing as the Materialists and Naturalists claim. Order and organization are purposefully and deliberately MADE; and, anything that was obviously made obviously has a Maker who designed it and made it. Physical matter was obviously made. The different quantum fields were obviously made. The different types of quantum particles were designed and made. For me personally, ALL of the order and organization that's being experienced and observed within our universe is one of the most convincing Scientific Proofs of God's Existence that we have. It explains why there is something rather than nothing.

In contrast, Creation from Nothing doesn't explain anything. It doesn't explain why there is something rather than nothing. Materialism, Naturalism, Nihilism, and Atheism are based exclusively on Creation Ex Nihilo or Creation from Nothing by Nothing. Creation from Nothing by Nothing has NO explanatory power. Creation Ex Nihilo can't explain why there is organized Physical Matter and organized Quantum Fields rather than chaos or nothing. However, the existence of the Organizer or the existence of God's Psyche and your psyche can indeed explain why there is something rather than nothing. By allowing Psyche and Quantum Mechanics in to play, we can literally explain everything that comes our way. Try it, you might like it!

Physical Matter Has a Short De Broglie Wavelength

https://www.youtube.com/watch?v=-IfmgyXs7z8

https://quantum-neuroscience.com/wp-content/uploads/2018/06/Is-Quantum-Tunneling-Faster-than-Light.zip

The Gods gave entropic physical matter a short De Broglie Wavelength which greatly limits (but does not completely eliminate) its innate quantum tunneling capabilities. In other words, the Gods gave entropic physical matter a very short quantum tunneling range by giving entropic physical matter a very short De Broglie Wavelength. Consequently, the physical atoms in your physical body aren't going to quantum tunnel away from you to the other side of the universe without God's permission to do so. God holds everything together which is why there is order and organization rather than complete chaos. Remember, anything that was obviously organized obviously has an Organizer who organized it.

The Gods gave entropic physical matter a very short De Broglie Wavelength in order to limit the distance or the range that physical particles can quantum tunnel or teleport. The atoms in your physical body are not going to quantum tunnel away from you, and you are not going to dissolve into thin air thanks to the physical limitations which God has imposed upon physical matter. However, God can give the physical matter in your body an infinite De Broglie Wavelength or an infinite velocity thereby making your physical body quantum tunnel as God sees fit. Quantum tunneling is experienced as time coming to a complete stop. You and your car are instantly someplace else with NO passage of time having taken place during the interim. It's a fascinating phenomenon to experience.

Everything can be explained once you realize that it is all comprised of raw energy and that energy can be FORMED or TRANSFORMED into anything that the Gods see fit to form it into. The order and organization come from God, particularly God's Psyche. Physical restrictions such as entropy were made, which means that entropic physical matter, entropy, and the other associated physical restrictions can be removed or disassembled back into raw energy anytime the Gods choose to do so. Choice is a function of Psyche or Intelligence. Our genes, physical matter, and "evolutionary history" do not make our choices for us. Our Human Psyche does. Physical restrictions and physical limitations such as entropy prevent our genes, evolution, and natural selection from making our choices for us. Genes and proteins don't make choices. They simply do what Nature's Psyche or God's Psyche tells them to do. They follow the physical laws unless God tells them to do otherwise. Our Human Psyche makes our choices for us and determines what we are going to do next.

Energy or psyche has innate superpowers or innate quantum mechanical capabilities that are limited, restricted, and dampened by physical laws, physical matter, physical constants, and physical restrictions. Physical matter, our genes, entropy, and the physical laws were made by the Gods to limit and restrict our innate superpowers or quantum mechanical capabilities. It works precisely as it was designed to work. The energy, or superpowers, or quantum mechanical capabilities are still there; but, we don't have access to them thanks to the mortal, fallen, physical, entropic nature of our physical bodies. The physical matter, physical restrictions, entropy, and physical laws are the veil that the Gods occasionally talk about. They dampen, limit, and restrict our superpowers. The Gods made entropy; and, the Gods can remove that entropy anytime they choose to do so. There's nothing sacrosanct or sacred about entropy. Entropy is a temporary construct.

That's the advantage of KNOWING that physical matter, the physical laws, the physical constants, and physical restrictions were MADE by the Gods. Instantly, you KNOW that the Gods can temporarily remove those physical restrictions as they see fit. Instantly, you can explain everything that has ever been experienced or observed. You KNOW that it's ALL made from raw energy; and, that organized energy can be changed, reorganized, rearranged, or even removed from this dimension as the Gods see fit. The FORM is never conserved. It can be changed whenever the Gods choose to do so. Choice is a function of

Psyche; and, God's Psyche is identified by the choices that it makes. God is identified by the things that He forms, organizes, and makes from raw energy.

You, your Psyche, are going to have to decide for yourself if I know what I'm talking about or not. That's something that no other psyche can choose to do for you. Your beliefs are chosen into existence by your own personal Psyche on both sides of the veil – both on the physical side and the spiritual side. ONLY your Psyche, or Life Force, or innate Intelligence has experiences and forms memories. Your physical body and spirit body don't have experiences and don't form memories while your Psyche is separated from them. This is what has been experienced and observed. Your Psyche is immortal and everlasting. Your Psyche or Core Energy was not made, and it cannot be destroyed. Your Psyche is conserved. Your Psyche makes your choices for you. So, choose well!

The Materialists, Naturalists, Darwinists, Nihilists, and Atheists don't want you to discover Psyche or Non-Local Consciousness and its innate ability to control and manipulate energy psychically and telekinetically at the quantum level. They know that if you discover Psyche then you will also discover God. Physicalism, Materialism, Naturalism, Darwinism, Nihilism, Behaviorism, Determinism, Physical Reductionism, and Atheism were designed to prevent us from making these kinds of scientific discoveries. They were designed to dumb us down and keep us in the dark. They work as advertised. I wasn't looking for any of this science and didn't know anything about any of this science while I was a Materialist, Naturalist, Nihilist, and Atheist. I simply believed that it didn't exist; and consequently, for me personally it didn't exist. No seeking, then no finding.

I have changed. I like the change. I like KNOWING what everything is and how it works. I like having a scientific explanation for everything that I encounter, experience, and observe. I like learning about Psyche, Non-Locality, Non-Physicality, Transdimensional Mechanisms, Supernatural Mechanisms, and Quantum Mechanics. My science and scientific understanding are infinitely superior to what they were before. I like it. By allowing Psyche and Quantum Mechanics in to play, we can literally explain everything that comes our way.

Quantum Tunneling, Psyche, Raw Energy, and Syntropy

https://phys.org/news/2015-05-physicists-quantum-tunneling-mystery.html

https://quantum-neuroscience.com/wp-content/uploads/2018/07/Physicists-Solve-Quantum-Tunneling-Mystery.pdf

https://en.wikipedia.org/wiki/Complex_number

https://quantum-neuroscience.com/wp-content/uploads/2018/07/Physicists-Solve-Quantum-Tunneling-Mystery.pdf

They discovered that quantum mechanics, quantum tunneling, psyche, raw energy, and syntropy are modeled by the "imaginary numbers".

The imaginary numbers or complex numbers model and describe the Quantum Realm or the Psyche Realm. Quantum Mechanics or Psyche really does go down the rabbit hole to infinity and beyond.

Atoms Are Sentient, Conscious, and Alive

https://phys.org/news/2015-10-zeno-effect-verifiedatoms-wont.html

https://quantum-neuroscience.com/wp-content/uploads/2018/07/Zeno-Effect-Verified-—-Atoms-Wont-Move-While-You-Watch-Them.pdf

The Quantum Zeno Effect tells us that the energy within atoms is sentient, perceptive, intelligent, responsive, conscious, and alive. Psyche is the innate intelligence within all the different forms of energy. Physical atoms are just one form of energy. Entropy is an emergent property of physical forms of energy. The form is NEVER conserved. Only the underlying energy or psyche is conserved. The energy is psychic, intelligent, conscious, and alive. Psyche is the innate intelligence within energy which gives that energy the inherent ability to understand, follow, and obey God's Laws and God's Commands.

Quantum Tunneling in Transistors and Computer Chips

https://phys.org/news/2014-06-long-range-tunneling-quantum-particles.html

https://quantum-neuroscience.com/wp-content/uploads/2018/07/Long-Range-Tunneling-of-Quantum-Particles.pdf

https://spectrum.ieee.org/semiconductors/devices/the-tunneling-transistor

https://quantum-neuroscience.com/wp-content/uploads/2018/07/The-Tunneling-Transistor.pdf

https://spectrum.ieee.org/nanoclast/semiconductors/devices/photonic-quantum-circuit-gets-photons-to-interact-for-quantum-computing

https://quantum-neuroscience.com/wp-content/uploads/2018/07/A-Photonic-Circuit-for-Quantum-Computers.pdf

https://spectrum.ieee.org/tech-talk/semiconductors/devices/tunnel-transistor-may-meet-power-needs-of-future-chips

https://quantum-neuroscience.com/wp-content/uploads/2018/07/Tunnel-Transistor-May-Meet-Power-Needs-of-Future-Chips.pdf

Quantum tunneling has been proven and verified to exist. Quantum Tunneling is a supernatural superpower. It's real; and, its very existence falsifies Materialism, Naturalism, Darwinism, Nihilism, and Atheism. The false is falsified by the truth; and, the truth is repeatedly experienced and observed. Scientific Naturalism, Physicalism, Darwinism, Nihilism, and Atheism have been falsified by the truth.

Nevertheless, you have to go out of your way to include Psyche or Quantum Mechanics into science because the Materialists, Naturalists, Nihilists, and Atheists deliberately try to eliminate it. These people actively try to hide Psyche or Quantum Mechanics from us. Scientific Naturalism, Physicalism, Darwinism, and Atheism were deliberately designed to prevent us from discovering Psyche, Action at a Distance, Syntropy, Spirituality, the Ultimate Law of Thermodynamics, the Quantum Law of Thermodynamics, Non-Locality, Quantum Mechanics, and God. They work as they were intended to work.

Quantum mechanics, spiritual mechanics, mystical mechanics, or supernatural mechanics are superpowers. Ask yourself why none of us have superpowers! The answer is obvious and clear. Our genes, our physical matter, physical restrictions, physical limitations, physical laws, physical constants, and the entropy within it all prevents us from obtaining and using superpowers or Quantum Mechanics. That's the way things really work. Is it not? That's the way things were designed to work. Physical matter and entropy prevent us from getting superpowers. While we are physical, fallen, mortal, entropic beings, God has to perform our miracles for us because physical matter and entropy were designed by the Gods to prevent us from performing miracles for ourselves.

If we could heal every illness at will, then everyone would be immortal, and nobody would die. Entropy or death was designed by the Gods to PREVENT US from becoming immortal, from developing perpetual motion machines, from evolving into something different, from getting superpowers, and from gaining access to the "unavailable energy" that the Gods don't want us to have access to. Remember, entropy and death were designed to prevent us from getting superpowers and becoming immortal at the physical level. When it comes to entropy or thermal equilibrium, ALL of the energy is still there. It's just not available for our use. It's anergy, entropy, or unavailable energy.

The greatest scientific discovery that we could make as fallen mortal physical beings would be to find a way to access all of that "unavailable energy", "psychic energy", or "spiritual energy" at will. The result would be immortality, perpetual motion machines, and superpowers. We would never run out of energy if we could gain direct access to the "Syntropy" or the "Conservation of Energy" that exists at the quantum level, and then use that energy at the physical level. According to the Quantum Law of Thermodynamics, heat death is impossible at the quantum level; and, psyche cannot die. All of the energy is still there waiting to be reused. The Quantum Realm or Psyche Realm is an Isothermal System which means that at the quantum level ALL of the energy is available for use all of the time. If we could figure out how to tap into that at the physical level, then entropy and death would cease to exist at the physical level; but, we will probably need God's help in order to be able to do so. Entropy or the second law of thermodynamics was designed to prevent us from doing so.

The Ultimate Law of Thermodynamics differentiates between what is being conserved and what is not conserved. Psyche, Life Force, Energy, Intelligence, and Syntropy are conserved; whereas, physical matter, death, and entropy are not conserved.

The Quantum Law of Thermodynamics states that ALL of the energy is available all of the time at the quantum level and the psyche level. In other words, anergy, entropy, unavailable energy, thermodynamics, heat death, death, and the second law of thermodynamics DO NOT EXIST in the Quantum Realm. This is what has been experienced and observed by Out-of-Body Travelers and Near-Death Experiencers. The BEST and most convincing Proof of Heaven is to go there and see it for yourself, or to choose to trust someone who has.

I found the truth. I found out how everything really works; and, that's all I wanted to know. It's simple – parsimonious. Of course, it's not good news for any physical, fallen, mortal, entropic being who desperately wants to have superpowers or immortality. The Gods designed physical matter, entropy, and death to prevent us from getting superpowers. Our superpowers are restored to us once we die – that reality has also been experienced and observed. For some strange reason, the Gods have decided that wicked, evil, and unproven individuals shouldn't be given access to superpowers while here in mortality on this physical planet. Go figure! I think it has something to do with absolute power corrupting absolutely.

By "removing" superpowers or quantum mechanics from our entropic physical reality, the Gods created the ultimate school-ground and the ultimate consensus reality where our choices have real and lasting consequences. The pain makes it real; and, we don't always get do-overs, especially when we choose to do something fatally stupid. We actually have skin in the game.

Mark My Words

Summarizing the Different Types of Syntropy

There's physical matter, and then there's everything else.

Entropy seems to be restricted exclusively to physical matter. Entropy is a function of physical matter. Everything else seems to be based upon Syntropy.

What a fascinating observation to make!

ALL of those invisible and intangible forces, fields, and quantum waves are based upon Syntropy. They NEVER burn out, and they NEVER wear out, because they are eternal and everlasting, without a beginning of days or an end of years.

There's physical matter or entropy; and then, there's everything else which is based upon Syntropy. There's classical physics, and then there's Quantum Mechanics or Transdimensional Physics. The one is based upon entropy, and the other is based upon Syntropy. This just might be the most interesting and useful scientific observation of all-time. Its explanatory power is through the roof, is it not? Syntropy has to exist, or entropy and physical matter would not exist.

It all depends upon the interpretation that you choose to give to Quantum Mechanics. Certain interpretations were designed to hide the truth from us; and, other interpretations were designed to reveal the truth to us. The false is falsified by the truth; and, the truth is repeatedly experienced and observed. If you successfully eliminate everything that is false, then only the truth will remain.

Remember, Syntropy has to exist, or we would not exist. Our very existence here in this physical universe in these physical bodies tells us that someone somewhere KNOWS how to do Syntropy or Organization where physical universes, genomes, proteins, eyes, brains, and physical bodies are concerned.

Organization is Syntropy. Psyche is Syntropy. Quantum Mechanics is Syntropy. The Priesthood Power of God is Syntropy. Spirit Matter is Syntropy. God is Syntropy. Quantum Mechanics or Syntropy is the Priesthood Power of God. Syntropy is a new and better way of doing science, explaining science, and understanding science. Syntropy is eternal and everlasting.

Syntropy means "without beginning of days or an end of years". Syntropy is Eternal Life. Syntropy or Eternal Existence is the opposite of creation ex nihilo. Creation ex nihilo or the catholic (universal) version of creation is Atheism – creation by nothing from nothing. I'm not a creationist. I'm definitely not that type of creationist. Creation ex nihilo is impossible. Not even God could do creation ex nihilo, because it's impossible. Creation ex nihilo is Atheism and magic. Creation ex nihilo is also unscientific, because it's impossible. Creation ex nihilo is the exact opposite of Syntropy. The one is magic, and the other is organization and science.

God is a scientist, and not a magician. God organized this physical universe from pre-existing Dark Matter or pre-existing Spirit Matter. Remember, it's the physical matter that is the anomaly – everything else is based upon Syntropy. The physical matter is visible to us with our physical eyes; and, everything else is not. We detect all the dozens of different invisible forces, fields, and laws by the effect that they have on physical matter. Physical matter or entropy is the anomaly. Everything else, or Syntropy, or Quantum Mechanics is the norm.

Syntropy concerns the things that are without a beginning of days or an end of years. Syntropy is the opposite of Materialism, Naturalism, Atheism, Entropy, Death, and Classical Physics. Syntropy is the original prime construct from which this physical universe and entropy were made.

Our physical universe had a beginning. It was ordered and organized by **someone** who has NO beginning and will have no end – Syntropy, Psyche, or God. Our physical universe or physical matter was built, or made, or organized from **something** that had NO beginning and will have no end – Spirit Matter, Dark Matter, Dark Energy, Quantum Waves, or Quantum Objects. Creation ex nihilo is falsified by Syntropy and all of that pre-existing pre-physical Spirit Matter or Dark Matter.

Observation or Premise: Anything that was obviously made obviously had a Maker or Creator who made it.

Observation or Premise: A genome was obviously made.

Logical Conclusion: Therefore, a genome obviously has a Maker or a Creator who designed it, programmed it, engineered it, field-tested it, fine-tuned it, made it, manufactured it, and deployed it.

In fact, evolution (genetic change or genetic drift), natural selection, and random mutations did not exist until after God designed, manufactured, and deployed the genomes in the first place. This is Logic 101.

To make or create something is to infuse it with Order or Syntropy – to reverse the entropy. Organization, Engineering, Manufacturing, and Creation are different types of Syntropy – a reversal of disorder or entropy. Syntropy is the opposite of spontaneous generation or magic. There is NO such thing as spontaneous generation, chemical evolution, creation ex nihilo, or macro-evolution. These things are prevented from happening by entropy or the second law of thermodynamics. These things are prevented from happening by physical matter. It requires some kind of Syntropy or Psyche in order to design and create and make things from scratch.

You know what this means, don't you?

It means that evolution can't happen. Evolution is impossible. The different types of evolution are prevented from happening by entropy and physical matter. Entropy and physical matter actually falsify the Theory of Evolution. Random mutations are entropy. Random chance or random diffusion is based upon entropy. Natural selection leads us towards entropy, death, and extinction. The chemical evolution of genes and proteins from atoms is prevented from happening by entropy or random diffusion. Spontaneous generation, abiogenesis, chemical evolution, and macro-evolution are prevented from happening by entropy, physical restrictions, and physical laws. Physical matter and genes impose a structure or an order that prevents evolution of any type from happening. Physical matter is the basis of the ultimate consensus reality.

It would be unpredictable chaos or evolution if it weren't for the order, structure, and limitations being imposed upon us by physical matter or by consensus. Natural evolution would not be a good thing. It would be chaos. That's why God, physical matter, genes, and entropy prevent any type of macro-evolution from happening. It's bad enough as it is that our genes are allowed to mutate. Think of all the problems which that causes! Now multiply that by a billion to imagine what it would be like if all the other types of evolution were allowed to happen.

Remember, classical physics or entropy prevents evolution. It was designed to do so. This just might be the most significant scientific discovery of all time.

It actually requires some type of Syntropy, Psyche, or Intelligence in order to be able to design, create, program, organize, engineer, field-test, fine-tune, manufacture, and bring physical things into existence. Some type of Syntropy is needed in order to bring life into existence, especially physical life. Psyche, Intelligence, or Quantum Non-Local Consciousness is LIFE, which means that it is Syntropy.

Remember, Syntropy has to exist, or we wouldn't exist. It's time that we, as scientists, finally start to use Syntropy, Psyche, and Quantum Mechanics to explain how things really work in every realm of existence, including both this physical realm and the non-local quantum realm. It's time for us to take Quantum Mechanics, Psyche, or Syntropy seriously, instead of ridiculing it, mocking it, and dismissing it as pseudo-science. It's time for us to upgrade our science to Science 2.0 and start allowing ALL of the evidence into evidence.

God's Psyche or God's Intelligence is Syntropy. It had no beginning and it will have no end. Your Psyche, or Intelligence, or Quantum Non-Local Consciousness is Syntropy. It had no beginning and it will have no end.

In contrast, entropy is death – physical death. It has an end. Physical matter or entropy is the anomaly. WE KNOW it is so, because everything else that we have encountered is based upon Syntropy – it NEVER ages, and it never wears out. Remember, Syntropy never ages and it never wears out. It is without a beginning of days or an end of years. Syntropy is eternal and everlasting. Compared with that, entropic physical matter is indeed the anomaly or the odd ball.

Physical bodies and physical universes begin and end thanks to both syntropy (the beginning of a physical universe and a physical body) and entropy (the ending of a physical universe and physical body); but, Syntropy or Quantum Mechanics or Psyche has no beginning and will have no end, because it is eternal and everlasting. Syntropy is light, life, love, psyche, intelligence, power, order, organization, and God. The Atonement of Christ is Syntropy. It has to be, because its purpose is to help us overcome entropy, corruption, sin, and death.

The obvious fact – that some type of Syntropy MUST exist or we would not exist – makes logical sense to me. The NEED for Syntropy is so glaringly obvious that Syntropy MUST exist; and therefore, its existence MUST be true. WE KNOW that some type of Syntropy exists, because entropy exists. We know that Syntropy exists because physical matter exists. There's entropic physical matter; and then, there's Syntropy. We have observed that everything else that we have encountered besides physical matter is based upon Syntropy – it never ages nor gets old. It's eternal and everlasting.

Like I said, this just might be the most significant and useful scientific discovery of all time. Its explanatory power is off the charts.

If entropic physical matter were the ONLY thing that exists, as the Materialists and Naturalists claim, then it would be physically impossible for our genes, proteins, eyes, brains, and physical bodies to exist. Entropy or random diffusion prevents physical matter from organizing spontaneously into functional genes, proteins, eyes, brains, and physical bodies. Entropy prevents abiogenesis. Evolution is entropy, and that means that the different types of evolution or the different types of entropy prevent the Theory of Evolution from becoming true. In contrast, psyche is the organizing force of the universe.

If entropy prevailed and there were no Order or no Syntropy, then the stars, planets, and galaxies would not form and would not exist. The Syntropy, the available energy, the zero-point field, the Higgs Field, the organizing force, or the love has to exist; or, there would be no genomes, no proteins, no planets, no stars, and no galaxies. There would be no physical matter and no physical life anywhere without some type of Syntropy or Order in this physical universe. This is rationally obvious to anyone like me who spends even the smallest amount of time thinking about it.

Thanks to Syntropy or Quantum Mechanics, we KNOW that Materialism, Naturalism, Darwinism, Nihilism, and Atheism are false and have been falsified. Thanks to Syntropy or Psyche, entropy is not the end. You will go on long after your physical brain is dead and gone. Thanks to Syntropy, your consciousness or psyche is eternal and everlasting. You have always existed, and you will always exist. Psyche is Syntropy. Relationships are Syntropy. Love is Syntropy. God is Syntropy. The best part of you is Syntropy, and it's capable of counteracting the effects of entropy and surviving the death of your physical brain.

Mark My Words

Physical Matter and Entropy Were Made

We are going to have to rethink entropy if we are going to make it match with what has been experienced and observed by Out-of-Body Travelers. Remember, God had to make physical matter, mass, entropy, and time because they don't exist at the quantum level nor at the psyche level. Entropy or the Second Law only exists at the physical level.

Technically, entropy doesn't exist as a person, place, or thing. Entropy doesn't exist in any useful, detectable, or measurable form. Entropy is defined as the non-existence of order (disorder), the non-existence of available energy (anergy), the non-existence of information (random chaos), the non-existence of life (death), the non-existence of thermal disequilibrium (heat death), or the non-existence of Psyche or Life Force (non-existence). It's physically impossible to make non-existence from physical matter, or from energy, or from anything else for that matter. It's also physically impossible to detect and to measure the non-existence of something, or to make the non-existence of something from physical matter. Instead, entropy is an emergent property of mass and the associated passage of time. The best that we can say where entropy is concerned is that entropic physical matter is MADE from many different FORMS of energy because entropy doesn't exist in any useful form nor in any measurable form. This is what has been experienced and observed.

This physical platform, this physical universe, and entropic physical matter were designed by the Gods to be temporary or entropic. Entropic means temporary. Entropy is a function of time. Entropy is an aging process based upon the passage of time. Entropy is the arrow of time. Entropy is death. Death wouldn't be possible without entropy. For example, Heat Death is impossible in the Quantum Realm or the Syntropy Realm because there is NO entropy, NO unavailable energy, NO heat flow, NO thermodynamics, and NO anergy in the Spirit World. ALL of the energy is always available for use in the Quantum Realm or the Psyche Realm. This is what has been experienced and observed.

You have to know and understand what the Materialists, Naturalists, Darwinists, Nihilists, Behaviorists, Determinists, and Atheists TEACH and BELIEVE before you can figure out what's wrong with it. These people teach and believe that entropic physical matter is the ONLY thing that exists. Do they not? That's indeed what they teach and believe!

Consequently, these people literally teach and believe that physical matter, heat death, death, and entropy are conserved. Many of these people have no clue that they have chosen to believe in the Conservation of Entropy, the Conservation of Physical Matter, the Conservation of Heat Death, and the Conservation of Death; or, they don't completely understand the full ramifications of their chosen beliefs if they do know. I KNOW because I didn't understand the full ramifications of my chosen beliefs when I was a Materialist, Naturalist, Nihilist, and Atheist. I hadn't given any of this much thought at the time.

The Physicalists and Naturalists teach and believe that physical matter and entropy are eternal and everlasting and that the whole universe is going to end in Heat Death. Entropy is death; and, these people teach and believe that death or entropy is conserved. They teach and believe that death is eternal and everlasting. This is what they teach and believe, is it not? I would guess that most of these people don't fully realize that they believe in the Conservation of Entropy, the Conservation of Death, the Conservation of Heat Death, and the Conservation of Physical Matter. Nevertheless, they do. That's the net result of choosing to believe that ONLY entropic physical matter exists. They end up believing in the non-existence of lots of different things. They end up believing in nothing.

Based upon $E = mc^2$, these people teach and believe that energy is synonymous with physical matter and that physical matter is conserved. Consequently, these people define the First Law of Thermodynamics as the "Conservation of Physical Matter". Based upon their Supremacy of Entropy Doctrine, these people teach and believe that entropy or death is conserved. Consequently, these people define the Second Law of Thermodynamics as the "Conservation of Entropy".

THEY ARE WRONG! These people got their science wrong. Read this carefully, because it's the KEY to everything else in Science.

The first law of thermodynamics is a version of the law of conservation of energy, adapted for thermodynamic systems. The law of conservation of energy states that the total energy of an isolated system is constant; energy can be transformed from one FORM to another but can be neither created nor destroyed.

https://en.wikipedia.org/wiki/First_law_of_thermodynamics

Once I fully understood and accepted this simple axiom or LAW, instantly everything became obvious and clear. I'm a scientist. My goal is to figure out what everything is and how it works. The First Law completely eliminates the Second Law at the quantum level.

Energy, Psyche, Intelligence, Life Force, or Syntropy can be neither created nor destroyed. However, Energy or Syntropy can be transformed from one FORM into another FORM. The FORM is never conserved. Entropic physical matter is comprised of many different FORMS of Energy, and the FORM is never conserved. Only the underlying Energy, Psyche, Life Force, Intelligence, or Syntropy is being conserved. This is the Ultimate Law of Thermodynamics which states that physical matter and entropy are NOT conserved. Remember, entropic physical matter is never conserved. The FORM is never conserved. Entropy is death; and ultimately, death is not conserved. This is what has been experienced and observed.

Energy can be transformed.

Transformed by whom? Who triggers this transformation? Who decides what FORM Energy will take? Someone Psyche has to make this decision, or nothing will happen, and physical matter will never be made. Physical matter, mass, entropy, and time don't exist at the quantum level which means that they have to be constructed or made by Someone

Psyche, or they won't exist. Physical matter, physical laws, physical restrictions, and entropy were made or caused to begin.

Made by whom? Made from what?

Well, it is obvious to me that entropic physical matter was made from different FORMS of Energy. The first person or first psyche who designed and made entropic physical matter from energy is obviously the person who first designed and made the stuff. Order and organization don't just spontaneously generate out of thin air. They are OBVIOUSLY designed, organized, made, and caused to begin. Psyche is the innate intelligence within all the different FORMS of Energy which gives that energy the inherent ability to understand, follow, and obey God's Laws and God's Commands. The different forms of physical matter were made.

Scientific Observation: Anything that was obviously made obviously had a Maker who made it.

Scientific Observation: Physical matter, physical laws, physical constants, physical restrictions, quantum fields, space-time, locality, time, and entropy were obviously made or caused to begin.

Scientific Conclusion: Therefore, it is logical and rational to conclude that physical matter, physical laws, physical constants, physical restrictions, quantum fields, space-time, locality, time, and entropy obviously have a Maker who made them or caused them to begin.

Watch the following PBS Space Time playlists on the "Origin of Matter and Time" and "Quantum Fields". If you do, you can see with your own eyes that entropic physical matter was designed, ordered, organized, structured, and made. The stuff is obeying LAWS. The Order, Organization, Structure, Rules, and LAWS are obvious to anyone who is willing to look and see. I find myself asking, "Who designed this stuff and made it work? Who made the quarks, electrons, bosons, and gluons? Who made the quantum fields? Who is making the gluons, bosons, quarks, electrons, and quantum fields OBEY these LAWS and these various rules or conventions?"

Order and organization, such as this, don't just spontaneously generate out of thin air from nothing. This stuff is deliberately, purposefully, and meaningfully DESIGNED and MADE. I can see the truth of it by watching the PBS Space Time videos.

https://www.youtube.com/channel/UC7_gcs09iThXybpVgjHZ_7g/videos

Mark My Words

The Origin of Matter and Time Playlist

https://www.youtube.com/playlist?reload=9&list=PLsPUh22kYmNCLrXgf8e6nC_xEzx dx4nmY

The Higgs Mechanism Explained

https://www.youtube.com/watch?v=kixAljyfdqU

Massless Speed-of-Light Particles or Massless Psyches USE mass to resist acceleration and to slow down to sub-light velocities. Mass produces the experience of time or the passage of time. Mass makes entropy possible. Mass is just a different FORM of energy. It's all energy, and energy is always conserved.

The most interesting science in this video is their discovery that, without the mass or the resistance to acceleration which the electron (and quark) gets from the Higgs Field and weak hyper-charge, the electron would be a massless particle and travel at the speed-of-light. In other words, the electron would have NO mass without the Higgs Field. The Higgs Field provides the mass, or resistance to acceleration, and slows the electron down.

Mass is resistance to acceleration. Mass slows particles down. Massless particles like photons travel at the speed-of-light, and they experience NO passage of time and therefore NO entropy. Photons and other Massless Particles are capable of being immortal and conserved. The Higgs Field produces mass, which slows quantum particles down to below the speed-of-light, which then produces the passage of time as a result. The passage of time makes entropy possible. The fundamental elementary components of the electrons and the quarks are Massless Speed-of-Light Particles; and, the Higgs Field gives the quarks and the electrons their small amount of mass and thereby slows them down to sub-light speeds.

Remember, mass produces resistance to acceleration and thereby slows Massless Particles down to sub-light speeds.

Most of the mass within protons and neutrons is produced by the Gluon Field or Strong Force, and NOT the quarks. The quarks only account for about 1% of a proton's mass and get their mass from the Higgs Field. The other 99% of the mass within a physical atom comes from the Energy or the Gluon Field that holds the quarks together inside the protons and neutrons. The elementary components of electrons and quarks are massless, and they would have NO mass and travel at the speed-of-light without the Higgs Field. Likewise, the elementary components or fundamental components of protons, neutrons, and atoms are Massless Particles; and, the protons and neutrons get most of their mass from the Gluon Field or Strong Force and NOT the quarks.

> **In theoretical particle physics, the Gluon Field is a four-vector field characterizing the propagation of gluons in the strong interaction between quarks. The Gluon Field plays the same role in quantum chromodynamics as the electromagnetic four-potential in quantum electrodynamics – the gluon field constructs the gluon field strength tensor.**

> **The strong nuclear force holds most ordinary matter together because it confines quarks into hadron particles such as the proton and neutron. In addition, the strong force binds neutrons and protons to create atomic nuclei. Most of the mass of a common proton or neutron is the result of the strong force field energy; the individual quarks provide only about 1% of the mass of a proton.**

I've always been interested in Particle Physics; therefore, it was extremely important to me to finally get this straight and get it right. Essentially, the Gluon Field or Strong Force

GLUES quarks together into protons and neutrons. In contrast, some Massless Speed-of-Light Particles USE the Higgs Field to slow themselves down thereby becoming electrons and quarks in the process. The electromagnetic force draws electrons to protons thereby forming physical atoms. As a consequence of the Pauli Exclusion Principle, only one fermion (quark or electron) can occupy a particular quantum state at any given time. The Pauli Exclusion Principle and Uncertainty Principle resist or counteract an electron's desire or urge to merge with a proton. Checks and balances!

Thanks to Materialism and Naturalism, Fermions are erroneously associated with matter and mass, whereas bosons are erroneously visualized as being massless force carrier particles. In truth, though, the elementary parts of BOTH fermions and bosons are in fact Massless Speed-of-Light Particles; and, it's the Quantum Fields or Force Fields associated with the Gluon Field and the Higgs Field that in FACT give the Massless Particles or Massless Psyches within atoms their mass or resistance to acceleration.

Massless Particles must travel at the speed-of-light, which means that time stops, or time is frozen for Massless Particles, which means that there is NO entropy at the quantum level or the psyche level. Each Massless Particle is some type of Psyche or Intelligence. According to the First Law of Thermodynamics, Massless Particles, Massless Energy, or Massless Psyches are the ultimate perpetual motion machine. Given the FACT that Energy is always conserved and given the FACT that the Second Law of Thermodynamics or Entropy does not exist at the quantum level and the psyche level, Massless Particles or Massless Psyches can keep going and keep working for ALL eternity because they are forever conserved. They can change form, assignment, or function; but, Massless Psyches or Massless Particles are always conserved.

The basic Quantum Field Theory equations for all the components of the atom leave them massless. This masslessness means that particles should travel only at the speed-of-light and experience no passage of time. Their clocks should be frozen. The quarks and the elections should be massless. But, quarks and electrons are distinctly not timeless. They evolve or change over time. The electron switches back and forth from left-handed chirality to right-handed chirality, which means that the electron does experience time, so it must have mass. The passage of time is produced by mass. Entropy is also produced by mass and the passage of time.

The photon is definitely massless. It travels at the speed of light and experiences its entire existence in an instant. A photon undergoes no internal evolution. A photon only changes if it bumps into something else and/or has to resolve its wave function. The photons and electrons are just excitations or vibrations in their own unique Quantum Fields. So, why does the electron have mass, and why doesn't the photon have mass? Why does the electron evolve or change over time? Why are photons massless and timeless?

There's something in the substrate of space everywhere that impedes and slows down the electron. It's the Higgs Field. The universe actually cares whether an electron has left-handed or right-handed chirality. ONLY left-handed electrons feel or experience the weak nuclear force. It's an open question why the universe cares which direction an electron is spinning. The universe cares so much that it won't let an electron flip from left to right unless it can ditch its weak hyper-charge; or flip back again unless it can pick some up. The Universe's Psyche or Nature's Psyche cares about these things. God is in the details.

Where does the weak hyper-charge come from, and where does it go to? The Higgs Field. The Higgs Field is really weird. It has a positive strength at all points in the universe. There's a little bit of Higgs in us everywhere. It's like the Light of Christ that proceeds forth from the presence of God to fill the immensity of space. The Higgs Field is an infinite source and sink of weak hyper-charge. On its own, the electron would travel at light speed, but trapped in this Higgs Field buzz, the electron feels mass and slows down.

The very existence of the Higgs Field falsifies Materialism, Naturalism, Nihilism, and even Atheism. The Higgs Field is an invisible, intangible, non-physical, and infinite ocean of some sort of field, energy, or charge that we've never heard of before, all invented so that electrons can be left-handed or right-handed at the same time. Nevertheless, something like this must be true. Quantum Field Theory is an incomplete theory without a mass-giving field. The discovery of the Higgs boson proved the existence of the Higgs Field.

All massless particles travel at the speed-of-light at the physical level. From their perspective, massless particles like the Photon experience no passage of time and effectively quantum tunnel to their destination. From our entropic and physical perspective, Photons travel at the speed-of-light. Without the electromagnetic field to restrict a Photon to the speed-of-light, a Photon could travel faster than the speed-of-light or simply quantum tunnel anywhere it wants to go. The various different Quantum Fields provide order, organization, stability, reliability, and velocity restrictions to ALL types of quantum particles or quantum waves. The Higgs Field might also provide an explanation for Dark Energy.

Technically, in $E = mc^2$, the "m" stands for mass and NOT matter. It's the energy or mass that's conserved and NOT the matter. Mass can be measured as a unit; but, entropic physical matter is MADE from many different FORMS of energy.

Mass is just bound or confined energy. Who binds it and confines it? Who made the rules, the laws, the forces, and the fields that bind energy and confine energy? In the case of the constituents of the atoms, mass comes from the Higgs Field. Who designed and made the Higgs Field? Some people call the Higgs Field, God's Field. It's obvious that God designed and made the Higgs Field. Such order, organization, structure, rule, and LAW doesn't just spring into existence from nothing. It is deliberately designed and made.

Quantum Field Theory describes the fundamental particles as excitations or energy waves in the quantum fields that fill our entire universe. Quantum Fields are omnipresent. An electron is an excitation or an energy wave in the electron field. Every point in the universe has a certain level of electron-ness. In "empty" space, that level hovers around zero. Space really isn't empty. Space is filled with many different Quantum Fields. Even in a vacuum, the electron field is there. Add some energy to that field at a particular spot, and it's like plucking a guitar string. The field vibrates, and that vibration is our electron. Every elementary particle is a vibration in its own unique Quantum Field; and, the vibrations and fields interact with each other transferring energy, momentum, and charge between the quantum particles and the Quantum Fields.

Naturalists and Atheists typically stop at the quantum level and refuse to take science to the psyche level or the God Level. However, in this video, he actually states that the Universe cares whether an electron is right-handed or left-handed. I choose to crawl down the rabbit hole and try to see how far down it goes. At the quantum level, Someone Psyche has to pluck the guitar string at a specific spot in a specific quantum field in order to produce a quantum wave packet of energy or a quantum particle. Someone Psyche has to deliberately add some energy to a quantum field or set the field vibrating at a specific spot in order to produce a quantum particle or a quantum wave packet of energy within a quantum field. Someone Psyche had to design and make the Quantum Fields. Someone Psyche has to maintain them and use them.

Thoughts and memories are quantum waves. Psyche or Life Force has its own unique quantum field. Psyche can produce, transmit, receive, and store quantum waves or thoughts. Psyche is an elementary particle – the elementary particle. Psyche can quantum tunnel just like any other elementary particle, unless God prevents it from doing so. Every elementary particle is a vibration in its own unique quantum field; and, these vibrations and fields interact with each other transferring energy, momentum, and charge between particles and fields. It's all made from energy, and Someone Psyche or Someone Intelligent made it all and controls it all. It is obvious. Every massless particle is some type of Psyche!

[Note: I don't always quote these videos verbatim. Without the diagrams that are presented in the video to enhance the concepts, I feel the need to paraphrase, adjust, adapt, and add to the concepts so that hopefully everything makes sense when I'm done. I also try to eliminate anything that has been falsified.]

The True Nature of Matter and Mass

https://www.youtube.com/watch?v=gSKzgpt4HBU

https://evolution-is-entropy.com/wp-content/uploads/2018/08/The-True-Nature-of-Matter-and-Mass.zip

One of the most noticeable features of entropic physical matter is that it has locality or physicality designed into it and programmed into it. It has been subjected to the effects of space-time. In contrast, massless elementary particles look like and ACT like Photons both at the quantum level and the physical level.

Resistance to acceleration or "Mass" is the word and the thing that we most tend to associate with matter and physicality. Therefore, massless particles are obviously by definition in principle spiritual, quantum, transdimensional, non-local, and non-physical in nature and origin. Massless elementary particles by themselves, on their own, have NO physical limitations and NO physical restrictions because they are NOT physical and have NO mass. Remember, Massless Particles have NO mass; and therefore, they have NO resistance to acceleration. Massless quanta are NOT subject to entropy because they experience NO passage of time and therefore do not age in their native original quantum format, spiritual format, non-local format, or supernatural format. Massless quantum particles are eternal and everlasting in nature. The Energy and the Psyche within them cannot be made, and it cannot be destroyed. It is always conserved. At the quantum level in the Quantum Realm, the Conservation of Energy or the Conservation of Psyche reigns supreme, and the Second Law of Thermodynamics or Entropy DOES NOT EXIST. This is what has been experienced and observed. Massless Particles do NOT experience the passage of time and therefore have NO entropy.

In this and a couple other videos, Matt states his belief that quarks and electrons are made up of massless elementary particles that are bound together by the Higgs Field. It's the Higgs field that provides the MASS or the resistance to acceleration, the element of time or the passage of time, and the effects of entropy. The different types of massless quantum particles experience NO TIME. In other words, massless quanta don't age; and, they must travel at the speed-of-light because they have no mass. These observations were the KEY that opened many doors for me and greatly increased my understanding of science in general. Massless quanta have no entropy because they experience NO passage of time, and entropy is a function of time. There is NO entropy at the quantum level and the psyche level because there is no aging process or passage of time for massless quanta in their native original psychic state. Technically, the word "massless" means non-physical. Massless elementary particles are quantum, supernatural, transdimensional, non-physical, and spiritual in nature and origin. Massless Particles are NOT subject to entropy because they experience NO passage of time. Massless Particles do not age; and consequently, the

180

energy within them is always conserved. In other words, Massless Particles are syntropic and NOT entropic. The Second Law of Thermodynamics DOES NOT EXIST at the quantum level and the psyche level. This is what has been experienced and observed.

Psyches use mass and control mass in order to get things done both at the quantum level and the physical level. Psyches are massless. Photons are massless. Anything without mass has to travel at the speed-of-light. From its perspective, it quantum tunnels to its destination and therefore experiences NO passage of time. From our limited perspective, it travels at the speed-of-light. There's something about mass that prevents an object from reaching the speed-of-light. Most of the mass within a physical atom comes from gluons that bind the quarks together into protons and neutrons. Mass is resistance to acceleration. Thanks to specific Quantum Fields, mass arises in the ensemble, where it doesn't exist in the massless parts – the massless quantum particles, the massless elementary particles, or the massless photons. Psyches or massless particles USE mass to slow themselves down to sub-light speeds. Psyches or massless particles also USE mass to form themselves or bind themselves into physical atoms. Every Massless Particle is some type of Psyche or Intelligence.

In this video, Matt says that communication between atoms is a speed-of-light interaction. In fact, communication between atoms and molecules is a psychic, telepathic, and instantaneous quantum mechanical phenomenon – instantaneous interaction at a distance or instantaneous communication at a distance. Telepathy is WiFi at the quantum level between photons, atoms, and molecules. Photons and atoms are NOT limited to the speed-of-light unless they choose to be – they can quantum tunnel unless God prevents them from doing so. God can limit photons to the speed-of-light and atoms to sub-light speeds; but, they had no such limitations in their original quantum environment. God deliberately slows things down so that we can live them, experience them, learn from them, and actually remember having done so. At the speed-of-light and during quantum tunneling, there is no passage of time and therefore nothing to experience and remember.

Quarks and electrons gain their tiny masses from a type of confinement via the Higgs Field. Take away the Higgs Field, and the components of the quarks and electrons are massless speed-of-light particles. Massless particles use the mass to slow down. It looks like everything with mass is composed of a combination of intrinsically massless light-speed particles that are prevented from streaming freely through the universe at the speed-of-light by Quantum Fields such as the Higgs Field and the Gluon Field, which confine those massless particles and limit those massless particles to sub-light-speed velocities.

Mass is not a fundamental property. Mass is the emergent result of massless particles bumping and sloshing around within confining and restricting Quantum Fields, resisting acceleration. The Higgs Field produces the mass within the electrons and quarks; and, the Gluon Field produces 99% of the mass within the protons and neutrons and atoms. The elementary particles within quarks and electrons are massless "quantum particles" or Photons. Inertial Mass and entropy are an emergent property of the effects being produced by the Higgs Field and the Gluon Field on massless quantum particles.

In theoretical particle physics, the Gluon Field is a four-vector field characterizing the propagation of gluons in the strong interaction between quarks. The Gluon Field plays the same role in quantum chromodynamics as the electromagnetic four-potential in quantum electrodynamics – the gluon field constructs the gluon field strength tensor.

https://en.wikipedia.org/wiki/Gluon_field

The strong nuclear force holds most ordinary matter together because it confines quarks into hadron particles such as the proton and neutron. In addition, the strong force binds neutrons and protons to create atomic nuclei. Most of the mass of a common proton or neutron is the result of the strong force field energy; the individual quarks provide only about 1% of the mass of a proton.

https://en.wikipedia.org/wiki/Strong_interaction

I've always been interested in Particle Physics; therefore, it was extremely important to me to finally get this straight and get it right. Essentially, the Gluon Field or Strong Force GLUES quarks together into protons and neutrons. In contrast, some Massless Speed-of-Light Particles USE the Higgs Field to slow themselves down thereby becoming electrons and quarks in the process. The electromagnetic force draws electrons to protons thereby forming physical atoms. As a consequence of the Pauli Exclusion Principle, only one fermion (quark or electron) can occupy a particular quantum state at any given time. The Pauli Exclusion Principle and Uncertainty Principle resist or counteract an electron's desire or urge to merge with a proton. Checks and balances!

Thanks to Materialism and Naturalism, Fermions are erroneously associated with matter and mass, whereas bosons are erroneously visualized as being massless force carrier particles. In truth, though, the elementary parts of BOTH fermions and bosons are in fact Massless Speed-of-Light Particles; and, it's the Quantum Fields or Force Fields associated with the Gluon Field and the Higgs Field that in FACT give the Massless Particles or Massless Psyches within atoms their mass or resistance to acceleration.

Massless Particles must travel at the speed-of-light, which means that time stops, or time is frozen for Massless Particles, which means that there is NO entropy at the quantum level or the psyche level. Each Massless Particle is some type of Psyche or Intelligence. According to the First Law of Thermodynamics, Massless Particles, Massless Energy, or Massless Psyches are the ultimate perpetual motion machine. Given the FACT that Energy is always conserved and given the FACT that the Second Law of Thermodynamics or Entropy does not exist at the quantum level and the psyche level, Massless Particles or Massless Psyches can keep going and keep working for ALL eternity because they are forever conserved. They can change form, assignment, or function; but, Massless Psyches or Massless Particles are always conserved.

In this and a couple other videos, Matt almost makes it all the way to a complete and total understanding of how everything works. This is the closest that I have ever seen anyone get. All he needs is a definition for Psyche and he is there.

The revelation that I received is that every massless elementary particle is a Psyche or Intelligence of some type. Massless Particles or Psyches look like and ACT like Photons or Pinpoints of Light, both at the quantum level and the physical level. Massless Particles are Psyches. This scientific discovery was a real game-changer for me. I realized that Psyche has been hiding in plain sight all along where nobody can see it nor find it because we aren't looking for it and have instead chosen to believe that it does not exist.

Any Massless Particle is some type of Psyche or Intelligence. There are different types of Psyche, each with its own unique God-given assignment, mission, purpose, or function. This truth is really simple, logical, parsimonious, obvious, and easy to understand which makes me wonder why nobody has ever discovered it or thought of it before.

Any massless "quantum particle" is a Psyche because it is psychic and intelligent and because it has NO physical limitations. A massless Photon is a type of Psyche. A Photon is psychic and intelligent. Psyche is intelligence and light. Psyches, whenever they are seen or observed, look like and ACT like Photons. Virtual Photons are Psyches because they are psychic and intelligent and because they have NO physical limitations. Photons are Psyches. Massless particles are non-physical Psyches or Intelligences of one type or another because they look like and ACT like Photons when they are on their own and when they are not using mass to slow themselves down and organize themselves into physical matter.

Remember, God gives each Psyche or Massless Particle a unique assignment, purpose, position, mission, station, and function in nature or reality – both at the quantum level and the physical level. Whenever they are seen or observed, Psyches or Massless Particles look like and ACT like Photons or Pinpoints of Light both at the quantum level and the physical level. This is what has been experienced and observed. Massless Particles, Photons, or Psyches have been experienced and observed, both at the quantum level and the physical level. This is Science. Science is observation and experience at any level of existence or reality.

The massless elementary components of quarks and electrons are Photons or Psyches. Any Massless Particle is some type of Psyche. The mass of the quarks and the electrons comes from the Higgs Field, which bind their internal massless Photons or Psyches together into massive quarks and mass-containing electrons. The quarks within protons and neutrons are bound together by the Strong Force or the Gluon Field. The elementary components of quarks and electrons are massless, timeless, ageless, syntropic, quantum, speed-of-light Photons or Psyches. Within quarks and electrons, the Higgs Field slows down the Photons or Psyches to sub-light-speed velocities so that these Psyches or Intelligences can then experience the passage of time, learn from it, and remember having done so. Massless Psyches or Massless Particles use mass to slow down. This is the way things work.

At the speed-of-light or faster, TIME STOPS; and, the massless Photons or Psyches experience NO passage of time. From their perspective, they simply quantum tunnel to their destination instead. Massless Psyches or Photons can quantum tunnel or travel at the speed-of-light unless God or Quantum Fields prevent them from doing so. Quantum Fields such as the Higgs Field and the Gluon Field slow down the Photons or Psyches by giving them mass and thereby making them subject to the passage of time. Psyches use mass to slow down. This explains everything that has ever been experienced and observed. It's the answer to life, the universe, and everything. This is how everything really works.

Mass is an emergent property of the interactions of massless particles. A single photon experiences no time, nor does any massless particle. Their clocks are frozen in time. Anything that has mass must experience the passage of time.

Inertial Mass is the degree to which an object resists being accelerated. Mass is a resistance to a change in speed. Mass is just one type of energy. There are many different types of energy. Most of the mass within atoms comes from the kinetic energy and binding energy of the gluons that are holding the quarks together in the protons and neutrons. The massless elementary parts of the quarks are like photons in that they don't age and don't experience the passage of time. It's the mass or the Higgs Field slowing them down – or slowing down the interaction of the internal massless components of the quarks and electrons – that is in fact causing them to FEEL and experience entropy or the passage of time.

Over ninety-nine percent of the mass of a proton is found in the vibrational energy and the binding energy of the gluon field. The quarks themselves have very little mass. The elementary parts of the quarks don't have any mass either – just the energy fields, binding the elementary parts of a quark into a quark, have the mass or binding energy that produces inertial mass. At the fundamental level, particles don't have mass and time. It's the energy fields or quantum fields binding the particles together that's producing the mass and slowing the particles down to sub-light speeds thereby producing the FEEL of time.

Do you ever get any sense that ALL of this stuff was deliberately designed and made? I do. I don't think it spontaneously generated out of thin air from nothing. Order and organization don't work that way – they don't just spontaneously generate out of thin air ex nihilo.

When you take a look at the particles that make up the proton – the three different quarks at the heart of them – their combined masses are only 0.2% of the mass of the proton as a whole. The way quarks bind into protons is fundamentally different from all the other forces and interactions we know of. Instead of the force getting stronger when objects get closer – like the gravitational, electric or magnetic forces – the attractive force goes down to zero when quarks get arbitrarily close. And instead of the force getting weaker when objects get farther away, the force pulling quarks back together gets stronger the farther away they get. This property of the strong nuclear force is known as asymptotic freedom, and the particles that mediate this force are known as gluons. Somehow, the energy binding the proton together, the other 99.8% of the proton's mass, comes from these gluons.

https://www.forbes.com/sites/startswithabang/2016/08/03/where-does-the-mass-of-a-proton-come-from/#7c2c1ba22e1d

https://psyche-ontology.com/wp-content/uploads/2018/09/Where-Does-The-Mass-Of-A-Proton-Come-From.pdf

The gluons are always portrayed as springs holding the quarks together. If the quarks try to get away from each other, the springs pull them back together.

Psyche has been observed by Out-of-Body Travelers as a pinpoint of light, a pinprick of light, a point particle of light, or a spark of light. That's precisely the way that the physicists describe the fermions – quarks, leptons, neutrinos, and electrons. These "particles" of light OBEY the Pauli exclusion principle. Obedience to Laws requires intelligence! The components of the quarks and the electrons are individual pinpricks of Psyche or Intelligence; and, the psyches within fermions USE the bosons, gluons, photons, quantum waves, and light to communicate with each other, interact with each other, and bind themselves to each other. The elementary parts of Fermions are Psyche or Intelligence. That would make Psyche or Intelligence the fundamental unit of reality and existence. Photons, gluons, and W and Z bosons are the four force-carrying gauge bosons of the Standard Model. There's NO mass in the elementary components of the fermions, the psyches, or the pinpricks of light. Over 99% of the mass within an atom is associated with the gluons. Psyche or Intelligence is driving the mass, controlling the gluons, and using the mass to slow down. It's all made from Energy which is always conserved.

From a different angle, we observe that the elementary components within quarks, fermions, electrons, neutrinos, leptons, gluons, and bosons are massless, timeless, ageless, syntropic "quantum particles," pinpoints of light, quantum waves, or Photons that we call Psyche, or Intelligence, or Quantum Non-Local Consciousness. Thoughts are quantum waves. Psyche or Intelligence is there in the models and the math just waiting to be identified as such. Every massless non-physical quantum particle is some type of Psyche.

Photons are Psyche or Intelligence, which means that the massless, timeless, ageless, syntropic elementary components or "elementary particles" within quarks, fermions, electrons, neutrinos, leptons, gluons, and bosons are in fact Photons, or Psyche,

184

or Intelligence, or Quantum Non-Local Consciousness. Psyche or Intelligence is Light and Truth. Psyche is a Photon; and, Photons are psychic. Psyche is observed by Out-of-Body Travelers as a pinpoint of light, a point particle of light, or a Photon. Each massless particle is some type of Psyche or Intelligence. It all fits together perfectly and makes logical sense.

Psyche is the innate intelligence within all the different forms of energy which gives that energy the inherent ability to understand, follow, and OBEY God's Laws and God's Commands. Obedience to Law requires some type of intelligence. This is what has been experienced and observed. Remember, any massless quantum particle is a Psyche. A Photon is a Psyche. The massless components of quarks and electrons are Psyches.

Photons or Psyches don't have to limit themselves to the speed-of-light. They choose to do so, if they do so. From their perspective, traveling Photons quantum tunnel to their destination every time anyway. Photons experience NO passage of time while they travel. Photons KNOW where they are going to land before they launch; otherwise, they don't launch. Photons are psychic. Psyche makes choices, and Psyche is psychic and intelligent. Psyches are Photons. Photons are Psyches. Every massless "quantum particle" is some type of sentient controlling Psyche, Photon, Spark, Life Force, or Intelligence.

Psyche has been experienced and observed by Out-of-Body Explorers and Near-Death Experiencers. The First Law of Thermodynamics tells us that Psyche or Energy is conserved, which means that it is eternal and everlasting. It cannot be made, and it cannot be destroyed. Psyche or Energy is syntropic which means that it is always conserved both at the quantum level and the physical level. Psyche looks like and acts like a Photon.

I prefer the scientific approach. I define science as observation and experience. Psyche has been experienced and observed; therefore, we KNOW that it exists. Quod erat demonstrandum! Our job as scientists is to explain what it is.

Law of Psyche: The Law of Psyche states that every massless quantum particle is some type of Psyche or Intelligence. Psyche is the innate intelligence within all the different forms of energy which gives that energy the inherent ability to understand, follow, and obey God's Laws and God's Commands. Every massless elementary particle is some type of Psyche because it is psychic, quantum, telepathic, makes choices, can self-generate or self-propagate, can think, can communicate, can transmute, can move, can phase-shift, can collapse its own wave function, is perceptive, is sentient, is omniscient, is omnipresent, is intelligent, is conscious, is alive, must travel at the speed-of-light from our perspective at the physical level, is prescient, and without any physical limitations imposed upon it can quantum tunnel at will in its native original environment at the quantum level. Remember, every massless quantum particle is some type of Psyche, or Intelligence, or Life Force. Psyche is seen by Out-of-Body Travelers as a pinpoint of light or a Photon. Each Photon is a type of Psyche or Intelligence. A Psyche is a Photon, a massless elementary particle. If it looks like a Photon and acts like a Photon, then it is some type of Psyche. Every massless quantum is a Psyche. Psyche USES mass, energy, or matter to get things done. This is what has been experienced and observed.

Psyche has been hiding in plain sight all the time where nobody can see it nor find it because they aren't looking for it. But, there's no great mystery there. Psyche is seen as a Photon BOTH at the quantum level and the physical level. Psyche or Intelligence IS what it appears to be, a Photon of Light or a Pinpoint of Light. No surprise there. Psyche is what it has been experienced and observed to be – a massless quantum of energy that looks like and acts like a Photon. Psyche, whenever it has been seen both at the quantum level and the physical level, looks like a Photon or a Pinpoint of Light.

Now, let's talk about time.

The passage of time is an emergent property of mass; and, mass or the slowing down of the particles is produced by the Higgs Fields. Entropy is an emergent property of the passage of time. Entropy is NOT one of the fundamental particles, nor is entropy a fundamental field, fundamental force, or fundamental law. Entropy is an emergent property of time, which in turn is an emergent property of mass, which in turn is an emergent property of the Higgs Field and other Quantum Fields. Mass itself is an emergent property of various different Quantum Fields. Mass is not a fundamental particle nor an elementary field either. Entropy, time, and mass are emergent properties of the different Quantum Fields. The speed-of-light from our perspective is quantum tunneling from the photon's perspective.

Both entropy and the passage of time are NOT essential parts of matter. Entropy, the aging process, and the passage of time can be left out anytime that God decides to do so. Temporarily remove the Higgs Field from your physical body or STOP the passage of time; and then, under the control of God's Spirit, your physical body can be teleported (or quantum tunneled) anywhere in the universe instantaneously at God's will. The Higgs Field is how God controls everything in the physical universe. It's how God slows everything down so that we can experience the passage of time and actually learn something from the experience. The Higgs Field is the braking mechanism. It slows everything down; and, God has His foot on the brakes. God can take His foot off the brakes anytime He chooses to do so because He made the brakes, or the Higgs Field and the other Quantum Fields, in the first place.

Quarks and electrons gain their mass from a type of confinement via the Higgs Field. Take away the Higgs Field, and they become massless speed-of-light particles; and then from their perspective without any mass, they experience NO passage of time and essentially quantum tunnel to their destination. It looks like everything with mass is composed of a combination of intrinsically massless light-speed particles that are prevented from streaming freely through the universe (or slowed down to sub-light velocities) by quantum fields that confine those particles and give them inertial mass or a resistance to acceleration. Mass is an emergent property of all those different confining quantum fields. Time is an emergent property of mass. Remove the effects of these quantum fields, and there is NO passage of time and the particle can then quantum tunnel at will.

In the physical realm, anything with mass has to quantum tunnel in order to travel faster than the speed-of-light. Anything without mass has to travel at the speed-of-light, while it is subject to the Natural Laws that God made. What is it about mass that prevents something from reaching the speed-of-light? God designed it that way so that people or psyches, who have a physical body, could experience the passage of time and actually learn something from the experience. TIME STOPS for photons traveling at the speed-of-light. From their perspective, they quantum tunnel to their destination every time. The photons experience NO passage of time while traveling to their destination. Mass or sub-light velocities are needed to produce the FEELING of time and the passage of time.

The Higgs Field is selective, which means that it is intelligent or psychic. The Higgs Field produces mass in certain Massless Particles such as electrons and quarks. In other words, the Higgs Field slows them down so that their occupants or psyches actually experience the passage of time. From their perspective, the particles don't quantum tunnel to their destination anymore. Instead, they FEEL TIME and pass through time thanks to the Higgs Field. Another way to look at this is from the perspective of the Massless Particles or Psyches who are intelligent and psychic, and actually reach out telepathically and use the Higgs Field to slow themselves down. When they slow themselves down, they experience the passage of time and become subject to entropy. Either way, it ALL fits together perfectly and is internally consistent, which tells me that it is true. Remember, every massless elementary particle is some type of intelligent conscious sentient Psyche.

When Time Breaks Down

https://www.youtube.com/watch?v=GguAN1_JouQ

https://evolution-is-entropy.com/wp-content/uploads/2018/08/When-Time-Breaks-Down.zip

Psyche has been experienced and observed by Out-of-Body Travelers; consequently, as scientists, we are obligated to try to figure out what it is and how it works. Science is observation and experience after all, or it should be. Our job as scientists is to explain the evidence – not to explain away the evidence. The Materialists, Naturalists, Darwinists, Nihilists, and Atheists take it upon themselves to explain away the evidence especially when it comes to Psyche or Quantum Non-Local Consciousness. This is unfortunate because with Quantum Mechanics and Psyche you can literally explain everything that comes your way.

Matter FEELS mass and the passage of time because of the energy of its internal moving parts. Our perception of time depends upon how much attention we pay to it. Perception of time is a function of our Human Psyche both in the physical realm and in the quantum spiritual realm. The psyche within matter literally FEELS mass and FEELS the passage of time because the Quantum Fields slow things down so that psyche can have experiences and FEEL the passage of time.

All of the atoms in a clock FEEL the same length of a second. The flow of time depends fundamentally on motion. The faster an object moves relative to you, the slower its clock appears to tick. A particle moving at the speed-of-light experiences no time. Its clock is frozen. The most elementary of particles are intrinsically timeless and massless. Mass produces the flow of time. Remove the mass and time stops. The familiar smooth flow of time only emerges as these particles are bundled into what we think of as matter. Time happens for the atom in a way that it doesn't happen for the atom's elementary parts. Light-speed particles are timeless. Massless particles are timeless. They FEEL or experience NO passage of time. Having mass and experiencing time are fundamentally connected.

"FEEL" is the word that he uses in this video to describe a particle's perception of time. It's the Psyche or the Intelligence within each particle who actually FEELS the passage of time. Perception is a function of Psyche. It's the FEELING or the perception of the passage of time that makes everything seem REAL to us. Psyche is the thing within us that CHOOSES to pay attention to what's happening around us. Feelings, perceptions, paying attention, and making choices are ALL functions of our Human Psyche.

The memories that survive the death of our physical brain are stored someplace else besides our physical brain. Some of them are stored within our Human Psyche. In contrast, our 360 degree after-death Life Review seems to be stored someplace else besides our Human Psyche. This is what has been experienced and observed.

In this video, in typical atheistic and materialistic fashion, he deliberately limits himself to the physical brain; but, I have trained myself to take it all the way down the rabbit hole until we reach The Elementary Particle, which is Psyche or Quantum Non-Local Consciousness. I'm not afraid to go where no man has gone before.

Out-of-Body Travelers and Near-Death Experiencers have observed that it is their Human Psyche, their immaterial viewpoint in space, who is having experiences and forming memories of those experiences – NOT their spirit body and definitely NOT their physical body. Your spirit body and physical body don't have experiences and don't form memories of any event, while your Human Psyche is offline or separated from them. It's the Psyche or Non-Local Consciousness who has experiences and then forms memories of those events.

At the physical level, the infusion of mass allows the psyche within a particle to FEEL the passage of time; otherwise, without the mass, the psyche within a particle FEELS no passage of time at the physical level. The passage of time or the FEELING of time is caused by an infusion of mass or a resistance to acceleration.

The photon clock, mentioned in this episode and the next episode, is probably my most favorite thought experiment in the whole of physics due to its massive explanatory power. In my humble opinion, Science should explain things rather than trying to explain them away as the Naturalists and Atheists often try to do.

Atoms are made up of things that would move at the speed-of-light, except that they are confined or limited by Quantum Fields. Quarks and electrons are confined by their coupling with the Higgs Field, and then again by the forces and fields binding them into atoms. All of these different forces and fields produce mass which then in turn produces the passage of time or the FEELING of time.

Just like what we observe in relationship to a photon clock, within an atom, the ticking corresponds to the interactions between its component particles and fields. The interactions that hold the atom together also produce the mass and the flow of time. The rate of these interactions, this ticking, represents the rate at which the atom will change from one arrangement or one state to the next. The confinement of light-speed particles gives matter mass. In fact, this confinement, this bundling of energetic moving parts, is what makes it physical matter. This same bundling of light-speed particles also gives matter time, or subjects physical matter to the passage of time. Atoms FEEL time in their internal evolution or internal revolutions.

Time is an emergent property of mass, and mass is produced by the Higgs Field and other Quantum Fields. The different Quantum Fields, including the Higgs Field, were designed, organized, and made by God. Psyches use mass to slow down.

The Origin of Matter and Time

https://www.youtube.com/watch?v=fHRqibyNMpw

https://scientific-proof-of-god.com/wp-content/uploads/2018/08/The-Origin-of-Matter-and-Time.zip

Real matter is comprised of massless light-speed components confined by interactions with other particles and force fields. Electrons and quarks are composites of massless particles confined or slowed down by the Higgs interaction. Clock ticks are interactions between the internal parts of our atoms, electrons, and quarks. Only the ensemble, the complete atom, can travel slower than the speed-of-light or be still. Its most elementary parts can't do that. They have to travel at light speed. Time, mass, and matter are emergent properties of the causal propagation of patterns of interactions between timeless massless parts. Most of the mass in an atom comes from the gluons that are binding the quarks together into protons and neutrons.

Mass and time and entropy are NOT the concrete things that we imagine them to be. Physical matter is NOT the concrete thing that we imagine it to be. It's all MADE from energy by Someone Psyche. Matter and time and entropy were obviously MADE. There is order and organization there – restrictions and limitations – which are obviously intentional. All of the different quantum fields and quantum particles were deliberately MADE and intended to function that way. Quantum Fields were also designed and intended to interact with each other in certain ways. It's ALL orderly rather than chaotic. The structure is obvious.

Things with mass experience the passage of time. As the result of mass, the psyche or intelligence within the massive particles experiences the passage of time. Remove the mass, and the particles have to travel at the speed-of-light, which means that from their perspective – from the perspective of the psyche within them – TIME STOPS, and they experience NO passage of time. From their perspective, massless particles like photons simply quantum tunnel to their destination.

Quantum tunneling is real. I experienced quantum tunneling when God teleported me and the car I was driving to safety. TIME STOPPED. The clocks froze. I FELT the whole of eternity fold into me. I could sense everything around me. Then instantly I was someplace completely different with the car stopped and the car turned off. I was like the photon. Time stopped, and from my perspective, I simply quantum tunneled to my destination instantaneously. What just happened to me? Did I do that? Can I do it again?

Those were some of the questions that went through my mind at the time. I didn't have a scientific explanation for it. It seemed quite miraculous to me. Of course, it wasn't anything that I did. I can't replicate it on demand. It was something that God did for me in order to save me and a bunch of other people.

I experienced it, so I KNOW that quantum tunneling physical matter is possible at the physical level. You just have to stop the flow of time or temporarily remove the effects of the Higgs Field. It can be done, even at the physical level. Quantum tunneling has been experienced and observed. From their perspective, Photons quantum tunnel to their destination, even though from our perspective they travel at the speed-of-light.

This video is cool. In this video, he explores the space-time diagram and world lines which give us a good feel for how time and space work. He demonstrates that causality is maintained or conserved at the quantum level even within an atom. If you allow yourself to do so, when it comes to mass and time, you can sense that this whole thing was deliberately designed and made. Anything that was obviously made obviously has a Maker who designed it and made it.

Quantum Field Theory

There are NO particles. There are only fields. The universe is made of quanta or packets of energy vibrating in quantum fields. Particles are epiphenomena arising from quantum fields. There is NO physical matter. Physical matter or particles are incompatible with quantum field theories. Particles or physical matter contradict relativity and quantum physics. Electrons, photons, and all the rest are excitations or waves in the fundamental fields. Quantum fields extend infinitely throughout the whole of space. Quanta are extended disturbances in space-filling fields. Space and the quantum vacuum contradict the particles view while confirming the field view. The Unruh effect and quantum nonlocality imply a field view. Many lines of reasoning contradict and falsify the "physical matter view" and confirm the "quantum fields view". Phenomena and theory lead to paradoxes if interpreted in terms of particles or physical matter, but they are comprehensible in terms of fields. Switching to an all-fields perspective solves the scientific problems caused by and created by Materialism, Naturalism, Darwinism, Nihilism, and Atheism. Psyche or Intelligence is a quantum, an energy packet, or a wave within the Psyche Field or the Thoughts Field. Thoughts are transmitted by Quantum Waves; and, memories are stored as Quantum Waves. Quantum Field Theory FALSIFIES Materialism, Physicalism, Naturalism, and Nihilism. When those go down, Darwinism, Behaviorism, Determinism, Physical Reductionism, and Atheism go down with them. The false is falsified by the truth; and, the truth is repeatedly experienced and observed.

https://arxiv.org/ftp/arxiv/papers/1204/1204.4616.pdf

http://quantum-law-of-thermodynamics.com/wp-content/uploads/2018/09/There-Are-No-Particles.pdf

https://pdfs.semanticscholar.org/c69f/09bcacd06bd41c72ef599a5f3b5adb8a524e.pdf

http://quantum-law-of-thermodynamics.com/wp-content/uploads/2018/09/There-Are-No-Particles-Only-Fields.pdf

Hobson, A. (2017). *Tales of the Quantum: Understanding Physics' Most Fundamental Theory*. New York: Oxford University Press.

I can see the Hand of God in the Quantum Fields and Quantum Field Theory. In fact, God seems to have a different finger and toe on each one of the Quantum Fields. Of course, it didn't used to be that way for me when I was a Materialist, Naturalist, Nihilist, and Atheist. I KNOW how it goes. I was completely oblivious to these things, didn't think about them, wasn't looking for them, didn't want them, and didn't know anything about them when I was in an atheistic frame of mind. Now I see things differently. In fact, for me personally, Quantum Fields and Quantum Field Theory are extremely convincing and undeniable Scientific Proofs of God's Existence.

Quantum Mechanics is Supernatural Mechanics or Spiritual Mechanics; and, I treat it as if it is real and truly exists because it has actually been experienced and observed.

I discuss Quantum Fields in other essays, so I just want to emphasize here that each Quantum Field was uniquely designed and made to fulfill a specific purpose. Quantum Fields are omnipresent and fill the immensity of space much like God's Influence. Quantum Fields were organized and made with a specific purpose in mind. Quantum Fields are the building blocks of our physical reality; and, they are invisible, intangible, non-physical, immaterial forces and fields. Their very existence FALSIFIES Materialism, Naturalism, Darwinism, Nihilism, and even Atheism which claim that invisible, intangible, non-physical forces and fields like Psyche and God's Psyche DO NOT EXIST.

Remember, Someone Psyche was NEEDED to design and make the Quantum Fields. Quantum Fields can't be made by entropic physical matter! In fact, Quantum Fields have to be designed and made BEFORE the Gods can then make entropic physical matter. The Quantum Fields are used by the Gods to produce and then sustain physical matter. That's just the way it is. It's obvious. First things first! Anything that is obviously made obviously has a Maker who made it. Quantum Fields were obviously designed and made. Therefore, Quantum Fields obviously have a Maker who designed them and made them. It's obvious.

PBS Space Time Quantum Field Theory Playlist

I'm a scientist; and, I had never heard about Quantum Fields until the summer of 2018. Why hadn't I ever heard of Quantum Fields? It's because we Materialists, Naturalists, Nihilists, and Atheists erroneously teach and believe that space is completely empty. We erroneously teach and believe that an atom is 99.999% empty space. We erroneously teach and believe that entropic physical matter is the ONLY thing that exists. Consequently, I wasn't looking for Quantum Fields because I didn't believe or know that such things exist. After fifty years of studying and admiring science, My Materialism, My Naturalism, My Nihilism, My Scientism, and My Atheism are still very much with me at my very core, even though I now know that they are false and have been falsified.

I wasn't looking for Quantum Fields because the Naturalists and Atheists had successfully hidden them from me. Why would they do such a thing? It's because immaterial, intangible, non-physical, spiritual, supernatural, non-local, omnipresent Quantum Fields FALSIFY Materialism, Naturalism, Darwinism, Nihilism, Behaviorism, Determinism, and Atheism which claim that psyche and the non-physical DO NOT EXIST. I had chosen to believe that things like Quantum Fields cannot exist and do not exist. I had chosen to believe that there could NEVER be any scientific proof of God's existence. I was wrong. My Psyche chose wrong; and, I got to suffer the consequences. I'm still suffering the consequences.

The Human Psyche can break its conditioning and brainwashing at will; but, it first has to acquire the desire or the will. It took me awhile to do so.

Each Psyche is identified by the CHOICES that it makes. Beliefs are chosen into existence by the Human Psyche, not our genes. Our genes have nothing to do with our chosen beliefs. Genes don't have beliefs, and neither does evolution or natural selection. Your genes aren't going to pass an IQ test either. Since our Psyche or Intelligence cannot be destroyed, our chosen beliefs, personality, and knowledge go forward with us into the next life after our physical brain and genes are dead and gone. Psyche is a quantum particle, The Elementary Particle; and, Psyche has its own unique Quantum Field that it uses to get things done.

Once I learned that Psyche has been repeatedly experienced and observed existing and functioning separately from its assigned spirit body, then I KNEW that we scientists have an obligation to define it and to try to figure out what it is and how it works. Psyche is the innate intelligence within all the different forms of energy which gives that energy the inherent ability to understand, follow, and obey God's Laws and God's Commands. Energy is psychic and telepathic. Telepathy or thought is WiFi at the quantum level and the psyche level. Psyche is the fundamental force of the multiverse within every dimension of reality and existence. Psyche is the Ultimate Cause and the ultimate causal agent.

Psyche controls and uses mass to get things done. Psyche or Life Force cannot be made, and it cannot be destroyed. Energy, and the Psyche within it, is syntropic or conserved. Psyche cannot die; consequently, Heat Death is impossible at the quantum level and the psyche level. Something Psyche or Something Intelligent within the energy KNOWS how to form into a specific type of Quantum Field and then KNOWS how to function as that type of Quantum Field on a universal scale; and/or, Someone Psyche KNOWS how to use Raw Energy to make and operate Quantum Fields on a universal scale at the quantum level. This reality explains why there is something rather than nothing. Once you choose to allow Psyche and Quantum Mechanics in to play, you can literally explain everything that comes your way. This is what has been experienced and observed.

The Order and Organization Is Obvious

I can see Order, Organization, Structure, LAW, and rules being obeyed ALL throughout this. When it comes to entropic physical matter and quantum fields, it is impossible for me to deny that there is Order and Organization throughout the whole of it. The Order and Organization is OBVIOUS! Entropic physical matter and quantum fields were obviously organized and made. Furthermore, Someone Psyche within the energy is CHOOSING to obey the associated LAWS that have been established by God for the production and maintenance of entropic physical matter and quantum fields. IT IS OBVIOUS.

Scientific Observation: Anything that is obviously organized obviously has an Organizer who organized it.

Scientific Observation: Entropic physical matter and quantum fields were obviously designed, structured, organized, made, and caused to begin.

Scientific Conclusion: Therefore, it is logical and rational to conclude that quantum fields, quarks, electrons, bosons, gluons, and entropic physical matter have an Organizer who designed them, organized them, made them, and deployed them. This is what we experience and observe.

$E = mc^2$ is LAWFUL. Wherever we encounter LAWS that are being followed and obeyed, we automatically encounter a Law Giver who designed and made those LAWS, a Law Enforcer who is teaching the LAWS and/or enforcing the LAWS, and Someone Psyche who is choosing to follow and obey those LAWS. Obedience to Law is deliberately chosen into existence by Someone Psyche or Someone Intelligent. IT IS OBVIOUS. This is Logic 101. First things first. I'm not going to apologize for figuring this stuff out. I'm a scientist. That's what we are supposed to do. We are supposed to explain the observational evidence rather than trying to explain away the observational evidence as the Naturalists, Darwinists, Nihilists, and Atheists typically try to do.

Atomic Orbitals Were Organized and Made

https://en.wikipedia.org/wiki/Atomic_orbital

https://syntropy.site/wp-content/uploads/2018/08/Atomic-Orbital.pdf

https://en.wikipedia.org/wiki/Electron

https://syntropy.site/wp-content/uploads/2018/08/Electron.pdf

https://en.wikipedia.org/wiki/Energy

https://syntropy.site/wp-content/uploads/2018/08/Energy.pdf

Can you sense any order, organization, structure, or law within the Atomic Orbitals or the Electron Orbitals? Why do the electron orbitals take on their characteristic, lawful, and predictable patterns and shapes rather than being totally random and chaotic like the Materialists, Darwinists, Naturalists, and Atheists say that everything should be? What's making the electrons do that? Where's the map, rules, and laws for electron orbitals being stored? How is this information or knowledge being accessed and used by the electrons?

It's as if the electrons are conscious, sentient, aware, purposeful, deliberate, and alive. They act differently depending upon where they are in the world or what assignment they have been given to perform. The Atomic Orbitals give structure, size, order, organization, law, and shape to atoms. Do they not? Physical atoms are comprised of different forms of energy.

Matter Is Made from Different Forms of Energy

Entropic physical matter is made from different FORMS of energy. The underlying energy is eternal and everlasting; but, the FORM of that energy is not conserved. The FORM is constantly changing. Remember, physical matter, the associated physical laws, and entropy are made from different FORMS of energy. Entropy is an emergent property of entropic physical matter; and, entropic physical matter was made from many different FORMS of energy. The FORM is never conserved. The FORM can be changed. The different "particles" or quanta are made from different waves of energy, or different forms of energy.

The following is the Science that I pulled from the internet. I eliminated the falsehoods and the falsified as I went along. It gives a much better picture of truth and reality than what we can get from the Materialists, Naturalists, and Atheists.

Most of an atom's mass is comprised of the kinetic energy and binding energy of the quarks, NOT the quarks themselves; but, even quarks are made from energy. It's ALL made from different FORMS of energy. Mass is just bound or confined energy. Mass is bound energy, frozen energy, confined energy, or condensed energy. Energy, and the psyche or intelligence within that energy, is the fundamental unit of reality. ALL energy is psychic, intelligent, conscious, alive, sentient, perceptive, and aware.

Quantum Field Theory defines "elementary particles" as excitations in the quantum fields that fill our entire universe. An electron is an excitation or a quantum wave in the electron field. They are NOT physical particles after all. They are quantum waves, excitations in fields, or quantum wave packets of energy called quanta. It's ALL energy. It's all comprised of different FORMS of energy. Quanta particlize when they reach a barrier or their destination. They are quantum waves of energy during their journey. Nature's Psyche within each quantum is all that's needed to collapse the wave function when it's time to do so.

Even in a vacuum, the electron field is there. Even in a planet or a star, the electron field is there. Fields are omnipresent. Add some energy to that field at a particular spot; and, it vibrates producing a quantum or a "particle". The vibration is your "particle", or the electron in the case of an electron field. "Particles" are energy vibrations or quantum waves of energy. Every "elementary particle" is a vibration in its own unique quantum field; and, these vibrations and fields interact with each other transferring energy, momentum, and charge between the different "particles" and fields. This explains how the quantum, the spiritual, or the sub-atomic really works. It's all comprised of intangible, invisible, non-physical waves, forces, fields, and energy.

In physics, a force is any interaction that, when unopposed, will change the motion of an object. A force is comprised of energy. A force is just one form of energy. There are many others.

A field is something that is present everywhere in space and time, and it can have waves in it. "Particles" are made from waves of energy rippling in a quantum field. "Particles" are quantum waves of energy within a quantum field of energy. All the different fields permeate the whole of space and time. They are highly organized energy. The very existence of these different invisible, intangible, non-physical forces and fields falsifies the claims of Materialism, Naturalism, and Atheism which claim that these invisible, intangible, non-physical forces and fields do not exist. Psyche is a force and a field within energy itself. For all we know, Psyche IS Energy, and Energy IS Psyche. ALL the different FORMS of energy are psychic, intelligent, conscious, sentient, perceptive, alive, and aware. Psyche is the intelligence within all the different forms of energy.

There are many different types of forces and fields; and, each one of them is made from energy. Each type of field was made with a different purpose, goal, or functionality in mind. In quantum field theory, what we perceive as "particles" are excitations of the quantum field itself. Each quantum field was made to function in a specific manner and to perform a specific function. The quantum fields co-exist with each other and interact with each other. They were obviously made.

The electric field is a part of nature that is found everywhere. At any given point in space, and at any particular time, you can measure it. If it's non-zero on average in some region, it can have physical effects, such as making your hair stand on end or causing a spark. The electric field can also have waves, in which the size of the field repeatedly becomes larger and smaller — visible light is such a wave, as are X-rays and radio waves, and all the other things we collectively call "electromagnetic waves". Magnets are invisible, intangible, non-physical waves of energy. Waves are made from energy, and so are quantum fields. They are all different FORMS of energy.

What is a particle? A "particle" is a quantum wave rippling in a quantum field. It's ALL energy. It's highly organized energy. A quantum field is comprised of organized energy; and, a quantum wave is also comprised of organized energy. They are all different forms of energy. A quantum or a "particle" is a quantum wave of energy or a packet of energy. Each massless quantum particle is some type of Psyche or Intelligence. Psyche or Intelligence uses matter, mass, and energy to get things done. This is what has been experienced and observed.

Photons are light waves. Einstein called these energy packets "photons", and these are now recognized as a fundamental particle. Quanta or "particles" are energy packets. Quanta are waves of energy in a quantum field. Mass-based quanta are caused or made or organized. Mass-based quanta have a beginning which means that they have someone or something that causes them to begin. In contrast, massless quanta or photons are some type of Psyche or Intelligence.

The least-intense possible wave that a quantum field can have is called a "quantum" or a "particle". A "particle" or "wave packet of energy" behaves in accordance with your typical idea of a "physical particle", functioning as a unit, moving in a straight line and bouncing indivisibly off of things, which is why we call it a "particle". A quantum or "particle" has a minimum energy intensity and therefore functions as a unit even though it is in truth comprised of quantum waves of energy propagating through a quantum field of energy.

A quantum is a packet of energy – it functions as a unit. A quantum is also a quantum wave rippling in a quantum field. A quantum is all-or-nothing. It functions as a unit or a whole even though it is comprised of energy waves within a quantum field. The wave is either there, or it isn't. A quantum wave functions as if it were a packet of energy or a "particle" of energy. Atoms, with their neutrons, protons, and electrons, are not particles at all but pure waves of energy. A "particle" is a quantum wave of energy propagating through a quantum field of energy. Every massless elementary "particle" is some type of Psyche.

In the case of the electromagnetic field, its particles are called "photons"; they represent the dimmest possible flash. Your eye can absorb light one photon at a time. Quanta or quantum waves function as a unit or a whole even though they are comprised of waves of energy rippling within a quantum field of energy. It's ALL comprised of energy.

Higgs bosons are ripples or waves of the Higgs Field. By finding the Higgs boson, we have evidence that the field itself exists – like seeing water waves and concluding there must be an ocean beneath. The Higgs Field is extremely important in particle physics. It's no exaggeration to say that it makes your existence possible. Without it, atoms would not exist; electrons would zip away from nuclei at the speed-of-light. The value of the Higgs Field determines what kinds of nuclei are stable and some of the differences between matter and antimatter. Almost all "particles" or quanta that we know of are affected in some way by the Higgs Field. The Higgs Field provides structure, stability, order, and organization to all the other fields. The Higgs Field shepherds the other fields. God is in the Higgs Field.

Organized fields of energy are made. Organized waves of energy or organized packets of energy are made or caused to begin. Quantum fields and quantum waves are made. Quanta or "particles" are made. Anything that was obviously made obviously has a Maker who designed it, made it, and caused it to begin.

A field is something that is present everywhere in space and time. The different fields are omnipresent. Fields are organized energy. They each have their own specific purpose and function. Quanta or "particles" are packets of energy or organized waves of quantum energy that are rippling through quantum fields of energy. Every massless elementary quantum is some sort of Psyche.

The Higgs Field (unlike most of the elementary fields of nature) has a non-zero average value throughout the entire universe. Because it does, many particles have mass, including the electron, quarks, and the W and Z boson particles of the weak interactions. The Higgs Field slows particles down to sub-light speed velocities producing entropy and the flow of time as emergent properties of those particles. If the Higgs Field's average value were zero, those particles would be massless or very light. That would be a disaster; atoms and atomic nuclei would disintegrate. Nothing like human beings, or the earth we live on, could exist without the Higgs Field having a non-zero average value. Our lives truly depend upon it. God made the Higgs Field have a non-zero value.

Quantum fields contain excited states that we observe as particles. These quantum fields can be divided into matter fields (whose particles are electrons, quarks, etc.) and force fields (whose particles are photons, gluons, etc.). The Higgs mechanism gives mass not only to weak particles, but also to electrons, quarks, and other fundamental particles. The more strongly a particle interacts with the Higgs Field, the more massive it is. It's important to note, however, that most of the mass in composite particles, like protons, nuclei, and atoms, does not come from the Higgs mechanism, but from the binding energy that holds these particles together.

Can you sense any order, organization, planning, intention, structure, purpose, or meaning in any of this? Or are you like the Materialists, Naturalists, Darwinists, Nihilists, and Atheists and simply believe that it all came about by random chance and blind luck? Anything that was obviously organized obviously has an Organizer who organized it. This is Logic 101.

The Materialists, Naturalists, Darwinists, Nihilists, Behaviorists, Determinists, Physical Reductionists, and Atheists are OBVIOUSLY WRONG whenever they claim that entropic physical matter is the only thing that exists, and that entropic physical matter is the fundamental unit of reality. The whole of Science and Quantum Field Theory proves that they are wrong.

Exergy is the energy that is available to be used. After the system and surroundings reach equilibrium, the exergy is zero. Determining exergy was the first goal of thermodynamics. Energy is neither created nor destroyed during a process. Energy changes from one form to another. In contrast, exergy is always destroyed when a process is irreversible, for example loss of heat to the environment. This destruction is proportional to the entropy increase of the system together with its surroundings. The destroyed exergy has been called anergy. Anergy is unavailable energy. For an isothermal process, exergy and energy are interchangeable terms, and there is no anergy in an isothermal system. In an isothermal system, ALL of the energy is available all the time. There is no anergy and no entropy in an isothermal system.

Energy is a conserved quantity; the law of conservation of energy states that energy can be converted in form, but not created nor destroyed. That would make energy, or the psyche within that energy, the fundamental unit of reality. Remember, any massless quantum particle is a Psyche. A Photon is a Psyche. The massless components of quarks and electrons are Psyches.

The above was adapted from Uncle Google. See in particular:

https://en.wikipedia.org/wiki/Energy

https://en.wikipedia.org/wiki/Quantum_field_theory

https://en.wikipedia.org/wiki/Exergy

https://profmattstrassler.com/articles-and-posts/the-higgs-particle/the-higgs-faq-2-0/

https://wikis.utexas.edu/display/utatlas/Higgs+boson+FAQ

https://www.youtube.com/watch?v=kixAljyfdqU

https://www.youtube.com/playlist?list=PLsPUh22kYmNBpDZPejCHGzxyfgitj26w9

Particles or quanta are constructs. They are instantiated or caused to begin. They are made. Technically, physical matter does not exist. It's an illusion. Physical matter is simply quantum waves rippling through quantum fields. It's all made from energy. It is the underlying energy that truly exists and is timeless and eternal. Physical matter is made from different FORMS of energy, which means that entropic physical matter can be disassembled and formed into something else instead. There's nothing eternal or everlasting about entropic physical matter. It was designed to be temporary and limited.

Psyche is the innate intelligence within all the different FORMS of energy which gives that energy the inherent ability to understand, follow, and obey God's Laws and God's Commands. Energy, and the psyche or intelligence within it, is eternal and everlasting. It cannot be made, and it cannot be destroyed. Energy of any kind is syntropic, sentient, conscious, perceptive, intelligent, psychic, alive, and aware. Energy, and the psyche within it, is always conserved.

We have the things that were made – quanta, particles, forces, and fields; and, we have the things that cannot be made and cannot be destroyed – energy, psyche, intelligence, and life force. Energy isn't made, it's TRANSFORMED; whereas, everything else is MADE from energy by Psyche. Energy or Psyche is the fundamental unit of reality – not entropic physical matter. Heat Death is impossible in the Quantum Realm, Psyche Realm, or Syntropy Realm because entropy is NOT conserved and entropy or anergy does NOT exist in the Spirit World. At the psyche level or the quantum level, ALL of the energy is available for use all of the time; consequently, death is impossible at the psyche level within the Quantum Realm. Your Psyche cannot die. Energy or Life Force cannot be created, and it cannot be destroyed. Energy, Psyche, Intelligence, or Life Force cannot die. Energy, Psyche, or Life Force is always conserved.

It all depends upon the interpretation that you choose to give to Quantum Mechanics. Certain interpretations were designed to hide the truth from us; and, other interpretations were designed to reveal the truth to us. The false is falsified by the truth; and, the truth is repeatedly experienced and observed. If you successfully eliminate everything that is false, then only the truth will remain.

Defining the Quantum Law of Thermodynamics

Exergy is the energy that is available to be used. After the system and surroundings reach equilibrium, the exergy is zero. Determining exergy was the first goal of thermodynamics. Energy is neither created nor destroyed during a process. Energy changes from one form to another. In contrast, exergy is always destroyed when a process is irreversible, for example loss of heat to the environment. This destruction is proportional to the entropy increase of the system together with its surroundings. The destroyed exergy has been called anergy. For an isothermal process, exergy and energy are interchangeable terms, and there is no anergy.

The Quantum Law of Thermodynamics states that heat death is impossible at the quantum level or the psyche level because entropy, anergy, or the second law of thermodynamics doesn't exist in the Psyche Realm or the Syntropic Realm. This is what has been experienced and observed. For an isothermal system, exergy and energy are interchangeable terms, and there is no anergy or no entropy. The Quantum Realm or Psyche Realm is an Isothermal System, which means that there is NO anergy or NO entropy within the Quantum Realm. The second law of thermodynamics doesn't exist in the Syntropic Realm. In an Isothermal System, ALL of the energy is available for use all of the time. The Quantum Law of Thermodynamics is Syntropy or Pure Exergy.

Syntropy is ZERO entropy. Syntropy is also Eternal Life or INFINITE LIFE. We all understand the mathematics behind ZERO and INFINITY; therefore, we KNOW that the math supporting the Quantum Law of Thermodynamics and Syntropy has already been discovered and observed. It's just sitting there waiting for us to find it and define it.

The Quantum Law of Thermodynamics or Syntropy states that death is impossible at the psyche level. Your psyche or intelligence has always existed and will always exist. It cannot be destroyed. It cannot die. Psyche is the innate intelligence within all the different forms of energy which gives that energy the inherent ability to understand, follow, and obey God's Laws and God's Commands; and, Energy is always conserved on both sides of the veil – both on the spiritual side and the physical side. Syntropy is the Conservation of Psyche, or the Conservation of Energy. This reality and truth explains everything that has ever been experienced and observed. The Quantum Law of Thermodynamics or Syntropy is the answer to life, the universe, and everything.

The Big Bang Singularity, if it existed, was an Isothermal System comprised of Pure Syntropy and made by the Gods. Only a God could make such a thing. By definition, in principle, there was NO entropy or NO anergy within the Big Bang Singularity. It was Pure Syntropy or Pure Exergy. The Quantum Law of Thermodynamics makes the Big Bang Singularity a realistic possibility because there is NO heat death and NO entropy at the quantum level or the psyche level. At the quantum level, ALL of the energy is available for use all of the time.

Do you see how that works? It's the answer to life, the universe, and everything. It explains everything that has ever been experienced or observed. It explains why there is something rather than nothing. It's all there hiding in plain sight where the Materialists, Nihilists, Atheists and Naturalists cannot see it nor find it because they refuse to go looking for it and don't want to find it. They prefer heat death, the end of life, and the end of our universe instead.

Rather than trying to explain away the evidence or destroy the evidence as the Naturalists and Atheists do, our job as scientists is to identify and explain everything that has ever been experienced and observed. I believe that I have now done so at the most fundamental level. Good enough! I've finally done my job.

Entropy and Time Were Made

Entropy and time were made, or more accurately, caused to begin. Entropy or the Second Law of Thermodynamics doesn't exist at the quantum level and the psyche level. Time is a construct. Entropy is a function of time, which means that entropy is a construct. Entropic physical matter is constructed from energy which means that it can be deconstructed and turned back into raw energy or available energy. Entropy is death. The way you convert entropic physical matter into syntropic physical matter is to remove the entropy or the death from the entropic physical matter. Resurrection from death has been experienced and observed. The way that resurrection works is that the entropy or the death is removed from the entropic physical matter; and thereby, that physical matter is converted into syntropic physical matter or immortal physical matter instead.

The Materialists, Naturalists, Darwinists, Nihilists, and Atheists will assure you that everything I have written here is wrong. They will tell you that I don't understand science. They define science as Materialism, Naturalism, Darwinism, Nihilism, Scientism, and Atheism. These people have told me many times that I don't understand science. They will tell you that there is nothing fundamental about energy, and that psyche does not exist. It's all over the internet. This is what these people teach and believe. These people will tell you that energy and consciousness is an emergent property of something much more fundamental. What's more fundamental than energy and consciousness? According to these people, physical matter is the fundamental unit of reality; and, energy or consciousness is just an emergent property of entropic physical matter. I think I understand their science quite well, don't you think? I KNOW what they teach and believe. I used to be a Materialist, Naturalist, Nihilist, and Atheist after all.

I eventually discovered that the Materialists, Naturalists, Darwinists, Nihilists, and Atheists are almost always wrong, especially when it comes to the fundamentals. The observations and experiences of the human race as a whole eventually convinced me that I was wrong. Nowadays, I define Science as observation and experience; and, I have observed that the observations and experiences of the human race FALSIFY Materialism, Naturalism, Darwinism, Nihilism, Atheism, Classical Physics, and their derivatives.

The fact that Materialism, Naturalism, Darwinism, Nihilism, Atheism, and their derivatives FALSIFY my current theories and ideas tells me that I'm finally on the right track and have finally found the truth. These people have a knack for finding the truth and then formally rejecting it. One of my greatest scientific discoveries came when I first realized that the truth is invariably the opposite of what the Materialists, Naturalists, and Atheists have chosen to believe. These people reject everything that has been experienced and observed by Someone Psyche, and then they preach the opposite instead as if it were true. The fact that Materialism and Naturalism FALSIFY my current theories and ideas is positive proof that I have finally found the truth.

The fatal flaw of the Materialists, Naturalists, Darwinists, Nihilists, and Atheists is that they teach and truly believe that ONLY entropic physical matter exists, which means that these people erroneously teach and believe that physical matter and entropy are conserved. They are wrong. These people deliberately misinterpret the First Law of Thermodynamics, which states that Energy or Syntropy is conserved; and, these people get the rest of us to misunderstand and misinterpret the First Law of Thermodynamics by successfully convincing us that only entropic physical matter exists and that consequently physical matter and entropy are being conserved. Since these people believe that ONLY entropic physical matter exists, these people literally and erroneously define the First Law of Thermodynamics as the "Conservation of Physical Matter" and the Second Law of Thermodynamics as the "Conservation of Entropy". Entropy is death; and, these people literally but erroneously teach that entropy or death is conserved. That's what they teach, is it not? We are going to have to rethink this thing if we are going to make it match with what has been experienced and observed by Out-of-Body Travelers.

One of my greatest scientific discoveries came when I first realized that physical matter and entropy are NOT conserved. The amount of physical matter, entropy, and space-time changes over time as God sees fit. Space, time, locality, entropy, spirit matter, and physical matter are different FORMS of energy. The FORM is never conserved. The FORM is constantly changing. Its amount or quantity is constantly changing. **It is the underlying Energy, Psyche, Life Force, or Syntropy that's being conserved according to the First Law of Thermodynamics and E = mc² —** NOT the physical matter and NOT the entropy. Physical matter and entropy are NOT conserved. This is the Ultimate Law of Thermodynamics.

The Ultimate Law of Thermodynamics differentiates between what is being conserved and what is not conserved. This scientific discovery is the answer to life, the universe, and everything. It explains everything that has ever been experienced and observed. Consequently, the greatest scientific discovery in applied science that fallen, mortal, physical beings like us could ever make would be to find a way to remove entropy from entropic physical matter and make all of that matter syntropic instead. That would indeed result in the answer to life, the universe, and everything. Would it not?

Remember, physical matter, entropy, locality, and space-time are NOT conserved. Their amount is constantly changing. Only the underlying Energy or Psyche is conserved.

Psyche is the innate intelligence within all the different forms of energy which gives that energy the inherent ability to understand, follow, and obey God's Laws and God's Commands. In contrast, entropic physical matter is made from different FORMS of energy. The FORM is never conserved. Energy's FORM can always be transformed into something else. Energy, and the psyche or the intelligence within that energy, is much smaller and more fundamental than physical matter. Physical matter is comprised of different forms of energy. This is the Ultimate Law of Thermodynamics. This is what has been experienced and observed. Physical matter and entropy are NOT conserved.

Physical matter, entropy, locality, and space-time are MADE, which means that they can be unmade or disassembled. With God's help or permission, we humans make brand new physical matter and entropy in our particle accelerators. We humans also destroy physical matter or unmake physical matter in our atomic bombs. The physical matter and the associated entropy are NOT conserved. Their quantity is constantly changing. Particles of physical matter are popping in-and-out of "existence" all the time. They don't literally cease to exist as the Materialists and Naturalists claim. Instead, they change FORM and switch dimensions; but, their underlying energy is always conserved. Electrons phase-shift and quantum tunnel both of which are quantum mechanical processes that have been experienced and observed. They don't cease to exist while they are phase-shifted into a different dimension. Their energy is always conserved.

There's NO such thing as Creation Ex Nihilo. The First Law of Thermodynamics and the Ultimate Law of Thermodynamics FALSIFY Creation Ex Nihilo.

Remember, Energy is constantly changing FORM while at the same time being conserved. Both syntropic physical matter and entropic physical matter are made from different FORMS of energy. That's the way that science really works contrary to what the Materialists, Naturalists, Darwinists, and Atheists claim to be true. The false is falsified by the truth; and, the truth is repeatedly experienced and observed. Energy or psyche is always conserved.

Physical matter, physical laws, physical constants, physical restrictions, space-time, space, locality, time, and entropy were MADE from energy which means that they have some kind of Maker. It also means that they can be disassembled or destroyed. They were made from Energy by God, which means that God can cause them to end or unmake them. They are different FORMS of Energy, and the FORM is never conserved. Only the underlying Energy, Psyche, or Syntropy is conserved. Entropic physical matter is comprised of different FORMS of Energy, which means that physical matter and entropy are never conserved. The FORM is never conserved. This is the Ultimate Law of Thermodynamics.

My purpose in life is to identify everything that is false, eliminate it from science, and replace it with the truth. The materialistic and naturalistic idea that physical matter and entropy are conserved is a false idea and a falsified idea. I have eliminated it from my science and replaced it with the truth – namely, that it's the underlying Energy, Psyche, or Syntropy that's actually being conserved and NOT the physical matter NOR the entropy. All I want is the truth because everything else will mess you up in the end. It's the underlying Energy or Syntropy that ends up being eternal and everlasting, not the physical matter nor the entropy. Remember, entropic physical matter is comprised of different FORMS of Energy, and the FORM is never conserved. Energy is Syntropy, and Energy is conserved. Syntropy is the First Law of Thermodynamics; and, it tells us that physical matter and entropy are NOT conserved.

Observation and experience have proven to us that the Ultimate Law of Thermodynamics is true. Physical matter, entropy, locality, time, space, and space-time are MADE, which means that they can also cease to exist anytime that God decides to put a stop to them. Only the underlying Psyche or Energy is conserved. It cannot be made, and it cannot be destroyed, which means that Energy or Psyche is conserved.

According to the Theory of Relativity, when a photon or tachyon travels at the speed-of-light or faster, TIME STOPS. Entropy is a function of time. Entropy measures the passage of time. Entropy is an aging process. Locality is associated with space. Entropy is associated with time. Entropy is the passage of time. According to the Theory of Relativity, when a photon or tachyon travels at the speed-of-light or faster, ENTROPY STOPS, or ENTROPY CEASES TO EXIST. From our entropic perspective, it took that photon 13.4 billion years to reach us. From the photon's syntropic perspective at the speed-of-light, it experienced NO passage of time; and, it arrived the very moment it launched. In other words, the photon and the tachyon didn't age. They didn't experience the passage of time. Entropy ceased to exist while they were traveling at the speed-of-light or faster. Entropy was NOT conserved. These objects literally quantum tunneled to their destination from their perspective. Remember, entropy is variable; and, entropy or the passage of time ceases to exist at the speed-of-light or faster. In other words, entropy or the passage of time is not conserved at the speed-of-light or faster. Instead, entropy ceases to exist; and, everything becomes syntropic instead at velocities and frequencies faster than the speed-of-light.

Once again, the Materialists, Naturalists, Darwinists, Nihilists, and Atheists erroneously teach that nothing can travel faster than the speed-of-light because these people erroneously teach that only physical matter and entropy exist. These people are right in that physical matter or entropic matter cannot travel faster than the speed-of-light; however, these people are wrong in that spirit matter, quantum matter, syntropic matter, spiritual matter, tachyons, and psyche can and do travel faster than the speed-of-light; and, they can also quantum tunnel at will. At velocities equal to or greater than the speed-of-light, TIME STOPS; and, entropy or the passage of time ceases to exist. There is no entropy in the Syntropic Realm. Objects existing at frequencies faster than the speed-of-light experience no entropy because they experience NO passage of time. Entropy ceases to exist at velocities and frequencies faster than or equal to the speed-of-light, according to the Theory of Relativity.

The Theory of Relativity and Entropy are limited and restricted. They ONLY apply to entropic physical matter. It has been observed that space is expanding faster than the speed-of-light or moving faster than the speed-of-light. How is that possible? It's because space isn't physical. Likewise, it has been experienced and observed that Psyche and Spirit Matter can move faster than the speed-of-light. How is that possible? It's because they aren't physical. In fact, it seems that EVERYTHING can move as fast as or faster than the speed-of-light, except for entropic physical matter. Likewise, entropy or the second law of thermodynamics only applies to entropic physical matter. According to the Quantum Law of Thermodynamics, there is NO heat death and NO entropy and NO unavailable energy in the Quantum Realm or Spirit World. We KNOW these things are true because they have been experienced and observed by Out-of-Body Travelers and Near-Death Experiencers. Entropy and relativity ONLY apply to entropic physical matter. These physical limitations and physical restrictions don't apply in the Quantum Realm or the Spirit World.

God deliberately created physical matter and entropy (the passage of time) for us in order to slow things down for us so that we can live them, observe them, experience them, learn from them, experiment with them, and remember having done so. At velocities equal to or faster than the speed-of-light, there is NO entropy because there is NO passage of time. Time stops. Time or entropy ceases to exist. Time, entropy, and the passage of time were made by God, which means that they can also cease to exist or come to a stop. I have experienced and observed that TIME STOPS and entropy (the passage of time) temporarily ceases to exist, just before you are quantum tunneled to a different location on this earth. God has complete control over Quantum Mechanics including quantum tunneling. From the photon's perspective, it quantum tunnels or teleports to its destination because it experiences no passage of time or no entropy during its voyage. From my perspective, time stopped, and I experienced no passage of time when God quantum tunneled me and the car I was driving to safety. God can quantum tunnel physical matter anywhere in this universe instantaneously simply by stopping the passage of time or temporarily removing the entropy from that physical matter.

Remember, there's nothing sacred about physical matter and entropy. It's the Energy, Psyche, or Syntropy that's being conserved. Energy can be organized into many different FORMS, including physical matter, spirit matter, space-time, physical laws, physical constants, space, time, locality, and entropy. Remember, it's the Energy or the Syntropy that's actually being conserved, not the physical matter nor the entropy. This is the Ultimate Law of Thermodynamics. It explains everything that we humans have experienced and observed. It corrects one of the fatal flaws in traditional science – namely, the erroneous belief that physical matter and entropy are being conserved. The Ultimate Law of Thermodynamics states that physical matter, entropy, locality, and space-time are NOT conserved. It's the underlying Energy or Syntropy that's being conserved, and NOT the physical matter nor the entropy. Entropic physical matter is just different FORMS of energy. Energy can be made to take on many different FORMS; and, the FORM is never conserved. It's the Energy, Psyche, or Syntropy that's actually being conserved and NOT the FORM. The Ultimate Law of Thermodynamics is the answer to life, the universe, and everything. It explains everything.

This Ultimate Law of Thermodynamics is one of the greatest scientific discoveries of all time; and, it's only made possible by eliminating Materialism, Naturalism, Darwinism, Nihilism, Behaviorism, Determinism, Physical Reductionism, and Atheism from science. If you successfully eliminate everything that is false and everything that has been falsified, then only the truth will remain. This is Logic 101. You start by eliminating Materialism, Naturalism, Darwinism, Nihilism, and Atheism from science; and then you see where that gets you. It's interesting to study and observe what remains after you have successfully removed Materialism, Naturalism, Darwinism, Atheism, and their derivatives from science. The experienced and the observed remain. The truth remains.

Remember, the Biblical God Jesus Christ has been experienced and observed both in the flesh and during our near-death experiences AFTER He rose from the dead. Science is observation and experience.

The Materialists, Naturalists, Darwinists, Nihilists, and Atheists define "science" as Materialism, Naturalism, Darwinism, Nihilism, and Atheism. Based upon all these different scientific discoveries that I have made during the past couple of years, I have upgraded my science to Science 2.0; and, I have redefined Science as "observation and experience". Observation and experience are a much better way to do science because the "non-existence of things" will never be experienced nor observed. It's impossible to experience and observe the "non-existence of something", including the non-existence of God. As scientists, we should go with what has been experienced and observed, not the wishful thinking of the Naturalists, Darwinists, and Atheists.

Physical matter and entropy were obviously made from Energy. Physical matter had physical limitations, entropy, locality, sub-light velocities, mass, time, space, and physical restrictions programmed into it. Physical matter is Organized Energy, and it was designed and organized by God. Since physical matter and entropy are never conserved, according to the Ultimate Law of Thermodynamics, God must of necessity exist in order to have organized, created, or made the first particles of physical matter from Raw Energy or Available Energy.

Scientific Observation: Anything that was obviously organized obviously had an Organizer who organized it.

Scientific Observation: Physical matter, physical laws, physical constants, space-time, physical restrictions, quantum fields, quanta, space, time, locality, and entropy were obviously organized and made from Energy. These things are made or brought into existence from different forms of Organized Energy. Physical matter is Organized Energy.

Scientific Conclusion: Therefore, it is logical and rational to conclude that physical matter, physical laws, physical constants, space-time, physical restrictions, quantum fields, quanta, space, time, locality, and entropy obviously have an Organizer who organized them, made them from Energy, and caused them to begin.

This truth has been experienced and observed. The false is falsified by the truth; and, the truth is repeatedly experienced and observed.

The Ultimate Law of Thermodynamics states that the FORM is never conserved, which means that the different forms of Energy such as entropic physical matter is never conserved. Their quantity or amount is constantly changing. "Particles" or quanta are being made and annihilated all the time. There's nothing sacred or inviolate about particles of matter. It's the underlying Energy or Syntropy that's always being conserved. It's the Energy, Syntropy, or Life Force that ends up being eternal and everlasting, not entropy or death. That is what has actually been experienced and observed.

Materialism, Naturalism, Darwinism, Nihilism, Behaviorism, Determinism, Physical Reductionism, and Atheism were designed to prevent us from discovering the Ultimate Law of Thermodynamics. Entropy is death; and, the Scientific Naturalists and Atheists erroneously teach that entropy or death is conserved. These people are wrong. Life after death has been experienced and observed.

The Psyche, Syntropy, Energy, or Life Force is always conserved; but, the entropic matter or physical matter is never conserved. Physical matter and entropy come and go as God sees fit. Physical matter and entropy are made or caused to begin, which means that they can be unmade, or disassembled, or brought to an end. In contrast, Psyche, Syntropy, Energy, or Life Force can be neither created nor destroyed. Psyche, Syntropy, Energy, and Life Force are eternal and everlasting, without a beginning of days or an end of years, because they are constantly being conserved. God's Psyche or God's Intelligence is being conserved. Your psyche or your intelligence is being conserved. In contrast, physical matter and entropy are never conserved. This is the Ultimate Law of Thermodynamics. The Ultimate Law of Thermodynamics is the answer to life, the universe, and everything. It has been experienced and observed.

Mark My Words

A New More Realistic Model for Entropy

We NEED a new, better, and more realistic model for entropy than the ones that we currently have. Death and Heat Death do not represent the true reality of the situation. Try this one on for size and see how you like it.

Let a blank CD or a blank DVD represent the beginning of our entropic physical universe and the physical matter within that universe. A blank CD or a blank DVD is at Maximum Syntropy. ALL of the space there is available for use to do work or to store information. The CD or DVD starts with Maximum Syntropy. Notice also that the blank CD or blank DVD was organized and made by Someone Psyche or Someone Intelligent with a specific purpose in mind. The system begins at Maximum Syntropy.

Now, let's write the history of our physical universe onto that CD or DVD and then close the disk when we are done writing. A closed CD or a closed DVD represents Maximum Entropy. There's NO space there available for use anymore. All that can be done with that CD or DVD has been done. The CD or DVD is now at Maximum Entropy.

What do we observe?

Even though that CD or DVD is now at Maximum Entropy, ALL of the Energy or ALL of the information on the DVD is still there just waiting for Someone Psyche, or Someone Intelligent, or Someone God-Like to come along, access it, use it, and do work with it. Even in Maximum Heat Death, our physical universe is still there, the quantum fields and quantum forces are still there, the information and the energy are still there. It's ALL there just waiting for Someone Psyche, Someone Intelligent, Someone Powerful, or Someone God-Like to come along, access it, use it, and do work with it. Everything is still there, even though our physical universe or DVD is currently at Maximum Entropy. It still can be used by Someone Psyche or Someone Intelligent to do useful work. ALL of the information or energy is still there even though we are currently looking at Maximum Entropy because Psyche, Energy, Intelligence, and the Information it contains is always conserved.

Now, in the extreme case that you tire of the CD or DVD and no longer want its contents, Someone Psyche, Someone Powerful, or Someone God-Like can take that CD or DVD, grind it down to its component parts, break it down into Pure Exergy or Pure Syntropy or Pure Energy, and then form it into a brand new blank CD or blank DVD that can then be used for a different purpose to store a different type of information. In other words, entropic physical universes that have reached Maximum Entropy can be recycled by Someone Psyche or Someone God-Like and become brand new, blank, Maximum Syntropy CDs or DVDs once again.

Now, there's a Model for Entropy that actually matches with the true reality of the situation.

Mark My Words

Was Thanos Right?

Let's turn to the comic books for a lesson in entropy and syntropy. Thanos' worldview or philosophy of life is based upon the Supremacy of Entropy Doctrine – the idea that the Second Law of Thermodynamics reigns supreme. Is he right, or is there something extremely important that he is missing?

What should we observe currently here on our earth if Thanos is right, and a population explosion is guaranteed to produce suffering, shortages, disease, starvation, wars, poverty, and death?

Well, we should observe worldwide famine, disease, poverty, death, mass extinction, misery, decline, shortages, hell, resource wars, suffering, and starvation as the population of our world increases. Is that what we observe? Now be honest with yourself. Is that what we really observe?

NO!

Why?

I think the following YouTube video is one of the most insightful ever made.

The Population Explosion – Was Thanos Right?

https://www.youtube.com/watch?v=qH_RE9HbNbA

https://evolution-is-entropy.com/wp-content/uploads/2018/09/Was-Thanos-Right.zip

What do we really observe as the population of our world increases?

Poverty rates have actually gone down as the population has grown. Prosperity has increased as the population has increased. Global food production has gone up and actually exceeds demand. Education rates and IQs have risen. Intelligence and knowledge have increased exponentially. Prices have gone down for basic commodities. Here we are with more people on the planet than we have ever had before, and not only is starvation not a problem, but obesity is becoming a worldwide epidemic instead. Clearly, the assumption "more people more problems" is not quite true. Something is obviously overriding and defeating the claim that Entropy Reigns Supreme. Something is clearly falsifying the claim that population explosions lead to misery, suffering, poverty, shortages, starvation, entropy, and death for all.

As the population of our earth has increased, rather than massive suffering and starvation, we have observed prosperity unlike anything that ever existed before. We have observed Syntropy and Economies of Scale that were NEVER dreamed of being possible on our limited finite earth. We observed that intelligent Human Psyches can actually introduce order, structure, purpose, meaning, syntropy, intelligence, solutions, and economies of scale to disorder, chaos, randomness, and entropy. We observed that intelligent Human Psyches can actually find ways to feed themselves and their pets. We observed that Syntropy, Psyche, or Intelligence reigns supreme. We observed the exact opposite of what we were supposed to observe if the Supremacy of Entropy Doctrine and the Second Law of Thermodynamics were absolutely true. We have observed the exact opposite of what Thanos predicts.

Why?

In this video, he accurately and perceptively identifies INTELLIGENCE as the thing that is overriding and overcoming the negative effects of entropy and population explosions.

All that's needed to eliminate Entropy from any system is Energy, Psyche, Syntropy, Life Force, Intelligence, Knowledge, Information, Planning, Foresight, Determination, Deliberation, Ordering, Organization, or Inspiration – all of which the Human Psyche provides in spades. With intervention from the Human Psyche, the whole ends up being much greater than the sum of the individual parts, which means that Thanos was wrong.

Misery, poverty, shortages, and starvation are NOT the inevitable end of population explosions. Psyche or Intelligence can override entropy, chaos, and disorder at will even in a fallen mortal physical world such as ours.

This is Good Science because it has actually been experienced and observed.

Entropy simply melts away and evaporates into thin air in the light of Psyche, Intelligence, Information, Knowledge, Energy, Cooperation, and God's Love. The Human Psyche can overcome and defeat entropy, disorder, and starvation at will. Likewise, God's Psyche can overcome and will overcome disease and death at will because He has the power, the energy, the knowledge, the information, and the love necessary to do so.

Entropy is NOT the fundamental law of the universe. Syntropy is. This is actually what has been experienced and observed. Syntropy or Psyche overcomes entropy at will bringing order, organization, and prosperity to our world and universe instead. The Human Psyche or Human Intelligence can actually pray to God and request help in overcoming the negative effects of entropy and population explosions, and actually receive the help or the syntropy that he or she requests. This reality and truth has been experienced and observed.

Mark My Words

Defining Everything in Terms of Science and Syntropy

I learn best through comparison and contrast by comparing what has been falsified with what has been proven to be true. The false is falsified by the truth; and, the truth is repeatedly experienced and observed. I define Science as observation and experience. I allow ALL of the evidence into evidence and pursue a preponderance of that evidence.

The Materialists, Naturalists, Darwinists, Nihilists, and Atheists teach and believe that entropic physical matter is the ONLY thing that exists, and that entropy or death is conserved. This is what they teach and believe. Is it not? Quantum Mechanics, Action at a Distance, Quantum Field Theory, and Quanta (Wave Packets of Energy) FALSIFY their belief that entropic physical matter is the only thing that exists. The observations and experiences of those who have had Near-Death Experiences (NDEs), Out-of-Body Experiences (OBEs), Shared-Death Experiences (SDEs), and After-Death Life reviews FALSIFY their belief that entropy or death is conserved. The First Law of Thermodynamics or Syntropy also FALSIFIES their belief that death or entropy is conserved.

They say that belief should wait on the evidence, which I take to mean that the evidence should take precedence over belief. I treat everything that has been experienced and observed on both sides of the veil as evidence.

Syntropy is the First Law of Thermodynamics or the Conservation of Energy.

What's really being conserved?

The first law of thermodynamics is a version of the law of conservation of energy, adapted for thermodynamic systems. The law of conservation of energy states that the total energy of an isolated system is constant; energy can be transformed from one form to another but can be neither created nor destroyed.

https://en.wikipedia.org/wiki/First_law_of_thermodynamics

Energy is a conserved quantity; the law of conservation of energy states that energy can be converted in form, but not created or destroyed.

https://en.wikipedia.org/wiki/Energy

Exergy is the energy that is available to be used. After the system and surroundings reach equilibrium, the exergy is zero. Determining exergy was the first goal of thermodynamics. Energy is neither created nor destroyed during a process. Energy changes from one form to another. In contrast, exergy is always destroyed when a process is irreversible, for example loss of heat to the environment. This destruction is proportional to the entropy increase of the system together with its surroundings. The destroyed exergy has been called anergy. For an isothermal process, exergy and energy are interchangeable terms, and there is no anergy.

https://en.wikipedia.org/wiki/Exergy

What did we just learn?

I learned that the Quantum Realm or the Psyche Realm is an isothermal system. There is NO anergy at the quantum level or the psyche level! The second law of thermodynamics doesn't exist at the quantum level because there is NO thermodynamics at the quantum level because the Quantum Realm or the Spirit World is an isothermal system. ALL of the Energy is ALWAYS available in an isothermal system; and, the Syntropy Realm or

Quantum Realm is an isothermal system or an exergy system. At the quantum level, both exergy and energy are always conserved. There is NO anergy or unavailable energy in the Quantum Realm or the Syntropic Realm. At the quantum level or the psyche level ALL of the energy is available for use all the time. There is NO anergy or entropy in the Quantum Realm or the Psyche Realm. Heat death is impossible at the quantum level or the psyche level. This is precisely what has been experienced and observed. It's there hidden in plain sight waiting for anyone to see it, find it, and understand it.

This is quite a revelation, isn't it? I bet you never learned anything about any of this in your college classrooms! Materialism, Naturalism, Darwinism, Nihilism, and Atheism were designed to hide these truths from us. They do what they were meant to do to hide the truth from you.

The Quantum Law of Thermodynamics states that there is NO anergy or NO entropy at the quantum level or the psyche level because the Quantum Realm is an Isothermal Realm. In the Quantum Realm or the Syntropy Realm, exergy and energy are always conserved. At the quantum level, ALL of the energy is available for use all of the time. This is what has been experienced and observed.

The Quantum Realm, Psyche Realm, or Syntropy Realm is an isothermal system which means that anergy, unavailable energy, entropy, and the second law of thermodynamics DO NOT EXIST at the quantum level or the psyche level. Heat death is impossible at the quantum level. This is so obvious and so obviously true that it makes you wonder why nobody has ever thought of it or discovered it before now; but, the reason why nobody has ever found nor seen this fundamental truth and reality before is because they didn't want to see it, find it, discover it, nor understand it. They don't want it to be true. Yet, it is consistent, holds together, and is obviously true. It explains why there is something rather than nothing. It has been experienced and observed. It's the answer to life, the universe, and everything. It explains how our physical universe could have begun in Syntropy. The explanatory power of Syntropy is huge!

Furthermore, according to the Ultimate Law of Thermodynamics, the Energy, Psyche, or Intelligence is being conserved NOT the form of that energy. Energy can always be transformed from one form to another. The FORM is never conserved. Only the underlying energy or psyche is conserved. In the equation $E = mc^2$, the left side of the equation (energy) is constantly being conserved; whereas; all the different stuff being constructed or made on the right sight of the equation (from energy) is NEVER conserved. If it is formed, constructed, or made, then it's NEVER conserved. The FORM is never conserved. Entropic physical matter is made from different FORMS of energy. Entropy is an emergent property of physical matter; and, entropic physical matter is NEVER conserved. This means that physical matter and entropy are NEVER conserved. This is the Ultimate Law of Thermodynamics. It also explains everything that has ever been experienced or observed.

Energy is the ultimate perpetual motion machine. Energy can be formed and transformed into anything, including physical matter and its associated entropy. Formed and transformed by whom? Who would have the power, knowledge, and ability to transform raw chaotic energy into organized physical matter and the associated physical laws or physical restrictions such as entropy? Organization of any kind requires an Organizer of some kind. This is Logic 101.

The physical laws and physical constants were made. That's why there is something rather than nothing. That's why there is order and organization rather than random chaos. Order and organization are made. They don't just spontaneously spring into existence from nothing as the Materialists and Naturalists claim. The physical laws and physical constants manifest as different FORMS of energy. The FORM is always formed or made. Entropic physical matter is made from different FORMS of energy. Mass is an accumulation of energy. It's ALL energy! Anything that was obviously formed or made obviously has a Maker who formed it or made it. Entropy is a physical law. The physical laws were made; therefore, entropy was made or caused to happen. Gravity, magnetism, the nuclear forces, quantum waves, visible light, radio waves, x-rays, microwaves, gamma rays, dark energy, dark matter or spirit matter, and physical matter were ALL made from different FORMS of Energy. It's ALL energy. In contrast, Entropy is just an emergent property of physical restrictions or physical laws. The physical laws and physical constants were made, and entropy is just one of the emergent results.

The Gods created entropy to prevent perpetual motion machines and immortality from being possible in a physical, fallen, entropic, mortal world such as ours. We have to fight against entropy in order to survive, learn, and grow. Entropy is death. It motivates us to learn. After use, the Energy is still there as heat within the environment. When it comes to entropy, the Energy is still there – it's just not available for usage by fallen, physical, mortal beings such as us. However, the Gods can access and use that "unavailable energy" at will because the underlying energy is always conserved. This is what has been experienced and observed. Jesus Christ removed the entropy, corruption, and death from His physical body and then rose from the dead. It's real. It truly happened.

Physical, fallen, mortal beings such us ourselves do NOT have access to the unavailable energy in this universe; but, the Gods who made that energy unavailable in the first place have direct access to the whole of it anytime they choose. The energy is still there waiting to be accessed and reused. The Gods have access to ALL of the energy within the environment, including the available energy as well as the energy that they made unavailable to us. That's the way things really work at all levels of existence. The Quantum Realm is an Isothermal Realm which means that anergy or entropy DOES NOT EXIST at the quantum level or the psyche level.

Entropy is an accounting – a function of time. Entropy is the arrow of time. Time and entropy are constructs. Through entropy the Gods keep track of the amount of energy that has been made unavailable to us. Entropy is a physical restriction that the Gods placed upon entropic physical matter; and, the Gods can remove that physical restriction anytime they choose to do so thereby converting entropic physical matter into syntropic physical matter instead. There's nothing sacrosanct about entropy or death. Entropy is an emergent property of physical matter and physical restrictions; and, physical matter is made from different forms of energy.

The various different physical laws and invisible forces and fields were made from raw energy; and, they can be disassembled and turned back into raw energy or chaotic energy anytime the Gods decide to do so. The energy is always conserved. The "unavailable energy" is still there in the system at the quantum level and is still available to the Gods. It's just not available to us. Remember, entropic physical matter is made from different FORMS of energy. Anything that is made is never conserved; but, the underlying energy is always conserved. This is what has actually been experienced and observed.

I wanted to find and know the truth. I used the Scientific Method and Negating the Consequent to FALSIFY Physicalism, Materialism, Naturalism, and Atheism. I used the observations and experiences of the human race to FALSIFY Nihilism, which is the belief that we cease to exist when we die. As a result, I discovered the Quantum Law of Thermodynamics which states that Heat Death is impossible in the Quantum Realm or the Syntropy Realm. There is NO second law of thermodynamics at the quantum level or the psyche level. Entropy or death doesn't exist at the psyche level. Psyche, Life Force, or Energy cannot be made, and it cannot be destroyed. It is always conserved. This is what has been experienced and observed.

Psyche, Life Force, or the Spark of Life has been experienced and observed by Out-of-Body Travelers as a pinpoint of light or a photon; therefore, we KNOW that it is real and truly exists. Our job as scientists is to explain what it is and how it works. Photons are Psyches. Every Psyche looks like a photon and ACTS like a photon both at the quantum level and at the physical level. Psyches are photons. Every massless "particle" or massless quantum is a Psyche of some type! Psyches or Massless Elementary Particles USE matter, mass, or energy to slow down and get things done. The massless elementary particles within quarks and electrons are Psyches or Intelligences. Every massless quantum is some type of Psyche. This is what has been experienced and observed.

Psyche or Intelligence has been experienced and observed. Psyche is seen or observed as a massless Pinpoint of Light or a Photon of Light. Psyche is experienced first-hand as an immaterial viewpoint in space. Every Massless Particle is some type of Psyche. There are different types of Psyche, each with its own God-given assignment, purpose, position, mission, or function. The Human Psyche's spirit body is a child of God – Heavenly Father and Heavenly Mother. A Human Psyche is a God. If you can read this, your surname is God. Your physical body is also descended from the Gods. Whenever a Human Psyche enters into a non-consensus reality in the Quantum Realm or Spirit World, that particular non-consensus reality orders itself or organizes itself to meet the demands and expectations of that Human Psyche. In any non-consensus reality at the quantum level, the Human Psyche is God. This is what has been experienced and observed.

Psyche has been hiding in plain sight all the way along where the Physicalists and Naturalists and Atheists will never see it nor find it because they aren't looking for it and have chosen instead to believe that it doesn't exist. But, their beliefs and wishful thinking are irrelevant because Psyche or Intelligence has been experienced and observed. Every Massless Particle is some type of Psyche; and, Massless Particles have been experienced and observed.

Psyche is the innate intelligence within all the different forms of energy which gives that energy the inherent ability to understand, follow, and obey God's Laws and God's Commands. Psyche and energy are syntropic which means that psyche, or the intelligence within energy, is always conserved. The Life Force is always conserved. Energy, Psyche, Intelligence, or Life Force can be neither created nor destroyed. It has always existed and will always exist. Life Force and Energy are conserved. Psyche and Energy are comprised of the same stuff and are effectively synonymous. It's ALL energy.

The Materialists, Naturalists, Darwinists, Nihilists, Behaviorists, Determinists, and Atheists erroneously teach that ONLY entropic physical matter exists. One error leads to another. Consequently, these people erroneously teach that physical matter and entropy are conserved. They erroneously teach that the whole thing is going to end in Heat Death and that we cease to exist when we die. That is what they teach, is it not? Entropy is death. These people erroneously teach that death or entropy is conserved. This is because these people erroneously equate energy with entropic physical matter; and then, they erroneously define the First Law of Thermodynamics as the "Conservation of Physical Matter" and the Second Law of Thermodynamics as the "Conservation of Entropy". These people erroneously teach that entropic physical matter is being conserved; whereas in truth, ONLY the underlying energy or psyche is being conserved. Physical matter and entropy are constructs which means that they can be deconstructed and turned into something else instead such as spirit matter and syntropy.

All I want is the truth. What has been experienced and observed?

The amount of entropy is constantly changing which means that entropy is NOT conserved. We humans make new physical matter in our particle accelerators; and, we destroy physical matter and the associated entropy in our hydrogen bombs; therefore, entropic physical matter is NOT conserved. This is what has been experienced and observed – physical matter and entropy are NOT conserved. This is the Ultimate Law of Thermodynamics.

By studying the errors that the Materialists, Naturalists, Darwinists, Nihilists, and Atheists make, I was able to discover and observe that the different FORMS of energy are never conserved. Entropic physical matter is comprised of different FORMS of organized energy. It's ALL energy. The FORM is never conserved. The FORM can always be transformed into a different form. Entropic physical matter is made from energy and then dissolved back into raw energy as the Gods see fit. Only the underlying energy or psyche is being conserved. These truths became the Ultimate Law of Thermodynamics. The Ultimate Law of Thermodynamics simply differentiates between what is conserved and what is not being conserved. Ultimately, that is what's most important to know. The Ultimate Law of Thermodynamics is like having a new toy to play with. Entropic physical matter is comprised of different forms of energy, and the FORM is never conserved.

Remember, entropic physical matter is comprised of different FORMS of Organized Energy. Anything that was obviously organized obviously has an Organizer who organized it. Anything that is obviously made or obviously organized obviously is NOT conserved. Physical matter and the different forces and fields within it are highly organized; therefore, it is obvious that entropic physical matter was organized, formed, and made. It didn't self-assemble from nothing out of thin air. Organization requires restrictions, rules, and laws carefully enforced and carefully obeyed. Organization and order require an intelligent Organizer and an intelligent Obeyer; otherwise, everything in this part of the universe would still be nothing but random chaos – matter unorganized or energy unorganized.

Entropic physical matter is made from different FORMS of energy. As mortal fallen scientists, the greatest and most impressive scientific discovery we could make would be to figure out how to remove the entropy from entropic physical matter thereby converting that entropic physical matter into syntropic physical matter instead. Immorality, Eternal Life, and Resurrection from Death would be the result. We would be like Adam and Eve in the Garden of Eden if we could remove all the entropy from our physical bodies. Nothing would ever wear out; and, we would be immortal. It has been experienced and observed; so, it's possible.

Psyche is Energy. Psyche, Life Force, or Energy is syntropic which means that it is always conserved on both sides of the veil – both the physical side and the spiritual side. Syntropy, or the Conservation of Psyche, ends up being a powerful defense of Quantum Mechanics. Psyche, or a Conscious Observer, or a Sentient Causer makes Quantum Mechanics possible in the first place. Quantum Mechanics is impossible without Psyche or Syntropy. Quantum mechanisms would have NO purpose or meaning without Psyche or Conscious Intent. The whole thing would be random chaos without God's Psyche to provide it with structure, order, organization, purpose, and meaning. Syntropy, or the Quantum Law of Thermodynamics, falsifies the second law of thermodynamics at a fundamental level. Something has to give, and entropy is ultimately defeated and destroyed by Syntropy, Psyche, Energy, or Intelligence. This is what has been experienced and observed.

I used to be a Materialist, Naturalist, Nihilist, and Atheist. The net effect of Entropy and the Second Law of Thermodynamics on my worldview or personal philosophy of life at the time is that heat death is inevitable, we cease to exist when we die, there's no purpose or meaning to life, existence is fleeting and pointless, and God does not exist. Entropy and the Second Law are the cornerstones of Nihilism. Entropy and the Second Law were designed to convince us that God does not exist, that life has no purpose or meaning, and that we cease to exist when we die. Ask yourself who first came up with the idea, and you will quickly discover its overall purpose. It does what it was meant to do.

"The Heat Death of the Universe Is Coming" is the mantra of the Materialists, Naturalists, Nihilists, and Atheists. They've made T-Shirts announcing and celebrating that universal belief. But, is it true?

The revelation that I received through all of this study and comparison of Syntropy and Entropy is that **heat death is impossible** at the psyche level or the quantum level. The Second Law of Thermodynamics does not exist in the Quantum Realm because there are NO thermodynamics or heat flow in the Quantum Realm. This reality and truth is the Quantum Law of Thermodynamics, or Syntropy. Syntropy means ZERO entropy. Syntropy also means Eternal Life or Infinite Life. In other words, your Life Force or your Psyche is always conserved on both sides of the veil – both in the physical realm and in the spirit realm. This is what has been experienced and observed by Out-of-Body Travelers and Near-Death Experiencers.

Remember, Psyche is the innate intelligence within all the different forms of energy which gives that energy the inherent ability to understand, follow, and obey God's Laws and God's Commands. Psyche is the organizing force in the multiverse at all dimensions of existence. Psyche, Intelligence, or Energy is always conserved. It cannot be made, and it cannot be destroyed. This reality and truth is the answer to life, the universe, and everything. It explains everything that has ever been experienced and observed.

Psyche is Energy. All you need is energy to remove entropy or heat death from any system. Psyche is Intelligence or Knowledge. All you need is knowledge or information to remove entropy or disorder from any system at will. Psyche is Eternal Life or Life Force. All you need is Eternal Life or Life Force to remove entropy or death from any system. Psyche is Syntropy. All you need is Syntropy to remove entropy from any system.

Syntropy is the Quantum Law of Thermodynamics which states that **heat death is impossible** at the psyche level or the quantum level; and technically, since **heat death is impossible** at the quantum level, it's also impossible at the physical level because the physical is ultimately based upon the quantum. **Ultimately, heat death is impossible and will not hold.** This is what has been experienced and observed.

My goal in life is to figure out what everything is and how it works. I believe that I have now done so. Good enough!

Mark My Words

—

Source Material for Defining Syntropy

I Am Not a Creationist: So What Am I?

https://www.amazon.com/dp/B071XTM8XY

References with a Similar Theme

God Is in the Light: God is light, and in Him is no darkness at all.

https://www.amazon.com/dp/B07168S37N

Quantum Mechanics from a Non-Physical Spiritual Perspective.

https://www.amazon.com/dp/B01J023TGU

Quantum Neuroscience: The Answer to Life, the Universe, and Everything.

https://www.amazon.com/dp/B079Z6QQQB

Science 2.0: I Upgraded My Science.

https://www.amazon.com/dp/B0771K6WTX

Scientific Proof of God's Existence: Finding God Where the Atheists Refuse to Look for Him

https://www.amazon.com/dp/B07B26CRHX

Symbolizing Syntropy

If I were to give you the task to produce the perfect symbol for Syntropy, what would you choose and what would we end up with? I gave myself such a task, and it wasn't easy at first.

Why?

It's because you have to have the right definition for Syntropy before you can find the perfect symbol to represent it.

So, what's the best definition for Syntropy?

The most common definition that I have encountered for Syntropy is that it's the opposite of entropy. According to this particular definition for Syntropy, we have to know what entropy is before we can define Syntropy.

So, what is entropy?

Entropy is typically defined as disorder, randomness, chaos, death, extinction, random mutations, corruption, cancer, and unavailable energy. By definition, in principle or practice, Materialism, Naturalism, Darwinism, Nihilism, Atheism, Behaviorism, Determinism, Scientism, Classical Physics, and the Theory of Evolution are based exclusively upon entropy. Materialism, Naturalism, Nihilism, and Atheism are defined as death or entropy.

Since we know that entropy exists, that means that Syntropy must exist as well. We can't have entropy without some kind of Syntropy to get it going in the first place.

Syntropy is supposed to be the opposite of entropy. Therefore, what do you visualize as being the opposite of these things? What do you visualize as being the opposite of entropy? Once we know what the opposite of entropy is, then we will have a good definition for what Syntropy is.

Syntropy is the opposite of entropy. Therefore, I visualize Syntropy as being order, organization, structure, function, purpose, teleology, eternal, everlasting, life, life force, psyche, intelligence, genetic engineering, programming, creation, incorruption, manufacturing, health, and endlessly available energy. Quantum Mechanics is Syntropy. Action at a Distance is Syntropy. Quantum Mechanics or Transdimensional Physics seems to be eternal and everlasting. Psyche, or Intelligence, or Consciousness, or Life is Syntropy. Syntropy is Supernatural. Quantum Mechanics is Supernatural and Eternal. Psyche is Supernatural and Eternal. Entropy is death. Syntropy is Eternal Life.

Syntropy or Quantum Mechanics is the opposite of Materialism, Naturalism, Darwinism, Nihilism, Atheism, Classical Physics, and the Theory of Evolution, which are based on entropy. Syntropy or Psyche or Intelligence produces order, organization, structure, purpose, meaning, teleology, life, genomes, proteins, life forms, planets, stars, galaxies, and machines. In contrast, random mutations, natural selection, evolution, and classical physics produce disorder, chaos, randomness, loss, disease, cancer, death, and extinction. Evolution, random mutation, or genetic change is entropy. Macro-evolution, chemical evolution, abiogenesis, and spontaneous generation are prevented from happening naturally by entropy. Evolution of any kind is entropy. It's physically impossible for entropy to design, create, and manufacture things. It can't happen, which means that it didn't happen.

A careful study of the scriptures that the Biblical God had a hand in writing and producing will reveal that Jesus Christ defines Syntropy as "without a beginning of days or an end of years". So, what about us is eternal, everlasting, and infinite?

Well, our Psyche or Intelligence or Quantum Non-Local Consciousness seems to be eternal and everlasting, without a beginning of days or an end of years. Furthermore, the spirit matter that comprises our spirit bodies seems to be without a beginning of days or an end of years.

However, the physical matter in our physical bodies had a beginning at the beginning of this physical universe and can therefore theoretically have an end. Our physical bodies definitely had a beginning and thanks to entropy will have an end. Entropy is a dead-end.

In mathematics, how do we symbolize things that have a beginning and an end, such as your physical body or mortal body? In geometry, a line segment is a part of a line that is bounded by two distinct endpoints, and it contains every point on the line between its endpoints. A line segment has both a beginning and an end.

—

Entropy is symbolized by a line segment.

So, what's the proper way to symbolize Syntropy, which is eternal, everlasting, and infinite? In mathematics, we use the Infinity Symbol. The Infinity Symbol represents Syntropy or Eternal Life. Infinity is also represented by a Circle, without a beginning or an end. Therefore, I chose the Infinity Symbol within a Circle as the site icon and logo for my websites about Syntropy. Syntropy must be represented by something that symbolizes Eternal Life.

As far as I can tell, the Infinity Symbol and a Circle are the best way to represent Syntropy or Completion, which means that the best way to represent entropy is as a line segment with a beginning and an end.

Symbols help us to see and understand what's happening and how things work. Entropy is a line segment. Syntropy is an Infinity Symbol or a Circle.

Other Informative Symbols

When it comes to Circles, Asymptotes, and Infinity, mathematics repeatedly tells us that such things must exist. Over and over again, math has proven to be a model for reality. Syntropy is the eternal nature of Psyche or Intelligence.

If entropy is modeled by a line segment, then how should Syntropy be modeled? Circles, Asymptotes, and the Infinity Symbol end up being the best way to model and explain Syntropy. However, there are other informative symbols that can be used to help us define and understand Syntropy.

The hexagram or the Star of David was designed to represent Syntropy or the union of opposites, as was the Chinese Yin and Yang symbol. Entropy exists to help us to learn to appreciate its opposite, Syntropy or God. Syntropy or God is the opposite of entropy or death. Syntropy is Eternal Life.

Hexagram or Star of David

The Star of David also represents Eternal Life. According to the Zohar, a medieval book of Jewish mysticism, the six points of the star represent the six male sefirot (attributes of God), in union with the seventh sefirah of the female (the center of the shape). Rosenzweig described two interlocking triangles — the corners of one representing creation, revelation, and redemption; and, the corners of the other representing Man, the World and God. — Uncle Google

It has also been suggested that the one triangle represents the three main offices of the Melchizedek Priesthood, and the other triangle represents the tree main offices within the Aaronic Priesthood. God's Priesthood or Quantum Power is also eternal and everlasting, cannot be made, and cannot be destroyed. It is the source of Resurrection or Immortality.

Yin and Yang

Yin and Yang represent the Eternal Life of male and female spirits. Yin and Yang also represent the Eternal Life of both good and evil Psyches or Souls.

Comparing Syntropy with Entropy

I have always loved science; and, ever since I got rid of My Materialism, My Naturalism, My Nihilism, My Scientism, and My Atheism, I have been a very open-minded scientist more than willing to think outside the box. I have developed a great love for Syntropy or Quantum Mechanics.

The best way to connect with these concepts related to Syntropy is through my many different books.

In all of my books, I make detailed comparisons between Syntropy and Entropy.

How does this work out in practice?

Well, it essentially means that I end up making detailed comparisons between Quantum Mechanics and Naturalism.

Quantum Mechanics, Spirit Matter, and Psyche are Syntropy. In contrast, Materialism, Naturalism, Darwinism, Nihilism, Atheism, and the Theory of Evolution are based upon Entropy, Chaos, Random Diffusion, Random Mutations, Physical Matter, and Classical Physics. There's a huge difference between the two. Syntropy is order and organization and life. Entropy is chaos, death, disease, cancer, and extinction. Evolution is entropy. Random mutations are entropy. Natural selection results in death, extinction, and entropy. Entropy cannot do order, organization, and life. It's physically impossible.

Physical Matter, Materialism, Naturalism, Darwinism, Nihilism, Behaviorism, Determinism, Classical Physics, and Atheism are based upon entropy. Entropy results in random mutations, death, and extinction.

Psyche, Consciousness, Thought, Memories, Quantum Waves, Spirit Matter, Order, Organization, Creation, Life, Supernatural Mechanisms, and Quantum Mechanics are based upon Syntropy. Syntropy is eternal and everlasting, without a beginning of days or an end of years.

See my Author Page on Amazon for books wherein I discuss these concepts and ideas in much greater detail.

https://amazon.com/author/science

As an open-minded scientist, I hope that you will welcome any comparison between Syntropy and Entropy.

Remember, Syntropy is the counterbalance to entropy. Syntropy has to exist, because entropy exists. Entropy couldn't exist without Syntropy.

Remember, Syntropy has to exist or we wouldn't exist; and, entropy or disorder would prevail across the universe.

It's time that we as scientists finally start to use Syntropy, Psyche, and Quantum Mechanics to explain how things really work in every realm of existence, including both the physical realm and the quantum non-local psyche realm.

It's time for us to upgrade our science to Science 2.0 and start allowing ALL of the evidence into evidence. Our very existence tells us that someone somewhere KNOWS how to do Syntropy where physical universes are concerned.

The Fleur de Lis

Finally, there is one other important point to make.

Somehow, I just instinctively KNEW that Syntropy or Unity or Oneness with God should be represented by the Fleur de Lis – the flower of light or the Light Bringers.

As a result of this inner feeling, the first symbol that I chose to present on my websites about Syntropy in order to represent Syntropy was in fact the Fleur de Lis.

Syntropy must be represented by something that symbolizes Eternal Life. A careful study of the Fleur de Lis reveals that it was designed to represent Eternal Life, Eternal Kingship, Kings and Queens, Gods and Goddesses, or Eternal Monarchy.

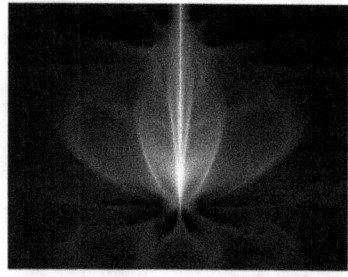

This image is a beautiful representation of Syntropy – the balance, the harmony, the peace, the eternity, the light, and the Glory of God contained within Psyche or Syntropy. God is called the Father of Lights for a reason. God is the Father of Psyches – in other words, God the Father and Heavenly Mother clothe our Psyches or Intelligences in spirit bodies, and later on within physical bodies. They give birth to us, thereby making us the Children of God. God is the Father of both our physical body and our spirit body, with our Psyche being eternal without a beginning of days or an end of years. Adam was a son of God, both spiritually and physically. The Flor de Lis symbolizes the Holy Trinity – Father, Son, and Holy Ghost who are the personification of Syntropy.

This individual says it well. This is beautiful material about Syntropy or Spirituality or God.

Flor de Lis is the symbol of the new times, of the new era, of what is being born and beginning to take place in our beloved planet.

The new ideas that we are giving birth to, recreating and co creating, putting into new words that include hidden and evident knowledge of humanity and of the Universe. These ideas have already been created in some space and time, in the evolution of Humanity and have now become Beings with pure hearts, pure energy.

The experience of "Sacredness" is essential in order to enter another stage of consciousness. This is an "emblem flower". There is an ancient and secret doctrine behind that emblem. There is a symbol in the signature of this sacred flower. An ancient symbol: a secret that only the initiated know about. The secret of "sacredness."

Each person perceives those things that belong to the highest level of consciousness to which he has access. The lips of Wisdom remain closed before the profane. But the flower is there, perfect and harmonious, to symbolize "sacredness", to grant us access to it. And she does not deny herself, she is all devotion. It's enough to look at her".

The Angel of the Paradise rescues the ancestral memory of the race, the yearning for perfection, it dresses up as a flower, descends to Earth and shows herself, imperceptibly but she does show herself... I am not a Fallen

Angel, I am an Elevated Being descending through a flower and I will show each what he can see... I will reveal what each can stand... because the truth is often unbearable. It is a blinding light. So, each one regulates the flow of energy and receives as much as he can absorb. The Lis doctrine is a secret that only the ancient hermetists know about.

Wisdom, experience in wisdom, this is the quality of the flower. It helps you to get closer to "sacredness". To get closer to a Greater Truth, to a conscious spirituality and to understand the Planetary Hermetic Wisdom.

You... you, who reads, you who listens... Do you still choose the duality? Or do you prefer to move forward?

Everything that the human being has experienced and that is registered in his genetic code has to be removed so that his inner conscience can be freed. It is there that the potential of this mutation is dynamited, transforming and irradiating in the adequate proportion and vibration for each member of the animal kingdom.

Apart from other meanings, the Flor-de-Lis is an esoteric symbol that has acted a lot from the subtle levels of consciousness, transmuting disharmonious situations that could have been even more tragic.

One of its tasks is to internally stimulate the formation of esoteric patterns used to capture the apparel of the new man (appearance of the 5th race).

The flor-de-lis teaches that the spirit has the duty to serve the spiritual world; the soul serves the intuitive world; and the outer "I" serves the world of shapes.

If we look at the flower we can see it has the shape of a compass, pointing to the right place, the north, the rectitude, the safe orientation. Seen from the side it looks like a flying bird, it symbolizes the Phoenix, the beginning of death and resurrection. The dove, the Holy Spirit.

It has six petals, three in the upper part and three in the lower part. These make two tetrahedrons that interweave and symbolize the Trinitarian correspondence between heaven and earth. In the center, where the upper and the lower parts meet, the flower has a ring that represents unity and fidelity between heaven and earth.

The central axis, forming a sword with two cutting edges, a double-edged weapon, reminds us how to use it and its consequences. The upper central petal represents heaven, the heavenly duties. The lower central petal represents the earth, the earthly duties.

The two upper side petals are God's eyes, with peripheral vision on alert, and the balance of opposed forces. The lower right petal represents the achievement of the balance of opposed forces and the one on the left reminds the service to our fellow men.

The three inferior leaves of the Flor-de-Lis refer to three fundamental states of Being, which the spiritual worker has to reach and integrate before reaching the three superior states of the Being represented by the three superior leaves of the Flor-de-Lis once it has gone through the central fringe.

CHRISTIC LOVE: The inferior left leaf represents Christic Love.

This relationship forms and consolidates a column of alignment and balance of the three inferior chakras, whose strength and stability are necessary to keep our heart open intensely and permanently. The consolidation of this "power column" takes place after our correct settlement and integration in the earthly and human reality, due to our knowledge of the vital energetic current that we permanently receive from Mother Earth and that allows us to keep alive.

Source:

http://www.peacewithinisworldpeace.com/flordelis/

https://syntropy.site/wp-content/uploads/2018/05/Flor-de-lis-symbol.pdf

Syntropy is all about getting God's Light into your psyche or light. The idea is to recharge your spiritual batteries once in a while.

Remember, every massless quantum or every massless elementary particle looks like and ACTS like a Photon or a Psyche. Every massless quantum particle is a Psyche. A Photon is a massless quantum particle or a massless elementary particle, which means that a Photon is one type of Psyche or Intelligence. The massless elementary particles within quarks and electrons are also a type of Psyche. Psyches have different assignments or different functions. The identity of Psyche has been hiding in plain sight all the way along; and, NONE of us realized what it is because we weren't looking for it and had chosen to believe that it doesn't exist. Psyche is some type of Photon or some type of massless elementary particle. This is what has been experienced and observed!

Psyche or Life Force is Syntropy or Eternal Life. Syntropy must be symbolized by something that represents Eternal Life because Syntropy IS Eternal Life. Syntropy is the eternal conservation of Psyche, Intelligence, Energy, and Life Force. It's all internally consistent, and it has all been experienced and observed, which means that it must be true.

Mark My Words

—

See the following websites for a discussion of these various different topics:

https://syntropy.site/

https://syntropy.website/

https://mypsyche.us/

https://quantum-neuroscience.com/

https://science-2-0.com/

https://scientific-proof-of-god.com/

https://god-is-light.com/

https://ldssoul.com/

http://ldssoul.com/forum/

Invisible Forces and Fields

The day science begins to study non-physical phenomena, it will make more progress in one decade than in all the previous centuries of its existence. — Nikola Tesla

Wind is an invisible force that we can't see directly with our physical eyes.

So, how do we know that it exists?

We know that the wind exists by the effect that it has on physical matter. We can feel that it exists by the effect that it has on our skin and hair.

Gravity and magnetism are a couple of invisible, immaterial, supernatural, non-physical forces or fields that we can't see directly with our physical eyes nor detect directly with our physical instruments. So, how do we know that they exist?

We know that gravity and magnetism exist by the effects that they have on physical matter. The effects that they have on physical matter is undeniable. However, the very existence of gravity and magnetism – the proven and verified existence of gravity and magnetism – falsifies the claims of Materialism and Naturalism which state that such supernatural, invisible, non-physical forces and fields do not exist.

Likewise, Dark Energy, Quantum Waves, Quantum Mechanisms, Action at a Distance, and the Human Psyche are invisible, immaterial, supernatural, non-physical forces or fields that we can't see directly with our physical eyes nor detect directly with our physical instruments. So, how do we know that they exist?

We know that they exit by the effects that they have on physical matter. Action at a Distance, Quantum Tunneling, and the effects of the Human Psyche have been caught in the act by the observable influence that they have on physical matter. The proven and verified existence of Quantum Mechanics or Transdimensional Physics FALSIFIES Materialism, Naturalism, Darwinism, and their derivatives.

Entropy is one more of those invisible forces or fields or phenomena that we can't see directly with our physical eyes, nor can we detect it directly with our physical instruments. So, how do we know that it exists?

We know that it exists by the effects that it has on physical matter. We see things running down all around us. The increasing disorder, corruption, and decay is impossible to deny.

Likewise, Syntropy is yet another of those invisible forces or fields or phenomena that we can't see directly with our physical eyes, nor can we detect it directly with our physical instruments. So, how do we know that it exists?

We know that Syntropy exists because entropy exists. Entropy wouldn't have been possible without a huge infusion of Syntropy and Order at the beginning of our physical universe. Logical. Simple. Parsimonious. True.

The Syntropy had to come from the Quantum Realm, or the Syntropy Realm, or the Psyche Realm because there doesn't seem to be any observable Syntropy within physical matter. Physical matter only seems capable of collecting entropy. Physical matter doesn't seem to be capable of producing Syntropy.

Instead, Syntropy seems to be an inherent part of Quantum Mechanics, Supernatural Mechanisms, Transdimensional Physics, Psyche, Non-Local Consciousness, Intelligence, Dark Matter or Spirit Matter, God, and Love – eternal and everlasting, without a beginning of days or an end of years. Syntropy seems to be the original primal state of existence; and, it's the physical matter which is the anomaly in all of this.

Gravity, Magnetism, the Strong Nuclear Force, the Weak Nuclear Force, the Zero-Point Field, Dark Energy, Quantum Waves, Quantum Mechanisms, Action at a Distance, and the Human Psyche ARE all manifestations of Syntropy. Even entropy is a manifestation of Syntropy. The only thing that doesn't seem to fit within this paradigm is physical matter. Physical matter seems to be the very manifestation of entropy. How fascinating is that?

Do you see what becomes possible by choosing to allow ALL of the evidence into evidence and then pursuing a preponderance of that evidence? We find ourselves approaching Truth or Syntropy.

Mark My Words

—

Source

Science 2.0: I Upgraded My Science

https://www.amazon.com/dp/B0771K6WTX

The Prime Event Proves that God Exists

Events must have causes.

There are many different theories for the origin of our physical universe – some more believable than others. Secular scientists are looking for an explanation that doesn't involve God's intervention. Some of them have proposed a big explosion or a big bomb to account for the origins of our physical universe and the current expansion of our physical universe.

However, there are serious philosophical repercussions with this Big Explosion Theory (Big Bang Theory). Explosions of any kind require someone to make the bomb, and then they require someone to trigger the bomb and make it explode. Anything that was obviously made obviously has a Maker who made it. Anything that was obviously organized obviously has an Organizer who organized it.

If you believe in the Big Explosion Theory, then you have to believe in God – the one who made the bomb and the one who triggered its explosion. You see, any kind of beginning requires Someone Psyche to cause it to begin which means that any kind of bomb requires Someone Psyche or Someone Intelligent to make it and then explode it. Events have causes. Someone Psyche is needed to cause events; otherwise, nothing would ever happen. The Prime Event, if it happened, definitely had a cause. All events have causes.

If the Big Explosion Theory could ever be proven to be true, then it ends up being Scientific Proof for God's Existence because someone had to make the bomb, and someone had to make the bomb explode. The secular scientists and unbelievers can't get away from God through the Big Explosion Theory because anything that was obviously made obviously has a Maker who made it. This is Logic 101.

In order to get away from God as the Maker of this physical universe, the safest and most effective course is simply to assume that our physical universe has always existed. If our physical universe has always existed, then it doesn't need a Maker to make it.

There are many complex problems and physical impossibilities associated with this Eternal Universe Theory, the biggest one being entropy. If our physical universe has always existed, then it would have ended in heat death an eternity or two ago. It is obvious that our current physical universe, physical sun, and physical earth have had some kind of beginning in the recent past because they still exist and haven't cooled off yet.

Anything that has a beginning obviously has Someone Psyche or Someone Intelligent who caused it to begin. Anything that was obviously made obviously has a Maker who made it. The physical matter within this physical universe was obviously made relatively recently. It is obvious. The physical matter within this physical universe was obviously made relatively recently, which means that it obviously has a Maker of some kind.

While speaking to their chosen prophets, the Gods frequently talk about some kind of beginning. What do they say began? The heavens and the ground. What's in the heavens and what is in the ground? Physical matter is found in the ground; and, physical planets, physical stars, and physical galaxies are found in the heavens or the universe. The Gods repeatedly state that they formed or made the planets and stars. Anything that was obviously made obviously has a Maker who made it. This is Logic 101. It is obvious that physical matter, physical planets, physical stars, and physical galaxies were made relatively recently, which makes it obvious that they have some kind of Maker who made them.

How do you make physical matter?

The construction of physical matter requires a massive infusion of raw energy; and, it also requires someone to organize that energy and make that energy follow and obey the physical laws that have been established.

Order and organization require Someone Psyche or Someone Intelligent to order it and organize it. Laws of any kind obviously require some kind of Law Maker and Law Enforcer. Law enforcement is essential, or laws are worthless. Anything that was obviously organized obviously has an Organizer who organized it. The physical matter in this physical universe was obviously organized from raw energy relatively recently because entropy hasn't caught up with it and taken it out of play yet.

The existence of usable physical matter is Scientific Proof for God's Existence. Anything that was obviously made obviously has a Maker who made it. Physical matter was obviously made relatively recently which means that physical matter obviously has some kind of Maker who made it. This is Logic 101.

Since our physical universe can't be infinitely old, that is scientific proof that our physical universe and the visible galaxies within it had a beginning relatively recently. Anything that has a beginning obviously has Someone Psyche or Someone Intelligent who caused it to begin. Physical matter, physical planets, and physical stars don't just spontaneously generate out of thin air from nothing. They have never been observed doing so. The laws of conservation prevent such a thing from happening. Physical matter is made, and physical matter had some kind of beginning, which means that physical matter has a Maker who began it. Physical matter, planets, stars, and galaxies were obviously made which means that they have some kind of Maker who made them.

When we start to map galaxies, we see order and organization rather than random diffusion. We see the signature of God. The galaxies were obviously ordered and organized into clusters surrounding voids. Anything that was obviously organized obviously has an

Organizer who organized it. Anything that was obviously put into place obviously has Someone Psyche or Someone Intelligent who put it into place. The visible galaxies within this physical universe were obviously made relatively recently and were obviously put into place in some kind of pattern or form relatively recently as well.

Finally, physical matter, planets, stars, and galaxies are born. They come into existence all at once. We know that this is so because there is no way to get expanding clouds of hydrogen and helium gas to coalesce into stars. When you force hydrogen or helium into a balloon or a canister, does it coalesce into a star? No! It expands the balloon and it also tries to expand the metal canister. Hydrogen molecules naturally repel each other, even under pressure and under the effects of gravity. Helium atoms naturally repel each other, even under pressure and under the effects of gravity. There is no physical mechanism in existence that will make expanding clouds of hydrogen and helium coalesce into planets and stars.

The planets and stars have to be born in place whole from the beginning, or they would never exist. In other words, the planets and stars have to be made. The planets and stars were obviously made. Anything that was obviously made obviously has a Maker who made it. The Gods admit that they made the planets and the stars.

While talking with their chosen prophets, the Gods frequently mention the beginning of our heavens and our earth.

What began? The planets, stars, and galaxies began. The physical matter began.

Anything that has a beginning obviously has Someone Psyche or Someone Intelligent who caused it to begin; otherwise, there would be nothing but unorganized mass or unorganized energy. It would be chaos. Organization of any kind obviously requires an Organizer of some kind. Physical matter, entropy, and time obviously began, which means that they have Someone Psyche or Someone Intelligent who caused them to begin. Physical matter, entropy, and time were obviously made which means that they obviously need Someone Psyche to make them. Since physical matter, entropy, and time were obviously made, that means that they can be unmade or disassembled back into raw energy by the Person who made them. Psyche is the Ultimate Cause, and Psyche is the ultimate causal agent.

These realities are obviously true because creation from nothing or spontaneous generation is physically impossible. The laws of conservation prevent it from happening.

The Gods can form energy into anything that they want it to be, including physical matter, physical laws, physical restrictions, velocity or momentum, mass, spirit matter (tachyons and dark matter), transdimensional laws (quantum mechanics), time, entropy, space, locality, waves, gravity, magnetism, photons, visible light, radio waves, microwaves, x-rays, gamma rays, the nuclear forces, and the vacuum energy or zero-point field (dark energy).

The Ultimate Law of Thermodynamics states that Energy, Syntropy, or Psyche is conserved; whereas, physical matter or entropy is never conserved. The Energy is conserved; but, the form of that Energy is never conserved. The Gods can form Energy into anything that they want it to be; and, that form is never conserved. The Gods can change the form at will. Only the Energy, Syntropy, or Psyche is being conserved. Entropic physical matter is comprised of different forms of energy. The form is never conserved. That means that the form is made, constructed, organized, or formed. That means that the form has an Organizer who organized it and formed it. Since entropic physical matter is comprised of different forms of Energy, entropic physical matter obviously has an Organizer who formed them and made them.

The whole of science tells us that this reality is true. It has been experienced and observed. Someone Psyche at the psyche level had to make the Syntropy, or the Big Bang Singularity, at a quantum level! Entropic physical matter couldn't have done the job because it didn't exist yet. Can you sense the truth of it? I can. It's obviously true.

Any way you choose to look at it, according to these observations, the chaos or unorganized energy has always existed and will always exist. Meanwhile, the Gods are actively spreading order and organization throughout the whole of it as time allows. A physical universe is simply a different form of energy, an organized form of energy. The fact that the expansion of our current physical universe appears to be accelerating is proof that the Gods continue to infuse energy between the galaxies and proof that the Gods are deliberately accelerating the expansion or the growth of this physical universe. This suggests that their creation of physical matter is ongoing and happening right now even as I write this. A physical universe is simply a different form of energy; and, the Gods are actively forming it at this very moment. The Energy, Syntropy, Psyche, Intelligence, or Life Force has always existed; but, its form is being constantly changed by the Gods as they see fit. New planets, stars, and galaxies are currently being formed.

The Gods have revealed to their chosen prophets that they organize or form everything spiritually first before giving it a physical form. The spiritual formation from spirit matter serves as the template or the blueprint for the later physical formation that will take place.

Physical matter is formed by infusing it with massive amounts of raw energy as well as order, organization, physical laws, physical restrictions, space, locality, entropy, time, meaning, and purpose. Since the visible physical matter within this physical universe was obviously formed relatively recently, it is obvious that this physical universe obviously has Someone Psyche or Someone Intelligent who formed it or made it relatively recently.

Physical matter is Scientific Proof of God's Existence whether we realize it or not. Usable physical matter wouldn't exist unless Someone Psyche or Someone Intelligent made it exist. A massive infusion of Energy or Syntropy is needed to make a particle of physical matter; otherwise, all of that subsequent entropy wouldn't be possible to begin with.

Human beings make new physical matter or entropy in their particle accelerators or atoms smashers; and, human beings destroy physical matter or entropy in their atomic bombs turning it back into energy. Physical matter or entropy is never conserved. Its amount is constantly changing as the Gods see fit. It's the underlying Energy or Syntropy that's being conserved, not the physical matter or the entropy. This is the Ultimate Law of Thermodynamics.

Mark My Words

—

Source

Scientific Proof of God's Existence: Finding God Where the Atheists Refuse to Look for Him

https://www.amazon.com/dp/B07B26CRHX

A Quantum Model for Origins

As physical, fallen, mortal beings our observations and experiences are constantly telling us that we are running out of energy and that entropic physical matter and entropy will eventually result in the heat death of our physical universe when there is no more available energy to use. The physical matter or entropy seems to be the REAL component in all of this while we slowly run out of energy over time. Entropy or the Second Law of Thermodynamics seems to reign supreme. Does it not?

[Due to] the deep impression that classical thermodynamics made upon me, [the Second Law of Thermodynamics] is the only physical theory of universal content which I am convinced will never be overthrown. — Albert Einstein

If your theory is found to be against the second law of thermodynamics, I can give you no hope; there is nothing for it but to collapse in deepest humiliation. — Arthur Eddington

And there you have it – the Supremacy of Entropy or the preeminence of the Second Law of Thermodynamics.

There is one major obstacle to the Supremacy of Entropy as it has been presented to us, though. If the ultimate end of all things is the heat death of our physical universe, then that end should have happened an eternity or two ago; and, there should have been NO big bang and NO massive infusion of Syntropy to reverse that entropy. Whether you believe in the Big Bang Theory or not, ALL of the observational evidence is telling us that our particular physical universe received a massively huge infusion of Syntropy or Available Energy approximately 13.8 billion years ago when our physical universe was made or caused to begin. This infusion of Syntropy happened relatively recently, or there would be no stars, planets, and galaxies to speak of. The Supremacy of Entropy would make a big bang impossible. Without a massive infusion of Syntropy somewhere sometime into our physical universe from someplace else or something else besides our physical universe, none of that subsequent entropy would have been possible; and, we wouldn't exist.

There's something fundamentally wrong with the Second Law of Thermodynamics or our understanding of the Second Law of Thermodynamics. The Second Law of Thermodynamics teaches that physical matter and entropy are conserved; but, observation and experience do NOT verify that claim nor prove it to be true.

Quantum Field Theory

https://www.youtube.com/watch?v=ATcrrzJFtBY

https://syntropy.site/wp-content/uploads/2018/07/The-First-Quantum-Field-Theory.zip

The first full formulation of quantum mechanics was Schrodinger's equation, and it couldn't account for light at all. The Schrodinger equation is totally incompatible with Einstein's relativity. The Schrodinger equation implicitly assumes that forces act instantaneously.

They can't see the value or usefulness of Instantaneous Action at a Distance on a Universal Scale because they don't want to. They don't want it to be true. They want entropic physical matter to be the fundamental unit of reality and existence. They don't value the Schrodinger equation because it doesn't support Heat Death, the Theory of Evolution, the Theory of Relativity, and their other pre-chosen conclusions.

They talk about the Schrodinger equation as if it is worthless because it doesn't match with classical physics and the theory of relativity. They literally declare everything that doesn't match with relativity as anathema, deprecated, inadequate, incomplete, worthless, and FALSE. They never once think or realize that things like the Schrodinger equation represent a TRUTH and a REALITY and an ETERNAL EPOCH that existed and came BEFORE God designed and created entropic physical matter, mass, time, entropy, locality, the physical laws, quantum fields, quantum particles, and various different physical restrictions such as the theory of relativity. Something like the Schrodinger equation is worthless to them because it doesn't match with and describe classical physics and the theory of relativity. They never once realize that the Schrodinger equation represents something much more primal and essential and fundamental than entropic physical matter, classical physics, and the theory of relativity. Entropic physical matter, classical physics, entropy, time, locality, and the theory of relativity are ADD-ONS – and NOT the fundamental units of reality. The Schrodinger equation represents something infinitely more fundamental and essential than entropic physical matter, classic physics, and the theory of relativity.

They don't realize it; but, the Schrodinger equation describes the Psyche State or Primal Quantum State BEFORE God imposed physical limitations, physical laws, physical restraints, classical physics, entropy, time, locality, and relativity onto our universe. Before God slowed everything down with physical matter, time, entropy, locality, and relativity, everything in the Quantum Realm or Psyche Realm was instantaneous, simultaneous, omnipresent, omniscient, infinite, and conserved. God had to slow events down for us so that we could live them, experience them, learn from them, and actually remember having done so. There was nothing worth remembering BEFORE God started to organize the thing and slow it down.

The Theory of Relativity is the pinnacle or the apex of classical physics. The Naturalists and Atheists don't value nor want their scientific discoveries unless they match with the Theory of Relativity. They reject any theory of gravity that doesn't limit and restrict gravity to the speed-of-light and the Theory of Relativity. They never once entertain the possibility that certain aspects of gravity and quantum fields function and ACT instantaneously across the whole universe as if the universe were one big Spiritual Entity or Quantum Entity. Instead, they automatically reject everything that doesn't match with the Theory of Relativity and Classical Physics. These people don't value nor want any math or theory that points them to something that is Purely Quantum, or Purely Spiritual, or Purely Psyche. Periodically, the math reveals to them something that is Purely Quantum, Purely Spiritual, Instantaneous Action at a Distance, or Forces that ACT Instantaneously across the universe; and, these people automatically reject these scientific discoveries and scientific observations because they don't match with the physical limitations and physical restrictions contained within the Theory of Relativity. The truth of the situation is hiding in plain sight where they can't see it nor find it because they don't want it to exist and don't want it to be true. Fascinating, is it not?

Quantum Field Theory teaches us that "particles" are omnipresent waves of energy or infinite vibrating quantum fields. The omnipresent, infinite, multidimensional aspect of energy waves or quantum fields, as well as the conservation of information aspect of quantum mechanics, directly contradicts the Second Law of Thermodynamics. They both can't be true simultaneously. Something has to give. **This video tells us that in particle interactions, particles are created and destroyed all the time! Paul Dirac's Second Quantization is all about creating and destroying particles.** In other words, particles or physical matter are NOT conserved! Physical particles as well as the associated entropy are created and destroyed all the time. According to Quantum Field Theory, physical matter and entropy are not conserved. Yet, the Second Law of Thermodynamics teaches us that physical matter and entropy are Supreme, which means that they are conserved. The Second Law of Thermodynamics actually teaches us that entropy is conserved, while at the same time Materialism, Naturalism, Atheism, and Classical Physics teach us that physical matter is conserved.

Is that true?

The Second Law of Thermodynamics is an attempt to convince us that entropic physical matter is conserved, namely, that the entropy within physical matter is conserved. That is FALSE. The amount of entropy changes over times, as does the amount of physical matter. There is something fundamentally wrong with the Second Law of Thermodynamics after all.

Can you sense what it is?

Entropy or the Second Law of Thermodynamics ONLY applies to entropic physical matter. If you eliminate all entropic physical matter, then there will be NO remaining entropy. Outside of entropic physical matter, entropy or the second law of thermodynamics no longer applies. Then we are looking at a superior law of thermodynamics, the Conservation of Energy. Entropy and physical matter are made, which means that they are NOT conserved, which means that the Second Law of Thermodynamics is false. Entropy or the Second Law of Thermodynamics is NOT conserved. It's NOT syntropic. Entropic physical matter is formed from raw energy. They are made, which means that they can be disassembled and formed into something else such as spirit matter, dark matter, or syntropic matter instead. There's nothing sacred or supreme about entropy and physical matter after all. They are constructs, which means that they can be deconstructed and formed into something else.

The First Law of Thermodynamics actually falsifies the Second Law of Thermodynamics. Quantum Field Theory and Dirac's Second Quantization falsifies the Second Law of Thermodynamics. The Conservation of Energy, or Syntropy, falsifies the Second Law of Thermodynamics. The Big Bang Theory or "the beginning of our physical universe filled full of syntropy" falsifies the Second Law of Thermodynamics. The Second Law of Thermodynamics only holds true when it comes to entropic physical matter – everything else is based upon Syntropy or the Conservation of Energy. Remove the entropy from entropic physical matter, and instantly it becomes syntropic physical matter, eternal and everlasting. That's the way that resurrection from the dead works. Adam and Eve had syntropic physical bodies – immortal and everlasting. The tree of knowledge introduced entropy into their physical bodies making them mortal. That was some apple! Entropy is death. Remove the entropy or the death from your physical body; and instantly, you are looking at Eternal Life or Syntropy. Syntropy falsifies the Second Law of Thermodynamics.

LIFE falsifies the Second Law of Thermodynamics. Entropy is death. Entropy or death cannot design and create. Entropy or death cannot produce life. The very existence of LIFE falsifies the Second Law of Thermodynamics. Psyche, Intelligence, or the Life Force falsifies the Second Law of Thermodynamics. The Materialists, Naturalists, Darwinists, Nihilists, Atheists, Behaviorists, Determinists, and Classical Physicists have made a prior commitment to the Supremacy of Entropy, the Supremacy of Death, and the Supremacy of Physical Matter; and, their pre-chosen commitment to Entropy and Physical Matter blinds them to all the other possibilities. These people enshrine the Second Law of Thermodynamics and ignore everything else, which makes them unable to see and understand anything else. I like to believe that there are always possibilities. Psyche,

Syntropy, or Life produces other possibilities. Psyche, Syntropy, or Life produces entropy, creates entropy, forms entropy, or makes entropy possible, which means that Psyche, Syntropy, or Life can eliminate entropy whenever it sees fit to do so. The experienced existence of Eternal Life or Immortality falsifies the Second Law of Thermodynamics.

Can you see the flaws in the Second Law of Thermodynamics? Most people can't because they have trained themselves not to be able to do so; but, the Conservation of Psyche, or the Conservation of Energy, or the Conservation of the Life Force falsifies the Second Law of Thermodynamics. Intelligence or Life falsifies the Second Law of Thermodynamics. The Big Bang Theory falsifies the Second Law of Thermodynamics. All that Syntropy at the beginning of our physical universe falsifies the Second Law of Thermodynamics. Quantum Mechanics or Supernatural Mechanics is syntropic. Quantum Mechanics or Transdimensional Physics falsifies the Second Law of Thermodynamics. The verified and proven existence of Syntropy falsifies the Second Law of Thermodynamics.

Remember, the Second Law of Thermodynamics **falsifies** or **contradicts** Eternal Life, Psyche, Syntropy, Choice, the Conservation of Psyche, the Conservation of Energy, Supernatural Mechanisms, God, God's Psyche, God's Priesthood Power, Intelligence, Miracles, Immorality, Atonement, Resurrection, Action at a Distance, and Quantum Mechanics. In contrast, the verified and proven existence of these things falsifies the Second Law of Thermodynamics. They are mutually exclusive. They both can't be true. Eternal Life and the Second Law of Thermodynamics are mutually exclusive. Syntropy and the Second Law of Thermodynamics are mutually exclusive. Something has to give; and, it's the Second Law of Thermodynamics that comes up with the shortest straw.

The Second Law of Thermodynamics is a temporary construct, which means that it can be reversed and eliminated whenever the Gods see fit to do so. Pull the entropy out of your physical body, and instantly you will be immortal or syntropic, as if you had been raised from the dead. There's nothing sacred or supreme about Entropy and the Second Law of Thermodynamics. It's a temporary construct and nothing more. Instead, it's the First Law of Thermodynamics, the Syntropy, or the Conservation of Psyche and Energy that ends up being sacred, inviolate, perfect, and supreme.

Contrary to the common sense of classical physics, the Syntropic Energy or "infinite quantum fields" end up being the REAL component in all of this while entropic physical matter and entropy end up being the ephemeral, limited, and temporary component. Entropy is a function of time and a product of time. Time is the temporal and ephemeral component; whereas, Syntropy or Eternal Life is the eternal and everlasting component. Find out what falsifies or overrides the Second Law of Thermodynamics, and you find the answer to life, the universe, and everything. The most unintuitive theory, Quantum Mechanics and Psyche, ends up overriding, overthrowing, and falsifying the most intuitive theory, Entropy and the Second Law of Thermodynamics.

Powerful!

Find out precisely what's being conserved in the First Law of Thermodynamics – along with what's not being conserved – and you will find the answer to life, the universe, and everything. That's a very powerful claim. These scientific observations are not going to sit well with the Materialists, Naturalists, Darwinists, Nihilists, Behaviorists, Determinists, Atheists, and Classical Physicists because these people have enshrined death or entropy. The falsification of the Second Law of Thermodynamics falsifies Materialism, Naturalism, Darwinism, Nihilism, Atheism, and even Classical Physics.

However, if you want to know the truth about everything, then you MUST figure out what is being conserved and what is changing over time. Everything else depends upon this crucial piece of information. You have to figure out what is eternal and everlasting, and what is ephemeral and temporary. The two are mutually exclusive. The whole of Science depends upon getting this straight and getting it right.

Quantum Mechanics is comprised of a wide variety of different invisible non-physical forces and fields. Energy or Syntropy can be enticed by God's Psyche to take on many different dimensions, phases, types, frequencies, and FORMS. According to the First Law of Thermodynamics, Energy or Syntropy is conserved, meaning that it is eternal and everlasting without a beginning of days or an end of years. Its overall quantity never changes. Energy or Psyche cannot be made, and it cannot be destroyed. Syntropy is the First Law of Thermodynamics or the Conservation of Energy. Psyche is the innate intelligence within all the different forms of energy which gives that energy the ability to understand, follow, and obey God's Laws and God's Commands.

In contrast, Physicalism, Scientific Naturalism, Darwinism, Nihilism, Atheism, Classical Physics, Entropic Science, and their derivatives ASSURE us that the non-physical DOES NOT EXIST. These people tell us that ONLY physical matter and entropy exist. Their purpose in life is to convince us that Psyche, Syntropy, Action at a Distance, and God do not exist. As a result, the Materialists and Naturalists teach that physical matter and entropy are conserved because physical matter and entropy are the only thing that exists.

They are wrong.

The different types of invisible non-physical forces and fields that we have discovered over the centuries are comprised of different types or different forms of Energy. Entropic physical matter is a FORM of energy; and, the FORM is never conserved because it can always be transformed into something else. Entropy is a Physical Law. The Physical Laws were made. Anything that is made is NOT conserved. Physical Laws require a Lawmaker, Lawgiver, and Law Enforcer. Anything that was obviously made obviously has a Maker who made it. Entropy was made which means that entropy is NOT conserved.

Do you see how that works? It's the answer to life, the universe, and everything.

Energy is Syntropy. Energy or Syntropy is quantum mechanical or supernatural in nature and origin. All of the different invisible non-physical forces and fields that we have discovered are in fact based upon Quantum Mechanics, Syntropy, or Energy. Syntropy or Energy is conserved according to the first law of thermodynamics. Quantum mechanisms are conserved. Psyche or Energy is the ultimate perpetual motion machine. It cannot be made nor destroyed. The proven and verified existence of Quantum Mechanics or invisible non-physical forces and fields falsifies Physicalism, Naturalism, Darwinism, Nihilism, and their derivatives.

$E = mc^2$. Energy equals mass times the speed-of-light squared. Matter is Organized Energy. Organized by whom?

Every type of organization requires an Organizer to organize it. There's no such thing as spontaneous generation or creation ex nihilo. We know this is so because Energy or Syntropy is conserved. The first law of thermodynamics falsifies spontaneous generation and creation ex nihilo, and even chemical evolution and macro-evolution. Things don't just spring into existence from nothing. Physical matter doesn't self-organize, but instead requires an Organizer or Syntropic Intelligence to make it organize.

Scientific Observation: Every type of organized system requires an Organizer to organize it; otherwise, chaos or random diffusion ensues. This truth has been experienced and observed.

Scientific Observation: Matter is Organized Energy. Physical matter is frozen energy or frozen light. Physical matter is highly organized. Entropic physical matter is made from raw energy. This is what has been experienced and observed.

Scientific Conclusion: Therefore, it is logical and rational to conclude that matter of any type was organized or made by an Organizer. Matter has an Organizer who made it or organized it from raw energy.

Quantum Mechanics is represented by the Energy side of $E = mc^2$. Energy can take on an infinite number of different FORMS, including spirit matter and physical matter. Whereas, the physical matter, physical laws, and entropy that we are familiar with are adequately represented by the Mass side of $E = mc^2$. The speed-of-light squared represents the Physical Laws or the conversion factor. In other words, God does something to Energy or Spirit Matter in order to convert it into physical matter. God imposes restrictions, order, organization, and physical laws onto it. God imposes locality and entropy (space-time) onto it. God imposes Mass or Organization onto it. That is how spirit matter is converted into physical matter. Energy or Syntropy is conserved; whereas, physical matter and entropy come and go as God sees fit.

Materialism, Naturalism, Darwinism, Nihilism, and Atheism were designed to prevent us from discovering these simple truths. Scientific Naturalism incorrectly teaches that only physical matter and entropy exist, and therefore that physical matter and entropy are conserved. These people are wrong.

Physical matter, locality, entropy, and space-time are MADE by God. They are different types of Energy or different FORMS of Energy. Spirit matter and physical matter are Organized Energy. Physical matter and entropy have a beginning, which means that they can be made to end or forced to end by being absorbed back into Syntropy or by losing their Mass or Organization to Syntropy.

Physical matter is Organized Energy, and so is spirit matter. Spirit matter and physical matter are just different types or different phases of Organized Energy. Physical matter has a lot more Mass or Organization than spirit matter. This means that physical matter has a lot more restrictions and limitations than spirit matter.

According to the First Law of Thermodynamics, Energy or Syntropy is the thing that is being conserved. Energy or Syntropy is eternal and everlasting, without a beginning of days or an end of years. Energy, or Syntropy, or Psyche, or God is indestructible. Materialism, Naturalism, Darwinism, Nihilism, and Atheism were designed and created to prevent us from discovering Syntropy. It works. Only a few people have ever heard of Syntropy. None of our scientists seem to know anything about Syntropy. It's always fascinating to find the truth hiding in plain sight where nobody else can see it.

The first law of thermodynamics is a version of the law of conservation of energy, adapted for thermodynamic systems. The law of conservation of energy states that the total energy of an isolated system is constant; energy can be transformed from one form to another but can be neither created nor destroyed.

https://en.wikipedia.org/wiki/First_law_of_thermodynamics

Energy or Syntropy is conserved according to the First Law of Thermodynamics; whereas, the physical matter and entropy come and go as God sees fit according to $E = mc^2$. This is what has been experienced and observed. Energy or Syntropy is eternal and everlasting; whereas, physical matter or entropy is ephemeral and temporary. This is the

Ultimate Law of Thermodynamics. Materialism, Naturalism, Darwinism, Nihilism, and Atheism were created to prevent us from discovering this simple truth.

We humans have produced particles of physical matter in our atom smashers or particle accelerators. We have converted energy into physical matter. We humans have also destroyed physical matter or converted physical matter back into Energy or Syntropy with our atomic bombs. Throughout it all, the Energy or the Syntropy is conserved; but, the physical matter or entropy comes and goes, or changes in quantity, according to God's Laws. Technically, the physical matter and the entropy ARE NOT CONSERVED because their quantity or amount is changing all the time; whereas, the amount of invisible non-physical Energy or Syntropy that is hidden within the system as a whole remains the same. The Energy or the Syntropy is conserved; whereas, the physical matter and entropy are not.

Do you see how that works? This is the KEY to everything else.

Energy or Syntropy, namely the conservation of Energy or Syntropy, is the answer to life, the universe, and everything. Energy or Syntropy is conserved. Energy or Syntropy cannot be created nor destroyed. It's eternal and everlasting, without a beginning of days or an end of years. However, Energy can be converted into many different FORMS, including invisible non-physical forces and fields, spirit matter or dark matter, and physical matter. The Energy, Syntropy, or Psyche is conscious, alive, and conserved; whereas, the different types of matter have been limited, restricted, ordered, organized, and restrained by different types of LAW. Psyche or Intelligence ACTS; and, the different types of matter were designed simply to REACT or to OBEY Psyche's commands and God's LAWS.

The verified and proven existence of invisible non-physical Magnetism, Gravity, Radio Waves, Microwaves, Dark Energy, Dark Matter, Invisible Light, Quantum Waves, Thoughts, Cognitions, after death Memories, Psyche, Quantum Mechanics, Quantum Tunneling, the Quantum Zeno Effect, and Action at a Distance FALSIFIES Materialism, Naturalism, Darwinism, Nihilism, Classical Physics, Behaviorism, Determinism, Physical Reductionism, and Atheism which state that these non-physical invisible forces and fields do not exist.

In this section, I attempt to develop a Quantum Model for the Origin of Physical Universes or a Syntropic Model for Origins to replace the materialistic, naturalistic, atheistic, and Darwinian models which have been FALSIFIED by scientific experiments, scientific observations, personal experiences, the scientific methods, and logical common sense. I chose to call it a Quantum Model for Origins in the title rather than a Syntropic Model for Origins because practically no one has ever heard of Syntropy. Energy is Syntropy. Psyche is Energy – an invisible non-physical force or field. Psyche is the intelligence within energy. Psyche is Syntropy meaning that psyche is syntropic or psyche is conserved. Energy or Syntropy is eternal and everlasting. Quantum Mechanics or Supernatural Mechanics is a type of Syntropy.

This Quantum Model for Origins is based upon the first law of thermodynamics which states that Energy or Syntropy is conserved, meaning that Energy or Syntropy is eternal and everlasting, without a beginning of days or an end of years. Syntropy or Energy has always existed and will always exist. Energy or Syntropy is God's Presence or God's Influence. It is immortal and omnipresent. Syntropy is Eternal Life.

Syntropy, Energy, Psyche, or Light is the default state or the Primal Construct; whereas, physical matter and entropy are the add-on or the exception to the rule. Remember, Conservation of Energy is the RULE; whereas, physical matter and entropy are the exception to the rule. Physical matter and entropy are NOT possible without a massive infusion of Energy or Syntropy during the creation of physical matter and entropy. Some type of Syntropy must exist, or all of that subsequent physical matter and entropy would not have been possible and would not exist.

ONLY physical matter is subject to entropy. The second law of thermodynamics ONLY applies to entropic physical matter. Everything else is Syntropy and has NO entropy, meaning that Syntropy of any type is NOT subject to the passage of time and therefore does NOT age. In other words, Energy or Syntropy is conserved. It doesn't experience entropy, and it doesn't age. Energy or Syntropy or Psyche will NEVER die. It is eternal and everlasting. Energy, or Syntropy, or God is indestructible. It has always existed and will always exist.

It's the Energy or the Syntropy that is conserved, and NOT the entropic state or the physical state. The physical matter comes and goes – begins and ends – according to God's will, God's Commands, and God's Law; whereas, the Energy or Syntropy is eternal and everlasting without a beginning of days or an end of years. God cannot create nor destroy Psyche or Energy. This is what has been experienced and observed.

Psyche is a type of Syntropy, or a type of Energy. Energy can take on many different FORMS. Energy can be formed into spirit matter. Spirit matter or dark matter is a type of Syntropy in that the energy within them is eternal and everlasting. Energy can also be formed into physical matter. Entropic physical matter is a type of entropic energy or entropic mass. Physical matter is highly organized Energy. Physical matter has been subjected to physical limitations such as entropy, the passage of time, the aging process, sub-light speed limits, mass, locality in space, and a limited or non-existent ability to quantum tunnel at will. Physical matter has been given a short De Broglie Wavelength which forces it to remain localized in space and time.

The following YouTube Video explains why physical matter cannot quantum tunnel long distances at will:

https://www.youtube.com/watch?v=-IfmgyXs7z8

https://quantum-neuroscience.com/wp-content/uploads/2018/06/Is-Quantum-Tunneling-Faster-than-Light.zip

A short De Broglie Wavelength limits and restricts a physical particle's ability to quantum tunnel or to be everywhere all at once. Huge amounts of mass limit and restrict a physical particle's ability to travel faster than the speed-of-light.

In contrast, give the physical matter within your physical body an infinite De Broglie Wavelength, and you would be instantly Omnipresent like God.

You have to think outside the box if you are going to figure out how everything really works, including the spiritual side or the psychic side of the equation. In order to explain everything that has ever been experienced or observed, you must include Psyche, Syntropy, Eternal Life, Non-Locality, the Non-Physical or the Transdimensional, Spirit Matter or Dark Matter, Invisible Intangible Forces and Fields such as Gravity and Dark Energy, Intelligence, Life Force, After-Death Life Reviews, Quantum Mechanics, Supernatural Mechanisms, Action at a Distance, and God into your explanations. There is no other way. You have to allow all of the evidence into evidence, or you are never going to be able to explain that evidence in a scientific manner.

I used to be a Materialist, Naturalist, Nihilist, and Atheist; but, I have changed. Nowadays, I'm primarily interested in finding a scientific explanation for all the phenomena that have been experienced and observed rather than trying to prove that Psyche, Spirit Matter, Psi, the Supernatural, or God does not exist. Claiming that something "does not exist" does absolutely nothing to explain what it is and how it works. I want to know what things are and how they work, including all those intangible, invisible, non-physical forces and fields that have been discovered such as Psyche, Quantum Waves, Dark Energy, and Dark Matter or Spirit Matter.

God put into place physical laws that prevent the atoms in your physical body from quantum tunneling away from you at will. Particularly, by imposing a very short De Broglie Wavelength on physical matter or entropic matter, the Gods basically localize physical atoms into space and time thereby limiting or even preventing their ability to quantum tunnel. If they do quantum tunnel, they can't go very far due to their very short De Broglie Wavelength. However, the Gods designed and created physical matter, including its assigned entropy and short De Broglie Wavelength, which means that the Gods can eliminate entropy and change the De Broglie Wavelength at will.

It has been observed that the Gods and resurrected Angels can teleport, or quantum tunnel, their physical bodies anywhere in the universe instantaneously at will. An easy and effective way to accomplish this task would be to temporarily give the atoms within your physical body an infinite De Broglie Wavelength and then simply ride the wave. Quantum tunneling has been documented and observed. Physical atoms can and do quantum tunnel. We KNOW it is possible. It has been documented and observed. Remember, physical atoms can and do quantum tunnel – just not very far thanks to their short De Broglie Wavelength. The Gods have deliberately given entropic physical atoms or fallen mortal beings a very short De Broglie Wavelength thereby greatly limiting and restricting their ability to quantum tunnel huge distances.

I'm particularly interested in documenting the transdimensional, spiritual, non-local, or non-physical side of Quantum Mechanics because most of the physicists have convinced themselves that the non-physical does not exist; consequently, they have done an adequate job explaining what happens on the physical side of the equation while simultaneously completely ignoring and rejecting the spiritual side of the equation.

Physicists spend a lot of time talking about and dreaming about FTL (faster-than-light) drives – a physical phenomenon. FTL will never happen. God put into place physical restrictions, physical limitations, and physical laws to prevent FTL travel and to greatly limit and restrict our ability to quantum tunnel. However, if the physicists were to spend one second and simply consider the spiritual side of Quantum Mechanics, they would quickly realize that by giving a physical atom an infinite De Broglie Wavelength (a spiritual process or a psychic process involving mind-over-matter), then that physical atom would be able to quantum tunnel or teleport anywhere in the universe instantaneously at will thereby completely eliminating any need for FTL drives based upon entropic matter, warping of space-time, fuel consumption, and propulsion.

Parsimony!

By using quantum tunneling or teleportation, a Blink Drive or a Jump Drive or a Quantum Drive would eliminate the need for fuel. Out-of-Body Travelers and Near-Death Experiences have observed that the spirit matter (or syntropic matter) within their spirit body can quantum tunnel or teleport, which means that spirit matter retains the long De Broglie Wavelength that is inherent in all quantum matter, or syntropic matter, or spirit matter.

Quantum Mechanics, Syntropy, Quantum Tunneling, and a long De Broglie Wavelength ARE the default state; whereas, the physical matter, sub-light speed limits, locality, entropy, space-time, a short De Broglie Wavelength, and the aging process are the exception to the rule. Physical matter and entropy were made by the Gods. The Gods deliberately add entropy, mass, and a short De Broglie Wavelength to physical matter or entropic fallen matter in order to localize it, limit it, and restrict it to space-time. Consequently, entropic physical matter is deliberately limited to sub-light velocities and prevented from quantum tunneling anything further than a very short distance. However, give that same physical atom a long or an infinite De Broglie Wavelength, and instantly it will be able to quantum tunnel or teleport anywhere you want it to go just like spirit matter.

Do you see how that works? It's got a great deal of explanatory power. In fact, it explains everything that has ever been experienced or observed. Does it not?

I'm a big fan of the quantum tunneling solution to FTL because I have experienced it. God teleported me and the car I was driving to safety preventing a lot of people from being hurt or killed. One second, I was on the side of the car that had pulled out in front of me with nowhere to go but into that car. Time stopped. Then instantly, I was someplace else in the middle of a lawn, headed north-east instead of north, with the car completely stopped, and the car turned off. It was like I had passed out and then woke up someplace else. The car was completely stopped because if I had kept going at the same speed that I was originally going but in that new direction, I would have gone over the embankment onto the freeway and into oncoming traffic. I went from 45 mph to 0 mph instantaneously.

I know that God can quantum tunnel entropic physical matter or ordinary physical matter because I experienced it. God could just as easily have stopped the passage of time and teleported me and the car to the opposite side of this universe instantaneously if He wanted to. According to the theory of relativity, at velocities faster than the speed-of-light, time stops; and, that's precisely what I experienced. I experienced quantum tunneling or teleportation. But, it was nothing that I did. It was something that God did for me in order to save me and a lot of other people.

Entropy is a function of time. Entropy is the result of the passage of time. Entropy is an aging process. ONLY physical matter has entropy. Only physical matter is subject to entropy, or the second law of thermodynamics, or the passage of time. Only physical matter has been localized into space-time or frozen into space-time. Physical matter and entropy are NOT conserved.

Everything else is Syntropy or Energy according to the first law of thermodynamics. Energy or Syntropy is conserved. It can neither be created nor destroyed. It has always existed and will always exist. It has been experienced and observed that Energy, Syntropy, Psyche, and Spirit Matter are able to phase-shift and quantum tunnel.

Physical matter and entropy have a beginning, which means that they can theoretically come to an end. Syntropy, Psyche, or Energy has NO beginning and therefore will have NO end. Syntropy or Energy is conserved! Syntropy is Eternal Life. Syntropy is the answer to life, the universe, and everything. Materialism, Naturalism, Darwinism, Nihilism, and Atheism were designed to prevent us from discovering Syntropy, Psyche, and God; and, they do just that! Practically everyone has heard of physical matter or entropy; whereas, hardly anyone has heard of Syntropy or the Counterforce to Entropy. Instead, the Physicalists, Naturalists, Darwinists, and Atheists ASSURE us that things like Psyche, Syntropy, Eternal Life, and God do not exist. These people tell us that entropy or death reigns supreme.

I used to be a Materialist, Naturalist, Nihilist, and Atheist. I've been a scientist all of my life. I didn't hear anything useful nor convincing about Syntropy or Psyche for the first fifty-five years of my life because the Physicalists, Naturalists, Darwinists, Nihilists, and Atheists had successfully convinced me that it doesn't exist. I believed them, so I never really went looking for Psyche, Syntropy, or God in earnest during the first five decades of my life. Self-deception works, and it works every time.

Physicalism, Scientific Naturalism, and Darwinism were designed to prevent us from discovering Syntropy; and, it works. Most scientists teach that Psyche or Syntropy or God does not exist; and, most of us have never heard of Syntropy as a result of what they have taught us all of our lives. Syntropy or Energy, as well as the Conservation of Syntropy or the Conservation of Energy, has been an integral and essential part of Science all the way along; but, the Materialists, Naturalists, Darwinists, Nihilists, and Atheists have successfully kept it hidden from us for millennia. It's what they do, and they are very good at what they do. Physicalism, Scientific Naturalism, Darwinism, and Atheism were designed to prevent us from discovering Syntropy. They were designed to prevent us from finding and knowing the truth.

Syntropy is the opposite of entropy. Entropy is death. The theory of evolution is based exclusively on entropy. Random mutations are entropy. Natural selection or survival of the fittest results in death. Natural selection doesn't touch our genes! The theory of evolution is Creation by Entropy or Creation by Death. Death cannot design and create. Death can only destroy. It requires some type of Syntropy in order to produce life. It requires some type of Psyche or Intelligence to collapse the wave function. This is what Quantum Mechanics or Syntropy is trying to teach us.

The idea that entropy or physical matter can design, create, produce, engineer, fine-tune, field-test, manufacture, make, and deploy life and genomes is our modern-day form of superstition. It's idolatry. Billions have chosen to believe in it. Creation ex nihilo, spontaneous generation, chemical evolution, and macro-evolution are the pinnacle of superstition. The Physicalists, Naturalists, Darwinists, and Atheists literally treat physical matter, entropy, and the different types of evolution as if they were supernatural beings, which they are not. Evolution is their god.

The Materialists, Naturalists, Darwinists, Nihilists, and Atheists teach that ONLY physical matter or entropy exists; consequently, they also teach that physical matter and entropy are conserved.

They are wrong!

It's the Energy or the Syntropy that is conserved; whereas, the physical matter or entropy comes and goes as God sees fit. The quantity of physical matter and entropy is constantly changing, which means that physical matter or entropy is NOT being conserved. Energy's FORM is never conserved. The FORM can change over time. This is the Ultimate Law of Thermodynamics.

This simple truth is the answer to life, the universe, and everything. It explains everything.

I belabor the point from a few different angles in this and other essays because most of us have never heard of Syntropy. Syntropy was one of my greatest scientific discoveries. Syntropy is the answer to life the universe and everything; and, I didn't even know it. Syntropy has been experienced and observed; whereas, the non-existence of Syntropy has not.

Quantum mechanical or supernatural phenomena have been experienced and observed; whereas, Creation by Rocks, or Creation by Entropy, or Creation by Death has not. Evolution is entropy. Evolution is death and extinction. Design and Creation by Evolution has NEVER been experienced nor observed because the different types of evolution cannot design and create. Truth and Science are defined by what has been experienced and observed; whereas, the Physicalists and Atheists define "science" as Materialism, Naturalism, Darwinism, Nihilism, Classical Physics, Creation by Physical Matter, Creation by Entropy, and Creation by Death. The one has been experienced and observed; whereas, the other has been falsified by a complete lack of observational support. The false is falsified by the truth; and, the truth is repeatedly experienced and observed.

The following essays in this Quantum Origins section were informed in part by the following YouTube Videos from Spike Psarris.

Science or Storytelling?

https://www.youtube.com/watch?v=gufYmnj0Gjw

https://www.youtube.com/watch?v=PwodpHGfI4k

What you Aren't Being Told about Astronomy

https://www.youtube.com/watch?v=CzyQbOQ0dv0

https://www.youtube.com/watch?v=E66409i-yn4

Creation Astronomy

https://www.youtube.com/watch?v=dPTcnZXN8WY

Defeating Atheism with Science

https://www.youtube.com/watch?v=sh1cXwhY1sE

Atheism, when applied consistently, makes Science impossible. – Spike Psarris

Spike Psarris was the one who introduced me to the first law of thermodynamics and the conservation of energy. Remember, it's the Syntropy or the Energy that is conserved; whereas, the physical matter and entropy come and go as God sees fit. There's NO such thing as the conservation of entropy or the conservation of physical matter because their quantity changes over time. Physical matter is born; and, physical matter or entropy can die and cease to exist. Physical matter and entropy are made or caused to begin, which means that they can be disassembled and formed into something else instead.

Entropic physical matter is born by infusing a particle of spirit matter with huge amounts Mass or Energy thereby slowing it down and making it subject to the passage of time or entropy thereby localizing it into space-time. A newborn particle of physical matter is localized into space-time; and, entropic physical matter has the passage of time or entropy enforced upon it. Remove all of that Mass or release all of that Energy, and the particle of physical matter dies or ceases to exist; and then, the resulting particle of spirit matter or residual particle of dark matter becomes syntropic and is NO LONGER subject to the passage of time or entropy, no longer has any physical limitations, and can quantum tunnel anywhere that it sees fit.

Do you see how that works?

The birth and death of physical matter or entropy is the answer to life, the universe, and everything. Conservation of Energy or Syntropy is the KEY to everything. Syntropy means eternal and everlasting, without a beginning of days or an end of years.

The birth of a new particle of physical matter entails phasing a particle of spirit matter into space-time and making that newly born particle of physical matter subject to locality and entropy. Locality is represented by space; and, entropy is represented by the passage of time. In other words, a newborn particle of entropic physical matter is subjected to space-time, or physical restrictions, or physical laws. Remove all of that Energy or Mass or Organization from a particle of physical matter, and it returns to being a particle of syntropic spirit matter or dark matter instead; and, it no longer has any physical limitations imposed upon it. Phase-shifting and Quantum Tunneling are quantum mechanical processes or supernatural processes.

Whenever a particle of physical matter is converted back into a particle of dark matter or spirit matter by having its MASS removed, it also has its entropy and its locality removed and is thereby no longer restrained by nor restricted by the various different physical laws and space-time. It becomes syntropic once again and becomes subject to quantum mechanics instead. The locality and entropy cease to exist when the MASS is removed from a physical particle and it becomes a syntropic particle or spirit matter once again. This also means that physical matter can be recharged by an infusion of Energy or Syntropy.

Entropy is NOT conserved. The amount of entropy changes over time. Physical matter is not conserved. The amount of physical matter changes over time. Matter in general is not conserved. The amount of matter and the types of matter can change over time. Matter is Organized Energy. Spirit matter and physical matter are different FORMS of Energy. The matter is not conserved. The Energy or Syntropy is conserved, but the different FORMS and quantities of Organized Energy can and do change over time. Spirit matter is a lot more syntropic and quantum in nature than physical matter or entropic matter.

These realities and truths have already been discovered and verified by Science. ONLY 4.6% of our universe is entropic physical matter or ordinary physical matter. The remaining 95.4% of our universe is syntropic in nature and origin. 22% of our universe is syntropic matter, spirit matter, or dark matter. 73% of our universe is Dark Energy or Syntropy. Energy or Syntropy is the Presence of God or the Glory of God. God converted ONLY 4.6% of our universe to physical matter, entropy, locality, and space-time when He made our physical universe. The rest of our universe remains as Syntropy. The rest of our universe is quantum mechanical and supernatural. Syntropy is indeed the answer to life, the universe, and everything. Syntropy explains everything.

Remember, entropy and physical matter ARE a function of time. Their quantity can and does change over time. Entropy and physical matter are MADE. They have a beginning. That means that physical matter or entropy can come to an end when it is phased back into Syntropy by having its Mass, Organization, Entropy, and physical restrictions removed. It's the underlying Syntropy or the Energy that is conserved. It's the Syntropy or the Energy that is timeless, and ageless, and eternal – not the physical matter and not the entropy. The physical matter and entropy come and go as God sees fit. This is the Ultimate Law of Thermodynamics.

When properly understood, the truth of all of this is found in the first and second LAWS of thermodynamics. These truths are axiomatic. They are LAW. They are the basic foundation of Physics. They also explain a Quantum Model for Origins.

Energy or Syntropy is versatile. It can take on many different FORMS. Psyche or Intelligence or Light is Syntropy. Energy is Syntropy. Psyche is the innate intelligence within all the different forms of energy. Quantum Mechanics is a type of Syntropy. Spirit matter is a lot more syntropic than entropic. Action at a Distance is syntropic. In contrast,

Materialism, Naturalism, Darwinism, Nihilism, Atheism, Classical Physics, and their derivatives are based exclusively on entropy and physical matter. These people erroneously teach that ONLY physical matter or entropy exists. Entropy is death. The theory of evolution is Creation by Entropy or Creation by Death. Entropy or death cannot design and create; but, Psyche or Syntropy certainly can. The Energy, or Syntropy, or Psyche is conserved; whereas, the physical matter or entropy comes and goes as God commands.

It's all there in plain sight, and the ONLY thing that prevents us from seeing it, understanding it, and accepting it is our firm belief in Physicalism and Scientific Naturalism which state that ONLY physical matter and entropy exist.

The First Law of Thermodynamics or the Conservation of Syntropy FALSIFIES Materialism, Physicalism, Naturalism, Darwinism, Nihilism, and even Atheism. The verified and proven existence of Quantum Mechanics and Action at a Distance FALSIFIES Physicalism, Naturalism, and their derivatives. Energy or Syntropy is the RULE; whereas, physical matter, physical laws, space-time, locality, and entropy are the exception to the rule. The Energy or Syntropy is conserved; whereas, the physical matter or entropy comes and goes as God sees fit. Entropy is NOT conserved. Entropy, locality, space-time, and the associated physical matter are born from Syntropy or made by God, which means that at some point along the way the physical matter, entropy, and physical properties can also die and come to an end by having their associated MASS or Organization removed and returned to Syntropy.

Do you see how that works? It's the answer to life, the universe, and everything.

Syntropy is the Conservation of Energy. Syntropy is Quantum Mechanics. Syntropy is the Presence of God. God is Syntropy.

The Syntropy, or the Conservation of Energy, or the Presence of God has been there in the Science all the way along; but, we never found it because we weren't looking for it and didn't want to find it. No seeking, then no finding. Our firm and solid commitment to Physicalism, Materialism, Naturalism, Darwinism, Nihilism, Behaviorism, Determinism, Physical Reductionism, Atheism, and the Supremacy of Entropy prevented us from discovering Syntropy, the Conservation of Energy, or the Presence of God. Naturalism, Darwinism, and Atheism were designed to convince us that it's the physical matter or entropy that's being conserved, and that Psyche, Syntropy, and God do not exist. Scientific Naturalism was designed to trick us, deceive us, and prevent us from finding and knowing the truth.

Do you see how that works?

I put some time and effort into this as the following essays will attest. As usual, you will have to decide for yourself if I meet my burden of proof or not. But, the first law of thermodynamics and the second law of thermodynamics have convinced me that what I write here is real and truly exists.

Remember, only physical matter is subject to entropy or the second law of thermodynamics because Energy or Syntropy is conserved according to the first law of thermodynamics. Only physical matter is subject to locality and entropy (space-time) because everything else is Syntropy. Energy or Syntropy is eternal and everlasting; whereas, physical matter or entropy comes and goes as God sees fit. Entropy is death and extinction; whereas, Syntropy is Eternal Life and the Presence of God. This is the Ultimate Law of Thermodynamics. I'm not going to apologize for finding the truth. That's what scientists are supposed to do; and, I'm a scientist.

Isn't it fascinating to realize that Materialism, Naturalism, Darwinism, Nihilism, and Atheism were designed to convince us that physical matter and entropy are conserved? Materialism, Naturalism, Nihilism, Atheism, and their derivatives were created to convince us that Psyche, Syntropy, and God do not exist. Darwinism and the theory of evolution were designed to convince us that physical matter or entropy can design and create at will. Materialism, Naturalism, Darwinism, Nihilism, and Atheism were designed and created to hide the truth from us while at the same time convincing us that a wide variety of different falsehoods are true. By figuring out what these people truly believe, we can then use Science to figure out why these people are truly wrong. The purpose of Physicalism, Scientific Naturalism, Darwinism, and their derivatives is to trick us, and deceive us, and convince us that Psyche, Syntropy, Eternal Life, Action at a Distance, and God do not exist. They were designed with a purpose in mind.

Matter of any kind is Organized Energy or Organized Syntropy. I have been a scientist all of my life; and, it took over fifty-six years of life before Syntropy was finally revealed to me. I never found it before because I wasn't looking for it before. Syntropy is the Presence of God. You will never find it until you actually go looking for it. I used to be a Materialist, Naturalist, Nihilist, and Atheist; and, I had been incorrectly taught that Psyche, Syntropy, and God do not exist.

Find out precisely what's being conserved in the First Law of Thermodynamics – along with what's not being conserved – and you will find the answer to life, the universe, and everything. It's staring us right in the face. Some part of the system is eternal, everlasting, indestructible, and conserved; and, some part of the system is temporary and ephemeral. God, Syntropy, Psyche, or Energy is being conserved; whereas, the physical matter, locality, and entropy are not being conserved but are constantly changing as God sees fit. That pretty much explains everything, doesn't it?

In Syntropy, or the Conservation of Energy, or the Conservation of Psyche, or the Presence of God, I have found the answer to life, the universe, and everything.

Mark My Words

The Causal Force Behind Choice

What is the causal force behind choice?

Can your genes make choices and determine your destiny as the Materialists, Naturalists, Darwinists, Nihilists, and Atheists claim? Can your genes really control cell requisition, cell differentiation, cell migration, protein synthesis, evolution, synaptogenesis, neurogenesis, synaptic pruning, and the construction of your physical body as the Physicalists and Scientific Naturalists claim is being done by your genes? If so, then how are your genes communicating with the cell and handling the communication between cells? How are the genes communicating with each other and the transcription enzymes? What motivates your genes to command and control everything within your physical body? How is that genetic information transferred from cell to cell during cell differentiation, cell migration, synaptogenesis, and synaptic pruning?

The teachings of the Physicalists, Naturalists, Darwinists, Evolutionists, and Atheists are internally inconsistent and therefore self-contradictory where this particular point is concerned. According to these people your genes DO NOT have a Psyche, Soul, or Mind and can therefore MAKE NO CHOICES whatsoever. Then in the next breath, these very same people will claim that your genes are controlling cell requisition, cell differentiation, cell migration, synaptic typing, synaptogenesis, protein synthesis, protein assembly, apoptosis, and a whole host of other physical processes that require some kind of CHOICE between two or more different options.

Are the genes making these choices, or are they not? The Materialists and Naturalists and Atheists want to have it both ways; but, their stance is logically incoherent, mutually exclusive, and therefore self-defeating. We have to bring in the Quantum or the Sub-Atomic if we want to explain how physical atoms, physical genes, and physical molecules are planning and controlling our destiny because these atoms and molecules have NO mechanism for doing so at the physical level.

Our genes CANNOT MAKE CHOICES, at least not at the physical level. Our genes DO NOT HAVE A PHYSICAL MECHANISM for communicating their choices to other genes, proteins, and other cells – at least not at the physical level. The genes have NO CHOICE but to do what they are told. So, who or what is making all of the different choices for our genes and cells at the quantum level or the psyche level, since choices cannot be made at the physical level?

In one of the most powerful and useful Scientific Observations ever made by the human race, Out-of-Body Explorers and Near-Death Experiencers have observed that BOTH our spirit body and our physical body DO NOT HAVE experiences, do not make choices, and do not form memories while our Human Psyche is separated from its assigned spirit body and physical body.

In other words, it is the Human Psyche or Human Intelligence who is having experiences, making choices, and forming memories – NOT our spirit body, and definitely NOT our physical body.

The causal force behind all CHOICE is Someone Psyche or Someone Intelligent. Our spirit body and physical body do NOT make choices, do NOT have experiences, and do NOT form memories while our Psyche is separated from them.

This is powerful Observational Science with a vast amount of explanatory power behind it.

Our physical genes do NOT make choices. Genes have no choice but to do what they are told to do. Our physical genes are NOT planning, controlling, and constructing our physical bodies, cell requisition, cell differentiation, cell migration, synaptogenesis, protein synthesis, cell construction, and our choices – not at the physical level at least. God knows what's going on at the quantum level or the psyche level where our physical instruments cannot reach.

Psyche is the Ultimate Causal Agent. Psyche is the Ultimate Cause. This reality and truth is the basis of a Psyche Ontology which is in fact the basis for the Ultimate Model of Reality. Within a Psyche Ontology, Psyche or Quantum Non-Local Consciousness is the fundamental unit of reality, existence, and choice.

When it comes to Science, this is where the tires really hit the pavement and we start to gain traction. Its explanatory power is through the roof to infinity and beyond.

Mark My Words

—

Source

The Ultimate Model of Reality: Psyche Is the Ultimate Cause

The Law of Psyche

Psyche has been experienced and observed by Out-of-Body Explorers and Near-Death Experiencers. The First Law of Thermodynamics tells us that Psyche or Energy is conserved, which means that it is eternal and everlasting. It cannot be made, and it cannot be destroyed. Psyche or Energy is syntropic which means that it is always conserved both at the quantum level and the physical level.

I prefer the scientific approach. I define science as observation and experience. Psyche has been experienced and observed; therefore, we KNOW that it exists. Quod erat demonstrandum!

Law of Psyche: The Law of Psyche states that every massless quantum particle is some type of Psyche or Intelligence. Psyche is the innate intelligence within all the different forms of energy which gives that energy the inherent ability to understand, follow, and obey God's Laws and God's Commands. Every massless elementary particle is some type of Psyche because it is psychic, quantum, telepathic, makes choices, can self-generate or self-propagate, can think, can communicate, can transmute, can move, can phase-shift, can collapse its own wave function, is perceptive, is sentient, is omniscient, is omnipresent, is intelligent, is conscious, is alive, must travel at the speed-of-light from our perspective at the physical level, is prescient, and without any physical limitations imposed upon it can quantum tunnel at will in its native original environment at the quantum level. Remember, every massless quantum particle is some type of Psyche, or Intelligence, or Life Force. Psyche is seen by Out-of-Body Travelers as a pinpoint of light or a Photon. Each Photon is a type of Psyche or Intelligence. A Psyche is a Photon, a massless elementary particle. If it looks like a Photon and acts like a Photon, then it is some type of Psyche. Every massless quantum is a Psyche. Psyche USES mass, energy, or matter to get things done. This is what has been experienced and observed.

Psyche has been hiding in plain sight all the time where nobody can see it nor find it because they aren't looking for it; but, there's no great mystery here. Psyche is seen as a Photon BOTH at the quantum level and the physical level because it is a Photon. Psyche or Intelligence IS what it appears to be, a Photon of Light or a Pinpoint of Light. No surprise there. Psyche is what it has been experienced and observed to be – a massless quantum of energy that looks like and acts like a Photon. Psyche, whenever it has been seen both at the quantum level and the physical level, looks like a Photon or a Pinpoint of Light.

I know that it's hard to believe because we have been trained, brainwashed, and conditioned all of our lives not to believe it; but, it is true, nonetheless. Psyche has been experienced and observed. Every massless quantum or every massless elementary particle looks like a photon and acts like a photon because is a Controlling Psyche who USES energy, mass, or matter to get things done both at the quantum level and the physical level.

Mark My Words

Experiential and Observational Proof of Psyche

The Materialists, Naturalists, Darwinists, Nihilists, Behaviorists, Determinists, Physical Reductionists, and Atheists teach and believe that there is NOTHING fundamental about energy or consciousness. I have even heard them say the words, "There is nothing fundamental about energy." These people teach us and tell us that Psyche or Non-Local Consciousness DOES NOT EXIST. These people erroneously teach and believe that ONLY entropic physical matter exists; consequently, these people teach and believe that energy or consciousness is an emergent property of entropic physical matter. Whether they realize it or not, these people also teach and believe that entropic physical matter and entropy are conserved. They got their priorities and science upside-down.

Psyche is the innate intelligence within all the different FORMS of energy which gives that energy the inherent ability to understand, follow, and obey God's Laws and God's Commands. Energy is psychic and intelligent. Energy is sentient, conscious, aware, and alive. Psyche is Energy, and Energy is Psychic. Energy or Psyche is the fundamental unit of reality because Psyche, Intelligence, Life Force, or Energy is always conserved. In contrast, physical matter, particles, and entropy are simply different FORMS of energy. The FORM is never conserved. This is what has been experienced and observed. Particles or quanta are popping in and out of existence all the time. However, they don't really cease to exist as the Materialists and Naturalists claim. They simply change state, form, phase, or dimension while their underlying Intelligence or Energy is always conserved. Particles or quanta are made and then destroyed. They are NOT conserved. Only the underlying energy, psyche, or intelligence is conserved. This is what has been experienced and observed. Entropic physical matter is NOT the fundamental unit of reality after all. Entropic physical matter is MADE from many different FORMS of energy; and, the FORM is never conserved. This is the Ultimate Law of Thermodynamics.

I learn best through comparison and contrast by comparing what has been experienced and observed with the philosophical speculation and wishful thinking of the Materialists, Naturalists, Darwinists, Nihilists, and Atheists.

Psyche has been experienced and observed. Psyche is experienced as an immaterial, non-physical, intangible viewpoint in space. Psyche is seen or observed by out-of-body travelers as a pinpoint of light, a point particle of light, or a spark of light.

As scientists, we have to expand, enhance, and upgrade Theoretical Physics and make it explain what has already been experienced and observed. Psyche is like a photon, or a particle of light, or a quantum of light, except for the fact that Psyche actually emits light, or emits quanta, or emits and stores quantum waves. Thoughts and memories are quantum waves. Psyche is like a permanent, infinitely stable, perfectly conserved particle of energy or quantum of energy. Psyche is NOT ephemeral – coming and going – transforming and being transformed – like the other elementary particles or quanta that have been discovered and observed. Psyche, and the energy within psyche, is conserved. It's stable. It's eternal and everlasting. Because Psyche is conserved, its FORM never changes. Psyche is always a pinprick of light, or a point particle of light, or a spark of light. Psyche is like a Starbase that instantiates, or starts, or makes the other quanta or quantum waves. The traveling quanta have to come from someplace, and they come from the Starbases that we call Psyche. Psyche makes and stores quantum waves. That's how Psyche is able to communicate with other Psyches – through quantum waves. Psyche ends up being the fundamental unit of reality, or the conserved unit of reality. Psyche is Energy, and Energy is Psychic.

The vacuum of space is NOT empty. It is comprised of many different fields of energy which are omnipresent in time and space. The various different Quantum Fields were MADE from energy. "Particles" or quanta are excitations or "packets of energy" within Quantum Fields. Psyche is the innate intelligence within ALL the different FORMS of energy which gives that energy the inherent ability to understand, follow, and obey God's Laws and God's Commands.

The **gluons** binding quarks together comprise 99% of the energy or mass in a proton and neutron. The quarks themselves comprise only 1% of the energy or mass in a proton and neutron. The massive energy of the **gluons** within a proton, as compared to the rest energy of the quarks alone in the QCD vacuum, accounts for almost 99% of the mass within a proton. It's ALL made from energy, not entropic physical matter! The gluons or energy are what give the neutrons and protons stability and make them last essentially for forever. Proton decay has NOT been observed. Why? It's because the underlying energy is always conserved. Gluons and quarks are just different FORMS of energy. **Gluons** are viciously strong attractive forces within the Quantum Chromodynamic Field. Who is making these quarks and gluons ACT this way rather than some other way? It has to be Someone Psyche. Therefore, Psyche or Consciousness ends up being the fundamental unit of reality – NOT quarks and gluons.

Protons and neutrons can be transformed into each other which means that they are NOT conserved; but, the underlying energy is always conserved. Remember, the FORM of the energy is NEVER conserved; but, the underlying energy is always conserved. If the Gods were to pull all of the **gluons** out of this physical universe into a different dimension or universe, the whole thing would dissolve back into the chaos or anarchy from whence it came. The gluons, quarks, forces, fields, laws, restrictions, order, and organization were MADE, which means that entropy and physical matter were also made. Anything that is made or organized is NOT being conserved. In contrast, the underlying Energy or Psyche or Intelligence was NOT made and it cannot be destroyed because it is always conserved. This is what has been experienced and observed.

I define Science as observation and experience; and, I have observed that the observations and experiences of the human race as a whole FALSIFY Materialism, Naturalism, Darwinism, Nihilism, Atheism, and their derivatives. Contrary to what these people claim, entropic physical matter is NOT the only thing that exists. Entropic physical matter is NOT the fundamental unit of reality. Materialism, Naturalism, Darwinism, Nihilism, and even Atheism are FALSIFIED by Quantum Field Theory and the Ultimate Law of Thermodynamics. Psyche or Quantum Non-Local Consciousness ends up being the fundamental unit of reality. A Psyche Ontology ends up being the Ultimate Model of Reality.

We have to expand, enhance, and upgrade our Theories of Physics to include Psyche or the Conserved; otherwise, we will NEVER be able to explain what has been experienced and observed.

Quantum Field Theory supports and predicts the Ultimate Law of Thermodynamics which states that physical matter and the other FORMS of energy such as entropy are NOT conserved. It's all made from energy, and entropic physical matter consists of different FORMS of energy. Quanta or "particles", including physical matter, are packets of energy. Quanta or particles are changing FORM all the time, which means that they are NOT conserved. Quantum Field Theory supports these ideas. The energy in the quantum fields is conserved; whereas, the "particles", or physical matter, or quanta come and go which means that they are NEVER conserved. You can see the truthfulness of these assertions with your own eyes by studying the Feynman Diagrams. There is NOTHING fundamental about physical matter. Everything is made from different FORMS of energy, including physical matter. Made by whom? Organized by whom? By Someone Psyche! Who else would be able to organize energy into its different FORMS at the quantum level or the psyche level?

Packets of energy, fields of energy, quarks made from energy, gluons made from energy, and the Psyche or Intelligence within all that energy ends up being the fundamental basis of reality. Psyche or Intelligence ends up being the fundamental unit of reality or the causal unit of reality – the conserved unit of reality. In contrast, entropic physical matter is MADE from many different quanta or packets of energy that we often call "particles"; and, when it comes to the electrons and the virtual photons in the Feynman Diagrams, it is obvious that they are NOT conserved. They are being transformed all the time! ONLY the underlying Energy, or Psyche, or Intelligence is conserved. This is what has been experienced and observed!

The Secrets of Feynman Diagrams

https://www.youtube.com/watch?v=fG52mXN-uWI

https://syntropy.website/wp-content/uploads/2018/08/Feynman-Diagrams.zip

http://hyperphysics.phy-astr.gsu.edu/hbase/Forces/feyns.html

Remember, Psyche is the person who makes the particles or the quanta and starts them moving in the first place. Psyche or Intelligence is the causal force or organizing force behind the quantum fields, quantum packets of energy, "particles", and the choices being made during the particle interactions that are diagrammed on the Feynman Diagrams. Psyche or Intelligence is the reason why the quarks, gluons, bosons, and electrons ACT the way they ACT rather than acting some other way or not acting at all. Psyche or Intelligence explains why there is something rather than nothing. Psyche is the thing that instantiates quanta, makes quanta, or causes the "particles" to begin in the first place. Psyche is the thing that chooses or decides what type of quantum a newly generated energy packet will be. Psyche is the Organizing Force behind everything else. Psyche is like a permanent Starbase – it generates, transmits, receives, and stores quantum waves or packets of energy. Psyche is conserved. Psyche is the fundamental unit of reality and existence. By allowing Psyche and Quantum Mechanics in to play, we can literally explain everything that comes our way. Try it, you might like it!

Due to the observations and the experiences of the human race as a whole, I upgraded my science to Science 2.0. According to Science 2.0, the BEST and FASTEST way to find and know the truth is to live it and experience it first-hand for yourself, or to choose to trust someone who has. The Human Psyche is KNOWN by living it and experiencing it, especially while your physical brain and physical body are dead and gone. Science 2.0 allows all of the evidence into evidence and then pursues a preponderance of that evidence. That's how I was able to identify and discover the Ultimate Law of Thermodynamics. That's how I was able to FALSIFY Materialism, Naturalism, Darwinism, Nihilism, and Atheism.

Phenomenology or the Lived Experiences of the human race had a very powerful influence in changing and altering my life, my science, my philosophy of life, and my worldview. The Phenomenologists, Near-Death Experiencers, and Out-of-Body Travelers changed the way that I look at the world. Thanks to these people, I'm no longer the same individual that I used to be. I'm no longer a Materialist, Naturalist, Nihilist, and Atheist.

In order to fully comprehend Quantum Mechanics and Syntropy, you must understand and accept the impact and influence of Quantum Non-Local Consciousness (or

Psyche) on quantum mechanisms. Without that knowledge and understanding, your comprehension of Quantum Mechanics will always be incomplete. You must find an interpretation of Quantum Mechanics and Syntropy that explains what the Human Psyche and Nature's Psyche are doing at the quantum level or psyche level in order to get things done for us at the physical level. Without this bridge between the quantum and the physical, your knowledge and understanding of science will always be incomplete and ineffective. You must know how Psyche or CHOICE fits into the picture, if you want to understand how science really works at every level of existence.

Here's a list of some of the observational experiences of Psyche or Intelligence that changed my life.

Psyche or Consciousness without a Spirit Body

Nowadays, I define Science as observation and experience; and, I treat experience as scientific evidence. After reading William Buhlman's account of how he left his physical body, asked to see his spirit body, and found that he was an immaterial viewpoint in space while looking at his spirit body, I started looking for corroborating evidence to verify what William Buhlman reported. I began looking for the people who had the experience of being an immaterial pinpoint of consciousness, a pinprick of light, or a spark on the ceiling instead of looking for themselves and seeing parts of their spirit body. I have been looking for the experience of Psyche and the observation of Psyche.

William Buhlman wrote the following:

Journal Entry, October 2, 1982

I hear the buzzing, engine-like sounds and will myself out-of-body. [He left his physical body behind.] **I step to the bedroom door and automatically request "Clarity now!" My vision improves, and I step through the door, into the living room. Still feeling a little out of sync, I verbally repeat my request with more emphasis, "Clarity now!" I feel my awareness and vision snap into place.**

My thoughts are clear, and I make a verbal demand, "I need to see the form I'm in now!" Instantly I feel an intense sensation of being drawn within myself. I'm suddenly different, weightless as though I'm floating in space. As I look forward I see a sparkling, bluish white form. For some reason, I seem to know that I'm looking at my nonphysical body from a different perspective. I stare in amazement at this form before me that shines and flows with energy and light. It looks like an energy mold created from a million tiny points of light; it radiates a bluish glow but appears to have a defined outer structure. The body of light before me is naked and is identical to my physical form. Even though my body looks firm, there is a noticeable energy motion and radiation present. I can see what appears to be an ocean of blue stars throughout my body. It's difficult to describe because the stars are stable yet moving at the same time; the light and energy of my [spirit] body appear to change and flow almost like the waves of an ocean.

As I stare at the body of light, it hits me that I must be in another body. Yet I can't perceive any form or substance; I'm like a viewpoint in space without shape or form of any kind. [He, his immaterial viewpoint in space, is pure psyche or intelligence or consciousness.] **As I reflect upon my new state of being, I feel a sensation of rapid motion and I'm instantly back within my physical body.**

Lying still and reviewing my experience, I'm struck by an inescapable conclusion: I must possess multiple energy-bodies. The form I just experienced was noticeably lighter (less dense) than even my second nonphysical body. I realize that the traditional view of our possessing two bodies — a physical body and a spiritual body — is far too simplistic; we are much more complex than this. Just as there are multiple nonphysical energy dimensions within the universe, each of us must consist of multiple energy-bodies or vehicles of expression.

Now I seriously wonder just how many nonphysical bodies or forms this involves. I suspect that there must be one within each dimension of the universe and that all of these are interrelated and connected, just as the physical body is connected to its first nonphysical (spiritual) body.

https://psyche-ontology.com/buhlman/

[See: Buhlman, W. L. (1996). *Adventures Beyond the Body: How to Experience Out-of-Body Travel*. New York: HarperCollins.]

For me personally, this was the turning point. This was the straw that broke the camel's back. After reading this book, and this journal entry in particular, I've KNOWN that Psyche or Intelligence is something completely different than one's spirit body. Since that moment, I've KNOWN that Psyche or Intelligence is an immaterial viewpoint in space – the part of us that has experiences and forms memories of those experiences. Ever since then, I've KNOWN that our spirit body and physical body don't have experiences and don't form memories while our Psyche is separated from them.

I treated this account as Scientific Evidence; and, I went searching for confirming evidence. Seek and ye shall find. I found what I was looking for, which tells me that it is true. Psyche is experienced as an immaterial viewpoint in space. Psyche is seen or observed as a point particle of light, a pinprick of light, or a spark of light. Psyche has been experienced and observed! That REALITY makes it Science because Science is observation and experience.

Science 2.0 allows ALL of the evidence into evidence, and then it pursues a preponderance of the evidence. The BEST and FASTEST way to find and know the truth is to live and experience it for yourself, or to choose to trust someone who has. If a phenomenon is real, then you should be able to find corroborating evidence; and I do, when it comes to Psyche or Quantum Non-Local Consciousness.

When people are separate from their spirit body and physical body, and they are viewing their physical body while they are an immaterial spark on the ceiling, they are experiencing their Psyche first-hand rather than their spirit body. It happens to some out-of-body travelers, but not all of them.

A more common out-of-body experience is to see parts of their spirit body – hands, arms, legs, and feet – but not their skull or eyes. If you close your eyes right now and try to sense where you are when your physical eyes are closed, you will sense that you (your psyche) is in the third eye position just behind your skull. That's the position from which you perceive things as well while inside your spirit body exploring the astral plane.

Psyche has been experienced and observed. Our job as scientists is to figure out what it is and how it works rather than trying to explain it away, dismissing it, and pretending that it doesn't exist. Denial of evidence and rejection of evidence makes for bad science. Feynman Diagrams map or portray the interaction of "particles" or quanta within Quantum Fields. Psyche is the thing that makes those particles and starts those particles moving in the first place.

One of my most interesting and useful Scientific Discoveries came to me when I first realized that Intelligences or Psyches, whenever they are seen or observed, manifest as a Pinpoint of Light, a Spark of Light, a Pinprick of Light, or a Photon of Light BOTH at the quantum spiritual level and at the macro physical level.

Direct experiential evidence of Psyche is rather rare; but, I have found a few experiential accounts of the nature and functionality of Psyche or Non-Local Consciousness while it is separated from its spirit body. Most of them were found on YouTube. Here are a few of them:

A Point of Consciousness in a Wonderful Black Void and Pinpricks of Light

https://www.youtube.com/watch?v=quU1xPeOtWs

She went looking for her body, hands, and feet and couldn't find them. "I was a point of consciousness," she said. She also saw other pinpricks of light that she recognized as souls that had passed on.

This NDE and OBE is probably the closest to what I experienced when I died. I remember floating in a crystal sharp, bright, black, obsidian void; and, for the first time in memory, I was completely and totally at peace. A part of me died. My five decades of fear and anxiety died. I experienced some kind of spiritual rebirth.

I didn't see my spirit body, didn't look for my spirit body, and didn't see anything else either that I remember. It was bright, black, and I was totally at peace. I had finally found rest and peace for my troubled soul.

Based upon my experience, the Atheists are right, there's nothing there when we die; however, the Atheists are wrong, because I was there when I died.

Each person who dies has a different experience and comes back with a different perspective, which is why it's important to allow ALL of the evidence into evidence, and then pursue a preponderance of the evidence.

I seemed to experience many of the benefits of a Near-Death Experience and none of the side-effects or bad-effects – without having any kind of astounding or amazing experience worth reporting. I was finally at peace; and, it stayed with me to this very day. The fear and anxiety were gone. I'm finally at peace. I see it as a gift from God.

A Unit of Consciousness

She talks about not being in a spirit body anymore but being a unit of consciousness. She went on a tour of the universe with the Being of Light at the speed of thought. The Human Psyche or Human Intelligence can leave its spirit body behind and separate from its spirit body.

https://www.youtube.com/watch?v=WUjP9kKoXfU

Lightning Strike Near-Death Experience – Dr. Anthony Cicoria

When Dr. Anthony Cicoria was struck by lightning, he died. He, his spirit body, was outside his physical body. As he walked up the stairs, his spirit body dissolved beneath him and transformed into an orb of light. He, his Psyche, was looking at his spirit body and later that orb of light from an immaterial third-person perspective or that Third Eye Perspective. This NDE is Scientific Evidence or Empirical Proof that Psyche, or Intelligence, is a different entity than our spirit body and our physical body.

https://www.youtube.com/watch?v=WUXzj0Tczz4

Consciousness without a Body Attached to It

https://www.youtube.com/watch?v=DmBnCTuQUOc

She Described Herself as Pure Consciousness While Out-of-Body

https://www.youtube.com/watch?v=V7xWffB2nH0

No Body – Just Pure Consciousness

https://www.youtube.com/watch?v=1kql9eD9qO4

Near Death Experience of Barbara Wilcox

In the next NDE, Barbara Wilcox explains not having a body, or spirit, or a soul but being Total Intelligence.

https://www.youtube.com/watch?v=JQzap_jFXT8

The Human Psyche or Human Intelligence is something completely different than our spirit body; and, our spirit body is something completely different than our physical body.

Intelligence or Psyche pre-dated the birth and organization of our Spirit Body. Our Psyche is described as an immaterial viewpoint in space, a point of consciousness, a spark on the ceiling, a pinprick of light, a speck of dust, a mote, or total intelligence.

Our spirit body seems to be malleable and can take on different shapes at will. In non-consensus realities, your spirit body can take on any shape that you want it to be. An orb of light is a common shape used for describing a spirit body. For the spirit children of God, our human shape is the most common shape that we see whenever our spirit body is viewed; however, Satan, the devils, and the evil spirits have denied and rejected their heritage as the spirit children of God, and these evil spirits can and do take on any shape that they want – serpents, dragons, alien grays, spacecraft, Nordic aliens, purple monstrosities, forked tail, pitchfork, red skin, and horns.

Our physical body is a part of the Ultimate Consensus Reality, which means that there's not a lot that we can do to change our physical body at will, because our physical body was designed to stay the same.

Both our physical body and our spirit body have NO memorable experiences and form NO new memories while the Human Psyche is separated from them. It's the Human Psyche who forms and stores new memories and has experiences worth remembering. Our spirit body is simply a convenient way to interface with the spirit realm; and, our physical body is an essential necessity for interfacing with this physical reality, which is the Ultimate Consensus Reality.

Susan Volt Talks about the Divine Spark within Each of Us

The word "spark" is used quite often by NDErs and others to describe the Human Psyche or the Human Intelligence. It's the spark of life. Psyche is life. Psyche is the Life Force.

Even in science fiction – something like the *Transformer* movies – they talk about the Spark or the AllSpark, which is the Life Force within us and within them.

The series *Earth: Final Conflict* did a great deal of discussion about Psyches and our spirit or ghost, often portraying them as two different things. There's a lot of talk about Quantum Mechanics within that series, because Psyche and Quantum Mechanics are integrated together and inseparable. In that series, Psyche or Consciousness has NO physical limitations, which means that it isn't limited by the speed-of-light. Entropy and the speed-of-light limitation applies only to physical matter. They don't apply to quantum mechanisms, spirit matter, and psyche. Syntropy is the natural state in the quantum realm, or the psyche realm, or the spirit world. There are NO physical limitations at the quantum level or the psyche level.

The Materialists and Naturalists have a hard time accepting and understanding Quantum Mechanics and Psyche, because Quantum Mechanics or Transdimensional Physics is supernatural in nature and origin. Quantum Tunneling or Teleportation is a supernatural phenomenon that happens automatically at the quantum level or the psyche level but is greatly limited and restricted at the physical level by the physical limitations that God has imposed on physical matter. You don't want your physical atoms within your physical body quantum tunneling away on you one at a time until you dissolve into thin air. God has imposed physical limitations on physical matter to prevent that from happening to you.

It's interesting to observe that whenever the Human Psyche is separated from its spirit body, it's the Human Psyche or Human Intelligence who is having the experiences and forming the memories, and NOT the spirit body. The Life or the Spark is something completely different than the spirit body and the physical body. It's always the Psyche or the Spark who has the experiences and forms the memories, not the spirit body and not the physical body.

Nicole Swann Is an Awareness Looking Out but No Body

It's our psyche, or intelligence, or non-local consciousness who has and experiences awareness. It's not our spirit body, and it's definitely not our physical body. Our physical body has no experiences and forms no memories while our psyche is separated from our physical body. The same can be said of our spirit body. It has no experiences and forms no memories while our psyche is separated from our spirit body.

Jessica Near-Death Experience of a Different Kind

In the following NDE, Jessica describes being a Point of Light, a Small Spec, and a Pinpoint of Light during her near-death experience. She, too, is describing Psyche or Non-Local Consciousness. Jessica also described the fact that we don't see everything and don't understand everything which is there to be seen and understood while out-of-body. Instead, there is ever-growing awareness and understanding, with still more to go that is never reached. Jessica could sense the presence of others, but she didn't engage with them during her NDE as much as others seem to do. Hers is the most "self-centered" NDE that I have encountered so far. Jessica's Psyche is very much the center of her universe.

There are many things about Jessica's NDE that are atypical, abnormal, weird, unusual, and strange – making some people believe that she's faking the whole thing. I include her in this short list, because she talks about being that immaterial spark during her NDE. I've never heard Jessica talk about her spirit body. She always describes herself as a spark; and because of my research, I define that "immaterial spark" or "viewpoint in space" as Psyche, Intelligence, or Non-Local Consciousness within my books.

It's interesting how they each describe something different, yet something similar as well.

Due to all the empirical evidence or observational evidence from NDEs and OBEs, I believe that I have finally found the Correct Model of Reality or the Ultimate Model of Reality.

"The Ultimate Model of Reality: Psyche Is the Ultimate Cause"

Psyche or Intelligence or Non-Local Consciousness is something completely different than a spirit body. The psyche and spirit body exist on different frequencies in different dimensions. The Psyche or Intelligence is eternal and everlasting without beginning of days or an end of years – PURE SYNTROPY.

In contrast, our spirit body had a point in time when it was organized by the Gods – a birth date. If you can read this, then your spirit body is a child of God, a child of your Heavenly Father and Heavenly Mother. Raw spirit matter is also PURE SYNTROPY, without beginning of days or an end of years. But, the birthdate of our spirit body or the organization of our spirit body did indeed have a beginning.

It's ALL matter and energy, all the way down to the most elementary of particles. If there is such a thing as a point particle, then Psyche or Intelligence is it. Psyche is some kind of smart dust or intelligent energy. Psyche or Intelligence is smaller than a string. Psyche is the thing that sets the frequency at which the string vibrates. Thoughts and memories are like quantum waves or vibrations within a string. The smaller can dwell within and control the larger. Psyche produces and controls the quantum waves or the vibrations within the strings. Psyche is like an infinite singularity.

Two particles of matter can occupy the same space at the same time if they are out of phase with each other, existing a different frequency or dimension from each other. Consequently, a spirit body and its assigned physical body as well as their Psyche can all occupy the same space at the same time because they are out of phase with each other. The quantum functionality known as phase-shifting allows the Psyche, the Spirit Body, and the Physical Body to occupy the same location or the same space at the same time. It works due to a feature known as Quantum Superposition. Different quantum waves can occupy the same space at the same time in a cumulative or additive fashion, forming a united whole, but also at the same time remaining completely separate from each other as well.

Due to the same Quantum Superposition feature, it's possible for your Psyche and your Spirit Body to be in two different places on opposite sides of our earth at the same time. Due to the fact that our physical body is part of the Ultimate Consensus Reality, this dual-presence feature doesn't seem to apply to our physical body. Although dual-presence is theoretically possible for a physical body thanks to Quantum Superposition, God seems to prevent that from happening where the physical body is concerned, because physical matter and physical bodies have been subjected to physical limitations and entropy by God in order to make our physical reality dependable, reliable, predictable, and controllable. There are NO physical limitations at the quantum level or the psyche level, so infinitely more is possible at the quantum level or the syntropy level.

Quantum Mechanics is infinitely more versatile and infinitely more powerful than Materialism, Naturalism, Nihilism, and Classical Physics. Quantum Mechanics has infinitely more explanatory power than Materialism and Naturalism do. Quantum Mechanics makes for better science and better philosophy than Materialism, Naturalism, and Classical Physics do.

WE KNOW that Psyche or Intelligence exists and that it is REAL because it has been observed as pinpricks of light and experienced first-hand from a first-person perspective. The existence of Psyche or Intelligence or Non-Local Consciousness has been VERIFIED by observation and experience.

In contrast, something like the Multi-Me Theory or the Parallel Worlds Theory has been FALSIFIED due to a complete lack of observation. Nobody has observed this phenomenon nor experienced it, outside of science fiction. Likewise, chemical evolution and the associated theory of evolution have been FALSIFIED due to a complete lack of observational evidence because chemical evolution of genes and proteins from atoms is physically impossible thanks to entropy or the second law of thermodynamics.

Do you see how that works?

Truths are repeatedly VERIFIED by being observed and experienced. In contrast, falsehoods and lies and fiction – like chemical evolution or the evolution of completely new and different life forms from random mutations and natural selection – are repeatedly FALSIFIED by a complete lack of verification or a complete lack of observational evidence.

According to Science 2.0, the BEST and FASTEST way to find and know the truth is to observe it, live it, experience it, and VERIFY it. The second-best way to find and know the truth is to use the *negating the consequent* version of the Scientific Methods to falsify and eliminate everything that is false so that only the truth remains. If you successfully eliminate everything that is false, then only the truth will remain. This is Logic 101.

Science 2.0: I Upgraded My Science

https://www.amazon.com/dp/B0771K6WTX

Materialism, Naturalism, Darwinism, Nihilism, and Atheism are FALSIFIED due to a complete lack of observational evidence, or a complete lack of supporting evidence, or a complete lack of verification. That's the same way that we have falsified the Multi-Me Theory or the Parallel Worlds Theory. Chemical evolution, spontaneous generation, abiogenesis, and macro-evolution have ALL been FALSIFIED due to a complete lack of observation. These things are physically impossible and prevented from happening by entropy or the second law of thermodynamics.

Every aspect of the Theory of Evolution has been FALSIFIED due to a complete lack of observation, or a complete lack of empirical evidence, or a complete lack of verification. Random mutations and natural selection have NEVER been caught in the act of designing and creating completely new and different life forms from the pre-existing life forms that God designed and created; and, evolution (genetic drift), random mutations, and natural selection did NOT exist until AFTER God designed and created the genes, genomes, proteins, eyes, brains, and life forms in the first place. These truths are so obvious that I sometimes wonder how I was able to overlook them for fifty years of my life.

Science is all about observation and experience. Go with the experienced and the observed; and then, get rid of all the rest – the philosophical speculation and wishful thinking.

Remember, Psyche or Quantum Non-Local Consciousness has been experienced and observed. Psyche or Intelligence is a massless quantum or a massless elementary particle; and, Psyche looks like and acts like a Photon whenever it is seen or observed, both at the quantum spiritual level and at the macro physical level. At every level of existence, Psyche or Intelligence uses mass, energy, or matter in order to get things done. Photons and other massless quantum particles have been experienced and observed which means that different types of Psyches or Intelligences have also been experienced and observed. The massless elementary particles within fermions, leptons, quarks, electrons, and neutrinos are Controlling Psyches or Controlling Particles who command and control the mass, energy, forces, and fields that surround them. This is what has been experienced and observed.

Mark My Words

My Time in the Void

I don't consider my experience to be an NDE. I don't think it happened when I was near-death during my suicide attempt, but sometime later during the withdrawal phase of my addiction when I was near death or wishing I could die. For me, withdrawal lasted six months after I decided to go cold turkey from all the different prescription drugs that they had me on and had gotten me addicted to. During this six-month withdrawal period, my sense of time was gone, my sense of reality was distorted, I was psychotic and delusional, my sense of right and wrong was gone, I was apathetic, I wanted to die, I couldn't remember how to turn on a computer or television, and I was in a lot of pain and confusion.

I was tripping the light fantastic. I found out later that one of the psychiatrists had diagnosed me as being schizophrenic. That was a misdiagnosis, because I was actually suffering from substance-induced psychosis and withdrawal symptoms; but, I didn't figure out any of this until years later. I didn't know what the hell was going on, but it was hell.

I didn't believe in God anymore. I was an Atheist. But, at a couple of the worst times during the withdrawal, I remember crying out, "God help me. I can't take this anymore." And apparently, He did. When the pain is severe enough, you will say and do anything to get it to stop.

At some point along the way, a part of me did die; and, for the first time in my existence, I finally achieved peace. I don't think it was a gift from my suicide attempt, but more of a gift from finally getting sober. Like I said, I lived for over six months without any real sense of time; and, I was very much in an atheistic frame of mind. I was delusional and hallucinating the whole time. I'm one of those who convinced himself that my NDE didn't happen – that I simply imagined it all – but the profound change in psyche, personality, psychology, and perspective was definitely real and life-changing.

After dreaming of the void or floating in the void completely at peace, my psychology completely changed from anxiety, dread, fear, paranoia, and anger to one of peace and compassion. It was a spiritual rebirth of some sort, and the peace has continued to last ever since.

I consider October 31, 2012, to be my birthday. By then I was back to the land of the living, and I fully realized how insane, psychotic, delusional, and paranoid had really been. By that day, I had finally started thinking rationally once again. All the addictive medications were out of my system. My brain had rebalanced or normalized, and the withdrawal period was coming to an end.

They wouldn't let me quit the Zoloft, but by October 31, 2012, they had had me tapering down the Zoloft to a maintenance dose for a month or more; and, I felt an infusion of happiness, peace, and joy. I was back, and better than before. Good enough.

The Blind Can See While Out-of-Body

Spiritual consciousness is like the internet. It continues to exist when the computer or the brain is turned off. Spiritual vision continues to exist, even though the physical brain was never able to see.

https://www.youtube.com/watch?v=gKyQJDZuMHE

https://www.youtube.com/watch?v=azIh8gsXVRg

Powerful Life Reviews

https://www.youtube.com/watch?v=gF_Dj6EduLY

Merging and Becoming One or Omniscient

Renee Pasarow - Near Death Experience

https://www.youtube.com/watch?v=qlFTanblpvg

https://www.youtube.com/watch?v=rSrHE8zkwYg

https://www.youtube.com/watch?v=xB-T78qgfHM

Eben Alexander: A Neurosurgeon's Journey through the Afterlife

https://www.youtube.com/watch?v=qbkqj5J91hE

Confirmed Out-of-Body Experiences

Saw and Described the Operating Theater

https://www.youtube.com/watch?v=J5_x8U7SR0I

Maria Sees a Tennis Shoe on a Ledge While Dead and Out-of-Body

https://www.youtube.com/watch?v=3gGqpxa32og

The Dead Person Describes the Operating Theater

https://www.youtube.com/watch?v=JL1oDuvQR08

A Shared-Death Experience (SDE)

https://www.youtube.com/watch?v=a1qGJfSZ_LQ

The Bubble of Protection, Quantum Tunneling, and Teleportation

https://www.youtube.com/watch?v=DmBnCTuQUOc

Near-Death Experience: Conversations with God

https://www.youtube.com/watch?v=Zrx8C2lxhJI

Rudolf Smit - The Self Does Not Die: 104 Cases of Verified NDEs

https://www.youtube.com/watch?v=2qJ5UGbBGhg

NDE Research Proves Afterlife Exists

https://www.youtube.com/watch?v=mcb2cQMTPRg

Dr. Jeffrey Long - God and the Afterlife - Science & Spirituality Collide

https://www.youtube.com/watch?v=SyhZV-LGtJ8

Dr. Gary Habermas - Near Death Experiences

https://www.youtube.com/watch?v=ac9pF32gRxU

Penny Satori – Documented Cases Near-Death Experiences

https://www.youtube.com/watch?v=yS2ITQzSPLk

https://www.youtube.com/watch?v=F6TRsTUj8WM

https://www.youtube.com/watch?v=MwXQaRwqUpk

https://www.youtube.com/watch?v=T_KJNQPPoZk

https://www.youtube.com/watch?v=n1JAPxxFkkA

https://www.youtube.com/watch?v=YkW0ikd8i7U

Shared-Death Experiences

https://www.youtube.com/watch?v=Z31cI73DI7M

https://www.youtube.com/watch?v=-0-R3nz0cdg

https://www.youtube.com/watch?v=lWjYjsh8i0w

https://www.youtube.com/watch?v=N8_2P8s77lc

William Buhlman Out-of-Body Experiences

William Buhlman has turned out-of-body experiences into a business, a science, and a way of life. There is no end to what he has available for free on the internet. What you won't get from William Buhlman is any contact with God or religion because he seems to have had no experience with either.

https://psyche-ontology.com/buhlman/

I used to be a Materialist, Naturalism, Nihilist, and Atheist; so, William Buhlman was the first person to get through to me and convince me that there's more to existence than this physical reality, because he treated his OBEs as a science and actually experimented with the phenomenon.

https://www.youtube.com/results?search_query=William+Buhlman

https://www.youtube.com/watch?v=v-Oa2lWrKOg

https://www.youtube.com/watch?v=FjbwXI2-0n8

https://www.youtube.com/watch?v=HtHEtWntLiw

https://www.youtube.com/watch?v=6HWGPSRLBTo

https://www.youtube.com/watch?v=4OR6Kiwlohw

https://www.youtube.com/watch?v=ZlZNmwCD1pA

https://www.youtube.com/watch?v=JEimkt3Wl98

https://www.youtube.com/watch?v=NaaUkJF2JMc

https://www.youtube.com/watch?v=Apw2WpuW60o

https://www.youtube.com/watch?v=Zy1-0wbRJpY

https://www.youtube.com/watch?v=IoQ9T7H4OrE

https://www.youtube.com/watch?v=gewT3DtsRWM

After Effects of NDEs

Alicia Fagan on some of the aftereffects of an NDE.

https://www.youtube.com/watch?v=71zXemHbHaE

Nancy Rynes on NDE aftereffects and adjustments.

https://www.youtube.com/watch?v=ii1UDGWi6Gk

The After Effects of My Personal Experiences

After my brush with death, spiritual death, and spiritual rebirth, I have experienced some of the positive effects of NDEs. I had fifty years of anxiety and fear. The fear is gone. For the first time in my remembered existence, I'm at peace. I'm not afraid of death.

I worked out a system with God to get answers to my questions. I want to know how things really work; and, whenever I ask God a question, He either tells me in my mind what book to read next where I find my answer, or He reveals to me the next morning while waking up the answer to my question if the answer is not easily available in one of the books that I own.

Sometimes, I'll spend a month or two writing a book based upon a hypnopompic flash of insight that came to me in answer to prayer while waking up the next morning. It can take months to put into words that single flash of insight, which I had months before in answer to my prayer.

I don't see dead people, and don't have visions or any of that. That doesn't seem to be my calling or gift. But, I do seem to get answers to my prayers whenever it comes to science, reality, and how things really work at all levels of existence or in every dimension of existence. Sometimes the answers have been quite surprising, and far outside the box of the natural sciences or the physical science.

Once I got rid of My Materialism, My Nihilism, and My Atheism, it opened me up to infinitely more. Materialism, Naturalism, and their derivatives are the greatest bane and hindrance to scientific discovery that mankind has ever created. Quantum Mechanics and Psyche are supernatural in nature and origin. The very existence of Psyche and Quantum Mechanics as well as verified proof of quantum mechanisms FALSIFIES Materialism, Naturalism, Darwinism, Nihilism, Behaviorism, Scientism, Determinism, and Atheism.

Quantum Mechanics and Naturalism are mutually exclusive because Quantum Mechanics is supernatural and Naturalism states that the supernatural does not exist. They both can't be true at the same time. If one of them is true, then the other has to be false. Quantum Mechanics has observational evidence, experimental evidence, and verified evidence supporting it; whereas, there is NO evidence and can NEVER be any evidence to support the claim of Naturalism which states that the supernatural does not exist. It's logically impossible to provide evidence of something's non-existence. It can't be done. It's physically impossible.

Quantum Mechanics is supernatural. Its very existence falsifies Naturalism, Materialism, and their derivatives such as Darwinism and Atheism. Quantum Mechanics and Psyche are true and verified; whereas, Materialism, Naturalism, and their derivatives are false and have been falsified. If you choose to follow the evidence and choose to allow ALL of the evidence into evidence, it is immediately clear that Psyche and Quantum Mechanics are true because they have been observed and experienced, and it's equally clear that Materialism and Naturalism are false because their claims are impossible to observe and experience. Observation trumps philosophical speculation every time, or at least it should.

Science is observation. Materialism and Naturalism are unscientific, because their hidden assumptions or major premises cannot be observed nor experienced. Psyche and Quantum Mechanics have been observed and experienced. Materialism and Naturalism have not been observed nor experienced because they can't be. You can't experience nor observe the non-existence of something, such as the non-existence of the supernatural or the non-existence of Quantum Mechanics. You cannot observe the non-existence of God and Psyche, either; but, you can definitely experience these things and observe these things for yourself or choose to trust someone who has.

Can you see the difference between the observed and the verified in comparison to the unobservable, unverified, and falsified? The one is true and the other is false. Quantum Mechanics and Naturalism are mutually exclusive. If one of them is true, then the other one is automatically false. Psyche, Intelligence, Life, Syntropy, the Supernatural, and Quantum Mechanics are true, which means that Materialism, Naturalism, Darwinism, and their derivatives are false. It's really simple to understand once a person chooses to do so.

How Quantum Physics and Psychology Affirm NDEs

https://www.youtube.com/watch?v=rBshmwf-iaw

https://www.youtube.com/watch?v=V9KnrVlpqoM

Since I'm a scientist by nature, I also turn to Quantum Mechanics in order to explain Quantum Non-Local Consciousness or Psyche, Out-of-Body Experiences, and Near-Death Experiences. Quantum Mechanics is vastly superior to classical physics when it comes time to try to explain what's happening to us at the quantum level or the psyche level. Quantum Mechanics is a proven and verified science, and it's time that we as a race start using Quantum Mechanics to explain our observational experiences while we are out-of-body exploring the Astral Plane.

Your consciousness, the Human Psyche, is what makes your reality in any plane of existence that we can possibly imagine. Wherever you go, there you are. You don't cease to exist when your physical body and physical brain die or go offline for an extended period of time.

Scientists Who Have Had NDEs

Life-long scientists have interesting and unique perspectives on NDEs, especially after they have had one. By best friend is a patent holding scientist. He is also a seer and has had NDEs and visions. He has been in the presence of Jesus Christ our Savior, and he has had a lot of other amazing and interesting experiences while out-of-body or in-the-spirit.

I have gone out of my way to collect NDEs from scientists who have had them; and, TED talks have proven to be a good source.

https://www.youtube.com/watch?v=EtbiUsX1klk

https://www.youtube.com/watch?v=Y8WIdDz4RxI

https://www.youtube.com/watch?v=MyaBeHeRK6M

https://www.youtube.com/watch?v=rbnBe-vXGQM

https://www.youtube.com/watch?v=mMYhqTgE6MU

There are others:

https://www.youtube.com/watch?v=PMICW2aaplA

https://www.youtube.com/results?search_query=Eben+Alexander

I'm a scientist. I define Science as observation and experience. As I see it, we scientists have to take these observations and experiences seriously if we want to find out what things really are and how they truly work. The BEST and FASTEST way to find and know the truth is to live it, experienced it, and observe it for yourself, or to choose to trust someone who has.

Encounters with God or Jesus Christ

Whenever the being of light and love identifies himself, it's always Jesus Christ whom NDErs encounter and experience when they have separated from their physical body. These various NDEs convinced me that Jesus Christ does in fact exist. Should you find yourself in hell, Jesus Christ will come to you and get you out of there just for the asking. That's probably the most valuable lesson that I have learned from all of this.

Howard Storm Saved from Hell by Jesus

https://www.youtube.com/watch?v=UPj4wci_bcI

https://www.youtube.com/watch?v=Y9AjcfM75gI

Taught by Jesus Christ: Ralph Jensen Shares NDE (Near Death Experience)

https://www.youtube.com/watch?v=uWshfNnyEQA

Lee Stoneking Addresses UN General Assembly

https://www.youtube.com/watch?v=FYt8sv4vzQs

George Ritchie

https://www.youtube.com/watch?v=DsQIU2dNY44

Richard Met God

https://www.youtube.com/watch?v=HAR5MpYNnRE

Erica Had Two Different Types of Life Review during Her Encounter with God

https://www.youtube.com/watch?v=xG_hEi8E4U8

Ian McCormack - NDE - former atheist - near death experience

https://www.youtube.com/watch?v=sTU7MfOgDKM

Dr. Mary Neal - Raised from the Dead

https://www.youtube.com/watch?v=DX473dF7ChY

https://www.youtube.com/watch?v=ULsl92H-Noc

https://www.youtube.com/watch?v=63wY2fylJD0

A Wide Variety of NDE Encounters with Jesus Christ

https://www.youtube.com/results?search_query=nde+jesus

A preponderance of the evidence tells us that Jesus Christ is the being of light and the being of love whom we encounter after we die. A preponderance of the evidence tells us that Jesus Christ exists and that He truly rose from the dead. Science is supposed to be about observation and experience, not philosophical speculation. Science is supposed to be phenomenological, not just hypothetical wishful thinking. The Materialists, Naturalists, Darwinists, Behaviorists, Determinists, Nihilists, and Atheists don't want God to exist, so these people deliberately reject, censor, block, ban, and destroy any evidence to the contrary. That's NOT science. That's the actions of religious dogmatism, fanatical extremism, wishful thinking, and blind faith. It's a perversion of rationality, logic, and science to block, ban, censor, ridicule, and destroy evidence. Materialism, Naturalism, Atheism, and their derivatives are based upon a refusal to look at evidence. *Refusing to look at evidence* is a logic fallacy.

Truth Is Known through a Preponderance of the Observational Evidence

According to Science 2.0, the BEST and FASTEST way to find and know the truth is to observe it, live it, and experience it for yourself, or to choose to trust someone who has. Through phenomenology or lived experiences, we can go directly to knowing the truth without having to run expensive and time-consuming science experiments. Furthermore, our physical science experiments definitely infer that the quantum level, quantum realm, and quantum mechanisms are real and truly exist; but, our physical experiments cannot provide direct evidence that the quantum level or psyche level is there nor explain what these things are truly like. When it comes to the quantum level, psyche level, or syntropy level, the BEST and FASTEST way to know that it is real and truly exists is to observe it, live it, and experience it for yourself or to choose to trust someone who has.

The preponderance of the evidence tells me that the NDE and OBE phenomenon is real, and that there really is an afterlife and a spirit world.

In contrast, we have the Materialists, Naturalists, Nihilists, and Atheists telling us that there is no such thing as psyche, God, non-locality, spirit, non-local consciousness, an afterlife, NDEs, OBEs, SDEs, Life Reviews, quantum mechanisms, supernatural mechanisms, and spiritual experiences.

How do they know?

They don't know. They can't know. It's impossible to know that something does not exist.

How do they prove that they are right?

The Materialists and Naturalists can't prove that they are right. There's NO way to prove that something doesn't exist. It's physically impossible, philosophically impossible, and logically impossible.

The major premises or primary assumptions of the Materialists, Naturalists, Nihilists, and Atheists are FALSIFIED by a complete lack of observational evidence or a complete lack of verification. You can't observe or verify that something doesn't exist.

ALL of the observational evidence, experiential evidence, empirical evidence, and eye-witness evidence tells us that the Materialist, Naturalists, Darwinists, Nihilists,

Behaviorists, Determinists, and Atheists are wrong. Materialism, Naturalism, and their derivatives are FALSIFIED by observational evidence and a preponderance of that evidence.

People choose to trust and believe the Materialists, Naturalists, Darwinists, Nihilists, Behaviorists, Determinists, and Atheists; BUT, these people haven't even observed, nor experienced, nor seen anything at all. It's impossible to observe that God does not exist. These people are flying on blind-faith and wishful thinking, and nothing more. When it comes to science, truth, and reality, you are much better served by choosing to trust and believe someone who has actually observed and experienced something.

Science 2.0 elevates observation and experience over the philosophical speculation of the Materialists, Naturalists, and Atheists. Science 2.0 allows all of the evidence into evidence and then pursues a preponderance of the evidence. Science 2.0 is the way that science should have always been done but wasn't. We should NEVER have let science be hijacked by the Materialists, Naturalists, Darwinists, Nihilists, and Atheists, because these people don't KNOW anything because they have never experienced anything and never observed anything.

Science 2.0: I Upgraded My Science

https://www.amazon.com/dp/B0771K6WTX

Remember, the BEST and FASTEST way to find and know the truth is to observe it, live it, and experience it for yourself, or to choose to trust someone who has. In contrast, know that Materialism, Naturalism, Nihilism, and Atheism are FALSIFIED due to a lack of observational evidence supporting their primary assumptions or major premises.

The truth is repeatedly observed, repeatedly experienced, and therefore repeatedly VERIFIED. In contrast, falsehoods like Materialism, Naturalism, and their derivatives are repeatedly FALSIFIED by a complete lack of observational evidence or a complete lack of verification. All you want is the truth, unless of course you are a Materialist, Naturalist, Darwinist, Nihilist, Behaviorist, or Atheist – then any old lie will do.

Mark My Words

—

Source

Putting Psyche Back into Psychology: Restoring Science to Consciousness

https://www.amazon.com/dp/B071NC987S

References

The Ultimate Model of Reality: Psyche Is the Ultimate Cause

https://www.amazon.com/dp/B071NC9JK6

Science 2.0: I Upgraded My Science

https://www.amazon.com/dp/B0771K6WTX

Quantum Neuroscience: The Answer to Life, the Universe, and Everything

https://www.amazon.com/dp/B079Z6QQQB

Theory of Mind and Intelligence

Siegler, R., DeLoache, J., & Eisenberg, N. (2010). *How Children Develop*, (3rd ed.). New York: Worth.

Mind is Psyche. Psyche is the innate intelligence within all the different forms of energy which gives that energy the inherent ability to understand, follow, and obey God's Laws and God's Commands. Someone Psyche designed and made the Quantum Fields upon which everything else is based.

The Growth of a Theory of Mind

Infants' and preschoolers' naïve psychology, together with their strong interest in other people, provides the foundation for the development of a theory of mind, an organized, integrated understanding of how psychological processes such as intentions, desires, perceptions, and emotions influence behavior. Preschoolers' theory of mind includes, for example, knowledge that beliefs often originate in perceptions, such as seeing an event or hearing someone describe it; that desires can originate either from physiological states, such as hunger or pain, or from psychological states, such as wanting to see a friend; and that desires and beliefs produce actions. One important component of such a theory of mind – understanding the connection between other people's desires and their actions – emerges by the end of the first year.

Although most 2-year-olds understand that *desires* can influence behavior, they show little understanding that *beliefs* are likewise influential.

By age 3 years, children show some understanding of the relation between beliefs and actions. Most 3-year-olds also have some knowledge of how beliefs originate. They know, for example, that seeing an event produces beliefs about it, whereas simply being next to someone who can see the event does not.

At the same time, 3-year-olds' understanding of the relation between people's beliefs and their actions, a key part of their theory of mind, is limited in important ways. These limitations are evident when children are presented with false-belief problems, in which another person believes something to be true that the child knows is false. The question is whether the child thinks that the other person will act in accord with his or her own false belief or in accord with the child's correct understanding of the situation. Studying such situations reveals children understand that other people's actions are determined by the contents of their own minds, rather than the objective truth of the situation or the child's understanding of it.

The 3-year-olds' responses show that they have difficulty understanding that other people act on their own beliefs, even when those beliefs are false. This finding is extremely robust.

(*How Children Develop*, pp. 267-268.)

https://books.google.com/books?id=OUgnEGbgFeYC&pg=PA267#v=onepage&q&f=false

Beliefs and desires are a function of Psyche – NOT our genes and NOT evolution or natural selection. This is what has been experienced and observed. Evolution, random mutations, natural selection, and our genes have NO beliefs and NO desires. The idea that genes can be selfish, and then spread their selfishness or motivations into our psyche, is nothing but science fiction. The genes, evolution, and natural selection have NO physical mechanism in place for influencing, choosing, and controlling our thoughts, desires, and beliefs. According to the Naturalists and Atheists, evolution, natural selection, and genes have NO soul, spirit, psyche, or mind. They don't exist as a person, place, or thing.

Young children learn to distinguish between psychological causes and physiological causes. The psychological causes are produced by Psyche; and, the physiological causes are related to one's physical body. They are completely different; and, young children learn to tell the difference when they develop a Theory of Mind.

Theory of Mind

Describe how children's theory of mind develops from age two to age three.

A theory of mind is a basic understanding of how another person's mind works and how the other person's mind influences their behaviors.

A study of the history of psychology reveals that Psychology was originally the scientific study of the human psyche. Then with the advent of Behaviorism in the 1920's, Psychology lost its mind. Study of mind or study of the human psyche has been banned from Psychology ever since. Therefore, I found it interesting when our textbook started talking about how young children develop a "theory of mind" or a "theory of psyche" as a natural and normal part of their developmental and growth process – a process that they subsequently disavow and reject twenty years later when they get their degree in Behaviorism, Materialism, Naturalism, Determinism, Nihilism, or Darwinism from their public college where all these childish things are brainwashed or programmed out of them.

Our textbook and lecture series suggest that children develop a theory of mind in stages; and, these stages overlap. There is no clear-cut age limit for each stage contrary to what this assignment suggests.

Starting in their first year and for the rest of their lives, their theory of mind starts with the development of joint attention skills, intersubjectivity skills, and the ability to understand the intentions of others.

Then, between age one and age two, children develop their first theory of mind – namely, the realization that another person's **desires** can influence the other person's chosen behaviors. Then, children continue to use this capability and refine this capability for the rest of their lives. In a sense, these children develop mind-reading skills. Our text suggests that most every human reaches this level of "theory of mind", namely that the other person's desires influence that other person's choices and actions.

By the age of three, some children start to develop the next level of theory of mind, which is the realization that another person's **beliefs** influence that other person's behavior, choices, and actions.

Our text emphasizes that most three-year-olds fail false-belief problems. False-belief problems are tasks that test a child's understanding that other people will act in accord with their own beliefs, even when the child knows that those beliefs are false. Our textbook states that a review of 178 studies revealed that 14% of three-year-olds get false-belief problems correct (probably by lucky guesses); whereas, 85% of 5-year-olds get false-belief problems correct. This demonstrates the discontinuous stage aspect of theory of mind in action.

Our text also emphasizes that most of the autistic children can't solve false-belief theory of mind problems even as teenagers. In other words, this last aspect or stage of theory of mind, the ability to get into another person's head and see things from their point of view, is not universal. Not everyone acquires the ability to solve false-belief problems, just like not everyone reaches Piaget's formal operations stage of development.

The question asks, "How does a child's theory of mind develop?" The "how" or the mechanism is in dispute.

The nativists, based upon the theory of evolution, tell us that evolution gave us a theory of mind module; and, this innate or inborn theory of mind module within our physical brain develops our theory of mind for us. Whether the scientists realize it or not, the theory of evolution has been falsified trillions of times in thousands of different ways – literally. Therefore, the nativists' theory of mind module has also been falsified, particularly the belief that evolution made this theory of mind module for us. Evolution is entropy, and entropy can't make anything.

The only way a "nativist" or "innate" point of view would work would be to take the Human Psyche into consideration. The Human Psyche or Human Intelligence is in fact all of these different "brain modules" that the nativists keep talking about. If there is anything "innate" or "there-from-birth" within our mental development or conceptual development, it comes from Psyche. God's Psyche or God's Intelligence made our genome and proteins. Nature's Psyche made and mapped our physical brain. And, the Human Psyche brings many innate capabilities with it from heaven. The question asked how a child's theory of mind develops, and Psyche is the most plausible explanation, especially given the fact that evolution or natural selection can't design and create anything from scratch. In fact, evolution (genetic drift), natural selection, and random mutations didn't even exist until after God designed, made, and deployed the genomes in the first place.

Of course, this Spiritual Nativist or Psyche Nativist view is not permitted into science by the Materialists, Naturalists, Darwinists, Nihilists, and Atheists because it falsifies their chosen beliefs. However, the empiricist point of view is allowed into science by these people.

The empiricists teach and believe that a child learns theory of mind through experiences and interaction with society. This empiricist model or point of view is actually more plausible than the belief that evolution made a theory of mind module for us. The empiricist point of view can be made to work on both sides of the veil. Out-of-Body Travelers and Near-Death Experiencers have observed that it is their human psyche, spark, or life force who has experiences and forms memories of those events, not their spirit body and not their physical body. Based upon these scientific observations or experiential evidence, we know that the Human Psyche or Human Intelligence has the innate or inborn capability to learn through experience; and, that ability is brought with us from heaven.

Therefore, the empiricists' model for theory of mind is more plausible and credible than the nativists' "evolution did it all" model for theory of mind, because the empiricist model actually works on both sides of the veil whereas the nativists' evolutionary model does not. Furthermore, the theory of evolution has been falsified which means that the nativists' evolutionary model for theory of mind has also been falsified.

So, in answer to the question, "How does a child's theory of mind develop?", the correct answer is that the human psyche or the child learns theory of mind through personal experience with the concept; but with one major caveat, a person's physical brain can restrict and limit the extent or the stages to which a Human Psyche can take theory of mind, as demonstrated by the autistic children who still can't solve false-belief problems in their teens and even into adulthood due to faulty development of their physical brain.

The different stages of theory of mind overlap; and, not everyone develops these stages on schedule by age two or by age three. The scientific observations mentioned in this essay describe how children's theory of mind develops from age two to age three, as well as describing how theory of mind develops before birth, before age two, and after age three. It's all part of the same process, and it's taken in stages. The human psyche or the child develops theory of mind through experience, subject to any limitations or restrictions from its physical brain. Evolution doesn't have anything to do with it. Evolution doesn't have a soul or a mind; therefore, evolution or natural selection doesn't have a theory of mind.

Mark My Words

Defining Intelligence

Intelligence can be viewed as one thing, several things, or many things. List the characteristics that you think are the most important components of intelligence and explain their relevance.

The LDS Church, Hindus, Muslims, Christians, and many other groups of people – including Out-of-Body Travelers and Near-Death Experiencers – teach and believe that the Human Psyche or Human Intelligence survives the death of its physical body and physical brain. According to these people, Intelligence or Psyche cannot be made, and it cannot be destroyed. It is eternal and always conserved. It is a single entity that has always existed and will always exist.

Likewise, some of the psychologists teach and believe that intelligence or g – what they call General Intelligence – represents one single entity. According to these people, since g correlates with every observable and measurable domain, intelligence must be one entity. This is what they believe. This single entity idea of g or General Intelligence matches perfectly with the idea that the Human Psyche or Human Intelligence is one single entity who survives the death of its physical body and physical brain. Therefore, this g factor or General Intelligence factor is one of the most important components of intelligence, because it actually survives the death of its physical body and physical brain according to the observational evidence that we have on hand as a race. At the quantum level or the psyche level, the Human Psyche, the Human Intelligence, or the g Factor is the most important component of intelligence. Psyche or Intelligence makes intelligent choices at every level of existence.

On the physical plane or at the physical level, however, if there is bad genetics or the physical development of the brain fails somewhere along the way, then IQ, g, General Intelligence, or Psyche won't be able to manifest its intelligence at the physical level through that broken, damaged, misfiring, or missing brain. At the physical level, a functional brain is absolutely essential for showing signs of intelligence. The Human Psyche, Human Intelligence, or g Factor is still in there; but, it can't work through nor get through the haze or nothingness that's being put up by its physical brain. At the physical level, a functional brain is the most important component of intelligence, because a person is not going to appear intelligent without it. The movie *Awakenings*, with Robert De Niro and Robin Williams, is a case study demonstrating these two essential aspects of intelligence.

These are the most important components of intelligence. When it comes to intelligence, the g Factor, Human Psyche, or Human Intelligence is the most relevant factor at the quantum level or the psyche level because it can function and exist without a physical brain. In contrast, when it comes to intelligence at the physical level, the physical brain is the most relevant and useful factor. This explains everything that has ever been experienced and observed in regard to human intelligence.

I have worked with kids who have Down Syndrome and Autism. Their g factor, psyche, intelligence, or unique personality is clearly in there and clearly does exist. Each one of them has a unique personality, a unique skill set, a unique history, and a unique set of chosen likes and chosen behaviors. The truth of it is obvious.

However, due to their bad genetics in the case of Down Syndrome and faulty brain development in the case of Autism, these people don't appear to be intelligent at the physical level. Instead, they appear to have intellectual handicaps at the physical level. Their g Factor, or Personality, or Preferences, Intelligence, or Psyche is still in there and clearly observable; but, these kids aren't going to do well on an IQ test because don't have the brain functionality necessary to do so.

How else is the g factor relevant?

Well, it's also relevant in the philosophical domain. The Materialists, Naturalists, Darwinists, Nihilists, and Atheists teach and believe that there is no "g thing", meaning that there is no Psyche or Non-Local Consciousness. These people teach that only entropic physical matter exists; therefore, there is no "g thing". Out-of-Body Travelers and Near-Death Experiencers have falsified their claim that there is no psyche or no g thing; but, the Materialists, Naturalists, and Atheists won't allow that evidence into evidence because that evidence falsifies their claims and their beliefs. For these people, their preferred beliefs take precedence over the observational evidence. That's the way they choose to roll. However, the observed, experienced, verified, and proven existence of Psyche or the g Factor ends up being scientific proof or observational proof that the Materialists and Naturalists have gotten their science wrong somewhere along the way. The false is falsified by the truth; and, the truth is repeatedly experienced and observed. Our chosen philosophy ends up being the foundation of our subsequent psychology and belief system, so it's very relevant to our psychology and intelligence in general. We are not going to learn certain concepts if we refuse to study them; and, our intelligence will be limited or restricted as a result. In other words, choice is an essential component of intelligence. Only psyche makes choices. Our genes, proteins, and molecules don't make our choices for us because that's physically impossible. They have no physical mechanism for making choices. Everything is determined by Someone Psyche at the physical level.

Finally, we must mention one of the other important and essential components of intelligence. A human psyche with a fully functional human brain is not going to do well on an IQ test if that person has had no physical experiences, no schooling, no language exposure, no food, and/or no training and education from the other psyches or persons surrounding them in their society. At the physical level, experiences, memories, and interactions with other psyches in some kind of social setting is absolutely essential if a person is to do well on an IQ test. Social interaction of some kind is one of the most important components of intelligence at any level of existence.

The Human Psyche or *g* Factor represents the **Psycho** portion of the multi-factorial BioPsychoSocial Model of intelligence and abnormal behavior. A person has to be conscious and alive, or have a psyche or life force, in order to do well on an IQ test at the physical level. Our genes and brain development represent the **Bio** portion of the multi-factorial BioPsychoSocial Model of intelligence and abnormal behavior. You must have a fully functional brain if you are going to do well on an IQ test at the physical level. There's no getting around it. Finally, all those other psyches, animals, and other people represent the environmental, cultural, or **Social** portion of the multi-factorial BioPsychoSocial Model of intelligence and abnormal behavior. If you have no interaction with society at the physical level, then you are not going to do well on an IQ test. At the physical level, all three of these components are absolutely essential if a person is to develop language skills, show signs of intelligence, and do well on an IQ test. If one of these factors is temporarily missing, damaged, hindered, or non-existent, then the person won't appear intelligent no matter what you try to do. He or she is not going to do well on an IQ test.

We could go on and talk about fluid intelligence, crystalized intelligence, practical intelligence, general capabilities of the individual, innate intelligence, physical skills, job training, experiences or experiential knowledge, long-term memories, available knowledge, mental abilities, genetics, developmental disabilities, developmental history, brain functionality, abnormal behavior, neurodegenerative diseases, cultural strengths and weaknesses, bad choices, government, the chosen behaviors or agency of the active child, their interests and preferences, their race, their home environment, their socioeconomic status, lack of schooling, poverty, and dozens of other factors that influence their IQ and their General Intelligence; but, all of these are sub-domains or sub-factors of the big three, which are BioPsychoSocial in nature and origin.

The purpose of this assignment was to discuss the most important components of intelligence and explain their relevance; and, I have now done so.

Mark My Words

What Has Been Experienced and Observed

Psyche or Non-Local Consciousness is experienced first-hand by Out-of-Body Explorers as an immaterial viewpoint in space. Whenever someone goes looking for their own Human Psyche while out-of-body on the astral plane, there's nothing there to be found because they ARE their Psyche. Your spirit body and physical body don't have experiences and don't form memories of events while your Human Psyche is separated from them. It's your Human Psyche who experiences events and then forms memories of those events – both at the quantum level and the physical level. This is what has been experienced.

Psyche is seen or observed by Out-of-Body Travelers as a pinpoint of light, a point particle of light, a spark of light, or a Photon. A Photon is one type of Psyche or Intelligence. This is what has been observed. Science is observation and experience. Remember, Photons are Psyches or Intelligences; and, Photons have already been experienced and observed on BOTH sides of the veil – both on the spiritual side and the physical side. Furthermore, each massless quantum particle or massless elementary particle that is discovered ACTS like a Photon and is therefore also a type of Psyche or Intelligence. Remember, ONLY Psyche or Intelligence can do command, control, and choice.

Mark My Words

—

Quantum Field Theory Playlist

My ultimate goal is to update Science and Quantum Mechanics and make them match with what has already been experienced and observed by Out-of-Body Travelers and Near-Death Experiencers. Science is observation and experience after all, or it should be. I start with what has been experienced and observed, and then I try to find a scientific explanation for it.

The following information was derived from the PBS Space Time Quantum Field Theory Playlist and then adapted and enhanced to include everything that has ever been experienced and observed:

https://www.youtube.com/playlist?list=PLsPUh22kYmNBpDZPejCHGzxyfgitj26w9

The Quantum Field Theory Playlist is also included within the PBS Space Time Quantum Mechanics Playlist, which I also took into consideration:

https://www.youtube.com/playlist?list=PLsPUh22kYmNCGaVGuGfKfJl-6RdHiCjo1

The mathematical descriptions provided by Quantum Mechanics and Quantum Field Theory reflect deep truths about reality. Quantum Mechanics points us directly to infinite possibilities. Through Quantum Mechanics or Spiritual Mechanics even our physical reality becomes possible. Quantum Field Theory is the best description we have of the fundamental workings of reality. Quantum Field Theory describes all elementary particles as vibrational modes in fundamental fields that exist at all points in space and time throughout the universe. Vibrating strings carry energy; likewise, vibrating Quantum Fields also carry energy. All "particles" or quanta are oscillations in different Quantum Fields. Every elementary particle or quantum has its own Quantum Field. Quantum Mechanics is Energy Mechanics or the way that energy works at the spiritual level or the quantum level rather than the physical level.

Quantum Fields, or Energy Fields, or Spiritual Fields, or Supernatural Fields are FUNDAMENTAL; and, quantum particles are just ways in which those fields vibrate or oscillate. A Quantum Field has some value everywhere throughout the whole universe; and, that value is usually zero. A Quantum Field is an energy web or energy membrane that has some value at every point throughout the entire universe. Quantum Fields are omnipresent. They are highly ordered and organized; and therefore, they are very dependable. Quantum Fields were the first thing that the Gods made; and thus, Quantum Fields would be the last thing to go should the Gods ever decide to throw in the towel and call it quits. Even if our observable physical universe were to suffer complete Heat Death, ALL of the Quantum Fields or Energy Fields will still be there waiting to be reused anytime the Gods decide to do so. Energy is always conserved.

Entropy and the theory of relativity are NOT the fundamental units of reality. They are ADD-ONS that came later when the Gods made entropic physical matter. Quantum Mechanics, Supernatural Mechanics, or Spiritual Mechanics is infinitely more fundamental and essential than entropic physical matter and the various different physical limitations such as the theory of relativity and entropy. When it comes to physics, Spiritual Fields or

Quantum Fields or Energy Fields are the fundamental building blocks of our universe, NOT physical matter and NOT the physical laws such as the theory of relativity or entropy.

Light is an energy wave in the electromagnetic field, NOT some type of point particle as the Materialists and Naturalists want it to be. Light only manifests as a particle when it reaches its destination and its wave function is forced to collapse. The electromagnetic field is a Quantum Field or a Spiritual Field. It's non-physical in nature and origin. We think of the electromagnetic field as existing everywhere in space, whether there is an electron or photon present or not. Quantum Fields can oscillate. Vibrations in the electromagnetic field are called photons, what we experience as light. An electron is just an excitation, a vibration, in a different field called the electron field. The electron and electromagnetic fields are interconnected. Vibrations in one can cause vibrations in the other.

Quantum Electrodynamics talks about the interaction of the electron field with the electromagnetic field, which means interactions between electrons, positrons, and photons. Anti-matter is represented as time-reversed matter. Electron, positrons, and everything with mass is affected by time. Time direction is irrelevant for photons. Photons experience NO passage of time. From their perspective, photons literally quantum tunnel to their destination every time. Photons aren't held back by the different quantum fields that introduce mass and the passage of time into the equations. The force between two colliding electrons is communicated through photons. The elementary particles communicate with each other through forces, through light, through photons, or through quantum waves. They often call these interactions between the elementary particles, "virtual photons", because they are more like thoughts than they are like actual particles. Thoughts and memories are quantum waves. Thoughts, intelligence, and memories are Light.

Virtual Particles aren't limited by the speed-of-light or the direction of time which means that they are truly quantum or truly psychic in every sense possible. Instantaneous Action at a Distance takes place through these Virtual Particles, or Psychic Particles, or Exchanges of Thought. Virtual Particles are the closest that the Feynman Diagrams get to allowing Psyche onto the map and into the equation; but, Psyche has many other uses as well, some of which are deliberately excluded and ignored because they FALSIFY Materialism, Naturalism, Darwinism, Nihilism, Behaviorism, Determinism, and Atheism. Feynman didn't limit particle velocity to the speed-of-light.

In Quantum Field Theory, each of the particle paths are infinite paths or infinite possibilities for particle momenta. We even have to consider "impossible" faster-than-light paths because at the quantum level or the psyche level, faster-than-light is totally possible and actually exists. Faster-than-Light Action at a Distance and Quantum Non-Locality are a verified and proven part of Quantum Mechanics. They have been experienced and observed.

Momentum is frequency for a matter wave or a wave function. Quantum particles are a superposition of waves that can be represented either in terms of position or momentum. Precision in one is actually constructed by the uncertainty in the other. A single perfectly localized particle is the same thing as an infinite number of momentum particles that occupy ALL of the locations in the universe. Photons are electromagnetic wave packets whose momentum is represented by their frequency. Any particle, or any wave function, can be represented as infinite locations in particle space or as a combination of infinite momenta in momentum space. Heisenberg's Uncertainty Principle represents Infinite Possibilities.

Photons are NOT affected by time; and, from their perspective, photons literally quantum tunnel to their destination "faster-than-light" because massless photons do NOT experience the passage of time like particles with mass do. It all fits together, and it all makes sense; but, only if you allow it to do so.

Conservation Laws limit the interactions that are possible between the electron field and the electromagnetic field. Energy and momentum conservation require that particles not just vanish or appear from nothing. Conservation Laws guarantee that if something goes in, then something else must come out. Charge must also be conserved.

As Feynman Diagrams demonstrate, there are INFINITE POSSIBILITIES whenever elementary particles interact. So, what do you get when you add Psyche into the equation? You get collapsing wave functions as INFINITE POSSIBILITIES literally collapse into a SINGLE CHOICE or a SINGLE CHOSEN ACTUALITY. Psyche makes CHOICES and thereby collapses wave functions – infinite possibilities consequently become one single chosen reality. Whether they realize it or not, Psyche also involves the CHOICE to make a quantum particle, or a quantum wave function, or a quantum wave packet of energy in the first place. Quanta are chosen into existence or made to exist. This reality and truth explains why there is something rather than nothing. Powerful Science! They never include Psyche into Feynman Diagrams as a separate entity; and, they probably should if they want a true and complete picture of what's really happening at the quantum level or the spiritual level.

Instead, they always pretend that Psyche doesn't exist because they don't want it to exist. They have observed that CHOICE doesn't exist at the physical level; so, in order to include CHOICE, they literally would have to include Psyche and a Psyche Level, which they refuse to do because CHOICE and Psyche FALSIFY Materialism, Naturalism, Darwinism, Nihilism, Behaviorism, Determinism, and Atheism which state that choice and psyche DO NOT EXIST. This is a case where their pre-chosen conclusions completely override common sense and scientific observations.

Psyche makes choices. Choices are a function of psyche. A specific psyche is identified by the choices that it makes. Psyche has been experienced and observed, but the Naturalists and Atheists deliberately exclude it because it FALSIFIES their religion. CHOICE has been experienced and observed, but they deliberately exclude it because it FALSIFIES their religion. Instead, they limit themselves exclusively to physical particles and let natural selection, abiogenesis, chemical evolution, creation ex nihilo, macro-evolution, or spontaneous generation take care of the CHOICES that are being made. They simply replace one God with a different god – a god that can't actually make CHOICES and do the work – a god that doesn't work. Natural selection doesn't touch our genes. Natural selection doesn't work as advertised. There's NO intelligence or psyche or soul within evolution and natural selection. Natural selection is an idol, a man-made god, and a false god; yet, these people are constantly telling us that natural selection is our Designer and Creator and God. They reify and deify natural selection even though it doesn't do anything constructive and doesn't touch our genes.

Fascinating, is it not?

I have observed that "Infinite Possibilities" is a better definition and a better use of Heisenberg's Uncertainty Principle than the word "uncertainty". "Uncertainty" points our minds towards the complete randomness of classical physics, entropy, and random diffusion. Quantum Mechanics is ALL about Infinite Possibilities. We are always dealing with Infinite Possibilities in Quantum Mechanics until Someone Psyche makes a CHOICE which then collapses those Infinite Possibilities into ONE single actual reality. This is the way things really work at the quantum level and the psyche level. It's all based upon Psyche or CHOICE in the final analysis. Nothing ever really happens, and nothing ever becomes actual and real, until some kind of CHOICE is made by Someone Psyche. This is just the way it is. Our observed mechanical reality or physical reality is in fact emergent from an underlying reality in which the quantum particles that are formed into physical matter arise from the combination of an infinity of possible properties. The vacuum itself is constructed from the sum of infinite possible particles and infinite possible fields. It should have been called Heisenberg's Infinite Possibilities Principle. According to Quantum Mechanics, Psyche or Consciousness CHOOSES from among infinite possibilities in order to produce ONE single actual reality. When it comes to Quantum Mechanics, Psyche or Consciousness is the BEST explanation.

The theory of relativity represents a physical limitation. In contrast, Quantum Mechanics and Psyche represent Infinite Possibilities, including the possibility of teleportation, quantum tunneling, or faster-than-light travel at the quantum level and the psyche level.

Quantum Tunneling Is Faster than Light

https://www.youtube.com/watch?v=-IfmgyXs7z8

https://syntropy.site/wp-content/uploads/2018/04/Is-Quantum-Tunneling-Faster-than-Light.zip

I have observed that the inclusion of Psyche or Intelligence into Science and Quantum Mechanics greatly enhances the explanatory power of our Science. Psyche or Consciousness explains why wave functions collapse. Psyche explains how and why wave functions or quantum particles are made. Psyche explains why we have finely-tuned Quantum Fields filling our universe rather than concrete. Psyche or Intelligence explains why there is something rather than nothing. Psyche explains why there is order and organization rather than randomness and chaos. Psyche explains why physical matter, physical constants, physical restrictions, and physical laws exist. Psyche explains Action at a Distance, Telepathy, Quantum Tunneling, Mind-Over-Matter Placebo Effects, and the Quantum Zeno Effect better than any other alternative. Psyche explains why the Planck Constant exists; and, the Planck Constant explains why we have Quantum Mechanics, Supernatural Mechanics, or Spiritual Mechanics. The purpose of Science is to explain what has been experienced and observed. Psyche or Intelligence explains it all. Does it not?

In terms of explanatory power, Psyche is vastly superior to the "nothing" that we get from the Materialists, Naturalists, Nihilists, and Atheists. Is it not? "Nothing" or "It Does Not Exist" doesn't explain Quantum Fields, Collapsing Wave Functions, the Production of Quantum Particles, the Planck Constant, Quantum Mechanics, Action at a Distance, Telepathy, Entropic Physical Matter, or anything else for that matter. There's NO explanatory power within Materialism, Naturalism, Darwinism, Nihilism, Behaviorism, Determinism, and Atheism because they rely exclusively on the creation of something from nothing by Nothing – creation ex nihilo. Nothing doesn't explain anything. Creation ex nihilo doesn't explain anything!

Psyche is the innate intelligence within all the different forms of energy which gives that energy the inherent ability to understand, follow, and obey God's Laws and God's Commands. Psyche has actually been experienced and observed by Out-of-Body Travelers and Near-Death Experiencers. In contrast, creation ex nihilo, design and creation by natural selection, abiogenesis, macro-evolution, chemical evolution, and spontaneous generation have NEVER been experienced NOR observed; and, they NEVER will be because they are physically impossible. They were falsified in 1859 by Louis Pasteur. This is what has actually been experienced and observed.

A Quantum Field is like an infinite array of springs. Each point in space can oscillate. The smallest possible oscillation above zero in the electromagnetic field is an indivisible packet of energy or quantum that we call a photon. There's no way in the universe that

272

physical matter can be the fundamental unit of reality as the Physicalists, Naturalists, and Atheists claim that it is. Oscillations in Quantum Fields produce the appearance of physical matter which would make the Quantum Fields more fundamental and more essential than the entropic physical matter that they produce.

Remember, there's NOTHING fundamental about entropic physical matter. In particle interactions, particles are created and destroyed all of the time. Paul Dirac's second quantization is all about creating and destroying particles. It works for ALL elementary particles. In other words, physical matter and entropy are NOT CONSERVED! Their quantity is changing all the time. Entropic physical matter is being created and destroyed all the time. This is the Ultimate Law of Thermodynamics; and, it has been experienced and observed. Physical matter and entropy are NOT conserved.

Photons can be stacked or combined infinitely at the same location; but, physical matter, quarks, and electrons cannot. Physical matter was deliberately limited and restricted by the Gods. The greater organization and the greater restrictions and limitations associated with physical matter introduce new and interesting possibilities. It also tells us that entropy and physical matter are MADE or caused to begin. Only the underlying energy is conserved. Anything that is made or caused to begin isn't being conserved. It was formed from something else more fundamental like Energy. Formed by whom? Formed by Someone Psyche or Someone Intelligent! Only Someone Psyche can manipulate energy at the quantum level, the psyche level, or the spiritual level. Physical matter can't do it from the physical level.

In these videos, Matt periodically states that the Schrodinger equation produced a version of Quantum Mechanics that NEEDED fixing. What needed to be fixed? The Schrodinger equation needed to be fixed by the Dirac equation and Feynman's path integral formulation so that the Schrodinger equation then matches with the theory of relativity and classical physics. They don't consider it to be a true Quantum Field Theory unless it matches with the theory of relativity. The Schrodinger equation is non-relativistic. The Schrodinger equation leads to infinite possibilities, untamed infinities, and infinite possible interaction events; and, it was a major challenge for physicists to FIX this so that Quantum Mechanics and Quantum Field Theory match with the limitations and restrictions associated with entropy, classical physics, and relativity.

The Schrodinger equation is totally incompatible with Einstein's relativity; but, this is a good thing because it points us to a fundamental truth and a fundamental reality that existed BEFORE God made entropic physical matter and the various different physical limitations such as entropy and relativity. The Schrodinger equation implicitly assumes that forces act instantaneously at a distance, which is precisely what they did BEFORE God imposed physical limitations onto them. God created physical matter to slow things down for us so that we can live them, experience them, learn from them, and actually remember having done so. At the speed-of-light, there is NO passage of time – time stops. From their perspective, objects traveling at the speed-of-light quantum tunnel to their destination instantaneously. They experience NO passage of time. God resolved that situation by creating physical matter, entropy, relativity, and the passage of time. Physical matter allows us to have experiences.

The Schrodinger equation suggests something infinitely more fundamental and essential than entropy and the sub-light velocity restrictions of relativity. Remember, there are hidden quantum states and quantum fields that exist, many of which we haven't discovered yet. The Naturalists and Atheists won't discover the spiritual states, spiritual fields, and psychic fields because these people aren't looking for these types of things and don't want them to exist. Remember, Quantum Mechanics is Supernatural Mechanics or

Spiritual Mechanics. It describes how energy really works at the quantum level or the spiritual level. It is BEST understood from a spiritual perspective or a psyche perspective.

We know that every elementary particle has an associated quantum field that fills all of space. These quantum fields are omnipresent. These energy fields or quantum fields are more like membranes, or webs, or lattices; and, they have a very definite energy level, usually zero. The elementary particles, quantum particles, or quantum waves are just regions where a quantum field has a bit more energy. This energy manifests as vibrations in the field. Anti-matter is a vibration in the same quantum field as its regular matter counterpart. All elementary particles have a quantum field; and, all have an anti-matter counterpart. Anti-matter particles have the same mass as their ordinary matter counterparts, but opposite charge. That mass is very real. It's all comprised of energy. This is Quantum Field Theory or Spiritual Field Theory. It is our best description of the underlying workings of reality.

The quantum fields that fill our universe also fluctuate due to the uncertainty principle resulting in what we know as vacuum energy; and, some quantum fields have an intrinsic non-zero zero-point energy before bringing Heisenberg into it. This leads to the famous Higgs mechanism and possibly also the phenomena of inflation and dark energy. To understand the universe, we need to understand how it behaves absent heat, absent light, and absent matter.

In summary, space itself is comprised of fundamental Quantum Fields, one for each elementary particle. Those fields oscillate or vibrate with different energies; and, those oscillations are the electrons, quarks, neutrinos, photons, and gluons that comprise the stuff of our universe. Quantum Fields can only be excited in quantized chunks of energy – integer multiples of some baseline energy – that we call quantum waves or quantum particles. When there are no quantum particles in a Quantum Field, we call this the vacuum state. The quantum vacuum is a seething ocean of activity. Space itself contains energy. Even in the complete Heat Death of our physical universe, these Quantum Fields and vacuum states will still exist waiting for Someone Psyche to come along and wake them up. Psyche or Intelligence is all that's needed to remove entropy or heat death from any system.

The "virtual particles" in Quantum Field Theory are like thoughts in the mind of God. They have real effects; but, we will never be able to get our hands on them.

Empty space is far from nothing. It's actually impossible to reduce any substance to absolute zero in temperature. There is a real minimum average kinetic energy called a zero-point energy. Furthermore, completely empty space has some real energy. It has vacuum energy. Dark energy may be vacuum energy. Vacuum energy also should produce gravitational effects. The whole thing is energy; and, the whole thing is supported and sustained by Quantum Fields.

The amount of vacuum energy is in dispute. It's called the Vacuum Catastrophe. One thing that is obvious, though, is that the Gods designed and made entropy or the second law of thermodynamics to prevent us from getting at and using all of this free vacuum energy that obviously exists. All of this vacuum energy is freely available and completely available to the Gods; but, as entropic physical beings, we can't get at it because of entropy, heat death, and the second law of thermodynamics. Even if a system contains a lot of energy, that energy may be inaccessible to fallen mortal entropic beings such as us who are an integral part of that system.

Even in an extremely energetic quantum vacuum, NONE of that energy is available for our use if it is in thermal equilibrium. Only the Gods can get at it because they are the ones who made that energy unavailable to us in the first place by creating the second law of thermodynamics and entropy. For us, the quantum vacuum contains NO useful energy if the vacuum has reached thermal equilibrium or maximum entropy. However, according to the Quantum Law of Thermodynamics, heat death and entropy are impossible at the quantum level and don't exist at the psyche level which means that the Quantum Realm is an Isothermal System and ALL of the energy within that realm is available for use ALL of the time by Someone Psyche. This reality and truth has been experienced and observed. However, only if we could persuade God to give us access to some of that unavailable energy would we ever be able to use it as a power source.

Meanwhile, there may be untapped sources of energy within ordinary physical matter that we could get at instead. Converting hydrogen into hydrinos or dark matter is one such method; and, so is Low Energy Nuclear Reactions (LENR).

https://brillouinenergy.com/

https://brilliantlightpower.com/

As far as I know, nobody has released a commercial or viable LENR product which tells me that they still have many technological and engineering hurdles to overcome before this tech becomes ready for the big time.

Remember, even in completely empty space totally void of any particles, the whole thing is still filled from one end of the universe to the other with a wide variety of different Quantum Fields. Quantum Fields are made from energy. Someone Psyche had to design, organize, and make the Quantum Fields; otherwise, physical matter couldn't exist. Quantum Fields obviously were NOT made by physical matter! Quantum Field Theory is in fact one of the most convincing Scientific Proofs of God's Existence. God's Psyche or God's Intelligence is needed to explain why we have Quantum Fields filling the immensity of space rather than chaos or nothing. Anything that is obviously organized and made obviously has an Organizer and Maker who organized it and made it. Quantum Fields were obviously organized and made. The truth of it is obvious. Quantum Fields can be used to make entropic physical matter; but, physical matter can't make Quantum Fields. Someone Psyche is needed to make Quantum Fields, and Quantum Fields are needed to make physical matter, which also means that Someone Psyche is needed to make physical matter. First things first!

As demonstrated by Charge Parity Violation, the Universe or Nature's Psyche does treat anti-matter differently than matter. That's why there is more matter than anti-matter in our visible universe. It doesn't have to be that way; but, Someone Psyche made that choice. Remember, Psyche is the innate intelligence within all the different forms of energy which gives that energy the ability to understand, follow, and obey God's Laws and God's Commands.

In terms of physics, Quantum Fields or Spiritual Fields are the fundamental unit of reality, NOT entropic physical matter. In fact, physical matter would be impossible to make without the Quantum Fields already in place to hold physical matter together at the quantum level or the spiritual level. First things first! Quantum Fields or Spiritual Fields are infinitely more fundamental and essential than entropic physical matter. Materialism, Naturalism, Nihilism, and Atheism were designed to HIDE this truth from us; and, they do precisely what they were designed to do. They skew our perspective and keep us in the dark.

Remember, by definition in principle, Quantum Fields are invisible, intangible, non-physical, supernatural, and spiritual in nature and origin. Quantum Fields, Energy Fields, Supernatural Fields, or Spiritual Fields are the things from which physical matter is MADE.

That would make Quantum Fields more fundamental and essential than entropic physical matter. Remember, the Gods made everything spiritually first before creating it physically. Quantum Fields tell us that it must be so. God also tells us that it is so.

Someone Psyche was needed to design and make the Quantum Fields in the first place. Order and organization don't just spring into existence from nothing as the Naturalists and Atheists claim. Order and organization are MADE at the quantum level or spiritual level by Someone Psyche or Someone Intelligent. This is just the way it is. Physical matter can't control the quantum level, the spiritual level, or the psyche level. The Quantum Fields or Spiritual Fields had to be MADE by Someone Psyche. There is no other way! Anything that was obviously made obviously has a Maker who made it. The various different Quantum Fields were obviously designed and made, each with a specific purpose and function in mind. It's obvious for anyone who is willing to look and see. From what I can see, the Psyche or Intelligence who designed and made the Quantum Fields, or Spiritual Fields, or Energy Fields would in fact be the fundamental and the most essential unit of reality and existence. This is Logic 101. We get to the truth through a process of elimination. If you eliminate everything that is false and everything that has been falsified such as Materialism, Naturalism, Darwinism, Nihilism, Behaviorism, Determinism, and Atheism, then ONLY the truth will remain.

Of course, the Materialists, Naturalists, Darwinists, Nihilists, and Atheists will tell you that everything I have written here is false; but, what do they know? They present us with NOTHING – things that they say DO NOT EXIST – which means that the know NOTHING. I'm going to go with what has been experienced and observed. I'm going to go with what is KNOWN to be true rather than going with nothing as the Naturalists and Atheists do. It's my choice, and you can choose to do otherwise if you want. That's the true nature of CHOICE. The Human Psyche can choose to believe that the Human Psyche doesn't exist. Our beliefs are chosen into existence by our Human Psyche; and, it can choose to believe anything it wants to believe. This is what has been experienced and observed.

When it comes to physics, Quantum Fields are the fundamental unit of reality, and NOT entropic physical matter. When it comes to reality and existence, including the design and production of Quantum Fields, Psyche or Intelligence is the fundamental unit of reality, NOT entropic physical matter. This is what has been experienced and observed. The very existence of Quantum Fields FALSIFIES Physicalism, Materialism, Naturalism, Nihilism, and even Atheism. Psyche is the ONLY thing that could make a Quantum Field at the quantum level and the psyche level. Entropic physical matter could NEVER do such a thing.

Mark My Words

The Foundations of Quantum Mechanics

Psyche has been observed by Out-of-Body Travelers as a pinpoint of light, a pinprick of light, a point particle of light, or a spark of light. That's precisely the way that the physicists describe the fermions – quarks, leptons, neutrinos, and electrons. These "particles" of light OBEY the Pauli exclusion principle. Obedience to Laws requires intelligence! The components of the quarks and the electrons are individual pinpricks of Psyche or Intelligence; and, the psyches within fermions USE the bosons, gluons, photons, quantum waves, and light to communicate with each other, interact with each other, and bind themselves to each other. The elementary parts of Fermions are Psyche or Intelligence. That would make Psyche or Intelligence the fundamental unit of reality and existence. Photons, gluons,

and W and Z bosons are the four force-carrying gauge bosons of the Standard Model. There's NO mass in the elementary components of the fermions, the psyches, or the pinpricks of light. Over 99% of the mass within an atom is associated with the gluons. Psyche or Intelligence is driving the mass or driving the gluons. It's all made from Energy which is always conserved.

Technically, the elementary particles or psyches that make up a gluon are massless like photons; but, the force mediated or controlled by the gluon has or produces a lot mass or resistance to acceleration. Likewise, the elementary particles or psyches that make up the quarks and the electrons are massless like photons; but, the Higgs Field that interacts with the quarks and the electrons gives them their small amount of mass. Massless particles are psyches.

From a different angle, we observe that the elementary components within quarks, fermions, electrons, neutrinos, leptons, and bosons are massless, timeless, ageless, syntropic "quantum particles," pinpoints of light, quantum waves, or Photons that we call Psyche or Intelligence or Quantum Non-Local Consciousness. Thoughts are quantum waves. Psyche or Intelligence is there in the models and the math just waiting to be identified as such.

Photons are Psyche or Intelligence, which means that the massless, timeless, ageless, syntropic elementary components or "elementary particles" within quarks, fermions, electrons, neutrinos, leptons, gluons, and bosons are in fact Photons, or Psyche, or Intelligence, or Quantum Non-Local Consciousness. Psyche or Intelligence is Light and Truth. Psyche is a Photon; and, Photons are psychic. Psyche is observed by Out-of-Body Travelers as a pinpoint of light, a point particle of light, or a Photon. It all fits together perfectly and makes logical sense.

Psyche is the innate intelligence within all the different forms of energy which gives that energy the inherent ability to understand, follow, and OBEY God's Laws and God's Commands. Obedience to Law requires some type of intelligence. This is what has been experienced and observed.

Remember, any massless quantum particle is a Psyche. A Photon is a Psyche. The massless components of quarks and electrons are Psyches.

Quantum Mechanics Playlist

There are many interesting videos in the PBS Space Time Quantum Mechanics Playlist:

https://www.youtube.com/playlist?list=PLsPUh22kYmNCGaVGuGfKfJl-6RdHiCjo1

In these essays, I mention and discuss the videos that present the points that I'm trying to make in this book. My goal is to use Psyche or Syntropy to defend and interpret Quantum Mechanics because this approach provides the best explanation and has the most explanatory power where Quantum Mechanics is concerned. I'm not going to apologize for figuring this stuff out because that's what scientists are supposed to do. I pursue the BEST explanation, which invariably ends up being whatever has been experienced and observed.

Quantum Entanglement and the Great Bohr-Einstein Debate

Things don't magically pop into and out of existence for no reason. Object permanence, the idea that the universe keeps existing when we're not looking at it, is a fundamental implied assumption behind all of classical physics. Most of science takes it for granted that the universe is real, whether or not we're looking at it. This notion that the universe exists independent of the mind of the observer is called Realism in physics.

Niels Bohr insisted that it was meaningless to assign reality to the universe in the absence of observation. Our experience of a well-defined material universe only has meaning at the moment of measurement. The Bohr camp eventually proposed Action at a Distance, Quantum Non-Locality, Entanglement, Faster-than-Light Interactions at the quantum level, and a Conscious Observer to collapse the wave function. Quantum Entanglement is an influence that could theoretically be transmitted instantly across any distance, and even back in time violating locality and possibly violating causality.

Albert Einstein insisted on an objective reality, a reality independent of our observation of it. Einstein backed Physical Realism. He insisted that the wave function, and by extension quantum mechanics, is incomplete. There must exist hidden variables that reflect a more physical underlying reality. According to Einstein's camp, in order to abandon Realism, you also had to abandon Locality and thereby violate the theory of relativity. Locality is the idea that each bit of the universe only acts on its immediate surroundings – there's NO action at a distance. This is fundamental to Einstein's relativity, which tells us that the chain of cause and effect can't propagate faster than the speed-of-light. Locality and relativity are classical physics. Einstein's camp thought that every point in the universe must be real and physical and defined by knowable quantities, local hidden variables that could affect each other no faster than the speed-of-light.

Scientific Experimentation and Scientific Observation revealed that Quantum Entanglement and Non-Locality are real and truly exist. Quantum particles can interact with each other faster than the speed-of-light. Instantaneous influence between entangled pairs has been observed over many kilometers of distance at this point. The wave function cannot have local hidden variables. The wave function is completely Non-Local in nature and origin.

So, does this confirm the Copenhagen Interpretation and kill both locality and realism? Do we live in a universe that vanishes into quantum abstraction when we aren't looking at it?

Einstein lost this debate. Yet, Physical Realism continues to exist and continues to be true as well. How's that possible? It's because each camp was missing an extremely important piece of information that most physicists completely overlook even to this very day – the identity of the observer who collapses the wave function.

Quantum Entanglement experiments do seem to violate local realism; but, that may mean a violation of realism, or just of locality. The violation of Bell's inequalities only disproved locality, NOT physical realism. Realism could be salvaged. Non-locality requires that entangled particles affect each other instantaneously. However, Non-Locality and Relativity can actually be perfectly consistent. The universe seems to conspire to avoid the paradox of information traveling faster than light, or backwards in time. The Copenhagen Interpretation remains consistent with all quantum observations. However, realist and hidden variable interpretations are also consistent as long as they abandon Locality. The Pilot-Wave Interpretation works by assuming Realism and Non-Local hidden variables.

Many people have claimed that the "act of observation" defines reality. Observation is NOT the thing that collapses the wave function. CHOICE is the thing that collapses the wave function – a CHOICE on the part of Nature's Psyche or the Universe's Psyche. Observations from Human Psyches don't do anything to collapse the wave function. It's Nature's Psyche who collapses the wave function according to the Orthodox Interpretation of Quantum Mechanics.

Years ago, I observed that Locality is Physicality; and, Non-Locality is Non-Physicality or Spirituality. Physical Realism can be salvaged and preserved by rejecting Locality and embracing Non-Locality or Non-Physicality or Spirituality instead. Can you visualize how this is done? It's the KEY to everything. It's the answer to life, the universe and everything!

Let's play around with this for a bit because it is important and foundational.

According to Einstein, "God doesn't play dice with the universe."

Actually, God does. The Gods are constantly experimenting with our universe and trying things out. The Gods are artists as well as scientists. The Gods are currently dumping HUGE amounts of Dark Energy into our universe at a universal scale.

The idea that God doesn't play dice with the universe is the same thing as saying that the physical universe is deterministic and that choice and free will do not exist in a physical universe. This is obviously false. The Human Psyche is making choices all the time changing the nature, purpose, and functionality of our physical universe. Human beings change their physical environment at will. Human beings play dice! It has been experienced and observed. Likewise, God's Psyche is making choices all the time as well. We are not the only ones who get to have some fun at the universe's expense.

Someone Psyche designed and made the Quantum Fields. We KNOW this is so because order and organization of any kind is always MADE by Someone Intelligent or Someone Psyche. Order doesn't just spring into existence from nothing. That's physically impossible. Order is DESIGNED and MADE. This also means that Someone Psyche designed and made Quantum Mechanics and entropic physical matter. Physical matter couldn't exist without the Quantum Fields and Quantum Mechanisms, both of which Someone Psyche MADE. First things first! Physical matter can't MAKE Quantum Fields and Supernatural Mechanisms; but, Quantum Fields and Spiritual Mechanisms can certainly be used to MAKE entropic physical matter. This is what has actually been experienced and observed. It's also logical common sense. Choice is a function of Psyche or Intelligence.

Einstein LOST this debate with Bohr. Quantum Entanglement and Quantum Non-Locality have been proven to exist through actual Science Experiments. Apparently, God does play dice with the universe, and choice is indeed an integral part of our physical universe. Go figure!

Early in my research into Quantum Mechanics, I realized that Non-Locality is the same thing as Non-Physicality, Spirituality, or Transdimensionality. Non-Locality means NOT located within our 3D physical space-time realm. Non-Locality means that a Quantum

Object is currently quantum, or spiritual, or transdimensional in nature. Non-Locality has actually been experienced and observed. Choice has actually been experienced and observed. Go with what has been experienced and observed, and NOT the wishful thinking of the Materialists, Naturalists, Darwinists, Nihilists, Behaviorists, Determinists, and Atheists. If you want the truth, then go with what has been experienced and observed.

Quantum Non-Locality and Quantum Entanglement are REAL and truly exist. I even wrote a book on the topic of Quantum Non-Locality:

Quantum Mechanics from a Non-Physical Spiritual Perspective

https://www.amazon.com/dp/B01J023TGU

Due to their compulsive NEED to eliminate psyche, choice, and responsibility from Quantum Mechanics and science in general, the Naturalists and Atheists introduce a bunch of other falsehoods into Quantum Mechanics that have to be dealt with and resolved if we are to make progress in science. This and other PBS Space Time videos touch on one of them in great detail – the identity of the observer who collapses wave functions. Who exists at the quantum level that can actually consciously choose to collapse a wave function at the quantum level?

The Materialists, Naturalists, Nihilists, and Atheists present Quantum Mechanics as a choice between "Physical Realism and Objective Reality and Determinism" or "Ghostly Spooky Action at a Distance and Subjective Reality and Random Chaos". This conundrum is created by refusing to allow Psyche or Choice into Quantum Mechanics.

Psyche or Choice resolves the situation perfectly. Psyche or Choice maintains Physical Realism or Objective Reality while at the same time allowing the Chooser or the Psyche to be non-local, non-physical, quantum, and spiritual in nature and origin. In other words, "The Universe" or Nature's Psyche makes the choices that collapses the wave function; and after the collapse of the wave function, Infinite Possibilities become ONE single real objective physical reality instead. The Observer Catastrophe or the Locality vs. Non-Locality Catastrophe is resolved instantly by abandoning Locality or Physicality and embracing Non-Locality or Spirituality instead. Because Quantum Non-Locality is at the foundation of everything else, Non-Locality works at ALL levels of existence – the spiritual levels and the physical levels – whereas Locality or Physicality does not. Locality or Physicality is limited exclusively to the physical level.

I don't know if they have an official name for the Observer Catastrophe in Quantum Mechanics; but, it's definitely an issue that needs to be resolved. Namely, who or what collapses the wave function? The Observer is in constant dispute in Quantum Mechanics.

This Locality vs. Non-Locality Catastrophe is also described by Einstein as "The Moon Doesn't Exist When No Human Is Looking at It" Catastrophe. It is the Observer Catastrophe, which states that things need a human observer to maintain their existence or which states that only a human can act as an observer. Einstein was rightfully trying to salvage Physical Realism or Objective Reality because it obviously exists; and, it obviously continues to exist when NO Human Psyche is observing it or looking at it. The contents of my hard drive and house continue to exist when NO human is looking at them. It's obvious. A Human Psyche or human observer isn't needed to maintain a physical object's existence.

In this video, Matt correctly states that Objective Reality or Realism can be salvaged and maintained by abandoning Locality and by embracing Quantum Non-Locality instead. That's precisely what Psyche does. Nature's Psyche or "The Universe" is a Non-Local Quantum Phenomenon that exists at the quantum level, which then produces Physical Realism or Objective Reality at the physical level by collapsing the necessary wave functions when it's time to do so. Objective Reality or Physical Realism remains in existence AFTER the wave function has been collapsed by Nature's Psyche whether there is any human psyche there to observe it or not. That's the nature of a physical consensus reality. It remains in place when nobody is looking at it. I don't have to keep looking at my essays and books on my hard drive to keep them in existence on my hard drive. Those wave functions have already been collapsed, and they will remain so until Someone Psyche starts messing with them. The "Things Blinking Out of Existence When Nobody Is Looking at Them" Catastrophe is resolved by introducing Psyche or Intelligence or Lasting Choices into the system – particularly Nature's Psyche or what Matt repeatedly calls "The Universe" in his videos.

Remember, according to the Orthodox Interpretation of Quantum Mechanics by Henry P. Stapp, it is Nature's Psyche or The Universe who collapses the wave function and NOT the Human Psyche. Paul Dirac defined the collapse of the wave function as a CHOICE on the part of Nature. In other words, it's Nature's Psyche or the Universe's Psyche who chooses and decides when to collapse the wave function and actually does the collapse of the wave function. This is important to get right because it resolves ALL of the catastrophes and controversies associated with "The Observer" in Quantum Mechanics.

It was a real Quantum Catastrophe when somebody decided that ONLY a Human Psyche, Human Consciousness, or Human Observer can collapse a wave function. The Human Psyche isn't consciously aware of any of these quantum mechanical processes; consequently, the Human Psyche can't be the entity who is collapsing the wave function. Everything in Quantum Mechanics finally makes sense when we observe that it is Nature's Psyche or what Matt calls "The Universe" who actually collapses the wave function and NOT the Human Psyche. By removing the Human Psyche as "the observer" who collapses the wave function and replacing him with The Universe or Nature's Psyche as the one who collapses the wave function, we also resolve "The Moon Doesn't Exist When No Human Is Looking at It" Catastrophe within Quantum Mechanics by allowing The Universe or Nature's Psyche to collapse the wave function rather than looking for some Human Psyche to do the job and keep doing the job.

Remember, the collapse of the wave function is a CHOICE that is made on the part of Nature's Psyche who decides when it is time to collapse the wave function. Technically, the Human Psyche doesn't have anything to do with the collapse of the wave function. The Human Psyche decides what it wants to do with its physical body and physical brain; and then, Nature's Psyche collapses the necessary wave functions and fires the necessary neurons in order to make the Human Psyche's requests a physical reality. The Human Psyche isn't consciously aware of any of these quantum mechanical processes. Nature's Psyche, The Universe, or God's Psyche handles it all for us.

Everything is resolved in Quantum Mechanics by figuring out that it is Nature's Psyche or The Universe's Psyche who collapses the wave function and NOT the Human Observer or Human Psyche. A Human Psyche is not needed to keep a physical consensus reality in existence. It does have to be Someone Psyche who is collapsing the wave functions because in principle entropic physical matter can't collapse wave functions at the quantum level nor the psyche level. We don't need to reject Realism once we realize that it is Nature's Psyche or The Universe's Psyche who collapses the wave function and NOT the Human Psyche. With this simple adjustment, we KNOW who collapses the wave function at the quantum level, and at the same time Objective Reality or Physical Realism is preserved. Nature's Psyche collapses the wave function at the quantum level thereby making the quantum particle physical, actual, and real at the physical level. Once a quantum particle is made physical, it remains physical until Someone Psyche starts messing with it once again.

All of this becomes possible by choosing to allow Psyche or Choice into Quantum Mechanics where it belongs – particularly Nature's Psyche or the Universe's Psyche. This is GOOD SCIENCE because it has been experienced and observed. It works because it is true. It explains everything that has ever been experienced and observed.

This is worth figuring out because the payoff is enormous. By allowing Psyche and Quantum Mechanics in to play, we can literally explain everything that comes our way.

ALL of these unsolvable problems, issues, controversies, and catastrophes in Science and Quantum Mechanics are easily resolved by choosing to include into Science everything that has ever been experienced and observed at the psyche level in the Quantum Realm or Spirit World. The explanatory power of Psyche, or Choice, or Intelligence is through the roof to infinity and beyond because it has actually been experienced and observed.

The Many Worlds of the Quantum Multiverse

https://www.youtube.com/watch?v=dzKWfw68M5U

http://quantum-law-of-thermodynamics.com/wp-content/uploads/2018/09/Many-Worlds-Interpretation-of-QM.zip

[Editor's Note: As usual, I adjusted this to make it match with what has actually been experienced and observed by Out-of-Body Travelers. Science is observation and experience after all, or it should be.]

One of the main goals in Quantum Mechanics is to find the right interpretation for the transition between the quantum and the classical worlds. In Quantum Mechanics, different possible superposed histories appear to converge on one final outcome; but, what causes that convergence? Who or what causes the collapse of the wave function? In the original Copenhagen Interpretation of Quantum Mechanics, the act of measurement was thought to collapse possibility space into a single reality at least with respect to the measured property. It collapses the wave function. That collapse signifies the transition between the quantum and classical realms. By the Copenhagen Interpretation, we might say that the universe chooses the final outcome of all of those infinite possibilities. The universe chooses an end result, such as a particle location on the screen or whether Schrodinger's cat is dead or alive. If a larger number of possible histories lead to a given result, then it's more likely that the universe will select that outcome. The Copenhagen Interpretation says that this selection happens in a fundamentally random way. The universe plays dice, even if the dice are weighted towards certain results. It is what we would call a nondeterministic interpretation because there's no underlying predictability behind the selection. The Universe or Nature's Psyche makes a CHOICE to collapse the wave function. It's truly a CHOICE in every sense of the word, a CHOICE made by The Universe or Nature's Psyche.

Without intending to do so, in this video, he demonstrated the vast superiority of the Copenhagen Interpretation over the Many Worlds Interpretation. This is possibly the BEST explanation of the Copenhagen Interpretation that I have encountered to date. Many people can't seem to see it because they don't want to see it and don't want it to be true; but, the Copenhagen Interpretation as presented in this video is vastly superior to the Many Worlds Interpretation.

The Materialists, Naturalists, Darwinists, Nihilists, Behaviorists, Determinists, and Atheists instinctively realize that CHOICE doesn't take place at the physical level. Since these people have chosen to believe that ONLY entropic physical matter exists, these people have also chosen to believe that Psyche or Choice DOES NOT EXIST.

They are wrong of course because Psyche or Choice exists at the quantum level, and choices are made by Someone Psyche at the psyche level. They are right, though, in that choices are NOT made at the physical level. Things at the physical level are DETERMINED by Someone Psyche at the quantum level and the psyche level. This is what has actually been experienced and observed.

I have observed that the Materialists, Naturalists, and Atheists always destroy the explanatory power of Quantum Mechanics whenever they try to turn it into Physical Mechanics or Classical Physics because Classical Physics or Physical Mechanics cannot be used to explain what Nature's Psyche is doing at the quantum level in order to get things done for us at the physical level.

The Many Worlds Interpretation and the Pilot-Wave Interpretation are an attempt to eliminate Psyche, Choice, Non-Locality, Spirituality, and God from Quantum Mechanics; and as a result, these interpretations are completely worthless when it comes time to explain what Nature's Psyche or The Universe is doing at the quantum level in order to get things done for us at the physical level.

The Many Worlds Interpretation breaks parsimony, is uneconomical, and is infinitely wasteful – just so that the Naturalists and Atheists can eliminate CHOICE, free will, responsibility, and the collapse of the wave function from Quantum Mechanics. Imagine it! Under the Many Worlds Interpretation, each time one of us makes a CHOICE a whole new universe is created just so that our choices don't really count, and we aren't responsible for our choices. Do you understand thrashing when it comes to a hard drive, when a hard drive is used as memory or cache? If the Many Worlds Interpretation were true, and we created a whole other universe every time one of us made a choice, there should be thrashing, and it should be noticeable. We observe gravity waves whenever black holes merge. Imagine what we would witness and observe every time the whole universe diverged into a completely separate universe. That would leave a mark! The Many Worlds Interpretation is a dumb idea just so that they can have a deterministic multiverse where Free Will, Choice, and Responsibility do not exist.

Remember, the Many Worlds Interpretation of Quantum Mechanics breaks parsimony at a universal scale. The Many Worlds Interpretation is worthless because it doesn't match with reality. We humans are making choices and taking responsibility for those choices all the time; and, our choices really make a difference determining whether buildings, houses, cars, bridges, planes, ships, factories, and computers will be made or not.

Why is the Copenhagen Interpretation vastly superior to the Many Worlds Interpretation?

It's because the Copenhagen Interpretation has actually been experienced and observed; whereas, the Many Worlds Interpretation has NOT. Observation and Experience trump philosophical speculation and wishful thinking every time, or they should. Science is observation and experience after all.

Pilot Wave Theory and Quantum Realism

https://www.youtube.com/watch?v=RlXdsyctD50

They talk about mysticism as if it were a bad thing, which it is if you are a Materialist, Naturalist, Nihilist, or Atheist because the proven and verified existence of mysticism or quantum mechanisms falsifies Materialism, Naturalism, Darwinism, Nihilism, Behaviorism, Determinism, Physical Reductionism, and Atheism.

This video portrays Pilot Wave Theory as an attempt to turn Quantum Mechanics into Physicalism, Materialism, Naturalism, Nihilism, and Classical Physics – an attempt that FAILS and is therefore FALSE. The creators and proponents of Pilot Wave Theory define physical matter as the only thing that is REAL. They define REAL as physical matter. Consequently, Pilot Wave Theory is falsified by the observation that the waves are non-physical and therefore NOT REAL. Pilot Wave Theory and Physicalism are self-defeating.

Pilot Wave Theory reasons that there is no need for quantum objects to transition in a mystical way between non-real waves and real particles. Why not just have real waves that push around real particles? This is Pilot Wave Theory. This real wave made of real stuff guides the motion of a real point-like particle that has a definite location at all times. Pilot Wave Theory maintains Locality and Physical Realism. A pilot-wave universe is completely deterministic. The initial motivation behind Pilot Wave Theory was to preserve the idea of REAL particles. It has a solid classical bias or physical bias. These people don't consider anything to be REAL unless it is physical. They beg the question and define REAL as physical. Pilot Wave Theory was also designed to eliminate psyche, quantum fields, non-physical quantum waves, choice, responsibility, and God from Quantum Mechanics thereby making Quantum Mechanics REAL or physical instead.

In case you missed it, the fundamental assumption here is that it's NOT REAL unless it is physical. That's a demonstrably FLAWED assumption. A quantum wave, thoughts, wind, magnetism, radio waves, microwaves, gravity waves, the fundamental forces, quantum fields, and a human psyche are very real, even though they are non-physical, non-local, intangible, invisible, transdimensional, spiritual, supernatural, and not detectable by human eyes. They beg the question and define REAL as physical. That's a logic fallacy.

Pilot Wave Theory requires a special pleading which is technically a logic fallacy. Pilot Wave Theory has hidden variables. In order to solve the problem, you have to switch from local hidden variables to non-local hidden variables; and, once you switch to something non-local or spiritual or supernatural then you no longer have a purely physical interpretation of Quantum Mechanics thereby defeating the purpose for which Pilot Wave Theory was made in the first place.

Remember, the wave function is not a wave in anything physical. Instead, the wave function is a wave in immaterial, non-physical, non-local, spiritual, intangible, supernatural, transdimensional Quantum Fields. As I see it, the proven and verified existence of Quantum Fields as well as a whole host of intangible, non-physical, mystical, supernatural forces FALSIFY Materialism, Naturalism, and Pilot Wave Theory. That observation is not going to sit well with the Naturalists and Atheists because it's not telling them what they want to hear. Nevertheless, experiments have demonstrated that some type of non-locality, non-physicality, or supernatural spirituality is REAL whether you believe in them or not.

His conclusion at the end of the video is that Pilot Wave Theory is certainly wrong because it doesn't account for relativity. At best, it's incomplete. Quantum

Field Theory matches with the Theory of Relativity; whereas, Pilot Wave Theory does not. Pilot Wave Theory is not consistent with Quantum Field Theory; and therefore, there isn't a complete relativistic formulation for Pilot Wave Theory yet.

In this and other videos, Matt seems surprised that Pilot Wave Theory is extremely popular on the internet. Of course, it is going to be popular because it is telling people what they want to hear and want to believe – namely, that psyche, choice, responsibility, and God DO NOT EXIST. That's the very same reason why the Many Worlds Interpretation has gotten a foothold in the science community – because it's telling people what they want to hear. Pilot Wave Theory goes out of its way to remain physical, naturalistic, materialistic, and atheistic. This is going to be popular among the people who beg the question and define Science as Materialism, Physicalism, Naturalism, Darwinism, Nihilism, and Atheism. That's just the way it is. In fact, a quantum interpretation isn't given a listing on the Wikipedia unless it is materialistic, naturalistic, physicalistic, and atheistic in nature and origin. Instead, it's listed as pseudo-science and given the Ig Nobel Prize instead.

I have no illusions. My current interpretations of Quantum Mechanics will never be popular on the internet and will never be accepted in mainstream science because they are based upon what has been experienced and observed rather than telling the Materialists, Naturalists, and Atheists what they want to hear. That's just the way it is. A lot of money can be made, and a lot of popularity can be gained by telling the Materialists, Naturalists, Darwinists, Behaviorists, Determinists, and Atheists what they want to hear. The most popular and best-selling "science" books of all time do just that. I KNOW how it goes because I used to be a Materialist, Naturalist, Nihilist, and Atheist until science itself convinced me that I was wrong. At the time, I wanted My Atheism to be true even though it was demonstrably false. We Materialists, Naturalists, Nihilists, and Atheists don't want Quantum Mechanics to be mystical, spiritual, or psychic in nature. We want it to be REAL which means that we want it to be physical.

A physical reality is the Ultimate Consensus Reality because it continues to exist even when it's not being observed. This is critical because it is obvious that the contents of my hard drive, my home, my bank account, my wife, my dog, and my car continue to exist even when they are unobserved. Physical configurations exist even when unobserved. This is what has been experienced and observed. A physical reality is reliable, dependable, predictable, and controllable because it was designed to be that way. What's not to like? The physical particles have been experienced and observed. Have they not? Once their wave function has been collapsed by The Universe or Nature's Psyche, physical particles continue to exist even when nobody is looking at them. Do they not?

There's no reason to insist that the "observation" must continue indefinitely in order for the physical particle to continue to exist. Once God or Nature's Psyche or The Universe has made the CHOICE that collapses the wave function and produces a physical particle, then the physical particle continues to exist until the Gods or humans decide to turn it into something else instead. Remember, the particles are REAL and continue to exist after Nature's Psyche, or The Universe, or God has collapsed the wave function.

The FLAW in all of this only comes about when the Naturalists and Atheists insist that entropic physical matter completely eliminates Psyche, God, Nature's Psyche, and The Universe from the equation thereby making them unnecessary and unreal and non-existent. Such a procedure assumes facts not in evidence. Furthermore, these people beg the question and axiomatically define REAL as "physical" which automatically makes anything non-physical like Quantum Fields or Psyche unreal, imaginary, and non-existent. Of course, the solution to all of this is to abandon the classical bias of Materialism, Naturalism, and Atheism and choose to go with Quantum Field Theory and Psyche or Non-Local Consciousness instead – which I have chosen to do.

A physical reality is the Ultimate Consensus Reality because it continues to exist even when it's not being observed. God designed it to be that way. God designed it to be REAL. When it comes to entropic physical matter, God deliberately slowed everything down so that we can observe it, live it, experience it, learn from it, and remember having done so. With the design and production of entropic physical matter, God removed Timelessness, Quantum Tunneling, and Teleportation as the fundamental laws of physics and replaced them with the passage of time instead at the physical level. These Quantum Laws or Quantum Freedoms continue to exist at the quantum level but are greatly limited and restricted at the physical level. The result was the Ultimate Consensus Reality, a physical reality. It's the perfect platform for learning and growing from experience. It's REAL because we actually have skin in the game. It's REAL because it hurts.

Remember, just because physical matter is REAL doesn't make Quantum Fields, Gravity, Magnetism, Photons, Quantum Waves, Thoughts, Psyche, Choice, and God any less real. Get real!

Mark My Words

Wiring the Brain

Materialists, Naturalists, Darwinists, Nihilists, and Atheists have defined axon growth and the arborization of dendrites along with synaptogenesis and synaptic pruning as "wiring the brain". Can you see any technical problems with their claims that our brains are wired together by axons, dendrites, and synapses?

Wired by whom?

Scientific Observation: We have observed that wired networks or networks of wires require some kind of Electrical Architect to design the schematic and some kind of Electrical Engineer, Wire Maker, and Cable Runner in order to implement and successfully wire that electrical network of wires.

Scientific Definition: The Materialists, Naturalists, Darwinists, Nihilists, and Atheists define axons and dendrites as "wires within a physical brain". These same people also define synapses as "wired connections between neurons in a physical brain".

Scientific Conclusion: Therefore, it is logical and rational to conclude that axons, dendrites, and synapses have some kind of Architect, Map Maker, Engineer, Manufacturer, and Cable Runner who decides and controls how synaptogenesis and synaptic pruning will take place. In order to have wires requires some kind of Wirer! Wires don't just spontaneously generate out of thin air!

If our brains are wired together by axons, dendrites, and synapses, then our brains need Someone Psyche or Someone Intelligent to do the wiring and to map-out the meaning and the purpose for each connection.

Clearly, our brains have been NETWORKED and MAPPED. MAPPED by whom?

Maps require some kind of Mapmaker! Networks need some kind of Network Engineer to plan them and make them.

There's another severe problem when it comes to the "wires within our brains" and the "normal wiring of our brain", which the Physicalists and Scientific Naturalists cannot see nor understand.

Do you know what that problem is?

There are NO physical wires within our brain!

Chemical Synapses are gaps in space – NOT physical connections. There are NO physical connections between neurons through Chemical Synapses, which means that axons and dendrites cannot function as physical wires within a physical brain.

The axons, dendrites, and synapses DO NOT FUNCTION as physical connections; and therefore, they do not function as wires and wired networks – not at the physical level at least! A physical synapse scrambles and randomizes everything that comes its way.

So, ask yourself how the different neurons are communicating with each other telepathically at a distance since they are NOT communicating with each other physically through their axons, dendrites, and synapses.

The materialistic and naturalistic claim that axons and dendrites form wires within our brains is deceptively false. Chemical Synapses are gaps in space – NOT physical connections. Technically, there are NO physical wires, physical connections, or physical cables within our brains connecting the neurons together into logic gates, transistors, and computer networks for computer processing and memory storage through our Chemical Synapses. If there is any computer processing or memory storage taking place within our physical brain, it's taking place at the quantum level or the psyche level because it's physically impossible at the physical level because chemical synapses are gaps in space. Synaptic clefts scramble or randomize everything that comes their way at the physical level, which means that communication between neurons has to be taking place at the quantum level or the psyche level since it's not taking place at the physical level.

This is what has been experienced and observed.

Mark My Words

—

Source Material

Quantum Neuroscience: The Answer to Life, the Universe, and Everything

https://www.amazon.com/dp/B079Z6QQQB

Rejecting Man-Made Gods

As a scientist, I have probably encountered at least a dozen different definitions for the word "entropy". Some of them are very confusing and technical, and others actually make sense.

Entropy is often defined as "disorder" and consists of the idea that disorder is increasing throughout our physical universe. Scientists teach that "maxim entropy" will result in the heat death of our physical universe. To me, "death" is the best and most explanatory definition for the word "entropy". Entropy is death.

The Physicalists, Naturalists, Darwinists, Nihilists, and Atheists literally define the Second Law of Thermodynamics as the Conservation of Entropy or the Conservation of Death; and then, they use the Second Law of Thermodynamics to falsify the First Law of Thermodynamics which is the Conservation of Psyche, or the Conservation of the Life Force, or the Conservation of Energy. For these people, entropy or death is God because in their minds entropy or death reigns supreme.

I have also seen entropy defined as a measure of the potential energy in the universe or a measure of the usable energy in the universe. The way this works is that low entropy means that there's still lots of usable energy in this physical universe; whereas, Maximum Entropy means there is "'no potential energy," or "no usable energy", or "heat death" where this physical universe is concerned. This definition for "entropy" gives the impression that we are running out of energy and that the amount of usable energy is decreasing. Most of the scientists that I have encountered teach that we are running out of energy and that someday our universe will end in Heat Death. They call it the Second Law of Thermodynamics – namely, the idea that we are running out of usable energy, someday there will be no more potential energy, and everything will end in Heat Death.

This definition for "entropy" has a demonstrable weakness or flaw. It violates the First Law of Thermodynamics. The Second Law of Thermodynamics, as presented to us by our scientists, violates the First Law of Thermodynamics. According to the First Law of Thermodynamics, energy is conserved. Energy can neither be created, nor destroyed. In other words, we can NEVER run out of usable energy or potential energy because energy is conserved. If the First Law of Thermodynamics is true, then this means that ALL of the energy is still there, and ALL of that energy is still usable despite the fact that high entropy makes it seem as if we have run out of usable energy or potential energy. In other words, entropy is just an illusion according to the First Law of Thermodynamics. Entropy makes it seem as if we have run out of potential energy or usable energy when in fact we have not. ALL of the energy is still there and still usable according to the First Law of Thermodynamics.

Concentrate on what the Second Law of Thermodynamics was designed to falsify and eliminate from science; and there, you will find the answer to life, the universe, and everything. The Second Law of Thermodynamics or the Conservation of Entropy was designed to eliminate Psyche, God, Syntropy, and the Conservation of Energy from science. Psyche is the innate intelligence within all the different forms of energy which gives that energy the inherent ability to understand, follow, and obey God's Laws and God's Commands. The Second Law of Thermodynamics was designed to eliminate the First Law of Thermodynamics from science – namely, the Conservation of Psyche or the Conservation of Energy. Syntropy is the First Law of Thermodynamics. The Second Law of Thermodynamics was designed to falsify and eliminate Syntropy from science. Explaining what it does and what it's meant to do ends up being a very powerful explanatory definition for the Second Law of Thermodynamics.

A favorite definition for "entropy" is the idea that entropy is the opposite of Syntropy. There's one major problem with this definition for entropy. The Materialists, Naturalists, Darwinists, Nihilists, Behaviorists, Determinists, Atheists, and Classical Physicists have chosen to believe that Syntropy or Psyche does not exist – Syntropy is Eternal Life, and these people have chosen to believe that Eternal Life does not exist. Syntropy is the First Law of Thermodynamics.

According to these people there is nothing opposite to entropy, and there is nothing opposing entropy. According to these people, ONLY entropy exists, and there is NO such thing as Syntropy. These people teach and preach the Supremacy of Entropy. These people literally teach that entropy or death is conserved. These people really do teach and believe that we are running out of usable energy, that our physical universe is running down, and that our physical universe will end in Heat Death. This is literally what they teach and believe. These people teach that entropy or death is being conserved in complete violation of the First Law of Thermodynamics which states that Energy or Psyche is conserved.

Something has to give. The Second Law of Thermodynamics (entropy) FALSIFIES the First Law of Thermodynamics (syntropy); and, Syntropy FALSIFIES entropy. They are mutually exclusive. They both can't be right. A correction or an amendment is in order.

How do the Materialists, Naturalists, Darwinists, Nihilists, and Atheists resolve this contradiction? They do so by stating that Syntropy and Psyche DO NOT EXIST. They do so by promoting the Supremacy of Entropy or the Supremacy of Death. They do so by stating that ONLY physical matter or entropy exists; whereas, Psyche or Syntropy DOES NOT EXIST.

How does stating that Syntropy and Psyche DO NOT EXIST explain what they are and how they work? It doesn't! It's bad science. Stating that something DOES NOT EXIST has no explanatory power.

I chose a different path once it was made clear to me that Psyche has been experienced and observed by Out-of-Body Travelers. Psyche has been experienced and observed which means that WE KNOW that it is real and truly exists. Now our job as scientists is to explain what it is and how it works. Psyche is the innate intelligence within all the different forms of energy which gives that energy the inherent ability to understand, follow, and obey God's Laws and God's Commands. Remember, Psyche has been experienced and observed. Now, it's left up to the scientists to figure out what it is and how it works.

The Materialists and Naturalists chose to promote the Supremacy of Entropy or the Supremacy of Death; and, I chose instead to promote the Supremacy of Syntropy, the Supremacy of Life, or the Supremacy of Energy and Psyche. The Materialists and Naturalists chose to use the Second Law of Thermodynamics to falsify the First Law of Thermodynamics; and, I chose to use the First Law of Thermodynamics to falsify the Second Law of Thermodynamics.

I call my version the Ultimate Law of Thermodynamics. The Ultimate Law of Thermodynamics states that Energy or Psyche is conserved; whereas, physical matter and entropy are NOT conserved. Entropy is temporary. Entropic physical matter is temporary. Entropic physical matter is made from raw energy which means that physical matter and entropy can be disassembled back into raw energy and then formed into something else instead. Energy or Psyche cannot be made nor destroyed which means that it is conserved. Energy, God's Psyche, and your psyche are eternal and everlasting. Energy or Psyche is syntropic which means that Energy or Psyche is conserved; but, the FORM of that energy is never conserved. Entropic physical matter is comprised of different forms of energy; and, the form is never conserved. Psyche is the innate intelligence within all the different forms of energy. This is the Ultimate Law of Thermodynamics. This is the answer to life, the universe, and everything. It explains everything that has ever been experienced or observed.

Entropy or death falsifies the existence of God's Psyche. The observed and proven existence of our resurrected Lord Jesus Christ falsifies entropy and death. Entropy or death is a man-made God. Jesus Christ is God. God's Psyche is syntropic; whereas, death is entropic. Entropy is temporary; and, Psyche, Energy, or Syntropy is eternal and everlasting. The two are mutually exclusive. Psyche has been experienced and observed;

therefore, our job as scientists is to figure out what it is and how it works. Our resurrected Lord Jesus Christ has been experienced and observed after He rose from the dead. Our job as scientists is to figure out what resurrection is and how it works. Entropy is not God. Entropy is not conserved. Entropy or death is temporary. The Ultimate Law of Thermodynamics states that entropy is temporary, meaning that entropy is NOT conserved.

The realization that entropy is temporary just might in fact be my greatest scientific discovery of all time. It explains everything that has ever been experienced or observed. Physical matter and entropy are made or caused to begin which means that physical matter and entropy are not conserved. Physical matter and entropy are constructs. Anything that is made is not being conserved! Entropy was made, the energy was made temporarily unavailable, which means that entropy is temporary. Entropic physical matter was made which means that entropic physical matter is temporary. The Heat Death of our physical universe will be temporary. It will NOT be conserved. The underlying Energy or Psyche is still there; and, the former Syntropy will regain supremacy once the entropy has been removed.

The Materialists and Naturalists preach and teach the Supremacy of Entropy or the Supremacy of Death. Observational evidence or scientific evidence preaches and teaches the Supremacy of Energy, the Supremacy of Psyche, or the Supremacy of the Life Force. They both can't be right. Something has to give.

Entropy is death. Syntropy is the opposite of entropy. So, what is the opposite of death? Eternal life is the opposite of death. Therefore, Syntropy is Eternal Life. Syntropy is the First Law of Thermodynamics. Syntropy is the conservation of Psyche, Life Force, Intelligence, Energy, and LIFE. God's Psyche is syntropic. It is conserved. It is eternal and everlasting. Psyche or Energy cannot be created, and it cannot be destroyed.

The Second Law of Thermodynamics was designed to falsify and eliminate the First Law of Thermodynamics. The Second Law of Thermodynamics was designed to falsify and eliminate Psyche, Life Force, Intelligence, Usable Energy, LIFE, Eternal Life, Action at a Distance, Quantum Mechanisms, Supernatural Mechanisms, and God from science. In contrast, the proven and verified existence of these different things FALSIFIES the Second Law of Thermodynamics. In other words, the First Law of Thermodynamics when properly understood FALSIFIES the Second Law of Thermodynamics. Entropy is death; and, Eternal Life or Immortality falsifies death or entropy. If the one is true, then the other is false.

The realization that the Second Law of Thermodynamics is fundamentally flawed and violates the First Law of Thermodynamics just might in fact be my greatest scientific discovery of all time. The First Law of Thermodynamics is the Conservation of Energy, the Conservation of Psyche, or the Conservation of the Life Force. The First Law of Thermodynamics is Syntropy or Eternal Life. Syntropy is the Conservation of Psyche or Energy. Syntropy falsifies entropy. Eternal Life falsifies death. The First Law of Thermodynamics FALSIFIES the Second Law of Thermodynamics.

In contrast, the Second Law of Thermodynamics is the Conservation of Entropy or the Conservation of Death. These people literally teach that entropy or death is conserved – namely, that entropy and death are eternal and everlasting. That's precisely what the Materialists, Naturalists, Darwinists, Nihilists, Behaviorists, Atheists, and Classical Physicists teach. Is it not? Entropy is a man-made God. These people teach and preach that Death or Entropy reigns supreme. These people literally teach and preach that entropy is God. They teach the Supremacy of Entropy Doctrine as an integral part of their religion.

The Second Law of Thermodynamics was designed to falsify and eliminate the First Law of Thermodynamics. Instead, I suggest for your consideration the idea that the First

Law of Thermodynamics – namely the Conservation of Psyche, the Conservation of Energy, or the Conservation of the Life Force – does in fact FALSIFY the Second Law of Thermodynamics. Entropy is temporary which means that entropy is NOT conserved! Since entropy is a temporary construct, the Gods can remove entropy and restore syntropy anytime they choose to do so. Entropy keeps score; but, the scoreboard can be removed anytime the Gods decide to do so. This is the Ultimate Law of Thermodynamics.

The Materialists and Naturalists teach that entropy or death is conserved, which means that they teach that entropy or death is eternal and everlasting. These people literally teach and promote the idea that entropy is God.

They are wrong.

According to the First Law of Thermodynamics, whenever we encounter a physical universe that has suffered Heat Death, ALL of the energy is still there and still usable. Remove the entropy, and that physical matter goes back to being Syntropy. The ultimate scientific discovery for human beings would be to discover how to remove the entropy from physical matter. Once we remove the entropy from entropic physical matter, then instantly it becomes syntropy physical matter or usable energy. It's possible. The Ultimate Law of Thermodynamics teaches us that entropy is temporary. Entropy was made, which means that entropy can be disassembled and turned back into raw energy or usable energy. Entropy is temporary, which means that entropy is NOT conserved. Remove all the entropy, and instantly we are looking at Syntropy. Remove the entropy from entropic physical matter, and instantly it becomes syntropic physical matter instead.

Adam and Eve had syntropic physical bodies while they were in the Garden of Eden. They were immortal, eternal, and everlasting. They ate the fruit from the Tree of Knowledge, and it introduced entropy into their physical bodies thereby converting their syntropic physical bodies into entropic physical bodies. They became mortal. If Adam and Eve could have gotten back to the Tree of Life, the fruit from the Tree of Life would have removed the entropy from their physical bodies converting their entropic physical bodies back into immortal syntropic physical bodies instead. The science is all there. The truth is there. We just have to figure out what it is and how it works. That's what scientists are supposed to do. The Gods are scientists. The greatest scientific discovery we could make would be to figure out how to remove the entropy from physical matter thereby converting it into syntropic physical matter instead.

Do you see what becomes possible by treating the Word of God or the Scriptures as scientific evidence? It tells us what's possible; and, it FALSIFIES the Second Law of Thermodynamics. The "conservation of entropy" or the "eternal and everlasting nature of entropy" is nothing but an illusion. The Materialists, Naturalists, Darwinists, Nihilists, and Atheists make NO effort to overcome entropy because they have chosen to believe that Entropy Is Supreme. Death or entropy is their God. Is it not?

Concentrate on what the Second Law of Thermodynamics was designed to falsify and eliminate from science, and there you will find the answer to life, the universe, and everything.

Mark My Words

Evolution Is a Man-Made God

The Theory of Evolution is the modern-day version of superstition or idolatry. Evolution is a man-made god. Evolution is entropy.

Materialism, Physicalism, Naturalism, Darwinism, Nihilism, Behaviorism, Determinism, Physical Reductionism, Classical Physics, the Theory of Evolution, and Atheism are based exclusively on entropy. Entropy is the driving force behind Materialism, Naturalism, and the Theory of Evolution. According to these people, ONLY physical matter or entropy exists. These people literally define "science" as Materialism, Naturalism, Darwinism, Nihilism, Atheism, Classical Physics, and Entropy. Whether they realize it or not, these people teach that Evolution or Entropy is God. These people worship entropy or the God of Death.

Entropy is death. Nihilism means to "cease to exist" when we die.

Materialism, Naturalism, and the Theory of Evolution are literally Creation by Entropy or Creation by Death. These people literally teach that only entropy or death exists and that Syntropy, Psyche, and Eternal Life DO NOT EXIST. This is what they teach and believe; but, they are wrong.

The different types of evolution or entropy have never been caught in the act of design and creation because evolution or entropy of any kind cannot design and create.

Natural selection does absolutely nothing to touch our genes or change our genes. Natural selection is absolutely worthless as a mechanism of change. The only time that natural selection or survival of the fittest comes into play is when you get selected against and die. Your death or entropy does NOT produce new and unique life forms from scratch. That's physically impossible. Random mutations are entropy. Random mutations slowly destroy our genome. Random mutations do not and cannot design and create new life forms from scratch. It's physically impossible. In fact, evolution (genetic change), random mutations, and natural selection didn't even exist until after God designed, programmed, engineered, field-tested, fine-tuned, manufactured, created, and deployed the proteins, their matching genes, the genomes, eyes, brains, and life forms in the first place.

Chemical evolution – the production of new functional proteins and the matching genes to go along with them – is prevented from happening by random diffusion and entropy. Chemical evolution is physically impossible. The First Law of Thermodynamics prevents creation ex nihilo from happening; and, the Second Law of Thermodynamics prevents proteins and their matching genes from spontaneously generating from atoms. Macro-evolution – members of one Family or Genus giving birth to an organism from a completely different Genus or Family – is prevented from happening by both genetics and random mutations. It's physically impossible for a sexually reproducing species to give birth to a genetically compatible Mr. and Mrs. Mutant at the same time in the same place generation after generation for millions of years of time. Macro-evolution is physically impossible.

Genes cannot do selection, meaning that genes cannot do choice. Genes have NO choice! Genes don't make any choices. Genes are programming code; and, Someone Psyche or Someone Intelligent is needed to determine when, where, how, and why that programming code is being used and will be used. Technically, the claim that our genes interact with each other and the environment is false. "Interaction" implies some kind of telepathic connection between the genes, their environment, and everything else. Technically, according to the Naturalists and Darwinists, genes have no intelligence, teleology, purpose, or goal in mind. By definition, the genes are incapable of selection or choice because they don't have a Psyche or a Mind according to the Naturalists, Darwinists, and Atheists. The theory of evolution or Creation by Genes is self-defeating because the genes by definition in principle cannot design and create anything from scratch. The proteins, genes, and enzymes simply do what they are told. Told by whom? The only logical answer is "Someone Psyche" because the atoms, molecules, and proteins don't have genes within them to tell them what to do – not at the physical level at least.

Choices are is made by the Human Psyche. If it can go either way – yes or no – then it is a choice that's being made by Someone Psyche. Our choices are NOT determined by our genes. Our genes have NO physical mechanism in place to make our choices for us. Genes cannot do selection or choice. ONLY Psyche can do selection or choice. Command and Control is a function of Psyche and NOT our genes.

Genes are NOT conscious, alive, psychic, telepathic, and in control of everything at the physical level. These kinds of capabilities that they normally attribute to our genes are ONLY possible at the quantum level, the syntropic level, or the psyche level. The physical atoms have NO physical mechanism in place to motivate them or force them to self-assemble into functional proteins and the matching genes to go along with those proteins. The assembly of functional proteins and their matching genes from atoms has to be taking place at the quantum level or the psyche level because chemical evolution is physically impossible at the physical level.

According to the Darwinists, Naturalists, Physicalists, and Atheists, our genes are NOT conscious and alive; but in the very same articles where these people make these claims, these people also talk about our genes and describe our genes as if they were conscious and alive. These people talk as if our genes are telepathically controlling everything within a living cell and interacting with our environment and planning our future telepathically, teleologically, and presciently as well. Their stance in relationship to our genes is internally inconsistent, contradictory, incongruous, and self-defeating. These people tell us that our genes are not psychic and not alive; yet, in all of their scientific explanations, these same people talk about our genes and describe our genes as if they are in fact intelligent, psychic, telepathic, cunning, insightful, selfish, interconnected, devious, predatory, motivated, prescient, teleological, purposeful, and alive. These people tell us that Psyche, Soul, Spirit, and Mind do not exist; but then, these people talk about our genes as if our genes have a Psyche, Spirit, Soul, or Mind. It is duplicity and hypocrisy. It can't possibly be true because it's self-contradictory.

Materialism or Physicalism is a philosophical belief or a religion that claims that the non-physical DOES NOT EXIST. Contrary to the scientific evidence and the observational evidence, these people claim that only entropic physical matter exists. Naturalism is the philosophical belief or the religion that states that Psyche, God, Spirit Matter, Dark Matter, Dark Energy, Quantum Fields, and Supernatural Mechanisms or Quantum Mechanisms DO NOT EXIST. Scientific Naturalism DOESN'T ALLOW anything invisible, intangible, quantum, transdimensional, spiritual, non-local, non-physical, or supernatural to exist. Naturalism claims that invisible, non-physical, and intangible forces and fields DO NOT EXIST. Atheism is a philosophical belief or a religion that claims that God DOES NOT EXIST. Behaviorism and Determinism are the philosophical belief or religion which states that Agency, Choice, and Free Will DO NOT EXIST. Only entropic physical matter exists. Nihilism is the philosophical or religious belief that there is NO afterlife and that we cease to exist when we die. Nihilism is the philosophical and religious belief which states that your Psyche, Mind, Spirit, or Soul DOES NOT EXIST. Darwinism or Abiogenesis is a form of Atheism which states that genes, proteins, and life forms spontaneously generated out of thin air from nothing. Darwinism is Creation by Nothing. The Theory of Evolution is Creation by Entropy or Creation by Death. There's literally NOTHING here for us to experience and observe! Science is observation and experience, which makes Materialism, Naturalism, Darwinism, Nihilism, Behaviorism, Determinism, and Atheism BAD SCIENCE or NOT SCIENCE. These falsified philosophies are PURE RELIGION. They have to be taken on blind faith as being real and true because there is literally NOTHING there for us to experience and observe.

Ironically, the Darwinists and Evolutionists put forth an invisible, intangible, non-physical, supernatural, spiritual, ethereal FORCE or FIELD which they call "natural selection" as our designer and creator. Go figure! These people claim that Natural Selection designed and made our genes, proteins, genomes, and physical cells. The ironic thing is that Natural Selection doesn't touch our genes and can't touch our genes! Natural Selection has NO physical mechanism for doing so because Natural Selection isn't a physical thing. Natural Selection is a philosophical concept – a figment of our imagination that we then imbue with God-like powers like prescience, fore-sight, cunning, planning, omniscience, omnipresence, telepathy, telekinesis, and teleology. Natural Selection doesn't do anything! In fact, evolution (genetic change), natural selection, and random mutations didn't even exist until AFTER God designed, engineered, created, field-tested, manufactured, and deployed our genes, proteins, genomes, and physical bodies in the first place. There's no way in the universe that Natural Selection could have been our designer and creator. It's physically impossible.

I used to be a Materialist, Naturalist, Nihilist, and Atheist until I actually started looking at and studying the scientific evidence or the observations and experiences of the human race. Then I realized that I was wrong and needed to up my game.

Entropy is death. The Ultimate Law of Thermodynamics distinguishes between what is being conserved and what is not conserved. According to the Ultimate Law of Thermodynamics, entropy is temporary which means that entropy is not conserved. According to the Ultimate Law of Thermodynamics, entropy or death is not conserved. ONLY Psyche, or Energy, or the Life Force is conserved. Psyche, Energy, Intelligence, or the Life Force is syntropic which means that it is conserved. Psyche or Energy cannot be made, and it cannot be destroyed. It is eternal and everlasting. Psyche is the innate intelligence within all the different forms of energy which gives that energy the inherent ability to understand, follow, and obey God's Laws and God's Commands. Psyche is conserved; and, entropy is not conserved according to the Ultimate Law of Thermodynamics. The Ultimate Law of Thermodynamics is the answer to life, the universe, and everything. Is it not? It explains everything that has ever been experience or observed. Does it not?

Evolution is entropy. Evolution is death. Entropy is temporary which means that the different types of evolution are temporal – they result in death, not life. Evolution produces death and extinction. That's never going to work for the Origin of Life.

The Materialists, Naturalists, Darwinists, Nihilists, Behaviorists, Determinists, Classical Physicists, and Atheists treat evolution, entropy, and death as if they are inviolate, sacrosanct, pure, untouchable, holy, and sacred. These people treat evolution, entropy, and death as if they were God. Do they not? Evolution or entropy is their God. These people teach that entropy or death is conserve, meaning that entropy or death is eternal and everlasting.

They are wrong.

One day it dawned on me that some type of God must exist in order to have done all of the Science which the different types of evolution or entropy could NEVER have done. Ever since then, I have KNOWN that God exists because it was the science that convinced me that God must exist.

While interacting with the Darwinists and Atheists online, one of them asked me which one of the man-made gods do I want him believing in? His implication was that all of the gods are man-made gods.

My recommendation at the time was the True and Living God – the one that has actually been experienced and observed by millions of different people. Nobody has ever seen Allah. However, thousands if not millions of different people have reported seeing and experiencing our resurrected Lord Jesus Christ either in the flesh or during their Near-Death Experiences. The only way to prove that God exists is through observation and experience. God has to find ways to reveal Himself to us, or He would remain forever concealed and unknown. Jesus Christ has done so. Jesus Christ isn't hiding from us. The BEST and FASTEST way to find and know the truth is to live it, observe it, and experience it for yourself or to choose to trust someone who has. Jesus Christ has been experienced and observed. Science is observation and experience, or it should be.

This person online was right in that the man-made gods such as Materialism, Naturalism, and the Theory of Evolution are absolutely worthless because they don't exist, have never been experienced nor observed, and can't do anything for us to help us or save us. There's nothing wrong with rejecting the man-made gods such as Evolution, Materialism, Darwinism, and Naturalism.

Evolution is a mental concept or philosophical construct that only exists within the psyche or the mind. Evolution is a figment of the imagination. By definition, in principle, there's no person called Evolution who can do things for us. Evolution will never come in person and retrieve you from hell; but, our Lord Jesus Christ will just for the asking. If you are going to go with a God, you want to go with the one that has actually been experienced and observed – the one that can actually come and get you out of hell. In all things, go with the one that has been experienced and observed. Science is observation and experience after all.

Evolution is entropy. Evolution is death. Evolution is a man-made god. Creation by Evolution is a figment of the imagination. Entropy is death. The theory of evolution is Creation by Entropy or Creation by Death. Evolution of any type cannot design and create. Evolution is a modern-day idol, and Darwinism is the modern-day form of idolatry. Evolution is a false god. Entropy and death cannot design and create. All of the different types of evolution cannot design and create. Evolution is worthless when it comes time to explain the origin of genomes, proteins, and life forms. Evolution will never come and get you out of hell, either; but, our Lord and Savior Jesus Christ can and will.

Should you ever find yourself in hell, remember that Jesus Christ can get you out of there just for the asking. Ask, and ye shall receive. Knock, and it shall be opened unto you.

Remember, Physicalism, Materialism, Naturalism, and the Theory of Evolution are based exclusively on entropy. According to these people, ONLY entropy or physical matter exists. There's nothing else. Entropy is death. The theory of evolution is Creation by Entropy or Creation by Death. Death cannot design and create, now can it?

As scientists, we NEED a much better Model for Origins than entropy, death, or evolution. Do we not?

I'm a scientist. Scientists are supposed to discover and present the truth rather than hide the truth from us. As a scientist, I have tried to develop better Models for Origins than entropy, death, or evolution.

Evolution of any type is a man-made god. I have formally rejected the man-made gods. I have formally rejected the false gods or the falsified gods such as Materialism, Naturalism, Darwinism, Nihilism, Atheism, Entropy, and the Theory of Evolution. I had to do so in order to discover and produce a Quantum Model for Origins.

I had to give up My Materialism, My Naturalism, My Nihilism, and My Atheism in order to make progress in Science. These falsified philosophies or falsified gods are anti-science and were designed to prevent discoveries in science. They were designed to hide scientific evidence from us. They were designed to hide observations and experiences from us. It works. I never discovered anything useful or interesting while I was a Physicalist, Naturalist, Nihilist, and Atheist; but, look at what I discovered once I decided to pursue the truth instead!

You have to understand and then reject the man-made Gods such as Spontaneous Generation, Creation Ex Nihilo, Chemical Evolution, Macro-Evolution, Entropy or the Second Law of Thermodynamics, Blind Luck, Wishful Thinking, and Natural Selection if you want to find and know the truth. I KNOW because I never made any progress while I was a Materialist, Naturalist, Nihilist, and Atheist. You can't find the truth while your "knowledge" is based exclusively upon a deception or a lie. The false is falsified by the truth; and, the truth is repeatedly experienced and observed. The truth has been "caught in the act" by Someone Psyche somewhere sometime. The truth has been experienced and observed.

In contrast, teaching that Psyche, Syntropy, Action at a Distance, the Non-Physical, and God DO NOT EXIST does absolutely nothing to explain what they are and how they work. The job of a scientist is to figure out what things are and how they work. I started to do my job after I got rid of My Materialism, My Naturalism, and My Atheism.

Mark My Words

Defining a Quantum Model for Origins

In the various different permutations of Syntropy, I have found the answer to life, the universe, and everything. Energy or Syntropy can be made to take on many different FORMS. Matter of any kind is Organized Energy; and, Energy or Syntropy is conserved. The amount of physical matter, or entropy, or time changes as God permits and deems fit. Therefore, the quantity of physical matter, or entropy, or time is NOT conserved. Location, entropy, or space-time is not conserved. These things can change over time.

It's the Energy or the Syntropy that's being conserved, meaning that Energy or Syntropy is eternal and everlasting without a beginning of days or an end of years. Psyche is Syntropy. Quantum Mechanics is a type of Syntropy. Spirit matter is syntropic in nature and origin. Only physical matter is subject to locality and entropy – space-time – or the physical laws. Everything else is Quantum Mechanics or Syntropy.

This is a different take on Reality and existence than the one that the Physicalists, Naturalists, Darwinists, Nihilists, and Atheists choose to give us. These people erroneously teach that physical matter and entropy are conserved, and that physical matter or entropy is the ONLY thing that exists. They are wrong. They deliberately choose to get their science wrong so that they can say that God does not exist. Materialism, Scientific Naturalism, Darwinism, Atheism, and their derivatives were deliberately designed to prevent us from discovering Syntropy, Psyche, and God.

I KNOW that the Science behind Syntropy is good, right, and true in large part because the Physicalists, Scientific Naturalists, and Atheists automatically reject it, dismiss it, and disagree with it BEFORE they actually look at any of the Science. These people have a talent for picking falsehoods and choosing to believe in them. These people teach that ONLY physical matter or entropy exists. They are wrong. There's so much more to life and reality than just physical matter.

I learn best by trying to put into words what I have experienced and observed. I can sometimes spend weeks or even months trying to put into words something that came in a single flash of insight; but, it's worth the effort because I learn the most while trying to explain it to someone else.

A quantum model or syntropic model for origins is based upon the following scientific observations and scientific concepts:

First, Energy is Syntropy. According to the first law of thermodynamics, Energy or Syntropy is conserved. That means that Energy or Syntropy is neither created nor destroyed. It is eternal and everlasting, without a beginning of days or an end of years. Quantum Mechanics or Syntropy is the Primal Construct. Syntropy is the NORM. Syntropy has NO entropy. Syntropy is the opposite of entropy. Entropy is death. Syntropy is Eternal Life. Energy is unorganized matter. Matter is organized energy or organized light. Physical matter is frozen or coalesced light. Energy or Syntropy is conserved, even during the mass-making or matter-making process. $E = mc^2$. Energy is what the Gods call "matter unorganized". Matter is organized energy or light. Matter and entropy are not conserved – their quantity or amount changes over time.

Second, entropy or the second law of thermodynamics ONLY applies to physical matter. Everything else is Syntropy. Everything else is spiritual in nature and origin. Everything else is eternal and everlasting. Physical matter or entropy is the exception to the NORM. Entropy is a function of time. Entropy is represented by the passage of time. Entropy is an aging process. This means that physical matter or entropy has a beginning. Physical matter or entropy is NOT the normal default state of existence. Physical matter or entropy is NOT the only thing that exists. Physical matter or entropy is the exception to the RULE, not the rule. Physical matter is subjected to locality in space. Physical matter can no longer quantum tunnel at will because it has been given a short De Broglie Wavelength and localized into space and time. Physical matter is subjected to space-time – locality and entropy. Everything else is Syntropy – eternal and everlasting. The first law of thermodynamics reveals to us the eternal nature of Energy or Syntropy; and, the second law of thermodynamics reveals to us the physical limitations, physical restrictions, entropy, and locality that have been placed upon physical matter by God. These physical limitations only apply to physical matter. Everything else is Syntropy. This pretty much explains everything, doesn't it? The truth has been staring us in the face all the way along; and, the ONLY thing preventing us from seeing it, understanding it, and accepting it is our deep and abiding belief in Materialism, Naturalism, Darwinism, Nihilism, and Atheism which state that the non-physical or the syntropic does not exist. These people have chosen to believe that only physical matter or entropy exists; and, their chosen belief prevents them from finding, seeing, and understanding anything else.

Third, the organization of spirit matter requires a small infusion of Energy, Order, Light, or Glory from an Intelligence, Psyche, or Consciousness. In other words, Psyche or Consciousness is required to collapse the wave function. This reality and truth is a fundamental and essential aspect of Quantum Mechanics. Psyche is a type of Energy or Syntropy – eternal and everlasting. Psyche or Consciousness is required to FORM energy into matter. Matter is Organized Energy, or Organized Light. Energy can be made to change FORM and thereby take on mass and become matter. In its native original format, Psyche or Intelligence has an infinite velocity or frequency and therefore experiences NO passage of time. Psyche is omnipresent in its native domain. God has to slow things down a bit in order for a Psyche or Intelligence to be able to experience the passage of time. Spirit matter has been slowed down a bit from Pure Energy, Pure Syntropy, or Pure Psyche. Spirit matter is non-local, meaning that it hasn't been located in our 3D physical space-time realm. Spirit matter has had order, organization, limitations, and restrictions placed upon it. Spirit matter or dark matter was made to respond to Psyche's commands. Psyche, spirit matter, and spirit bodies in their native original environment can quantum tunnel anywhere in the universe instantaneously at will until God places some limitations, restrictions, consensus, order, and laws upon them.

Fourth, the conversion of spirit matter into physical matter requires a massive infusion of Energy, Mass, Order, Structure, Syntropy, Light, Law, or Glory from Psyche, Intelligence, or God in order for physical matter to come into existence. Physical matter has space-time enforced upon it. Space is represented by locality – physical matter can no longer quantum tunnel away from us at will. Time is represented by entropy or an aging process. Physical matter is slowed down by God so that we can live it, experience it, learn from it, manipulate it, and remember having done so. Physical matter is limited to sub-light velocities and frequencies. Physical matter is the basis for the Ultimate Consensus Reality. A physical reality is highly dependable, reliable, predictable, controllable, limited, lawful, and organized. In a physical reality, I can depend upon this essay being there tomorrow on my hard drive and cloud drive when I go looking for it. My physical hard drive is not going to quantum tunnel away from me at will because it has had locality or space-time enforced upon it by the Gods. That's the main benefit of a physical reality. You can depend upon it being there tomorrow when you wake up. A physical reality is not ephemeral like a non-physical spiritual non-consensus reality can be and typically is.

Fifth, there is NO such thing as a coalescing cloud of gas. Out in space, gas does not coalesce. Gas expands and dissipates into space. This is what has been experienced and observed. This Observational Reality or Scientific Reality means that planets, stars, and galaxies are born whole in situ. This is what has been experienced and observed. The galaxies, planets, and stars come into existence whole or are born all at once. We observe NO galactic-sized coalescing clouds of gas. Gas clouds out in space do not coalesce. They disperse under the effects of gas pressure and random diffusion. Coalescing gas clouds would require some kind of external force or God in order to make them coalesce. When it comes to physical planets, physical stars, and physical galaxies, they are born whole in place all at once. That is what has been observed through our telescopes. The birth of physical planets, stars, and galaxies is done by Quantum Mechanics and NOT classical physics. That is why this is called a Quantum Model for Origins. Dark Matter or Spirit Matter served as the template or the blueprint for this physical creation; and, Dark Energy is the Power by which this physical universe was made.

Sixth, Quantum Mechanisms such as Quantum Tunneling, Phase-Shifting, Telepathy, the Quantum Zeno Effect, and Action at a Distance have been experienced and observed. The Gods organized the planets, stars, and galaxies spiritually before organizing them physically. A "spirit body" serves as the template or the blueprint for the forthcoming physical body. A planet, star, or galaxy is born physically by being phased into existence or quantum tunneled into existence whole by the Gods. It requires a massive infusion of Energy or Syntropy from the Gods in order to convert a spirit planet into a physical planet. When a planet or star is born, it phases or quantum tunnels whole from the spirit realm into this physical realm. It changes dimensions and frequencies. It phases into its physical existence. It is born. That is why this is called a Quantum Model for Origins – it's based upon Quantum Mechanics and NOT classical physics. This Syntropic Model for Origins is based in large part upon Quantum Tunneling and Phase-Shifting both of which have been experienced, observed, and verified to exist. This Quantum Model for Origins is also based upon the Conservation of Energy or Syntropy, or the first law of thermodynamics. The Syntropy or Energy is conserved; whereas, the physical matter or entropy comes and goes as God sees fit. Physical matter is highly organized Energy. The spirit earth is still there; but, it has been added upon by a massive infusion of Syntropy or Energy, and that added Mass made it physical instead of just spiritual. When our earth was born, it phase-shifted from the spirit realm to our physical realm. It was added upon. It had locality and entropy (space-time) placed upon it by the Gods when it became physical.

Seventh, our earth, star, and galaxy were organized spiritually before being converted into physical matter. The Gods infused a massive amount of Energy or Syntropy into each particle of spirit matter within this earth thereby converting this whole earth into physical matter; and thus, our physical earth was born and made subject to space-time. Our physical earth was phased into existence or quantum tunneled into existence. The spirit body or spirit earth is still there. It served and serves as the template or the blueprint for the physical body or physical earth. Our earth was simply added upon when it experienced its physical birth. The same process happened with our sun and our galaxy. Under this Syntropic Model for Origins, new galaxies will be born as our physical universe continues to expand. It has been observed that our physical universe continues to expand and that that expansion is accelerating under the effects of what the astrophysicists call Dark Energy, which means that the Gods are continuing to infuse Energy or Syntropy into this physical universe in the form of what we call Dark Energy in order to make this universe expand. In other words, God's creation of this physical universe is ongoing, right here, right now. In the future, additional galaxies will be born or phased into a physical existence and receive a physical life when the time is right. The only way that the expansion of this universe could be currently accelerating is if the Gods are currently infusing massive amounts of Syntropy or Energy into this physical universe right now from someplace else besides this physical universe.

Conclusion: All of this is made possible by the fact that Energy or Syntropy is conserved – it never ages, and it never dies. Syntropy or Energy or Quantum Mechanics is eternal and everlasting, without a beginning of days or an end of years. It's ONLY physical matter that experiences a beginning or a birth. It's ONLY physical matter that is subjected to locality and entropy, or what we typically call space-time.

This ends up being a Quantum Model for Origins rather than a model for origins that's based exclusively upon classical physics and entropy. Quantum Mechanics has infinitely more explanatory power than Materialism, Naturalism, Classical Physics, and Entropy. This Quantum Model for Origins or Syntropic Model for Origins is much more likely to be true because Quantum Mechanics, Phase-Shifting, Quantum Tunneling, and Action at a Distance have actually been experienced and observed whereas coalescing gas clouds have not. Expanding clouds of gas do NOT coalesce without some kind of external intervention from God. In open space, when left on its own, gas expands, dissipates, and diffuses into space. It does not coalesce. That is what has been experienced and observed.

Energy, Syntropy, Psyche, or God is conserved. It's eternal and everlasting without a beginning of days or an end of years. In contrast, the physical matter or entropy comes and goes as God sees fit. The quantity of physical matter or entropy is changing all the time. It's not being conserved. A huge infusion of Syntropy, Energy, Organization, or Mass produces physical matter, locality, and entropy. Remove that Mass and Entropy and Physical Restrictions, and that particle of physical matter goes back to being spirit matter or syntropic matter instead.

This obvious truth has been staring us in the face all the way along; but, our deep and abiding commitment to Physicalism, Naturalism, Classical Physics, Darwinism, Nihilism, and Atheism has prevented us from seeing it and understanding it. Materialism, Naturalism, Darwinism, Nihilism, and Atheism were designed to prevent us from discovering Psyche, Syntropy, Quantum Mechanics, Supernatural Mechanisms, Invisible Non-Physical Forces and Fields, Quantum Tunneling, Phase-Shifting, the Quantum Zeno Effect, the Zero-Point Field of Light, Action at a Distance, and God.

Self-deception works, and it works every time. No seeking, then no finding.

We have known all the way along that Energy or Syntropy is conserved, and that physical matter or entropy comes and goes and is therefore not conserved. Physicalism, Naturalism, Darwinism, Nihilism, and Atheism were designed to prevent us from discovering, understanding, and accepting this simple truth. These people erroneously teach that physical matter or entropy is conserved, and that physical matter or entropy is the only thing that exists. They are wrong. Only the Syntropy or Energy is conserved. The amount of physical matter and entropy changes over time and is therefore NOT conserved.

Mark My Words

Tachyons

Kaku, M. (2008). *Physics of the Impossible: A Scientific Exploration into the World of Phasers, Force Fields, Teleportation, and Time Travel*. New York: Anchor Books.

Even though Michio Kaku is a materialist, naturalist, Darwinist, big banger, evolutionist, and agnostic skeptic supporting the party-line, he is not hateful, mean-spirited, and caustic like the New Atheists are. Even though Michio Kaku still hasn't gotten the message that Darwinism and Materialism are FALSIFIED and DEAD, he ironically supports and promotes theories and ideas that have NO physical support nor any materialistic naturalistic support to sustain them – intangible and unseen things like String Theory, M-Theory, Extra-Dimensionality, and Cosmic Music which are typically labeled as "not science" or "unscientific" by the Materialists, Naturalists, and Evolutionists at-large.

Michio Kaku is my favorite materialist and evolutionist to study and read because I'm fascinated by string theory, m-theory, cosmic consciousness, and extra-dimensionality. He also produces and promotes a lot of science that is invisible, immaterial, and non-physical in nature and origin.

I used to be a Materialist, Naturalist, Nihilist, and Atheist until the scientific evidence and observational experiences of the human race convinced me that I was wrong. Eventually Science led me to the Non-Physical, the Quantum Mechanical, or the Supernatural; and, my life, chosen beliefs, and worldview began to change.

In his book, *Consciousness Beyond Life: The Science of the Near-Death Experience*, Pim van Lommel convinced me that in Quantum Non-Locality or the Non-Local Spirit Realm, the Quantum Objects or Particles of Matter that reside there can theoretically exist at infinite velocities much faster than the speed-of-light.

Based upon that evidence, I concluded that spirit matter is the same thing as physical matter, EXCEPT for the fact that spirit matter exists at velocities greater than the speed-of-light and physical matter exists at velocities less than the speed-of-light. Physical matter has a lot of mass; whereas, spirit matter does not. Remove all the mass, and an infinite velocity becomes possible. Then I eventually concluded that immaterial Psyche exists at an infinite velocity or an infinite frequency capable of being instantaneously and simultaneously everywhere while in its native original format.

God marries our Psyche or Intelligence to spirit matter and later physical matter in order to slow it down, so that we Psyches can have Lived Experiences, learn from our experiences, and remember having done so. There is NO passage of time at velocities faster than the speed-of-light according to the theory of relativity. We have to exist or reside at velocities or frequencies less than the speed-of-light in order to experience entropy or the passage of time. We can't have experiences and form new memories without the passage of time. Where our physical realm is concerned, entropy or time is a localized physical construct that allows us to experience the passage of time and then remember having done so.

I eventually realized that matter existing at different velocities or existing in different phases can actually occupy the same space because they are out-of-phase with each other. This is the Quantum Law of Superposition. Therefore, Psyche, spirit matter, and physical matter can occupy the same space at the same time because they have different phases of existence or exist at different velocities or frequencies in different dimensions.

I eventually was taught and realized that a Quantum Object is a Particle of Matter, and that matter has two main states of existence according to the Quantum Law of Complementarity – a spiritual state and a physical state. Physical matter exists at velocities less than the speed-of-light because of all the MASS or resistance to acceleration that physical matter has; whereas, spirit matter with little or no mass exists at velocities greater than the speed-of-light.

We will NEVER be able to accelerate physical matter faster than the speed-of-light because God put physical limitations into place to prevent us from doing so. We would have to phase the physical matter into a quantum state first, and then we would teleport or quantum tunnel that matter anywhere in the universe we want it to go before phasing it back into a physical reality. This is how the Gods teleport their physical bodies anywhere in the universe at will. They have complete control over Quantum Mechanics. We do not.

In time, I came to realize that Dark Matter is in fact Spirit Matter.

For a very long time, I thought I was the ONLY one on the planet who realized that Spirit Matter exists at velocities faster than the speed-of-light and is therefore undetectable by physical instruments directly.

Last night 29JUL2017, I was reading in my most favorite introduction to Popular Physics or Speculative Physics – a book entitled *Physics of the Impossible* by Michio Kaku – the following description for Tachyons:

> **Tachyons live in a strange world where everything travels faster than light. As tachyons lose energy** [mass]**, they travel faster, which violates common sense. In fact, if they lose all energy** [or mass]**, they travel at infinite velocity. As tachyons gain energy** [or mass]**, however, they slow down until they reach the speed-of-light. (p. 280).**

That's the exact SAME definition for Spirit Matter and Dark Matter that I have been giving in all of my books for the past year or so.

Tachyons ARE Spirit Matter!

Of course, Michio Kaku doesn't realize that Tachyons are Spirit Matter because as a Naturalist and Physicalist he isn't looking for Spirit Matter; but, I instantly recognized the definition for Spirit Matter that I have been providing for the past year in my books, within the definition that Michio Kaku provided for Tachyons.

Cool, huh?

Well, at least I thought it was cool.

Tachyons live in a strange world that we have been calling the Non-Local Realm, or the Spirit Realm, or the Quantum Realm. As tachyons or spirit matter lose energy, their mass or resistance to acceleration decrease, which makes perfect sense to me. In fact, if the tachyons or spirit matter were to lose ALL of their mass, they would become immaterial Psyche and exist at an infinite velocity or frequency being instantaneously and simultaneously everywhere in their native original format.

Remember, ZERO MASS = Infinite Velocity. MASS or Inertia is resistance to acceleration. No resistance to acceleration results in an infinite velocity. It's always made perfect sense to me, ever since I learned about it. I first learned about spirit matter and its potentially infinite velocity from Pim van Lommel in his book, *Consciousness Beyond Life: The Science of the Near-Death Experience*.

Remember, Matter is a Quantum Object, and a Quantum Object has two mutually exclusive states of existence – a spiritual state faster than the speed-of-light and a physical state slower than the speed-of-light.

As God inputs more mass or energy into a Tachyon or particle of Spirit Matter or Dark Matter, its velocity slows down until it reaches the speed-of-light. Then, by consciously inserting additional energy or mass or order into that Tachyon or particle of Spirit Matter, God slows that particle of matter to sub-light velocities thereby converting that particle of Spirit Matter into a particle of Physical Matter.

There in Michio Kaku's paragraph, I found independent confirmation of what I have been trying to teach my readers for the past year of my life.

Non-Local Matter, or Spirit Matter, or Quantum Matter, or Dark Matter exists in the Non-Local Realm at velocities or frequencies greater than the speed-of-light. Spirit Matter has little mass or no mass and consists of Tachyons that exist at velocities greater than the speed-of-light. According to the Quantum Law of Superposition, spirit matter exists in a different phase, dimension, or frequency than physical matter; and, Psyche exists in a different phase, dimension, or form than spirit matter.

Mass is one FORM of Energy. This gets tricky. Very low mass, or a different type of mass, or a different organization of the Energy results in spirit matter, low entropy or a different type of time stream, non-locality, and the ability to travel faster than the speed-of-light. High amounts of mass or energy result in physical matter, entropy, the passage of time, locality in space, physical limitations, and NO ability to travel faster than the speed-of-light. It's hard to tell when spirit matter phases into physical matter. Should a Quantum Object ever shed ALL of its mass, then it ceases to be matter and becomes Psyche or Non-Local Consciousness instead. Psyche, or Quantum Non-Local Consciousness, IS the fundamental unit of reality. This is a Psyche Ontology, or the Ultimate Model of Reality.

Psyche ACTS; whereas, matter of any kind REACTS to psyche's command. Psyche is the ACTOR, because in its native original format it exists at an infinite velocity which means that it is instantaneously and simultaneously everywhere in its native original format. God marries our psyches or intelligences to spirit matter, and eventually physical matter, in order to slow us down so that we are subject to the passage of TIME or ENTROPY and can therefore have experiences and remember having had them. Each time that God introduces limitations into our lives, bodies, and/or psyches, God also introduces new and interesting opportunities to learn and grow.

Tachyons ARE Spirit Matter because they exist at velocities greater than the speed-of-light. I also have reasons for believing that Dark Matter is Spirit Matter. From my perspective, Science itself has PROVEN the existence of Spirit Matter, in the form of Tachyons or Dark Matter. Science is observation and experience. Spirit matter, dark matter, or tachyons have also been experienced and observed during our Near-Death Experiences (NDEs), Shared-Death Experiences (SDEs), and Out-of-Body Experiences (OBEs). We KNOW that they exist because they have been experienced and observed. Science is observation and experience, or it should be.

If we are going to take Quantum Mechanics, Quantum Non-Locality, Quantum Non-Local Consciousness (Psyche), Supernatural Mechanisms, Action at a Distance, and Syntropy seriously, then some major adjustments need to be made to Science in order allow for it. Our materialistic and naturalistic physical sciences as they currently stand make NO accommodation for Quantum Mechanics in any way, shape, or form. That has to change if we want to upgrade our Science to something that is more real, actual, useful, and true.

What Is Being Conserved?

Energy is Syntropy. Syntropy is the Conservation of Energy. Syntropy is the First Law of Thermodynamics. Energy, Psyche, Syntropy, or Life is conserved.

WE KNOW for a fact that it is the Syntropy or the Energy that is being conserved and not the amount of matter or entropy. The amount of matter as well as the amount of entropy comes and goes as God sees fit; but, the amount of the underlying Energy or Syntropy is always conserved. Physical matter, locality, entropy, time, and space are NOT conserved. Their quantity or their amount changes over time. We human beings create new physical matter in our atom smashers or particle accelerators. We human beings destroy physical matter in our atomic bombs. Physical matter or entropy is NOT conserved.

Understanding this Reality and Truth is the KEY to everything.

The first law of thermodynamics is a version of the law of conservation of energy adapted for thermodynamic systems. The law of conservation of energy states that the total energy of an isolated system is constant; energy can be transformed from one form to another but can be neither created nor destroyed.

https://en.wikipedia.org/wiki/First_law_of_thermodynamics

They are talking about Syntropy here. Syntropy is the First Law of Thermodynamics. Energy is Syntropy. Syntropy is eternal and everlasting, without a beginning of days or an end of years. Syntropy can neither be created nor destroyed. It has always existed and will always exist. Energy or Syntropy is conserved. Meanwhile, physical matter or entropy is NOT conserved. Its quantity changes. Entropy and physical matter are different FORMS of Energy. The FORM changes. The FORM is never conserved! Entropy is death. Physical matter, entropy, and death are NEVER conserved. This is the Ultimate Law of Thermodynamics. It explains everything.

The Materialists, Naturalists, Darwinists, Nihilists, and Atheists erroneously teach that physical matter and entropy are conserved. They teach that ONLY physical matter or entropy exists, which means that they teach that physical matter and entropy are conserved; and, they ruin and destroy Science in the process by introducing a lie into Science. Entropy is death; and, these people teach that death is conserved. They teach that death is eternal and everlasting. They are wrong.

303

The Physicalists and Scientific Naturalists are always wrong when it comes to Psyche, Syntropy, Quantum Mechanics, Supernatural Mechanisms, Spirit Matter, Action at a Distance, and what's being conserved. If you want to find and know the truth, then pick the exact opposite of Materialism, Naturalism, Darwinism, Nihilism, Atheism, and the Theory of Evolution. It works! It's reliable, replicable, and dependable. You pick the exact opposite of everything that has been FALSIFIED if you want to find and know the truth. Materialism, Naturalism, Darwinism, Atheism, and their derivatives have been falsified trillions of times in thousands of different ways. This is one of the most useful Scientific Observations that you can make. If you successfully eliminate everything that is false and everything that has been falsified, then only the truth will remain. If you want to find and know the truth, you start by eliminating Materialism, Naturalism, Darwinism, Nihilism, Behaviorism, Determinism, Physical Reductionism, and Atheism which have been falsified by the observations and the experiences of the human race in general.

I have observed that Syntropy is Conservation of Energy or the First Law of Thermodynamics. Energy is Syntropy. Conservation of Energy is Syntropy. Psyche, Intelligence, or Quantum Non-Local Consciousness is the fundamental unit of Syntropy, Existence, and Reality. According to the First Law of Thermodynamics, Energy, Psyche, or Syntropy is conserved; whereas, the different FORMS of Energy are NOT conserved. The FORM is never conserved. Space-time is a FORM of Energy. Physical matter and spirit matter are different FORMS of Energy. Entropy and locality are different FORMS of Energy. All of the different invisible, intangible, non-physical forces and fields that we have discovered are different FORMS of Energy or Syntropy. Physical matter, space-time, entropy, and locality are NOT conserved. The FORM is never conserved. ONLY the underlying Psyche, Syntropy, or Energy is conserved. This is the Ultimate Law of Thermodynamics. It explains how everything really works.

The amount of entropy is constantly changing, which means that entropy is NOT conserved. It also means that the entropy can be completely eliminated from physical matter because entropy is not conserved. In fact, the Apostle Paul suggests in 1 Corinthians 15 that a resurrected physical body is a "spiritual body" meaning that it has had its entropy or corruption removed. A resurrected body or a "spiritual body" is tangible like our physical body but it is syntropic, or eternal and everlasting. Entropy is like an in-built clock that measures the passage of time and provides a theoretical expiration date. Remove that clock or that entropy from physical matter, and it will become some type of syntropic matter instead. Meanwhile, all the way along, the Energy or the Syntropy is conserved. The entropy or the FORM is never conserved; but, the Energy or the Syntropy is.

Energy or Syntropy can take on many different FORMS. There are many different types of syntropic matter – spirit matter, dark matter, exotic matter, phase-shifted matter, tachyons, translated matter, and resurrected matter. ONLY physical matter is entropic matter or baryonic matter. Remove the entropy and the locality (space-time) from physical matter, and it becomes some kind of syntropic matter or quantum matter instead. It switches from classical physics to Quantum Mechanics.

Many scientists have observed that protons do NOT decay. Why don't they decay? They should decay if Materialism, Naturalism, Darwinism, Nihilism, and Atheism are true. The protons don't decay because they are syntropic in nature and origin – they are being conserved. The Energy or the Syntropy is always conserved no matter how God chooses to FORM it or ORGANIZE it. Entropy is only one FORM of Energy. The FORM is never conserved; but, the underlying Energy or Syntropy is. Energy, Psyche, or Syntropy is eternal and everlasting, without a beginning of days or an end of years. God can change its FORM; but, the Energy or the Syntropy is always conserved. The Energy or Syntropy can neither be made nor destroyed; whereas, physical matter and entropy can be made and destroyed.

Do you see how that works? It explains everything that we have experienced and observed, which means that it's Good Science. It's already been verified and proven to be true.

Energy, Psyche, or Syntropy can neither be created nor destroyed; whereas, the FORM is never conserved. That's what the First Law of Thermodynamics is trying to teach us; and, it's really simple to understand once you have eliminated Physicalism and Scientific Naturalism from your philosophy or worldview. Materialism, Naturalism, Darwinism, Nihilism, and Atheism teach that ONLY physical matter or entropy exists, which means that these people erroneously teach that physical matter and entropy are being conserved. They are wrong. Entropic physical matter is a FORM of Energy; and, the FORM is never conserved.

Just remember, physical matter, locality, entropy, and space-time are NEVER conserved. Physical matter and entropy come and go as God sees fit; but, the underlying Energy, Psyche, or Syntropy is always conserved. This is the Ultimate Law of Thermodynamics.

Here we have the answer to life, the universe, and everything.

The Bible says that God is Light, which means that God is Energy, Psyche, or Syntropy. It also means that God is conserved. God is eternal and everlasting, without a beginning of days or an end of years. That's made possible by the fact that God is Energy or Syntropy, and Energy or Syntropy is always conserved.

Entropy is a "FORM" of energy in that entropy is a measure of energy's transition from a useful form to an unavailable form. God's Psyche can make unavailable energy available again because God designed and made the rules which made that energy unavailable in the first place. Space is a FORM of Energy. Time and locality are a FORM of Energy. Space-time is a FORM of Energy. Physical matter is a FORM of Energy. The FORM of the Energy is never conserved. Energy or Syntropy can take on many different FORMS. Psyche, Intelligence, or Life is a FORM of Energy or Syntropy. So are spirit matter and physical matter. In fact, Psyche or Intelligence IS Energy or Syntropy; and, Psyche or Syntropy FORMS the basis for everything else. In contrast, entropy is death. Entropy or death is never conserved.

The Ultimate Law of Thermodynamics states that entropy or death is NEVER conserved, while the underlying Psyche or Syntropy is always conserved. This Reality explains everything that has ever been experienced or observed, does it not?

Remember, the FORM of the Energy is never conserved, while the underlying Psyche, Syntropy, or Energy is always conserved. Quantum Non-Local Consciousness, Psyche, Spirit Matter, Physical Matter, Physical Limitations, Locality, Entropy, Space, Time, and Space-Time are different FORMS of Energy. The FORM is never conserved; but, the underlying Psyche, Intelligence, Syntropy, or Energy is always conserved. The ORGANIZATION or the FORM of the Energy constantly changes; but, the quantity of the Energy is conserved. Energy, Psyche, or Syntropy is eternal and everlasting; whereas, physical matter, entropy, and death are NOT. Physical matter, entropy, and death are temporary. This is the Ultimate Law of Thermodynamics. It takes everything into consideration.

Let's Analyze Some of the Corollaries

Matter of any kind is Organized Energy or Organized Syntropy. Physical matter is more organized or more limited and restricted than spirit matter, which means that physical matter has more Energy, Syntropy, Order, Organization, or Mass infused into it than spirit matter. Remember, physical matter has more limitations or restrictions than spirit matter. Physical matter is more organized than spirit matter.

Tachyons are spirit matter.

Spirit matter or dark matter is a lot more syntropic than physical matter. Physical matter has had severe physical limitations placed upon it so that your physical body can't quantum tunnel away from you at will. God has forced a huge amount of Syntropy, Energy, Mass, Organization, Locality, and Entropy into physical matter. Physical matter or entropic matter has been bound to space-time (locality and entropy); whereas, spirit matter has not. Although spirit matter can be made to function within a time-stream, within its native, original, syntropic dimension, spirit matter has NO entropy, experiences NO passage of time, doesn't age, is omnipresent, has a long or infinite De Broglie Wavelength, and has NO physical limitations meaning that spirit matter can theoretically Quantum Tunnel anywhere in the universe at will and/or travel at an infinite velocity.

Remember, when it comes to the First Law of Thermodynamics, it's the Energy or the Syntropy that is being conserved and NOT the physical limitations, physical state, entropic state, or physical matter. The amount of physical matter, spirit matter, and entropy can and does change over time. Matter, entropy, and locality are NOT conserved. Space-time is NOT conserved. Space-time comes and goes as God sees fit. Entropy comes and goes as God sees fit. Entropy is NOT conserved. Entropy is not eternal and everlasting, as the Materialists and Naturalists claim that it is.

The amount of Energy or Syntropy remains constant; whereas, the amount of physical matter and entropy are constantly in flux and constantly changing. Remember, we human beings have created physical matter in our particle accelerators or atom smashers; and, we have destroyed physical matter in our atomic bombs. Physical matter is NOT conserved, as the Materialists and Naturalists claim that it is. Physical matter or entropy is NOT the only thing that exists.

Syntropy is the answer to life, the universe, and everything. It explains everything. Physical matter or entropy is the exception to this syntropic RULE, rather than being the rule.

Our scientists have observed that parts of our universe are currently expanding faster than the speed-of-light and will soon disappear from view. The expansion of our universe is currently accelerating. Our expanding universe and Dark Energy are proof positive that our physical universe is NOT a closed system. Our physical universe continues to receive a massive infusion of Energy or Syntropy from the Gods to this very day.

Is that infusion of Syntropy or Dark Energy centrally located forcing our whole universe to expand from the center, or is that infusion of Syntropy universal in nature across the whole universe causing the whole universe to expand equally? Is it our physical universe that's expanding its boundaries into nothing; or, is it the order, the organization, and the Syntropy that's expanding our universe as a whole into formerly unorganized chaos or space? Which one we are currently experiencing and observing? Either inference explains the observational evidence, so it's impossible to know which explanation is true unless God were to reveal it to us.

Despite all the unanswered questions that remain, I found this Science extremely useful and interesting. I'm going with the Science on this one rather than the wishful thinking of the Materialists, Naturalists, and Atheists who state that Dark Matter, Spirit Matter, and Tachyons do not exist because these syntropic particles of matter are invisible, immaterial, intangible, and non-physical. I prefer the science over the philosophy. That is my right.

I have chosen to believe that Spirit Matter, Dark Matter, or Tachyons do in fact exist. Our beliefs are chosen into existence by our Psyche, on both sides of the veil; and, that's what I have chosen to believe. You are free to choose otherwise. I have observed that Spirit Matter, Dark Matter, Tachyons, Quantum Mechanics, Dark Energy, and Syntropy have infinitely more explanatory power than Materialism, Naturalism, Darwinism, Nihilism, Atheism, and Classical Physics. Syntropic matter or spirit matter is Good Science – its explanatory power is through the roof to infinity and beyond; but, you are going to have to decide for yourself whether you think so or not.

The false is falsified by the truth; and, the truth is repeatedly experienced and observed. Science is observation and experience, and NOT the wishful thinking and the philosophical speculation of the Materialists, Naturalists, Darwinists, Nihilists, and Atheists. All you want is the truth. Everything else is worthless and absurd.

Mark My Words

—

Source

God Is in the Light: God is light, and in Him is no darkness at all.

> https://www.amazon.com/dp/B07168S37N

Science 2.0: I Upgraded My Science.

> https://www.amazon.com/dp/B0771K6WTX

References

Quantum Neuroscience: The Answer to Life, the Universe, and Everything.

> https://www.amazon.com/dp/B079Z6QQQB

Quantum Mechanics from a Non-Physical Spiritual Perspective.

> https://www.amazon.com/dp/B01J023TGU

> https://www.amazon.com/dp/1521132380

Dark Matter and Dark Energy

The Physicalists, Naturalists, Darwinists, Nihilists, Classical Physicists, Behaviorists, Determinists, Physical Reductionists, and Atheists have NO scientific explanation for Invisible Non-Physical Matter or Dark Matter or Spirit Matter, Dark Energy or Syntropy, Near-Death Experiences, Out-of-Body Experiences, After-Death Memories during Life Reviews, Invisible Non-Physical Forces and Fields such as Psyche, the Quantum Zeno Effect, Teleportation or Quantum Tunneling, and Action at a Distance such as Telepathy and Telekinesis.

How do they explain these things scientifically?

Their "scientific explanation" for Psyche, Dark Mater, Dark Energy, Syntropy, Non-Physical Invisible Forces and Fields, Quantum Mechanisms, Supernatural Mechanics, and Action at a Distance is to state that these things DO NOT EXIST. How does the "non-existence of Quantum Mechanics" explain all the science behind Quantum Mechanics? It doesn't! It can't! The "non-existence of Psyche" does NOT explain what it is and how it works. Materialists and Scientific Naturalists have NO scientific explanation for all the different invisible non-physical forces and fields that we have encountered throughout our lives; but, the Quantum Physicists certainly do, if they are willing to use it.

Dark Energy is a type of Syntropy. Energy in any form is conserved. Syntropy is the conservation of energy. Dark Energy is an invisible non-physical force or field. Energy can be made to take on many different FORMS. Spirit matter is a form of organized energy. Spirit matter or dark matter is a type of organized Energy or organized Syntropy. Physical Matter is a form of organized energy that has had even greater restrictions and limitations placed upon it, in order to keep it from quantum tunneling away from us at will.

It's fascinating to observe and study the psychology of the Materialists, Naturalists, Darwinists, Nihilists, Behaviorists, Determinists, and Atheists. They are in denial most of their lives; and, they suffer gravely from selective amnesia. They are motivated by their extreme desire to keep God and His Commandments out of their lives; and, they get their wish. However, it results in a great deal of hypocrisy and inconsistency, especially when it comes to their science.

For example, these people can typically accept the existence of gravity, even though they can't see gravity with their own eyes. Nobody has ever seen gravity with their physical eyes; but, we KNOW that it exists by the effect that it has on physical matter. They can accept gravity, but they refuse to accept other invisible non-physical forces and fields that they don't like and don't want to exist such as God's Psyche, the Human Psyche, Spirit Matter, and the Light of Christ or the Zero-Point Field of Light. These people are selective about which of the invisible non-physical forces and fields that they will accept. They consistently reject the non-physical forces and fields that prove that God exists, that prove that there will be some type of after-life, and that prove that there will be some type of Final Judgment. These people are afraid of the knowledge that someday God will judge them for their sins and call them to account.

Selectively accepting the existence of certain invisible non-physical forces and fields while at the same time selectively rejecting others that they don't like is called insincerity, pretense, two-facedness, falseness, deception, hypocrisy, and duplicity. It's a double standard! Yet, these people have the audacity to call it "science". These people literally call their "rejection of observational evidence" and their "rejection of experiential evidence" Science.

What other invisible non-physical forces and fields are these people willing to accept in the name of science?

Have you ever played with a magnet? Nobody can see magnetic waves or quantum waves with their physical eyes; but, we KNOW that they exist by the effect that they have on physical matter. Have you ever seen the wind? NO! Technically, you have never seen the wind with your physical eyes. However, you have definitely seen the effect that wind has on physical matter.

Have you ever seen the strong and weak nuclear forces? No, you have not. But, we KNOW that they exist by the effect that they have on physical matter. Have you ever seen a Psyche or a Spirit? Some have while out-of-body or in-the-spirit; but, technically it is physically impossible to see a Psyche or a Spirit with your physical eyes. Just because you can't see it with your physical eyes doesn't mean that it doesn't exist. The Materialists and Naturalists don't make any significant scientific discoveries because they conclude *a priori* before doing any actual science that if you can't see it with your physical eyes, then it doesn't exist.

I have looked at pond water under a microscope, and the little buggers do indeed exist; but, I can't see them with the naked eye.

Visible light is an infinitesimally small portion of the electromagnetic spectrum; and, there are many different types of light, energy, or quantum waves besides the electromagnetic variety. Energy or Syntropy can be made to take on many different FORMS, including different types of light!

The visible spectrum is the portion of the electromagnetic spectrum that is visible to the human eye. Electromagnetic radiation in this range of wavelengths is called visible light or simply light. A typical human eye will respond to wavelengths from about 390 to 700 nm.

https://en.wikipedia.org/wiki/Visible_spectrum

https://en.wikipedia.org/wiki/Electromagnetic_spectrum

Just because you can't see radio waves, microwaves, x-rays, and gamma rays with your physical eyes, that reality doesn't mean that they don't exist.

Furthermore, there are many different types of waves or "light" that are NOT on the electromagnetic spectrum. There are quantum waves. Thoughts and memories are quantum waves. We KNOW because thoughts and memories survive the death of our physical body and physical brain according to the empirical evidence from Near-Death Experiences (NDEs), Out-of-Body Experiences (OBEs), Shared-Death Experiences (SDEs), and our after-death Life Reviews. If a thought or memory survives the death of your physical brain, then it's truly a memory in every sense of the word.

Thanks to Quantum Superposition, many different types of quantum waves (energy or light) and matter can exist in the same space at the same time as long as they are out-of-phase with each other. That's how your physical body, spirit body, and psyche are able to exist in the same space at the same time – they are out of phase with each other. It works thanks to Quantum Superposition. This means that radio waves, visible light, microwaves, x-rays, magnetism, psyche, quantum waves, spirit matter, physical matter, and psyche can ALL exist at the same time in the same space because they exist at different frequencies and many of them are also out-of-phase with each other. That's why an fMRI machine doesn't scramble your memories or your thoughts – they are out-of-phase with each other.

Nobody has ever seen Dark Matter and Dark Energy with their physical eyes. Dark matter is a type of spirit matter. There are many different types and frequencies of spirit matter. Dark Energy like Gravity is a type of quantum wave – an invisible non-physical force or a field. Dark Energy and Gravity are a different type of energy or light. Just because we can't see it with our physical eyes doesn't mean that it doesn't exist. We KNOW that Dark Energy, Dark Matter, and Gravity exist by the effects that they have on physical matter. Energy is Syntropy. Most of the time we can't see Energy or Syntropy with our physical eyes; but, we know that it exists because of the effects that it has on physical matter.

The very existence of these supernatural forces, fields, invisible wavelengths of light, quantum waves, quantum mechanisms, psyche, intelligence, consciousness, spirit matter, dark matter, dark energy, gravity, magnetism, and wind FALSIFIES the claims of Materialism, Naturalism, Behaviorism, Determinism, and Atheism which claim that these different types of invisible, non-physical, supernatural forces, fields, waves, energy, and light DO NOT EXIST.

However, EVERYTHING FALSIFIES Materialism, Naturalism, Darwinism, Nihilism, and Atheism – even physical matter. Physical matter is also composed of a lot of different types of invisible energy, forces, fields, waves, mass, and light. Physicist David Bohm has told us that physical matter is frozen light. What happens when you freeze water? It converts into ice. What happens when you freeze invisible light? It converts into physical matter and becomes visible to your physical eyes. Everything falsifies Materialism and Naturalism, including physical matter.

Scientific Observations: Anything that was obviously made obviously has a Maker or Manufacturer who made it. Anything that was obviously organized obviously has an Organizer who organized it. Anything that was obviously fine-tuned obviously has a Fine-Tuner who fine-tuned it. Anything that obviously began, obviously has Someone Psyche or Someone Intelligent who caused it to begin. The truthfulness of these scientific observations is obvious.

Scientific Observations: Physical matter obviously had a beginning. Physical matter was obviously made. The Physical Constants, Physical Laws, Physical Restrictions, and Physical Limitations have obviously been fine-tuned. The physical galaxies were obviously organized. Physical stars and physical planets obviously began and were obviously organized or made. Physical genomes and functional proteins were obviously made and obviously had a beginning. They have obviously been fine-tuned.

Scientific Conclusion: Therefore, it is logical and rational to conclude that physical matter, physical universes, physical galaxies, physical stars, physical planets, physical genomes, physical brains, and physical bodies obviously have a Maker, Organizer, Fine-Tuner, Someone Psyche, and Someone Intelligent who made them, organized them, fine-tuned them, and caused them to begin.

That's what the Science is trying to tell us, is it not? The "doing of science" requires some kind of intelligent psyche or intelligent person to do that Science – does it not? Otherwise, nothing would ever get done. Science is observation and experience – NOT the wishful thinking of Materialists and Naturalists who assure us that the science or the observations do not exist. Science in general, and Psychology in particular, have infinitely more explanatory by choosing to include Psyche into the mix. By choosing to allow Psyche and Quantum Mechanics in to play, we can literally explain everything that comes our way.

The Big Bang Theory is in dispute. Raw Energy or Available Energy is "matter unorganized" or "potential matter". A pre-existing energy universe, that was subsequently organized into physical galaxies and that is now currently expanding thanks to an ongoing infusion of Dark Energy from some other dimension, can explain the observational evidence just as well as the Big Bang Theory. It's very possible that the Big Bang didn't happen. Most of the scientists are willing to admit that the Big Bang has NO explanatory power when it comes Time 0 or the Origin of our Physical Universe. The Big Bang Theory simply explains what might have happened AFTER the organization or creation of physical matter, time, space, entropy, and locality.

Scientists have observed that our physical universe is expanding under the influence of Dark Energy. The galaxies furthest away are moving away from us the fastest. At the furthest reaches of the observable universe, space is expanding faster than the speed-of-light. Such a huge universe-wide expansion would require a massive infusion of dark energy at a universal scale. That energy has to be coming from some other dimension besides our physical universe because our physical universe is actually losing available energy to entropy. Energy is always conserved; and, it's as if the energy that's being "lost" through entropy is being fed back to us as dark energy. Currently, our universe is gaining infinitely more from dark energy than what it is "losing" to entropy.

Scientists have observed that our physical universe is expanding. So, they rationally take that expansion and run it backwards in time. Then, they take a leap of faith and keep reversing that expansion and keep running it backwards until they have our whole universe residing in a single point that they call the Big Bang Singularity. Now, think about this process logically and rationally. While rewinding everything and moving backwards in time towards that hypothetical Big Bang Singularity, YOU CAN STOP anywhere along the way and still produce the same results that we see today when you start running the clock forward again. In other words, you don't have to run the clock all the way back to the Big Bang Singularity and still end up with the expanding universe that we see around us today. You could just as easily and more logically run the clock back to the time where the Gods started making physical matter, planets, stars, and galaxies from Raw Energy or Available Energy; and, it would still produce the same observable universe that we have today as long as the Gods start infusing huge amounts of dark energy on a universal scale somewhere along the way. The Big Bang Singularity is unnecessary to produce what we see today. You can stop your rewind anywhere along the way and still produce the results that we see around us today. Remember, you don't have to rewind it all the way to produce the results that we see today.

So, what about the cosmic background radiation?

Well, the scientists have proven that it is impossible to reach absolute zero at the physical level due to Quantum Mechanics or Energy Mechanics. Through the process of making planets, stars, and galaxies from Raw Energy or Available Energy, it's going to leave its mark on the "absolute zero" of space. There's always going to be some residual heat or radiation because absolute zero is impossible to achieve at the physical level.

Absolute Cold

https://www.youtube.com/watch?v=OvgZqGxF3eo

https://evolution-is-entropy.com/wp-content/uploads/2018/08/Absolute-Cold.zip

The quantum fields that fill our universe also fluctuate due to the uncertainty principle resulting in what we know as vacuum energy; and, some quantum fields have an intrinsic non-zero zero-point energy before bringing Heisenberg into it. This leads to the famous Higgs mechanism and possibly also the phenomena of inflation and dark energy. To understand the universe, we need to understand how it behaves absent heat, absent light, and absent matter.

Furthermore, the Science tells us that the production of hydrogen molecules is what produced the Cosmic Microwave Background Radiation in the first place. That's when the CMB began, when the Gods produced physical matter from energy or formed physical matter from energy. Therefore, all you have to do is rewind the expansion of our universe back to the point where the Gods made or produced the physical matter in the first place, and you automatically end up with the cosmic background radiation that we see today when you start running the clock forward again. You don't have to run the clock backwards all the way to a Big Bang Singularity in order to have the Cosmic Microwave Background Radiation that we see today. You just have to rewind our universe to the point where the Gods started making physical matter out of Raw Energy or Available Energy.

I love science. There is always more than one way to explain what we observe.

Many scientists believe that the Big Bang Singularity was a huge black hole. I'm sorry to inform you, but it would take a God to make a black hole explode. Black holes were definitely designed not to explode. They can evaporate, but not explode.

https://en.wikipedia.org/wiki/Horizon_problem

https://en.wikipedia.org/wiki/Flatness_problem

https://en.wikipedia.org/wiki/Inflation_(cosmology)

The Big Bang Theory is successfully falsified by the Horizon Problem, and by the "fix" to the Horizon Problem which they call Inflation. It would take some kind of God to start and stop Inflation on a universal scale. Would it not? To solve the Flatness Problem requires Fine-Tuning by the Hand of God. As I see it, the Big Bang Singularity would have had to have been a White Hole – a complete and total reversal of entropy – Pure Syntropy. It would take a God to make such a thing.

https://www.youtube.com/watch?v=S4aqGI1mSqo

https://scientific-proof-of-god.com/wp-content/uploads/2018/08/White-Holes.zip

https://www.youtube.com/watch?v=nhy4Z_32kQo

https://science-2-0.com/wp-content/uploads/2018/08/Are-You-a-Fluctuation-in-Entropy.zip

The Big Bang Theory breaks parsimony. It's anything but simple. It violates the Second Law of Thermodynamics with a massive reversal of entropy or a massive infusion of syntropy at the beginning of the process and/or throughout the process. Only a God could reverse entropy. Only a God could infuse Syntropy or Dark Energy into the system at a universal scale.

Furthermore, if the newborn universe was dense enough to form stars and primordial black holes, then it would have been dense enough to collapse back into a black hole. It would take a God to make something that dense to begin with; and, it would take massive amounts of intervention from God to prevent the newborn universe from collapsing into a black hole. It would take a God to finely tune and drive the inflation and then the subsequent expansion. It would take a God to force portions of that expanding cloud of hydrogen and helium to coalesce into planets, stars, and galaxies.

https://www.youtube.com/channel/UC7_gcs09iThXybpVgjHZ_7g/search?query=Big+Bang

There are NO Population III stars comprised ONLY of Hydrogen and Helium; and, they should exist in ALL sizes and in great abundance if the Big Bang truly happened. Add to that the FACT that expanding clouds of hydrogen and helium gas are NEVER going to coalesce into planets and stars without some kind of direct intervention from God; and, I'm left with the feeling that the Big Bang never happened.

However, if you choose to believe that the Big Bang happened (and most scientists do), then you have to believe in God anyway, because Someone Psyche or Someone Intelligent was needed to form the bomb and then make it go bang. Furthermore, Someone Psyche or Someone Powerful was needed to drive inflation, stop inflation, and then billions of years later start and drive the accelerating expansion of our physical universe through massive infusions of Dark Energy from some other dimension. Someone Psyche or God-Like is needed to keep varying the Cosmological Constant or the amount of Dark Energy on a universal scale, and to control the timing of it all. It would take a God to vary the Cosmological Constant at a universal scale for Inflation and Universal Expansion, would it not?

If the Big Bang truly happened, then it is Scientific Proof of God's Existence. Someone Psyche has to be driving the bus. God had to make the Big Bang Singularity, and God had to trigger it and make it go bang. Anything that was obviously made obviously has a Maker who made it. The Big Bang Singularity, if it ever existed, was obviously made. It's not going to spring into existence from nothing through a random fluctuation in entropy. That would violate the First Law of Thermodynamics. Furthermore, the only reason we have Laws is because we have a Law Maker. God's Psyche is needed to explain why there is something rather than nothing. Creation Ex Nihilo, or creation from nothing by Nothing, can never explain why there is Dark Energy filling our universe rather than concrete filling our universe. You are going to have to think logically about this stuff if you are ever going to figure out how it works.

Raw Energy is unorganized matter. From my perspective, it is clear and obvious that our observable physical universe received a massive infusion of Syntropy, or a massive infusion of Energy, or a massive Organization 13.8 billion years ago, which would represent the beginning of our physical universe as it currently exists. There had to be a massive infusion of Syntropy 13.8 billion years ago or some kind of infusion of Physical Law 13.8 billion years ago, or all of that subsequent entropy and physical organization would not have been possible.

Whether we are talking about the Big Bang or the Big Organization, it took place 13.8 billion years ago in the beginning of our Physical Universe and the beginning of the physical matter in our universe. According to the Big Bang Theory, over 13 billion years ago physical matter was organized or made, and then that physical matter was subsequently organized into planets, stars, and galaxies. Anything that was obviously organized obviously requires an Organizer to organize it. That reality applies directly to both the Big Bang Singularity and the organization or formation of physical matter and our physical universe. The very existence of physical matter is Scientific Proof of God's Existence. Physical matter is organized energy. The very existence of planets, stars, and galaxies is Scientific Proof of God's Existence. Planets, stars, and galaxies are organized physical matter. There would be NO order and organization without God. The whole thing would still be unorganized energy or "matter unorganized" without God's direct intervention. The whole thing would still be chaos.

Anything that was obviously organized obviously has an Organizer who organized it. In an entropic physical universe, order and organization don't just spontaneously generate out of thin air from nothing as the Materialists and Naturalists claim. An entropic physical universe requires an Organizer, or nothing would ever happen. The very existence of entropy as well as the very existence of order and organization is Scientific Proof of God's Existence.

The very existence of organized planets, stars, and galaxies is Scientific Proof of God's Existence. There's no way in the universe that clouds of hydrogen and helium gas are ever going to coalesce into planets, stars, and galaxies without some kind of external force from God making them do so.

Anything that was obviously made obviously has a Maker who made it. In an entropic universe, order and organization don't just spontaneously generate out of thin air from nothing as the Materialists and Naturalists claim. They are made!

The very existence of a functional information-rich genome is Scientific Proof of God's Existence. Your genome is God's Signature. Such a thing is never going to spontaneously generate out of thin air from nothing as the Materialists, Naturalists, Darwinists, and Atheists claim happened. The genome along with the associated proteins that it codes for was in fact the first convincing Scientific Proof of God's Existence that I encountered in my journey to discover God.

Because Dark Matter, Magnetism, Gravity, Psyche, Quantum Waves, the Nuclear Forces, and Dark Energy are intangible, non-physical, invisible, and spiritual in nature and origin, they have to be defined in terms of what they do rather than in terms of what they are. However, Science tells us precisely what they are, nonetheless. They are different forms of ENERGY. The First Law of Thermodynamics tells us clearly and conclusively that they are different forms of Energy. In fact, Psyche is the innate intelligence within all the different forms of energy which gives that energy the inherent ability to understand, follow, and obey God's Commands and God's Laws.

The first law of thermodynamics is a version of the law of conservation of energy, adapted for thermodynamic systems. The law of conservation of energy states that the total energy of an isolated system is constant; energy can be transformed from one form to another but can be neither created nor destroyed.

https://en.wikipedia.org/wiki/First_law_of_thermodynamics

I used to be a Materialist, Naturalist, Nihilist, and Atheist. I was holding out for Scientific Proof of God's Existence. I demanded Scientific Proof for God's Existence thinking that I would never find it. I truly believed that I was safe and that there would never be any convincing Scientific Proof of God's Existence. I was wrong. Once it was revealed to me, I had no choice but to believe in God's existence if I were going to remain true to myself and the commitment that I had made.

The very existence of stars made from hydrogen and helium as well as your genome are two of the most-convincing Scientific Proofs of God's Existence that I have found so far; and, there are others. The very existence of usable energy within this entropic physical universe is Scientific Proof of God's Existence. The very existence of Organized Energy or Physical Matter is Scientific Proof of God's Existence. Anything that was obviously organized obviously has an Organizer who organized it. Order and organization of any kind is Scientific Proof of God's Existence. Someone Psyche had to organize raw energy into entropy and physical matter, or they wouldn't exist.

Anything that was obviously made obviously has a Maker who made it. Proteins and their matching genes were obviously made. Planets, stars, and galaxies were obviously made. The fact that there is something rather than nothing is Scientific Proof of God's Existence. Someone Psyche had to do the Science, or none of this would exist. Science itself is Scientific Proof of God's Existence. There would be no Science if God did not exist. The verified and proven existence of Dark Energy and Dark Matter is Scientific Proof of God's Existence. Someone Psyche is needed to make that Dark Matter and Dark Energy work, and to build it or organize it into useful forms. It would take some sort of God to organize Dark Matter or Spirit Matter into planets, stars, and galaxies on a universal scale.

Would it not? The First Law of Thermodynamics, Syntropy, or the Conservation of God's Energy and God's Psyche is Scientific Proof of God's Existence. The Fine-Tuning associated with all the different Physical Laws and Physical Constants is very convincing Scientific Proof of God's Existence – an endless supply of proof that God exists.

I'm not going to apologize for finding the truth. That's precisely what scientists are supposed to do; and, I'm a scientist. Materialism, Naturalism, Darwinism, Nihilism, Behaviorism, Determinism, Physical Reductionism, and Atheism were designed to prevent us from discovering these truths. They work as advertised. I had to be willing to abandon My Materialism, My Naturalism, My Nihilism, and My Atheism BEFORE I was finally able to find, know, and understand these scientific truths. My Atheism was preventing me from finding and knowing these truths. Self-deception works, and it works every time.

Mark My Words

Quantum Fields

If I were a God tasked with bringing order to a small portion of the chaos which is our overall Greater Universe, I would start by designing and creating a network or a lattice of various different Quantum Fields. In other words, I would ORGANIZE space at the quantum level or the spiritual level. I would organize everything spiritually or quantum mechanically BEFORE trying to bring physical matter into existence. Physical matter wouldn't work without this underlying network of Quantum Fields to hold it together.

These Quantum Fields of Energy are fundamental and essential – infinitely more fundamental than entropic physical matter. If I were dealing with unorganized matter and unorganized energy – chaos I would bring order to it first by establishing functional and useful Quantum Fields within it at the quantum level or the spiritual level.

You see, our observable physical universe as a whole is a quantum wave, quantum particle, or quantum object – a blip of order and organization within an overall chaos that's completely hidden away from us outside our cosmic horizon. Our organized universe with its Quantum Fields is currently expanding into chaos bring order and organization to the chaos as our physical universe expands. The Gods continue to organize and create things spiritually or quantum mechanically beyond the cosmic horizon with the intent to eventually organize them physically as well. It all starts by making Quantum Fields or organizing the space at the quantum level or the spiritual level.

Many of the physicists can't see this nor understand it because they don't want to. They want entropic physical matter to be the fundamental unit of reality, NOT Psyche, Energy, Quantum Fields, and God. Nevertheless, even if our own unique observable physical universe were to suffer complete Heat Death, ALL of the Quantum Fields will still be there waiting to be reused anytime the Gods choose to do so. In other words, the Quantum Fields will be CONSERVED even after the Heat Death of our physical universe. According to the Quantum Law of Thermodynamics, heat death is impossible at the quantum level and the psyche level. In the Quantum Realm or Spirit World, everything is CONSERVED including the Quantum Fields that have been made. All the Gods have to do to start the whole process all over again is to bring some more entropic physical matter, or syntropic physical matter, into existence from all of that Dark Matter and Dark Energy that's lying around – anytime they choose to do so. This would be easy for them to do because ALL of the Quantum Fields are still there even if our physical universe appears to have suffered

Heat Death at the physical level. Remember, there is NO heat death or entropy at the quantum level or the spiritual level because everything there is CONSERVED.

Contrary to what the Naturalists and Atheists are trying to teach us, Quantum Fields are infinitely more fundamental than entropic physical matter. There's really nothing fundamental about entropic physical matter – it's an after-the-fact ADD-ON. It's the Psyche, Energy, and Quantum Fields that ARE fundamental, NOT the entropic physical matter. According to the Ultimate Law of Thermodynamics, physical matter and entropy are NOT conserved, which means that the CONSERVED (Psyche, Energy, Intelligence, Life Force, Quantum Fields) ends up being more fundamental and essential than anything that's not being conserved.

Do you see how that works? It's the answer to life, the universe, and everything. It explains everything that has ever been experienced and observed.

Remember, physical matter is impossible without the Quantum Fields or the Spiritual Fields to sustain it and hold it together. Quantum Fields or Spiritual Fields are infinitely more important, essential, and fundamental than entropic physical matter. Psyche is the innate intelligence within all the different forms of energy which gives that energy the inherent ability to understand, follow, and obey God's Laws and God's Commands. This is what has been experienced and observed. Science is observation and experience after all.

The lowest energy state is called the vacuum state. It's ALL comprised of energy or psyche, including the void or the vacuum of space.

Quantum Fields are Organized Webs of Energy. They fill the immensity of space.

Organized by whom? Who made them? Who is using them? Inquiring minds want to know.

https://en.wikipedia.org/wiki/Quantum_field_theory

https://en.wikipedia.org/wiki/Feynman_diagram

The particle creation and particle annihilation shown on Feynman Diagrams is made possible by the Conservation of Energy. Quantum Field Theory and the Feynman Diagrams are scientific proof that physical matter is NOT conserved. Physical particles are created and destroyed all the time. ONLY the underlying energy is conserved. Quantum Field Theory is scientific proof that the Ultimate Law of Thermodynamics is true. The Ultimate Law of Thermodynamics states that physical matter and entropy are not conserved. Only the underlying Energy or Psyche is conserved. The FORM is never conserved. Energy can always be transformed into a different form anytime the Gods see fit to do so.

According to the Copenhagen Interpretation of Quantum Mechanics, Someone Psyche is needed to start the electron's or fermion's initial state (or to start and instantiate a photon's initial state) thereby causing the quantum, quantum wave, or "particle" to exist and begin moving through the quantum field in the first place. In the Feynman Diagrams, they deliberately ignore the Psyche or the Intelligence who creates the quantum particles and then starts them moving though their respective quantum fields. That's what Materialists, Naturalists, and Atheists do. These people just assume that these quantum particles come into existence from nothing and return to nothing when they are done doing their thing. These people assume Creation Ex Nihilo as the fundamental foundation of their science rather than the Conservation of Energy or the Conservation of Psyche. These atheistic people literally use "nothing" to explain everything. In contrast, WE KNOW that Someone Psyche or Someone Intelligent is needed to explain why there is something rather than nothing.

One of my greatest and most useful scientific discoveries came when I first realized that the Materialists and Naturalists teach, preach, and believe Creation Ex Nihilo, Abiogenesis, Spontaneous Generation, Conservation of Entropy, Conservation of Death, the Supremacy of Entropy, and the Conservation of Physical Matter in direct violation of the First Law of Thermodynamics because these people teach and believe that ONLY entropic physical matter exists and that ONLY entropic physical matter is conserved. These people also teach and believe that Psyche, Life Force, or Consciousness DOES NOT EXIST. They are wrong.

By studying the Space Time Quantum Field Theory playlist below, you can see with your own eyes that Quantum Fields are highly organized webs of Energy that fill the immensity of space. The order and organization behind it all is obvious and clear for anyone who has eyes and a willingness to see and understand. The production of Quantum Fields and the Initialization of Quanta, or the Creation of Quanta, or the Initial States of Quanta are Scientific Proof of God's Necessity. God's Psyche is needed to explain why there is something rather than nothing, especially at the quantum level or the psyche level. Quantum Fields are NOT made by entropic physical matter. Instead, Quantum Fields facilitate the creation and movement of quanta, wave packets, energy packets, or "particles" of matter by Someone Psyche.

The First Quantum Field Theory

https://www.youtube.com/watch?v=ATcrrzJFtBY

https://syntropy.site/wp-content/uploads/2018/07/The-First-Quantum-Field-Theory.zip

"Particles" are quanta – wave packets of energy. Quantum Field Theory describes all elementary "particles" or quanta as vibrations, or ripples, in the Fundamental Quantum Fields that exist at all points in space and time throughout the universe. Quantum Fields of Energy fill the immensity of space. Space isn't empty after all. Photons are "field quanta", "light quanta", energy packets, or wave packets moving through a Quantum Field or the Electromagnetic Field.

These Quantum Fields of Energy seem to be the fundamental structure of space-time, and therefore the fundamental unit of our physical universe and physical reality.

As I see it, the Psyche or Intelligence who made and organized the Quantum Fields ends up being the most fundamental or the most essential unit of them all, when it comes to Reality as we know it. Organization doesn't spring whole from nothing.

Quantum Fields were obviously organized and obviously made. When it comes to Quantum Fields and Quantum Field Theory, we can clearly see the order, organization, deliberation, intelligent planning, structure, restrictions, and laws behind it all.

Anything that was obviously organized obviously has an Organizer who organized it. Anything that was obviously made or obviously fine-tuned obviously has a Maker who designed it, made it, and fine-tuned it. Quantum Fields were obviously designed, made, organized, and fine-tuned. There is structure, order, organization, and intelligent planning there – rather than nothing. Psyche is the innate intelligence within all the different forms of energy which gives that energy the inherent ability to understand, follow, and obey God's Laws and God's Commands. Psyche, in particular God's Psyche or God's Intelligence, explains why there is something rather than nothing. To discover what things are, you must discover who made them and why they exist. "Nothing" is not an adequate explanation for the existence of something.

The Materialists and Naturalists claim that space is empty and that only entropic physical matter exists. They are wrong. Quantum Fields of Energy are scientific proof that entropic physical matter is NOT the fundamental unit of reality nor the ONLY unit of reality as the Materialists and Naturalists claim that it is. Most people can't see it because they choose not to see it and understand it; but, Quantum Field Theory and Quantum Fields FALSIFY Materialism, Naturalism, Darwinism, Nihilism, and even Atheism. It's obvious and clear for anyone who is willing to look and see.

Quantum Mechanics, Quantum Field Theory, and Action at a Distance FALSIFY Materialism, Naturalism, Darwinism, Nihilism, Behaviorism, Determinism, Physical Reductionism, Atheism, and Classical Physics. That's why the Materialists and Atheists keep calling Quantum Mechanics "unintuitive". Quantum Mechanics can't be explained by Materialism, Naturalism, Darwinism, Nihilism, Atheism, nor Classical Physics.

In contrast, Quantum Mechanics is perfectly intuitive to anyone who KNOWS that the Non-Physical, the Non-Local, the Transdimensional, the Intangible, the Spiritual, and all those different Invisible Forces and Fields do indeed exist in spades throughout the whole of space and time. In fact, Psyche and Quantum Mechanics are essential if we want to understand what these things are and how they work. The verified and proven existence of these quantum mechanical realities FALSIFIES Materialism, Naturalism, Darwinism, Nihilism, Atheism, and their derivatives; and, it ends up being Scientific Proof of God's Existence instead.

Mark My Words

Entropic Physical Matter

We need a brief introduction to Physical Matter before we can explore Dark Matter.

The following playlist and video from Space Time does a good job of explaining the origin of matter and time from the perspective of Quantum Field Theory. Of course, we need something like God's Psyche to explain the origin of Quantum Fields, Quantum Mechanics, and Physical Laws. Any type of law, rule, restriction, order, or organization requires an Organizer of Law Giver, as well as Law Enforcer and Law Followers. Psyche is the innate intelligence within all the different forms of energy which gives that energy the ability to understand, follow, and obey God's Laws and God's Commands. Someone Psyche is needed for the construction of the Quantum Fields and Physical Laws in the first place. Creation Ex Nihilo has NO explanatory power. It cannot explain why there is order rather than chaos. "Creation from nothing by Nothing" cannot explain why our universe is filled with Dark Energy or Quantum Fields and why there are halos of Dark Matter surrounding and shepherding galaxies keeping them from flying apart.

The Origin of Matter and Time

https://www.youtube.com/playlist?list=PLsPUh22kYmNCLrXgf8e6nC_xEzxdx4nmY

The Higgs Mechanism Explained

https://www.youtube.com/watch?v=kixAljyfdqU

https://syntropy.site/wp-content/uploads/2018/08/The-Higgs-Mechanism-Explained.zip

Massless Speed-of-Light Particles or Massless Psyches USE mass to resist acceleration and to slow down to sub-light velocities. Mass produces the experience of time or the passage of time. Mass makes entropy possible. Mass is just a different FORM of energy. It's all energy, and energy is always conserved.

The most interesting science in this video is their discovery that, without the mass or the resistance to acceleration which the electron (and quark) gets from the Higgs Field and weak hyper-charge, the electron would be a massless particle and travel at the speed-of-light. In other words, the electron would have NO mass without the Higgs Field. The Higgs Field provides the mass, or resistance to acceleration, and slows the electron down.

Mass is resistance to acceleration. Mass slows particles down. Massless particles like photons travel at the speed-of-light, and they experience NO passage of time and therefore NO entropy. Photons and other Massless Particles are capable of being immortal and conserved. The Higgs Field produces mass, which slows quantum particles down to below the speed-of-light, which then produces the passage of time as a result. The passage of time makes entropy possible. The fundamental elementary components of the electrons and the quarks are Massless Speed-of-Light Particles; and, the Higgs Field gives the quarks and the electrons their small amount of mass and thereby slows them down to sub-light speeds.

Remember, mass produces resistance to acceleration and thereby slows Massless Particles down to sub-light speeds.

The Origins of Mass

https://www.youtube.com/watch?v=x8grN3zP8cg

https://science-2-0.com/wp-content/uploads/2018/08/The-Origins-of-Mass.zip

Mass is really just the energy or the gluons that bind the quarks together.

Time stops at the speed-of-light. Real matter or physical matter is comprised of massless light-speed components confined by interactions with other particles and force fields. 99% of the mass of an atom is trapped in and produced by the gluons and quantum fields that bind the elementary particles together; and, it's that mass or field that feels, causes, or experiences the passage of time, and not the massless timeless light-speed components or quantum particles within the physical matter. The passage of time is built into the quantum fields, not the massless timeless elementary quantum particles that comprise physical matter.

Everything with mass is comprised of massless, timeless, speed-of-light particles that are confined, limited, and prevented from streaming freely at the speed-of-light by the various different quantum fields that confine those particles and limit them to sub-light speeds. In other words, the mass or resistance to acceleration is found in the quantum fields and NOT the quantum particles themselves. The mass in quarks and electrons is produced by the Higgs Field. The mass in the protons and neutrons is produced by the Strong Force or Gluon Field that binds the quarks together.

Mass is energy, NOT entropic physical matter. Don't you get the sense from Quantum Mechanics and Quantum Field Theory that Energy or Psyche is the fundamental unit of reality and NOT entropic physical matter? I certainly do. When it comes to the universe, the whole thing is comprised of Energy or Psyche, and NOT physical matter. Technically, entropy and physical matter do not exist, certainly not as any kind of mass or stuff. The majority of the mass or resistance to acceleration is in the force fields, or the quantum fields, or the energy fields binding the quarks together – NOT in the quarks, or the "matter", or the "particles" themselves.

In theoretical particle physics, the Gluon Field is a four-vector field characterizing the propagation of gluons in the strong interaction between quarks. The Gluon Field plays the same role in quantum chromodynamics as the electromagnetic four-potential in quantum electrodynamics – the gluon field constructs the gluon field strength tensor.

https://en.wikipedia.org/wiki/Gluon_field

The strong nuclear force holds most ordinary matter together because it confines quarks into hadron particles such as the proton and neutron. In addition, the strong force binds neutrons and protons to create atomic nuclei. Most of the mass of a common proton or neutron is the result of the strong force field energy; the individual quarks provide only about 1% of the mass of a proton.

https://en.wikipedia.org/wiki/Strong_interaction

I've always been interested in Particle Physics; therefore, it was extremely important to me to finally get this straight and get it right. Essentially, the Gluon Field or Strong Force GLUES quarks together into protons and neutrons. In contrast, some Massless Speed-of-Light Particles USE the Higgs Field to slow themselves down thereby becoming electrons and quarks in the process. The electromagnetic force draws electrons to protons thereby forming physical atoms. As a consequence of the Pauli Exclusion Principle, only one fermion (quark or electron) can occupy a particular quantum state at any given time. The Pauli Exclusion Principle and Uncertainty Principle resist or counteract an electron's desire or urge to merge with a proton. Checks and balances!

Thanks to Materialism and Naturalism, Fermions are erroneously associated with matter and mass, whereas bosons are erroneously visualized as being massless force carrier particles. In truth, though, the elementary parts of BOTH fermions and bosons are in fact Massless Speed-of-Light Particles; and, it's the Quantum Fields or Force Fields associated with the Gluon Field and the Higgs Field that in FACT give the Massless Particles or Massless Psyches within atoms their mass or resistance to acceleration.

Massless Particles must travel at the speed-of-light, which means that time stops, or time is frozen for Massless Particles, which means that there is NO entropy at the quantum level or the psyche level. Each Massless Particle is some type of Psyche or Intelligence. According to the First Law of Thermodynamics, Massless Particles, Massless Energy, or Massless Psyches are the ultimate perpetual motion machine. Given the FACT that Energy is always conserved and given the FACT that the Second Law of Thermodynamics or Entropy does not exist at the quantum level and the psyche level, Massless Particles or Massless Psyches can keep going and keep working for ALL eternity because they are forever conserved. They can change form, assignment, or function; but, Massless Psyches or Massless Particles are always conserved.

The True Nature of Matter and Mass

https://www.youtube.com/watch?v=gSKzgpt4HBU

https://evolution-is-entropy.com/wp-content/uploads/2018/08/The-True-Nature-of-Matter-and-Mass.zip

The Real Meaning of E = mc2

https://www.youtube.com/watch?v=Xo232kyTsO0

The Speed-of-light Is Not About Light

https://www.youtube.com/watch?v=msVuCEs8Ydo

When Time Breaks Down

https://www.youtube.com/watch?v=GguAN1_JouQ

https://evolution-is-entropy.com/wp-content/uploads/2018/08/When-Time-Breaks-Down.zip

The Origin of Matter and Time

https://www.youtube.com/watch?v=fHRqibyNMpw

https://scientific-proof-of-god.com/wp-content/uploads/2018/08/The-Origin-of-Matter-and-Time.zip

These videos from Space Time and Fermilab make it obvious and clear that entropic physical matter is NOT the fundamental unit of reality as the Materialists and Naturalists claim that it is. In fact, these videos seem to imply that Quantum Fields are the fundamental structure of reality, not physical matter. Of course, I like to ask the question, "Who made the Quantum Fields?" Scientists are supposed to ask questions. It seems to me that whoever made the Quantum Fields or Quantum Structures would in fact be more fundamental or essential than the Quantum Fields. Interacting Quantum Fields, each filling the whole of space and time, implies order and organization at the grandest scale. Who would be capable of making a Quantum Field at the universal scale, determining what it does and how it's going to work? The only answer that I can come up with would be Someone Psyche. Only Psyche would be able to manipulate energy and change it into different forms at the quantum level. It seems to me that Psyche would in fact be the fundamental unit of reality, or the most essential unit for existence and reality.

Quantum Field Theory and Quantum Mechanics in general FALSIFIES Materialism, Naturalism, Darwinism, Nihilism, and Atheism which claim and state that ONLY entropic physical matter exists, that entropic physical matter is the fundamental unit of reality, and therefore that physical matter and entropy are conserved. The false is falsified by the truth; and, the truth is repeatedly experienced and observed. Physical matter is NOT the fundamental unit of reality. Entropic physical matter is NOT the only thing that exists. Physical matter and entropy are NOT conserved. Instead, Science tells us that Energy or Psyche is the fundamental unit of reality, and NOT entropic physical matter.

Do you see how that works? It's the answer to life, the universe, and everything. Psyche or Life Force is the innate intelligence within all the different forms of energy which gives that energy the inherent ability to understand, follow, and obey God's Laws and God's Commands. This reality explains everything that has ever been experienced and observed.

Dark Matter

Nobody knows what Dark Matter is. For all we know, Dark Matter could in fact be one of the many different types of Spirit Matter. That idea is no less crazy than the other ideas that have been proposed. In fact, Spirit Matter has actually been experienced and observed, whereas the other explanations for Dark Matter have not been observed.

https://www.youtube.com/watch?v=1pq9hovXI44

http://origin-science.org/wp-content/uploads/2018/09/Dark-Matter.zip

I've learned to go with what has been experienced and observed. Spirit Matter has been experienced and observed by Out-of-Body Travelers; therefore, I'm running with the idea that Dark Matter is some type of Spirit Matter until somebody comes up with a better explanation or idea. Dark Matter certainly looks like and acts like Spirit Matter. So, that's good enough for now. It gives us a frame of reference to work with.

The Materialists and Naturalists and Atheists erroneously define entropic physical matter as the fundamental unit of reality and the only unit of reality; so, it shouldn't be too surprising that these people will also end up being wrong about Dark Matter or Spirit Matter.

The Materialists, Naturalists, Darwinists, Nihilists, Atheists, Behaviorists, Determinists, Physical Reductionists, Classical Physicists, and Atheists ASSURE us that Spirit Matter DOES NOT EXIST. However, if Dark Matter exists, then it is some type of Spirit Matter; and, many of these same people are willing to allow Dark Matter to exist.

Dark Matter is the invisible, intangible, non-physical, spiritual stuff that has been gathered into haloes around galaxies to keep those physical galaxies from flying apart.

We know for a fact that expanding clouds of hydrogen and helium gas will NEVER coalesce into planets, stars, and galaxies without some kind of EXTERNAL FORCE making them do so. The very existence of planets, stars, and galaxies is Scientific Proof of God's Existence.

A number of scientists have stated that Dark Matter forces hydrogen and helium gas to coalesce into planets, stars, and galaxies. That seems like a semi-plausible explanation for star and galaxy formation; but, it begs the question. Who is forcing the Dark Matter to form into planets, stars, and galaxies? There's NO such thing as spontaneous generation at any level of existence, especially the physical level. Entropy prevents spontaneous generation and chemical evolution from happening at the physical level. It can't be done which means that it wasn't done. Hydrogen molecules naturally repel each other; and, helium naturally repels everything. There's no way in the universe that clouds of hydrogen and helium gas are ever going to coalesce into planets, stars, and galaxies which means that planets, stars, and galaxies have to be made or they wouldn't exist.

https://ldssoul.com/the-birth-of-our-earth/

https://ldssoul.com/forum/viewtopic.php?f=103&t=325

Big Bang Theorists repeatedly state that it was the influence of Dark Matter that formed hydrogen and helium into stars and galaxies. If so, then who is driving the Dark Matter? Someone Psyche or Someone Intelligent has to be forming that Dark Matter into galaxies and stars so that the Dark Matter can then shepherd or form ordinary physical matter into planets, stars, and galaxies. Any way that we choose to look at it, it ends up being Scientific Proof of God's Existence.

The scientists in general have proven that Dark Matter exists. Dark Matter is some type of Spirit Matter or Intangible Matter. Who has the ability to manipulate and organize Spirit Matter? It has to be Someone Psyche. Dark Matter and Baryonic Matter were made from raw energy. Matter of any kind is Organized Energy. Anything that was obviously made obviously has a Maker who made it. Physical matter was obviously made. Anything that was obviously organized obviously has an Organizer who organized it. If stars and galaxies really did coalesce from expanding clouds of hydrogen and helium gas as the Big Bang Theorists claim, then Someone Psyche obviously organized the Dark Matter or Spirit Matter into galaxies and stars, in order to shepherd or form physical matter into galaxies and stars. The existence of organization and the existence of matter of any kind is Scientific Proof of God's Existence. Order, Organization, Physical Matter, and Physical Laws don't just spontaneously generate out of thin air from nothing as the Materialists and Naturalists claim. There is NO such thing as Creation Ex Nihilo. The First Law of Thermodynamics or the Conservation of Energy FALSIFIES Creation Ex Nihilo.

Physical matter was organized from raw energy. Dark Matter, Spirit Matter, and Physical Matter have obviously been organized into planets, stars, gas clouds, and galaxies. Anything that was obviously organized obviously has an Organizer who organized it. That's why there is something rather than nothing. The very existence of planets, stars, and galaxies is Scientific Proof of God's Existence. There's no way in the universe that expanding clouds of hydrogen and helium gas will ever coalesce into planets, stars, and galaxies without some kind of outside intervention from God. God is needed to do the physically impossible. That's just the way it is.

Defining Dark Matter:

Dark matter is a hypothetical type of matter distinct from baryonic matter (ordinary matter such as protons and neutrons), neutrinos, and dark energy. Dark matter has never been directly observed; however, its existence would explain a number of otherwise puzzling astronomical observations. The name refers to the fact that it does not emit or interact with electromagnetic radiation, such as light, and is thus invisible to the entire electromagnetic spectrum. Although dark matter has not been directly observed, its existence and properties are inferred from its gravitational effects such as the motions of visible matter, gravitational lensing, its influence on the universe's large-scale structure, on galaxies, and its effects on the cosmic microwave background.

The standard model of cosmology indicates that the total mass – energy of the universe contains 4.9% ordinary matter, 26.8% dark matter and 68.3% dark energy. Thus, dark matter constitutes 84.5% of total mass, while dark energy plus dark matter constitutes 95.1% of total mass – energy content. The great majority of ordinary matter in the universe is also unseen. Visible stars and gas inside galaxies and clusters account for less than 10% of the ordinary matter contribution to the mass-energy density of the universe.

Dark matter is a theorized form of matter that is believed to account for approximately 80% of the matter in the universe, and about a quarter of its total energy density. The majority of dark matter is thought to be non-baryonic in nature, possibly being composed of some as-yet undiscovered

subatomic particles. Dark matter has not been directly observed, but its presence is implied in a variety of astrophysical observations, including gravitational effects that cannot be explained unless more matter is present than can be seen. For this reason, most experts believe dark matter to be ubiquitous in the universe and to have had a strong influence on its structure and evolution. The name dark matter refers to the fact that it does not appear to interact with observable electromagnetic radiation, such as light, and is thus invisible (or 'dark') to the entire electromagnetic spectrum, making it extremely difficult to detect using usual astronomical equipment.

The primary evidence for dark matter is that calculations show that many galaxies would fly apart instead of rotating, or would not have formed or move as they do, if they did not contain a large amount of unseen matter.

From different versions of the Wikipedia on this topic.

https://en.wikipedia.org/wiki/Dark_matter

They call it Dark Energy and Dark Matter because it isn't visible to our physical eyes and isn't directly detectable with our physical instruments. These things are quantum, non-physical, and supernatural in nature and origin. Dark Matter is one of the different types of Spirit Matter. It's invisible and non-physical (non-baryonic) in nature and origin. Dark Matter or Spirit Matter is a different FORM of Energy or Syntropy. Energy or Syntropy can be made to take on many different FORMS by Psyche or Intelligence.

Whether the Materialists and Naturalists realize it or not, the demonstrable and proven existence of Dark Matter and Dark Energy FALSIFIES Materialism and Naturalism.

Think about it logically.

Dark Matter is invisible non-physical "matter" that cannot be detected with our physical instruments. Such a thing is supposed to be physically impossible according to the Materialists and Naturalists; yet, there's Dark Matter in your living room all around you right now that you can't see with your physical eyes. Invisible non-physical matter is the very definition of Spirit Matter. Dark Matter or Spirit Matter is inferred to exist by the gravitational and organizational effect it has on the physical matter, ordinary matter, regular matter, or baryonic matter that's suspended within it.

Dark Energy or Syntropy is also an invisible non-physical force or field.

What's the difference between dark matter and dark energy? Is it valid to think about dark matter and dark energy in the same way? Are they two states of the same dark "thing"? Are they interchangeable?

The short answer to your question is that we don't know if dark matter and dark energy are manifestations of the same dark "thing". We know they both must exist to explain certain phenomena, but we still know very little about their make-up, so we cannot assume they are linked. For now, we think of them as separate, and we believe the cosmos to be composed of roughly 0.03% heavy elements (anything other than hydrogen and helium), 0.3% neutrinos, 0.5% stars, 4% free hydrogen and helium, 25% dark matter, and 70% dark energy.

Here is how we define them separately:

Dark matter must exist to account for the gravity that holds galaxies together. If the only matter in the universe was matter we could directly detect, galaxies would not have had enough matter to have ever formed. The galaxies we observe today would fly apart because they wouldn't have enough matter to create a strong enough gravitational force to hold themselves together. Dark matter is also responsible for amplifying small fluctuations in the Cosmic Microwave Background back in the early universe to create the large-scale structure we observe in the universe today.

Dark energy, which also goes by the names of the cosmological constant or quintessence, must exist due to the rate of expansion we observe for our universe. Not only is the universe expanding, but this expansion is also accelerating, so the unknown 'anti-gravity' force at work is termed 'dark energy'.

http://curious.astro.cornell.edu/about-us/108-the-universe/cosmology-and-the-big-bang/dark-matter/658-what-s-the-difference-between-dark-matter-and-dark-energy-intermediate

Technically, the proven existence of invisible forces and fields such as quantum waves, light waves, electromagnetic waves, the strong force, the weak force, gravity waves, and dark energy waves FALSIFIES the major premises or primary assumptions of Materialism and Naturalism which claim that the non-physical or the supernatural does not exist. ALL of these different types of waves, forces, and fields are immaterial, non-physical, and supernatural in nature and origin.

When it comes to all of these different waves, their wave function can supposedly collapse into something that's detectable or inferable through the effect that it has on the physical world. The quantum can influence and control the physical. The different waves, forces, and fields can have physical effects. The smaller can dwell within and control the larger.

Energy or Syntropy can be made to take on different FORMS.

https://en.wikipedia.org/wiki/Elementary_particle

Light waves reduce to photons. Magnetic waves and electric waves are also tied in with photons, although they could theoretically reduce to an electron. Photovoltaic cells were designed to convert photons into electrons, thereby converting light into electricity. Electrons are always half-in and half-out – spending half their time in the non-physical quantum realm and half their time in the physical realm.

The strong force reduces to gluons. The weak nuclear force reduces to bosons. Quantum waves reduce to thoughts, memories, and cognitions. Gravity waves reduce to the hypothetical graviton. Dark energy waves would reduce to some sort of hypothetical darkon. It's going to be some type of boson when they finally figure out what it is.

The problem is that these particular bosons (cognitions, gravitons, and darkons) are purely hypothetical in nature and will probably never have a verifiable physical component being purely supernatural, or non-physical, or quantum in nature and origin. At some point along the continuum, the physical has to cease to exist and the spiritual or the quantum has to take over and continue onward from there. The quantum or the supernatural is the core construct. The physical is simply a different phase or a different dimension of the quantum or the supernatural. The physical is a very small sub-set of the quantum or the supernatural. Energy is Syntropy, eternal and everlasting; whereas, physical matter or entropy is the exception to the rule. Physical matter and entropy began, and they can theoretically come to an end when they are absorbed back into Syntropy from whence they came.

Syntropy is supernatural. WE KNOW this is so, because only physical matter is subject to entropy. Quantum Mechanisms are supernatural in nature, origin, and function. Quantum Mechanisms are Syntropy in action. The very existence of Quantum Mechanics FALSIFIES Materialism and Naturalism.

The very existence of quantum waves – thoughts and memories – that survive the death of our physical brain is evidentiary proof that FALSIFIES Materialism and Naturalism. WE KNOW that our thoughts, memories, and dreams are non-physical, non-local, transdimensional, and spiritual in nature and origin because they cannot be detected nor recorded by our physical instruments.

It's clear to me and to many other people that Dark Matter is in fact Spirit Matter. Invisible Matter or Spirit Matter is not supposed to exist according to the Materialists, Naturalists, Darwinists, Nihilists, and Atheists; yet, there it is. It's all around us, and we can't even see it nor detect it with our physical instruments; but, we KNOW it exists by the effects that it has on physical matter.

Remember, there are different types of spirit matter, different levels of spirit matter, different frequencies of spirit matter, different dimensions of spirit matter, and different types of reality comprised of spirit matter. In the spirit world, there are consensus realities comprised of communities of spirit beings; and, there are non-consensus realities that mold themselves to the demands of your Psyche whenever you enter into them.

According to the Law of Quantum Superposition, two particles of matter can exist at the same time in the same location as long as they are out of phase with each other. Spirit matter and physical matter occupy the same space at the same time. This is possible because they are out of phase with each other or exist in different dimensions at different frequencies. Quantum Superposition makes this reality and truth possible. Quantum waves are additive, and they can also be separated from each other, too. It works both ways thanks to Quantum Superposition. There are also different levels or frequencies or dimensions of physical matter. They all exist in the same space or location simultaneously, whether we realize it or not.

Clearly, Dark Matter is some type of Spirit Matter. What else would you call invisible matter that isn't detectable directly by our physical instruments, but has to be inferred to exist instead by the influence that it has on physical matter? They call it Dark Matter because it isn't visible and isn't detectable by our physical instruments. The same truth and reality applies to Spirit Matter.

The quantum can influence and control the physical. Dark Matter or Spirit Matter is quantum in nature and origin; but, this Dark Matter shepherds or organizes or controls the physical matter, according to the science we have on hand.

Dark matter must exist to account for the gravity that holds galaxies together. If the only matter in the universe was matter we could directly detect, galaxies would not have had enough matter to have ever formed. The galaxies we observe today would fly apart because they wouldn't have enough matter to create a strong enough gravitational force to hold themselves together.

Who organized this Spirit Matter or Dark Matter into Galaxies, so as to draw the associated physical matter into a galactic shape or form?

Even the Big Bang Theory points to these truths. According to the Big Bang Theory, our physical universe should be nothing more than one huge homogeneous ball of gas with every physical particle moving away from every other physical particle. So, who put the Dark Matter into place so as to force the ordinary physical matter to form into galaxies, and later to force some of that same ordinary physical matter to form into stars and planets? It is Dark Matter or Spirit Matter that is currently holding our galaxies together; and, it's Dark Matter that allegedly brought our galaxies, stars, and planets together in the first place! Who created everything spiritually, before He created it physically? Who could do such a thing? It had to be Someone Psyche, Someone Powerful, and Someone Intelligent. It would have to be a God.

If it weren't for Dark Matter or Spirit Matter, every physical particle in this physical universe would be moving away from every other physical particle. Who do we know of that can control and manipulate and organize Spirit Matter or Dark Matter? Well, it's certainly NOT evolution. Evolution of any kind is restricted to or limited to the physical level. Evolution can't touch the quantum level or the psyche level. So, who do we know of that operates on the quantum level or the dark matter level? It has to be Someone Psyche. Who was moving behind the scenes before the first particle of physical matter was organized and made? Who was moving behind the scenes before this physical universe was brought into existence? It wasn't evolution as the Darwinists and Naturalists claim because evolution and physical matter didn't exist yet.

Who created everything spiritually, before He created it physically? The Biblical God Jesus Christ tells us that He created everything spiritually before He created it physically.

Moses 3: 4-5:

And now, behold, I say unto you, that these are the generations of the heaven and of the earth, when they were created, in the day that I, the Lord God, made the heaven and the earth, and every plant of the field before it was in the earth, and every herb of the field before it grew.

For I, the Lord God, created all things, of which I have spoken, spiritually, before they were naturally upon the face of the earth. For I, the Lord God, had not caused it to rain upon the face of the earth. And I, the Lord God, had created all the children of men; and not yet a man to till the ground; for in heaven created I them; and there was not yet flesh upon the earth, neither in the water, neither in the air.

Go figure!

God organized everything spiritually before He organized it physically, just as the Dark Matter Scenario suggests! Dark Matter is Spirit Matter. God organized the Dark Matter first and then used it as a template or blueprint to organize the Physical Matter. Organized Dark Matter or Spirit Matter served and serves as the template or blueprint for this physical universe. Truth is consistent with truth. Truth supports truth. Truth FALSIFIES Materialism, Naturalism, Darwinism, Nihilism, Atheism, Classical Physics, and the Theory of Evolution.

By choosing to allow ALL of the evidence into evidence, we can answer every question that comes our way with an answer that makes logical scientific sense. Quantum Mechanics, Syntropy, and Quantum Non-Local Consciousness (Psyche) have a vast amount of explanatory power that Classical Physics, Entropy, Materialism, and Naturalism simply lack.

Both Quantum Mechanisms and Dark Matter require some kind of Chooser or Actor to organize them and trigger them; otherwise, they aren't going to do anything intelligent or useful or productive. Otherwise, it would be chaos. In the Realm of Syntropy, Psyche or Quantum Non-Local Consciousness is that Chooser or Actor that Quantum Mechanisms and Dark Matter require. Remember, some type of Psyche or Syntropy is required to collapse wave functions; otherwise, nothing happens.

This is a Quantum Model for Origins and not a classical physics model for origins. It's based upon Quantum Mechanics and not classical physics.

Dark Energy

Most of the Materialists, Naturalists, Darwinists, Nihilists, Atheists, Behaviorists, Determinists, Physical Reductionists, Classical Physicists, and Atheists ASSURE us that Dark Energy, or the Light of Christ, or the Quantum Sea of Light, or the Zero-Point Field of Light does not exist.

Are they right; or, did they get their science wrong?

Defining Dark Energy:

In physical cosmology and astronomy, *dark energy* is an unknown form of energy which is hypothesized to permeate all of space, tending to accelerate the expansion of the universe. Dark energy is the most accepted hypothesis to explain the observations since the 1990s indicating that the universe is expanding at an accelerating rate.

Assuming that the standard model of cosmology is correct, the best current measurements indicate that dark energy contributes 68.3% of the total energy in the present-day observable universe. The mass - energy of dark matter and ordinary (baryonic) matter contribute 26.8% and 4.9%, respectively, and other components such as neutrinos and photons contribute a very small amount.

The density of dark energy ($\sim 7 \times 10\text{--}30$ g/cm3) is very low, much less than the density of ordinary matter or dark matter within galaxies. However, it comes to dominate the mass - energy of the universe because it is uniform across space.

Two proposed forms for dark energy are the cosmological constant, representing a constant energy density filling space homogeneously, and scalar fields such as quintessence or moduli, dynamic quantities whose energy density can vary in time and space. Contributions from scalar fields that are constant in space are usually also included in the cosmological constant. The cosmological constant can be formulated to be equivalent to the zero-point radiation of space i.e. the vacuum energy. Scalar fields that change in space can be difficult to distinguish from a cosmological constant because the change may be extremely slow.

https://en.wikipedia.org/wiki/Dark_energy

Dark Energy is the expanding force of the universe. Dark Energy permeates the whole universe and is evenly distributed throughout the whole universe, much like gravity.

Dark Energy has been called Gravity's "opposing force". Dark Energy or Syntropy is the Power by which this physical universe was made.

It's hard to know what to do with Dark Energy. I have observed that EVERYTHING UNEXPLAINABLE that we have come across in science, mysticism, new age, religion, physics, and astrophysics seems to get dumped into what the astrophysicists call Dark Energy. Dark Energy is some sort of quantum mechanism or supernatural mechanism. Dark Energy is non-local, transdimensional, non-physical, quantum, syntropic, and supernatural in nature and origin. Dark Energy is a type of Syntropy. Dark Energy is an invisible non-physical force or field. Dark energy is omnipresent.

Dark Energy has been called the Vacuum Energy of Free Space, the Zero-Point Field, the Zero-Point Field of Light, the Quantum Sea of Light, Anti-Gravity, Quintessence, the Ether, Essence, Soul, Spirit, the Quantum Construct, the Music of the Spheres, the Cosmic Internet, the Universal Broadcast, the Light of Christ, and the Holy Spirit. I'm beginning to believe that Dark Energy is a Quantum Field or is the result of interactions between different Quantum Fields. It's all comprised of Energy. Dark Energy is just one of the many different types of Energy.

Dark Energy is some type of syntropy or some type of quantum mechanism. Dark Energy permeates the whole of time and space. Dark Energy is the stretching force in this universe. Dark Energy is an invisible non-physical force or field. Dark Energy has the opposite effect of Gravity. Dark Energy could be used by the Gods to do levitation. Gravity is an invisible tractor beam; and, Dark Energy is an invisible repulsor beam. They are invisible and non-physical because they exist at the quantum level. In fact, the proven and verified existence of something like Gravity or Magnetism or Invisible Light or Dark Energy actually falsifies the claims of Materialism, Naturalism, Darwinism, Nihilism, and Atheism which claim that invisible non-physical forces and fields do not exist.

Psyche is the innate intelligence within all the different forms of energy which gives that energy the inherent ability to understand, follow, and obey God's Laws and God's Commands. The Materialists, Naturalists, Darwinists, Nihilists, Behaviorists, Determinists, Physical Reductionists, and Atheists teach the Conservation of Entropy or the Conservation of Death as the Second Law of Thermodynamics thereby FALSIFYING the First Law of Thermodynamics which teaches Syntropy, Eternal Life, the Conservation of Psyche, the Conservation of the Life Force, the Conservation of Intelligence, and the Conservation of Energy instead. These atheistic people use the Second Law of Thermodynamics to falsify and eliminate the First Law of Thermodynamics.

Matt makes this point in the following PBS Space Time video:

https://www.youtube.com/watch?v=UwYSWAlAewc

https://syntropy.website/wp-content/uploads/2018/07/Anti-gravity-and-the-True-Nature-of-Dark-Energy.zip

In this video, he states:

> **It takes work to expand the volume of the universe, and the new energy turns into dark energy. But, who's doing that work? Where does the energy come from? The answer is maybe nowhere. This is where intuition goes completely out the window. See, the law of conservation of energy no longer applies in an expanding universe. This law is a property of a Newtonian universe, in which space and time are fixed static dimensions. In a universe governed by general relativity, this is no longer true. Energy**

can be forever lost or gained from nothing within an expanding curved spacetime.

Ouch!

He uses Creation Ex Nihilo and his Atheism to FALSIFY the First Law of Thermodynamics. His materialism, naturalism, and atheism are messing him up. It's causing him to go so far as to deny the truthfulness of the First Law of Thermodynamics and to replace Conservation of Energy with Creation Ex Nihilo instead.

He is suggesting that Dark Energy is being made from nothing as our universe expands. I don't buy it! That's magic, not science. Something has to be fundamental; otherwise, our universe wouldn't exist, we wouldn't exist, and chaos would reign supreme. ALL of the observational evidence is telling us that the Conservation of Energy is the Fundamental Law of the Multiverse – NOT Creation Ex Nihilo, Blind Luck, nor Random Fluctuations in Entropy. Energy, and the psyche or intelligence within it, is the fundamental substance of the Multiverse – NOT nothing. Dark Energy is another Quantum Field; and, Quantum Fields are made from energy, NOT nothing from nothing by Nothing. Energy is something, NOT nothing; and, Energy cannot be made, and it cannot be destroyed, which means that Dark Energy cannot be made from nothing. Dark Energy is made from energy, which means that Dark Energy is just another form of energy.

Creation from nothing by Nothing cannot explain why Nothing is producing Dark Energy rather than concrete throughout the whole of our universe. In contrast, Creation by Intelligent Psyche does indeed explain why there is something rather than nothing. Psyche or Intelligence explains why Dark Energy is being used to expand our universe rather than concrete. Do you see how that works? An influx of concrete would indeed expand our universe, but then that would violate God's Anthropic Goals and make life as we know it impossible. We must use a little philosophical common sense when trying to concoct causal explanations for the physical phenomena that we are observing and experiencing.

His causal explanation for Dark Energy violates the whole of Physics and destroys the whole of Science. He invokes Creation from nothing by Nothing. He invokes magic. He denies the truthfulness of the First Law of Thermodynamics. The truth as it has been experienced and observed is the exact opposite of what he said here!

This quote demonstrates what I have been saying all the way along – the Materialists, Naturalists, and Atheists preach and teach the Conservation of Death, the Conservation of Heat Death, and the Conservation of Entropy in complete violation of the First Law of Thermodynamics which is the Conservation of Psyche or the Conservation of Energy.

Ultimately, the First Law of Thermodynamics and the Second Law of Thermodynamics are mutually exclusive. They both can't be right because they falsify each other. The Materialists, Naturalists, and Atheists use entropy and the Second Law of Thermodynamics to falsify the existence of God; whereas, the First Law of Thermodynamics verifies and confirms the existence of God's Psyche or God's Energy. "Entropy proves that God does not exist" was the take-away message that I got from the Second Law of Thermodynamics when I was a Physicalist, Naturalist, Nihilist, and Atheist. That's the message that we all get from entropy or the Second Law of Thermodynamics. It's a message that many of us want to hear. We literally use the Conservation of Entropy and the Second Law of Thermodynamics to FALSIFY the Conservation of Psyche or the Conservation of Energy which is the First Law of Thermodynamics.

His is a materialistic and naturalistic explanation of Dark Energy; and, I have observed that the Materialists and Naturalists are always wrong. The Materialists,

Naturalists, and Atheists preach the Supremacy of Entropy, the Conservation of Death, the Conservation of Entropy, Creation Ex Nihilo, Spontaneous Generation, Abiogenesis, Chemical Evolution, Macro-Evolution, the non-existence of Psyche, and the non-existence of God. These people teach and preach that entropy or death is conserved while at the same time teaching and preaching that Energy or Psyche is NOT conserved or is non-existent. They use the Second Law of Thermodynamics to FALSIFY the First Law of Thermodynamics. They use entropy, death, or the Second Law of Thermodynamics to falsify the existence of God. These people deliberately give preeminence to entropy or the Second Law of Thermodynamics because that is the one they want to be true because that is the one that falsifies the existence of Psyche and God. Find out what the Second Law of Thermodynamics, the Theory of Evolution, and Creation Ex Nihilo were designed to falsify and eliminate from science; and there, you will find the answer to life, the universe, and everything.

By his own admission, his chosen beliefs violate parsimony because they are unintuitive and violate common sense. This is one of the most fascinating quotes I have come across. It demonstrates precisely and clearly that we Naturalists and Atheists are more than willing to reject the First Law of Thermodynamics and embrace Creation Ex Nihilo instead just so that we can continue to believe that God does not exist. I KNOW how these thought processes work. I used to be a Materialist, Naturalist, Nihilist, and Atheist until the science and the observations of the human race convinced me that I was wrong.

He is a brilliant individual, but his explanation violates the First Law of Thermodynamics. His explanation for expansion and dark energy is Creation Ex Nihilo and states that "energy can be gained or lost from nothing". Creation Ex Nihilo is magic. Creation Ex Nihilo is unpredictable, unreliable, non-demonstrable, non-replicable, and uncontrollable. What's to guarantee that you are going to get energy from nothing? You could just as easily get nothing from nothing. You could just as easily get concrete from nothing. Creation Ex Nihilo is BAD SCIENCE. It completely lacks explanatory power. Why is he going to get dark energy from nothing rather than new physical matter or new gravity waves? With Creation Ex Nihilo, he can't explain why nothing is producing dark energy rather than producing something else like spirit matter or entropy. Creation Ex Nihilo violates the First Law of Thermodynamics. By his own admission, his explanation for the causal origin of Dark Energy violates the First Law of Thermodynamics or the Conservation of Energy. The observed and experienced science, or the truth, FALSIFIES what he says here at the end this video in this quote.

Creation by nothing from nothing doesn't really explain anything. Now does it? Creation from nothing by nothing doesn't explain why there is something. Creation from nothing by nothing doesn't explain how the Physical Constants got their precise finely-tuned values rather than some other random value. Creation Ex Nihilo fails to satisfy. It doesn't explain what things are and how they work. Creation Ex Nihilo can't explain why there is something rather than nothing.

What if the truth is the exact opposite of everything that he said? Wouldn't that be something?

The Materialists, Naturalists, Darwinists, Nihilists, Behaviorists, Determinists, and Atheists have a knack for finding and embracing everything that is false and everything that has been falsified by observational evidence and experiential evidence. "The exact opposite of everything he said" can also be used to explain the causal origin of Dark Energy in a satisfactory way that actually makes logical sense.

His explanation for dark energy and expansion derives from his prior commitment to his chosen belief that God does not exist, that religion sucks, and that God cannot be the source of that new energy and work – a commitment and belief that requires a monumental leap of blind faith into nothing. He is more than willing to reject the First Law of Thermodynamics and embrace Creation Ex Nihilo instead just so that he can continue to believe that God does not exist. Self-deception works, and it works every time. However, the only thing he got right in all of this is his claim that his intuition has gone completely out the window. There's always another way to look at things.

Spontaneous Generation, Chemical Evolution, Abiogenesis, or Creation Ex Nihilo is physically impossible. It requires Syntropy, Psyche, Energy, Intelligence, and Supernatural Mechanisms in order to produce genomes, physical life, physical constants, physical matter, and physical laws.

Science of any kind requires a Scientist to do that science. Work of any kind requires a Worker to do that work.

"Who is doing that work?"

God is doing that work.

"Where does the energy come from?"

It comes from the presence of God to fill the immensity of space. How do we know that this Dark Energy or new energy is coming from the presence of God? It's because God revealed it to us.

I prefer God's explanation for Dark Energy over the one given in this video. When it comes to the causal origin of Dark Energy, I think this video is wrong and that God is right. However, I link to this video and include this video in my archive because it's one of the best explanations of Dark Energy that I have ever encountered. He does an excellent job explaining what Dark Energy is and how it works despite the fact that he got its causal origin wrong.

God revealed to us that Dark Energy and Quantum Fields come from the Presence of God.

The Biblical God Jesus Christ talks about Dark Energy and Quantum Fields; but obviously, He doesn't use the words "dark energy" or "quantum field" to describe the phenomenon. He calls it the Light of Christ. It's the same thing in the end. Energy or Syntropy can take on many different purposes and FORMS, including quantum fields, dark energy, dark matter, spirit matter, and physical matter. Dark Energy, Syntropy, or the Light of Christ is the Primal Construct according to Jesus Christ.

Doctrine and Covenants 88: 3-13:

Wherefore, I now send upon you another Comforter, even upon you my friends, that it may abide in your hearts, even the Holy Spirit of promise; which other Comforter is the same that I promised unto my disciples, as is recorded in the testimony of John. This Comforter is the promise which I give unto you of eternal life, even the glory of the celestial kingdom; which glory is that of the church of the Firstborn, even of God, the holiest of all, through Jesus Christ his Son — he that ascended up on high, as also he descended below all things, in that he comprehended all things, that he might be in all and through all things, the light of truth; which truth shineth. This is the light of Christ. As also he is in the sun, and the light of the sun, and the power thereof by which it was made. As also he is in the moon, and is the light of the moon, and the power thereof by which it was made; as also the light of the stars, and the power thereof by which they were made; and the earth also, and the power thereof, even the earth upon which you

332

stand. And the light which shineth, which giveth you light, is through him who enlighteneth your eyes, which is the same light that quickeneth your understandings; which light proceedeth forth from the presence of God to fill the immensity of space — the light which is in all things, which giveth life to all things, which is the law by which all things are governed, even the power of God who sitteth upon his throne, who is in the bosom of eternity, who is in the midst of all things.

This sounds like Dark Energy or Quantum Fields to me. The Light of Christ fills the immensity of space. Dark Energy or the Light of Christ is the Power by which the earth, moon, stars, and galaxies were made. This Dark Energy was used to phase-shift or to quantum tunnel the planets, stars, and galaxies into existence whole. It has been observed that galaxies form whole, in situ or in place. Galaxy formation is the result of Quantum Mechanics or Syntropy, and NOT classical physics. There's NO such thing as a coalescing cloud of gas out in expanding space. Hydrogen and helium gas do not coalesce. Gas expands and disperses. In order to get gas to coalesce, it would require some kind of external intervention from God in order to make it happen.

https://ldssoul.com/the-birth-of-our-earth/

https://ldssoul.com/forum/viewtopic.php?f=103&t=325

The Light of Christ (or Dark Energy) is some kind of quantum force by which everything was made. It proceeds forth from the presence of God to fill the immensity of space. It permeates everything in this physical universe. It's the stretching force or the expanding force of the universe. It's dark energy or invisible energy. It's the Priesthood Power of God. Dark Energy is Syntropy. Dark Energy is Quantum Mechanics. Dark Energy is the Power by which the physical worlds and stars were made or phased into our physical reality.

Quantum Fields are energy fields; and, Quantum Fields fill the immensity of space. Quantum Fields are also the Light of Christ. Quantum Fields are the implementation of the Light of Christ. In fact, many scientists believe that Dark Energy is just another Quantum Field – just like Gravity is a Quantum Field. I believe they are right. Quantum Fields proceed forth from the presence of God to fill the immensity of space. Each one was designed and made with a different purpose and function in mind. The Light of Christ is the Organizing Force of the universe. Anything that was obviously organized obviously has an Organizer who organized it. Our physical universe has obviously been organized. At the most fundamental level, it is orderly rather than chaotic.

Dark Energy, Quantum Fields, and Gravity are subsumed somewhere and somehow within the Light of Christ. Energy can take on many different FORMS. Energy or Syntropy is conserved; but, Energy or Syntropy can take on many different FORMS such as spirit matter, physical matter, dark energy, gravity, magnetism, quantum fields, quantum waves, entropy, time, and psyche. A huge infusion of Syntropy, Energy, or Mass will convert a particle of spirit matter into a particle of physical matter; and, physical matter is slowed down and made subject to space-time, or locality and entropy, by an infusion of Order, Energy, Structure, or Mass. We will NEVER be able to accelerate physical matter faster than the speed-of-light because an infusion of Energy or Mass is what made it physical matter in the first place.

The Energy, Syntropy, Mass, Entropy, Passage of Time, and Locality have to be temporarily removed from physical matter in order to quantum tunnel that physical matter someplace else in this universe. Adding more energy or mass to physical matter will only slow it down by making it more massive in the end. According to the Theory of Relativity, at velocities faster than the speed-of-light, time stops and there is no passage of time – there is no entropy or no aging process. In order to quantum tunnel physical matter instantaneously to the opposite side of the universe, the God's stop the passage of time or temporarily remove the entropy from that physical matter. Entropy was put into entropic physical matter by the Gods to prevent that physical matter from quantum tunneling away from us at will. Entropy puts the brakes on. Does it not?

It has been observed that the expansion of our universe is accelerating. That means that the Gods are currently infusing massive amounts of Dark Energy or Syntropy into our physical universe in preparation for the next phase of its existence. This observational truth and reality is proof positive that our physical universe is NOT a closed system. There's currently Energy coming into this physical universe from someplace else besides this physical universe.

Fascinating, is it not? Well, I think it is.

I learned in the end that God knows what He's talking about because He designed, organized, created, and made these things; and, He's currently making good use of them to accomplish His purposes and goals.

The Materialists, Naturalists, Darwinists, Nihilists, Behaviorists, Determinists, and Atheists use the Second Law of Thermodynamics or the Conservation of Entropy to FALSIFY the First Law of Thermodynamics, the Conservation of Energy, or the Conservation of God's Psyche. These people use entropy to falsify syntropy. The Materialists and Naturalists erroneously teach and preach that entropy or death is conserved. Do they not?

In contrast, God uses the First Law of Thermodynamics, Syntropy, the Atonement of Christ, Resurrection from Death, the Conservation of Energy, and the Conservation of Psyche to FALSIFY the Conservation of Entropy, the Conservation of Death, or the Second Law of Thermodynamics.

The First Law of Thermodynamics and the Second Law of Thermodynamics are mutually exclusive. They both can't be true simultaneously because they falsify each other. Either the Conservation of Entropy is true, or the Conservation of Energy is true; but, they both can't be true simultaneously because they contradict each other.

God states that Psyche, Intelligence, Life Force, or Energy is conserved. God tells us that the First Law of Thermodynamics is true, which means that the Conservation of Entropy, the Conservation of Death, or the Second Law of Thermodynamics is false.

Doctrine and Covenants 93: 29-30:

Man was also in the beginning with God. Intelligence, or the light of truth, was not created or made, neither indeed can be. All truth is independent in that sphere in which God has placed it, to act for itself, as all intelligence also; otherwise there is no existence.

Psyche, Intelligence, or Energy cannot be created, and it cannot be destroyed. It is eternal and everlasting. It is always conserved, which means that Psyche or Energy is syntropic. It has always existed, and it will always exist. Psyche is the innate intelligence within all the different forms of energy which gives that energy the inherent ability to understand, follow, and obey God's Laws and God's Commands. God states that Psyche or Energy or Intelligence is conserved, which means that entropy and entropic physical matter are NOT conserved. The First Law of Thermodynamics FALSIFIES the Second Law of Thermodynamics; and vice versa, the Second Law of Thermodynamics FALSIFIES the First Law of Thermodynamics. They both can't be true. Something has to give.

Study what the Second Law of Thermodynamics was designed to falsify and eliminate from science; and there, you will find the answer to life, the universe, and everything. The Materialists, Naturalists, and Atheists use the Second Law of Thermodynamics to falsify the First Law of Thermodynamics. God uses the First Law of Thermodynamics or Syntropy to falsify Materialism, Naturalism, Darwinism, Nihilism, Atheism, and the Second Law of Thermodynamics. You will have to decide for yourself who got their science right and who did not; but, I'm going with God on this one and not the Naturalists and Atheists. I used to be a Materialist, Naturalist, Nihilist, and Atheist until the science convinced me that I was wrong. Science is observation and experience. The observations and experiences of the human race convinced me that I was wrong and that the revelations from God are true.

So, what is God's purpose, work, or plan?

Moses 1: 37-39: **And the Lord God spake unto Moses, saying: The heavens, they are many, and they cannot be numbered unto man; but they are numbered unto me, for they are mine. And as one earth shall pass away, and the heavens thereof even so shall another come; and there is no end to my works, neither to my words. For behold, this is my work and my glory — to bring to pass the immortality and eternal life of man.**

God's work, purpose, and glory is to bring to pass your physical immortality and eternal life in His presence. Of course, He needs your cooperation to do so.

God isn't hiding from us. He's telling us precisely what He is doing and why He is doing it. Good enough!

As I see it, you've got to figure out what everything is before you can figure out what to do with it; and, you are NEVER going to figure out anything if you limit yourself exclusively to Materialism, Naturalism, Darwinism, Classical Physics, and Entropy. Materialism, Naturalism, and their derivatives will NEVER help you to figure out what things are and how they really work because Materialism and Naturalism of any kind are based upon a refusal to look at evidence and a refusal to accept the evidence that God has given us. You will NEVER find the truth by refusing to look at evidence. It's physically impossible!

Materialism, Naturalism, Darwinism, Nihilism, Behaviorism, Determinism, and Atheism were designed to prevent us from discovering Psyche, Syntropy, Invisible Intangible Non-Physical Forces and Fields, Spirit Matter or Dark Matter, the Light of Christ or Dark Energy, and God. The Second Law of Thermodynamics was designed to falsify the First Law of Thermodynamics.

A Quantum Model for Origins or a Syntropic Model for Origins actually explains Dark Matter and Dark Energy, and it explains the need for Dark Matter and Dark Energy; whereas, Materialism, Naturalism, Darwinism, Classical Physics, and Entropy DO NOT. The Dark Matter or Spirit Matter serves as the template or the blueprint for this physical creation; and, Dark Energy or the Light of Christ is the Power by which everything physical was made or brought into a physical existence. Physical organization is an upgrade or a phase-shift from being exclusively spirit matter into having physical matter as well. The physical planets, stars, and galaxies are born or made; and, Dark Energy or the Light of Christ is the Power by which they are born and become physical.

When it comes to the origins of this physical universe, the Biblical God Jesus Christ knows what He is talking about because He was there and made it happen. Someone Psyche is required to collapse the wave functions; or, nothing happens in the first place. This is what has been experienced and observed.

This ends up being a Quantum Model or a Syntropic Model for Origins because it based upon Quantum Mechanics, Phase-Shifting, Action at a Distance, Quantum Tunneling, and the First Law of Thermodynamics which states that Energy or Syntropy is conserved. The Dark Energy or Syntropy has always existed and will always exist because it is conserved. Syntropy or Dark Energy is the power by which our physical universe, physical matter, physical earth, physical sun, and physical galaxy were made. By definition, in principle, only a God would have access to and control over Syntropy or Dark Energy. We physical beings can't get our hands on such a thing – not at the physical level at least.

You are going to have to decide for yourself who got their science right and who got their science wrong. I KNOW what the observations and experiences of the human race as a whole are telling me, though. God got His science right; and, the Naturalists and Atheists did not.

Dark Energy Expands Planets, Stars, and Universes

This section attempts to discover what Dark Energy is and what Dark Energy does. Why does it exist, and how is it used?

How Will the Universe End?

This is the ultimate question, is it not? Science attempts to predict how everything will turn out or how everything will end.

From one episode to the next, the host of PBS Space Time – Matt O'Dowd – does an excellent job presenting and explaining the materialistic, naturalistic, nihilistic, and atheistic religion or worldview. If you want to know what these people truly believe, then this is the show for you. It will give you something to falsify too, if you are so inclined.

How Will the Universe End?

https://www.youtube.com/watch?v=Qg4vb-KH5F4

https://god-is-light.com/wp-content/uploads/2018/08/How-Will-the-Universe-End.zip

How will our universe end? **Heat Death** is Matt's ongoing prediction from one episode of PBS Space Time to the next because he doesn't believe in God and thinks that religion is silly. Matt truly believes in Heat Death. It's his religion or belief system. He's even had T-shirts made to celebrate the fact that **The Heat Death of the Universe Is Coming**.

https://store.dftba.com/collections/space-time/products/heat-death-shirt

The Materialists, Naturalists, Darwinists, Nihilists, Behaviorists, Determinists, Physical Reductionists, Classical Physicists, and Atheists teach and believe that Heat Death is unavoidable; but, are they right? Is Entropy or the Second Law of Thermodynamics the fundamental LAW of the universe, or is there something more fundamental that these people are deliberately overlooking and ignoring?

In this and other episodes of Space Time, Matt repeatedly preaches and teaches the fundamental religion or the naturalistic and atheistic paradigm. He teaches and believes that the Second Law of Thermodynamics is the dominant LAW of the universe, that all men must die, that physical matter is the only thing that matters, that when physical matter ends life ends, that life or consciousness is an emergent property of physical matter, that psyche and an after-life do not exist, that heat death is the dominant law of physics, and that heat death or entropy is ultimately conserved. He takes a stand. But is he right? The Biblical God Jesus Christ repeatedly tells us that he is wrong. So, who is right, and who got their science wrong? If Matt is right, then Heat Death is coming and therefore your extinction is coming. If he is wrong, Eternal Life is coming. So, which one is actually coming, and which one is science fiction?

In other episodes Matt tells us that massive amounts new Dark Energy is being made from **nothing** as a part of expanding space-time. His science is ultimately based upon nothing. It's Creation Ex Nihilo – the creation of something from nothing by Nothing. He actually tells us that the Conservation of Energy no longer applies and no longer holds true in a universe based upon relativity and expanding space-time – the Dark Energy literally comes from nothing as space expands. So, why is space expanding? "Well, nothing is making it expand." Why is nothing making it expand rather than contract? "Well, it just is. Nothing has decided that it must be so." So, why is nothing filling our universe full of Dark Energy rather than concrete? "Well, it just is. Nothing knows what it is doing."

That's no explanation! That's religion and blind faith, not science! In fact, it's been falsified by Science. Abiogenesis, Spontaneous Generation, Creation Ex Nihilo, Creation from Nothing, Creation by Entropy, Creation by Death, Creation by Random Fluctuations in Entropy, Creation by Quantum Fluctuations, Materialism, Naturalism, Darwinism, Chemical Evolution, and the Theory of Evolution were FALSIFIED in 1859 by Louis Pasteur. They've always been false and continue to be false because they are physically impossible.

Creation Ex Nihilo has no explanatory power; and in many of his videos, Matt actually uses Creation Ex Nihilo to falsify the First Law of Thermodynamics. He uses **nothing** to falsify everything. He has literally placed all of his faith into nothing. However, my faith in God is no different than his faith in nothing, except that I believe that my God will conquer his nothing anytime that my God chooses to do so.

Creation Ex Nihilo is magic; and, even magic (if it were real) has to come from somewhere and from some type of Magician; otherwise, nothing would ever happen in the first place. If you are a scientist, you have got to find some way to explain why there is something rather than nothing. They can't see it nor understand it because they don't want to. Someone Psyche or Someone Intelligent has to be behind it all; otherwise, nothing would ever happen. God explains why we have Dark Energy filling our universe right now rather than concrete. Nothing doesn't explain it. Nothing can't explain it. Nothing doesn't explain anything.

I've learned to see things differently. I've broken my fifty years of brainwashing and conditioning. I'm no longer a Materialist, Naturalist, Nihilist, and Atheist. Science itself convinced me that I was wrong. Science convinced me that God exists. Science is observation and experience. Observation and experience explain why there is something rather than nothing. The Biblical God Jesus Christ has been experienced and observed after He rose from the dead. Psyche has been experienced and observed by Out-of-Body Travelers and Near-Death Experiencers. Science is observation and experience, not philosophical speculation and wishful thinking.

The ongoing idea – that the expansion of our physical universe is accelerating along with the observation that our physical earth seems to be expanding or growing – suggests that some type of invisible energy or intangible energy is currently being infused into our physical universe and into our physical earth from someplace else besides our physical universe. These scientific observations suggest that Dark Energy is transdimensional or quantum in nature and origin rather than having some kind of physical origin or physical cause. With Dark Matter and Dark Energy, there's more than meets the eye.

I believe that the Conservation of Energy or the Conservation of Psyche is the Fundamental Law of Reality. Given the fact that our universe is currently expanding faster than the speed-of-light and given the fact that an infusion of Dark Energy is driving that expansion, imagine how many planets, stars, and galaxies the Gods will be able to make out of all that brand-new Dark Energy once the Gods finally choose to do so.

The people who made all of this stuff that we see around us have other plans for it besides heat death. It's impossible for it to end in heat death thanks to the Conservation of Energy or the First Law of Thermodynamics. After the New Heaven and the New Earth – after the resurrection – there shall be no more death.

<u>Revelation 21: 1-7</u>:

And I saw a new heaven and a new earth: for the first heaven and the first earth were passed away; and there was no more sea. And I John saw the holy city, new Jerusalem, coming down from God out of heaven, prepared as a bride adorned for her husband. And I heard a great voice out of heaven saying, Behold, the tabernacle of God is with men, and he will dwell with them, and they shall be his people, and God himself shall be with them, and be their God. And God shall wipe away all tears from their eyes; and there shall be no more death, neither sorrow, nor crying, neither shall there be any more pain: for the former things are passed away. And he that sat upon the throne said, Behold, I make all things new. And he said unto me, Write: for these words are true and faithful. And he said unto me, It is done. I am Alpha and Omega, the beginning and the end. I will give unto him that is athirst of the fountain of the water of life freely. He that overcometh shall inherit all things; and I will be his God, and he shall be my son.

God tells us that these words are true and faithful, which means that they are guaranteed. The Revelator saw it in vision. It has happened before, and it will happen again. The contrast between the New Heaven and heat death couldn't be more stark or extreme. They are mutually exclusive. The one falsifies the other. So, which one is true?

Entropy is death. After the New Heaven and the New Earth – after the resurrection – there shall be no more death. That includes heat death. No more heat death! No more entropy! The only way that's physically possible is if the Conservation of Energy or the Conservation of Psyche is the dominant LAW of the multiverse. The only way to end death, including heat death, is to eliminate entropy or the second law of thermodynamics with some type of Psyche, Intelligence, Information, Energy, Life Force, Structure, Organization, or Syntropy.

According to the Quantum Law of Thermodynamics, heat death or entropy doesn't exist at the quantum level in the Quantum Realm or the Psyche Realm, which means that heat death technically doesn't exist at the physical level either because the physical level is based upon and made from the quantum level. Heat death or entropy is simply an illusion or a figment of the imagination. Entropy or death is temporary, meaning that entropy or death is temporal rather than quantum or spiritual. Entropic physical matter is not conserved. This is what has been observed.

In the Quantum Realm, we observe that Syntropy, Psyche, Life Force, or the Conservation of Energy reigns supreme. Syntropy, Psyche, or Energy makes the end-of-death possible, including the end of heat death or the end of entropy, because Psyche and Energy are always conserved.

God says that He is going to put an end to death. That means that He is going to put an end to entropy or heat death. I believe God can do it. The Law of the Conservation of Energy tells me that He can do it. The Ultimate Law of Thermodynamics and the Quantum Law of Thermodynamics tell me that He can do it. Syntropy tells me that He can do it. Psyche or Intelligence tells me that He can do it. All of that new Dark Energy currently being infused into our physical universe from someplace else tells me that He can do it. God says that He can do it. He would know. Entropy is death. God made entropy or death, so He can get rid of it at will anytime He wants to. Physical matter and entropy are NOT conserved, which means that death is NOT conserved.

According to God, that massive ongoing infusion of Dark Energy from some other dimension is in preparation for something wonderful and new in our little corner of space-time. Heat death is impossible at the quantum level, which means that heat death is also technically impossible at the physical level. It's not going to hold! Even at maximum entropy, all of the energy is still there waiting to be reused by Someone Psyche. Psyche is Intelligence. Intelligence, information, or knowledge is all that's needed to overcome entropy. Psyche is Energy. Energy is all that's needed to overcome entropy. Psyche or Life Force is Syntropy. Life Force is all that's needed to overcome entropy or death. Syntropy is all that's needed to overcome entropy. It's not going to end in heat death.

Heat death is impossible at the quantum level which means that heat death is ultimately impossible at the physical level. That's the revelation that I received. We need to make a T-shirt that says that "Heat Death Is Impossible" because that is what has actually been experienced and observed.

Even under their Heat Death Model, the universe, space, and the quantum fields continue to exist. The space or the vacuum is still there full of quantum fields from one end of the universe to the other. Quantum fields are organized energy. Quantum fields are the fundamental construct; and, they are constructed from energy by Someone Psyche or Someone Intelligent. Quantum fields are maintained by Someone Psyche. Psyche is the innate intelligence within all the different forms of energy which gives that energy the inherent ability to understand, follow, and obey God's Laws and God's Commands. Even at maximum entropy, ALL of the energy is still there and so is the psyche within it. Ultimately heat death cannot prevail so long as Energy or Psyche is conserved.

Mark My Words

A Psyche Ontology

A Psyche Ontology or the Ultimate Model of Reality states that Psyche, Intelligence, or Non-Local Consciousness is the fundamental unit of Reality, Existence, Personality, and Individuality. Psyche is Energy; and, Energy is psychic. God is Light; and, God is in the light. The glory of God is intelligence, or light and truth.

The Law of Psyche or a Psyche Ontology states that Psyche, Intelligence, or Life Force is the fundamental unit of reality, the fundamental force in the universe, and the organizing force of the universe at all levels of existence. Psyche is the innate intelligence within all the different forms of energy which gives that energy the inherent ability to understand, follow, and obey God's Laws and God's

Commands. Psyche or Life Force explains why there is something rather than nothing.

Psyche or the Life Force seems to have its own unique quantum field or network through which transpersonal communication or psyche-to-psyche communication is made possible. It's as if each individual Psyche has its own unique telephone line or web site on this psychic network or psychic quantum field. Telepathy is WiFi at the quantum level. All of this has been experienced and observed. Science is observation and experience after all.

Psyche is Energy; and, Energy is psychic. Your Psyche or Intelligence is eternal and everlasting. Like Energy, your Psyche cannot be made, and it cannot be destroyed. Your Psyche, Intelligence, Life Force, or Personality is always conserved. Remember, Psyche or Life Force is the Organizing Force of the Universe. This is what has been experienced and observed. Psyche or Energy is always conserved.

I start with what has been experienced and observed; and then, I try to find a scientific explanation for all of it. After the death of your physical body comes the Eternal Life of your Intelligence or Psyche. This is what has been experienced and observed. After the heat death of our physical universe comes the Eternal Life of Our Universe, or the Conservation of Energy and Psyche, assuming that the heat death of our physical universe actually comes. According to God and the Atonement of Jesus Christ, after death comes resurrection and immortality, and this reality applies on a universal scale. The Gods can raise anything from the dead anytime they choose to do so.

How is that possible?

$E = mc^2$ works both ways! The Materialists, Naturalists, Darwinists, Nihilists, and Atheists erroneously teach and believe that $E = mc^2$ is a one-way process and that entropy, heat death, death, physical reality, and entropic physical matter are ultimately conserved. These people are wrong. According to this equation, the Gods can make spirit matter, dark matter, syntropic physical matter, spiritual matter, entropic physical matter, and resurrected physical matter from Dark Energy anytime they choose to do so. Furthermore, according to this equation, the Gods can turn entropic physical matter back into Raw Energy, or Available Energy, or Dark Energy anytime they choose to do so! That's how things really work at the quantum level.

$E = mc^2$ is the Conservation of Energy or the Conservation of Psyche; and, it ultimately falsifies and defeats the second law of thermodynamics, death, heat death, and entropy. That's GOOD SCIENCE because it has actually been experienced and observed. The Conservation of Energy is the Fundamental Law of Reality and Existence. Psyche is Energy; and, Energy is psychic. This is what has been experienced and observed. Psyche or the Life Force has been experienced and observed by Out-of-Body Travelers and Near-Death Experiencers. Psyche is experienced as an immaterial viewpoint in space; and, Psyche is observed or seen as a pinprick of light, a point particle of light, or a spark of light.

According to the Ultimate Law of Thermodynamics, physical matter and entropy are NOT conserved. This is what has actually been experienced and observed. Their quantity is changing all the time! Entropic physical matter is made and destroyed all the time. Entropic physical matter is made from different forms of energy; and, the FORM is never conserved.

According to the Quantum Law of Thermodynamics, heat death doesn't apply to your psyche or life force; and, heat death doesn't apply to God's Psyche or God's Life Force either.

According to the Quantum Law of Thermodynamics, heat death is impossible at the quantum level in the Quantum Realm. Entropy or the second law of thermodynamics doesn't exist at the psyche level or the quantum level in the Quantum Realm, Psyche Realm, or Spirit World which means that entropy won't prevail at the physical level either. This is what has been experienced and observed. Resurrection from death has been experienced and observed.

God made entropy and entropic physical matter; and, He can get rid of them anytime He chooses to do so. The Atonement of Christ overcomes, defeats, and eliminates entropy, death, and heat death at will. That's why the Atonement of Christ exists. This is what has been experienced and observed.

Do you see how that works?

This is GOOD SCIENCE! It's the answer to life, the universe, and everything. It explains everything that has ever been experienced and observed. Does it not?

It's a bit more satisfying and explanatory than Materialism, Naturalism, Darwinism, Nihilism, Atheism, Creation Ex Nihilo, and the Heat Death of our Universe. Is it not? It explains why there is something rather than nothing.

According to the "Heat Death of Our Universe Model" or Entropy Model, our physical universe shouldn't exist in the first place; yet it does exist, which means that there's something fundamentally wrong with the Heat Death of Our Universe, Entropy, and the Second Law of Thermodynamics. This is logical common sense and is based upon what has already been experienced and observed. Syntropy, the Law of Psyche, the Atonement of Christ, the Resurrection, the First Law of Thermodynamics, the Conservation of Energy and Psyche, the Ultimate Law of Thermodynamics, and the Quantum Law of Thermodynamics FALSIFY and defeat heat death, entropy, death, and the second law of thermodynamics.

Entropy and entropic physical matter couldn't exist and wouldn't exist without some kind of Syntropy or Organizing Force in this universe. Psyche or Intelligence is the fundamental unit of Reality and Existence as well as the Organizing Force of the Universe on both sides of the veil – both on the physical side and on the spiritual quantum side. Psyche, Intelligence, Energy, or Life Force can overcome disorder and entropy at will. This has already been experienced and observed.

Whenever a Human Psyche (one of the Gods) enters into a non-consensus realm in the Spirit World, the spirit matter and energy within that realm organizes according to the demands, commands, and expectations of that Human Psyche. The psyche within spirit matter and energy reads the mind of the Human Psyche and then organizes itself accordingly. Psyche really is the Organizing Force of the Universe at every level of existence. This is what has been experienced and observed.

I just provided you with the answer to life, the universe, and everything. Did I not? What has been mentioned here explains everything that has ever been experienced and observed.

Mark My Words

Quantum Fields: The Real Building Blocks of the Universe

https://www.youtube.com/watch?v=zNVQfWC_evg

Quantum fields forever! I want to meet the guy who made, organized, and maintains the quantum fields. Then I think we'll finally be getting somewhere. Building blocks need a Builder; otherwise, nothing will ever be made from them. Building blocks also need a Maker; otherwise, they would never exist in the first place. I want to meet the Builder and the Maker of the real building blocks of the universe. I want to meet the guy who maintains them and is keeping them going. That's the guy who knows how to do science for real. Remember, the Gods are scientists, not magicians. The Gods know how to do science. They can science anything they want, including an end to heat death.

Mark My Words

The Growing Universe

https://en.wikipedia.org/wiki/Expansion_of_the_universe

https://en.wikipedia.org/wiki/Observable_universe

The expansion of our universe is not in dispute. The scientists are in complete agreement that our physical universe is growing and expanding. The thing that is in dispute is the mechanism behind the expansion or the growth. Where's all of that extra energy coming from? The expansion of our universe isn't slowing down – it's accelerating. The only way that's logically possible is if our physical universe as a whole is currently receiving a massive infusion of energy or mass from someplace besides our physical universe. This is just logical common sense.

Of course, contrary to rational common sense, the Naturalists and Atheists on PBS assure us that the new dark energy is coming from nothing and is just an emergent property of expanding space-time. They assure us that the conservation of energy doesn't apply in expanding space-time and that the new dark energy is being made from nothing as space expands. These people use Creation Ex Nihilo to falsify the First Law of Thermodynamics or the Conservation of Energy. Anything but God! Creation Ex Nihilo is magic. If you buy the magic, then you're definitely an atheist. I don't buy it. I believe that the First Law of Thermodynamics or the Conservation of Energy falsifies Creation Ex Nihilo. To each their own. The best I can do is to present the evidence and then let you make up your own mind what you want to believe. Your beliefs are chosen into existence by your Human Psyche on both sides of the veil – both the physical side and the spiritual side.

The Growing Earth

https://www.youtube.com/watch?v=ydFItltGzUk

https://www.youtube.com/watch?v=7kL7qDeI05U

https://www.youtube.com/watch?v=VYSSIpP3r9w

https://www.youtube.com/watch?v=V840anEvGPw

The Theory of Plate Tectonics is in dispute. Obviously, there are crustal plates that we call continents; and equally obvious, crust floats on mantle. The part of Plate Tectonics that is in dispute derives from the fact that there is NO subduction. The Theory of Plate Tectonics relies upon subduction for the formation of new crust. There is NO subduction. There's just the constantly stretching and growing earth taking place towards the center of the Pacific and Atlantic Oceans. As the earth expands and grows, the crust buckles or settles thereby forming mountains – not along the coastlines but inward towards the middle of the continents as we observe. Buckling of the crust forms mountains, not subduction. There is no subduction. That is what has been experienced and observed. The Growing Earth Theory actually has more explanatory power than Plate Tectonics; and, the Growing Earth Theory actually fits better with observational evidence than Plate Tectonics.

I define Science as observation and experience; so, the observations have had an influence on me in recent years whereas the philosophical speculation and wishful thinking not so much. Most of our scientists define science as Physicalism, Materialism, Naturalism, Darwinism, Nihilism, Behaviorism, Determinism, Physical Reductionism, and Atheism. These things have never been experienced nor observed. They are in dispute. They are pure philosophy – nothing but wishful thinking. These people do science by Community Consensus, Groupthink, and Popular Vote – not observation and experience. Instead, I choose to go with the observations and the experiences that falsify Materialism, Naturalism, Darwinism, Atheism, and their derivatives. I personally find the observations, and the experiences, and the scientific experiments more convincing than the wishful thinking and philosophical speculation of the Physicalists, Darwinists, Atheists, and Scientific Naturalists.

Of course, the Expanding Earth Theory is also in dispute. It happens so slowly that the claim has been made that it has never been experienced nor observed. I do find the circumstantial evidence compelling, nonetheless. It would explain why we still have a magnetic field acting as a force field to protect us from the sun after billions of years of time.

Planets and Stars Are Made

Dark energy or the light of Christ is the power by which the planets and stars were made. It takes a massive infusion of dark energy, the light of Christ, or vacuum energy to give birth to a physical planet, or a physical star, or a physical galaxy. The raw energy is there waiting to be used. We look through our telescopes and we observe that physical galaxies are born all at once, and then they start to wind up from there. It takes a massive infusion of dark energy, the light of Christ, or vacuum energy to stretch and expand a physical universe as is currently happening to our physical universe. It takes a constant infusion of dark energy, the light of Christ, or vacuum energy to make our physical earth expand and grow. The magnetic field within the earth is often used to explain how physical matter is made within our earth. Dark energy could indeed be another source of energy for the production of physical matter within our earth especially during the times when our earth doesn't have a strong magnetic field. The explanatory power of dark energy or the light of Christ is impressive. It proves that our physical universe is not a closed system but is currently being made and formed by the Gods.

Since it appears that the Gods are currently infusing new raw energy into this physical universe from someplace else besides this physical universe, the Gods can form that new energy into new planets, stars, and galaxies essentially for all eternity which means that our physical universe could be sustained and kept in existence essentially for forever constantly growing and expanding forever out into the chaos that preceded the formation of our physical universe in this region of reality.

Add to that the Ultimate Law of Thermodynamics which states that Energy, Psyche or Syntropy is conserved – not physical matter and not entropy; and, we find ourselves looking at the answer to life, the universe, and everything. The Ultimate Law of Thermodynamics explains everything that we have ever experienced or observed. Physical matter, spirit matter, physical laws, physical constants, gravity, magnetism, visible light, photons, the nuclear forces, invisible intangible forces and fields, quantum waves, space-time, space or locality, time or entropy, and the vacuum energy or zero-point field of light are different FORMS of energy. According to the Ultimate Law of Thermodynamics, the FORM is never conserved. The Gods can change one FORM into a completely different FORM anytime they choose to do so. Only the underlying Psyche or Energy is being conserved. Psyche or Energy is syntropic which means that it is eternal and everlasting, which means that it's always being conserved.

In contrast, the different FORMS of energy were formed by the Gods, including physical matter and entropy, which means that they can be eliminated by the Gods and formed into something else instead. The Gods can form energy into anything, including physical matter and entropy, which means that they can transform physical matter and entropy into something else or turn it back into raw unorganized energy anytime that they see fit to do so. The FORM is never conserved. The underlying psyche or energy is conserved. Psyche is the intelligence within all the different forms of energy that have ever been experienced or observed. Psyche, Intelligence, Consciousness, or Energy is conserved. Syntropy is the conservation of energy. This is the Ultimate Law of Thermodynamics.

Physical matter and entropy are never conserved. Their amount is constantly being changed by the Gods. Physical matter and the associated entropy are made or caused to begin by the Gods; and, these things can and do cease to exist whenever the Gods convert them back into raw syntropic energy or into some other FORM of energy. Human beings, the children of the Gods, create new physical matter and entropy in their particle accelerators or atom smashers. Human beings also destroy physical matter and entropy in their atomic bombs by converting them back into raw energy. Physical matter and entropy are never conserved. It's the underlying Energy or Psyche that's being conserved.

The Gods can't make energy or psyche, and they can't destroy energy or psyche, which means that your psyche and the choices that your psyche makes are being conserved. Your psyche is syntropic which means that it has always existed and will always exist because it cannot be destroyed because it is constantly being conserved.

The Ultimate Law of Thermodynamics is the answer to life, the universe, and everything. Physical matter and entropy are never conserved; whereas, the underlying energy or psyche is syntropic which means that it is always conserved. This explains everything that has ever been experienced and observed.

The intelligent and purposeful construction or production of physical matter from energy is a lot more scientific, demonstrable, parsimonious, credible, and believable than the "spontaneous generation" of Materialism and Naturalism or the "creation ex nihilo" of Atheism. Construction of physical matter from energy doesn't violate the First Law of Thermodynamics; whereas, creation ex nihilo does. Atheism is creation from nothing by nothing. Atheism is creation ex nihilo. Atheism violates the First Law of Thermodynamics; whereas, the intelligent and purposeful organization of energy into different forms of energy such as physical matter and entropy does not violate the First Law of Thermodynamics.

Furthermore, Naturalism, Darwinism, Materialism, and Physicalism are spontaneous generation or abiogenesis, which violates the Second Law of Thermodynamics. Chemical evolution and macro-evolution are prevented from happening by entropy. However, the telekinetic organization or intelligent organization of physical matter into proteins and genes does not violate the Second Law of Thermodynamics. Intelligent beings are able to form physical matter into different things. It has been experienced and observed; whereas, spontaneous generation, abiogenesis, chemical evolution, and macro-evolution have never been experienced nor observed.

Materialism, Physicalism, Naturalism, Darwinism, Nihilism, Behaviorism, Physical Reductionism, and Atheism were designed to prevent us from discovering the Ultimate Law of Thermodynamics. These people erroneously teach that physical matter or entropy is the only thing that exists, which means that these people erroneously teach that physical matter and entropy are being conserved. Entropy is death; and, these people literally teach that death is being conserved. That's what they really teach, is it not? These people deny the existence of Psyche or Syntropy or Intelligence or Life Force; and, they instead teach that physical matter or entropy is the only thing that exists, and that physical matter and entropy are being conserved.

They are wrong.

Materialism, Naturalism, Darwinism, Nihilism, and Atheism are falsified by the truth; whereas, the truth is repeatedly experienced and observed.

The falsified philosophies of Naturalism, Darwinism, and Atheism were designed to prevent us from discovering Psyche, Syntropy, Intelligence, Eternal Life, Immortality, Quantum Mechanics, Non-Locality, God, and the Ultimate Law of Thermodynamics.

Mark My Words

Birth of Our Earth

The birth of our earth has been seen in vision. Some of the prophets of God have been shown in vision how it was done.

Enoch wrote:

I had a vision from God, and I saw and felt the birth of our physical earth in this vision.

It was an oh-wow type of an experience. So, that's how it was done!

It was fast, like an infusion of energy from the universe while at the same time being phase-shifted from the spiritual into the physical. I could see and feel the earth being lifted up from the spiritual into the physical; and, I could also see and feel rays of energy coming in from all directions converging on our earth and making it physical. Our earth was born. Our earth was made. It was a quantum mechanical process, not a physical process. In other words, it was a spiritual process and not a physical process. It was a birth.

The revelation was that physical matter, physical genomes, physical proteins, physical planets, physical stars, physical galaxies, and physical universes are made by the Gods. They are made whole all at once. They don't coalesce out of expanding clouds of gas. They wouldn't coalesce out of expanding energy either. Expanding clouds of gas don't ever coalesce naturally under the effects of gravity. That's the revelation or the insight that I received.

Physical matter is made from energy. Physical matter is just a different form of energy. The energy is eternal. Energy cannot be made nor destroyed. However, physical matter is made, which means that it can also be destroyed or come to an end. Energy is syntropic; whereas, physical matter is entropic. Syntropy is eternal life; whereas, entropy is death. Birth and death are the signs of entropy. Eternal life is the surest sign of syntropy. Raw energy is unorganized mass or unorganized matter. Raw energy hasn't been organized yet. It hasn't been given a physical form. Energy is matter unorganized. The Gods can form energy into anything including physical matter; but, the Gods can't create nor destroy energy. The energy has always existed meaning that the energy is conserved.

Psyche or intelligence is the innate life force within all the different forms of energy. This psyche or life force gives energy the ability to obey God's commands and God's laws. This is a different way of looking at science that is totally consistent with the first law of thermodynamics. It also allows or permits the second law of thermodynamics or entropy. It teaches that entropy only applies to physical matter. Entropy and physical matter are made by the Gods. In contrast, energy is syntropic which means that energy is conserved thereby making energy eternal and everlasting.

The idea that we are star stuff is actually false. A lot of people love that idea even though it is physically impossible. There's no way in the universe that clouds of gas, expanding out into space after a super nova, will ever coalesce into a planet or another sun. In fact, our whole universe is currently expanding faster under the influence of dark energy than it is coalescing or converging under the influence of gravity. God actually has to place huge amounts of spirit matter or dark matter around galaxies to keep them from flying apart into space. The idea that planets, stars, and galaxies coalesce out of expanding clouds of gas and physical matter is nothing but science fiction. That's not the way planets, stars, and galaxies are made. Gravity simply is not an adequate mechanism for getting

expanding clouds of gas and physical matter to coalesce into planets, stars, and galaxies. There has to be a better way.

Now granted, the Gods have sufficient power and ability to take expanding clouds of gas and force them to coalesce into planets and stars. That's one way that the creation of planets and stars can be done; however, that's not the way it was done in the vision that I had. The physical matter was made in place whole during the birth of our earth. It was fast. It was a transition or a phase shift from the spiritual to the physical. There wasn't any cloud of gas. Our earth didn't coalesce out of gas. It was too fast for those incoming rays of energy to have been physical matter. Physical matter cannot move at the speed-of-light. The dark energy, infusion of mass, or rays of energy came in straight from all directions converging on our earth all at once producing the physical matter for our earth instantaneously.

It was dark. There wasn't any sunlight. There wasn't any starlight. The newborn earth was dark and void. It was just an infusion of raw mass or raw energy from all directions all at once. It was a birth. The physical matter for our earth came into existence all at once. It was nothing like the way we scientists typically visualize planets and stars slowly coalescing out of a huge swirling cloud of gas under the effects of gravity and mutual attraction between atoms and rocks, all lit up by the newborn sun in the center of that swirling cloud of gas. It was a birth.

It was a vision. It was seen and felt; and, it was completely unexpected. It had all the hallmarks of a vision.

I know what I saw and felt in the vision. What I don't know is if I have successfully described the event in terms of today's scientific knowledge and vocabulary. I did the best I could, though, given what I know of science.

https://ldssoul.com/the-birth-of-our-earth/

https://ldssoul.com/forum/viewtopic.php?f=103&t=325

Our earth was made. It didn't coalesce out of a swirling cloud of gas. Dark energy is stronger than gravity; therefore, expanding clouds of gas will never coalesce into planets, stars, and galaxies. Furthermore, hydrogen molecules and helium atoms naturally repel each other. They will never coalesce into planets, stars, and galaxies under the influence of gravity.

Entropic physical matter is made from energy. Proteins and genes are made from physical matter. Planets, stars, and galaxies are made – either through birth into physical mortality by quantum processes or through the forceful concentration of expanding clouds of gas by quantum processes. They don't just spontaneously generate out of thin air; and, they don't coalesce out of expanding clouds of gas through natural processes. They are made. Anything that is obviously made obviously has a Maker who made it. Anything that has a beginning has Someone Psyche or Someone Intelligent who caused it to begin. This is the way things work.

Only Energy, Psyche, or Intelligence is eternal and everlasting without a beginning or an end. Only Energy, Psyche, or Intelligence is conserved. Psyche is the innate intelligence or life force within all the different forms of energy. It is syntropic which means that it is conserved. The fact that physical matter and entropy are not conserved whereas Psyche and Energy are conserved ends up being the answer to life, the universe, and everything. It explains everything that we humans have ever experienced or observed.

Things that Act and Things that Are Acted Upon

According to the Biblical God Jesus Christ, there are things that act and things that are acted upon.

2 Nephi 2: 11-16, 25-29:

For it must needs be, that there is an opposition in all things. If not so, my firstborn in the wilderness, righteousness could not be brought to pass, neither wickedness, neither holiness nor misery, neither good nor bad. Wherefore, all things must needs be a compound in one; wherefore, if it should be one body it must needs remain as dead, having no life neither death, nor corruption nor incorruption, happiness nor misery, neither sense nor insensibility. Wherefore, it must needs have been created for a thing of naught; wherefore there would have been no purpose in the end of its creation. Wherefore, this thing must needs destroy the wisdom of God and his eternal purposes, and also the power, and the mercy, and the justice of God. And if ye shall say there is no law, ye shall also say there is no sin. If ye shall say there is no sin, ye shall also say there is no righteousness. And if there be no righteousness there be no happiness. And if there be no righteousness nor happiness there be no punishment nor misery. And if these things are not there is no God. And if there is no God we are not, neither the earth; for there could have been no creation of things, neither to act nor to be acted upon; wherefore, all things must have vanished away. And now, my sons, I speak unto you these things for your profit and learning; for there is a God, and he hath created all things, both the heavens and the earth, and all things that in them are, both things to act and things to be acted upon. And to bring about his eternal purposes in the end of man, after he had created our first parents, and the beasts of the field and the fowls of the air, and in fine, all things which are created, it must needs be that there was an opposition; even the forbidden fruit in opposition to the tree of life; the one being sweet and the other bitter. Wherefore, the Lord God gave unto man that he should act for himself. Wherefore, man could not act for himself save it should be that he was enticed by the one or the other. Adam fell that men might be; and men are, that they might have joy. And the Messiah cometh in the fulness of time, that he may redeem the children of men from the fall. And because that they are redeemed from the fall they have become free forever, knowing good from evil; to act for themselves and not to be acted upon, save it be by the punishment of the law at the great and last day, according to the commandments which God hath given. Wherefore, men are free according to the flesh; and all things are given them which are expedient unto man. And they are free to choose liberty and eternal life, through the great Mediator of all men, or to choose captivity and death, according to the captivity and power of the devil; for he seeketh that all men might be miserable like unto himself. And now, my sons, I would that ye should look to the great Mediator, and hearken unto his great commandments; and be faithful unto his words, and choose eternal life, according to the will of his Holy Spirit; and not choose eternal death, according to the will of the flesh and the evil which is therein, which giveth the spirit of the devil power to captivate, to bring you down to hell, that he may reign over you in his own kingdom.

This description of reality from 600 B.C. matches with observational reality or the scientific evidence that we now have on hand as a race.

From what I have been able to tell, Psyche or Intelligence CHOOSES or ACTS; whereas, the different types of matter simply REACT. In other words, Energy is Syntropy, eternal and everlasting without a beginning of days or an end of years; whereas, the different types of matter or different forms of matter are MADE and therefore have some type of beginning. Energy or Syntropy is conserved according to the first law of thermodynamics; but, the amount of physical matter or entropy changes according to the second law of thermodynamics.

There's also corruption and incorruption when it comes to matter. Entropic matter or physical matter gets filled with corruption or entropy; whereas, syntropic matter or spirit matter is incorruptible, eternal, and everlasting in nature. Entropic matter or physical matter has been made subject to the passage of time and localized into space; whereas, syntropic matter, dark matter, or spirit matter is ageless, timeless, eternal, and based upon quantum mechanics.

According to the Biblical God Jesus Christ, this is how things really work when it comes to this universe and all the different objects within it. I choose to believe that God knows what He is talking about, whereas the Naturalists and Atheists do not.

Conclusions Regarding Dark Matter and Dark Energy

Materialism, Physicalism, Naturalism, Darwinism, Nihilism, Classical Physics, and Atheism formally claim that invisible non-physical forces and fields such as Dark Energy and Psyche DO NOT EXIST.

Are these people right; or, did they get their science wrong?

The scientists who discovered Dark Matter and Dark Energy knew for a fact that the discovery of Dark Matter and Dark Energy falsifies the claims of Physicalism and Scientific Naturalism which state that invisible non-physical forces and fields such as Dark Matter and Dark Energy DO NOT EXIST. The scientists who discovered Dark Matter and Dark Energy didn't care that these different types of Organized Energy or Organized Syntropy falsify Materialism, Naturalism, and their derivatives. These scientists were only interested in finding and knowing the truth.

Materialism, Darwinism, and Naturalism are Bad Science. How does claiming that Dark Matter and Dark Energy DO NOT EXIST explain scientifically what they are and how they work? It doesn't! It can't! Materialism, Naturalism, Darwinism, and Atheism completely lack explanatory power whenever it comes time to explain how Psyche, Consciousness, After-Death Memories, Non-Locality, Quantum Mechanics, Supernatural Mechanisms, Syntropy, Action at a Distance, and Invisible Non-Physical Forces and Fields work, what they are, and why they can't be seen with the naked eye.

British TV Documentary, "Edge of the Universe", 2008, Season 1 Episode 3, "Final Frontier".

> **Space can be twisted out of shape by the massive, unseen force of gravity.**
>
> **There's nowhere near enough matter in all these galaxies to pull the universe into a curved shape. But more alarmingly, there is only a fraction of the matter required to keep it flat. To make sense of his unexpected**

discovery, Brian Boyle has invoked two of the most mysterious forces in the universe: dark matter and dark energy.

Although the galaxy survey has only found one third of the mass required to make the universe flat, we now believe that the remaining two thirds of the matter energy in the universe is contained in this so-called dark energy.

Astronomers have long speculated that, in addition to the familiar matter of stars and planets, there is a vast proportion of the universe that is invisible, made up of dark matter and dark energy. But where's the proof? These forces are just theories. Until now.

In one of the most remarkable discoveries of recent years, the evidence has been found in the microwave background radiation. This radiation, which has travelled through space since the beginning of time, has confirmed the existence of these invisible cosmic forces. In proving that the universe is flat, the microwave background has also confirmed that astronomers were right to assume the existence of dark energy and dark matter. Everything we see is just a small fraction of the total cosmic mass. Most of the universe is invisible.

Astronomers have discovered that galaxies are not only rushing apart from each other, they are gathering speed. It seems that four billion years ago, the universe changed up a gear and ever since has been expanding faster and faster. But what is the force driving this cosmic acceleration?

Now if it's true that the rate of expansion is picking up, that suggests that there's something like an antigravity force, something opposing the normal force of gravitation, the pulling force, something pushing galaxies apart faster and faster. This antigravity force is dark energy. Astronomers now believe the universe must be buzzing with this invisible energy. Space, it seems, is far from empty. We just can't see what's out there. The unseen forces of the universe, gravity and dark energy, are locked in the ultimate battle. At stake is the destiny of the universe. Dark energy will defeat gravity and succeed in stretching the universe into oblivion.

This model we have requires dark energy that we've never seen, and dark matter which also we have never seen. And the atoms that we know about are making up only 5% of the universe.

Somebody's got their foot on the gas. The rate of Dark Energy infusion increased noticeably four billion years ago. Will our universe expand forever as the Naturalists and Atheists claim? ONLY if the God who has his foot on the gas keeps applying the gas. The observed fact that He put the pedal to the metal four billion years ago tells us that He can take his foot off the gas and apply the brakes anytime He wants to; and, if He runs out of gas, it will have the same effect as well.

Anything that we KNOW to exist, which is unseen or unseeable, will in fact be spiritual in nature or non-physical in nature. Whenever Scientists use the words "unseen", or "force", or "field" to describe something, that particular "something" will in fact be immaterial, non-physical, and PROVE beyond a shadow of a doubt that Materialism or Scientific Naturalism is FALSE.

The tangible stuff that we can get our hands on makes up only 5% of this universe! Materialism or Scientific Naturalism only covers 5% of the universe; and, the Materialists and Naturalists formally reject all the rest. Not only does that make Materialism incomplete and FALSE, but that also makes Materialism or Scientific Naturalism BAD SCIENCE.

I used to be a Materialist, Naturalist, Nihilist, and Atheist. I never understood Quantum Mechanics until after I got rid of My Materialism, Naturalism, Nihilism, and Atheism. Physicalism, Naturalism, Darwinism, Nihilism, Classical Physics, and Atheism were designed to prevent us from understanding Quantum Mechanics or Syntropy because Quantum Mechanics is supernatural in nature and origin.

The Materialists, Naturalists, Darwinists, Nihilists, Classical Physicists, and Atheists will assure you that I don't understand Science. They have told me many times that I don't understand Science. These people define "science" as Physicalism, Naturalism, Darwinism, Nihilism, Atheism, and the Theory of Evolution. When it comes to Science, are these people right, or have they gotten their science wrong?

According to the Physicalists, Naturalists, Darwinists, Nihilists, and Atheists, dark matter and dark energy DO NOT EXIST. According to these people, Psyche, Syntropy, Quantum Mechanisms, Supernatural Mechanisms, and Action at a Distance DO NOT EXIST. According to these people, invisible non-physical forces and fields do not exist. ONLY entropy or physical matter exists. Are they right, or have the gotten their science wrong?

In order to be consistent, the Materialists and Naturalists must also insist that invisible non-physical forces and fields such as magnetism, gravity, the strong nuclear force, the weak nuclear force, quantum waves, psyche, radio waves, microwaves, gamma rays, and x-rays DO NOT EXIST. The fact that these different invisible non-physical forces and fields have been proven to exist is proof positive that the Materialists and Naturalists have gotten their science wrong.

Entropy, heat death, or thermal equilibrium is invisible, intangible, and technically non-physical; and, according to the Materialists and Naturalists, entropy does in fact exist.

So, what gives?

These people are hypocrites. They allow the existence of the invisible and non-physical forces and fields that they like while at the same time denying the existence of the invisible and non-physical forces and fields that they don't like. These people deny the existence of any invisible non-physical force or field that proves that God exists or that falsifies Materialism and Naturalism. Materialism and Naturalism are self-defeating because of all the internal contradictions, inconsistencies, discrepancies, hypocrisy, and falsified philosophy that are built into Materialism, Naturalism, Darwinism, Nihilism, and Atheism.

Physicalism, Naturalism, Darwinism, Nihilism, Atheism, and their derivatives were designed to hide the truth from us. They were designed to hide the observations and the experiences of the human race from us. Physicalism and Naturalism were designed to hide the observational evidence, scientific evidence, experiential evidence, and eye-witness evidence from us. They were designed to prevent us from making scientific discoveries that contradict them and prove them false. They were designed to hide the science experiments that falsify Materialism, Naturalism, and Darwinism. Physicalism and Naturalism were designed to contradict the observational evidence and experiential evidence that we have on hand as a race. They were designed to hide the evidence and the truth from us so that we will never find it unless we deliberately go looking for it.

I have observed that the production of physical matter requires a massive infusion of Energy, Syntropy, or Mass – on both sides of the veil. This massive infusion of Syntropy, Energy, Order, or Mass slows the particle down, localizing it in space and making it subject to entropy or the passage of time. A massive infusion of Energy, Syntropy, or Mass forces a particle of spirit matter to be converted into a particle of physical matter and forces that newly born particle of physical matter to be subject to space-time limitations.

By definition, in principle, it has been observed that entropic physical matter is made by infusing a huge amount of Mass or Energy into a particle of spirit matter or a particle of dark matter thereby converting that particle of spirit matter into physical matter and locating it into space-time consequently making it subject to space-time, or locality and entropy. Space results in locality; and, time results in entropy. Entropy is a function of time. Entropy is the result of a huge infusion of Energy, Syntropy, or Mass. There is NO entropy in the Syntropic Realm or Quantum Realm where everything is everlasting, timeless, ageless, eternal, and conserved.

The Energy or Syntropy is conserved; whereas, the physical matter or entropy comes and goes as God sees fit. Physical matter or entropy began, which means that it can theoretically come to an end when it is absorbed back into Syntropy or when it has all of that mass, or entropy, or energy removed. Physical matter is produced by a massive infusion of Energy or Mass. Remove all of that Energy, Mass, Structure, or Entropy, and that particle of physical matter will go back to being spirit matter instead. This is the way things work in real life. It explains everything, doesn't it?

Energy or Syntropy or the First Law of Thermodynamics is the ANSWER to life, the universe, and everything. Energy or Syntropy is conserved; and, a massive infusion of Energy, Syntropy, or Mass slows a particle of spirit matter down changing it into a particle of physical matter thereby making that newly created particle of physical matter subject to space-time, or locality and entropy. This explains everything, doesn't it?

Syntropy is the answer to life, the universe, and everything.

It is the amount of Energy or Syntropy that is being conserved or that remains constant; whereas, the amount of entropy and physical matter changes over time. We actually create new physical matter in particle accelerators and atom smashers by converting energy into physical matter. The amount of physical matter does NOT remain constant and is therefore NOT conserved. Likewise, the amount of entropy or the age of a physical particle is changing all the time. Only entropic physical matter is subject to entropy, locality, sub-light speed limitations, the aging process, the passage of time, quantum tunneling restrictions, and other physical limitations. Everything else is Syntropy. Everything else is timeless and ageless. Everything else is conserved.

Energy or Syntropy is the RULE; and, physical matter or entropy is the exception to the rule. Energy or Syntropy is conserved. Energy or Syntropy is eternal and everlasting, without a beginning of days or an end of years. Energy or Syntropy can change FORM; but, it is always conserved. Energy or Syntropy can be formed into spirit matter; and, a huge massive infusion of Energy or Syntropy or Mass can change a particle of spirit matter into a particle of physical matter. Remove all of that Energy, Entropy, or Mass, and that particle of physical matter will return to being an invisible non-physical particle of spirit matter or dark matter instead. This is the answer to life, the universe, and everything. It explains how everything works.

The Materialists, Naturalists, and Atheists will assure you that I got my science wrong and that I don't understand science. Are they right? Or is there something extremely important that these people are missing?

I take Quantum Mechanics or Syntropy seriously; and, the Materialists, Naturalists, and Atheists do not. That's the main difference between us. Instead, the Physicalists, Naturalists, and Atheists deny the existence of Psyche, Syntropy, and Supernatural

Mechanics. Quantum Mechanics is supernatural in nature and origin; and, the Scientific Naturalists deny this fact.

Instead, the Materialists, Naturalists, Darwinists, and Atheists assure us that dark energy and dark matter DO NOT EXIST. These people laugh at and ridicule the Light of Christ, the Quantum Sea of Light, the Zero-Point Field of Light, or Dark Energy. These people laugh at and ridicule Spirit Matter or Dark Matter. These people laugh at and ridicule Psyche or Syntropy. Are they right, or did they get their science wrong?

Science is observation and experience. Science is knowledge. The Materialists and Naturalists are motivated to deny our observations and experiences so that they can say that God does not exist. In other words, these people are motivated to deny the Scientific Evidence proving that Psyche, Syntropy, Action at a Distance, and God do exist, just so that they can convince themselves that God does not exist. They have an agenda. Materialism, Naturalism, Darwinism, Nihilism, and Atheism are based upon a denial of Reality, a rejection of observational evidence, and a refusal to look at any evidence that falsifies Materialism, Naturalism, and Darwinism. These people deliberately hide and destroy evidence in order to make their case and prove that Materialism and Naturalism are true. These people are in denial all of their lives. Self-deception works, and it works every time. I KNOW because I used to be a Materialist, Naturalist, Nihilist, and Atheist.

How does saying that "Dark Matter and Dark Energy DO NOT EXIST" explain what they are and how they work? It doesn't! Materialism and Naturalism are Bad Science in that they cannot explain dark matter and dark energy – what they are and how they work. Materialism and Naturalism lack explanatory power when it comes to all the different invisible non-physical forces and fields that we scientists have discovered over the years. Psyche or Consciousness is an invisible non-physical force or field. So is dark energy, magnetism, and gravity. The Materialists and Naturalists can't explain Psyche or Consciousness in any way that makes logical sense. Instead, these people assure us that Dark Energy, Dark Matter, Psyche, Syntropy, and Consciousness DO NOT EXIST.

How does saying that "Psyche, Syntropy, Quantum Mechanics, and Action at a Distance DO NOT EXIST" explain what they are and how they work? It doesn't! Materialism and Naturalism are worthless because they can't explain the observations and experiences associated with Psyche, Syntropy, Quantum Mechanics, Quantum Tunneling, Phase-Shifting, the Quantum Zeno Effect, Collapsing Wave Functions, Supernatural Mechanisms, and Action at a Distance.

In contrast, once you choose to allow Psyche, Syntropy, and Quantum Mechanics in to play, you can literally explain everything that comes your way. The explanatory power of Quantum Mechanics and Syntropy is through the roof to infinity and beyond.

Mark My Words

—

Source Material

God Is in the Light: God is light, and in Him is no darkness at all.

https://www.amazon.com/dp/B07168S37N

Science 2.0: I Upgraded My Science.

https://www.amazon.com/dp/B0771K6WTX

References

Matter Is Organized Light or Organized Energy

$E = mc^2$.

It's ALL matter. It's ALL energy. It's ALL mass. It's ALL light.

Energy is light. Mass is matter. Energy is mass.

It's the same difference.

Mass or matter is organized energy or organized light.

Energy, or psyche, or light is disorganized matter in its most primal form.

These things progress from disorder or chaos to ever increasing orders of organization under the command and control of Someone Psyche or Someone Intelligent. The more organized it becomes, the more laws or restrictions that are put into place to keep it organized.

Anything that is obviously organized obviously has an Organizer who organized it.

Restrictions or Laws always have some sort of Law Giver and Law Enforcer. Laws are worthless unless they are enforced. The physical laws and physical constants would be worthless if they weren't enforced by the Law Giver. Order and organization come from Laws and Restrictions meticulously enforced. This is the way our physical universe works. Someone has to be enforcing the physical laws, or this universe would be nothing but chaos, and we wouldn't exist as physical beings because physical matter wouldn't exist. Entropic physical matter is organized energy.

Mass and energy are conserved according to the first law of thermodynamics. They have always existed, and they will always exist. They are interchangeable but indestructible. Conservation of mass and energy is a type of Syntropy. It's eternal and everlasting; but, energy can change FORM, or change from one phase of existence to a different state of existence.

Technically, it is the Energy or the Syntropy that is conserved, and NOT the physical state. The physical state comes and goes as needed. Entropy comes and goes as God sees fit. The state, the phase, or the FORM of the Energy or Syntropy changes; but, the Syntropy or Energy is conserved.

Based upon the Second Law of Thermodynamics and the Supremacy of Entropy Doctrine, the Materialists, Naturalists, Darwinists, Nihilists, Behaviorists, Determinists, Atheists, and Classical Physicists erroneously teach that only physical matter or entropy exists. These people deny the existence of everything else. According to these people physical matter and entropy are conserved. Entropy is death; and, these people teach that

354

death or entropy is conserved. They teach that death and entropy are eternal and everlasting. That's what they teach, is it not? These people literally teach that entropic physical matter is conserved and that it is all going to end in heat death.

They are wrong.

I know that God exists because physical matter and entropy convinced me that He exists. Entropic physical matter is comprised of different FORMS of Energy. The FORM is never conserved. The FORM is made, organized, or formed. Anything that is obviously made obviously has a Maker who made it. Physical matter and entropy were obviously made – relatively recently in fact. The existence of visible stars and galaxies is scientific proof that they were made relatively recently. They haven't had time to burn out yet. Anything that is obviously organized obviously had an Organizer who organized it. It is obvious from our visible universe and the existence of usable physical matter that raw energy was relatively recently organized into physical matter and entropy by the Gods.

Anything that is formed obviously has Someone Psyche or Someone Intelligent who formed it; otherwise, it would be nothing but chaos. Physical matter or entropy – organized energy or organized light – is Scientific Proof of God's Existence whether we realize it or not. The existence of visible galaxies and our sun is scientific proof that the physical matter within them was made by God relatively recently because entropy hasn't had time to catch up with them and burn them out yet. Physical matter or entropy was made which means that it can be disassembled back into raw energy whenever God chooses to do so. It is Energy, Syntropy, or Psyche that's being conserved, not the physical matter and not the entropy.

The science makes it obvious that God exists.

Determine what the Second Law of Thermodynamics was designed to falsify; and, you automatically discover what is eternal, syntropic, and being conserved. The Second Law of Thermodynamics **falsifies** or **contradicts** Syntropy, Psyche, Non-Local Consciousness, the Conservation of Psyche, God's Psyche, the Conservation of Energy, the First Law of Thermodynamics, Non-Locality, Immortality, Eternal Life, Intelligence, the Life Force, Intangible Non-Physical Forces and Fields, Supernatural Mechanisms, Action at a Distance, Telepathy, Telekinesis, Quantum Tunneling, Teleportation, Levitation, and the Perpetual Motion Machinery associated with Quantum Mechanics. In contrast, the verified and proven existence of these things falsifies the Second Law of Thermodynamics.

People can't see nor understand this concept because they don't want to. Due to their prior commitment to the Supremacy of Entropy and Physical Matter, they lose the ability to falsify the Second Law of Thermodynamics. They can't see the various different ways to falsify the Second Law of Thermodynamics because they don't want to. They don't want it to be true. Nobody is immune. Self-deception works, and it works every time.

[Due to] the deep impression that classical thermodynamics made upon me, [the Second Law of Thermodynamics] is the only physical theory of universal content which I am convinced will never be overthrown. — Albert Einstein

If your theory is found to be against the second law of thermodynamics, I can give you no hope; there is nothing for it but to collapse in deepest humiliation. — Arthur Eddington

These people can't see that the Second Law of Thermodynamics is limited and flawed because they don't want to see it. They don't want it to be true. These people go out of their way to prevent themselves from making these kinds of scientific discoveries. I KNOW because I used to be a Materialist, Naturalist, Nihilist, and Atheist. I've always been a scientist; but for over 55 years, I didn't find any weakness in the Second Law of Thermodynamics because I wasn't looking for it. You'll never find it unless you go looking for it.

Entropy is death. Eternal Life falsifies death or the Second Law of Thermodynamics. Syntropy of any kind falsifies the Second Law of Thermodynamics. The verified and proven existence of Syntropy or Psyche falsifies entropy or the Second Law of Thermodynamics. Psyche has been experienced and observed by Out-of-Body Travelers. It's real and truly exists. The eternal or syntropic nature of Psyche and Energy falsifies the Second Law of Thermodynamics. Psyche is the innate intelligence within all the different forms of energy which gives that raw energy the ability to understand, follow, and obey God's Laws and God's Commands. Syntropy is the conservation of energy or the conservation of psyche. Syntropy, Psyche, Eternal Life, the Life Force, Intelligence, and Energy falsify the Second Law of Thermodynamics. The First Law of Thermodynamics falsifies the Second Law of Thermodynamics. Somewhere in there we find the answer to life, the universe, and everything.

I'm not embarrassed for discovering that the Second Law of Thermodynamics is flawed, that entropy is NOT conserved, and that **Entropy Is Temporary**. I haven't lost hope because I successfully falsified the Second Law of Thermodynamics. I'm not embarrassed or humiliated for finding the truth. That's what scientists are supposed to do; and, I'm a scientist. In order to get at the truth, we have to adjust our scientific explanations in order to make them match with what has been experienced and observed. Psyche has been experienced and observed by Out-of-Body Travelers and Near-Death Experiencers. Our job as scientists is to figure out what it is and how it works.

My ultimate goal in life is to figure out what things are and how they work. That can't be done without taking the observations and experiences of other people seriously and treating their non-local observations or spiritual experiences as scientific evidence. How else is Psyche and/or Spirit Matter to be discovered and verified besides observing it and experiencing it for yourself or choosing to trust someone who has? Science is observation and experience after all. Claiming that Psyche or Spirit Matter DOES NOT EXIST does not explain what it is and how it works. Now does it? You can't explain what something is and how it works by denying its existence. It doesn't work.

The Physicalists, Materialists, Naturalists, Darwinists, Nihilists, Behaviorists, Determinists, Atheists, and Classical Physicists present us with an Entropic Model for Origins or a Death-Based Model for Origins. Evolution is entropy. Evolution is based exclusively on entropy or death. Death cannot design and create. The Theory of Evolution is Creation by Entropy or Creation by Death because these people teach that only entropy or death exists. These people teach that only entropic physical matter exists.

I'm proud to announce and to know that God the Father has presented us with a Syntropic Model for Origins – the Atonement of Jesus Christ, the Plan of Salvation, and Resurrection from Death. As I see it, a Syntropic Model for Origins is vastly superior to an Entropic Model for Origins; and, the Syntropic Model for Origins has the added benefit of actually being possible and true. There's no way in the universe that entropy or death is ever going to be caught in the act of producing Syntropy and Life. It can't. It's physically impossible. Entropy or death cannot produce genomes and life. Entropy or death isn't going to produce planets and stars either.

Remember, matter of any kind is organized energy. Matter is organized light. Organized by whom? Matter has Order or Laws placed upon it to restrict it, and limit it, and organize it. Brand new physical particles and entropy are made from energy in super colliders, atom smashers, or particle accelerators. Anything that is made is NOT conserved.

Anything that is obviously made obviously has a Maker who made it. Anything that was caused to begin has Someone Psyche or Someone Intelligent who caused it to begin. Physical matter and entropy are obviously made or caused to begin, which means that they have a Creator who designed them and made them. They are the way they are because they were made to be that way. Without a Maker or an Organizer, everything would be nothing but random chaos, and there would be no physical matter or entropy.

Energy is disorganized matter, or disorganized mass. Explode an atomic bomb and all of that physical matter and entropy seems to disappear into thin air and becomes invisible and intangible as its energy dissipates throughout the environment. The Energy is conserved but the FORM is not. Entropic physical matter is comprised of different FORMS of energy which means that they are NOT conserved. The FORM is never conserved. Only the underlying energy is conserved. Energy or Psyche cannot be made, and it cannot be destroyed. In this comparison, only physical matter and entropy can be made and destroyed. The FORM is made, which means that the FORM can be destroyed or disassembled back into raw energy. The FORM is never conserved. The FORM is temporary. This is the Ultimate Law of Thermodynamics. It explains everything that has ever been experienced and observed.

$E = mc^2$.

Matter is the physical side of the equation, and energy is the spiritual side or the quantum side of the equation. The speed-of-light or a physical law provides the conversion factor. It's really simple to understand once a person chooses to do so. Mass or matter is energy that has had physical laws and physical restrictions imposed upon it, including sub-light velocity restrictions, space, locality, entropy, and time. Entropy is an aging process or the passage of time. Entropy is like a timer or a clock. Stop-watches and clocks are made. At velocities faster than the speed-of-light, time stops and entropy or the passage of time ceases to exist. In other words, the entropy is not conserved! The underlying Energy is always conserved; but, the FORM of that energy is never conserved. The Gods can change the FORM of Energy anytime they decide to do so. The Gods can't make Energy, Psyche, or Syntropy; and, they can't destroy it either. But, they can change its FORM any time they want to. The FORM is never conserved. Physical matter and entropy are never conserved because entropic physical matter is made from different FORMS of energy; and, the FORM is never conserved. Entropy is temporary which means that entropic physical matter is also temporary. This is the Ultimate Law of Thermodynamics. It explains everything that has ever been experienced or observed.

The very existence of quantum waves, quantum fields, invisible non-physical forces and fields, radio waves, magnetic fields, gravity, dark energy, dark matter, and invisible energy falsifies the claims of Materialism, Naturalism, Darwinism, Nihilism, and Atheism which state that these things DO NOT EXIST. Syntropy of any kind falsifies the Second Law of Thermodynamics or the Supremacy of Entropy.

Quantum fields and matter particles are just two sides of the same coin. A particle is a knot in a quantum field. A particle of matter is just a knot in the energy or the light. A quantum field or a quantum wave is energy or light. Energy or psyche is syntropic which means that it is always conserved. It is eternal and everlasting. Syntropy is Eternal Life.

Dark energy and dark matter are just two sides of the same coin. They both are different types or different states of mass; and, they both are different types or different states of energy or light. Dark matter is spirit matter, or invisible non-physical matter. Dark Energy permeates the whole of space. Dark matter or spirit matter tends toward locality. Dark Energy is omnipresent. The Dark Energy is more quantum or syntropic than Spirit Matter or Dark Energy. Do you see how that works?

The Materialists, Naturalists, Darwinists, Nihilists, Behaviorists, Determinists, Physical Reductionists, Classical Physicists, and Atheists state that ONLY physical matter or entropy exists. Clearly, they are wrong.

There has to be some type of eternal and everlasting Syntropy behind all of this, or all of that subsequent physical matter and entropy wouldn't have been possible in the first place. This is logical common sense.

Entropy is a physical law. The Physical Laws were made. Physical Laws require some type of Law Giver. More importantly, physical laws require some type of Law Enforcer. Through entropy and the other Physical Laws, God is preventing the atoms within your physical body from quantum tunneling away from you at will. The physical laws or physical restrictions that God put into place greatly limit and restrict the quantum mechanical capabilities that are an innate part of matter and energy. Without God to enforce these physical laws, the atoms in your physical body, your car, your computer, your house, your spouse, and your dog would quantum tunnel away from you into different parts of this universe and you would cease to exist as a physical being. This is what the physical laws and physical constants are trying to teach us.

Scientific Observation: Physical Laws and Physical Constants require some type of Law Giver and Law Enforcer. They also require a Psyche intelligent enough to choose to follow those Laws. Laws are no good without a Law Follower. Laws or Restrictions bring order and organization; but, only if they are followed and obeyed.

Scientific Observation: Our physical consensus reality has Physical Laws that are restraining it, controlling it, restricting it, limiting it, ordering it, organizing it, and forcing it to be reliable, predictable, dependable, and trustworthy.

Scientific Conclusion: Our physical universe and the physical matter within that universe have some type of Law Giver and Law Enforcer who is making the physical matter within this physical universe obey the Physical Laws and Physical Constants of this universe. Psyche is the innate intelligence within all the different forms of energy which gives that energy the inherent ability to understand, follow, and obey God's Laws and God's Commands.

These scientific observations are in fact Scientific Proof of God's Existence. Our physical universe would be disorder and chaos without the Physical Laws and the Enforcement of those Physical Laws. Enforcement requires some kind of Enforcer. The physical laws would be absolutely worthless without the enforcement of those physical laws. It's the physical laws and the enforcement of those physical laws by God that's actually preventing the atoms within your physical body from quantum tunneling away from you at will. The physical laws were designed in large part to PREVENT quantum tunneling or teleportation from happening at the physical level. Fascinating, is it not?

Science is Observation and Experience

Science is all about observation and experience. Go with the experienced and the observed; and then, get rid of all the rest – the philosophical speculation and wishful thinking.

Remember, Psyche or Quantum Non-Local Consciousness has been experienced and observed. Psyche or Intelligence is a massless quantum or a massless elementary particle; and, Psyche looks like and acts like a Photon whenever it is seen or observed, both at the quantum spiritual level and at the macro physical level. At every level of existence, Psyche or Intelligence uses mass, energy, or matter in order to get things done. Photons and other massless quantum particles have been experienced and observed which means that different types of Psyches or Intelligences have also been experienced and observed. The massless elementary particles within fermions, leptons, quarks, electrons, and neutrinos are Controlling Psyches or Controlling Particles who command and control the mass, energy, forces, and fields that surround them. This is what has been experienced and observed.

Concentrating Energy

$E = mc^2$.

Do you have any idea how much energy has to be organized and concentrated in order to produce one atom of physical matter? Now imagine the amount of concentrated energy that would be required to produce the estimated 10^{78} to 10^{82} atoms in the known observable universe in their prime original state.

Energy doesn't just spontaneously concentrate into physical atoms; and, physical atoms don't just spontaneously congregate into planets, stars, and galaxies. We humans use huge particle accelerators or atom smashers that cost billions of dollars to produce single particles of physical matter from energy. So, what did God use to produce the 10^{82} atoms in the known observable universe? That was one huge atom smasher, was it not?

Physical atoms don't just spontaneously generate out of thin air ex nihilo. They were deliberately made! Organized Energy and the Physical Laws were obviously made!

Scientific Observation: Any type of Law that is conserved or preserved requires both a Law Maker and a Law Enforcer.

Scientific Observation: The various different physical constants and physical laws are obviously preserved and conserved universally.

Scientific Conclusion: Therefore, it is logical and rational to conclude that the physical laws and physical constants have some kind of Law Giver and Law Enforcer who is preserving them or conserving them universally where this physical universe is concerned.

Physical atoms don't just spontaneously generate out of thin air ex nihilo. They were deliberately made! Someone Psyche, Someone Intelligent, and Someone Powerful had to make the 10^{82} atoms in the known observable universe before we humans came along and started making a couple of them in our atom smashers or particle accelerators from time to time. That's one huge atom smasher is it not?

Scientific Observation: Anything that was obviously made obviously had a Maker who made it.

Scientific Observation: Every physical atom was obviously made. Physical matter is Organized Energy. Physical matter is Concentrated Energy.

Scientific Conclusion: Therefore, it is logical and rational to conclude that every physical atom has a Maker or an Organizer who made it, concentrated it, caused it to be made, and/or allowed it to be made.

The proven and verified existence of one particle of physical matter is Scientific Proof of God's Existence. The gathering of enough physical matter into one place so as to make a planet or a star is Scientific Proof of God's Existence whether we realize it or not.

Scientific Observation: Anything that was obviously made obviously had a Maker who made it.

Scientific Observation: Every physical atom, physical planet, physical star, physical galaxy, and physical universe was obviously made.

Scientific Conclusion: Therefore, it is logical and rational to conclude that everything physical has a Maker who made it or brought it into existence in the first place.

When it comes to origins, follow the Energy, or follow the Syntropy. It takes a huge amount of Organized Energy to produce one single particle of physical matter. Matter of any kind is concentrated Energy, or concentrated Syntropy. Physical atoms don't just spontaneously generate out of thin air. They are made! They have to be made, or they wouldn't exist.

The following YouTube video is based upon many of these concepts.

Scientific Proof that God Exists

https://www.youtube.com/watch?v=5b9PssoJfLg

https://scientific-proof-of-god.com/wp-content/uploads/2018/06/Scientific-Proof-that-God-Exists.zip

All of that concentrated energy, within the 10^{82} atoms in the known observable universe, is Scientific Proof of God's Necessity and therefore Scientific Proof of God's Existence especially given the fact that physical matter is NOT conserved! The quantity and type of physical matter changes over time! The Energy or the Syntropy is conserved; but, the physical matter or entropy is not.

Five million tons of physical matter are converted into energy every second within our sun. Quadrillions of tons of physical matter are being converted to energy all throughout this physical galaxy every second. So, imagine the amount of energy that the Gods had to concentrate in order to form the initial particles of physical matter in the first place. That's a lot of Power! There's no way that we human beings could produce such a thing. It would require a God. It would require complete control over Quantum Mechanics or Supernatural Mechanics, which we human beings don't have.

Another fascinating aspect of all of this is that the fusion of hydrogen atoms into helium atoms produces tons of energy every second; yet, particles of physical matter (helium) remain in existence after the fusion process has been completed. That reality demonstrates a high level of Law, Order, and Organization where Organized Energy or physical matter is concerned.

Scientific Observation: Laws require Law Makers and Law Enforcers. Organization requires an Organizer.

Scientific Observation: It's obvious that physical matter is highly organized energy. It is obviously that physical matter is somehow being forced to follow physical laws. Physical matter is highly organized and highly concentrated Energy.

Scientific Conclusion: Therefore, it is logical and rational to conclude that physical matter of any kind has an Organizer who organized it as well as a Law Giver and a Law Enforcer who is making that physical matter obey the different Laws that were used to make it in the first place.

The Organized Energy, the very existence of Physical Matter, the Syntropy behind it all, the Organization of Physical Matter into planets and stars, Physical Genomes, and the Physical Laws are Scientific Proof of God's Existence. Your genome is God's Signature. None of this would exist without the direct intervention of Someone Powerful, Someone Intelligent, and Someone Psyche. Organization of any kind requires an Organizer; otherwise, everything would be disorder and chaos, and we wouldn't exist.

Matter is Organized Energy. Logic and common-sense Science tell us that there has to be Someone Psyche, Someone Syntropic, Someone Intelligent, and Someone Powerful who organized all of that Energy into each particle of physical matter; or, that physical matter would not exist in the first place. Organized Energy requires an Organizer to organize it. Particles of physical matter don't just spontaneously generate out of thin air ex nihilo as the Materialists, Naturalists, and Atheists claim. Physical atoms were organized or made, which means that they require some kind of Organizer or Maker to organize them and make them.

I'm not the first person to have figured this out; and, I won't be the last.

Mark My Words

Learning and Understanding the Physical Laws

The important thing to KNOW is what has been experienced and observed. You absolutely MUST gain an understanding of basic Physics if you want to find Scientific Proof for God's Existence.

These YouTube Videos from Spike Psarris do an excellent job in that regard.

Science or Storytelling?

https://www.youtube.com/watch?v=gufYmnj0Gjw

https://www.youtube.com/watch?v=PwodpHGfI4k

What you Aren't Being Told about Astronomy

https://www.youtube.com/watch?v=CzyQbOQ0dv0

https://www.youtube.com/watch?v=E66409i-yn4

Creation Astronomy

https://www.youtube.com/watch?v=dPTcnZXN8WY

Entropy is confusing and initially counter-intuitive. In popular language, people talk about entropy increasing, or disorder increasing, or death increasing as one of the others increases.

The second law of thermodynamics states that the entropy of an isolated system never decreases. Such systems spontaneously evolve towards thermodynamic equilibrium, the state with maximum entropy. Non-isolated systems may lose entropy, provided their environment's entropy increases by at least that amount so that the total entropy increases. Entropy is a function of the state of the system, so the change in entropy of a system is determined by its initial and final states. In the idealization that a process is reversible, the entropy does not change, while irreversible processes always increase the total entropy.

https://en.wikipedia.org/wiki/Entropy

This definition for entropy has the entropy increasing and moving towards a state with maximum entropy, or maximum disorder, or maximum heat death.

However, in the following definition for entropy, they have the entropy moving towards zero as the physical object moves towards absolute zero or heat death. They seem to have the entropy decreasing as the disorder or heat-loss increases.

The third law of thermodynamics is sometimes stated as follows: "The entropy, of a perfect crystal of any pure substance, approaches zero as the temperature approaches absolute zero." At zero temperature the system must be in a state with the minimum thermal energy. The entropy of a system approaches a constant value as the temperature approaches zero. The constant value (not necessarily zero) is called the residual entropy of the system.

https://en.wikipedia.org/wiki/Laws_of_thermodynamics

In some descriptions and definitions for entropy, they have the entropy increasing towards maximum entropy and maximum disorder. In other definitions for entropy, they have the entropy decreasing towards zero as the physical system moves towards absolute zero. The net effect of entropy is the SAME in both definitions for entropy – the physical object moves towards disorder and heat death. However, the way that entropy is defined and the direction of its movement – increasing or decreasing – is different within these two conflicting definitions for entropy.

I don't know why these inconsistencies in the definitions for entropy exist – maybe it's to trick us and deceive us. But, one thing that I do KNOW for sure is that the Laws of Thermodynamics prove the existence of some kind of Law Giver or God. These physical laws would NOT exist without some type of Law Giver and Law Enforcer. Laws require some type of Law Giver, or they will never exist in the first place.

$E = mc2$. The law of conservation of energy: This states that energy can be neither created nor destroyed. However, energy can change forms, and energy can flow from one place to another. A particular consequence of the law of conservation of energy is that the total energy of an isolated system does not change.

https://en.wikipedia.org/wiki/Laws_of_thermodynamics

The first law of thermodynamics, $E = mc^2$, or "conservation of energy and mass" FALISIFIES creation ex nihilo and spontaneous generation, thereby providing proof of God's necessity. In other words, the energy that went into the transformation of this physical universe from spirit matter into physical matter had to come from someplace else besides this physical universe because there is NO such thing as creation ex nihilo or spontaneous generation.

Energy is Syntropy. It has no beginning and will have no end because it is conserved. Its quantity does NOT change over time.

ENERGY CAN CHANGE FORMS. Energy can be made into spirit matter; and, with an even greater infusion of energy or mass, spirit matter can be made into physical matter.

The first law of thermodynamics makes it obvious and clear to me that our physical universe is NOT an isolated system. Our physical universe is receiving infusions of Energy or infusions of Syntropy from time to time. Dark Energy has been called "The Stretching Force of this Universe". That invisible non-physical Stretching Force, or that Dark Energy, or that Syntropy has to be coming from someplace else besides this physical universe; and, the amount of Dark Energy is currently on the increase according to scientific observations. This observed reality demonstrates that our physical universe is NOT a closed system. There's Syntropy or Dark Energy coming into this physical universe from someplace else besides our physical universe.

Based upon the first law of thermodynamics, I have concluded that the invisible non-physical Dark Matter that has been inferred to exist is in fact Spirit Matter. Spirit Matter or Dark Matter is just a different FORM of energy or light. I take these Physical Laws seriously, and I see what falls out when I start to examine them in detail.

The second law of thermodynamics PREVENTS chemical evolution, macro-evolution, spontaneous generation, and abiogenesis from happening. In other words, the second law of thermodynamics PREVENTS the theory of evolution from becoming true. Entropy is death. Materialism, Naturalism, Darwinism, Nihilism, Atheism, Classical Physics, and their derivatives are based exclusively on entropy; and, these philosophies or religions emphatically state that Syntropy does not exist.

The theory of evolution is Creation by Entropy or Creation by Death. Since the second law of thermodynamics PREVENTS the theory of evolution from becoming true, God must of necessity exist in order to have done all of the Order, Organization, Law, Syntropy, and Science which the different types of evolution could never have done. Evolution, or entropy, or death cannot design and create. In fact, evolution (genetic change), random mutations, and natural selection didn't even exist until AFTER God designed, programmed, engineered, field-tested, fine-tuned, produced, manufactured, created, and deployed the proteins, genes, genomes, eyes, brains, and life forms in the first place. This is logical common sense.

Design, creation, organization, programming, engineering, and life require the intervention or the infusion of some type of Syntropy or Psyche or Intelligence in order to become actual and real. Since energy is conserved, and since our physical universe and physical matter are based exclusively on entropy, all of that Syntropy has to be coming from someplace else besides this physical universe.

This is logical common sense. This is what the Laws of Physics are trying to tell us and teach us.

Energy within Matter Is Conserved

The first law of thermodynamics is extremely powerful science.

E = mc2. The law of conservation of energy: This states that energy can be neither created nor destroyed. However, energy can change forms, and energy can flow from one place to another. A particular consequence of the law of conservation of energy is that the total energy of an isolated system does not change.

https://en.wikipedia.org/wiki/Laws_of_thermodynamics

Energy is conserved. Energy can change FORM, but it is always conserved. The Energy or Syntropy has always existed and it will always exist. The Energy or Syntropy can be transformed, but it cannot be created nor destroyed. This is what the first law of thermodynamics is trying to teach us.

Let's study the ramifications of this Axiom or LAW.

Energy has always existed, and it will always exist, which means that Energy or Syntropy does NOT require any type of Creator, Organizer, or Maker in order to come into existence. However, Energy or Syntropy can change FORM. In other words, energy can be made into spirit matter; and, with an additional infusion of energy or mass, spirit matter can be converted into physical matter. Energy or Syntropy has always existed and will always exist; but, spirit matter and physical matter can be FORMED or MADE from energy, which means that spirit matter and physical matter can begin, and they also theoretically can come to an end.

Remember, only physical matter is subject to entropy or the second law of thermodynamics because Energy or Syntropy is conserved according to the first law of thermodynamics. The amount of entropy and physical matter changes over time; but, the amount of Energy or Syntropy does not. Energy or Syntropy is timeless and eternal, without a beginning of days or an end of years.

Psyche, Intelligence, or Quantum Non-Local Consciousness is a type of Energy, Syntropy, or Light. It is living light or conscious light. Psyche is a type of Syntropy or a type of Energy, which means that it is eternal and everlasting. It has always existed, and it will always exist. Psyche has NO creator or maker who caused it to begin.

Scientific Observation: Anything that obviously began obviously had a Maker, or a Creator, or an Organizer who caused it to begin.

Scientific Observation: Our physical universe, physical matter, the physical laws, and entropy obviously began.

Scientific Conclusion: Therefore, our physical universe, physical matter, the physical laws, and entropy obviously have a Maker, Creator, or Organizer who caused them to begin.

Energy or Syntropy cannot be created nor destroyed. Energy has always existed and will always exist. Energy or Syntropy is the Primal Construct. Physical matter or entropy is the exception to the rule. There is NO conservation where entropy is concerned. Its quantity changes over time. Physical matter or entropy has a beginning, which means that it can also have an end. Entropy or the second law of thermodynamics applies only to physical matter. Everything else is Syntropy.

Physical matter really isn't created ex nihilo. Creation ex nihilo is impossible. Spontaneous generation is impossible. It can't be done which means that it wasn't done. The first law of thermodynamics, or conservation of energy and mass, FALSIFIES creation ex nihilo, spontaneous generation, or the theory of evolution. The physical matter and entropy were made from some type of Energy, Syntropy, Psyche, or Light. Physical matter is a FORM of energy or light that has been restricted, limited, ordered, and organized by entropy, locality, the passage of time, sub-light speed limits, and other physical laws. The Energy or Syntropy is eternal and everlasting; whereas, the physical limitations within physical matter have a beginning and can theoretically come to an end. These observed and experienced Realities are the answer to life, the universe, and everything.

Physical matter is a different form of energy or a different type of energy. The main difference between physical matter and the syntropic FORMS of energy is that physical matter has been localized into space and made subject to entropy or the passage of time. Physical matter has been made subject to the aging process or the passage of time, which means that it can take on entropy. Physical matter has been localized into space, which means that physical atoms can no longer quantum tunnel at will and actually need God's permission and intervention in order to do so. In summary, physical matter has been subjected to the restrictions of space-time; whereas, spirit matter and the other types of Syntropy have not.

So, the birth of physical matter from spirit matter really isn't any type of creation – it's a re-organization or a re-ordering. It's an infusion of energy or mass. It's a state change or a phase shift. Yes, there was a beginning – a beginning of physical matter and physical restrictions; but, physical matter was organized or made or constructed from something that already existed in the first place, namely all of that pre-existing spirit matter or dark matter.

Remember, only physical matter is subject to entropy or the second law of thermodynamics because Energy or Syntropy is conserved according to the first law of thermodynamics.

Psyche or Intelligence ACTS. Matter of any type REACTS to Psyche's commands.

Was there a time when spirit matter or dark matter was made by Someone Psyche? It's possible, according to the first law of thermodynamics. Energy can change FORM. Anything that phase-shifts or changes form has someone or something that causes it to change its dimension, phase, or form. This is the answer to life, the universe, and everything. It explains everything that we know about, doesn't it?

Physicalism and Naturalism Lack Explanatory Power

Over time, I eventually learned that Materialism, Naturalism, Atheism, and their derivatives contradict Observational Science. If the Observational Science is true, then Atheism is not. Picture it! If the Science is true, then Atheism is false. If the observational experiences of the human race are true, then Atheism, Materialism, Naturalism, and Darwinism have been falsified.

Materialism, Naturalism, Darwinism, Nihilism, and Atheism CANNOT EXPLAIN Syntropy nor the eternal and everlasting nature of Energy. All that these people can do is blindly assure us Syntropy, Psyche, Spirit Matter, and God DO NOT EXIST. These people are wrong. Science itself proves that they are wrong.

How wrong are they?

Our universe currently consists of 5% physical matter, 23% spirit matter or dark matter, and 72% dark energy or invisible light.

Overall, dark energy is thought to contribute **73 percent** of all the mass and energy in the universe. Another **23 percent** is dark matter, which leaves only **4 percent** of the universe composed of regular matter, such as stars, planets, and people. — Uncle Google

By claiming that ONLY physical matter exists, the Materialists, Naturalists, Darwinists, Nihilists, and Atheists are WRONG about 96% of this universe, which is in fact immaterial and non-physical in nature and origin. This physical universe was energy and spirit matter BEFORE a very small part of it was made into physical matter. This universe was Pure Syntropy or Pure Energy BEFORE parts of it were organized into physical matter, having entropy and locality forced upon it or infused into it by the Gods.

Remember, only physical matter is subject to entropy or the second law of thermodynamics because Energy or Syntropy is conserved according to the first law of thermodynamics.

The entropy ONLY exists in physical matter. Entropy is an add-on, meaning that physical matter is an add-on. Energy or Syntropy is the default state or the Primal Construct. This is really simple to understand once a person chooses to do so.

Physical matter is assigned a location in space, so that it can no longer quantum tunnel away from us at will. In other words, when spirit matter is converted into physical matter, space-time is forced upon that particle of spirit matter thereby converting it into a particle of physical matter. This is really simple to understand once a person chooses to do so.

According to $E = mc^2$, God can convert dark energy or invisible light into dark matter or spirit matter; and, that also of course means that God can create physical matter by infusing a lot of dark energy, or power, or energy, or light into spirit matter. The more energy or mass that you infuse or force into a particle of spirit matter, the more physical or massive it becomes.

Our universe is comprised of only 4.6% physical matter.

This scientific observation is telling us that God converted only 4.6% of our universe into physical matter; and, the rest of this universe God left as dark matter and dark energy. God left over 95% of our universe as spirit matter and energy. The Materialists and Naturalists are wrong about 95% of our universe.

This scientific observation is also telling us that there might in fact be an infinite number of different types of light or an infinite number of different ways that light can be organized or formed. Energy or light can be converted into matter or mass. Spirit matter, dark matter, or quantum matter can be converted into physical matter by infusing a huge amount of energy or mass into a single particle of spirit matter.

To make or create physical matter requires a massive infusion of energy or light. Physical matter is highly organized or highly structured. Physical matter has been ordered and severe limitations have been placed upon it. These limitations and restrictions, which have been placed upon physical matter by the Law Giver and the Law Enforcer, are what we call the Physical Constants and the Physical Laws. Physical matter has been ordered and organized. Laws provide order and organization by imposing restrictions and limitations on the energy or the mass. There has to be some type of Law Enforcer, or the LAWS would be meaningless and ineffective.

Energy is unorganized matter. Energy is disorganized mass. Energy is matter unorganized. Energy is unformed matter. In contrast, matter is organized energy. Matter is organized light. There are many different ways to say the same thing.

The more restrictions and laws that are placed upon Energy or Light, the more material and physical it becomes. The more ordered or restricted or organized that a particle becomes, the more physical it becomes because of all the additional restrictions, orders, or laws that are placed upon it. Law provides order and organization by providing restrictions and limitations. This is the way things really work in this physical universe.

Remember, only physical matter is subject to entropy or the second law of thermodynamics because Energy or Syntropy is conserved according to the first law of thermodynamics.

These observations prove that some type of Syntropy or Psychic Force must exist, or all of these different physical laws, physical constants, entropy, and physical matter would NOT exist. The Physical Laws could not exist without some type of Law Giver and Law Enforcer. Laws of any type are worthless without some type of enforcement. Where matter, laws, order, and organization are concerned, enforcement implies some type of Psychic Force or Psychic Enforcer who is doing all of that enforcement. Something besides the matter and the laws are making the matter obey the laws. This is logical common sense.

As the result of Physical Laws and Physical Constants, physical matter loses its ability to quantum tunnel without some kind of external help or intervention from God. The result is the Ultimate Consensus Reality, a physical reality. Physical matter is highly reliable, dependable, predictable, and controllable because of all the different limitations that have been placed upon it by God. The atoms within your physical body are not going to quantum tunnel away from you at will because of all the restrictions that God has placed upon them. In other words, God prevents the atoms in your physical body from quantum tunneling away from you, through the physical laws that He put into place and is forcing the physical atoms obey. God has forced "locality in space" or a short De Broglie Wavelength onto the physical atoms so that they can no longer quantum tunnel without His permission. This means that physical atoms cannot travel faster than the speed-of-light without God's permission.

This is really simple to understand once a person chooses to do so. The TRUTH is always simple to understand. It's the lies and deceptions that require a ton of explanation in order to get you to buy into them. For example, they created saltation and punctuated equilibrium in order to make the theory of evolution palatable; but, saltation and punctuated equilibrium have gone the way of the dinosaur as well.

Energy or Syntropy is conserved. It has always existed, and it will always exist.

Energy is light. Energy is Syntropy. There are many different types, phases, forms, and dimensions of light or energy. Matter is a type of energy or light. Matter is a form of energy or light.

Spirit matter is organized light. It has structure or order. Things can be made from spirit matter by Psyche, Intelligence, or Consciousness. Psyche ACTS. Matter REACTS. Spirit matter has NO physical limitations. Spirit matter doesn't experience entropy or the passage of time, meaning that it doesn't age. Spirit matter seems to be eternal and everlasting once it is formed from energy or light. Furthermore, spirit matter can be used to make things such as spirit bodies and spirit worlds.

Physical matter is spirit matter that has been organized and restricted even further by physical laws, locality, and entropy or the passage of time. Unlike spirit matter, physical matter is restricted to space-time. It ALL makes sense so long as you choose to allow ALL of the evidence into evidence.

Remember, only physical matter is subject to entropy or the second law of thermodynamics because Energy or Syntropy is conserved according to the first law of thermodynamics.

I'm trying to give you everything that I have found so far concerning Organization or Origins, especially the stuff that I know to be real and true. I got tired of all the falsehoods, deceptions, and lies that the Materialists, Naturalists, and Darwinists were trying to force upon me throughout the six decades of my mortal life. The non-existence of things cannot be experienced and observed. The parts of Materialism, Naturalism, Darwinism, Nihilism, Determinism, and Atheism – which claim that certain things do not exist – cannot be experienced and observed.

Science is observation and experience NOT philosophical speculation and wishful thinking. The first law of thermodynamics has been experienced and observed on BOTH sides of the veil – both on the spiritual side and the physical side. Energy or Syntropy is conserved, meaning that it has always existed and will always exist.

Entropy or the second law of thermodynamics ONLY applies to physical matter. Entropy is a function of time. Locality is a function of space. Physical matter has been made subject to space-time; whereas, everything else has not. In other words, physical matter has been made subject to locality and entropy. Entropy is an aging process that is based upon the passage of time.

Remember, when it comes to the first law of thermodynamics, it's the Energy or the Syntropy that is conserved and NOT the entropy, physical limitations, and physical matter. Physical matter and the associated entropy are MADE, which means that they had a beginning and can therefore theoretically come to an end when they are absorbed back into Syntropy. In contrast, it is the Energy or the Syntropy that is eternal and everlasting, without a beginning of days or an end of years.

Syntropy is the answer to life, the universe, and everything. In other words, the conservation of Energy or Syntropy – the first law of thermodynamics – is the answer to life, the universe, and everything. Energy or Syntropy is the Rule. Consequently, the entropy or physical matter is the exception to the rule rather than being the rule.

The Materialists, Naturalists, Darwinists, Nihilists, and Atheists assure us that Syntropy or Psyche does not exist. These people tell us that ONLY physical matter or entropy exists. They are wrong. The Science tells us that they are wrong. There's more to life than physical matter.

Mark My Words

Models for Origins

There are as many different models for origins or creation as there are people on this planet. We each have our own pet theories, and they can't all be right because they tend to contradict each other. I've found and developed a few Models for Origins that were helpful, intriguing, interesting, and even believable. I present different models for the origin of this physical universe in different books and essays that I have written. I haven't settled upon one or chosen one because my mind remains open to all possibilities. In this essay, I present my Quantum Mechanical Model for the Origin of Our Physical Universe.

I used to be a Materialist, Naturalist, Nihilist, and Atheist. I'm a scientist, among other things. I was a practitioner of Scientism. The scientific evidence, experiential evidence, and observational evidence finally convinced me that I was wrong. The evidence eventually convinced me that Materialism, Naturalism, and their derivatives are based exclusively on entropy, which means that Naturalism, Darwinism, and Physicalism are Creation by Entropy or Creation by Death. I now know for a FACT that Materialism, Naturalism, Darwinism, Nihilism, and Atheism are physically impossible because entropy or death cannot design and create. I no longer subscribe to the philosophical speculation, blind faith, and wishful thinking of the Materialists, Naturalists, and Atheists.

Nowadays, I have chosen to let science, observation, experience, and logical common sense reveal the truth to me instead.

Scientific Observation: Anything that obviously began obviously had Someone Intelligent or Someone Psyche who caused it to begin.

Scientific Observation: The physical matter in our physical universe obviously began.

Scientific Conclusion: Therefore, it is logical and rational to conclude that physical matter has Someone Psyche or Someone Intelligent who caused it to begin.

The idea that physical matter had someone who caused it to begin makes infinitely more logical sense to me nowadays than the Atheistic and Christian claim that physical matter and this physical universe sprang into existence from nothing. Science itself has proven that there's no such thing as creation ex nihilo or spontaneous generation. The repeated verification of Conservation of Energy, or the first law of thermodynamics, has successfully falsified creation ex nihilo and any version of the Big Bang Theory that claims that our universe came into existence from nothing.

This is good stuff. This is Good Science; but, only if you choose to take advantage of it by choosing to believe that it is real and true.

The Gods are scientists, not magicians. God can't do creation ex nihilo or spontaneous generation. The Gods do science, order, organization, and manufacturing instead. The Gods talk about our earth being formed. Energy can be made to change FORM. The Energy or Syntropy is eternal and everlasting – it is conserved – but energy can be made to take on different FORMS or STATES including psyche, spirit matter, and physical matter. This is the answer to life, the universe, and everything.

I have chosen to go with the science and the observations on this one, when it comes to Origins, rather than the philosophical speculation and wishful thinking of the Materialists, Naturalists, Darwinists, Nihilists, and Atheists. Even though I used to be a Materialist, Naturalist, Nihilist, and Atheist, the science and the observational evidence eventually convinced me that I was wrong. Since then, I have tried to upgrade my Science and make it better than it was before. I have redefined "science" from Materialism and Naturalism TO observation and experience. "Observation and Experience" is an infinitely better definition for Science than Materialism, Naturalism, Darwinism, Nihilism, and Atheism.

Said another way, some people define "science" as classical physics. I have redefined Science from classical physics TO Quantum Mechanics and Action at a Distance. I have also refined, clarified, and perfected the LAWS of Thermodynamics.

Remember, only physical matter is subject to entropy or the second law of thermodynamics because Energy or Syntropy is conserved according to the first law of thermodynamics. Conservation of Energy would be impossible if entropy or change applied to Energy or Syntropy. That's why entropy ONLY applies to physical matter. The first law of thermodynamics applies to Energy or Syntropy; whereas, the second law of thermodynamics applies to Physical Matter. Once this distinction is made, suddenly everything becomes obvious and clear. Syntropy is the answer to life, the universe, and everything.

I have also concluded based upon the available evidence that the Biblical God Jesus Christ was there during the organization of this physical universe from spirit matter and saw how it was really done. Therefore, His observational statements and experiences in relationship to the origin of the heavens and the earth now take precedence in my Science over the philosophical speculation and wishful thinking of the Materialists, Naturalists, Darwinists, Nihilists, Behaviorists, Determinists, Physical Reductionists, Classical Physicists, and Atheists who by their own admission weren't there and have no memory of how it was really done. I think the Biblical God Jesus Christ knows more about the subject than they do because He was there, saw how it was done, and even helped to get it done.

Let me share a small part of what I have found so far in relationship to this topic.

God Organized Everything Spiritually First

The Gods don't need physical technology and don't use physical technology because the Gods are syntropic and the Gods have direct access to and control over Quantum Mechanics. Quantum Mechanics or Spiritual Mechanics is vastly superior to Classical Physics. The Gods can quantum tunnel their syntropic physical bodies anywhere in the universe at will. The Gods can manipulate energy and form it into anything they want it to be, at the quantum level or the sub-atomic level, including the formation of physical matter and entropy from raw energy. The Gods are eternal and everlasting, meaning that the Gods are syntropic.

The Gods don't need technology to escape the death of our physical universe. They can simply create another one at will or recharge this one as they see fit. The Gods can raise physical bodies, physical planets, and physical stars from the dead. A resurrected star would be indistinguishable to us from an entropic mortal physical star. They would both function the same, except the resurrected star would be syntropic and the mortal star would be entropic. Syntropic physical matter is vastly superior to and a lot more versatile and durable than entropic physical matter. Syntropy means eternal and everlasting. Syntropy is immortality and Eternal Life. Energy is syntropic. It never runs out. Syntropic physical matter would never wear out. These scientific discoveries result from taking Psyche, Syntropy, Action at a Distance, Dark Matter or Spirit Matter, and Quantum Mechanics seriously and treating them as Science.

The Gods organized everything spiritually before organizing them or creating them physically. The Gods organized bodies, planets, stars, galaxies, and universes spiritually from spirit matter BEFORE organizing them further into physical bodies made of physical matter.

Spirit Organization

See also Creation; Man, Antemortal Existence of

- every plant of the field before it was in the earth, Gen. 2:5 (Abr. 5:5).
- you are children of the most High, Ps. 82:6 (Rom. 8:16).
- spirit shall return unto God who gave it, Eccl. 12:7.
- he that giveth … spirit to them that walk therein, Isa. 42:5.
- Ye are the sons of the living God, Hosea 1:10.
- formeth the spirit of man within him, Zech. 12:1.
- we are the offspring of God, Acts 17:29.
- in subjection unto the Father of spirits, Heb. 12:9.
- First spiritual, secondly temporal, which is the beginning of my work, D&C 29:32.
- temporal in the likeness of that which is spiritual, D&C 77:2.
- All spirit is matter, but it is more fine or pure, D&C 131:7.
- created all things … spiritually, before they were naturally, Moses 3:5.
- beheld the spirits that God had created, Moses 6:36.
- I made the world, and men before they were in the flesh, Moses 6:51.
- spirits … have no beginning; they existed before, Abr. 3:18.
- he stood among those that were spirits, Abr. 3:23.

See also Num. 16:22; Job 32:8.

A spirit or a ghost is a psyche and spirit body combination. Spirit is a combination of intelligence and spirit body. A psyche is something completely different than a spirit body. A soul has often been defined as a spirit and physical body combination. When you give up the ghost, your spirit leaves, and your physical body dies.

God organized everything spiritually before making it physical. This reality also applies to human beings.

The Antemortal Existence of Man

See also Council in Heaven; Foreordination; Man, a Spirit Child of Heavenly Father; Spirit Creation

- God of the spirits of all flesh, Num. 16:22 (27:16).
- all the sons of God shouted for joy, Job 38:7.
- the spirit shall return unto God who gave it, Eccl. 12:7.
- Before I formed thee in the belly I knew thee, Jer. 1:5.
- Lord … formeth the spirit of man within him, Zech. 12:1.
- who did sin, this man, or his parents, that he was born blind, John 9:2?

- poets have said, "For we are also his offspring", Acts 17:28.
- For whom he did foreknow, he also did predestinate, Rom. 8:29.
- chosen us in him before the foundation of the world, Eph. 1:4.
- subjection unto the Father of spirits, Heb. 12:9.
- angels which kept not their first estate, Jude 1:6.
- Michael and his angels fought against the dragon, Rev. 12:7.
- called and prepared from the foundation of the world, Alma 13:3.
- bringeth them back into the presence of the Lord, Hel. 14:17.
- third part of the hosts of heaven turned he away, D&C 29:36.
- seraphic hosts of heaven, before the world was made, D&C 38:1.
- man, according to his creation before the world, D&C 49:17.
- Man was also in the beginning with God, D&C 93:29.
- choice spirits who were reserved to come forth, D&C 138:53.
- before they were born ... received their first lessons, D&C 138:56.
- in heaven created I them; and there was not yet flesh upon the earth, Moses 3:5.
- he beheld the spirits that God had created, Moses 6:36.
- intelligences that were organized before the world was, Abr. 3:22.
- he stood among those that were spirits, Abr. 3:23.
- took his spirit ... and put it into him, Abr. 5:7.

See also Prov. 8:22-31; John 1:2, 14; 8:58; 16:28; 17:5, 24; 2 Tim. 1:9; Titus 1:2; 3 Ne. 1:13; 26:5; Ether 3:16.

How would you organize Intelligences or Psyches? How would you organize Quantum Non-Local Consciousness? How would you raise Psyches to a higher order or level of existence?

First, by placing Psyche into some type of spirit body; and second, by eventually placing that Spirit or Ghost into a physical body. God organized everything spiritually before organizing it physically.

A spirit or a ghost is a psyche and spirit body combination. Spirit is a combination of intelligence and spirit body. A psyche is something completely different than a spirit body. Psyche is a different FORM of light or energy. Energy, Psyche, Syntropy, or Light is conserved according to the first law of thermodynamics; whereas, the amount of physical matter or entropy changes over time according to the second law of thermodynamics.

A soul has often been defined as a spirit and physical body combination. Man became a living soul when God place his spirit into his physical body.

Man Is a Spirit Child of Heavenly Father

See also Man, Antemortal Existence of; Spirit Body; Worth of Souls

- God of the spirits of all flesh, Num. 16:22.
- Ye are the children of the Lord your God, Deut. 14:1.
- there is a spirit in man, Job 32:8.
- breath of the Almighty hath given me life, Job 33:4.
- Ye are gods … children of the most High, Ps. 82:6.
- the spirit shall return unto God who gave it, Eccl. 12:7.
- he that giveth breath … and spirit to them that walk, Isa. 42:5.
- Ye are the sons of the living God, Hosea 1:10.
- Have we not all one father, Mal. 2:10.
- Be ye therefore perfect, even as your Father, Matt. 5:48 (3 Ne. 12:48).
- Our Father which art in heaven, Matt. 6:9 (3 Ne. 13:9).
- we are the offspring of God, Acts 17:29.
- Spirit itself beareth witness … we are the children of God, Rom. 8:16.
- One God and Father of all, Eph. 4:6.
- be in subjection unto the Father of spirits, Heb. 12:9.
- he hath created his children, 1 Ne. 17:36.
- spirits … taken home to that God who gave them life, Alma 40:11.
- from God, for the benefit of the children of God, D&C 46:26.
- inhabitants thereof are begotten sons and daughters unto God, D&C 76:24.
- spirit of man in the likeness of his person, D&C 77:2.
- your Father, who is in heaven, knoweth, D&C 84:83.
- spirit and the body are the soul of man, D&C 88:15.
- I may testify unto your Father, and your God, D&C 88:75.
- man is spirit, D&C 93:33.
- God had created all the children of men, Moses 3:5.
- I made the world, and men before they were in the flesh, Moses 6:51.
- intelligences that were organized before the world, Abr. 3:22.

    See also Luke 23:46; Acts 7:59; Eph. 3:14–15; 5:1.

You are one of the Gods. Your surname is God. You, your spirit, helped to create or organize this physical universe and this physical earth in one way or another.

A spirit body is something completely different than an Intelligence or a Psyche. A spirit or a ghost is a spirit body and psyche combination. A soul is often defined as a physical body and ghost combination.

Thanks to the LAW of Quantum Superposition, a psyche, spirit body, and physical body can occupy the same space at the same time because they are out-of-phase with each other.

Do you see how that works?

By choosing to allow Psyche, Syntropy, and Quantum Mechanics in to play, you can literally explain everything that comes your way.

Spirit Body

See also Man, Antemortal Existence of; Man, a Spirit Child of Heavenly Father; Spirit Creation

- God of the spirits of all flesh, Num. 16:22 (27:16).
- let this child's soul come into him again, 1 Kgs. 17:21.
- there is a spirit in man, Job 32:8.
- spirit shall return unto God who gave it, Eccl. 12:7.
- spirit indeed is willing, but the flesh is weak, Matt. 26:41 (Mark 14:38).
- if a spirit or an angel hath spoken to him, Acts 23:9.
- glorify God in your body, and in your spirit, 1 Cor. 6:20.
- subjection unto the Father of spirits, Heb. 12:9 (D&C 129:3).
- body without the spirit is dead, James 2:26.
- preached unto the spirits in prison, 1 Pet. 3:19.
- form of a man ... it was the Spirit of the Lord, 1 Ne. 11:11.
- bodies and the spirits of men will be restored, 2 Ne. 9:12.
- my immortal spirit may join the choirs above, Mosiah 2:28.
- their spirits uniting with their bodies, Alma 11:45.
- which ye now behold, is the body of my spirit, Ether 3:16.
- until my spirit and body shall again reunite, Moro. 10:34.
- long absence of your spirits from your bodies, D&C 45:17 (138:50).
- spirits of men kept in prison, whom the Son visited, D&C 76:73.
- spirit of man in the likeness of his person, D&C 77:2.
- spirit and the body are the soul of man, D&C 88:15.
- celestial spirit shall receive the same body, D&C 88:28.
- spirits of men who are to be judged, D&C 88:100.

- spirit and element, inseparably connected, D&C 93:33.

- Every spirit of man was innocent in the beginning, D&C 93:38.

- spirits of just men made perfect … not resurrected, D&C 129:3.

- Holy Ghost … is a personage of Spirit, D&C 130:22.

- All spirit is matter, but it is more fine or pure, D&C 131:7.

- spirits … have no beginning; they existed before, Abr. 3:18.

- took his spirit (that is, the man's spirit), Abr. 5:7.

   See also Luke 1:47; 1 Cor. 2:10–11; 2 Cor. 4:16.

ALL of these things have been experienced and observed by Someone Psyche or Someone Intelligent. Science is observation and experience. The truth bears record of the truth. The false is falsified by the truth.

Once I decided to accept this scientific evidence or observational evidence as being real and true, then I chose to go into it all the way.

If you believe that the Bible is true, then you simply KNOW that the Book of Mormon: Another Testament of Jesus Christ, the Doctrine and Covenants, and the Pearl of Great Price are true because they ALL are saying the same thing or teaching the same principles. You simply KNOW that the Biblical God Jesus Christ is the source for all these different observations and ideas because they are consistent with each other and back each other up. The truth is always consistent with all the other the truths.

Adam was the son of God.

Luke 3: 38: **Which was the son of Enos, which was the son of Seth, which was the son of Adam, which was the son of God.**

When it comes to Adam, both his spirit body and his physical body were the son of God. His spirit body and his physical body were made by the Gods and/or descended from the Gods. You are one of the Gods. Your spirit body and physical body are descended from the Gods. Your surname is God.

All of this is possible because spirit matter and energy are conserved. Matter and energy are just different phases, states, dimensions, velocities, and/or frequencies of the same stuff. It all fits together like a well-designed machine.

The false is falsified by the truth; and, the truth is repeatedly experienced and observed. The first law of thermodynamics has been experienced and observed on both sides of the veil. Energy or Syntropy is conserved. Matter and energy are two different sides of the same coin. Energy or light tends to be on the spiritual side or quantum side of the equation; and, matter or mass tends to be on the physical side of the equation. This is what has been experienced and observed.

Since everything is comprised of energy or light, technically, there really is only one substance in this universe with that energy or light being in different states, types, categories, forms, phases, dimensions, velocities, and frequencies of existence. The different types of quantum waves or invisible light are non-physical energy; whereas, the different types of matter represent different states or different types of organized light or organized energy which has had restrictions and limitations placed upon it by Psyche or by

God. It's ALL energy. It's ALL matter. It's just different types of energy or light in different states or phases of existence or being.

The Human Psyche

The Human Psyche or Human Intelligence is also matter and energy. Matter and energy are interchangeable according to $E = mc^2$. The Energy or Syntropy is always conserved according to the first law of thermodynamics; but, the Energy or Syntropy can change FORM according to the second law of thermodynamics. Do you see how that works? It's the answer to life, the universe, and everything.

Psyche or Intelligence has been experienced by Out-of-Body Explorers as an immaterial viewpoint in space. Psyche or Intelligence has been seen by Out-of-Body Explorers and Near-Death Experiencers as a pinprick of light or a spark.

I think the Human Psyche is an infinite singularity in its energy form and a point particle in its "physical form".

The math keeps telling us that some type of infinite singularity does in fact exist. If there is such a thing as an infinite singularity, Psyche is it. If there is such a thing as a point particle or ultimate elementary particle, Psyche is it. This is a Psyche Ontology, wherein Psyche or Intelligence is the fundamental unit of reality.

It is Psyche or Consciousness who collapses the wave function thereby changing spirit matter into physical matter. It is Psyche or Consciousness who collapses the wave function thereby making our physical choices or physical behaviors actual and real. This is the Ultimate Model of Reality. Psyche ACTS. Matter or non-Psyche types of energy REACT. Matter or energy is acted upon by Psyche or Intelligence.

Energy is Syntropy. Energy or Syntropy is conserved. Something is syntropic if it is eternal and everlasting. Psyche or Intelligence is syntropic. It has always existed, and it will always exist. In contrast, entropy or physical matter or physical limitations began which means that they theoretically can come to an end by being absorbed back into Syntropy.

Intelligence or Psyche

See also God, Glory of; God, Intelligence of; God, Omniscience of; Knowledge; Light [*noun*]; Light of Christ; Man, Antemortal Existence of; Understanding; Wisdom

- intelligence cleaveth unto intelligence, D&C 88:40.
- Intelligence ... was not created or made, D&C 93:29.
- All truth is independent ... intelligence also, D&C 93:30.
- glory of God is intelligence, D&C 93:36.
- Whatever principle of intelligence we attain unto in this life, D&C 130:18.
- The Lord is more intelligent than they all, Abr. 3:19.
- I rule ... over all the intelligences thine eyes have seen, Abr. 3:21.
- shown unto me ... the intelligences that were organized, Abr. 3:22.

See also Isa. 55:8–9.

Intelligence or Psyche is something completely different than matter or organized energy. Intelligence or Psyche organizes matter into useful things. Psyche or Intelligence converts energy into matter.

- Gods organized and formed the heavens and the earth, Abr. 4:1.
- Gods went down to organize man, Abr. 4:27.

The Gods are scientists. The Gods do organization, NOT creation ex nihilo! Creation ex nihilo or spontaneous generation has been falsified by science. There's no such thing as spontaneous generation or creation ex nihilo.

The following YouTube videos from Spike Psarris do an excellent job falsifying spontaneous generation or creation ex nihilo.

Science or Storytelling?

https://www.youtube.com/watch?v=qufYmnj0Gjw

https://www.youtube.com/watch?v=PwodpHGfI4k

What you Aren't Being Told about Astronomy

https://www.youtube.com/watch?v=CzyQbOQ0dv0

https://www.youtube.com/watch?v=E66409i-yn4

Creation Astronomy

https://www.youtube.com/watch?v=dPTcnZXN8WY

These videos from Spike Psarris do a really good job falsifying Creation Ex Nihilo or spontaneous generation from nothing. Spike Psarris is like me in that he used to be an atheist, and it was the Science, scientific evidence, and science experiments that convinced us that God exists.

The Materialists, Naturalists, Darwinists, Nihilists, and Atheists told me that there is NO scientific proof for God's existence and that there will NEVER be any Scientific Proof of God's Existence; and, I believed them. Consequently, I never found any convincing proof of God's existence while I believed that there was NO such proof to be had. These people are right. If you choose to define "science" as Materialism, Naturalism, Darwinism, Nihilism, Scientism, Behaviorism, Determinism, Physical Reductionism, and Atheism, then there will NEVER be any scientific proof for God's existence. Materialism and Naturalism cannot be used to prove that God exists because these people assume on blind faith that God does NOT exist.

I was holding out for Scientific Proof of God's Existence before I was going to believe in God, firm in the belief that such proof would NEVER be forthcoming. And, I never found any convincing proof for God's existence until I actually started to study and understand the scientific evidence that proves that God exists. Do you see how we Atheists and Naturalists paint ourselves into a corner and prevent ourselves from finding evidence for God's existence? We convince ourselves that no such evidence exists; and as a result, we NEVER go looking for any such evidence to begin with. No seek, then no finding. Self-deception works, and it works every time.

Rather than defining "science" as Physicalism and Naturalism, if you choose to define Science as observation and experience instead, instantly you find yourself looking at an

377

infinite amount of Scientific Proof for God's Existence. Rather than labeling God as non-existent as the Materialists, Naturalists, and Atheists do, if you choose to define Science as observation and experience, suddenly you find yourself looking at an endless supply of observations and experiences that demonstrate and prove that the Biblical God Jesus Christ does in fact exist and that He did in fact rise from the dead.

We get what we seek and no more than that. Seek and ye shall find. Knock and it shall be opened unto you. Sow and ye shall reap. In contrast, stick your head in the sand, and you are going to get dust in your eyes. If you go looking for nothing, you will find nothing. It works! It's all dependent upon what you, your Human Psyche, chooses to do. Beliefs of any kind are chosen into existence by the Human Psyche on both sides of the veil; and, the truth is always found through observation and experience, not philosophical speculation and wishful thinking.

The truth is always in conformity with other truths. The false is falsified by the truth; and, the truth is constantly experienced and observed. Once I decided that I was in, I went in ALL the way.

If you believe the Bible to be true, then you simply KNOW that the Book of Mormon: Another Testament of Jesus Christ, the Doctrine and Covenants, and the Pearl of Great Price are true because they are saying the same thing in different ways. That's what the evidence is telling us. I'm trying to give you everything that I have found so far concerning Organization or Origins, especially the stuff that I know to be real and true. I got tired of all the falsehoods, deceptions, and lies that the Materialists, Naturalists, and Darwinists were trying to force upon me throughout the six decades of my mortal life. The non-existence of things cannot be experienced and observed.

My best friend experienced phase-shifting. I experienced Quantum Tunneling. Each one of us on this planet is given a piece of the puzzle. You have been given a piece of the puzzle that I don't have. Phase-Shifting and Quantum Tunneling have been experienced and observed, which means that these quantum phenomena are real and truly exist.

Action at a Distance and the Quantum Zeno Effect have also been experienced and observed. Science is observation and experience, or it should be. Quantum Mechanics, or Spiritual Mechanics, or Supernatural Mechanics is real and truly exists. WE KNOW because it has been experienced and observed.

I have observed that there is absolutely NO conflict whatsoever between Revealed Religion and Science as long as Science is defined as observation and experience. However, there is massive amounts of disagreement between Science and Religion if Science (or Religion) is defined as Materialism, Naturalism, Darwinism, Nihilism, Behaviorism, Determinism, Physical Reductionism, Classical Physics, and Atheism. Materialism, Naturalism, and their derivatives destroy everything they encounter including Science and Religion. They were designed to do so.

The man-made philosophies or religions such as Materialism, Naturalism, and Atheism will always conflict with Science, Observational Experiences, and Revealed Religion in one way or another because they are designed to do so. Materialism, Naturalism, and their derivatives were designed to conflict with and contradict our observations and our experiences. In contrast, observations and experiences falsify Materialism, Naturalism, Darwinism, Nihilism, and Atheism because these falsified philosophies incorrectly state that these various different observations and experiences do not exist. The non-existence of things cannot be experienced and observed. It takes a monumental blind leap of faith to believe that Materialism, Naturalism, Darwinism, Nihilism, and Atheism are true.

We can use the scientific methods to falsify our theories. We can even use science experiments to verify our theories; but, we cannot go into all dimensions of the multiverse and see that something does not exist. There's no way to experience and observe the non-existence of something.

The false is falsified by the truth; and, the truth is repeatedly experienced and observed. There is absolutely no conflict between the Revealed Religion which was repeatedly revealed to us by the Biblical God Jesus Christ and the Science that has been repeatedly revealed to us through all of our different scientific observations and experiences as a race, because they both are presenting to us the truth in one form or another. Truth is self-consistent; whereas, falsehoods are self-defeating. Materialism, Naturalism, Creation by Entropy, Nihilism, Atheism, and their derivatives are self-defeating because there's no way that they can possibly be true because there's no way that their hidden assumptions or major premises can be experienced and observed.

Remember, it's scientifically and logically impossible to experience and observe that something doesn't exist; and, that's the fatal flaw of Materialism, Naturalism, Nihilism, Determinism, Atheism, and their derivatives because they each claim that something does not exist. These philosophical ideas or falsified religions are fatally flawed because their major premises or hidden assumptions cannot be experienced and observed. If you choose to define Science as observation and experience, then Science itself falsifies Materialism, Naturalism, Nihilism, Determinism, Behaviorism, Classical Physics, and Atheism.

It's impossible to experience and observe that something doesn't exist. The best we can do as scientists is to verify our scientific theories with observations and experiences. Materialism, Naturalism, Nihilism, Determinism, and Atheism are automatically FALSIFIED due to a complete lack of observational evidence supporting their major premises or hidden assumptions which state that certain things do not exist.

If you successfully eliminate everything that is false such as Materialism, Naturalism, Darwinism, Atheism, Nihilism, and their derivatives then ONLY the truth will remain. This is Logic 101.

All you want is the truth because everything else is a lie.

Mark My Words

Phase-Shifting or Quantum Tunneling into the Physical Realm

When it comes to this physical universe, God organized everything spiritually before He made it physical.

Visualize this!

Visualize a solar system like ours that has been made from spirit matter. Then God speaks the Word of Command or Sings the Song of Creation; and, there comes a massive infusion of energy and law into every particle of spirit matter within that solar system. You can literally watch a world like our earth phase shift into a physical existence as the spirit matter particles within that world are all converted into physical matter simultaneously. It's a phase shift. That matter or that energy has been organized or limited even further. It has become physical. Physical planets, physical stars, and physical galaxies are quantum tunneled or phase-shifted into existence whole all at once by the Gods. This is what the

Quantum Model for Origins is trying to tell us and teach us. This is also what the scientific observations or astronomical observations are trying to tell us and teach us.

Galaxies come into existence whole all at once; and, then they begin to wind up under the pressure from physical laws. We observe whole galaxies, or we observe nothing at all. There are NO clouds of gas coalescing into galaxies. Such a thing has NEVER been experienced nor observed. Gas pressure makes the gas disperse into empty space. Because of gas pressure, the gas clouds can NEVER coalesce into planets, stars, and galaxies without some kind of external intervention from God.

Physical planets, physical stars, and physical galaxies are organized in situ from spirit matter by a massive infusion of energy or mass into each and every particle of spirit matter thereby converting that spirit matter into physical matter, and locating or phasing that newly made world into our physical 3D space-time reality, and making that world subject to entropy (the passage of time), and making that world subject to physical limitations such as sub-light speed limits – no more quantum tunneling at will, no more omnipresence, and a definite specific predictable location in space-time or physical reality.

All of this science has already been discovered, is already in place, and it just makes sense. It has been experienced and observed.

That newly created physical matter has been infused with a huge amount of energy or mass. The net effect of all that new mass is that the physical matter is slowed down to sub-light velocities, forced to become localized into our 3D space-time dimension or phase, thereby being made subject to entropy (the passage of time) and the theory of relativity (speed-limits or locality). That spirit matter has been phase shifted from non-locality (a spiritual or quantum dimension) to physical locality (a physical or localized 3D space-time dimension) thereby becoming physical matter instead.

A spirit world such our earth literally phases or phase-shifts into physical existence instantaneously in situ through a massive infusion of energy or mass into each and every particle of spirit matter within that world thereby slowing it down or giving it mass and thereby making it subject to entropy, physical limitations, and physical laws. It's as if the quantum mechanical or non-local is phased out or taken out of that spirit matter and replaced by classical physics, locality, physical laws, entropy, the passage of time, sub-light velocities, and physical limitations instead when that spirit matter is converted into physical matter by God's Word of Command.

We are erroneously taught by the Materialists, Naturalists, and Atheists that the physical matter has always existed, and that physical matter is the only thing that exists. They are wrong. Physical matter is made. Physical matter is highly organized or highly limited energy or light. Physical matter or entropy has a beginning which theoretically means that it can have an end when it is absorbed back into Syntropy.

Physical matter is concentrated light or frozen light according to physicist David Bohm. Physical matter is a limited and restricted FORM of Energy or Light. Physical matter has a beginning, which means that it can theoretically have an end when it is absorbed back into Syntropy or back into the Light. Spirit is light. Light is spirit. Spirit matter is just a different form of Energy or Syntropy. The very existence of energy, spirit, or light falsifies the claims of Materialism, Naturalism, Darwinism, Nihilism, and Atheism which state that spirit, energy, psyche, or light does not exist. These people are wrong at a very fundamental level. Their purpose is to convince us that God's Spirit, or God's Psyche, or God's Light, or God's Glory does not exist which is why these people are determined to be wrong. These people are motivated to be wrong and are therefore eager to promote and preach false doctrine, false dogma, false philosophies, false religions, and false ideas.

Energy, Syntropy, Psyche, Intelligence, or Light is the thing that has always existed and will always exist. The first law of thermodynamics tells us that the energy or the light has always existed and will always exist – it's called conservation of energy. It's the matter or the mass that gets organized or made into something else besides what it was before. Energy or light is unorganized matter. Matter of any type is organized energy or organized light.

Spirit matter is syntropic. It has NO physical limitations. The Energy or Syntropy within it has always existed, and it will always exist, according to the first law of thermodynamics. Energy or Syntropy is conserved. Matter is a different FORM of energy, and the energy is always conserved. It's the FORM, or the shape, or the dimension, or the phase of the Energy that changes and is NOT conserved according to the second law of thermodynamics. In is primal unorganized non-consensus state, spirit matter can quantum tunnel anywhere in the universe instantaneously at will, at least until the Gods organize it or limit it into spirit bodies, spirit planets, spirit stars, spirit galaxies, and spirit universes by converting it into a consensus reality or by organizing it.

The Gods create or organize everything spiritually before they bring it into existence physically or phase it into existence physically.

I've visualized this. I've seen it in my mind's eye, as a whole world like our earth phases into existence from a spiritual state into a physical state. You can just feel the massive infusion of energy, mass, order, organization, and stability as that thing converts or phases from a spiritual world into a physical world. It solidifies.

When our earth phased into physicality, it was dark and void. It was just raw physical matter. Then the Gods set to work organizing it further and beautifying it.

Who were these Gods?

You were one of them. We are the children of the Gods – both spiritually and physically. Adam was a son of God. That means that both his spirit body and his physical body descended from the Gods.

One of the reasons the Gods took billions of years to organize our earth or terraform our earth so that it would be capable of sustaining life is to give each one of us the opportunity to participate in the process and learn from the process. The Gods deliberately slowed down physical matter to sub-light speeds so that we could live it, experience it, manipulate it, control it, learn from it, and remember having done so. At the speed-of-light or faster, there is NO passage of time. There is NO entropy. The theory of relativity tells us that at the speed-of-light or faster TIME STOPS. This is what the science is trying to tell us.

The Energy or Syntropy is conserved; but, Energy can be changed into different FORMS over time. Syntropy is the answer to life, the universe, and everything. Energy or Syntropy is eternal and everlasting without a beginning of days or an end of years; however, the amount of entropy or physical matter changes over time according to the second law of thermodynamics.

Phase-Shifting Planets and Stars

Study the science fiction and the Blink Drive in something like the television series *Dark Matter*. They phase out of this physical reality, quantum tunnel to a different location in the universe, and then phase back into this physical reality. It is a Blink Drive or a Jump Drive that is based upon Phase-Shifting and Quantum Tunneling.

The reason I mention this is because both Phase-Shifting and Quantum Tunneling have been experienced and observed. By best friend experienced Phase-Shifting as an elk passed through his truck unphased; and, I experienced Quantum Tunneling when God teleported me and the car I was driving to safety into the middle of a lawn. I was traveling 40 mph and about to hit a car that had pulled out in front of me unexpectedly. There was nowhere to go, except into that car. I felt time stop. Everything froze. Then instantly I was in the middle of a lawn with the car stopped and the car turned off. I blinked, or quantum tunneled, just as I was approaching a freeway overpass. If I would have kept going at 40 mph in that new direction of travel, I would have gone down the embankment and ended up in the middle of the freeway. I was supernaturally calm. There hadn't been enough time for my adrenaline to start pumping. However, it was nothing that I did. It was something that God did for me in order to save me and a bunch of other people as well.

With this Quantum Model for Origins or Spiritual Model for Origins that I'm presenting here, instead of a ship blinking into a physical existence from a quantum existence or a spiritual existence, we have whole planets, stars, and galaxies blinking into physical existence or phasing into physical existence all at once.

The scientists or astrophysicists have observed that physical galaxies seem to come into existence all at once in situ, or in place, or in position; and, they do. Last I heard, our Milky Way Galaxy and the Andromeda Galaxy have been in existence long enough to have rotated once or twice, and that's it. The rate of galaxy rotation is in constant dispute due to the effects of dark matter. However, it has been estimated by some that it takes a billion years for a galaxy to rotate. That would mean that our galaxy and the Andromeda Galaxy are a couple of billion years old. It would also mean that our "official" radiometric dating estimates are wrong.

We look out into the universe, and we DO NOT observe huge clouds of gas. Instead, we observe whole galaxies. Science is observation and experience, NOT philosophical speculation and wishful thinking. Galaxies come into existence all at once, whole. Galaxies do NOT coalesce into existence from one huge gas cloud. They have NEVER been observed doing so. Galaxies come into existence or blink into existence whole. A barred galaxy is one of God's Signatures in the heavens. Such a thing could never coalesce out of a gas cloud. It's physically impossible.

Coalescing gas clouds and coalescing planets and stars from gas clouds are physically impossible. Gas clouds DO NOT coalesce. They expand away from themselves and dissipate into empty space. Coalescing of stars, planets, and galaxies from gas clouds cannot be done, which means that it wasn't done that way. Spike Psarris does a good job explaining this in the YouTube videos that I linked to above.

In contrast, quantum-tunneling and phase-shifting are the NORM. Quantum tunneling, or phase-shifting, or blinking into a physical existence from a former spiritual existence IS the way that planets, stars, and galaxies are made or brought into a physical existence. Blinking into physical existence from a former quantum existence is the way things seem to work in reality. That's literally what we see when we look out into our physical universe. Whole galaxies have blinked into existence all at once. They are just there, or they are not there at all. Galaxies are NOT half in and half out. Galaxies do NOT coalesce out of gas clouds. That's prevented from happening by gas pressure, entropy, and random diffusion.

Electrons are still half in and half out, as well – they are blinking in and out of our physical existence all the time. In contrast, the quarks, neutrons, and protons seem to be locked into our physical existence all of the time. The quarks have been infused with a huge amount of mass or energy localizing them into our physical reality; and, a physical atom as a whole has also been infused with a massive amount of space, locality, entropy, restrictions, order, and LAW which we call space-time. Entropy is a function of time. Entropy is the passage of time. There's a lot of space between the nucleus of an atom and its electron shells, which are being maintained in orbit by all of that energy or mass in the nucleus of the atom. It has been observed that if you infuse even more energy into a physical atom, the electrons will achieve an even higher orbit for a while. That's just the way that these things seem to work.

Temporarily remove the energy, mass, space-time, or entropy from a particle of physical matter; stop the passage of time; temporarily make it spirit matter; and you can then quantum tunnel or teleport that particle of physical matter anywhere in the universe instantaneously before the physical limitations are restored. By temporarily STOPPING THE PASSAGE OF TIME, the Gods could theoretically teleport or quantum tunnel whole planets at a time. Quantum Mechanics is based upon quantum tunneling and phase-shifting; and, these things have been experienced and observed.

The Theoretical Expansion Zone in the Center

Under this Quantum Model for Origins, there seems to be a construction zone or expansion zone within the center of a physical universe where new galaxies are periodically being phased into existence, while the furthest galaxies in our physical universe are eventually lost from view. Everything seems to be moving away from the center or expanding away from the center which gives us the impression that all of the newly organized galaxies are being infused into the center, or phased into the center, or born into the center of this ever-expanding universe. Since it takes so long for the light to reach us from the outskirts, it seems as if all of those distant galaxies are young and newborn from our central and relatively new perspective on the scene. In fact, the Gods seem to time each new planting or each new expansion so that as we look back into the history of our universe all of the galaxies that we can see seem to be young and just starting to wind up.

God phases or blinks the galaxies into existence towards the center of this physical universe so that we don't observe galaxies blinking into existence or phasing into existence in the distant outskirts of this physical universe. Meanwhile, God is continuously and constantly infusing Dark Energy or the Stretching Force of this universe into this universe so that our physical universe keeps expanding outward from its central construction zone.

Our physical universe is expanding. Our earth and some of the planets and moons within our solar system seem to be expanding too. There seems to be an endless supply of YouTube Videos documenting and demonstrating our expanding earth or growing earth.

https://www.youtube.com/results?search_query=Expanding+Earth

https://www.youtube.com/results?search_query=Growing+Earth

According to these YouTube Videos, our earth is alive and it's growing.

Most of the scientists are in agreement that our physical universe is stretching or expanding. The expanding earth is in dispute.

If our earth really is growing, then this would seem to imply that there is an infusion of energy taking place both in the center of our physical universe and also in the center of our earth. As time goes by, a phase shift takes place and that energy eventually becomes matter. With Quantum Mechanics, Science becomes possible, and the physically impossible becomes possible.

$E = mc^2$.

Remember, energy or light is the spiritual side or the quantum side of the equation – it's mostly invisible and non-physical; whereas, matter or mass is the physical side of the equation.

The verified existence of invisible non-physical forces, fields, quantum waves, light, psyche, gravity, magnetism, and dark energy FALSIFIES Materialism, Naturalism, Darwinism, Nihilism, Behaviorism, Determinism, Physical Reductionism, Classical Physics, and Atheism which state that these invisible non-physical forces and fields do not exist. The verified and proven existence of Quantum Mechanics or Action at a Distance FALSIFIES Materialism, Naturalism, Darwinism, and their derivatives which state that Action at a Distance is impossible and does not exist. The Materialists, Naturalists, Darwinists, and Atheists are demonstrably wrong.

Is our physical universe really 13.8 billion years old with light becoming visible 13.2 billion years ago; or, is 13.2 billion years the furthest back that our telescopes can currently see; or is it the fact that the dark energy permeating and infusing into our rapidly expanding universe has our universe expanding so fast that all the galaxies older than 13.2 billion years have been lost from view by now? There's really NO way to know for sure how old our physical universe really is. If parts or all of our physical universe is expanding faster than the speed-of-light, then our particular physical universe could be trillions of years old with the vast majority of what's out there having been lost from view by now. Did all of those galaxies that we cannot see die in a massive heat death; or, did God raise them from the dead with a new infusion of Syntropy?

There's no way to know for sure.

A newly resurrected syntropic galaxy would look just like a new-born entropic galaxy, so there would be no way to tell the difference between them from our perspective as long as they exist in the same phase or dimension of reality. The new-born entropic physical galaxy would be made subject to entropy or the passage of time; but, the resurrected syntropic galaxy would be without entropy, timeless, and eternal. Other than that, there would be no difference between them that we could observe.

Fascinating, is it not?

For all we know, some of the galaxies that we can see through our telescopes may have already been raised from the dead.

From what we can see, though, ALL of the evidence is telling us that galaxies blink into existence, or quantum tunnel into existence, or phase into a physical existence all at once whole. We either see complete galaxies, or we see nothing at all. We don't see huge galactic-sized coalescing clouds of gas. We don't detect stars between galaxies either.

Scientific Observation: Anything that was obviously made obviously has a Maker who made it.

Scientific Observation: Physical galaxies are obviously made.

Scientific Conclusion: Therefore, physical galaxies obviously have a Maker who made them.

Under this Quantum Model for Creation or Quantum Model for Origins – where planets, stars, and galaxies blink into existence or phase into existence whole – God's creation or God's construction project is ongoing within the center of this physical universe rather than having taken place all at once in some distant past. I feel strongly about the truthfulness of this Quantum Model for Origins because I saw it in vision or experienced in my mind's eye as if I were there and saw it happen where this earth is concerned. I felt it and saw it phase into this physical reality. It was dark and void. There was still a lot of construction and terraforming to be done.

Looking back at my life, I can now see that God designed my life so that I would eventually be led to discover and present this Quantum Model for Origins or Syntropic Model for Origins. This is one of my reasons for being. My hope is to set people free, as I have been set free. If I can prevent just one person from going through the misery, frustration, confusion, and hell that I went through during my life, then my mission or my purpose for living will have been achieved.

There's NO Limit to Philosophical Speculation and Guesswork

There are as many different scenarios for origins as there are people on this planet, or so it seems. Each person thinks that he knows how it was done. What's interesting about all of this is that it's literally possible to concoct an infinite number of different scenarios for origins all of which can be made to match with the observational evidence after the-fact. ONLY God knows which one of them is true or if they are all false.

We do KNOW, however, that the coalescing gas-cloud theory for origins is science fiction. Planets, Stars, and Galaxies don't coalesce from gas clouds because they can't. Gas pressure exceeds gravity. Gas clouds expand. They don't contract. That is what has been experienced and observed.

If the Big Bang Theory were 100% true, then our whole universe would be nothing but one huge homogenous ball of gas with every atom traveling away from every other atom. The fact that we actually have planets, starts, and galaxies is proof positive that there's something fundamentally wrong with Materialism, Naturalism, Classical Physics, and our Big Bang Model. In fact, there's observational evidence suggesting that our universe is expanding faster now than it was expanding four or five billion years ago. Our physical universe is growing again preparing for the next phase of its existence.

A growing earth and/or a growing physical universe would require a massive infusion of Dark Energy or Syntropy or Invisible Light from someplace else besides this physical universe. Under this Quantum Model for Origins, eventually some of that Spirit Matter or Dark Matter will be phased into this physical reality and become physical planets, physical stars, and physical galaxies.

Remember, stars, planets, and galaxies do NOT coalesce out of gas clouds. They can't. It's physically impossible. It's prevented from happening by gas pressure, entropy, and random diffusion.

Galaxies are born all at once whole. That's what has been observed! We observe whole galaxies while looking out into the universe, NOT galactic-sized coalescing clouds of gas.

Galaxies phase-shift, blink, or quantum tunnel into a physical existence from a former spiritual existence or quantum existence when God infuses a huge amount of energy or mass into each particle of spirit matter thereby converting it into physical matter. In that manner, a galaxy is born.

Quantum Tunneling and the intricately related Phase-Shifting Phenomenon have been experienced, observed, verified, and proven to exist. Quantum Tunneling and Quantum Mechanics are proven and verified science, even at the physical level. Quantum Mechanics is the way that science is done. Quantum Mechanics is also the way that the Gods do science. Quantum Mechanics is the way that the Gods make or produce physical planets, physical stars, physical galaxies, and physical universes. We KNOW this is so because random diffusion, gas pressure, and entropy prevent gas clouds from coalescing into stars, planets, and galaxies.

Energy or Syntropy is conserved; whereas, physical matter and entropy are "born" and begin to exist. Quantum Mechanics and the first and second laws of thermodynamics have been repeatedly experienced, observed, verified, and proven to exist. This Quantum Model for the Origin of Physical Universes is based upon Quantum Mechanics and the first and second laws of thermodynamics. It also takes Dark Energy and Dark Matter into account. This Syntropic Model for Origins fits with and explains the observational evidence just as well if not better than the other models for origins that I have encountered over the years. Good enough. It's a viable theory for origins. That's what I was going after in the first place. I wanted something that is based upon observation and experience. All I really want is the truth.

At some future time, newer physical galaxies will be phased into existence as the older ones disappear from view. Throughout it all will remain the cosmic background radiation that represents a residual leftover from all of that new energy that is being infused into the system each time a new galaxy is born or blinks into existence and begins radiating energy or light into the universe. That's why the cosmic background radiation looks lumpy. The galaxies were born one after the other rather than all at once. As our universe as a whole continues to expand under the influence of dark energy and an infusion of energy from the quantum realm or spirit realm, distant galaxies will disappear from view if any of that expansion is taking place faster than the speed-of-light. The cosmic background radiation could also be the remnant of our expanding universe as energy has been periodically infused into our universe from the Quantum Realm or the Syntropic Realm.

Wouldn't it be cool to witness a whole galaxy blink into existence or quantum tunnel into existence nearby? One day it's not there; and, the next day it is. Suddenly, there is a massive infusion of space, mass, energy, entropic potential, law, and time nearby as the spiritual or the quantum instantly becomes physical and is subjected to space-time, entropy, locality, and the physical laws. I think it would be cool.

Since each one of us is descended from the Gods and each one of us has a Psyche or Intelligence that is eternal and everlasting, chances are good that many of us if not all of us have actually witnessed physical earths, physical stars, and physical galaxies being phased into a physical existence all at once. That's the way things are done because it's physically impossible to coalesce a galaxy, planets, and stars from clouds of gas.

In the beginning of our physical galaxy, there was a blink or a phase-shift when our galaxy switched from being spiritual to being physical. In the beginning, our physical galaxy quantum tunneled or phased into our physical reality and began its physical existence. In the beginning of our physical galaxy, our physical galaxy was born.

Mark My Words

Explanatory Power

Science of any kind requires an intelligent Scientist in order to make it work and get it done; otherwise, this whole universe would be nothing more than random diffusion, Brownian motion, and one huge homogenous ball of gas expanding forever into the void. This is what has been experienced and observed.

Science is worthless unless it explains what's happening in the world around us. Science is worthless unless it explains what has been experienced and observed.

Quantum tunneling, phase-shifting, the quantum zeno effect, mind-over-matter, telepathy, consciousness, thought, and action at a distance have been experienced and observed. What's the scientific explanation for these things?

The Materialists, Naturalists, Darwinists, Nihilists, Behaviorists, Determinists, Physical Reductionists, Classical Physicists, and Atheists ASSURE us that the non-physical DOES NOT EXIST. What kind of scientific explanation is that? How does the non-existence of something explain how it works? The claim that invisible non-physical forces and fields DO NOT EXIST is worthless. It has NO explanatory power. It's Bad Science because it doesn't explain what's happening in the world around us. It's a cop-out and a dodge. It's a denial of reality – all so that they can say that God does not exist.

These people define "science" as Materialism, Naturalism, Darwinism, Nihilism, and Atheism. It's worthless because it lacks explanatory power, especially when it comes to all the different invisible non-physical forces and fields that we see and experience all around us every day. Materialism and Naturalism are a denial of reality. Something is said to be scientific if it matches with Reality. Physicalism, Naturalism, and their derivatives are UNSCIENTIFIC because they deny the Realties that we observe and experience everyday all around us.

How does denying the existence of the non-physical and the invisible explain magnetism, gravity, invisible light, radio waves, microwaves, x-rays, gamma rays, thought, consciousness, after-death memories, dark matter or spirit matter, dark energy, the zero-point field of light, the strong and weak nuclear forces, quantum waves, quantum phenomena, and action at a distance? It doesn't!

Materialism, Naturalism, Darwinism, Nihilism, Atheism, and their derivatives are absolutely worthless because they can't explain scientifically all the different invisible and non-physical forces and fields that we experience and observe all around us every day of our lives.

Remember, Science is worthless if it can't explain what's happening in the world all around us right now. Physicalism, Naturalism, Darwinism, and their derivatives are Bad Science. Materialism and Naturalism are a Bad Definition for science because they can't explain all the different invisible and non-physical forces and fields that are being experienced and observed right here, right now, in our everyday lives.

Mark My Words

—

Source

Quantum Mechanics from a Non-Physical Spiritual Perspective

NATURE vs. NURTURE vs. NIRVANA: An Introduction to Reality

I Am Not a Creationist: So What Am I?

Summary Of: The Theory of Evolution Proves that God Exists

The Second Comforter: Supping with Our Resurrected Lord Jesus Christ

Origin Science

We are going to have to rethink Science if we are going to make it match with what has been experienced and observed by Out-of-Body Travelers.

Materialism, Naturalism, Darwinism, Nihilism, Behaviorism, Determinism, Physical Reductionism, Classical Physics, and Atheism cannot explain origins nor the origin sciences in a way that makes logical sense without invoking dumb idols and magic such as spontaneous generation, creation ex nihilo, macro-evolution, chemical evolution, creation by random chance, collapsing clouds of hydrogen and helium gas, design and creation by natural selection, and creation by entropy – all of which are physically impossible and all of which have been falsified by observations, personal experiences, scientific experimentation, and scientific evidence. Materialism, Naturalism, Darwinism, Nihilism, and Atheism have been falsified by Science – they have been falsified by observations and the experiences of the human race as a whole. They have been falsified by the Scientific Method.

The only scientific explanations that make any logical and rational sense are those which are based upon transdimensional physics or quantum mechanics – the ones that involve intelligence or psyche, syntropy or the conservation of energy, transpersonal communication between atoms and molecules wirelessly at a distance, intelligent command-and-control from a distance or from the quantum realm, quantum blueprints and schematics and maps, an intelligent transdimensional genome maker, protein maker, map maker, and fine tuner, as well as someone intelligent or someone psyche who is handling the logistics for all of this wirelessly at a distance from the quantum realm or the syntropic realm. Quantum Mechanics or transdimensional physics is the only scientific explanation that makes logical sense when it comes time to explain what we are actually experiencing and observing in the physical world.

Quantum mechanics or transdimensional physics has been experienced and observed. Psyche has been experienced and observed. In contrast, spontaneous generation, creation from nothing, macro-evolution, chemical evolution, design and creation by natural selection, creation by entropy, and collapsing clouds of expanding hydrogen and helium gas have NEVER been experienced nor observed and NEVER will be because they are physically impossible. There isn't a physical explanation for the origin of life and this physical universe that makes logical and rational sense and that doesn't rely upon dumb idols or magic.

The Ultimate Law of Thermodynamics

Time and entropy were made by God, which means that God can change them, manipulate them, and even destroy them. Entropy is like a clock. It keeps track of the passage of time within physical matter or entropic matter. A clock is made, and a clock can be destroyed or come to an end. Entropy was made, which means that entropy can be disassembled and turned back into raw energy. Physical matter was made, which means that physical matter can be disassembled and turned back into raw energy. Entropy and physical matter are NOT conserved.

Physical matter, space-time, space or locality, and time or entropy were made by God, which means that God can change them, manipulate them, and even eliminate them.

It's the underlying Energy or Syntropy that are being conserved, not the physical matter nor the entropy. Physical matter, spirit matter, entropy, space, and time are simply different forms of energy. The form is never conserved. Physical matter and entropy are never conserved. They are temporary. The form is constantly being changed by God as God sees fit. This is the Ultimate Law of Thermodynamics. The Ultimate Law of Thermodynamics differentiates between what is being conserved and what is not conserved. Psyche, Syntropy, or Energy is conserved. Physical matter and entropy are NOT conserved.

It's the Energy, or the Syntropy, or the Psyche that cannot be made nor destroyed. God cannot make Energy, Syntropy, or Psyche, and God cannot destroy Energy, Syntropy, or Psyche. It's the Energy, or the Syntropy, or the Intelligence that's eternal and everlasting without a beginning of days or an end of years. It's the Energy, or the Syntropy, or the Psyche who is being conserved. However, God can transform raw energy or unorganized energy into anything He wants it to be, including physical matter, physical bodies, planets, stars, galaxies, spirit bodies, forces, fields, space, time, locality, and entropy. The form is never conserved. The form can change as God sees fit. It's the underlying Energy, Syntropy, or Psyche that's being conserved.

This is called the Ultimate Law of Thermodynamics. It explains everything that we human beings have ever encountered.

It's one eternal round.

Time or entropy, space or locality, physical constants, physical laws, physical matter, physical genomes, physical proteins, physical planets, physical stars, physical galaxies, and physical universes are made and disassembled by God as He sees fit. Only the underlying Psyche, Syntropy, or Energy is being conserved. The energy cannot be made, nor can it be destroyed. The energy or psyche has always existed and will always exist. It is syntropic, without a beginning of days or an end of years. It is eternal and everlasting. God's Psyche has always existed and will always exist. Your psyche or your intelligence has always existed and will always exist. Energy or psyche is syntropic which means that it is always being conserved; whereas, the form of that energy is never conserved. The form is constantly being changed by God as God sees fit.

This is called the Ultimate Law of Thermodynamics. It is the answer to life, the universe, and everything.

The Ultimate Law of Thermodynamics is hidden within the First Law of Thermodynamics which states that energy is conserved.

The first law of thermodynamics is a version of the law of conservation of energy, adapted for thermodynamic systems. The law of conservation of energy states that the total energy of an isolated system is constant; energy can be transformed from one form to another but can be neither created nor destroyed.

https://en.wikipedia.org/wiki/First_law_of_thermodynamics

Energy can neither be created nor destroyed. Psyche can neither be created nor destroyed. Psyche is the intelligence or the life force within energy that gives energy the ability to obey God's Laws and God's Commands. Psyche or Energy is conserved. The First Law of Thermodynamics tells us that the energy is constant, and that the energy is conserved – not the FORM of that energy. Energy can be transformed from one form to another, which means that the FORM is never conserved.

Transformed by whom?

Transformation of any kind requires Someone Psyche or Someone Intelligent to instigate, originate, initiate, and start that transformation. Any type of beginning requires Someone Psyche to cause it to begin; otherwise, random chaos would continue to reign supreme.

The Ultimate Law of Thermodynamics emphasizes that the FORM is never conserved, which means that the different FORMS of energy such as physical matter, entropy, space, locality, and time are never conserved. Whenever something is formed or made or organized, that means that its FORM is not being conserved. Only Psyche or Energy is syntropic or conserved. Energy or Psyche cannot be created, and it cannot be destroyed. However, energy can be transformed into many different things by the Gods or by Someone Psyche. The FORM is never conserved. Only the underlying energy is conserved.

Remember, the Gods can form energy into anything. The form is never conserved because the Gods can always form that energy into something else whenever they choose to do so. The Gods have complete control over everything at the quantum level or the psyche level, except for your choices. Your psyche or your energy is being conserved, which means that your choices and your ability to choose is being conserved. It's eternal and everlasting.

These realities and truths explain everything that we human beings have ever experienced or observed.

Mark My Words

An Introduction to Quantum Neuroscience

The physical brain is an enigma.

Why?

It's because there is no physical explanation for how the brain works or why it works. A brain shouldn't work according to classical physics and common-sense logic; yet, it does.

Don't believe me? Then examine the evidence! The evidence is clear.

Brain Functionality Is Physically Impossible

Each neuron in your brain is separated from every other neuron in your brain by gaps called synapses. Synapses are gaps, not connections.

The claim that your brain is wired together into logic gates, transistors, memory storage units (RAM), and computer processors (CPUs) is nothing but a myth and a lie. There are no wires within a physical brain. There's no RAM or CPU in a neuron either. The scientists erroneously call a chemical synapse a "connection". They are wrong. Chemical synapses are gaps, not connections. There's no way to send bytes of information through a gap at the physical level. Gaps prevent the transmission of information at the physical level. In short, there's no way for neurons to communicate with each other at the physical level.

The scientists claim that neurons are communicating with each other across synaptic gaps or synaptic clefts with neurotransmitters. Neurotransmitters are molecules – not wires, cables, CPUs, or random-access memory (RAM).

It's physically impossible to load bytes of data, information, memories, or programming code into a neurotransmitter, which means that it's physically impossible to transmit a message from one neuron to the next through neurotransmitters. It can't be done which means that it isn't being done.

Furthermore, neurotransmitters are released randomly into the synaptic gap or the synaptic cleft. Therefore, even if it were possible to store bytes of information within a neurotransmitter, any message stored within your neurotransmitters would be completely scrambled and randomized whenever your neurotransmitters are dumped randomly into the synaptic cleft. It's physically impossible to transmit a message consisting of bytes of organized information or bytes of programming code through a synaptic cleft. It can't be done which means that it isn't being done.

Finally, what are all those neurotransmitters doing to the post-synaptic neuron when they finally reach and dock with the post-synaptic neuron's dendrites?

They are changing the post-synaptic neuron's voltage level. They are simply changing the post-synaptic potential (PSP) up or down. That's it.

There can be as many as 15,000 synapses on a neuron, and all of that synaptic input gets reduced to one BIT of information – the voltage level of the neuron or the post-synaptic potential of the neuron. The PSP determines whether the neuron turns ON (fires) or stays OFF. A neuron is nothing more than a switch. A neuron is either ON or OFF. That's it. At the physical level, a neuron represents nothing more than one BIT of information – ON or OFF. It's physically impossible to transmit bytes of information, memories, data, and programming code through a single hardware BIT or a single switch that can only be ON or OFF. It's physically impossible to transmit bytes of information through neurons, which means that it's physically impossible to store bytes of information within neurons. Only one BIT of information is processed through a neuron – ON or OFF.

Sensory-input neurons are registers. They register physical input – ON or OFF. Motor neurons are switches – ON or OFF. That's it. There isn't even a single byte of memory storage capacity or computer processing power within a physical brain at the physical level. Every type of synapse simply determines whether the neuron turns ON or stays OFF. Neurons are simply ON or OFF; and, ON or OFF represents one BIT of information at the physical level. The neurons are not wired together into logic gates, transistors, bytes of information, CPUs, RAM, data, memory storage, or programming code. The synapses prevent that from happening.

The scientists are convinced that neurons are communicating with each other; but, the only way that they could be communicating with each other is wirelessly at a distance at the quantum level. Communication at a distance, action at a distance, or telepathy is a quantum mechanical process and not a physical process. In fact, the proven and verified existence of Action at a Distance is scientific proof that materialism, naturalism, and physicalism are false.

God made it physically impossible for the neurons to communicate with each other at the physical level; so, how are the neurons communicating with each other? There's only one logical answer that science provides to us. Brain functionality, its memory storage capacity, and all of its "computer processing" has to be taking place at the transdimensional level (quantum level) because it's physically impossible at the physical level.

Your brain is simply the physical interface – the registers and switches – between your physical body and the invisible intangible non-physical transdimensional computer (or quantum computer) that God's Psyche or Nature's Psyche has made for your Human Psyche to access and use.

Nature's Psyche (or God's Psyche) maps your physical brain at the quantum level and thereby determines what each neuron means and what each neuron does at the physical level. Then Nature's Psyche uses those quantum maps of physical functionality which it has made at the quantum level in order to get things done for us at the physical level. It's Nature's Psyche who collapses the wave function and turn's the neurons on. Nature's Psyche operates the quantum computer that it has made at the quantum level in order to control our brain at the physical level.

The quantum computer which Nature's Psyche has made and Nature's Psyche process all of the information that registers onto our sensory input neurons. The Human Psyche also has access to some of that information. Nature's Psyche determines the meaning and the purpose of each sensory input neuron and creates a quantum map or quantum register to explain it all, access it all, and use it all.

Nature's Psyche also maps the motor neurons and decides what each motor neuron triggers or activates. These maps are made at the quantum level by Nature's Psyche (or God's Psyche), and then used by Nature's Psyche at the quantum level to get things done for us at the physical level.

Energy and Psyche are synonymous. Whenever God interacts with raw energy directly, God is interacting directly with the psyche or the consciousness within that energy. Psyche or energy is always conserved, which means that it cannot be made, and it cannot be destroyed. It has always existed, and it will always exist. In contrast, entropic physical matter is made or formed from raw energy, which means that the physical matter or entropy is never conserved because entropic physical matter can be formed into something different anytime the Gods decided to do so. This is the Ultimate Law of Thermodynamics. It explains everything that we have ever been experienced or observed.

Whenever the Human Psyche chooses to do something with its physical body, then Nature's Psyche decides if it is physically possible or physically lawful; and then if possible and lawful, Nature's Psyche collapses the necessary wave functions and triggers the appropriate motor neurons in order to make the Human Psyche's requests a physical reality. It's Nature's Psyche who collapses the wave function and/or triggers a specific motor neuron. It isn't your genes. Your genes have no physical mechanism for doing so.

Out-of-body travelers and near-death experiencers have noticed that both the physical body and the spirit body have no experiences and make no memories while the Human Psyche is separated from them. It's the Human Psyche who is having experiences and forming memories, not the spirit body and not the physical body. The Human Psyche is experienced first-person as an immaterial viewpoint in space. Whenever the Human Psyche is separated from the spirit body looking at its spirit body, there's nothing to be seen whenever a person goes looking for his Psyche, Intelligence, or Consciousness. The Human Psyche is an immaterial viewpoint in space. The Psyche is seen or viewed as a pinprick of light, a point particle of light, or a spark of light.

At the physical level, the physical brain is the physical interface between the Human Psyche and its physical body. Nature's Psyche makes some kind of map or quantum computer at the quantum level which determines what each neuron, synapse, and neurotransmitter means at the physical level. It's Nature's Psyche who runs the physical brain at the quantum level because the Human Psyche isn't consciously aware of any of these quantum processes that Nature's Psyche uses to control our physical brain and trigger our motor neurons.

A trigger requires someone intelligent or someone psyche to pull the trigger, or nothing happens. Motor neurons are triggers or switches. Someone psyche has to pull the trigger or throw the switch; or, your physical body would never move.

The Human Psyche simply decides what it wants to do with its physical body; and then, Nature's Psyche collapses the necessary wave functions and/or turns ON the necessary neurons in order to get the job done. It's Nature's Psyche who maps purpose and meaning onto each and every neuron, synapse, and neurotransmitter operating within your physical brain. It's Nature's Psyche who decides how to construct your physical brain, where to place the synapses, what type of neurotransmitter will pass through each synaptic cleft, and what it all means. It's Nature's Psyche who creates a memory or a map of it all as it goes along. Nature's Psyche is keeping track of all of this at the quantum level so that we don't have to at the physical level. Besides, all of this is physically impossible at the physical level. Quantum mechanical processes and physical atoms cannot be commanded and controlled from the physical level by our proteins and our genes. Quantum mechanical processes and physical atoms cannot be commanded and controlled from the physical level by neurons and synaptic gaps either.

The only thing that can be done at the physical level is to turn a neuron ON or to keep it OFF. A neuron was designed and programmed to trigger an action potential automatically once the post-synaptic potential (PSP) reaches a certain voltage level. Not a lot can be done at the physical level. If a neurosurgeon stimulates or triggers a neuron at the physical level, either false sensory input is generated, or a motor neuron is triggered, and a part of your physical body moves. The purpose of each neuron was mapped into it by Nature's Psyche, not the neurosurgeon and not our genes. Neurons are registers and switches at the physical level. All of the processing, mapping, purposing, meaning, and memory storage for the physical brain takes place at the quantum level because it's physically impossible at the physical level.

A physical brain is solid physical proof that God's Psyche or Nature's Psyche must exist because there's no physical explanation for how a physical brain works because a physical brain can't do computer processing and memory storage at the physical level, which means that all of this has to be taking place at the transdimensional level instead.

The Brain Has Been Mapped

Clearly, the physical brain has been mapped at the quantum level to provide specific physical functions at the physical level.

Mapped by whom? Maps of any kind obviously require some kind of Map Maker. Maps don't just spontaneously generate out of thin air from nothing. That's impossible.

The Map Maker cannot be the Human Psyche because the Human Psyche isn't consciously aware of any of these quantum mechanical processes. The Map Maker cannot be our genes because our genes by definition in principle or practice cannot make choices and therefore cannot produce maps.

There's no physical mechanism for transferring sensory information into our genes; and, there's no physical mechanism within our genes for making choices and then transferring that information out of out of our genes to the neurons and the cells within our physical body and physical brain. Our genes are not controlling our physical brain. Our genes and proteins are not making and controlling the quantum maps, the quantum memory storage, and the quantum computers within our physical brain either.

Our genes code for proteins, not blueprints, schematics, and maps. Even if there were brain maps encoded within our genes, our genes have no physical mechanism for transferring those maps and those neural assignments into a physical brain. Our genes code for proteins, so if there are maps, blueprints, and schematics within our genes, then they exist at the quantum level because they are clearly non-existent at the physical level.

Your genome is nothing but a complex map for making proteins. Genes map for proteins, and nothing else.

Who made the map? Maps of any kind require some kind of Map Maker.

The most obvious map in the physical brain is the motor cortex. The motor cortex is the region of the cerebral cortex involved in the control and execution of voluntary movements. The motor cortex has obviously been mapped to provide control over specific physical movements.

Voluntary movements require some kind of choice which means that they require some kind of intelligent Chooser. The Human Psyche chooses what it wants to do with its assigned physical body and physical brain. Your genes don't control your physical body. They have no mechanism for doing so. Your genes can't make choices. Your genes simply do what they are told to do when they are told to do it.

Told by whom? Who commands and controls the transcription enzymes that read your genes? Who tells a transcription enzyme, a molecule, which gene to read? It isn't the Human Psyche because the Human Psyche isn't consciously aware of this process. It can't be your genes because your genes have no physical mechanism in place for commanding and controlling the transcription enzymes. Command and control of the physical brain has to be taking place at the quantum level or the transdimensional level because it's physically impossible at the physical level.

Your Psyche controls your physical body. Your Psyche chooses what it wants to do with its physical body. The Human Psyche does all of the command and control, all of the choosing and decision-making, where its physical body is concerned. Your genes simply code for proteins. Your genes are not involved in command and control.

Someone intelligent is controlling your enzymes, proteins, amino acids, and genes telling them where to go, how to act, what to do, and what to become when they get there. It isn't the Human Psyche and it isn't your genes, so it has to be Nature's Psyche or God's Psyche.

Who is the Map Maker for all of these different maps?

It has to be God's Psyche or Nature's Psyche. There is no other logical explanation because it definitely can't be our genes. Our genes code for proteins and not blueprints, schematics, and maps.

The synapses or synaptic gaps within your physical brain as well as synaptic typing were mapped by Someone Psyche or Someone Intelligent during synaptogenesis. Someone Psyche has to determine which neurotransmitters and neuropeptides to dump into the synaptic gap and to receive at the opposite side of the synaptic gap. The only way to map the meaning and the purpose of a physical gap existing at the physical level is from the quantum level or the transdimensional level. This is logical common-sense. There is no way to map meaning and purpose onto a physical gap at the physical level. It requires Action at a Distance which is only possible at the quantum level or the spiritual level. All of this mapping of physical functionality has to be taking place at the quantum level or the psyche level because it's physically impossible at the physical level.

Your eyes and vision were obviously mapped to your occipital lobe and the visual cortex. The neurons in your visual cortex register input from the neurons and sensors in your eyes.

Mapped by whom? Maps of any kind require an intelligent Map Maker.

Your ears and hearing were obviously mapped to your auditory cortex. The neurons in your auditory cortex register the input from the neurons and sensors in your ears.

The axons from olfactory receptor neurons located in your nose were obviously mapped and guided to the neuronal registers within your olfactory system.

Your skin, your sense of touch, and your taste buds were obviously mapped onto your somatosensory cortex by Nature's Psyche (or God's Psyche) in order to register physical input from your skin and your tongue.

Whenever your sensory input neurons are malfunctioning or firing randomly on their own, it results in hallucinations or false sensory input because those neurons were designed and mapped by Nature's Psyche to produce and register some kind of sensory input into the Human Psyche at the quantum level. Certain drugs produce malfunctions, delusions, and hallucinations. The physical machinery has to be working right in order for correct input to arrive at the Human Psyche and Nature's Psyche. That means that the physical machinery has to be mapped right, assigned correctly, functioning properly, and operated properly by Nature's Psyche in order to avoid delusions, hallucinations, and false sensory input.

Other parts of your brain were mapped for specific physical functionality by Nature's Psyche (or God's Psyche). Without all of this mapping at the quantum level, your physical brain wouldn't work right at the physical level. This is the only explanation for brain functionality that makes any logical sense. The idea that your genes are controlling it all at the physical level is nothing but science fiction. Genes, molecules, atoms, and proteins cannot do command and control at the physical level. They simply do what they are told to do by Nature's Psyche or God's Psyche.

The most obvious map in the physical brain is the motor cortex. The motor cortex is the region of the cerebral cortex involved in the control and execution of voluntary movements. The motor cortex has obviously been mapped to provide control over specific physical movements.

Mapped by whom?

Mapping of any kind requires some kind of Map Maker. Mapping requires intelligence, intention, observation, purpose, record-keeping, targeting, fine-tuning, planning, organization, manufacturing, and deliberation.

Who assigns that specific motor neuron to trigger the movement of your left index finger? In other words, who maps that specific motor neuron in your physical brain to turn ON the movement of your left index finger? Who makes all of these assignments or does all of this mapping? Who makes that specific neuron's axon grow directly to the muscles that are attached to your left index finger? In other words, who does the targeting for each and every axon within your physical brain? Someone Intelligent has to do the targeting. Targeting can't take place by random processes. Axon growth requires targeting. Synaptogenesis requires targeting. Targeting of any kind requires Someone Intelligent or Someone Psyche to do the targeting; otherwise, the brain would be nothing but random chaos and nothing would work right within the physical body.

So, who is doing the targeting for cell migration, axon growth, dendritic growth, and synaptogenesis? Who is doing the targeting for the transcription enzymes and telling each transcription enzyme which gene to read and where that gene is located?

It can't be the Human Psyche because the Human Psyche isn't consciously aware of any of these processes. It definitely can't be our genes. Our genes code for proteins, not mapping and targeting. Proteins are used for making cells not targeting axons, dendrites, and synapses through space at a distance.

Who decides where to place each and every synapse within your brain; and, who decides which neurotransmitter and which neuropeptide to send across the synaptic gap or synaptic cleft from the pre-synaptic neuron to the post-synaptic neuron? Who decides what each synapse means and what each neuron does? Who maps the physical functionality, purpose, typing, and meaning onto each and every neuron within your brain and every cell within your body? Who maps the physical functionality, purpose, and meaning onto each and every synapse within your brain? Who makes your physical brain and makes your physical brain work? Who is making all of these decisions and choices? Who is operating your physical brain? Who is collapsing the wave functions and thereby turning ON specific motor neurons within your physical brain?

It can't be the Human Psyche because the Human Psyche isn't consciously aware of any of these processes. It definitely can't be your genes because molecules such as proteins and genes cannot make decisions and choices. Molecules, proteins, and genes simply do what they are told.

Told by whom? Who can command and control atoms, molecules, amino acids, enzymes, proteins, and genes? If I gave you the assignment to command and control physical atoms, how would you go about that task?

Well, you are not going to be doing it at the physical level. You are going to have to do it at the quantum level or the psyche level. Action at a distance, or telepathy and telekinesis, is a transdimensional process and not a physical process.

Maps require some kind of intelligent Map Maker.

Decisions and choices require some kind of intelligent Psyche. Selection or choice is a function of Intelligence or Psyche. By definition in principle, the atoms and the molecules are incapable of making choices. Physical matter cannot do selection or choice. Instead, physical matter simply does what it is told to do.

Told by whom? Who or what would have the capability of telling atoms, molecules, proteins, enzymes, and genes where to go and what to do when they get there?

Who is doing all of this targeting for the atoms, molecules, proteins, enzymes, amino acids, axons, and synapses within your physical cells? Targeting requires intelligence, planning, fine-tuning, and foresight. Fine-tuning of any kind requires an intelligent Fine-Tuner. According to the materialists, naturalists, and physicalists, there is no Intelligence or Psyche within a gene or any other molecule or atom. According to these people, Psyche or Syntropy does not exist. Obviously, they are wrong. Someone intelligent is doing the targeting and the synaptogenesis for each and every neuron, axon, dendrite, and synapse within your physical brain; otherwise, the whole thing would be nothing but chaos. Someone is obviously constructing the physical maps within your physical brain, operating them, and making them work right when it is done.

It can't be the Human Psyche; and, it definitely is not your genes. Genes cannot do quantum mechanics or transdimensional physics. Only someone psyche can. Genes are physical molecules which means that genes cannot make decisions and choices at the quantum level or the psyche level; and, genes have no physical mechanism for doing so at the physical level either.

The only logical answer for all of these questions is Nature's Psyche or God's Psyche. Only someone psyche or someone intelligent could tell an atom where to go and what to do when it gets there. Only someone intelligent or someone psyche could tell transcription enzymes, messenger RNA, transfer RNA, amino acids, codons, and proteins where to go and what to do when they get there. Only someone psyche could tell cells where to migrate, axons and dendrites where to grow, and neurons where to place their synapses. Action at a distance, or telepathy and telekinesis, can only be done by someone psyche because there is no physical mechanism at the physical level for producing this type of functionality. This is what we have experienced and observed.

Synaptic gaps and neural mapping require action at a distance, telepathy, psyche, telekinesis, quantum mechanics, psyche, and intelligence in order to transmit information between neurons at the quantum level because synaptic gaps make it physically impossible to transfer bytes of information between neurons at the physical level. It's also physically impossible to store bytes of information within a neuron at the physical level because a neuron is one single hardware BIT which is either ON or OFF. The only information stored within a neuron at the physical level is ON or OFF.

Neurons have to be mapped at the quantum level by Nature's Psyche or God's Psyche because the synaptic gaps prevent the neurons from being mapped and handled at the physical level. Transcription enzyme targeting, protein synthesis, axon growth, axon targeting, dendritic growth and targeting, synaptogenesis, and cell migration targeting require intelligence and action at a distance which are transdimensional processes because there is no physical mechanism that can do action at a distance or targeting with atoms and molecules at the physical level. By definition in principle, atoms, molecules, enzymes, proteins, and genes cannot do selection or choice. They have no physical mechanism for doing so. They simply do what they are told to do by Nature's Psyche or God's Psyche. Psyche commands, controls, and makes choices. Physical matter simply reacts and does what it is told to do by someone psyche.

This is the way things really work when it comes to our physical brain.

Quantum Neuroscience

Transdimensional Neuroscience (quantum neuroscience) is the scientific study of how the Human Psyche interfaces with and controls its physical brain through Quantum Telepathy, Quantum Tunneling, and Quantum Telekinesis.

Since there are no wires, CPUs, and RAM within our neurons and within our physical brain, we must look to the transdimensional realm and transdimensional physics for the source of our brain's computer-like functionality. Out-of-body travelers, while separated from their spirit body and looking at their spirit body, have observed that it is their immaterial intangible psyche or spark who is having experiences and forming memories of those experiences – not their spirit body and not their physical body. The brain's quantum functionality is not to be found within the physical matter and our genes, which means that we need a better explanation for brain functionality than genes and physical matter. Psyche and Transdimensional Physics (Quantum Mechanics) are the best explanation when it comes to a physical brain's computer-like functionality.

Quantum Mechanics and Quantum Neuroscience are really easy to understand because they are based exclusively upon observational evidence, experiential evidence, empirical evidence, experimental evidence, and scientific evidence. The quantum or the transdimensional has been experienced and observed. Quantum Mechanics or Spiritual Mechanics is a proven science. It has been verified and proven to be real through scientific experimentation and practical application.

However, you can spend dozens of years explaining how Quantum Mechanics works with all the different physical processes, and then providing evidentiary and logical Proof of Concept, by trying to study and demonstrate how Quantum Mechanics applies to all of the different ions, molecules, nanomachinery, synapses, neurons, glial cells, cerebral fluid, and neurotransmitters within the physical brain.

Obviously, Queen Mary was interested in all of this and tried to take it to the next level; but, I won't go into any of that here and now. I simply wanted to demonstrate that there is no logical explanation for how a physical brain works at the physical level which means that we must turn to the transdimensional level or the quantum level in order to find a logical explanation for how it's being done.

The physical brain is the physical manifestation of a quantum mechanical machine or mechanism. Your physical brain is an interface between your Psyche and your physical body. Your Psyche controls your physical body through your physical brain. That's the way things were designed and that's the way things work.

Different Parts of Your Brain Are Specialized, Coded, and Mapped

Certain parts of the brain are specialized for coding particular types of spatial information; and, this coding or specialization occurs in both hemispheres.

https://books.google.com/books?id=OUgnEGbqFeYC&pg=PA279#v=onepage&q&f=false

https://quantum-neuroscience.com/wp-content/uploads/2018/08/Your-Brain-Was-Specialized-and-Coded-and-Mapped.png

Coded or MAPPED by whom? Specialized by whom? Who codes or maps a physical brain? Who assigns neurons and synapses to specific functions? Who handles the logistics and typing for brain cell migration, brain cell differentiation, and synaptogenesis? Who is doing the targeting and target acquisition? Who really makes and runs your physical brain? It can't be evolution or natural selection because evolution doesn't have any physical mechanism for coding, typing, differentiating, or mapping a physical brain. It can't be our genes because our genes code for proteins and NOT brain maps or synapses.

Causes are necessary for events to occur. Who is the Ultimate Cause behind the different events associated with brain development and brain functionality? It can't be evolution and it can't be our genes, so who is it? Who can make Quantum Maps of Physical Functionality at the quantum level and then use those maps at the quantum level in order to get things done for us at the physical level?

Code of any kind requires an intelligent Programmer. Specialization of any kind requires an intelligent Specialist. Mapping of any kind requires an intelligent Mapmaker. Operating machinery like a physical brain requires an intelligent Operator. There is NO intelligence within evolution or natural selection. A rock has more intelligence than evolution or natural selection. Anything that is obviously made obviously has a Maker who made it. A physical brain was obviously made. It didn't burst into existence from nothing. Each physical brain was obviously made and obviously mapped, which means that each physical brain obviously has a Maker of some kind who made it and then mapped it. A synapse is a gap-in-space. Who can map purpose and meaning onto a synapse or a gap-in-space at the quantum level? Certainly NOT our genes; and, definitely NOT evolution and natural selection. By definition, in principle, there is NO intelligence within evolution and natural selection.

Intelligence of any kind requires Someone Psyche; and, evolution has NO soul and NO mind.

Mark My Words

Evolution Made Your Brain

https://evolution-is-entropy.com/evolution-made-your-brain/

One science article after another states, "Over millions of years of time under the unrelenting pressure of natural selection on your genes, evolution made your brain."

https://www.google.com/search?q=%22evolution+made+your+brain%22

This is science fiction. Natural selection doesn't do anything. Natural selection doesn't touch our genes! Natural selection is an immaterial, non-physical, intangible, incorporeal, ethereal, insubstantial thought or idea. Natural selection isn't an entity or a person. Natural selection doesn't have hands or a mind. The Naturalists, Darwinists, and Atheists talk about natural selection as if it were a God; but, natural selection doesn't change nor touch our genes. Natural selection has NO physical mechanism for doing so.

In almost every science textbook that I own, the authors state that evolution made your brain. They never explain how. They simply say that evolution did it. It's kind of like saying that God did it. Of course, if you try to pursue the details, you eventually get down to the claim that random chance or blind luck produced your brain. Evolution is synonymous with blind luck and random chance.

Random chance doesn't explain anything. Now does it? Random chance doesn't explain how life was made and how life originated. There's no explanatory power in random chance. Random chance is worthless as a scientific explanation for Origins because random chance is not replicable, reliable, dependable, demonstrable, nor scientific. We can't rely on random chance or blind luck to produce anything for us.

Materialism, Physicalism, Naturalism, Darwinism, Behaviorism, Atheism, Genetic Mutations, the Theory of Evolution, and even Natural Selection are based exclusively on blind luck or random chance. Random chance is nothing but entropy. Entropy can't design and create.

Random chance is worthless as a scientific explanation for Origins because random chance is not replicable, reliable, dependable, nor scientific. Random chance is unscientific. It's physically impossible for functional proteins and their matching genes to develop from atoms by random chance or blind luck. It can't happen which means that it didn't happen. It's prevented from happening by entropy or the second law of thermodynamics.

Spontaneous generation or Creation by Luck was falsified in 1859 by Louis Pasteur – the same year that Charles Darwin published "On the Origin of Species by Means of *Random Chance*". We've known since the very beginning of the Theory of Evolution that it is false; but, most of our scientists have chosen to ignore, dismiss, and eliminate the scientific evidence that falsifies the Theory of Evolution.

Proteins and their matching genes were made which means that they have some kind of Maker who made them.

Many scientists on this planet claim that evolution or natural selection made your brain. That's physically impossible. Transdimensional Physics or Quantum Mechanics doesn't make "evolution by random chance" possible either. By definition in principle, there is no psyche or soul within evolution or natural selection. Your genes and proteins cannot do selection or choice at the physical level. Genes and proteins simply do what they are told to do by Nature's Psyche or God's Psyche.

Quantum Mechanics is Transdimensional Physics or Non-Local Mechanisms. Transdimensional means "not located in our physical reality". Transdimensional is synonymous with Non-Locality or Non-Physicality. Transdimensional is also synonymous with the word "Quantum". Transdimensional Physics or Quantum Mechanics is the study of how spirit matter and psyche function and work in the Transdimensional Realm, Non-Local Realm, Quantum Realm, or Spirit World in order to get things done for us here in the physical realm.

As their major premise, primary assumption, and fundamental axiom, the Materialists, Naturalists, and Atheists state that the Transdimensional, or the Non-Local, or the Non-Physical, or the Quantum, or the Spiritual does NOT exist. By axiom and fiat, these people define the Transdimensional or the Non-Local out of existence.

The Materialists and Naturalists are wrong, and they are proven wrong through observational evidence, experiential evidence, empirical evidence, scientific experimentation, spiritual experiences, and first-hand eye-witness phenomena or events known as Lived Experiences. Through Lived Experiences, a Human Psyche can go directly to KNOWING the truth, which is a lot faster than trying to use the Scientific Methods and Science Experiments to find the truth through a process of elimination or through a process of trial and error.

Remember, the Darwinists, Evolutionists, Naturalists, Materialists, Atheists, and Abiogenesists literally want you to believe that genomes, proteins, ribosomes, mitochondria, living cells, eyes, synapses, neurotransmitter systems, physical brains, and the complex designs and programming associated with each of these can spontaneously generate out of thin air from nothing; but, such a thing is physically impossible. Abiogenesis, Spontaneous Generation, or Macro-Evolution has never been observed nor experienced in real life, in the lab, or in the field. It does not exist. It's physically impossible! We KNOW that it does not exist because Abiogenesis or Macro-Evolution has NEVER been experienced nor observed, and it never will be. It's physically impossible!

Remember, Materialism, Naturalism, Darwinism, Macro-Evolution, Abiogenesis, and Spontaneous Generation were falsified by Louis Pasteur in 1859 – the very same year that Charles Darwin published "On the Origin of Species".

In almost all of the college textbooks that I own, they claim that the unrelenting pressure from natural selection made you, your mind, and your brain. There's no explanatory power in natural selection when it comes to the origin of our genes and genomes. Natural selection doesn't touch our genes. It can't. It's physically impossible.

Can you feel that unrelenting pressure from Natural Selection designing, and planning, and controlling your every plan, action, choice, encounter, the destiny of your whole life, and the destiny of the whole human race?

No?

Well, neither can I!

Remember, the processes of Evolution or Genetic Drift, Random Mutations and Natural Selection do NOT provide some kind of unrelenting pressure organizing, designing, programming, planning, manufacturing, and upgrading your genome from scratch. Mutations and Selection are blind physical processes – like a rock rolling down the hill. Can you control the outcome of that?

No!

When it comes to Random Mutations and Natural Selection, the Darwinists *over-sell* it, which is a logic fallacy. These people *attribute* to Mutation and Selection the capability of intelligent design, planning, foresight, teleology, purpose, meaning, and creation; whereas, the influence of natural selection in your life is relatively insignificant. Natural selection doesn't touch your genes; and, random mutations scramble or randomize your genes making them non-functional. Remember, the Materialists, Naturalists, and Atheists are always wrong – demonstrably wrong.

Random Mutations functioned in your life in a "creative fashion" only when your sperm and your egg were made by your father and mother. A random shuffling of your parent's two chromosomes takes place, and those two shuffled chromosomes become one chromosome that goes either into your sperm or your egg. That's the only thing that Random Chance or Random Mutations did for you during your whole life. After that, any Random Mutations that have taken place in the cells of your physical body have brought you disease, cancer, death, and eventually the extinction of the whole human race if it goes on long enough through a process called Genetic Entropy.

The ONLY time that Natural Selection does anything for you or to you during your life is when you get selected against and die. Natural selection is death or entropy. Natural Selection doesn't design and control and plan your whole future and the whole destiny of the human race, as the Materialists, Naturalists, and Darwinists claim. Sure, your death will be a significant event for you personally, but it really doesn't have anything to do with the rest of the human race. Remember, Natural Selection only comes into play when somebody gets selected against and dies. It's not a panacea controlling the destiny of the whole human race, as it's typically portrayed as being by the Materialists and Naturalists. Natural Selection is not God. Remember that!

Evolution or Genetic Drift, Random Mutations, and Natural Selection do not function as advertised – they don't design, plan, organize, and create. It's physically impossible for them to do so.

Evolution didn't make your brain. Evolution or natural selection can't even make a protein or its matching gene. In fact, evolution (genetic drift), random mutations, and natural selection didn't even exist until after God designed and created the first proteins, genes, genomes, eyes, brains, and life forms to begin with. The genomes were made first by God; and, then we started to experience random mutations and natural selection afterwards.

Mark My Words

Mapping Language Skills onto the Physical Brain

Scientists have observed that there are many different maps within a physical brain. Clearly, the different parts of our physical brain have been mapped to produce and provide specific physical functionality. Mapped by whom? Anything that was obviously mapped obviously has a Mapmaker who mapped it. Maps don't just spontaneously generate out of thin air. Maps are made, which means that maps have some kind of Mapmaker who made them. That is what has been experienced and observed. Is it not?

According to the Orthodox Interpretation of Quantum Mechanics by Henry P. Stapp, the Human Psyche (process 1) chooses what it wants to do with its physical body; and then, Nature's Psyche (processes 2 and 3) collapses the necessary wave functions and turns on or fires the necessary neurons in order to make the Human Psyche's choices a physical reality.

Psyche makes choices. Process 1 is represented by the choice of the observer or the choice made by the Human Psyche. The Human Psyche chooses what it wants to do with its physical body. Process 2 is the collapse of the wave function or the physical manifestation of the Human Psyche's choices. Nature's Psyche collapses the wave functions and turns on the appropriate neurons according to Orthodox Quantum Mechanics. Nature's Psyche also maps the physical brain, synaptogenesis, neurons, and axon growth at the quantum level and then uses that quantum map to get things done for us at the physical level. The Human Psyche isn't consciously aware of any of these quantum mechanical processes. Process 3 is the choice of the feedback (or the choice of the outcome) – what Dirac called: "a choice on the part of nature". Nature's Psyche chooses whether to comply with or to reject the Human Psyche's requests. If I were to choose to put my fist through a brick wall, the results would be completely different if Nature's Psyche responded "Yes" rather than responding "No".

Based upon the Orthodox Interpretation of Quantum Mechanics, it is Nature's Psyche who maps our physical brain thereby providing physical functionality to our physical brain. It is Nature's Psyche who makes our physical brain and does synaptogenesis for our physical brain deciding what each synapse means and does. It's Nature's Psyche who does cell migration and axon growth. Nature's Psyche is the Mapmaker who maps our physical brain. These maps are constructed and stored at the quantum level and then used by Nature's Psyche to provide us with physical functionality at the physical level. This reality and truth explains everything that we have ever experienced or observed when it comes to our physical brain. This is the basis for Quantum Neuroscience.

Mapping Language Skills onto the Brain

Siegler, R., DeLoache, J., & Eisenberg, N. (2010). *How Children Develop*, (3rd ed.). New York: Worth. The following quotes are in the third edition, but they were mostly pulled from the online second edition.

It is obvious that our brains have been mapped to provide specific physical functions. It is also obvious that maps require some kind of Mapmaker to make them or map them. There are signs of this map-making process all throughout the whole of the psychological sciences and neuroscience. It can't be avoided because it is obviously real and truly exists.

This time around, I'm going to use "How Children Develop" second and third editions as the source material to demonstrate the necessity and the existence of this invisible, intangible, non-physical, quantum map-making process that is taking place within each and every physical brain on this planet. These are situations in which Psyche is the best explanation; and, our genes are the worst explanation. By definition in principle, our genes cannot make choices, cannot do selection, and cannot perform quantum mechanical processes. Genes are passive, not active. Genes are "read" in order to produce RNA and proteins. Genes don't have command and control capabilities; and, genes don't have target acquisition capabilities either. Genes don't have a Psyche or a Soul according to the Materialists, Naturalists, Darwinists, and Atheists. Genes don't contain maps or blueprints for cellular organelles, synaptogenesis, synaptic typing, synaptic pruning, cell migration, cell typing, synaptic typing, a physical cell, or a physical body. Genes simply code for proteins. All of the data or information for the various different 3D maps and blueprints for our physical body are stored someplace else besides our genes. In such a situation, Psyche and Quantum Mechanics becomes the best explanation, and not genetics and classical physics.

A major landmark in early language development is achieved when children start combining some words into sentences, an advance that enables them to express increasingly complex ideas. The degree to which children develop syntax, and the speed with which they do it, is what most distinguishes their language abilities from those of nonhuman primates. (*How Children Develop*, p. 241.)

https://books.google.com/books?id=WMEz44QTjN8C&pg=PA236&lpg#v=onepage&q&f=false

In other words, there are NO evolutionary precursors for syntax and sentence formation within the animal kingdom. The animals cannot write sentences, and they cannot speak sentences. Syntax and sentence construction did not come from our genes. Our genes do NOT learn our language, which is proven by the fact that our genes do not pass on our language to the next generation. Our genes, and therefore evolution, have no mechanism for learning our language and then passing our language on to the next generation. Therefore, we must look for some other source or cause for language, syntax, and sentence construction besides evolution and our genes. This is logical common sense.

If language development and sentence construction are not a product of our genes and evolution, then there's only one other logical explanation for their cause and their source – the Human Psyche. In fact, the preponderance of the evidence points to the Human Psyche as the source and the cause of language development within human beings.

Out-of-Body Explorers and Near-Death Experiencers have observed that the Human Psyche continues to think and communicate telepathically long after their physical brain is dead and gone. Ultimately, it is the Human Psyche who acquires a language, and then uses that language to communicate with other human psyches and God's Psyche.

However, the Human Psyche is not consciously aware of all these quantum mechanical map-making processes, which also suggests that Nature's Psyche or God's Psyche plays a significant role in the development of language functionality within the human being and the human brain.

The Orthodox Interpretation of Quantum Mechanics by Henry P. Stapp suggests that although the Human Psyche does indeed learn the language, it is Nature's Psyche who actually maps those language skills onto the physical brain by developing and constructing the physical brain as well as the quantum maps associated with that specific physical brain. Language development is a complex process that involves interaction between the Human Psyche and Nature's Psyche – a process that has absolutely nothing to do with evolution or our genes. Our physical genes code for physical proteins – NOT our language skills, NOT synaptogenesis, NOT language understanding, and NOT the quantum maps of physical functionality that Nature's Psyche produces for our physical brain both at the quantum level and the synaptic level.

A synapse is a gap-in-space. You can't map meaning and purpose onto a gap-in-space at the physical level. It can only be done at the quantum level or the psyche level. You can't map language skills onto synapses at the physical level. It can only be done at the quantum level. The maps are made at the quantum level by Nature's Psyche and then used by Nature's Psyche in order to get things done for us at the physical level. I call these things Quantum Maps of Physical Functionality.

The different types of evolution consist of different physical processes, many of which have proven to be physically impossible. Genes are physical machines that are used by Nature's Psyche to produce and deploy physical proteins. There's nothing within genes or evolution that can do Quantum Mapping. By definition, evolution of any type cannot touch the quantum, the spiritual, the non-local, or the non-physical. Evolution is restricted to physical matter and entropy. In principle, genes cannot produce Quantum Maps. They have no mechanism for doing so. Genes map for or code for proteins at the physical level. Therefore, by process of elimination, Nature's Psyche (or God's Psyche) ends up being the source and the cause of Quantum Maps of Physical Functionality because the Human Psyche isn't consciously aware of these quantum mechanical map-making processes.

This is logical common sense derived through a process of observation, experience, and the elimination of everything that doesn't work. Evolution and our genes do not work at the quantum level; but, the Human Psyche and Nature's Psyche certainly do. I define science as observation and experience. The Human Psyche has been experienced and observed; therefore, we KNOW that it is real and truly exists.

I'm choosing to go with what has been experienced and observed – NOT the wishful thinking and selective ignorance of the Materialists, Naturalists, Darwinists, and Atheists who assure us that Psyche and Syntropy do not exist. The goal is to find the truth – not to hide the truth from the rest of the world. Materialism, Naturalism, Darwinism, Nihilism, Behaviorism, Determinism, Physical Reductionism, and Atheism were designed to prevent us from discovering Nature's Psyche, the Human Psyche, and God's Psyche. They were designed to hide the truth from us. Consequently, I prefer to go with what has been experienced and observed.

I used to be a Materialist, Naturalist, Nihilist, and Atheist until science (observations and experiences) convinced me that I was wrong. Science convinced me that God exists. Science also convinced me that Psyche exists. Psyche is the innate intelligence within all the different forms of energy that have ever been experienced or observed.

By definition, there is no intelligence or psyche within any type of evolution. The Darwinists or Evolutionists assure us that psyche or intelligence does not exist. There's no way in the universe that evolution could ever do Quantum Mapping; so, we have to look for some other source or cause of the Quantum Maps that have been mapped onto our physical brains – besides evolution or our genes. Nature's Psyche is the best and most logical explanation because it could actually do the job that evolution and our genes could never do.

A third theme that recurs throughout the chapter is *individual differences*. As you will see, there is great variability in the timing of most aspects of language development. For any given milestone, some children will achieve it much earlier, and some much later, than others. The *active child* theme also puts in repeated appearances here. Infants and young children pay close attention to language and a wide variety of symbolic artifacts, and they work hard at figuring out how to use them to communicate with other people. (p. 217.)

https://books.google.com/books?id=WMEz44QTjN8C&pg=PA213&lpg#v=one page&q&f=false

It's the child, the individual Human Psyche, who is working hard to figure out how to communicate with other people – not our genes, not evolution, and not natural selection. Our genes don't learn our language; the Human Psyche does. How do we know? Language skills survive the death of our physical brain and therefore the death of our genes. Our genes don't learn our language. If they did, then our language skills would be automatically inherited from our parents and our psyche wouldn't have to learn our language.

The individual, personality, or self is comprised of the Human Psyche. Individual differences, psyche, memories, acquired language skills, and intelligence persist long after the physical body and physical brain are dead and gone. Psyche or energy is conserved; and, psyche is the intelligence within all the different forms of energy.

The Human Psyche within infants and young children chooses what to pay attention to. Choice is a function of psyche. Choices are produced by Psyche.

The Active Child, Targets of Attention, Language Acquisition, and Individual Differences are in fact the "psycho" part of the BioPsychoSocial Model in action. Our genes have no mechanism for making our choices for us. Genes code for proteins, not decisions and choices. Our choices have nothing to do with our genes, evolution, and natural selection. In fact, our genes can't do choice or selection according to the Materialists, Naturalists, Darwinists, Nihilists, and Atheists. Choice is impossible at the physical level according to the Behaviorists. Our genes and proteins simply do what they are told to do.

What is Required for Language?

What does it take to be able to learn a language in the first place? Full-fledged language is achieved only by humans, but only if they have experience with other humans using language for communication.

A Human Brain

The key to full-fledged language development is in the human brain. Language is a *species-specific* behavior, in that only humans acquire language in the normal course of development in their normal environment. Furthermore, it is *species-universal* in that virtually all young humans learn language. It takes highly abnormal environmental conditions or relatively severe cognitive impairment to disrupt children's language development.

In contrast, no other animals naturally develop anything approaching the complexity or generativity of human language, even though they can communicate with each other. (p. 219.)

Once again, they are wrong. The Human Psyche continues to think and communicate long after its physical brain is dead and gone. The Human Psyche is the essential requirement for language, not the physical brain! The children of God, human beings, inherited the ability for language from their parents, Heavenly Father and Heavenly Mother. Brains are made from proteins, and there's no significant difference between the proteins in the human brain and an ape's brain. It's all physical matter. The proteins and the genes are NOT learning our language. The Human Psyche is!

Yes, there seems to be a correlation between the human brain and human language, but correlation is not causation!

The Human Psyche learns the language, and Nature's Psyche maps the associated physical functionality onto the human brain as needed. The human brain gives Nature's Psyche something to work with when it comes time to do mapping and synaptic pruning at the physical level.

What is required for language?

The Human Psyche, not a human brain! The human brain is only useful for physical functionality, not the telepathic communication between Psyches taking place wirelessly at a distance in the spirit world. Technically, your physical brain doesn't learn a language. Psyche does! A physical brain isn't needed to communicate in the spirit world.

There is one thing that they did get right, though. There are NO evolutionary precursors for human language. Evolution didn't produce human language, and neither did our genes and proteins. The existence of human language falsifies Materialism, Naturalism, Darwinism, Nihilism, and Atheism.

Out-of-Body Explorers and Near-Death Experiencers have observed first-hand that a physical brain isn't needed for human language and communication between Human Psyches in the spirit realm; however, a Human Psyche certainly is.

Nature and nurture – genes, proteins, and a physical environment – are needed to produce the physical aspects and physical manifestations of our language skills; however, a psyche is needed to produce the quantum mechanical, the non-local, and the psychic aspects of our language skills.

The Materialists, Naturalists, Darwinists, Nihilists, Behaviorists, Determinists, and Atheists will tell you that evolution or natural selection made your brain. They are wrong. Nature's Psyche made your brain. A physical brain is mapped by Nature's Psyche, and synaptic pruning is done by Nature's Psyche during that mapping process. Evolution and natural selection cannot do the quantum mapping that's taking place within your physical brain. Nature's Psyche can. Nature's Psyche maps your physical brain at the quantum level

406

and then uses those Quantum Maps of Physical Functionality at the quantum level in order to get things done for us at the physical level. This is how things really work; and, it has nothing to do with evolution and natural selection. Evolution or natural selection doesn't touch our spirit, mind, psyche, or soul.

Physical matter is made by the Gods with physical limitations. Physical limitations cannot do quantum mapping. Evolution and natural selection are based exclusively on random chance and blind luck. Random chance and blind luck cannot do quantum mapping.

Evolution (genetic change), natural selection, random mutations, Materialism, Naturalism, Darwinism, Nihilism, Behaviorism, Determinism, Physical Reductionism, Classical Physics, and Atheism are based exclusively on entropy. Evolution is entropy. Entropy is death. The theory of evolution is Creation by Entropy or Creation by Death. Entropy and death cannot do quantum mapping. Entropy and death cannot produce life. Evolution or death cannot produce life. Therefore, evolution, natural selection, random mutations, entropy, and death cannot do quantum mapping.

Do you see what becomes possible once a person chooses to think about these things critically, logically, and rationally? The farther down the rabbit hole that you go, the more obvious it becomes that Psyche is the best and most parsimonious scientific explanation that can be given to the scientific evidence that we have on hand as a race.

In most people, language is primarily represented in the left hemisphere of the cerebral cortex. (p. 221.)

In other words, in most people, language has been mapped to the left hemisphere of the cerebral cortex by Nature's Psyche. However, Nature's Psyche could just as easily have chosen to map a person's language skills to his or her right hemisphere instead. It's Nature's Psyche who does the mapping, therefore it is Nature's Psyche who decides where to do that mapping. Nature's Psyche makes the map and does the mapping both at the quantum level and the physical level. The Human Psyche isn't consciously aware of these quantum mechanical processes.

According to the Orthodox Interpretation of Quantum Mechanics by Henry P. Stapp, the Human Psyche chooses what it wants to do with its physical body and physical brain; and, it's Nature's Psyche who collapses the necessary wave functions and triggers the necessary neurons needed to make the Human Psyche's requests a physical reality. Nature's Psyche makes the maps; and, then Nature's Psyche uses those maps at the quantum level in order to get things done for us at the physical level.

Remember, language in the human brain can be mapped to the other hemisphere by Nature's Psyche who constructs and then maps our physical brain!

Remember, genes and proteins cannot produce quantum maps. Only Nature's Psyche can. Nature's Psyche handles neurogenesis, synaptogenesis, brain mapping, and synaptic pruning.

Critical Period for Language Development

A considerable body of evidence has given rise to the hypothesis that the early years constitute a critical period during which language develops readily. (p. 221.)

Why the early years?

It's because that's the period of time when Nature's Psyche is mapping language functionality onto the human brain and doing synaptic pruning to remove the synapses that Nature's Psyche hasn't mapped to anything. A synapse is a gap-in-space. How would you map meaning, purpose, and specialized functionality onto a gap-in-space at the physical level? It can't be done, which means that it isn't being done. Synaptic typing, synaptic mapping, synaptic meaning, and synaptic purposing are only possible at the quantum level. How else are you going to map meaning, purpose, and functionality onto a gap-in-space? It can't be done at the physical level by our genes; so, it has to be done at the quantum level or the psyche level.

This critical period for language development takes place while the brain is being mapped by Nature's Psyche and the synapses are being pruned by Nature's Psyche. After Nature's Psyche has done its work and located its quantum maps within the physical structures of the physical brain, it's a lot harder to re-map the brain than it is to map the brain correctly in the first place.

There's a critical period for language development. If damage to a physical brain takes place during neurogenesis and synaptogenesis, then Nature's Psyche has nothing to work with at the physical level when it comes to language development. If synaptogenesis has already taken place but mapping of the physical brain and synaptic pruning have not yet taken place; then, if the language portions of the brain are destroyed, Nature's Psyche can simply map that physical functionality someplace else within the physical brain.

After synaptic pruning, physical mapping, and quantum mapping have taken place, then damage to specific areas of the brain can produce long-lasting effects; and, it takes a concentrated effort on the part of the Human Psyche to get Nature's Psyche to remap that physical functionality to some other part of the brain. Stroke victims can and do recover if they are determined to do so and live long enough for Nature's Psyche to remap their brains. There's intelligence and life force there on both sides of the aisle. Sever the major axons, though, and the physical functionality seldom comes back. There's nothing there for the brain to be remapped onto if the axons or nerves have been cut.

When it comes to language development and physical development, Nature's Psyche actually provides the best explanation for what's really happening both at the quantum level and the physical level. The mapping being done by Nature's Psyche explains everything that has ever been experienced and observed.

Not only does the Human Psyche learn its native language; but, Nature's Psyche seems to learn it as well so that Nature's Psyche can then map the associated physical functionality onto specific parts of the physical brain. This is fascinating science. Its explanatory power is through the roof to infinity and beyond.

It is obvious that our physical brains have been mapped. There are dozens of maps within our brains. They have been experienced and observed. Maps of any kind obviously require some kind of intelligent Mapmaker; or, you end up with random chaos when you are done.

Ironically, all of this is hidden within the scientific explanations that are typically given by the Naturalists and Atheists simply waiting to be observed and found. Let me show you what I mean.

Left-hemisphere damage produces aphasia in deaf signers just as it does for users of a spoken language. This suggests that the left hemisphere is actually specialized for the kind of analytic, serial processing required for language, not for the specific modality (spoken words or signs) in which it is expressed. (p. 221.)

Specialized by whom? Specialization requires a Specialist! Nature's Psyche specializes in mapping the physical brain both at the quantum level and the physical level. Remember, genes and proteins can't map your brain at the quantum level. It requires a Specialist in order to get that job done. The genes have NO physical mechanism for choosing which neurons to fire and which neurons to leave off. The genes aren't running your physical brain as the Naturalists and Atheists claim. Nature's Psyche is.

Although the most advanced nonhuman communicators combine symbols in utterances, there is little evidence for syntactic structure, which is a defining feature of language. In short, only the human brain acquires a communicative system with the complexity, structure, and generativity of language. (p. 220.)

This is wrong!

Your genes and your proteins are NOT learning your language. If your genes actually learned your language, then your language would be passed onto your children through your genes. In other words, your physical brain is NOT learning your language. The physical brain doesn't learn anything. The Human Psyche does.

Out-of-Body Travelers and Near-Death Experiencers have observed that while their Psyche is separated from their spirit body looking at their spirit body, it is in fact their Psyche who is having the experience and forming memories – not their spirit body. Likewise, while the Psyche is separated from a physical body, a physical body doesn't have experiences and doesn't form memories. It has been observed that language skills survive the death of our physical brain. Only the Human Psyche acquires language skills. It's a spiritual gift or a psychic inheritance and has nothing to do with our genes. Clearly, our genes don't learn and pass on our language. The Human Psyche does.

Your brain doesn't learn your language. Your brain simply does what Nature's Psyche tells it to do. Consequently, the language that your Human Psyche learns goes with you into the spirit realm after your physical body and physical brain are dead and gone. It's your psyche that learns your language, not your brain. Your brain doesn't learn anything. Instead, your brain is mapped by Nature's Psyche to provide specific physical functions. Psyche learns. Your brain does not. Psyche learns. Your genes do not.

However, these people are right in that there are NO evolutionary precursors for syntactic structure and the generativity of language that these people have chosen to associate with our brain. In other words, syntactic structure and language generativity did not evolve through natural selection. Evolution, natural selection, and our genes do not learn our language. The Human Psyche does.

Remember, there are NO evolutionary precursors to syntactic structure, sentence construction, writing skills, and human language! Evolution didn't produce it!

Psyche is identified by choice. Any time you catch choice in the act, you have in fact caught some kind of psyche in the act.

Brain-Language Relations

A vast amount of research has examined brain-language relations. One thing that is clear is that language processing involves a substantial degree of functional localization. At the broadest level, there are hemispheric differences in language functioning. For the 90% of people

who are right-handed, language is primarily represented and controlled by the left hemisphere of the cerebral cortex.

Listening to speech is associated with greater electrical activity in the left hemisphere. Thus, the left hemisphere shows some specialization for language (or language-like stimuli) at a very early age, with the degree of hemispheric specialization for language increasing over time. Specialization for language is also evident *within* the left hemisphere. (p. 220.)

Specialization requires a Specialist! Random chance can't do specialization nor mapping. Maps of any kind require an intelligent Mapmaker to make them and then use them.

The physical brain of human beings is mapped by Nature's Psyche to facilitate language functionality. For 90% of the people who are right-handed, language skills are mapped to the left hemisphere by Nature's Psyche; but, for 10% of the right-handed, Nature's Psyche chooses to map language functionality to the right hemisphere instead.

It's a choice! Nature's Psyche (or God's Psyche) makes this choice. Nature's Psyche makes the choice as to where to map or specialize language functionality within the human brain. Obviously, anything that was mapped has an intelligent Mapmaker who mapped it. Maps don't just spontaneously generate out of thin air from nothing. Obviously, the intelligent Mapmaker who mapped language functionality onto your physical brain chose where to map that language functionality into your physical brain. The intelligent Mapmaker chooses where to do the mapping.

It's as if Nature's Psyche programs or maps a quantum computer or a quantum brain and then makes that quantum computer match with and control your physical brain. If a part of your physical brain gets damage, Nature's Psyche can be persuaded to map that functionality someplace else if given time enough and incentive enough to do so. Stroke victims can recover by getting Nature's Psyche to remap their brain by getting Nature's Psyche to remap desired functionality onto a different part of the brain.

Choice is a sure sign of Psyche or Intelligence. Mapping is a sure sign of Psyche or Intelligence. Language in the Human Brain can be mapped to either hemisphere by Nature's Psyche, who constructs and then maps our physical brain.

Do you see how these scientists have the Human Psyche hidden within all of their scientific descriptions where the Atheists and Naturalists will never see it nor find it? The Materialists, Naturalists, Darwinists, Nihilists, and Atheists talk about Psyche in all of their publications without ever once realizing that they are actually talking about Psyche. Psyche is hidden in plain sight where the Naturalists and Atheists will never see it nor find it because they refuse to look for it.

Remember, mapping of any kind requires an intelligent Mapmaker to make the map. Mapping also requires an intelligent User to use the map. Maps don't just spontaneously generate out of thin air as the Naturalists and Atheists claim. Maps are made. Anything that was obviously made obviously has a Maker who made it.

A classic study demonstrated *fast mapping* – the process of rapidly learning a new word simply from hearing the contrastive use of a familiar word and the unfamiliar word.

When the experimenter said, "Show me the blicket," the children mapped the novel label to the novel object, the one for which they had no name."

As you can see, infants and young children have a remarkable ability to learn new words as object names. Interestingly, they are equally able to learn nonlinguistic "labels" for objects. Infants between 13 and 18 months of age map gestures or nonverbal sounds (e.g. squeaks and whistles) onto novel objects just as readily as they map words. By 20 to 26 months, however, they accept only words as names. (pp. 237, 238, 241.)

There are dozens of different maps within our physical brain. A genome is a map. A genome maps the production sequences for all the different proteins that are used in a living cell. The scientists are in universal agreement that each physical brain on this planet is being mapped. They understand mapping and how it works. But, there is no universal agreement when it comes to the identity of the Mapmaker.

The Materialists, Naturalists, Darwinists, Nihilists, and Atheists will tell you that it is evolution or natural selection who is mapping our brains even though their hypothetical concepts don't have a physical mechanism for doing so. There is NO physical mechanism that will do quantum mapping of a physical brain and assign purpose, meaning, information, and knowledge to the gaps-in-space that we call synapses. The quantum mapping of purpose, meaning, teleology, intention, and goals can only be done at the quantum level by Someone Psyche or Someone Intelligent. Our genes have no mechanism for doing so.

By definition, evolution and natural selection are dumb and blind without a soul or a mind. That means that evolution and natural selection have NO mechanism for learning our language and then passing that language on to our children through our genes.

Here in this quote, they indirectly state that it is the children who are doing the mapping – their Human Psyche – not their genes, not their environment, and not their society. Psyche or Intelligence does the mapping, not our physical genes! The truth of it is hidden in plain sight where the Naturalists and Atheists will never see it nor find it.

Adults who learned a second language at 1 to 3 years of age show the normal pattern of greater activity in the left hemisphere during a test of grammatical knowledge. Those who learned the second language later show less localized activity, including more right-hemisphere activity. (p. 222.)

In this scientific research, it is observed that the mapping of the brain and synaptic pruning by Nature's Psyche, which took place later in the development of bilingual skills, produced completely different results than the results produced by early development of bilingual skills.

From birth to age three, Nature's Psyche maps both the first and the second language to the same hemisphere during the normal process of synaptic mapping and synaptic pruning. If a second language is learned later in life, however, then Nature's Psyche maps a lot of that physical functionality into the right hemisphere that's not being used for language functionality until a second language is actually learned. The genes are precisely the same in both instances. What differs is the timing when Nature's Psyche is doing the mapping of the synapses and the pruning of the synapses.

Remember, our genes and our proteins are NOT learning our language. Our language doesn't get passed on to our children through our genes. It's Psyche – the Human Psyche and Nature's Psyche – who are learning the language, not our genes. It has been observed that our language skills survive the death of our physical brain and our genes. That's only possible because it's the Human Psyche who learns our language, and not our genes. Nature's Psyche (or God's Psyche) also learns our language because it is Nature's Psyche who is actively mapping language functionality onto our physical brains.

411

Your brain isn't learning your language. Your brain is being mapped by Nature's Psyche instead. It is Nature's Psyche who fires the neurons or turns on the physical brain whenever the Human Psyche starts thinking about its first language or its second language. Nature's Psyche maps the brain at the quantum level, and then Nature's Psyche uses those Quantum Maps of Physical Functionality at the quantum level in order to get things done for us at the physical level. The mapping of Quantum Maps onto our physical brain has nothing to do with our genes, evolution, or natural selection. It requires Someone Psyche or Someone Intelligent to function as a Mapmaker in order to produce Quantum Maps of Physical Functionality and then use those maps at the quantum level in order to get things done for us at the physical level. This is logical common sense. It actually explains what's going on at every level of existence.

When it comes to scientific theories, what good is it if it can't be used to explain how things really work? The explanatory power of Quantum Mechanics and Psyche is vastly superior to anything that the Materialists, Naturalists, Darwinists, Nihilists, Behaviorists, and Atheists are able to produce. Is it not? By choosing to allow Quantum Mechanics and Psyche in to play, we can literally explain everything that comes our way. It's all hidden in plain sight where the Naturalists and Atheists will never see it nor find it. They literally prevent themselves from doing so.

This is good stuff, but only if you choose to see it as such. It's complete nonsense to anyone who has chosen to reject it. I KNOW because I used to be a Materialist, Naturalist, Nihilist, and Atheist until the scientific evidence convinced me that I was wrong.

We thus see that infants begin life equipped with the two basic necessities for acquiring language: a human brain and a human environment. So long as they do not suffer from serious brain injury or developmental disorders or grow up in conditions of extreme social deprivation, they will acquire their native language. (p. 225.)

These scientists deliberately overlook the Third Necessity; but, there's no way that they can exclude it. The Third Necessity is hidden within everything that they say, write, study, examine, or do.

What is the Third Necessity?

Human infants begin life equipped with a Human Psyche which is necessary for acquiring human language skills. Human infants bring their Human Psyche with them from heaven, and they take their Human Psyche with them into the afterlife after their physical body and physical brain are dead and gone.

A human body and human brain are indeed necessary for physical language functionality in a physical realm. The human society or human environment is also necessary in order to provide a specific language to learn. However, our genes, our proteins, and our brains do not learn our language. The Human Psyche does. We KNOW this is so because it has been observed that the languages that we learn go with us after our physical brain is dead and gone. Memory is a function of the Human Psyche, not our genes! If a thought or memory survives the death of your physical brain, then it's truly a memory in every sense of the word. Is it not?

The Third Basic Necessity for language acquisition is a Human Psyche.

Out-of-Body Travelers and Near-Death Experiencers have observed that while their Psyche is separated from their spirit body looking at their spirit body, it is in fact their Psyche who is having the experience and forming memories – not their spirit body.

Likewise, while the Psyche is separated from a physical body, a physical body doesn't have experiences and doesn't form memories.

It has been observed that we are born with innate language acquisition capabilities almost as if we had learned a language before we were born here in mortality.

> **One difference between infants' and adults' distinction among speech sounds, however, is that young infants actually make *more* distinctions than adults do.**

> **Infants can distinguish between phonemic contrasts made in all the languages of the world – about 600 consonants and 200 vowels.**

> **This research reveals an ability that is both innate, in the sense that it is present at birth, and independent of experience, because infants can discriminate between speech sounds they have never heard before.**

> **Infants are born with the ability to discriminate between speech sounds in any language, but they gradually begin to specialize, retaining their sensitivity to sounds in the language they hear every day – their native language – but becoming increasingly less sensitive to nonnative speech sounds. This change begins as early as 7 months for some infants.** (pp. 227-228.)

Think about it! Before Nature's Psyche starts mapping and pruning a child's brain, a child can distinguish between all of the languages in the world. They are born with this ability; and, they didn't get it from their genes or society. Our genes do NOT learn our language. Our genes have no mechanism for doing so. Universal phonemic differentiation is present at birth before any physical experiences take place. How's that possible? It didn't come from our genes, which means that it didn't come from evolution and natural selection.

This physical ability to distinguish between all the different languages of the world takes place before and while the brain is being made, differentiated, and mapped by Nature's Psyche! It doesn't come from our genes. Our genes have no mechanism for hearing and learning our spoken languages; but, Nature's Psyche and the Human Psyche do. Language acquisition is one of the smoking guns that clearly reveals the existence of Psyche or Intelligence. Our genes do not learn and pass on our language. The Human Psyche does.

This reality is obviously true; and, it shows up in the scientific research that has been done. However, these people will tell you that evolution and natural selection produced this ability to discriminate between speech sounds that infants have never heard before. In other words, the Naturalists and Darwinists are claiming that evolution and natural selection have learned all the languages of the world, and then pass on the ability to differentiate between them to newborn infants through their genes. It takes an infinite amount of blind faith to believe that it is so; but, that's precisely what these people choose to believe whenever they start talking about the innate capabilities that infants are born with. In this book, these scientists attribute these innate capabilities to evolution, natural selection, and our genes. These people teach that ONLY physical matter exists, which means that these people teach that ALL of our innate capabilities are coming from our genes thanks to the mystical, magical, and non-physical endowments of evolution and natural selection. The theory of evolution is internally inconsistent and self-contradictory; but, that doesn't matter. Their motto – anything but God – overrides every other consideration.

The Human Psyche is a better explanation – a lot more parsimonious and logical than to claim that the mindless and random processes of evolution and natural selection are learning our language and then passing our language on to our children through our genes.

To each their own!

Remember, the infant (the Human Psyche) is born with innate knowledge, innate capabilities, and innate functionality. Gradually, the infant (the Human Psyche) begins to specialize. Based upon this physical input, Nature's Psyche begins to map purpose and meaning onto the synapses (gaps-in-space) and neurons within the physical brain. As this mapping takes place, as specialization takes place, and as Nature's Psyche does synaptic pruning, the physical brain slowly loses its ability to facilitate the differentiation between speech sounds.

Over time, the physical brain gets mapped by Nature's Psyche to discriminate or identify specific sounds while at the same time ignoring other sounds that have no purpose or meaning. Brain mapping by Nature's Psyche has powerful explanatory power. It effectively explains everything that we have ever experienced and observed. Attention is a function of psyche, not our genes! Choice is a function of psyche, not our genes! Our genes don't pay attention to what we are seeing and hearing; and, our genes do not learn our language. The Human Psyche does.

Specialization of any kind requires a Specialist. Genes can't function as a psychic, quantum mechanical, intelligent Specialist; but, the Human Psyche certainly can. Genes can't learn a language; but, the Human Psyche can. This means that evolution and natural selection can't learn a language and don't learn a language; but, the Human Psyche does.

Nine-month-old American infants attend longer to lists of words that follow this pattern than to words in which the second syllable is stressed.

Thus, the infants used recurrent sound patterns to fish words out of the passing stream of speech. Probably the most salient regularity in what infants and toddlers hear is their own name. As early as 5 months of age, they show the 'cocktail party effect,' attending selectively to the sound of their own name among a stream of speech sounds they are hearing. (p. 229.)

Another smoking gun where the Human Psyche is concerned is "selective attention". Selective attention is a foundational cornerstone of Psychology. When it comes to "attention", we are in fact talking about selection or choice. The Human Psyche is the BEST explanation for this particular capability.

The Human Psyche CHOOSES what to pay attention to – not our genes, and not our environment. There's nothing within our genes that's capable of paying attention to our external environment. Even if our genes were capable of paying attention to our environment, they have no physical mechanism in place for making our choices for us.

Our society (those other psyches) can force us to do things against our will; but then, it's no longer a choice. Where a human physical brain and human physical body are concerned, CHOICE is a function and a product of the individual Human Psyche – not our genes. Our genes can't make our choices for us. Our genes can only enhance or limit our options. The same can be said for our society. Our society either presents opportunities, or blocks access to our opportunities. CHOICE is done by the individual Human Psyche where a human society is concerned.

Attention is chosen into existence by the Human Psyche – not our genes!

Love, friendship, loyalty, mercy, belief, compassion, hatred, and duty are chosen into existence by the Human Psyche. They don't exist until they are chosen into existence or believed into existence by the Human Psyche. Selection or choice is a function and a product of the Human Psyche. You can tear the whole universe down to individual atoms and molecules; and, you will NEVER find a single atom of mercy, kindness, friendship, and love. These things don't exist at the physical level. They are psychic phenomena or quantum mechanical phenomena.

Ironically, natural selection, random mutations, and evolution cannot do selection or choice. These are different forms of entropy, and entropy has NO CHOICE! Imagine it! They built a whole science on a concept that can't do selection or choice – natural selection. Natural selection can't do selection or choice! Natural selection is entropy or death. Natural selection doesn't touch our genes. Natural selection doesn't make any choices where our lives are concerned. Natural selection, or survival of the fittest, only comes into play when we get selected against and die. Remember, natural selection is entropy or death. Entropy is death. The theory of evolution is Creation by Entropy or Creation by Death. Please explain to me how that's supposed to work! You can't do it because it doesn't work. Entropy and death cannot design and create. Entropy and death cannot do CHOICE! That is what has been experienced and observed.

The false is falsified by the truth; and, the truth is repeatedly experienced and observed. The truth is predictable and dependable. Evolution, natural selection, and random mutations reliably predict entropy, death, and extinction. Entropy, death, and extinction cannot produce life. That's what's been experienced and observed.

When it comes to the Human Psyche, they can't see it because they don't want to see it; but, it's there hidden within their words, nonetheless. Within all their scientific explanations, the Human Psyche is there waiting to be found. By their own admission, it's the child or the human psyche who learns the language; and, not our genes. There's no mention of genes learning a language because our genes can't learn our language. They have no physical mechanism for doing so.

The truth is hidden within the scientific observations, science experiments, and scientific research; but, it's only found when we actually go looking for it and are willing to receive it when we finally find it. If you successfully eliminate everything that is false, then only the truth will remain. We start our search for truth by eliminating Physicalism, Materialism, Naturalism, Darwinism, Nihilism, Behaviorism, Determinism, Physical Reductionism, and Atheism. If you eliminate everything that is false, then only the truth will remain. It's fascinating to study and observe what remains after you have eliminated Scientific Naturalism and its derivatives from the equation. The experienced and the observed remain. The truth remains. This is what I have experienced and observed. The truth is recognized by the fact that it has been experienced and observed by Someone Psyche somewhere sometime.

Psyche is experienced first-hand as an immaterial viewpoint in space. Whenever near-death experiencers and out-of-body travelers are outside of their spirit body looking at their spirit body, there's nothing there or nothing to be found whenever they go looking for their psyche, intelligence, or consciousness. These people have observed that it is their Psyche who makes choices, has experiences, and forms memories – not their spirit body and not their physical body. While out-of-body, other Psyches are viewed, seen, or observed as being a pinprick of light, a point particle of light, or a spark of light. Psyche has been experienced and observed. Science is observation and experience; and, I want my science to match with what has been experienced and observed.

415

Successful communication also requires *intersubjectivity*, in which to interacting partners share a mutual understanding. The foundation of intersubjectivity is *joint attention*, which is established by the parent's following the baby's lead, looking at and commenting on whatever the infant is looking at.

Once infants can recognize recurrent units from the speech they hear, the state is set for them to make a truly major advance. They are ready to address the problem of *reference*, to start associating words and meaning. (pp. 231, 232.)

The Human Psyche is the BEST explanation for intersubjectivity, joint attention, reference, choice, and meaning – not our genes. Our genes aren't paying any attention to what we are saying, thinking, and doing. Genes simply code for proteins. Genes don't code for blueprints, maps, subjectivity, joint attention, reference, meaning, or human language! Your genes don't learn your language. If they did, then we would inherit our language skills from our parents; and then, there would be only ONE language on this planet. The Human Psyche is the BEST explanation for human language – the best explanation for what we experience and observe.

Not only does the Human Psyche learn its native language; but, the Human Psyche also creates languages. Children who are born deaf will create their own unique sign language in order to communicate with other people. It has nothing to do with their genes, and everything to do with their Psyche or Intelligence.

Attention is a function of Psyche, and not our genes. Our genes only pay attention to transcription factors and transcription enzymes, not our human language and not the meaning and purposes being mapped into our synapses (gaps-in-space) by Nature's Psyche. Only Nature's Psyche is capable of mapping purpose and meaning onto a gap-in-space that we call a synapse. There is NO physical mechanism for doing so! Meaning, purpose, teleology, subjectivity, reference, and attention are products of Psyche, not our genes.

It's two Human Psyches who are in fact interacting and sharing mutual understanding on both sides of the veil – not their genes! The genes in one person have no physical means for interacting with the genes in another person thereby sharing mutual understanding between their genes. Genes have no physical means for interacting with and sharing mutual understanding at the quantum level or the psychic level either. Psyche does. Psyche can function on both sides of the veil – both at the quantum level and the physical level. Our genes cannot. By definition in principle, our genes are limited to the physical level or the entropic level. Our Psyche is not. Our Psyche doesn't have physical limitations. Our Psyche is syntropic – eternal and everlasting, without a beginning of days or an end of years. Psyche is the innate intelligence within all the different forms of energy which gives that energy the ability to follow and obey God's Laws or God's Commands.

This is what has been experienced and observed.

Notice the process! Notice that I have gone out of my way to put Psyche back into Psychology. You too can do this same thing with the whole of psychology – go out of your way to include Psyche in Psychology. Just look for the psychic explanation or the quantum mechanical explanation for every science experiment and all the scientific research that has ever been done. Psyche is a good fit when it comes to Psychology. Invariably, Psyche ends up being the BEST explanation whenever we are dealing with Psychology. Psychology was originally the scientific study of the Human Psyche; and therefore, Psyche is still the BEST explanation when it comes to the different theories and models that we find associated with Psychology.

Theoretical Models for Language Development

Siegler, R., DeLoache, J., & Eisenberg, N. (2010). *How Children Develop*, (3rd ed.). New York: Worth.

There seem to be as many different models for Language Development and Language Origins as there are scientists on this planet. This is yet another sure sign that we are dealing with something psychic or something quantum mechanical. Whenever the Physicalists, Naturalists, Darwinists, Nihilists, Behaviorists, and Atheists can't figure out how something works, invariably they are dealing with Psyche or Quantum Mechanics. These people don't have the scientific means whereby to explain the quantum, the intangible, the psychic, the non-corporeal, the spiritual, the non-local, or the non-physical because these people have formally rejected the science behind these things.

No seeking, then no finding.

These people deliberately limit themselves to nature and nurture – their physical genes and their physical environment. They deliberately exclude everything else.

Therefore, these people will never discover that it is in fact Nature's Psyche who maps language skills onto the physical structures of each physical brain.

Nativist Views

From Plato to Kant to their modern intellectual descendants, nativists have argued that language is too complex to arise from experience alone, so there must be some preexisting, innate structures that enable young humans to acquire it. The most influential modern version of this point of view was advanced by Noam Chomsky, a linguist who argues that using a language requires knowledge of a set of highly abstract, unconscious rules.

These rules, which Chomsky referred to as *universal grammar*, are common to all languages, and they are what makes it possible for people to learn particular languages. Children are assisted in acquiring their native language by their inborn knowledge of the general form that languages can take. Because of this inborn knowledge, children require only minimal input to trigger language development; simply hearing other people use language is enough to get it going. (pp. 247-248).

Inborn knowledge of language – their Psyche brings it with them from heaven where they were communicating with other Psyches before they came here. Our genes don't learn our language at the physical level, nor do they pass it on at the physical level.

Any time the scientists mention the word "unconscious", they are in fact talking about Psyche, particularly Nature's Psyche. The Human Psyche isn't consciously aware of what Nature's Psyche is doing.

Nature's Psyche produces the innate structures within a physical brain. Nature's Psyche maps physical functions, meaning, and purpose onto the physical brain. Anytime we are dealing with a pre-mortal existence or innate capabilities, we are in fact dealing with Psyche. Psyche is syntropic, which means that it is conserved. Psyche is syntropic, which means that Psyche is eternal and everlasting, without a beginning of days or an end of years. Psyche is conserved!

The Materialists, Naturalists, Darwinists, Nihilists, Behaviorists, and Atheists assure us that all of this innate knowledge comes to us from evolution and natural selection through our genes. They are wrong. We don't get our language skills and syntax capabilities from our genes. If we did, then the apes would have the same capabilities that we do – having acquired their language skills from their genes. They have most of the same genes that we do, after all. Our genes don't learn our language. If they did, then we would simply inherit our language from our parents, and there would then in fact be only one language on this planet. Our genes don't learn our language. The human psyche does. We KNOW this is so because it has been observed by out-of-body travelers that our language skills survive the death of our physical body and physical brain. In fact, the physical body and physical brain have NO experiences and form NO memories while the Human Psyche is separated from the physical body. It's the Human Psyche who has experiences and forms memories, not our physical body and not even our spirit body.

When it comes to language acquisition, the Human Psyche ends up being the BEST explanation for its cause and its origin. Psyche is the most elementary of particles and the fundamental unit of reality. This is a Psyche Ontology.

When it comes to science and psychology, I have chosen to take Quantum Mechanics and Psyche seriously. Most scientists don't, which makes it seem as if I have made scientific discoveries that none of the other scientists have ever discovered before. That's what happens when you eliminate the elementary particle or the fundamental unit of reality from your science. That's what happens when you eliminate Psyche. Instantly, you have no scientific explanation for the psychic, the non-local, the intangible, the non-physical, the spiritual, the invisible, the transdimensional, or the quantum mechanical. Your explanatory power goes down the toilet whenever you remove Psyche, Action at a Distance, Syntropy, Quantum Mechanics, and Intelligence from your arsenal.

Inborn knowledge cannot be explained by our genes because our genes have no physical mechanism for passing that knowledge on to the Human Psyche or human consciousness. Our genes don't learn our language, and then pass that language on to the next generation.

> **According to the nativist view, the cognitive abilities that support language development are highly specific to language. As Steven Pinker (1994) describes it, language is "a distinct piece of the biological makeup of our brains . . . distinct from more general abilities to process information or behave intelligently".**

> **This claim is taken one step further by the *modularity hypothesis*, which proposes that the human brain contains an innate, self-contained language module that is separate from other aspects of cognitive functioning. The idea of specialized mental modules is not limited to language. Innate, special-purpose modules have been proposed to underlie a variety of functions, including perception, spatial skills, and social understanding.**

> **Nativist views of language development are supported by the fact that virtually all children exposed to full-fledged language acquire it and by the fact that no other animals do. Demonstrations of critical periods for language development, as well as specific links between brain structures and language abilities, are generally taken as supportive of the nativist position.**

Probably the strongest support for the idea that children come equipped with some fundamental knowledge of linguistic rules is the research on the *invention* of sign language by groups of deaf children, with no linguistic input from adults. The fact that these children spontaneously imposed grammatical structure onto their simple sign systems suggests preexisting structural knowledge, especially because aspects of the grammatical system that some of these children invented match that of existing spoken languages.

Nativist views have been criticized for many things, especially for the idea of a universal grammar common to all languages, as well as for the idea that language acquisition is supported by special language-specific mechanisms. They have also been criticized for focusing almost exclusively on syntactic development while ignoring the importance of the communicative role of language.

Connectionism is a type of information processing theory that emphasizes the simultaneous activity of numerous interconnected processing units. These programs learn from experience, gradually strengthening certain connections among units. This account is based on computer models of neural networks that have the capacity to modify themselves as a result of input. These models can learn from experience, sometimes displaying patterns of acquisition that are surprisingly similar to those of young children. Connectionist models are always open to criticism regarding the features that were built into the models in the first place. (pp. 248-251.)

https://books.google.com/books?id=WMEz44QTjN8C&pg=PA244#v=onepage&q&f=false

Notice what the Nativists immediately do. They immediately and erroneously assign our inborn language capabilities and **inborn knowledge** to our biology, our genes, and our brains because they have trained themselves to believe that ONLY physical matter exists. One error leads to another which leads to another. Garbage in, then garbage out! They can't help it because they have brainwashed themselves into believing that ONLY physical matter or entropy exists. They believe it because they want to believe it. I KNOW because I used to be a Materialist, Naturalist, Nihilist, and Atheist until the science convinced me that I was wrong. I believed it because I wanted to believe it – not because it was true.

According the materialistic and naturalistic point of view, our native inborn capabilities must derive from our genes because there is no other mechanism to produce them. However, our genes do NOT learn our language, and our genes do NOT pass on our language. Therefore, we must find some other explanation for the "preexisting structural knowledge" that we have been born with and have subsequently incorporate into our lives and our interaction with others.

Once again, the Human Psyche provides the BEST and most parsimonious explanation.

Abductive reasoning (also called abduction, abductive inference, or retroduction) is a form of logical inference which starts with an observation or set of observations then seeks to find the simplest and most likely explanation.

https://en.wikipedia.org/wiki/Abductive_reasoning

Abductive reasoning falls back on the BEST and most parsimonious explanation. The Human Psyche is the BEST and most parsimonious explanation for ALL of the innate inborn capabilities that we cannot get from our genes and our society. This is logical common sense through a process of eliminating the impossible from consideration.

How often have I said to you that when you have eliminated the impossible [and the false]**, whatever remains, however improbable, must be the truth? — Sherlock Holmes.**

Remember, if you successfully eliminate everything that is impossible and false, then ONLY the truth will remain.

It's physically impossible for our genes to learn our language and then pass on our language skills to the next generation. It can't be done, which means that it isn't being done. There is NO physical explanation for the language skills that we are born with. Apes have most of our genes; but, they don't have language skills, writing skills, and syntax skills. We didn't get our language skills from our genes; otherwise, we would all have the same language, and the apes would have the same language and language skills as well.

The Darwinists, Naturalists, Evolutionists, and Atheists automatically assign our unique language abilities to "a distinct piece of the biological makeup of our brains". These people tell us that evolution or natural selection made our brains. They are wrong.

Evolution and natural selection don't learn and pass on our languages. It's the Human Psyche who invents and then uses languages, including sign languages. Our genes don't learn and then pass on sign languages. Evolution and natural selection don't learn and then pass on sign languages. If they did, then there would be only one universal sign language, and everyone would know it from the very moment they were born. There's NO such thing as genetic memory in the physical realm. Genes don't have a psyche or a soul that's permitted by God to pass information on to the Human Psyche. From the perspective of the Human Psyche, the genes are dumb and blind without a soul or a mind. When it comes to fallen mortal beings, there's NO transpersonal communication or telepathic communication between the Human Psyche and its genes.

Evolutionists, Naturalists, Physicalists, and Atheists credit **inborn knowledge** to evolution and our genes. I credit it to the Human Psyche, because our genes and evolution have NO physical mechanism for learning and then passing on our **inborn knowledge**. Genes and evolution don't learn our language. The Human Psyche does!

Inborn knowledge and language have nothing to do with our brains or our genes because we continue to communicate and use language long after our physical brain and genes are dead and gone, according to the scientific evidence or observational evidence from Out-of-Body Experiences and Near-Death Experiences. **Inborn knowledge** is a function and a product of the Human Psyche, and it goes with us into the next world when our physical body and physical brain die and cease to exist at the physical level.

Inborn knowledge is brought with us from heaven through the Human Psyche, where we were communicating with each other and using language long before we came here to this fallen physical mortal existence. Our genes don't learn our language at the physical level, nor do they pass it on at the physical level.

There are NO "interconnected processing units" within a physical brain, because synapses are gaps in space not wired connections. The connectionist model or computer model doesn't work where the physical brain is concerned. There are no wires or connections with a physical brain at the physical level. If these connections, neural networks, and maps exist, then they exist at the quantum level because they are physically impossible at the physical level where the physical brain is concerned. There are NO CPUs, connections, or wires within a physical brain. A physical brain is NOT a computer and doesn't work like a computer.

Clearly our physical brain has been mapped, programmed, or modularized. The **modularity hypothesis** proposes that the human brain is comprised of CPUs or Modules, and the **modularity hypothesis** is based upon computer science. Alas, there are NO wires within a physical brain connecting the neurons together into functional logic gates, transistors, CPUs, and RAM. There are NO modules or CPUs within our brain at the physical level. If they exist, they are mapped or made at the quantum level by Nature's Psyche (or God's Psyche) because the Human Psyche isn't directly aware of these quantum mechanical processes, mechanisms, maps, and machinery.

If your brain has a computer inside it consisting of wires, transistors, logic gates, CPUs, and RAM, then that computer exists at the quantum level because it's physically impossible for it to exist at the physical level. Neurons are stand-alone switches that are either ON or OFF. Neurons are single hardware BITS. It's physically impossible to connect neurons or switches into functional CPUs, RAM, and BYTES of code and memory without any wires to connect them. The synapses or gaps-in-space prevent the neurons from functioning as anything but single stand-alone hardware BITS. If there are any BYTES within your physical brain, they exist at the quantum level, not the physical level. They are Quantum Modules, not physical modules.

The truth is that Nature's Psyche maps or makes these modules at the quantum level, and then uses these Quantum Maps of Physical Functionality to get things done for us at the physical level. It's Nature's Psyche who makes our brain and synapses, and then decides what meaning or purpose to assign to each neuron and synapse. It's Nature's Psyche who maps our brain. It's Nature's Psyche who maps the different physical functions, purposes, and meaning onto the neurons and synapses of our brain. We KNOW this is true, because according to the Orthodox Interpretation of Quantum Mechanics, it is Nature's Psyche who collapses the wave function, and it is Nature's Psyche who chooses which motor neurons to trigger or to fire in order to make our physical bodies move. The Human Psyche isn't consciously aware of any of these quantum mechanical processes. The Human Psyche simply decides and chooses what it wants to do with its physical body and physical brain; and then, Nature's Psyche collapses the necessary wave functions and fires the necessary neurons in order to make the Human Psyche's choices actual, physical, and real.

Clearly, our physical brain has been mapped, networked, programmed, or modularized; but, that process is physically impossible at the physical level because there are NO wires or connections within a physical brain. Synapses are gaps-in-space not physical wires. You can't make physical computers out of gaps-in-space. Physical computers require wires to connect everything together into a functional whole. There are NO wires within a physical brain!

The brain is mapped by Nature's Psyche to facilitate the production and use of language both at the quantum level and the physical level. Nature's Psyche can choose to map language skills and inborn knowledge to either hemisphere of the physical brain. Quantum Maps of Physical Functionality made and used by Nature's Psyche provides us with the BEST and most parsimonious explanation for what we have experienced and observed on both sides of the veil. Clearly the physical brain has been mapped. Maps of any kind obviously require some kind of intelligent Mapmaker in order to make them and map them.

The "specific links between brain structures and language abilities" has to take place at the quantum level or the psyche level because it's physically impossible at the physical level. Synapses are gaps-in-space, not wired connections! The ONLY way to map purpose and meaning onto a gap-in-space is at the quantum level or the psychic level because it can't be done at the physical level. The specific links between brain structures and language abilities ends up being Quantum Mapping at its very finest. To explain Quantum Mapping logically and scientifically, we must invoke a Quantum Agent such as Nature's Psyche or the Human Psyche. ONLY Psyche can do Quantum Mapping. Our genes and proteins cannot do Quantum Mapping. Our genes and proteins simply do what they are told to do by God's Psyche or Nature's Psyche. Our genes and proteins don't plan and choose our every move. Our Human Psyche does!

Deliberation, planning, design, purposing, teleology, goals, intentions, foresight, mapping, programming, creation, and science are functions of Psyche or Intelligence. Our genes cannot do these things.

Clearly, our brains have been mapped to provide specific physical functions. Remember these "specific links" or Quantum Maps are made by Nature's Psyche and not our genes. Our genes don't learn our language, or they would be passing on our language to the next generation; and then, everyone would have the same language. Human language is learned by the Human Psyche; and, then the physical functionality for all of that is mapped onto the human brain at the quantum level by Nature's Psyche. When the brain lights up and the neurons fire, it is Nature's Psyche who is collapsing the wave functions and turning on the specific neurons that it needs to activate at the quantum level in order to make the Human Psyche's choices actual and real at the physical level. This is how things really work when it comes to our physical brain.

The explanatory power of Quantum Mechanics and Psyche is through the roof to infinity and beyond. There's no limit to what you can explain when you finally chose to bring Psyche and Quantum Mechanics into the game.

Attention is chosen into existence by our Human Psyche. It's the Human Psyche who chooses what to do with its physical body and physical brain. The Human Psyche chooses what to pay attention to – not our genes. According to the Behaviorists, our genes are incapable of selection or incapable of making choices. Our genes can't make choices at the physical level. Choice doesn't exist at the physical level. Our genes, molecules, atoms, and proteins simply do what they are told to do by Nature's Psyche or God's Psyche. Choice is a quantum mechanical process that only Psyche can make, or do, or perform. Our genes can't make our choices for us. Choice is a function and a product of Someone Psyche. ONLY Psyche or Intelligence can do choice or make choices. Out-of-Body Explorers have observed that ONLY the Psyche has experiences and forms memories while that Psyche is separated from its spirit body and its physical body. That means that ONLY Psyche can do choice because ONLY Psyche can have experiences and then remember having had those experiences. Your spirit body and physical body don't have experiences and don't form memories while your Psyche is separated from them. This is what has been experienced and observed. Science is observation and experience after all.

Remember, our physical genes are NOT wired together into CPUs, Modules, Transistors, and RAM. The DNA within our genes is the result of chemical bonding. In other words, invisible, non-physical, intangible, non-local, transdimensional, quantum forces and fields act at a distance through space to bind the individual atoms and DNA molecules into the functional information that we call a gene. Individual atoms and molecules interact with each, target each other, and communicate with each other wirelessly at a distance at the quantum level because they have NO physical mechanism in place for doing so at the physical level. The individual atoms and molecules have NO wires tying them together into functional modules, transistors, logic gates, CPUs, and RAM at the physical level! Likewise, synapses are gaps in space. There are NO wires tying the synapses together into functional modules, transistors, logic gates, CPUs, and RAM. There are no wires, CPUs, and RAM within a physical brain at the physical level. They have never been experienced nor observed. If such things exist, they exist at the quantum level or the psyche level because they don't exist at the physical level.

Once again, the Human Psyche and Nature's Psyche end up being the BEST and most parsimonious explanation that we can give for what we have experienced and observed as a race. By allowing Psyche and Quantum Mechanics in to play, we can literally explain everything that comes our way. Remember, the BEST and FASTEST way to find and know the truth is to live it, observe it, and experience it first-hand for yourself, or to choose to trust someone who has. The false is repeatedly falsified by the truth; and, the truth is repeatedly experienced and observed.

Design and creation by "something that does not exist" has NEVER been experienced nor observed. It's impossible to experience and observe that "something does not exist". You simply have to operate on blind faith that it does not exist, and then go from there. Blind faith is religion, not science. Materialism, Naturalism, Darwinism, Nihilism, Behaviorism, Determinism, and Atheism are religions – not science – because they are based upon nothing but blind faith. Scientific Naturalism has to be taken on blind faith as being true. It also requires that we dismiss and ignore mounds of scientific evidence of a psychic nature or quantum nature. In contrast, Psyche has been experienced and observed by Out-of-Body Travelers and Near-Death Experiencers. Psyche is experienced as an immaterial viewpoint in space; and, Psyche is observed as a pinprick of light, a spark of light, or a point particle of light. You can't prove that Psyche does not exist, especially after it has been experienced and observed.

Furthermore, it's impossible to prove that the Biblical God Jesus Christ does not exist, especially since He has been experienced and observed after He rose from the dead. Thousands, if not millions, of us have seen Him and touched Him both in the flesh and during our Near-Death Experiences.

Science is observation and experience. The fact that Jesus Christ has been experienced and observed after He rose from the dead is scientific proof that Materialism, Naturalism, Darwinism, Nihilism, Behaviorism, Determinism, and Atheism are false. The false is repeatedly falsified by the truth; and, the truth is repeatedly experienced and observed. That's just the way things work.

Children take giant steps in mastering the phonology, semantics, syntax, and pragmatics of their native language. This remarkable achievement is made possible by the joint prerequisites of a human brain, [a human psyche]**, and exposure to human communication.**

Although language is our preeminent symbol system, humans have invented a wealth of other kinds of symbols to communicate with each other. Virtually anything can serve as a nonlinguistic [non-physical] **symbol so long as someone** [psyche] **intends it to stand for something other than itself.**

The list of symbols you regularly encounter is long and varied, ranging from the printed words, numbers, graphs, photographs, and drawings in your textbooks to thousands of everyday items such as TV, movies, computer icons, maps, clocks, and so on. (pp. 251-252.)

Notice carefully that it's the children (the Human Psyche) who learn their native language – not their genes. It is humans (the Human Psyche) who invent – not their genes! Genes don't invent! Genes and proteins simply do what Nature's Psyche tells them to do. Remember, it's the Human Psyche who learns symbols and then passes these symbols on to others – not our genes. Our genes have NO physical mechanism in place for learning our symbols and our native language. Think about it. Maps provide information. Information is a product of Someone Psyche or Someone Intelligent. Maps don't just spontaneously generate out of thin air from nothing. Maps are made. Symbols are made. Anything that is obviously made obviously has a Maker who made it. This is Logic 101.

Symbols are mental constructs or psychic constructs. Symbols come from a mind. Our symbols are NOT invented by our genes, nor are our symbols passed onto the next generation by our genes.

Evolution, natural selection, and genes have NO physical mechanism for doing the Quantum Mapping of a physical brain. Evolution, natural selection and our genes do NOT learn our native language. Evolution, natural selection, and genes cannot map purpose and meaning onto synapses or gaps-in-space at the physical level either. Such mapping of purpose and meaning has to take place at the quantum level or the psyche level because it can't be done at the physical level. The truth is hidden in plain sight where the Naturalists and Atheists will never see it, find it, nor understand it. Materialism, Naturalism, Darwinism, Nihilism, and Atheism were designed to hide these truths from us and to prevent us from discovering them.

The Materialists, Naturalists, Darwinists, Nihilists, Behaviorists, and Atheists deliberately leave out the Human Psyche from their science and thereby damage and destroy the explanatory power of their science. We must include Quantum Mechanics and Psyche into science, if we want science to be able to explain everything that we have ever encountered, observed, or experienced as a race. The explanatory power of Quantum Mechanics and Psyche is through the roof to infinity and beyond. Naturalism and Atheism? Not so much! Naturalism and Atheism can't explain Psyche and Action at a Distance. They have no way of doing so. Instead, they deny the existence of Psyche and Action at a Distance and leave it at that. Denying it doesn't explain it! Our goal as scientists is to try to explain how everything works on both sides of the veil. Atheism, Naturalism, Materialism, Darwinism, Nihilism, Determinism, Behaviorism, and Classical Physics are absolutely worthless when it comes time to explain how Quantum Mechanics, Telepathy, Telekinesis, Quantum Tunneling, Phase-Shifting, Supernatural Mechanisms, or Action at a Distance works.

Human language is made possible through the interaction of the Human Psyche with Nature's Psyche on both sides of the veil – both the quantum side and the physical side. The physical functionality for human language is uniquely mapped onto each human brain at the quantum level by Nature's Psyche, and then used at the quantum level by Nature's Psyche to get things done for us at the physical level.

This is the way things really work when it comes to language acquisition and the human brain. The Orthodox Interpretation of Quantum Mechanics tells us that it is true. Nature's Psyche collapses the necessary wave functions and triggers the specific neurons needed to make the Human Psyche's choices a physical reality. Orthodox Quantum Mechanics ends up being the answer to life, the universe, and everything.

Psyche and Orthodox Quantum Mechanics explain everything that we have ever experienced or observed; whereas, Naturalism and Atheism do not. Instead, Atheism and Naturalism deny the existence of psyche and the supernatural or the quantum mechanical. Stating that something doesn't exist will NEVER explain what it is or how it works. Stating that Psyche, Quantum Mechanisms, Supernatural Mechanisms, Syntropy, Action at a Distance, and God "do not exist" will NEVER explain what they are and how they work.

Language ability is species-specific. The first prerequisite for its full-fledged development is a human brain, areas of which are specialized for the comprehension and production of language. Researchers have succeeded in teaching nonhuman primates remarkable symbolic skills but not full-fledged language. (p. 256.)

"Areas of which are specialized for the comprehension and production of language." Specialized by whom? Who is mapping or specializing our physical brain? Specialization requires a Specialist!

The Human Psyche wants to communicate with and interact with other Psyches. This desire on the part of the Human Psyche to use its brain to communicate with other psyches is what motivates Nature's Psyche to map language functionality onto the physical brain at the quantum level so that the Human Psyche and Nature's Psyche can then use those Quantum Maps at the quantum level in order to get things done at the physical level.

The KEY component for language acquisition is in fact the Human Psyche, not the human brain. A language, once acquired, continues onward with us into the next life after our physical body and physical brain are dead and gone. This reality and truth has been experienced and observed.

Young infants are actually better than adults at discriminating between speech sounds not in their native language. As they learn the sounds that are important in their native language, their ability to distinguish between sounds in other languages declines. (p. 256.)

The BEST and most parsimonious explanation for these phenomena comes from the fact that it is the Human Psyche who learns its native language, and NOT its genes. The Human Psyche has inborn knowledge that makes it capable of learning and then passing on a language. The Human Psyche has learned other languages before it came here to this earth and learned its native language on this earth.

When children are born, their brains have not been mapped, limited, and restricted yet. As Nature's Psyche maps physical functionality for the native language onto the physical brain and does synaptic pruning removing the synapses that no longer have purpose or meaning or use, then the Human Psyche's ability to distinguish and hear the difference between sounds in foreign languages is slowly lost over time.

Once again, Psyche provides the BEST and most parsimonious explanation for the phenomena that we have experienced and observed. Our genes and proteins do NOT learn our native language. If they did, then they would pass it on to our children; and then, our planet and all the animals on our planet would have only one language, and that language would be passed onto our children through our genes. Quod erat demonstrandum!

Since our genes and proteins obviously don't learn our language and pass it on to the next generation, that means that evolution and natural selection don't learn our language either. Genes, evolution, and natural selection are the worst possible explanation for where our inborn knowledge is coming from because our genes, evolution, and natural selection don't acquire knowledge at the physical level. According to the Darwinists and Evolutionists, genes, evolution, and natural selection are dumb and blind without a soul or a mind, thereby making the Theory of Evolution self-defeating and impossible to implement at the physical level.

These realities and truths explain everything that we have ever experienced or observed on both sides of the veil – both on the psyche side and the physical side. Good enough!

My ultimate goal in life is to explain what things are and how they work. I can do so, once I choose to allow Psyche, Quantum Mechanics, and Action at a Distance into science where they belong. The explanatory power of Psyche, Quantum Mechanics, and Action at a Distance is infinite and unlimited. We can't say the same thing about Materialism, Naturalism, Darwinism, Nihilism, Behaviorism, Determinism, Physical Reductionism, Classical Physics, and Atheism. Denying the existence of something that has been experienced and observed will NEVER explain what it is and how it works. Denying the existence of the True and Living God, who has been experienced and observed, will NEVER explain who He is and how He works. This is logical common sense.

John 17: 3: **And this is life eternal, that they might know thee the only true God, and Jesus Christ, whom thou hast sent.**

There it is plain as day. Syntropy is Eternal Life. Syntropy is the conservation of Energy, or Psyche. Syntropy is the First Law of Thermodynamics. The Gods are syntropic. They are eternal and everlasting. Psyche or Intelligence is conserved. Psyche is the innate intelligence within all the different forms of Energy, which gives that energy the ability to understand, follow, and obey God's Commands and God's Laws. Psyche is the innate intelligence within the different forms of Energy which gives that energy the ability to KNOW God. By allowing Psyche and Quantum Mechanics in to play, we can literally explain everything that comes our way. This is what I have experienced and observed.

When it comes to God the Father and his son Jesus Christ, the only way to go where they have gone is to learn to do what they have done. It's going to be your Psyche who learns to do what they have done – not your spirit body and not your physical body. Your spirit body and your physical body already have a direct connection with God's Psyche and Nature's Psyche. Your Human Psyche does not. Spirit matter, physical matter, and Nature's Psyche simply do what God commands them to do because they have covenanted to do so. The same reality does not apply to your Human Psyche, unless you have actually made covenants with God and have chosen to keep those covenants or obey those covenants.

Our beliefs are chosen into existence by our Human Psyche – not our genes, not our spirit body, and not our physical body. Choice is a function and a product of Psyche. Choice is physically impossible at the physical level. Physical matter, genes, and proteins simply do what Nature's Psyche or God's Psyche tells them to do. Physical matter and physical reactions are determined by what Nature's Psyche tells them to do. Physical matter or Nature's Psyche simply follows God's Laws, God's Commands, and God's Psyche. Only Psyche can do choice, which means that choice is only possible at the quantum level or the psyche level under the command and control of Someone Psyche. Out-of-Body Travelers have observed that even spirit matter, in the non-consensus realities, conforms itself to the demands and commands of their Psyche or Intelligence. Their Psyche can even give their spirit body a different form than the human form that they inherited from the Gods. Psyche or Intelligence rules supreme. This is the way things really work at all levels of existence.

Energy and the associated intelligence are always conserved; but, the form of that energy is never conserved. Energy is syntropic. Energy is intelligent and alive. Energy is psychic. Energy is transdimensional. Energy is conserved. Energy is eternal and everlasting.

The Ultimate Law of Thermodynamics differentiates between what is being conserved and what is not being conserved. The form is never conserved. Entropic physical matter is comprised of different forms of energy. Physical matter and entropy are NOT conserved. The form is never conserved. Spirit matter and physical matter can be formed into many

different things. The form is never conserved. Only the underlying energy is conserved.
Psyche is the innate intelligence within all the different forms of energy which gives that
energy the ability to follow, understand, and obey God's Laws and God's Commands.
Energy and Psyche are syntropic, which means that they are conserved. They are eternal
and everlasting. They are alive. However, the form of that energy is never conserved. The
form is temporary. Physical matter and entropy are temporary. Their form is not being
conserved. This is the Ultimate Law of Thermodynamics. It explains everything that has
ever been experienced and observed.

I find the explanatory power of Psyche and Quantum Mechanics simply amazing.
They can explain everything that you will ever encounter on both sides of the veil. I'm
passionate about this because I have finally found the truth that I have been searching for
all of my life, for over fifty years of my life. I have found the answer to life, the universe,
and everything. This is good stuff – good science; but, you are going to have to decide for
yourself whether I know what I'm talking about or not. That's something that nobody else
can do for you. That's something that you are going to have to do for yourself or go
without. I wish you well.

Mark My Words

—

Source

Quantum Mechanics from a Non-Physical Spiritual Perspective

https://www.amazon.com/dp/B01J023TGU

Source Material

***Scientific Proof of God's Existence: Finding God Where the Atheists Refuse
to Look for Him***

https://www.amazon.com/dp/B07B26CRHX

The Appearance of Design

Biology is the study of complicated things that have the appearance of having been designed with a purpose. – Richard Dawkins.

They have the appearance of having been designed because they were designed. There's NO other logical explanation for their origin because entropy or physical matter cannot design and create anything. It's NEVER been caught in the act of doing so. It can't, which means that it didn't.

Spontaneous generation, creation by entropy, macro-evolution, abiogenesis, creation ex nihilo, chemical evolution, OR the theory of evolution was FALSIFIED in 1859 by Louis Pasteur. Biology appears to have been designed because it was designed.

Now try this philosophical proof of God's existence. It's an extremely powerful syllogism, because it is also a Scientific Proof of God's Existence or an OBSERVED Proof of God's Existence.

First Observation or Premise: Anything that was obviously made obviously had a Maker or Creator who made it. This is Logic 101.

Second Observation or Premise: A genome was obviously made. Such a thing doesn't just spontaneously generate out of thin air. It took some planning, programming, science, and manufacturing to get the job done.

Conclusion: Therefore, a genome obviously has a Maker or a Creator who designed it, programmed it, engineered it, field-tested it, fine-tuned it, made it, manufactured it, and then deployed it.

These observations convinced me that God does in fact exist, because I'm a scientist and I believe in the scientific evidence that has been experienced and observed. In other words, I KNOW that the premises are true; therefore, I KNOW that the conclusion MUST be true as well.

Some type of Syntropy, Psyche, or God MUST exist; or, physical matter, entropy, genomes, proteins, and this physical universe would NOT exist. It's elementary.

Consequently, since WE KNOW for a fact that God does indeed exist, the next task is to figure out who He is. I choose to go with our resurrected Lord Jesus Christ, because He has been EXPERIENCED and OBSERVED both in the flesh and during our Near-Death Experiences (NDEs), after He rose from the dead. I choose to go with the ONE who has been experienced and observed in real life by real people like you and me.

Mark My Words

—

Source

Syntropy in Defense of Quantum Mechanics: The Answer to Life, the Universe, and Everything

https://www.amazon.com/dp/B07BPT3W8R/

The Scientific Method Is Based Upon Affirming the Consequent

When it comes to the Philosophy of Science, Science, Personality Theory, Psychology, and the Scientific Method, I discovered that studying and learning the difference between *affirming the consequent* and *negating the consequent* is the most interesting and most useful concept that one can study and learn about. It is the concept that paid the most dividends in the end. All of the Psychologists and Personality Theorists turn to Joseph Rychlak for this information because he pioneered Humanism, and he was twice a president of the APA's division of Theoretical and Philosophical Psychology.

I used to be a practitioner and promoter of Scientism – the philosophical belief that Science and the Scientific Method is the only way to find and know the truth.

The Scientific Method is a good way for coming to know the truth; but, most people do not realize that the Scientific Method is not a foolproof way of knowing the truth nor is it the best way for finding and knowing the truth.

Like me, most people are shocked when they first realize or are first taught that the Scientific Method is based upon a logic fallacy called "affirming the consequent".

It took an honest philosopher and scientist to reveal this truth to me because many people like me have been brainwashed into believing that the Scientific Method is an infallible god. The people who are part of the religion called Scientism actually worship Science and treat Science as if it were God. These people pin all of their hope and faith on the Scientific Method, never once realizing that there is a fatal flaw at the heart of the Scientific Method which has the power to mess them up every time.

I finish this essay in the following book; and, it's also available online.

BioPsychoSocial: Including Psyche or Light into our Theoretical Models

https://www.amazon.com/dp/B0713NDHVW

Mark My Words

—

References

Rychlak, J. F. (1981). *A Philosophy of Science for Personality Theory* (2nd ed.). Malabar, FL: Robert E. Krieger Publishing Company.

Rychlak, J. F. (1988). *The Psychology of Rigorous Humanism* (2nd ed.). New York: New York University Press.

Rychlak, J. F. (1991). *Artificial Intelligence and Human Reason: A Teleological Critique*. New York: Colombia University Press.

Rychlak, J. F. (1994). *Logical Learning Theory: A Human Teleology and Its Empirical Support*. Lincoln, NE: Nebraska University Press.

Rychlak, J. F. (1997). *In Defense of Human Consciousness*. Washington DC: American Psychological Association.

Psyche Is Separate from the Physical Brain

When it comes to the Philosophy of Science, Science, Personality Theory, Psychology, and the Scientific Method, I discovered that studying and learning the difference between *affirming the consequent* and *negating the consequent* is the most interesting and most useful concept that one can study and learn about.

It's the concept that paid the most dividends in the end. All of the Psychologists and Personality Theorists turn to Joseph Rychlak for this information because he pioneered Humanism, and he was twice a president of the APA's division of Theoretical and Philosophical Psychology.

Social psychology studies our thinking, influence, and relationships by asking questions that have intrigued us all. Here are some examples:

How Much of Our Social World Is Just in Our Heads? **As we will see in later chapters, our social behavior varies not just with the objective situation but also with how we construe it. Social beliefs can be self-fulfilling.** (*Social Psychology*, p. 4.)

This is mind-over-matter that these psychologists and scientists are talking about. Our mind or our psyche constructs our reality, on both sides of the veil. Construing reality or constructing a worldview or a belief system is a function of the Human Psyche.

George Kelly taught his students that the Human Psyche construes reality. Joseph Rychlak was one of his students. My philosophy of science and the models of reality that I ascribe to are descended from this line of thought.

https://en.wikipedia.org/wiki/George_Kelly_(psychologist)

If every psychological event (every thought, every emotion, every behavior) is simultaneously a biological event, then we can also examine the neurobiology that underlies social behavior. What brain areas enable our experiences of love and contempt, helping and aggression, perception and belief? How do brain, mind, and behavior function together as one coordinated system? What does the timing of brain events reveal about how we process information? Such questions are asked by those in *social neuroscience.*

Social neuroscientists do not reduce complex social behaviors, such as helping and hurting, to simple neural or molecular mechanisms. Their point is this: To understand social behavior, we must consider both under-the-skin (biological) and between-skins (social) influences. Mind and body are one grand system. Stress hormones affect how we feel and act. Social ostracism elevates blood pressure. Social support strengthens the disease-fighting immune system. *We are bio-psycho-social organisms.* We reflect the interplay of our biological, psychological, and social influences. And that is why today's psychologists study behavior from these different levels of analysis. (*Social Psychology*, p. 9.)

Earlier in his career, Rychlak tended to define "science" as Materialism, Naturalism, Scientism, and Atheism – just like the rest of us. He also tended to define Psychology as Humanism, which is a type of Atheism. However, he adjusted his beliefs and definitions as more information and research became available.

The Materialists, Naturalists, Darwinists, Nihilists, Behaviorists, Determinists, Physical Reductionists, and Atheists tell us that psyche DOES NOT EXIST. They are wrong. They don't know what they are talking about. Every massless elementary particle or every massless quantum is some type of Psyche or Intelligence. A Photon is a type of Psyche or Intelligence. Psyche has been experienced and observed! Therefore, we KNOW that it is real and truly exists.

After decades of searching for the mind within the physical brain, the father of direct brain stimulation, Neurosurgeon Wilder Penfield, concluded that the mind or psyche is separate from the physical brain, which would be the case if my Ultimate Model of Reality or my Psyche Ontology is correct and true.

Wilder Penfield (Penfield, 1975, *The Mystery of the Mind: A Critical Study of Consciousness and the Human Brain*) **is the father of direct brain stimulation. In 1933 he stumbled onto the remarkable effects of electrically stimulating the brain. During operations on patients with epilepsy, in which abnormally firing portions of the cortex were severed or removed (Penfield, p. 13), Penfield inserted an electrode administering 60 pulses per second into various portions of the cortex to see whether brain functions could be induced. These operations are done with the use of local analgesics, so that the patient feels no discomfort when electrodes are inserted.**

After years of such research, Penfield concluded that all he ever dealt with was the physical hardware of the brain and that he never influenced the mind per se. Thus, he could say with confidence that "there is no place in the cerebral cortex where electrical stimulation will cause a patient to believe or to decide" (Penfield, p. 77). In fact, quite the opposite suggestion leaps out at us from Penfield's clinical examples.

Thus, when he would stimulate a patient's motor cortex, causing a movement in the hand, Penfield would ask the patient about this motion. Invariably, the patient's response was "I didn't do that. You did." (Penfield, p. 76). Sometimes, the patient would reach over with the other hand and stop the motion entirely (Penfield, p. 77). When made to vocalize, the patient would add "I didn't make that sound. You pulled it out of me" (p. 76). All such behaviors are consistent with delimiting oppositionality.

Penfield framed a dualistic theory of brain functioning that held that "the mind may be a distinct and *different essence"* **from the brain (Penfield, p. 62). Elaborating on this point, Penfield tells us that "a man's mind, one might say, is the person. He walks about the world, depending always upon his private computer [i.e., brain], which he programs continuously to suit his ever-changing purposes and interests" (Penfield, p. 60). The mind directs, and the [brain]-mechanism executes. The mind is like the program writer, relying on the memory bank of the brain. The brain is like the hardware, carrying out the directions of the higher [mind] source.**

As Penfield suggests, the brain can be considered an instrumentality rather than the originator of thought. It is in the active patterning of taking a position that thought occurs. To think is to take a position in the light of a certain predication relating to a targeted referent, as well as the meaning-extension that follows. Directed thought is relevant, well-focused position taking and meaning-extension.

(Rychlak, *Logical Learning Theory: A Human Teleology and Its Empirical Support*, 1994, pp. 235-236).

Belief, choosing to believe, purpose, teleology, planning, deciding, evaluation, predication, cognition, designing, creation, choice, and negation are integral parts of the human psyche; and, these functions are not found in the physical brain. Things like friendship, compassion, mercy, forgiveness, and love do not exist until after they are chosen into existence or believed into existence by the human psyche. That's Death's ultimate lesson in *The Hogfather*. You can tear the whole universe down to its component parts, and you will never find a single atom or molecule of mercy, justice, compassion, or love because these things do not exist in physical matter. These things are chosen into existence or believed into existence by the human psyche.

I was a computer programmer for a decade, and I know for a fact that computers cannot program themselves. They are not self-aware. Computers have no purpose, plans, desires, or goals. Computers can't do science, designing, manufacturing, teleology, final cause, ultimate cause, or anything else that requires planning, meaning, programming, and some kind of ultimate goal. Computer are like rocks. Computers can only do what we make them do. They are not going to come alive on us and suddenly become self-aware and self-directing. They can't because there is no psyche, or soul there – no ghost in the machine. If computers ever take over our nuclear arsenal and bomb us to death, it will be because some human or external psyche programmed them to do so. They aren't going to do so on their own. Computers can't do thought, predication, meaning-extension, position-taking, deciding, choosing, willing, or desiring.

(There's no sense re-inventing the wheel. See Rychlak, 1991, *Artificial Intelligence and Human Reason: A Teleological Critique*, for a more thorough development of this reality. Rychlak covered all the bases needed to support this Ultimate Cause Model of Reality, including the reality that an AI will never become a human psyche and will never survive bodily death like the human psyche does. Pull the plug, and that's the end of an AI. Not so with the human psyche! The human psyche goes on, after brain death and bodily death according to the Science of NDEs and OBEs. Other books from Rychlak on the same basic theme are *Discovering Free Will and Personal Responsibility*, and *In Defense of Human Consciousness*).

The Psyche, the Self, or the "I" is separate from the physical brain. Once that truth is understood and fully accepted, then a whole new world of reality opens-up to one's view. Once you find the truth, you will notice that it is consistent with all other truths; and at the same time, the truth will expose the lies and deceptions of Materialism and Naturalism for what they are – unscientific, incompatible with reality, and incompatible with the truth.

I will never apologies for seeing through the lies to the truths that the Materialists and Naturalists have been trying to hide from us. Once I got rid of My Materialism and My Atheism, I didn't want to go back. In fact, I can't go back, because now I know better. Materialism is an extremely poor, limited, and incomplete model of reality; and, I eventually started looking for something better.

Synaptogenesis and Synaptic Pruning

A detailed study of synaptogenesis and synaptic pruning reveals a separation of duties between Nature's Psyche and our genetic inheritance.

Who or what decides where to place a synapse?

Who decides what type of synapse it is going to be?

How is that synaptic typing information transmitted to both the pre-synaptic neuron and the post-synaptic neuron preceding the construction of the synapse?

Who decides the shape and the size of each synapse before it is built? Who is overseeing and controlling the construction process?

It can't be our genes because our genomes are only capable of storing 750 megabytes of information; and, the quadrillion synapses that a child is born with would require petabytes of data to code for 3D location addressing and synaptic typing. Furthermore, genes code for proteins and NOT synapses.

The command and control, location addressing, and typing information for synaptogenesis is being stored and processed someplace else besides our genes.

So, where is this information being stored, how is it being accessed, and how is it being transmitted telepathically at a distance to the axons and dendrites at the site of synaptogenesis? Who is overseeing all of this, deciding where each synapse will be built and what type and shape of synapse it will be?

It can't be the Human Psyche because we are NOT consciously aware of synaptogenesis or synaptic pruning.

It has to be Nature's Psyche or God's Psyche who is doing all of this synaptogenesis and synaptic pruning for us. It's Nature's Psyche who collapses the wave function according to the Orthodox Interpretation of Quantum Mechanics.

It has been theorized that Learning is NOT the process of forming synaptic connections but of eliminating, or removing, or the dying away of excessive synapses or the pruning of an overabundance of synapses. Synaptic pruning occurs at different times and different areas of the brain as a result of development and the chosen experiences of the active child.

Synaptic pruning poses some very interesting and challenging philosophical questions as well.

Who decides which synapses to prune away or kill? Why are those synapses chosen for culling and others are left in place?

Once again, it can't be the Human Psyche because the Human Psyche isn't consciously aware of any of these synaptic pruning processes. The ONLY thing the Human Psyche does is choose what it wants to do with its physical body and physical brain. Nature's Psyche handles the collapse of the necessary wave functions and the triggering of the necessary neurons in order to make the Human Psyche's choices and requests a physical reality. Therefore, it has to be Nature's Psyche who is choosing which synapses have no meaning or purpose and need to be pruned away instead.

When it comes to the cause and the source of synaptic pruning, it definitely can't be our genes because our genomes have nowhere near enough information storage capacity to code for and handle synaptogenesis and synaptic pruning. Furthermore, there is NO known physical mechanism in place that would allow genes to command and control synapses. The genes code for proteins, and that's it. Someone else besides our genes, Someone Psyche, is requisitioning each protein and then deciding where each requisitioned protein is going to be deployed or used. Once again, we are looking at Nature's Psyche or God's Psyche for the source and the cause of protein synthesis, synaptogenesis, and synaptic pruning because there is nowhere near enough information storage capacity within our genome for such a task.

Finally, who or what is deciding what each synapse means? Synapses literally have NO meaning whatsoever at the physical level. A synapse is a gap between neurons and NOT a direct wired connection. Technically, a gap in space has no physical meaning and no physical purpose, now does it? There are NO direct physical wires connecting our neurons together into logic gates and transistors either. It would literally take Someone Psyche to assign a meaning and a purpose to a gap in physical space.

We know that Someone Psyche is deciding what each synapse means at the quantum level or the psyche level because synapses have no meaning at the physical level. Therefore, the same Psyche who is deciding what each synapse means must also be the same Psyche who is deciding which synapses have NO meaning or purpose and need to be pruned away.

When it comes to a physical brain, it is my theory that Nature's Psyche is in fact creating a Quantum MAP of Physical Functionality at the quantum level and then using that MAP to get things done for us at the physical level. In other words, Nature's Psyche is building for each one of us a quantum computer within our brain area at the quantum level; and, then using that quantum computer or that Quantum MAP of Physical Functionality at the quantum level to fulfill the Human Psyche's commands and choices at the physical level.

It is clear at the physical level that our brains have been MAPPED. Mapped by whom? It is obvious that maps require some type of intelligent Map Maker. It can't be our genes who are doing this mapping because our genes don't have any physical mechanism in place for commanding, controlling, and mapping our synapses. The telepathic, wireless, or WiFi command and control of our different cellular processes is NOT being handled by our genes either – not at the physical level at least. Who knows what the genes and proteins are doing at the quantum level? Quantum Maps of Physical Functionality was one of my most useful and most interesting scientific discoveries.

I call this new science Quantum Neuroscience. Quantum Neuroscience is an attempt to ascertain what Nature's Psyche is doing for us at the quantum level or the psyche level in order to get things done for us at the physical level.

Mark My Words

—

Source

The Ultimate Model of Reality: Psyche Is the Ultimate Cause

https://www.amazon.com/dp/B071NC9JK6

Reference Materials

Myers, D. G. (2010). _Social Psychology_ (10th ed.). New York: McGraw-Hill.

Rychlak, J. F. (1994). _Logical Learning Theory: A Human Teleology and Its Empirical Support_. Lincoln, NE: Nebraska University Press.

Using the Scientific Method to Falsify Theories

Ask yourself, "What hidden assumptions do the Materialists, Naturalists, Darwinists, Nihilists, and Atheists make at the beginning of each of their scientific arguments or scientific proofs, every time they try to do science?" A lot can be learned by choosing to study people's hidden assumptions.

Ask yourself how these people convince themselves that the Theory of Evolution is true, because they successfully do so.

Ask yourself what logic fallacies these people use to verify and prove that the Theory of Evolution is true.

The Scientific Method as the Materialists, Naturalists, Darwinists, Nihilists, Behaviorists, Determinists, and Atheists typically use the Scientific Method is based upon the *begging the question* and the *affirming the consequent* logic fallacies. These people rely upon these logic fallacies in order to make their case for the Theory of Evolution; therefore, if you are a real scientist, it's extremely important to know what these logic fallacies are and how they work.

This is Philosophy of Science 101.

This topic is important to know about and understand. Science is supposed to be all about falsifying and eliminating everything that is false. Science is supposed to be about observation and experience. Traditional science is not about these things. Traditional science, based upon Materialism and Naturalism, is all about trying to trick us and deceive us into believing that science is Materialism and Naturalism. It's a hoax and a scam that millions of us have fallen for.

Begging the Question

Begging the question is the process of using your pre-chosen conclusion as your hidden assumption, major premise, or primary assumption in all of your scientific arguments, as evidence or proof that your pre-chosen conclusion is true. These people literally start every science experiment with the pre-chosen conclusion that the Theory of Evolution is true; and then, these people use the Truthfulness of Evolution as their first primary assumption or major premise, as evidentiary proof that the Theory of Evolution is true. It's *circular reasoning*, another logic fallacy. Their pre-chosen conclusion is used axiomatically as the hidden assumption or first premise in every scientific argument they make and every science experiment that they do. That's the way these people do science. It's posturing, smoke and mirrors, and deception. It's cheating.

Begging the question consists of making your conclusion one of the axiomatic premises, and effectively results in *jumping to conclusions*. The Materialists and Darwinists do this all the time, without realizing it.

Affirming the Consequent

Let's discuss how Science can go wrong. We can't use Science and the Scientific Methods to verify and know the Truth. Scientific Methods cannot be used to produce Knowledge.

Due to the *affirming the consequent* logic fallacy, and the *jumping to conclusions* logic fallacy, and the *category error* logic fallacies which are built into the scientific methods, it is impossible to use Science and the Scientific Methods to know the truth and to prove the truth.

If your ultimate goal in life is to KNOW THE TRUTH and prove the truth, then Science and the Scientific Methods are in fact one of the worst ways for accomplishing that task.

This is how scientific verification works in practice.

The following logical argument outlines the basic approach that has been taken by traditional scientists throughout the history of science:

Scientific Hypothesis: If Theory X is true, then we will observe Y.

Scientific Observations: We observe Y.

Scientific Conclusion: Therefore, Theory X is true.

This sort of thinking, however, reflects a logical fallacy called *affirming the consequent*. Here's a comparable example to demonstrate:

We hypothesize: If Sally's pet is a cat, it will have a tail.

We observe: Sally's pet has a tail.

We conclude: Therefore, Sally's pet is a cat.

We can easily see that this logic is fallacious. Just because we observe Y (a pet with a tail) that does not mean that our theory X (the pet is a cat) is true. After all, dogs, lizards, birds, and mice have tails too.

Yet, **this IS the scientific method**, and this is exactly how traditional scientists use the scientific methods to demonstrate and prove the truth. The whole enterprise is based upon a logic fallacy or two.

By *affirming the consequent*, you can prove anything to be true – anything – including Materialism, Naturalism, Darwinism, Nihilism, Atheism, and the Theory of Evolution. By *affirming the consequent*, you can prove that Sally's pet is a cat, even though it is in fact a bird. By *affirming the consequent*, you can prove that the theory of evolution is true, even though it is in fact a dog. Do you see how that works?

We hypothesize: If the Theory of Evolution is true, it will match with the fossil record and Darwin's tree of life.

We observe that the Theory of Evolution matches with the fossil record and Darwin's tree of life.

We conclude: Therefore, the Theory of Evolution is true.

This is how they prove that the theory of evolution is true. In fact, I see it on almost every page of my college textbooks where the Evolutionists are simply amazed at how miraculously all that evidence supports the truthfulness and predictions of the Theory of Evolution. These people never realize that they are *affirming the consequent*. They never realize that Darwin's tree of life and the theory of evolution were carefully designed and tailor-made to match with any and all evidence – past, present, and future. But, their arguments are fallacious, and they don't even know it.

Why is this argument fallacious?

It's because some version of Darwin's tree of life can also be made to match with Intelligent Design Theory or Christianity; and, the different types of Christianity and Islam can be made to match the fossil record as well. The whole thing is based upon *circular reasoning*, *begging the question*, or *affirming the consequent*. It's a logic fallacy.

Affirming the consequent is how the Materialists, Naturalists, Darwinists, Nihilists, Behaviorists, Determinists, and Atheists use the Scientific Method to prove that the Theory of Evolution is true. They cheat; and, most of them don't even realize that they are doing so, because they have never studied the Philosophy of Science.

Let's try it and see how it works in practice.

Jumping to conclusions and *affirming the consequent* is precisely what the Darwinists and Atheists DO each and every time they publicly declare that Science has proven the Theory of Evolution true.

Let's demonstrate their error by *affirming the consequent*:

If the Theory of Evolution is true, the fossil record will match with Darwin's Tree of Life.

We observe that the fossil record matches with Darwin's Tree of Life.

Therefore, the Theory of Evolution and Darwinism are true.

This is *affirming the consequent*. This is how the Darwinists and Naturalists prove that the Theory of Evolution is true. They do it every day. You can go online, and you will find them online right now *affirming their conclusions* or *jumping to conclusions*.

Most of the scientists can't see anything wrong with this argument. They use it every day to prove to themselves that the Theory of Evolution is true. They've been taught to do so in school by their science professors.

The problem with this argument, besides the fact that it is *affirming the consequent*, comes from the fact that Intelligent Design Theory, the Spaghetti Monster Theory, and even Christianity match with the fossil record, or can be made to match with the fossil record. Anything can be made to match with the fossil record and the future fossils that we dig out of the ground, including the Theory of Evolution and Darwin's Tree of Life.

Online, we see the Darwinists, Evolutionists, Materialists, Naturalists, and Atheists telling us every day that the fossil record is real and that the fossil record proves that the Theory of Evolution is true. "I've seen the fossils myself. I've seen evolution in action," they say. They are *affirming the consequent* and don't even realize that they are doing so. They have convinced themselves that the fossil record proves that the Theory of Evolution is true. Self-deception works, and it works every time.

Begging the question and *affirming the consequent* are two of the ways by which these people convince themselves that the Theory of Evolution is true. It works, because it was designed to work that way.

The few scientists, who fully understand that *affirming the consequent* can be used and is used to prove that anything is true, will correctly state that the Scientific Methods cannot be used to prove anything true. But, these same people do not realize that that truth and reality applies directly to the Theory of Evolution as well. When it comes to the Theory of Evolution, the very same people employ a *special pleading,* attempting to grant Darwinism, Materialism, and Naturalism immunity from the rules of science. They cheat and grant Darwinism an exemption and make it an exception. The *special pleading* logic fallacy is alive and well where Materialism, Naturalism, Darwinism, Nihilism, and Atheism are concerned. They cheat. Their first premise is allowed to be an Axiom or a Given Truth, which doesn't require any evidence, verification, or proof of its truthfulness; and, their first premise is invariably their pre-chosen conclusion. It's called *begging the question*, and it's a logic fallacy.

> **Scientific Hypothesis: If the Theory of Evolution is true, we will observe that the Theory of Evolution is true.**

> **Scientific Observation: We observe that the Theory of Evolution is true.**

> **Scientific Conclusion: Therefore, the Theory of Evolution is obviously true.**

This is precisely the scientific argument that I watched a couple of the New Atheists using the other day. They were totally convinced that the Theory of Evolution is true. Go figure! They take a blind leap of faith and go straight to their pre-chosen conclusion. They use their pre-chosen conclusion as proof that their pre-chosen conclusion is true.

You can't lose when it comes to *begging the question* and *affirming the consequent.* It is "heads I win, tails you lose". It makes you a winner every time! But, that's exactly the way that these people do science, by *affirming the consequent* with every word that they speak.

Proving the Theory of Evolution True

Due to the *affirming the consequent* logic fallacy, and the *jumping to conclusions* logic fallacy, and the *category error* logic fallacies which are built into the traditional scientific methods, it is impossible to use Science and the Scientific Methods to know the truth and to prove the truth. This is Philosophy of Science 101; and, almost everyone is completely oblivious to how often they use *affirming the consequent* to make their case and to prove their point. Scientists and PhDs are particularly vulnerable to this logic fallacy because this is the way they were taught to do science while they were in school.

This is how *affirming the consequent* works in practice. The following logical argument is the traditional Scientific Method in action. The following logical argument outlines the basic approach that has been taken by scientists to prove that the Theory of Evolution is true:

> **Scientific Hypothesis: If Theory X is true, then we will observe Y.**

> **Scientific Observations: We observe Y.**

> **Scientific Conclusion: Therefore, Theory X is true.**

This IS the Scientific Method in action! This is the way that the Scientific Method is traditionally used to do science. This sort of thinking, however, reflects a logic fallacy called *affirming the consequent.*

Now, let's do as the Materialists, Naturalists, Darwinists, and Atheists do – let's use *affirming the consequent* to prove that the Theory of Evolution is true.

If the Theory of Evolution is true, then we will observe natural selection.

We observe natural selection.

Therefore, the Theory of Evolution is true.

We see the Materialists, Naturalists, Darwinists, and Atheists everyday online and within their research papers using natural selection to prove that the theory of evolution is true; and, these people don't stop there.

If the Theory of Evolution is true, then we will observe random mutations.

We observe random mutations.

Therefore, the Theory of Evolution is true.

We see the Materialists, Naturalists, Darwinists, and Atheists everyday online and within their research papers using random mutations to prove that the theory of evolution is true.

These people literally use the existence of natural selection and random mutations as scientific proof that Evolution happens and as scientific proof that the Theory of Evolution is true. That's how these people prove that the Theory of Evolution is true; and, millions have fallen for the deception or the ruse. They can't see anything wrong with it.

If Chemical Evolution is true, then we will observe genes and proteins.

We observe genes and proteins.

Therefore, Chemical Evolution is true, and the Theory of Evolution is true.

We see the Materialists, Naturalists, Darwinists, and Atheists everyday online and within their research papers using Chemical Evolution and the existence of genes and proteins as proof that the Theory of Evolution is true. Clearly, evolution made those genes and proteins, because nothing else could have. That's *begging the question* and *affirming the consequent*.

These people literally use the existence of genes and proteins as scientific proof that Chemical Evolution happened and as scientific proof that the Theory of Evolution is true.

These people are using the *affirming the consequent* logic fallacy to prove that the Theory of Evolution is true; and, the unwary, the uneducated, and the scientists themselves don't even realize that they are doing so. We are being tricked and deceived, and we don't even know it.

We see these arguments being used every day online; and, nobody is aware of the logic fallacy that is contained within them.

I have seen this argument hundreds of times, and so have you.

Definition: Evolution is natural selection and random mutations. Evolution is genetic change.

Obviously, we observe natural selection, random mutations, and genetic change every day on this planet.

Therefore, the Theory of Evolution is obviously true; and, natural selection, random mutations, and genetic change obviously designed and created all of the life forms on this planet.

By *affirming the consequent*, these people *jump to the conclusion* that the Theory of Evolution is true without even realizing that they are doing so. It makes logical sense to them. They simply know that the Theory of Evolution is true, and never once do they question it. Many of them promote it with a blind-faith and zeal that rivals anything that can be found from an Evangelical Christian or a Muslim Fanatic.

Of course, the Theory of Evolution is true because natural selection and random mutations have been observed.

These people don't see anything wrong with their argument. In their minds, they have proven that the Theory of Evolution is true.

Of course, Evolution is true. It has been observed. Random mutations and natural selection have been observed. Of course, the Theory of Evolution is true. It's obvious. It's been observed.

These people *affirm the consequent* and *jump straight to the conclusion* that the Theory of Evolution is true. This is how they do science, and they can't see anything wrong with it.

The following is one of my favorites. I see it being used all the time by PhD scientists online, in their research papers, and everywhere else.

Scientific Hypothesis: If Materialism, Naturalism, Darwinism, Nihilism, Atheism, and the Theory of Evolution are true, then we will observe entropy.

Scientific Observation: We observe entropy all around us all the time.

Scientific Conclusion: Therefore, Materialism, Naturalism, Darwinism, Nihilism, Atheism, and the Theory of Evolution are true.

This IS the Scientific Method in action – the way that the Scientific Method is traditionally used to do science. Most people can't see anything wrong with it; but, it's based upon a logic fallacy called "affirming the consequent".

There's nothing wrong with *affirming the consequent*; is there? As long as it gets us the truth, it should be okay; shouldn't it?

Well, let's demonstrate what's wrong with *affirming the consequent* and *jumping to conclusions*.

If the Intelligent Design Theory is true, then we will observe natural selection.

We observe natural selection.

Therefore, the Intelligent Design Theory is true, and God clearly designed living organisms to be influenced by natural selection and sexual selection.

Notice how I first *affirmed the consequent*, and then I *jumped to the conclusion* that I wanted to be true.

By *affirming the consequent* and *jumping to conclusions*, we can prove anything to be true.

If the Christianity is true, then we will observe random mutations.

We observe random mutations.

Therefore, Christianity is true, and God clearly designed and created our genomes so that random mutations would be possible and would be observed.

Notice how I first *affirmed the consequent*, and then I *jumped to the conclusion* that I wanted to be true.

By *affirming the consequent* and *jumping to conclusions*, we can prove anything to be true; and, that's why *affirming the consequent* and *jumping to conclusions* are logic fallacies. By using logic fallacies, we can prove anything to be true.

If Materialism and Naturalism are true, then we will observe physical matter.

We observe physical matter.

Therefore, Materialism and Naturalism are true.

By *affirming the consequent*, Materialists, Naturalists, and Scientists prove that Materialism and Naturalism are true. This is the traditional Scientific Method in action. You see them online each and every day within their scientific arguments and research papers proving that Materialism, Naturalism, and the Theory of Evolution are true by *affirming the consequent* and *jumping to conclusions*. That's the way these people do science. We observe them doing so every day and for centuries at a time.

It's easy to prove anything true by *affirming the consequent*. These people do it all the time, and they don't even know it.

Scientific Hypothesis: If Darwinism and the Theory of Evolution are true, then we will observe DNA, genes, genomes, proteins, eyes, brains, and living life forms.

Scientific Observations: We observe DNA, genes, genomes, proteins, eyes, brains, and living life forms.

Scientific Conclusion: Therefore, Darwinism and the Theory of Evolution are true; and, Evolution clearly designed and created our DNA, genes, genomes, proteins, eyes, brains, and physical bodies.

Notice how I first *affirmed the consequent*, and then I *jumped to the conclusion* that I wanted to be true. It works every time, even though it is a logic fallacy.

We see the New Atheists using this argument all the time to make their case and prove that the Theory of Evolution is true. These people are *affirming the consequent* and *jumping to conclusions*, and they don't even know it.

I see these very same arguments in almost ALL of the college textbooks that I own and online every day. By *affirming the consequent* and *jumping to conclusions*, these scientists prove that the Theory of Evolution is true. They are completely oblivious and

441

unaware of the logic fallacies that they used to make their case. They were taught in college that *affirming the consequent* is the way that they are supposed to use the Scientific Method to do science. They were taught wrong; but, most of them don't even know it.

Falsifying Theories by Negating the Consequent

Now, let's use the Scientific Method for what it was designed for. Let's falsify some theories. Unlike *affirming the consequent*, *negating the consequent* or *falsifying a theory* is philosophically and logically sound. Here's how *falsifying a theory*, or *negating the consequent*, works in principle using the Scientific Method:

Scientific Hypothesis: If Theory X is true, then we will observe Y.

Scientific Observations: We don't observe Y.

Scientific Conclusion: Therefore, Theory X is false and has been falsified by the Scientific Method.

Here's how *negating the consequent* or *falsifying a theory* works in practice:

Scientific Hypothesis: If the Theory of Evolution, Materialism, Naturalism, and Darwinism are true, then we will observe the rocks and physical reactions designing, creating, and manufacturing genomes and life forms from scratch. If the Theory of Evolution is true, we will observe chemical evolution, abiogenesis, and spontaneous generation all around us all the time in real time.

Scientific Observations: We have NEVER observed the rocks and physical reactions designing and creating genomes and life forms; and, we NEVER will. They can't. The chemical evolution, abiogenesis, or spontaneous generation of proteins and genes from atoms is physically impossible and has NEVER been observed.

The Scientific Conclusion: Therefore, the Theory of Evolution, Materialism, Naturalism, and Darwinism are false and have been falsified by the Scientific Method or Scientific Observations.

I just successfully falsified Materialism, Naturalism, Darwinism, and the Theory of Evolution; and, I used the Scientific Method to do so! Best of all, my argument is philosophically and logically sound. I have in fact **falsified** Materialism, Darwinism, Naturalism, Creation by Rocks, and the Theory of Evolution for REAL! I have PROVEN them false! It's that simple! I successfully used the Scientific Methods for what they are good for – falsifying theories which are false. I wish I would have known how to do that forty years ago. It would have saved me a lot of confusion, frustration, time, and grief.

Materialism, Naturalism, Atheism, and their derivatives are falsified by a complete lack of observational evidence supporting their Major Premises or Hidden Assumptions. That's what falsifying a theory means – eliminating every premise, theory, and conclusion that has NEVER been experienced nor observed.

Now, let's run Atheism through the Scientific Method by *negating the consequent* to falsify it:

Scientific Hypothesis: If Atheism is true, we will observe "nothing" designing, creating, and manufacturing everything, including genomes and

life forms. It would be chaos, but that's exactly what we should be observing if Atheism were true.

Scientific Observations: Obviously, we have NEVER observed "nothing" designing and creating anything; and, we NEVER will. It's patently absurd.

The Scientific Conclusion: Therefore, Patent Absurdity or Atheism is false and has been falsified by the Scientific Method and Scientific Observations.

I just falsified Atheism for REAL; and, it's philosophically and logically sound. I have in fact falsified Atheism, using the Scientific Method and Scientific Observations to do so!

Materialism, Naturalism, Atheism, and their derivatives are falsified by a complete lack of observational evidence supporting their Major Premises or Hidden Assumptions. That's what falsifying a theory means – eliminating every premise, theory, and conclusion that has NEVER been experienced nor observed. Materialism, Naturalism, Darwinism, and Atheism have been falsified or proven false. There's no way to observe that something does not exist. There's no way to observe that the non-physical does not exist. There's no way to observe that the supernatural or the quantum does not exist. There's no way to observe that an after-life does not exist. There's no way to observe that life has no purpose and no meaning. There's no way to observe that God does not exist. There's NO way to use *affirming the consequent* to prove the truth.

However, you can definitely use *negating the consequent* to prove that your favorite theories are false. Such a course of action is philosophically valid and scientifically sound.

Remember, there's only one way to use the Scientific Method "to prove the truth" and that is to use *negating the consequent* and the Scientific Methods to eliminate everything that is false so that ONLY the truth or ONLY the observed remains. It's fascinating to observe what remains after eliminating everything that you know to be false or everything that has been proven false, such as Materialism, Naturalism, Darwinism, Nihilism, Atheism, and their derivatives.

Falsifying the Theory of Evolution

Now, let's use the Scientific Method for what it was designed for. Let's falsify some of people's favorite theories. *Negating the consequent* or *falsifying a theory* is philosophically valid and logically sound. *Negating the consequent* is NOT a logic fallacy. *Negating the consequent* is the way that Science should be done and the way that the Scientific Methods should be used.

Sherlock Holmes taught me how this works in principle or practice.

How often have I said to you that when you have eliminated the impossible [and the false]**, whatever remains, however improbable, must be the truth? — Sherlock Holmes.**

Sherlock Holmes and the Philosophy of Science taught me how to use the Scientific Method to falsify the Theory of Evolution, Materialism, Naturalism, and Atheism.

Here's how *falsifying a theory*, or *negating the consequent*, works in principle using the Scientific Method:

Scientific Hypothesis: If Theory X is true, then we will observe Y.

Scientific Observations: We don't observe Y.

Scientific Conclusion: Therefore, Theory X is false and has been falsified by the Scientific Method.

This scientific argument is philosophically valid and logically sound. It works. Theory X has been proven false by the Scientific Method. Theory X has been falsified by a complete lack of observational evidence supporting the hypothesis. That's how you falsify a theory! It's dependable. It works.

One of the primary tenets of Science 2.0 is the observation that if we successfully eliminate everything that is false or everything that has been falsified, then ONLY the truth will remain. Under Science 2.0, the truth is found by eliminating everything that is false, everything that is impossible, and everything that has been falsified.

Observe that *negating the consequent* or *falsifying a theory* requires careful consideration of what one is really trying to prove. In other words, *negating the consequent* requires a great deal more intelligence and precision than *affirming the consequent*.

Let's use *negating the consequent* and the Scientific Method to falsify a theory or two. This can be a great deal of fun to do.

Scientific Hypothesis: If Chemical Evolution is true, then we will observe atoms spontaneously generating into DNA, genes, genomes, and proteins.

Scientific Observations: We have NEVER observed atoms spontaneously generating into DNA, genes, genomes, and proteins. It's physically impossible. It's prevented from happening by entropy or the second law of thermodynamics.

Scientific Conclusion: Therefore, Chemical Evolution is false and has been falsified by the Scientific Methods. In other words, Chemical Evolution is falsified by a complete lack of observational support. Chemical Evolution or spontaneous generation has NEVER been observed.

Thanks to entropy or the second law of thermodynamics, it's physically impossible for atoms to spontaneously generate into functional proteins and the matching genes to go along with those proteins. Chemical Evolution is physically impossible. It's prevented from happening by entropy. That's the REAL science. Chemical Evolution has been falsified by a complete lack of supporting evidence. Since Chemical Evolution is false and impossible, the Theory of Evolution is false and impossible.

Let's try another one by using the *negating the consequent* version of the Scientific Method:

Scientific Hypothesis: If Macro-Evolution is true, then we will observe cats occasionally giving birth to dogs, and we will observe the occasional ape giving birth to humans.

Scientific Observations: We have NEVER observed one species giving birth to a completely different species. It's physically impossible. It's prevented from happening by genetics and random mutations.

Scientific Conclusion: Therefore, Macro-Evolution is false and has been falsified by the Scientific Method. In other words, Macro-Evolution is falsified by a complete lack of observational support.

Thanks to entropy or the second law of thermodynamics, it's physically impossible for random mutations to produce a functionally compatible or genetically compatible Mr. and Mrs. Mutant at the very same time in the very same place. It can't be done, which means that it wasn't done. Thanks to random mutations or entropy, it's physically impossible for chimp-like ancestors to give birth to genetically compatible male and female chimpanzees, or genetically compatible male and female human beings. It can't be done, which means that it wasn't done. That's the REAL science.

Macro-Evolution is false because it is prevented from happening by entropy, random mutations, and genetics. Macro-Evolution has NEVER been experienced nor observed; and, it NEVER will be, because it is physically impossible.

Scientific Hypothesis: If the Theory of Evolution is true, then we will observe natural selection and random mutations producing new and unique life-forms in real-time. It should be happening here and now, and not in some distant unseen past.

Scientific Observations: We have NEVER observed one species giving birth to a completely different species. It's physically impossible. It's prevented from happening by genetics and random mutations.

Scientific Conclusion: Therefore, the Theory of Evolution is false and has been falsified by the Scientific Method. In other words, the Theory of Evolution is falsified by the fact that random mutations and natural selection have NEVER been caught in the act of designing and creating new and unique life-forms from scratch.

These are powerful observations, which just happen to be true. The Theory of Evolution is falsified by a complete lack of observational support for its major claim or primary premise, which states that natural selection and random mutations designed and created ALL of the life forms on this planet. In truth, evolution (genetic change), random mutations, and natural selection didn't even exist until AFTER God designed, created, and manufactured the genes, proteins, genomes, eyes, brains, and physical life forms in the first place. Your genome is God's Signature. This is Logic 101; and, it falsifies the Theory of Evolution. Evolution, natural selection, and random mutations didn't exist and couldn't exist until after Someone Psyche or Someone Intelligent designed and created the genes and the genomes in the first place.

Furthermore, random mutations and natural selection cannot do computer programming or genetic programming. A genome is programming code and computer hardware. Thanks to entropy, it's physically impossible for random mutations to design, program, engineer, create, and manufacture functional genomes from scratch. Random mutations are entropy, NOT designers and creators.

The false is falsified by the truth. The false is falsified by observation and common sense; whereas, the truth is repeatedly experienced and observed by Someone Psyche somewhere sometime.

Let's falsify a couple other theories by using the *negating the consequent* version of the Scientific Method:

Scientific Hypothesis: If the claims of Materialism and Naturalism are true, then we will observe atoms spontaneously generating into DNA, genes, genomes, and proteins.

Scientific Observations: We have NEVER observed atoms spontaneously generating into DNA, genes, genomes, and proteins. In fact, spontaneous generation was falsified in 1859 by Louis Pasteur.

Scientific Conclusion: Therefore, Materialism and Naturalism are false and have been falsified by the Scientific Method. In other words, Materialism and Naturalism are falsified by a complete lack of observational evidence supporting their hidden assumptions or major premises which claim that the non-physical, the supernatural, or the quantum does not exist. It's physically impossible to observe that something does not exist.

The Materialists and Naturalists *jump to the conclusion* that only physical matter exists and therefore atoms must spontaneously generate into DNA, genes, proteins, genomes, and life forms. They are wrong. Spontaneous Generation, Abiogenesis, Materialism, Naturalism, Darwinism, and the Theory of Evolution were FALSIFIED by Louis Pasteur in 1859, the very same year that Charles Darwin published "On the Origin of Species". We knew the truth from the very beginning but chose to ignore it.

Once we have successfully falsified Materialism and Naturalism, then the Theory of Evolution and Atheism go down with them as fruit from the poisoned tree. Atheism – the idea that "Nothing" designed and created everything – is patently absurd. Atheism is creation ex nihilo – creation from nothing by nothing. Atheism or creation ex nihilo is obviously false to any rationally thinking person who comes across it. "Nothing" cannot design and create anything.

Jumping to conclusions, the Materialists, Naturalists, Darwinists, Nihilists, and Atheists define science as Materialism, Naturalism, Darwinism, Nihilism, and Atheism. That's the way these people do science. It's *begging the question* – using their pre-chosen conclusion as scientific evidence that their pre-chosen conclusion is true. It's *circular reasoning*.

The Materialists and Naturalists define science as Materialism and Naturalism, thereby making Materialism and Naturalism axiomatically true and thereby proving that Materialism and Naturalism are true. That's *begging the question*. By *affirming the consequent*, *begging the question*, and *jumping to conclusions*, you can prove anything to be true – even Materialism, Naturalism, Atheism, and the Theory of Evolution. It works, because logic fallacies can be used to prove anything true. It's called sophistry. Self-deception works, and it works every time.

In contrast, *negating the consequent* or *falsifying a theory* involves careful examination of the observational evidence and careful examination of the hidden conclusions that these people are jumping to.

I wish I would have known about *negating the consequent* or *falsifying theories* forty years ago when I was first in college; but, these are concepts that you will NEVER learn about from the Materialists, Naturalists, and Atheists who control and run our public schools. This type of information is banned from our public schools and science labs because it falsifies Materialism, Naturalism, and Darwinism. This type of information is forbidden in our public schools. The purpose of our public schools is to indoctrinate you in Materialism, Naturalism, Darwinism, Scientism, Nihilism, Behaviorism, Determinism, Classical Physics, and Atheism. Yet, these falsifiable and falsified philosophies or religions

446

are in fact FALSIFIED by *negating the consequent*. The truth can be found through a process of elimination by falsifying and eliminating from science anything that is false.

This is the way that science should be done – by falsifying and eliminating theories. This is the way that science should have been done but wasn't. Instead, by defining science as Materialism, Naturalism, Darwinism, Nihilism, and Atheism, falsified theories are retained as the core foundation of the physical sciences and classical physics. In other words, these people cheat and use logic fallacies to make their case and prove that Materialism, Naturalism, and the Theory of Evolution are true.

When it comes to Science and my understanding of the Scientific Methods, the eight months I spent studying the Philosophy of Science, *affirming the consequent*, and *negating the consequent* paid the most dividends in the end. It is powerful stuff and very interesting too. By using the *negating the consequent* version of the Scientific Method, I have successfully FALSIFIED Materialism, Naturalism, Darwinism, Atheism, and the Theory of Evolution many different times in many different ways.

In order to find and know the truth, we must successfully eliminate everything that is false, starting with Materialism, Naturalism, Darwinism, Nihilism, Atheism, and their derivatives.

Negating the consequent or *falsifying theories* was really cool once I discovered how it truly works. Now that I KNOW how *negating the consequent* works and how it can be used to prove theories false, there's NO going back to *affirming the consequent* and the falsehoods and lies that it is being used to produce. Cool, huh?

Using the Scientific Method to falsify Materialism, Naturalism, Darwinism, Atheism, and the Theory of Evolution just might be the greatest scientific discovery that I have made during my lifetime. I certainly find it powerful and useful. I wish I would have known how to falsify these false philosophies or false religions forty years ago when I was first in college. It would have saved me a ton of time, confusion, and grief. All you want is the truth, because falsehoods and falsified theories don't do anyone any good, unless you are making a living off the lies.

What do you think?

Is my discovery of the difference between *affirming the consequent* and *negating the consequent* one of my greatest scientific discoveries of all time; or, am I deluding myself? Self-deception works, and it works every time. Am I deceiving myself; or, do you think I have a point?

Obviously, I'm not the first person to discover *affirming the consequent* and *negating the consequent*; but, as far as I know, I am the first person to apply them directly to Materialism, Naturalism, Darwinism, Nihilism, Atheism, and the Theory of Evolution.

Ever since I successfully falsified Materialism, Naturalism, and their derivatives, I HAVE KNOWN that they are false because I HAVE KNOWN why they are false. That knowledge is particularly valuable to me because I used to be a Materialist, Naturalist, Nihilist, and Atheist. In my ignorance, I couldn't see anything wrong with the arguments and scientific proofs that were being presented to me by these people. I fell for it – hook, line, and sinker.

But now, I have seen the light.

Science 2.0 Is the Way that Science Should Have Been Done

Science 2.0 deliberately allows ALL of the evidence into evidence, and the goal is to pursue a preponderance of the evidence. Like the forensic sciences, Science 2.0 relies upon a Burden of Proof Methodology when it comes to the non-physical and the non-replicable; and, a Burden of Proof Methodology is dependent upon a preponderance of the evidence just like in a court of law. The goal is to get beyond a reasonable doubt.

The cool thing about Science 2.0 is that it allows into evidence EVERYTHING that we have observed and experienced as a race concerning the physical realm, as well as EVERYTHING that we have lived and experienced and observed in respect to the quantum realm or the supernatural realm.

Science 2.0 defines "science" as observation and experience. Under Science 2.0, nothing is off-limits or out-of-bounds. ALL of the evidence is allowed into evidence, as science should be done. When it comes to the quantum realm, or the supernatural realm, or the transdimensional realm, or the spirit world, Science 2.0 relies upon the observed, verified, and proven evidence from Quantum Mechanics, or Transdimensional Physics, or Supernatural Physics to explain what's happening in the quantum realm. It works, if allowed to work.

Science 2.0 demands that we find an Interpretation of Quantum Mechanics that explains in detail what the Human Psyche and Nature's Psyche are doing at the quantum level in order to get things done for us at the physical level. I found two such Interpretations for Quantum Mechanics, and I discuss them in detail in my books about Quantum Mechanics:

"Quantum Neuroscience: The Answer to Life, the Universe, and Everything".

"Quantum Mechanics from a Non-Physical Spiritual Perspective".

How often have I said to you that when you have eliminated the impossible [and the false], **whatever remains, however improbable, must be the truth? — Sherlock Holmes.**

This is Science 2.0 in action. The goal of Science 2.0 is to eliminate the impossible, the false, and the falsified while at the same time keeping and promoting the experienced and the observed.

If we successfully eliminate everything that is false, then only the TRUTH will remain. It's fascinating to observe what remains after Materialism, Naturalism, Darwinism, Nihilism, Behaviorism, Determinism, Scientism, Atheism, and the Theory of Evolution have been FALSIFIED and ELIMINATED from science.

The truth remains.

The VERIFIED, the OBSERVED, and the EXPERIENCED remain. The eye-witness evidence, the experiential evidence, the observational evidence, and the empirical evidence remain. Science is observation, or it should be. Science 2.0 defines science as Observation and Experience. The TRUTH is repeatedly verified, repeatedly experienced, and repeatedly observed by everyone who is willing to observe and learn. Science 2.0 allows ALL of the evidence into evidence and then pursues a preponderance of the evidence. Science 2.0 uses a Burden of Proof Methodology, which is based upon a preponderance of the observed evidence. Science 2.0 is how science should have been done but wasn't.

According to Science 2.0, if we successfully eliminate everything that is false, then ONLY the TRUTH will remain. According to Science 2.0, the BEST and FASTEST way to find and know the truth is to observe it, live it, and experience it for yourself, or to choose to trust someone who has. According to Science 2.0, science is observation, and NOT philosophical speculation and wishful thinking. According to Science 2.0, Materialism, Naturalism, Darwinism, Nihilism, Atheism, and the Theory of Evolution are FALSIFIED due to a complete lack of observational evidence supporting their hidden assumptions or major premises which claim that the supernatural or the quantum does not exist.

Quantum Mechanics is Supernatural Mechanics – it's non-physical in nature and origin. Quantum Mechanisms or Psychic Mechanisms in their natural state or their original non-local non-physical state are NOT limited by physical laws nor physical restrictions. Quantum Mechanisms are Syntropy, which means that they are not subject to entropy in their original primal state. In a physical state, however, Quantum Mechanisms, Supernatural Mechanisms, or Psyche Mechanisms are dampened, limited, and restrained by physical matter and entropy. This is the way things really work, at both the quantum level and the physical level.

Science 2.0 chooses to give precedence to quantum mechanisms or supernatural mechanisms, because they have been observed, experienced, and verified. Likewise, Science 2.0 eliminates Materialism, Naturalism, and Darwinism from science because they are falsifiable and have been falsified due to a complete lack of observational evidence supporting their hidden assumptions and hidden conclusions which claim that the supernatural or the quantum does not exist.

Under Science 2.0, the psyche level or the quantum level or the supernatural level remains, because it has been OBSERVED, EXPERIENCED, and repeatedly VERIFIED by lived experiences and scientific experimentation. Quantum Mechanics is our best-proven, most-verified, and most-used science that we have. Quantum Mechanics is Supernatural Mechanics. Likewise, the physical level remains under Science 2.0, because the physical level has been OBSERVED, EXPERIENCED, and repeatedly VERIFIED by lived experiences and scientific experimentation.

In contrast, Materialism and Naturalism are falsified by Science 2.0, because the non-existence of the supernatural or the non-existence of the quantum has NEVER been observed, experienced, nor verified. Materialism, Naturalism, and their derivatives are FALSIFIED by observational evidence to the contrary. The observation and experience of Quantum Mechanisms, or Psychic Mechanisms, or Supernatural Mechanisms FALSIFIES Materialism, Naturalism, and their derivatives.

Under Science 2.0, design and creation by Intelligent Beings remains because it has been OBSERVED, EXPERIENCED, and repeatedly VERIFIED by lived experiences and scientific experimentation. Science 2.0 retains what is TRUE because it retains what has been observed, experienced, lived, and remembered. Likewise, Science 2.0 consistently eliminates everything that is false and has been falsified like Materialism, Naturalism, Darwinism, Nihilism, Scientism, Behaviorism, Determinism, Atheism, and the Theory of Evolution.

Science 2.0 is the way that science should have been done but wasn't.

Mark My Words

Source

Using THE SCIENTIFIC METHOD to Eliminate the Usual Suspects and to Prove the Truth.

https://www.amazon.com/dp/B01J6STHP0

https://www.amazon.com/dp/1521133581

References

Rychlak, J. F. (1981). *A Philosophy of Science for Personality Theory* (2nd ed.). Malabar, FL: Robert E. Krieger Publishing Company.

Slife, B. D. & Williams, R. N. (1995). Science and Human Behavior. In *What's Behind the Research? Discovering Hidden Assumptions in the Behavioral Sciences*, (pp. 167–204). Thousand Oaks, CA: SAGE Publications.

http://mypsyche.us/science/

Science 2.0: I Upgraded My Science.

https://www.amazon.com/dp/B0771K6WTX

Quantum Neuroscience: The Answer to Life, the Universe, and Everything.

https://www.amazon.com/dp/B079Z6QQQB

Quantum Mechanics from a Non-Physical Spiritual Perspective.

https://www.amazon.com/dp/B01J023TGU

https://www.amazon.com/dp/1521132380

Analysis of Falsification

What did we just learn from this exercise?

We learned that the Scientific Methods can be used to prove theories false or to falsify theories. Technically, according to the laws of logic, the Scientific Method cannot be used to prove anything true, due to the *affirming the consequent* logic fallacy; however, the Scientific Methods can definitely be used to falsify theories and thereby prove that they are false. Falsifying a theory means proving that a theory is false. The Scientific Methods are used to prove theories false; and, it works. In contrast, *affirming the consequent* can be used to trick us and deceive us. Self-deception works, and it works every time.

What else did we learn from this exercise in falsification?

We learned that a lack of observation FALSIFIES theories. That's what we really learned, isn't it? We learned that a lack of observation or a lack of observational evidence is the foundational hallmark of Materialism, Naturalism, Darwinism, Nihilism, and Atheism. These philosophical theories are FALSE because there is NO observational evidence supporting their major premises or primary assumptions which claim that the quantum, or the supernatural, or the non-physical does not exist.

We learned that there is no evidence and can be no evidence for the claim that Psyche does not exist. We learned that there is no evidence and can be no evidence for the claim that the Non-Local or the Non-Physical does not exist. We learned that there is no evidence and can be no evidence supporting the materialistic and naturalistic claim that the quantum level or the psyche level does not exist. We learned that ALL of the observational evidence and experimental evidence that we have on hand as a race FALSIFIES Materialism, Naturalism, Darwinism, Nihilism, Behaviorism, Scientism, Determinism, Physical Reductionism, and Atheism.

We learned that science begins by allowing ALL of the evidence into evidence. We learned that whenever science is done right, by *negating the consequent* or by falsifying and eliminating theories that are false, we are pointed to the truth as a result. We learned that the truth is discovered by falsifying and eliminating Materialism, Naturalism, Darwinism, Atheism, and their derivatives. We will learn that it can be extremely fascinating and informative to study what remains after everything that is false has been deliberately eliminated.

We learned that if you successfully eliminate everything that is false, then only the truth will remain. If you successfully eliminate everything that has NEVER been experienced nor observed by anyone, then ONLY the truth will remain. If you successfully allow ALL of the evidence into evidence, then ONLY the truth will remain. We learned that science is personal experience and observation.

We learned that we don't have to run any science experiments in order to find and know the truth. We can go straight to KNOWING the TRUTH simply by observing it, experiencing it, and living it for ourselves. We learned that the BEST and the fastest way to find and know the truth is to live it and experience it for yourself, or to choose to trust someone who has. We learned that the second-best way to find and know the truth is to eliminate everything that is false so that only the truth remains. We learned that the discovery of truth starts by eliminating Materialism, Naturalism, Darwinism, Nihilism, Atheism, and their derivatives.

Powerful, is it not?

We learned that there is NO empirical evidence and will NEVER be any observational evidence supporting the major premises or primary assumptions of Materialism, Naturalism, Darwinism, and Atheism. We learned that the Scientific Method FALSIFIES Materialism, Naturalism, Darwinism, and Atheism. We learned that there is NO evidentiary support for Materialism, Naturalism, Darwinism, Atheism, and the Theory of Evolution. There are NO observations and can NEVER be any observations supporting the claims of Materialism, Naturalism, Darwinism, and Atheism, because these things have to be taken on blind faith as being real and true, because they can't be experienced nor lived nor learned from.

We learned that there is NO phenomenological support or observational support for Materialism, Naturalism, Darwinism, and their derivatives. These things are pure hypothesis, pure philosophy, pure fiction, pure speculation, pure dogma, pure religion, pure pseudo-science, and the very pinnacle of wishful thinking.

We learned that Materialism, Naturalism, Darwinism, Nihilism, Atheism, and the Theory of Evolution are FALSE because there is no observational evidence supporting them. We learned that science is OBSERVATION. We learned that you can get a lot of mileage out of the truth, especially when that truth is experienced and observed.

We learned that most of us have been brainwashed and conditioned in our public schools not to be able to see, understand, nor accept these realities and truths. You will NEVER learn anything about any of this from your materialistic, naturalistic, and atheistic professors and teachers in our public schools because this kind of information has been deliberately banned, blocked, censored, and eliminated from the curriculum. The purpose of our public schools is to indoctrinate us into Materialism, Naturalism, Darwinism, Nihilism, Scientism, Behaviorism, Determinism, and Atheism.

Fascinating, is it not?

It explains what has been done to us. It answers a lot of unasked questions, doesn't it? Instantly, the first fifty years of my life made sense to me, as it finally dawned on me how I had been tricked and deceived all of my life. Ye shall know the truth, and the truth shall set you free. My hope is to set a lot of people free, as I have been set free.

Ask yourself, "What hidden assumptions do the Materialists, Naturalists, Darwinists, Nihilists, and Atheists make at the beginning of each of their scientific arguments or scientific proofs, every time they try to do science?"

These people always start with the hidden assumption that Psyche, Spirit Matter, Quantum Non-Local Consciousness, Syntropy, God, Action at a Distance, Quantum Non-Locality, and Quantum Mechanisms do NOT exist. *Begging the question*, these people always start with these hidden assumptions; and therefore, these people always come to the very same conclusions – there is NO such thing as Syntropy, God, Quantum Non-Locality, Action at a Distance, Psyche, Choice, Spirit Matter, Consciousness, or Eternal Life. Garbage in, then garbage out. These hidden assumptions are inserted at the beginning of every argument that these people make, whether they realize it or not. Consequently, these people always come to the same conclusions. It's unavoidable. If you start your science with false assumptions, you are going to end up with false conclusions when you are done. Simple. Logical. True.

There's no way possible to observe that Psyche and Syntropy do not exist. It's physically impossible to prove that Psyche and Syntropy do not exist. There's no way to observe that spirit matter does not exist. There's no way to observe that God does not exist. There is no way to observe that the quantum realm, the transdimensional realm, the spirit realm, or the non-local realm does not exist. In fact, Materialism, Naturalism, Darwinism, Nihilism, and Atheism completely fall apart and fail each and every time someone sees, talks with, and OBSERVES our Resurrected Lord Jesus Christ. Observation destroys Materialism, Naturalism, Darwinism, and their derivatives.

Each time we OBSERVE or experience Syntropy, God, Action at a Distance, Resurrected Beings, Consciousness, Quantum Mechanics, Psyche or Choice, Thoughts and Memories, and the Spirit World, those observations and experiences FALSIFY Materialism, Naturalism, Darwinism, Nihilism, Behaviorism, Scientism, and Atheism. That's the way evidence works! Evidence falsifies everything that is false. Science is observation, or it should be. Observation falsifies Materialism, Naturalism, and their derivatives. Observations should be given precedence and priority over philosophical speculations and wishful thinking, especially when it comes to science.

The very existence of Quantum Mechanics proves that the Materialists and Naturalists are wrong. Quantum Mechanics is Syntropy. The very existence of Quantum Mechanics, Syntropy, Action at a Distance, Quantum Non-Locality, and Psyche or Quantum Non-Local Consciousness FALSIFIES Materialism, Naturalism, Darwinism, and their derivatives.

Quantum Mechanics is very difficult to falsify, due to the fact that it exists at the quantum level or the psyche level. However, Quantum Mechanics is constantly and repeatedly VERIFIED each and every time we get smart enough as a race to design an experiment that will indeed verify it or use it. Quantum Mechanics is verified by the effects that it has on physical matter. Syntropy, Quantum Mechanics, Quantum Non-Local Consciousness, Intelligent Intervention, Action at a Distance, or Transdimensional Physics is very REAL; and, Quantum Mechanics is in fact our most-verified, best-proven, and most-used science on the planet right now.

WE KNOW that Materialism, Naturalism, Darwinism, Nihilism, Behaviorism, Scientism, and Atheism are FALSE, because there is NO observational evidence supporting their hidden assumptions, major premises, or primary assumptions. Materialism, Naturalism, Darwinism, and their derivatives are unscientific because there is NO observational evidence supporting their major premises or primary assumptions. That's the way science works; or at least, that's the way science is supposed to work.

Anything that has been successfully falsified should be eliminated from science as being unscientific. If you want to find and know the truth, then you must eliminate everything that is false and everything that has been falsified, starting with Materialism, Naturalism, Darwinism, Nihilism, Behaviorism, Scientism, Atheism, and the Theory of Evolution. There is no other way to find and know the truth. The truth is what remains after everything that is false has been eliminated.

Remember, entropy degenerates genomes. It doesn't improve and enhance them. It can't. It's physically impossible. Entropy cannot do order, organization, design, programming, engineering, fine-tuning, field-testing, and manufacturing. Entropy makes chemical evolution, abiogenesis, spontaneous generation, and macro-evolution physically impossible. Random mutations are entropy. Natural selection is a dead-end, death, entropy, or the extinction of species. These things destroy and end genomes. They do NOT enhance and improve genomes. Evolution of any kind is entropy. It can't design and create.

Begging the Question

One of my most important and significant discoveries in Science and Philosophy is the discovery that Materialism, Naturalism, Darwinism, Nihilism, Atheism, and their derivatives are based exclusively on *Begging the Question*, which is a logic fallacy. Materialism, Naturalism, and Atheism have been based upon the *Begging the Question* logic fallacy since the beginning of recorded history.

So, what is *Begging the Question*?

Begging the Question is a logic fallacy wherein they take their pre-chosen conclusion and insert their pre-chosen conclusion as evidence and proof that their pre-chosen conclusion is true. In other words, these people are using their pre-chosen conclusions as their Major Premises or Hidden Assumptions in every scientific argument, scientific interpretation, scientific inference, and scientific proof that they make.

So, what are these Hidden Assumptions or Pre-Chosen Conclusions that these people are using to make their case and prove that they are right? They have been doing so for thousands of years. Wouldn't you want to know?

Materialism is based upon the pre-chosen conclusion that the non-physical does not exist. Therefore, these people (often unknowingly) insert the **non-existence of the non-physical** as their first Major Premise or Hidden Assumption; and thereafter, these people ONLY consider physical matter, and they completely and deliberately exclude and refuse to do research into all the different invisible, non-physical forces, fields, and entities which have been experienced, observed, and proven to exist.

Naturalism is based upon the pre-chosen conclusion that the supernatural or the quantum does not exist. Naturalism is also based upon the pre-chosen conclusion that Psyche, Spirit Matter, and Quantum Non-Local Consciousness do not exist. These people deny the existence of Intelligence, Psyche, or the Life Force. It's philosophically impossible to verify, or to prove, or even to falsify such a pre-chosen conclusion; but, they don't care. These people deny the existence of Supernatural Mechanisms or Quantum Mechanisms; and therefore, these people insert the **non-existence of the supernatural** or the **non-existence of the quantum** as their first Major Premise or Hidden Assumption in every scientific argument, scientific proof, scientific interpretation, and scientific inference that they make. Consequently, these people ONLY consider nature or classical physics, and they completely and deliberately exclude the quantum, the supernatural, the psyche, or Syntropy from consideration when it comes to their science and their philosophy.

Nihilism is the philosophical belief that there is no after-life for the Psyche or the Soul, and therefore no such thing as a Quantum Realm, Syntropy Realm, Psyche Realm, or Spirit Realm. These people choose to believe that there is no such thing as the Non-Local Realm, Non-Physical Realm, or Transdimensional Realm. These people use the **non-existence of psyche**, and the **non-existence of spirit matter**, and the **non-existence of syntropy** as scientific proof that these things do not exist. They *beg the question*. These people insert the non-existence of these things into every scientific argument, scientific proof, scientific interpretation, and scientific inference that they make as evidence or proof that their pre-chosen conclusions are true. These people use the non-existence of these things as evidence or proof that these things do not exist. It's illogical and irrational; but, that's the nature of logic fallacies such as *begging the question*. Most of them don't even realize that they are *begging the question* and *affirming the consequent*. They do it automatically because they have been trained to do so in college by their college professors who were also Materialists, Naturalists, Darwinists, Nihilists, and Atheists.

Atheism is based upon the pre-chosen conclusion that God or God's Psyche does not exist. It's philosophically impossible to verify, or to prove, or even to falsify such a pre-chosen conclusion. Atheism is the design and creation of something by nothing. Atheism is irrational, illogical, and patently absurd. There's no logic to it whatsoever. These people deny the existence of a Maker, even when it comes to things that were obviously made, such as proteins and genomes; consequently, these people insert the **non-existence of God** or the **non-existence of Makers** as their first Major Premise or Hidden Assumption into every scientific argument, scientific proof, scientific interpretation, and scientific inference that they make. As a result of this pre-chosen conclusion, these people convince themselves that proteins, genes, genomes, eyes, brains, and life forms spontaneously generate out of thin air all the time. Such a belief or claim has NEVER been experienced nor observed. In fact, Louis Pasteur falsified spontaneous generation in 1859, the very same year that Charles Darwin published "On the Origin of Species". It's ironic, but Louis Pasteur falsified Spontaneous Generation, Abiogenesis, Creation Ex Nihilo, Materialism, Naturalism, Darwinism, Chemical Evolution, Macro-Evolution, and the Theory of Evolution in 1859; and, the Naturalists have been denying that fact ever since.

Darwinism is a carefully designed combination of Materialism, Naturalism, Nihilism, and Atheism. Darwinism or the Theory of Evolution is based upon the **non-existence of God**, the **non-existence of the non-physical**, the **non-existence of the supernatural**, the **non-existence of psyche**, and the **non-existence of the quantum realm** or transdimensional realm. These people quietly insert the non-existence of these things as their Major Premise or Hidden Assumption into every science argument, science proof,

454

interpretation of data, and scientific inference that they make. These people use their pre-chosen conclusions as evidence and proof that their pre-chosen conclusions are true.

This is how these people do Science and Philosophy, by excluding a priori anything and everything that falsifies Materialism, Naturalism, Darwinism, Nihilism, and Atheism. They cheat. They *beg the question*! They declare their pre-chosen conclusions to be Axioms, Laws, or Given Truths, and then insert those Axioms or Given Truths (their pre-chosen conclusions) as their Major Premise into every scientific argument, scientific proof, scientific interpretation, and scientific inference that they make. Most of them don't even realize that they are doing so; but, they are. It's what they do. It's what they have been trained to do in college.

These people also make a *special pleading* and adjust the rules of science and philosophy by axiomatically declaring that their First Premise in any argument that they make is allowed to be a Given Truth or an Axiom, and therefore doesn't have to be seen, verified, nor proven to be true. That's how they get around the rules of logic and the rules of science, by unilaterally declaring that their First Premise in any argument that they make is allowed to be an Axiom or a Given Truth. That's how these people trick us and deceive us. That's how they trick and deceive themselves. They cheat. They *beg the question* and *affirm the consequent*. They use *circular reasoning* to make their case and prove their point.

Because I understand the *begging the question* and the *affirming the consequent* logic fallacies, whenever I do philosophical proofs and scientific arguments, I demand that ALL of my premises as well as my conclusion be experienced and observed by Someone Psyche somewhere sometime, or I consider my philosophical arguments and scientific proofs to be worthless. I upgraded my science to Science 2.0. I don't allow Axioms or Given Truths to be used as the Premises in any of my scientific arguments or philosophical proofs. Everything – the premises and the conclusion – has to be based upon some kind of observational evidence, experiential evidence, or empirical evidence; or, I'm liable to be deceived. Using their pre-chosen conclusions as their First Premise, Given Truth, or Axiom is precisely how the Materialists, Naturalists, Darwinists, Nihilists, and Atheists deceive themselves. Self-deception works, and it works every time.

By using their pre-chosen conclusions as evidence or as proof that their pre-chosen conclusions are true, these people easily and convincingly prove to themselves and to others that their pre-chosen conclusions are indeed true. You can't miss by *begging the question*! By *begging the question* and *affirming the consequent*, you can prove anything to be true, including Materialism, Naturalism, Darwinism, Nihilism, and Atheism. *Begging the question* is cheating. It's a logic fallacy. But, these people don't care, because *begging the question* and *affirming the consequent* proves that their theories, philosophies, and ideas are true; and, that's all they really care about – winning the argument. And, these people have won millions if not billions over to their side.

When it comes to Materialism, Naturalism, Darwinism, and Atheism, the whole thing is a sham or a scam; and, most people don't even know it. Fascinating, is it not? Well, I think so. I used to be a Materialist, Naturalist, Nihilist, and Atheist. I'd been tricked and deceived just like everyone else; and, I didn't even know it. Philosophy (or religion and blind faith) has a definite impact in our lives, especially when it comes to Science. By using philosophy and *begging the question*, we can prove anything to be true.

This is the benefit of studying the Philosophy of Science and the various different logic fallacies that the Materialists and Naturalists use to trick us and deceive us. You can actually figure out how they are tricking you and deceiving you. You can also take steps to correct the problem.

It's time that we upgrade our science to Science 2.0.

Begging the question, traditional science is defined as Materialism, Naturalism, Darwinism, Behaviorism, Scientism, Determinism, Physical Reductionism, Nihilism, and Atheism. It's time for us to re-define Science as observation and experience. Let's get rid of the philosophical speculation and the *begging the question* logic fallacy, and switch over to observational evidence, experiential evidence, eye-witness evidence, and empirical evidence for our proof. Science 2.0 allows ALL of the evidence into evidence and pursues a preponderance of the evidence. Let's demand that every premise and every conclusion be experienced and observed by Someone Psyche somewhere sometime; or consider our scientific proofs and philosophical arguments to be falsified if there is no observational evidence supporting one of our premises or our pre-chosen conclusion. The different types of evolution have NEVER been caught in the act of design and creation; therefore, they are false and have been successfully falsified. They should be eliminated from consideration. That's the way science should be done but isn't.

Remember, Materialism, Naturalism, Darwinism, Nihilism, and Atheism are FALSIFIED by a complete lack of observational evidence supporting their Major Premises or Hidden Assumptions. These things should have been falsified and eliminated long ago, but they weren't because their proponents cheat and *beg the question* in order to make their case and prove their point. These falsified philosophies should have been eliminated from science, but they weren't. That's the way science should have worked but didn't. Materialism, Naturalism, Darwinism, Nihilism, and Atheism should be eliminated from science; but, they aren't because they have been declared axiomatically true and are then used as Axioms or Given Truths as the First Premise in every scientific argument, scientific inference, scientific interpretation, and scientific proof that these people make.

By *begging the question* and *affirming the consequent*, these people successfully prove that the Theory of Evolution, Materialism, Naturalism, Nihilism, and Atheism are true; and, millions if not billions have fallen for the ruse because they don't understand the Philosophy of Science and how things really work.

If you tear something down, you are obligated to try to build something better in its place, which I have also tried to do.

Mission Statement

After successfully using the Scientific Method and Logic to falsify Materialism, Naturalism, Darwinism, and Atheism, then I felt a compelling need to re-envision and re-do Science and Philosophy in order to bring them into line with the millions of Near-Death Experiences (NDEs) and Out-of-Body Experiences (OBEs) which are on record and that have been experienced and observed first-hand by real human beings or human psyches. In order to do so, I had to switch Science and Philosophy from a platform of speculation and wishful thinking based upon Materialism and Naturalism over to a new paradigm based upon the observations and verified proof that have been obtained from Quantum Mechanics or Supernatural Mechanics. In other words, I had to bring the quantum or the supernatural or the non-physical into Science and Philosophy in order to upgrade them, fix them, and bring them into the modern age.

As a result, Science 2.0 and the Ultimate Model of Reality were born and brought into existence.

Mark My Words

The Fruits of These Truths

Science 2.0 was the result of these science discoveries. If you study this essay carefully, you will notice that Science 2.0 follows naturally from *negating the consequent* and using the Scientific Method the way it was meant to be used. The Scientific Method was supposed to be used to falsify and eliminate everything that is false; but, it is seldom used that way by the Materialists, Naturalists, and Darwinists because the Scientific Method and *negating the consequent* falsifies Materialism, Naturalism, and their derivatives. Science 2.0 is discussed in much greater detail in the following book.

Science 2.0: I Upgraded My Science.

https://www.amazon.com/dp/B0771K6WTX

The Ultimate Model of Reality and Ultimate Cause were also the result of using the Scientific Method and Logic to successfully falsify Materialism, Naturalism, Darwinism, and Atheism. The purpose of the Ultimate Model of Reality, a Psyche Ontology, a Psyche Epistemology, and Ultimate Cause is to add a necessary Fifth Cause to Aristotle's Four Causes which were originally based upon Materialism, Naturalism, and Atheism. In other words, I added Syntropy, Psyche, Non-Locality, Quantum Mechanics, the Non-Physical, or the Supernatural to Aristotle's Four Causes so as to bring Philosophy and the Philosophy of Science into the modern age or the quantum age. The Ultimate Model of Reality is discussed in much greater detail in the following books.

The Ultimate Model of Reality: Psyche Is the Ultimate Cause

https://www.amazon.com/dp/B071NC9JK6

Putting Psyche Back into Psychology: Restoring Science to Consciousness

https://www.amazon.com/dp/B071NC987S

BioPsychoSocial: Including Psyche or Light into our Theoretical Models

https://www.amazon.com/dp/B0713NDHVW

NATURE vs. NURTURE vs. NIRVANA: An Introduction to Reality

https://www.amazon.com/dp/B01JWRCSVA

https://www.amazon.com/dp/1521132615

Source

Science 2.0: I Upgraded My Science.

https://www.amazon.com/dp/B0771K6WTX

Using THE SCIENTIFIC METHOD to Eliminate the Usual Suspects and to Prove the Truth.

https://www.amazon.com/dp/B01J6STHP0

https://www.amazon.com/dp/1521133581

Reference Materials

*Using **THE SCIENTIFIC METHOD** to Eliminate the Usual Suspects and to Prove the Truth*.

https://www.amazon.com/dp/B01J6STHP0

https://www.amazon.com/dp/1521133581

The Theory of Evolution Proved to Me that God Exists: Why I Am No Longer an Atheist and Why I No Longer Believe in the Theory of Evolution.

https://www.amazon.com/dp/B01HZYBZ7K

https://www.amazon.com/dp/1521131228

The Scientific Method Proves That the Theory of Evolution Is False.

https://www.amazon.com/dp/B01IAAIRT2

https://www.amazon.com/dp/1521133611

Summary Of: The Theory of Evolution Proves that God Exists.

https://www.amazon.com/dp/B01GQCWED6

https://www.amazon.com/dp/1521130485

Quantum Mechanics from a Non-Physical Spiritual Perspective.

https://www.amazon.com/dp/B01J023TGU

https://www.amazon.com/dp/1521132380

Quantum Neuroscience: The Answer to Life, the Universe, and Everything.

https://www.amazon.com/dp/B079Z6QQQB

Rychlak, J. F. (1981). *A Philosophy of Science for Personality Theory* (2nd ed.). Malabar, FL: Robert E. Krieger Publishing Company.

Slife, B. D. & Williams, R. N. (1995). Science and Human Behavior. In *What's Behind the Research? Discovering Hidden Assumptions in the Behavioral Sciences*, (pp. 167–204). Thousand Oaks, CA: SAGE Publications.

http://mypsyche.us/science/

—

What did we learn from this?

We learned that the dude wrote some books, and he wouldn't mind if you were to take some time to buy them and read them.

Philosophy of Science

I had to develop a new and better way of looking at the world and reality, because the one that I had gotten from my society wasn't working for me. It was working against me.

I don't have a lot of respect for philosophy and haven't seen much value in it, because philosophy is typically used to deceive people. Materialism, Naturalism, Darwinism, Nihilism, Behaviorism, Determinism, Scientism, and Atheism are philosophies or dogma, NOT science. They are religions, NOT science. They have to be taken on blind faith as being true, because there can NEVER be any observational evidence supporting their major premises or hidden assumptions which claim that the non-physical does not exist. In contrast, every observation and experience of the non-physical FALSIFIES Materialism, Naturalism, Darwinism, Nihilism, Atheism, and their derivatives.

Science is observation, or it should be.

There is NO observational evidence supporting the major premises or hidden assumptions of Materialism, Naturalism, and their derivatives. These philosophies and their philosophical arguments are used to trick us and deceive us; consequently, I have typically found philosophy to be unreliable and unsound because their arguments usually don't match with reality and are seldom based upon observational evidence.

An argument is *valid* if the conclusion necessarily follows from the premises.

An argument is *sound* if it is valid and the premises are true. Falsified premises cannot produce a *sound* argument.

Materialism, Naturalism, and their derivatives will NEVER be logically sound and will NEVER be philosophically sound because they will NEVER have any observational evidence supporting their major premises which claim that the quantum, or the spiritual, or the supernatural does not exist. Every observation and experience of the quantum, the spiritual, the supernatural, the transdimensional, and the non-physical FALSIFIES Materialism, Naturalism, Darwinism, Nihilism, Atheism, and their derivatives. In other words, repeated verification of Quantum Mechanics, Action at a Distance, or Supernatural Mechanisms FALSIFIES Materialism, Naturalism, and their derivatives. Materialism, Naturalism, Darwinism, Nihilism, Behaviorism, Determinism, Scientism, and Atheism cannot be true, because their premises are demonstrably false.

When it comes to philosophy, some have claimed that it is best left ignored. However, I see it differently. When it comes to philosophy, you absolutely MUST KNOW how these people are using philosophy to trick you and deceive you. If you don't, then you are going to be tricked and deceived. Billions of people have been tricked and deceived by the Materialists, Naturalists, Darwinists, Nihilists, and Atheists; and, they don't even know it. I used to be one of them. I used to be a Materialist, Naturalist, Nihilist, and Atheist.

Remember, when it comes to philosophy, your premises have to be flawless and perfect, or your conclusion is going to be wrong.

The Essential Philosophy of Science

The fundamental concept that you must understand about the Philosophy of Science is to know what *Affirming the Consequent* is, how it works, and what it does to the Scientific Method.

Affirming the Consequent is a logic fallacy, and it is the logic fallacy upon which Traditional Science, Materialism, Physicalism, Naturalism, Darwinism, and Atheism are based. *Affirming the Consequent* is the way that these people do science; and, they don't even know it, because they don't understand the Philosophy of Science.

By *affirming the consequent*, you can prove anything to be true, including Materialism, Naturalism, Atheism, and the Theory of Evolution; but, the process is nothing more than trickery and deception, which billions have fallen for over the millennia.

In my humble opinion, *affirming the consequent* is the most interesting and useful thing to know about and understand, when it comes to Science and the Philosophy of Science.

I discuss *Affirming the Consequent* in great detail in most of my books because its antithesis, *Negating the Consequent*, is the core foundation of my science and the way that I use the scientific methods. Verification uses *affirming the consequent*. Falsification uses *negating the consequent*. *Negating the Consequent* is philosophically and logically sound; whereas, *Affirming the Consequent* is not. Consequently, I use *negating the consequent* in all of my scientific arguments and while using the Scientific Method in order to eliminate everything that is false, so that I can increase my chances that my conclusions and interpretations of the evidence will actually be true.

Remember, if you successfully eliminate everything that is false, then ONLY the truth will remain. I use the scientific methods and *negating the consequent* to eliminate everything that is false, in the hope that ONLY the Truth will remain.

Mark My Words

—

Source

1. *The Ultimate Model of Reality: Psyche Is the Ultimate Cause*

   https://www.amazon.com/dp/B071NC9JK6

2. *Science 2.0: I Upgraded My Science*

   https://www.amazon.com/dp/B0771K6WTX

3. *Syntropy in Defense of Quantum Mechanics: The Answer to Life, the Universe, and Everything*

   https://www.amazon.com/dp/B07BPT3W8R/

4. Web Page:

   https://philosophy-of-science.com/philosophy-of-science/

5. The Official Website:

   https://philosophy-of-science.com/

References

1. Slife, B. D. & Williams, R. N. (1995). Science and Human Behavior. In *What's Behind the Research? Discovering Hidden Assumptions in the Behavioral Sciences*, (pp. 167–204). Thousand Oaks, CA: SAGE Publications.

   https://mypsyche.us/wp-content/uploads/2017/04/Science.pdf

   https://philosophy-of-science.com/wp-content/uploads/2018/04/Science.pdf

2. Gantt, E. (2014). Logical Arguments. In *Psychology 353 – LDS Perspectives in Psychology*, (pp. 8-11). Provo, UT: Brigham Young University.

   https://philosophy-of-science.com/wp-content/uploads/2018/04/Logical-Arguments.pdf

3. Gantt, E. (2014). Leveling the Playing Field – Why Science is Not a Trump Card. In *Psychology 353 – LDS Perspectives in Psychology*, (pp. 50-58). Provo, UT: Brigham Young University.

   https://philosophy-of-science.com/wp-content/uploads/2018/04/Verification-vs-Falsification.pdf

4. Rychlak, J. F. (1981). *A Philosophy of Science for Personality Theory* (2nd ed.). Malabar, FL: Robert E. Krieger Publishing Company.

5. **My Amazon Page:**

   https://amazon.com/author/science

6. *Science 2.0: I Upgraded My Science*

   https://www.amazon.com/dp/B0771K6WTX

Fixing Philosophy with the Ultimate Model of Reality

Rychlak, J. F. (1981). *A Philosophy of Science for Personality Theory* (2nd ed.). Malabar, FL: Robert E. Krieger Publishing Company.

Rychlak, J. F. (1970). The Human Person in Modern Psychological Science. *British Journal of Medical Psychology*, 43(3), 233–240.

https://mypsyche.us/wp-content/uploads/2016/12/Rychlak.pdf

Aristotle's Four Causes NEED a Fifth Cause

A couple of my college classes introduced me to Joseph Rychlak and Aristotle's four causes.

I found the whole thing severely disturbing.

Why?

Traditional philosophy is completely missing the Ultimate Cause or the Quantum Mechanics. The deliberate omission was glaringly obvious to me.

Traditional philosophy has limited itself exclusively to Materialism, Naturalism, Nihilism, and Atheism for millennia. Traditional philosophy is incomplete because it's lacking the Quantum Mechanics, Psyche Mechanisms, Supernatural Mechanisms, Non-Locality, Action at a Distance, Quantum Tunneling, Dark Matter or Spirit Matter, Dark Energy, the Strong and Weak Nuclear Forces, and even things like Gravity and Magnetism. Traditional philosophy is missing the Psyche, Syntropy, or the Ultimate Causal Agent. The verified and proven existence of Dark Energy, Dark Matter, and Quantum Mechanics FALSIFIES traditional philosophy, because traditional philosophy is based exclusively on Materialism, Naturalism, Darwinism, Nihilism, Behaviorism, Determinism, Physical Reductionism, Atheism, and Classical Physics.

I realized that it was finally time for one of us to seriously upgrade and fix traditional philosophy and bring it into the modern age. In order to do so, I had to add a fifth cause to Aristotle's four causes – the Ultimate Cause. Psyche is the Ultimate Cause. Psyche is the ultimate causal agent. Psyche or Quantum Non-Local Consciousness is the fundamental unit of reality.

These adjustments to traditional philosophy resulted in what I call a Psyche Ontology and a Psyche Epistemology, wherein Psyche is the fundamental unit of reality, Psyche or experience is the best way to know reality, Psyche is the ultimate causal agent, and Psyche is the Ultimate Cause. A Psyche Ontology is the Ultimate Model of Reality.

Mark My Words

—

Source

Syntropy in Defense of Quantum Mechanics: The Answer to Life, the Universe, and Everything

https://www.amazon.com/dp/B07BPT3W8R/

—

The books in which I develop and present the Ultimate Model of Reality, a Psyche Ontology, and the Ultimate Cause.

The Ultimate Model of Reality: Psyche Is the Ultimate Cause

https://www.amazon.com/dp/B071NC9JK6

Putting Psyche Back into Psychology: Restoring Science to Consciousness

https://www.amazon.com/dp/B071NC987S

BioPsychoSocial: Including Psyche or Light into our Theoretical Models

https://www.amazon.com/dp/B0713NDHVW

NATURE vs. NURTURE vs. NIRVANA: An Introduction to Reality

https://www.amazon.com/dp/B01JWRCSVA

https://www.amazon.com/dp/1521132615

Fixing the Scientific Method with the Philosophy of Science

Rychlak, J. F. (1981). *A Philosophy of Science for Personality Theory* (2nd ed.). Malabar, FL: Robert E. Krieger Publishing Company.

Slife, B. D. & Williams, R. N. (1995). Science and Human Behavior. In *What's Behind the Research? Discovering Hidden Assumptions in the Behavioral Sciences*, (pp. 167–204). Thousand Oaks, CA: SAGE Publications.

https://mypsyche.us/wp-content/uploads/2017/04/Science.pdf

Science 2.0: I Upgraded My Science.

https://www.amazon.com/dp/B0771K6WTX

Science NEEDS to Be Fixed and Upgraded

After reading "Science and Human Behavior," I was severely disturbed by their message.

Why?

I learned that Science and the Scientific Method have been based exclusively upon the *affirming the consequent* logic fallacy since the very beginning of Science and the Scientific Method. The net result is that Science was axiomatically defined as Materialism, Naturalism, Darwinism, Nihilism, Scientism, Determinism, Physical Reductionism, and Atheism BEFORE any actual observations were done. Our modern-day scientists automatically censor, ban, and delete anything and everything that falsifies Materialism, Naturalism, and the Theory of Evolution. That's dogma, extremism, and religion in action – NOT science. Science is supposed to be observation and experience, and NOT the hiding and the elimination of evidence.

By *affirming the consequent*, you can prove anything to be true, even Materialism, Naturalism, and the Theory of Evolution. *Affirming the consequent* is a logic fallacy, and it is being used to trick us, deceive us, and lie to us in the name of science.

I realized that conventional modern-day Science is in severe need of an upgrade. Science needs to be upgraded to include Quantum Mechanics, Action at a Distance, Dark Matter or Spirit Matter, Dark Energy, the Strong and Weak Nuclear Forces, Invisible Non-Physical Forces and Fields, Quantum Waves, Psyche, and Syntropy into the mix. It's simply wrong to limit Science exclusively to physical matter and entropy. It's wrong to limit science exclusively to Materialism, Naturalism, Darwinism, Nihilism, Determinism, Physical Reductionism, Atheism, Entropy, and Classical Physics. It's wrong to define science as Materialism, Naturalism, Darwinism, Nihilism, and Atheism. The verified and proven existence of Quantum Mechanics FALSIFIES Materialism, Naturalism, and their derivatives such as Darwinism and Atheism.

It's time to upgrade Science to Quantum Mechanics, Action at a Distance, Non-Locality, Quantum Non-Local Consciousness, Transdimensional Physics, Psyche, and Syntropy. It's time that we scientists allow ALL of the evidence into evidence and start to pursue a preponderance of that evidence.

It's time that we scientists learn our trade and actively switch from *affirming the consequent* over to *negating the consequent*, where the Scientific Methods are concerned. It's time that we learn how to falsify and eliminate our theories, instead of wasting all of our time trying to verify theories that have already been falsified.

In the summer of 2017, I upgraded my science to Science 2.0; and now, I allow ALL of the Quantum Evidence into evidence and actively pursue a preponderance of that evidence. I now use the *negating the consequent* version of the Scientific Method to eliminate everything that is false. Remember, if you successfully eliminate everything that is false and everything that has been falsified – such as Materialism, Naturalism, Nihilism, Atheism, Darwinism, Behaviorism, Determinism, Physical Reductionism, and the Theory of Evolution – then ONLY the truth will remain.

It's fascinating to study and observe what remains after you have successfully eliminated everything that is false. The observed, the experienced, and the verified remains. The Quantum Mechanics, or Supernatural Mechanisms, or Psyche Mechanisms remain. The truth remains.

Mark My Words

—

Source

Syntropy in Defense of Quantum Mechanics: The Answer to Life, the Universe, and Everything

https://www.amazon.com/dp/B07BPT3W8R/

—

I fix and upgrade Science, the Philosophy of Science, and the Scientific Method in the following books:

1. *Science 2.0: I Upgraded My Science*.

    https://www.amazon.com/dp/B0771K6WTX

2. *Using THE SCIENTIFIC METHOD to Eliminate the Usual Suspects and to Prove the Truth*.

    https://www.amazon.com/dp/B01J6STHP0

    https://www.amazon.com/dp/1521133581

3. *The Scientific Method Proves That the Theory of Evolution Is False*.

    https://www.amazon.com/dp/B01IAAIRT2

    https://www.amazon.com/dp/1521133611

4. **Quantum Neuroscience: The Answer to Life, the Universe, and Everything**.

    https://www.amazon.com/dp/B079Z6QQQB

5. **Quantum Mechanics from a Non-Physical Spiritual Perspective**.

https://www.amazon.com/dp/B01J023TGU

https://www.amazon.com/dp/1521132380

6. *Scientific Proof of God's Existence: A Primer*

https://www.amazon.com/dp/B071713NNL

https://www.amazon.com/dp/1521325170

7. *Scientific Proof of God's Existence: Finding God Where the Atheists Refuse to Look for Him*

https://www.amazon.com/dp/B07B26CRHX

8. *The Ultimate Model of Reality: Psyche Is the Ultimate Cause*

https://www.amazon.com/dp/B071NC9JK6

9. *Putting Psyche Back into Psychology: Restoring Science to Consciousness*

https://www.amazon.com/dp/B071NC987S

10. *BioPsychoSocial: Including Psyche or Light into our Theoretical Models*

https://www.amazon.com/dp/B0713NDHVW

11. *NATURE vs. NURTURE vs. NIRVANA: An Introduction to Reality*

https://www.amazon.com/dp/B01JWRCSVA

https://www.amazon.com/dp/1521132615

12. *Syntropy in Defense of Quantum Mechanics: The Answer to Life, the Universe, and Everything*

https://www.amazon.com/dp/B07BPT3W8R/

Science is extensive, so I have tried to fix it and upgrade it from every angle or discipline that I can possibly imagine. I've been busy; and, it's taken some time. All you want is the truth, because lies and half-truths are what got us into this mess in the first place.

Mark My Words

Philosophical Assumptions and Logic Fallacies

The thing I do love to study about philosophy is ALL of the different logic fallacies, falsified premises, and hidden assumptions that the Materialists, Naturalists, Darwinists, Nihilists, and Atheists use to trick us and deceive us. Clearly, these people are able to trick us and deceive us, and I always wanted to know how they do it; and, now I know.

Some of these people know precisely what they are doing, and they are deliberately trying to trick us, deceive us, and lie to us. Most of them, though, don't have a clue, when it comes to the Philosophy of Science and Logic Fallacies. The clueless have been tricked and deceived by their college professors, right along with the rest of us.

During my research, I have found the following list of logic fallacies useful:

https://en.wikipedia.org/wiki/List_of_fallacies

https://philosophy-of-science.com/wp-content/uploads/2018/04/List-of-Fallacies.pdf

They keep adding to this list as they find more. Within my books, I have created and added a couple of my own, which aren't on this list.

Scientific inferences are logic fallacies because they involve *jumping to conclusions*, *begging the question*, *circular reasoning*, and typically *assume facts not in evidence*. *Scientific inferences* are correlational rather than observational. Correlation is NOT causation; yet, the Materialists and Naturalists have created whole "science" disciplines that are based exclusively on *scientific inferences* which are correlational but that they treat as being causational – the theory of evolution being one of the first to come to mind.

The whole theory of evolution is correlational because chemical evolution, spontaneous generation, creation ex nihilo, abiogenesis, or macro-evolution has NEVER been experienced nor observed in the wild. In fact, these things are physically impossible and prevented from happening by entropy or the second law of thermodynamics. The real science falsifies them and prevents them from happening. The chemical evolution of proteins and the matching genes to go along with them is prevented from happening by entropy or the second law of thermodynamics. Random mutations are entropy. Death or natural selection is entropy. Entropy cannot design and create anything.

Nowadays, when it comes to the obligatory Shrine to Evolution that is found in every college textbook on psychology and biology, I can author whole books debunking it, demonstrating how it's false and why it is false. Many of the logic fallacies that these people use to make their case and prove their point are glaringly obvious to me, because I have trained myself to detect them and debunk them.

Materialism, Naturalism, Darwinism, Nihilism, Behaviorism, Determinism, Scientism, and Atheism ARE philosophies or religions, which means that they are easily defeated with philosophy and religion, because they have NO observational evidence and can have NO observational evidence supporting their major premises or hidden assumptions which claim that the quantum or the supernatural does not exist. The verified existence of anything non-physical, such as forces and fields, immediately falsifies the claims of Materialism, Naturalism, and their derivatives. The proven existence of gravity waves, quantum waves, magnetic waves, dark energy, the strong nuclear force, the weak nuclear force, and dark matter falsifies Materialism, Naturalism, and their derivatives. Quantum Mechanics or Supernatural Mechanisms falsify Materialism and Naturalism. When those go down, Darwinism and Atheism go down with them as fruit from the poisoned tree.

Thoughts and memories are quantum waves. We KNOW this is so, because our thoughts and memories survive the death of our physical brain according to the empirical evidence or observational evidence obtained from Near-Death Experiences (NDEs), Out-of-Body Experiences (OBEs), Shared-Death Experiences (SDEs), and our after-death Life Reviews. If a thought or memory survives the death of your physical brain, then it's truly a memory in every sense of the word. Is it not?

All you have to do is think logically and rationally to defeat Materialism, Naturalism, Nihilism, Atheism, and the Theory of Evolution because there is NO observational evidence supporting their major premises or hidden assumptions which state that the non-physical does not exist. The different forces and fields, which I have already listed above, clearly and obviously exist. We KNOW that they exist by the effect that they have on physical matter. Their very existence FALSIFIES Materialism, Naturalism, and their derivatives such as Darwinism, Nihilism, and Atheism.

Remember, the demonstrated and proven existence of non-physical forces and fields falsifies Materialism, Naturalism, and their derivatives. Psyche or your Intelligence is a non-physical force or field that has been experienced and observed. Clearly you have Intelligence, or you wouldn't be reading any of this.

If you successfully eliminate everything that is false and everything that has been falsified, then ONLY the truth will remain. This is Logic 101. You start by eliminating Materialism, Naturalism, Darwinism, Nihilism, Behaviorism, Determinism, Scientism, and Atheism; and then, you go on from there and eliminate everything else that has NEVER been experienced, observed, nor verified. It's fascinating to observe what remains after you have successfully eliminated everything that is false. The experienced and the observed remain. The verified truths remain. Quantum Mechanisms or Supernatural Mechanisms remain. Dark matter or spirit matter remains. Psyche or Intelligence remains. You continue to think and remember, long after your physical brain is dead and gone.

Mark My Words

—

References

I discuss these concepts in much greater detail in the following books:

1. *Science 2.0: I Upgraded My Science*.

   https://www.amazon.com/dp/B0771K6WTX

2. *Quantum Neuroscience: The Answer to Life, the Universe, and Everything*.

   https://www.amazon.com/dp/B079Z6QQQB

3. *Scientific Proof of God's Existence: Finding God Where the Atheists Refuse to Look for Him*.

   https://www.amazon.com/dp/B07B26CRHX

4. *Quantum Mechanics from a Non-Physical Spiritual Perspective*.

   https://www.amazon.com/dp/B01J023TGU

5. *Scientific Proof of God's Existence: A Primer*.

I used to be a Materialist, Naturalist, Nihilist, and Atheist. Obviously, I learned not to be afraid of God, religion, spirituality, psyche, fringe science, and quantum mechanics. I had nothing to lose and everything to gain. I encourage you to join me.

Developing the Ultimate Model of Reality

Website: https://ultimate-model-of-reality.com/

The more research that I did, the clearer it became to me that both science and philosophy are broken. They were destroyed by Materialism, Naturalism, Darwinism, Nihilism, Scientism, Behaviorism, Determinism, Physical Reductionism, and Atheism.

Therefore, my goal became an attempt to re-make and to fix both philosophy and science so that they are brought into line with the observational evidence being obtained from Quantum Mechanics or Supernatural Mechanics, Action at a Distance, Syntropy, Near-Death Experiences (NDEs), Out-of-Body Experiences (OBEs), Shared-Death Experiences (SDEs), Spiritual Experiences, Mystical Experiences, Lived Experiences, Phenomenology, Psyche, and our after-death Life Reviews. Science and philosophy should be based exclusively on observational evidence and the lived experiences of the human race, and NOT the philosophical speculation, confirmation biases, blind-faith, and wishful thinking of the Materialists, Naturalists, Darwinists, Nihilists, and Atheists.

As I see it, both traditional science and conventional philosophy need to be repaired, upgraded, overhauled, and fixed so that they allow ALL of the evidence into evidence, rather than restricting them exclusively to Materialism and Naturalism and defining them as Materialism, Naturalism, Darwinism, Nihilism, and Atheism. It has been my observation that defining science and philosophy as Materialism and Naturalism greatly limits science and philosophy, and eventually devastates them in the end. It kills their explanatory power. A Philosophy of Science based upon Materialism and Naturalism, and restricted exclusively to Darwinism and Atheism, ruins and destroys both science and philosophy. Materialism and Naturalism makes everything lose its explanatory power, because Materialism and Naturalism cannot explain Quantum Mechanics, Supernatural Mechanisms, Action at a Distance, Quantum Tunneling, Syntropy, Spirit Matter or Dark Matter, Dark Energy, and Psyche.

Fixing Science

In my attempt to fix science, I upgraded my science to Science 2.0, which allows ALL of the evidence into evidence and pursues a preponderance of the evidence. Science 2.0 defines science as Observation and Experience, rather than Materialism and Naturalism. Science 2.0 allows Quantum Mechanics, Syntropy, Action at a Distance, Supernatural Mechanisms, Dark Matter or Spirit Matter, Dark Energy, Transdimensional Explanations, Non-Locality or Non-Physicality, Quantum Non-Local Consciousness, Near-Death Experiences, Out-of-Body Experiences, Shared-Death Experiences, Life Reviews, and Psyche into evidence; whereas, Materialism and Naturalism do not. Science 2.0 is a new and better way of doing science. Science 2.0 has infinitely more explanatory power than Materialism and Naturalism ever did. Science 2.0 is the way that science should have always been done but wasn't.

Science 2.0: I Upgraded My Science

Fixing Philosophy

In my attempt to fix or repair philosophy, I added an Ultimate Cause to Aristotle's four causes; and, I developed the Ultimate Model of Reality. The Ultimate Model of Reality is a Psyche Ontology wherein Psyche, Intelligence, or Quantum Non-Local Consciousness is the fundamental unit of reality. Psyche is the Ultimate Cause; and, Psyche is the ultimate causal agent. Adding an Ultimate Cause to Aristotle's four causes completes Philosophy bringing it in-line with observed reality. A Philosophy of Science based upon Ultimate Cause has infinitely more explanatory power than a science or a philosophy that has been limited exclusively to Materialism, Naturalism, Darwinism, Nihilism, and Atheism.

The Ultimate Model of Reality, Ultimate Cause, and a Psyche Ontology are developed in the following books:

The Ultimate Model of Reality: Psyche Is the Ultimate Cause

Putting Psyche Back into Psychology: Restoring Science to Consciousness

BioPsychoSocial: Including Psyche or Light into our Theoretical Models

NATURE vs. NURTURE vs. NIRVANA: An Introduction to Reality

Fixing Neuroscience

Finally, pursuant to Science 2.0 and my decision to allow ALL of the evidence into evidence, I moved Neuroscience from a materialistic, naturalistic, atheistic, and classical physics foundation over to a Quantum Mechanical, Psychical, or Supernatural foundation based upon Syntropy and Transdimensional Physics; and, I developed a new science called Quantum Neuroscience. Quantum Neuroscience has infinitely more explanatory power than traditional Neuroscience, which is based exclusively upon Materialism, Naturalism, and Classical Physics.

Quantum Neuroscience: The Answer to Life, the Universe, and Everything.

Quantum Mechanics from a Non-Physical Spiritual Perspective

I know that it is audacious, impudent, and daring to try to fix everything that is wrong with modern-day science and philosophy; but, I believe that I have succeeded in doing so. However, I am under no illusions. I don't expect any of this to be used in practice or embraced during my lifetime. It's sufficient enough that it's available to me during the remainder of my life; and, that will have to do.

Obviously, I haven't fixed **everything** in science and philosophy that were ruined, limited, broken, stifled, or destroyed by Materialism and Naturalism; but, I've made a lot of progress in that direction, as the books which I have written will attest. I hope that someone will find my efforts useful and liberating. My ultimate goal was to set people free.

Only YOU can decide if I have succeeded in my efforts or not. Nobody else can do that for you.

Mark My Words

—

My Amazon Author Page:

https://amazon.com/author/science

—

Source

1. *Syntropy in Defense of Quantum Mechanics: The Answer to Life, the Universe, and Everything*

https://www.amazon.com/dp/B07BPT3W8R/

References

1. Rychlak, J. F. (1981). *A Philosophy of Science for Personality Theory* (2nd ed.). Malabar, FL: Robert E. Krieger Publishing Company.

2. Slife, B. D. & Williams, R. N. (1995). Science and Human Behavior. In *What's Behind the Research? Discovering Hidden Assumptions in the Behavioral Sciences*, (pp. 167–204). Thousand Oaks, CA: SAGE Publications.

https://mypsyche.us/wp-content/uploads/2017/04/Science.pdf

3. *Science 2.0: I Upgraded My Science*.

https://www.amazon.com/dp/B0771K6WTX

4. *The Ultimate Model of Reality: Psyche Is the Ultimate Cause*

https://www.amazon.com/dp/B071NC9JK6

5. *Putting Psyche Back into Psychology: Restoring Science to Consciousness*

https://www.amazon.com/dp/B071NC987S

6. ***BioPsychoSocial: Including Psyche or Light into our Theoretical Models***

https://www.amazon.com/dp/B0713NDHVW

7. ***NATURE vs. NURTURE vs. NIRVANA: An Introduction to Reality***

https://www.amazon.com/dp/B01JWRCSVA

https://www.amazon.com/dp/1521132615

8. **Quantum Neuroscience: The Answer to Life, the Universe, and Everything**.

https://www.amazon.com/dp/B079Z6QQQB

9. **Quantum Mechanics from a Non-Physical Spiritual Perspective**.

https://www.amazon.com/dp/B01J023TGU

https://www.amazon.com/dp/1521132380

10. Rychlak, J. F. (1970). The Human Person in Modern Psychological Science. *British Journal of Medical Psychology*, 43(3), 233–240.

https://mypsyche.us/wp-content/uploads/2016/12/Rychlak.pdf

Syntropy: The Answer to Life, the Universe, and Everything

Psyche Is Syntropy

Ultimately, Psyche or Syntropy is the pin prick of light at the center of everything else. This reality is based upon a Psyche Ontology wherein Psyche or Quantum Non-Local Consciousness is the fundamental unit of reality. In every picture that tries to model Syntropy or Quantum Mechanics, Psyche is the center point of everything else; and, Psyche is the causal agent behind everything else. Psyche is Syntropy. It is without a beginning of days or an end of years. It has always existed and will always exist.

Psyche Is the Ultimate Cause

Psyche is the Ultimate Cause; and, Psyche is the ultimate causal agent.

Introducing a Psyche Ontology and a Psyche Epistemology wherein Psyche is the fundamental unit of reality. Psyche is the Ultimate Cause; and, Psyche is the ultimate causal agent. Those who have seen Psyche or Quantum Non-Local Consciousness describe it as a pinprick of light or a point of light.

Psyche transmits, receives, and stores Quantum Waves. Thoughts and memories are quantum waves. While out-of-body, Psyche or Intelligence is experienced as an immaterial viewpoint in space.

Quantum Waves

Catch the wave!

Thoughts and memories are quantum waves. They survive the death of our physical body according to the empirical evidence from Near-Death Experiences (NDEs), Out-of-Body Experiences (OBEs), Shared-Death Experiences (SDEs), and our after-death Life Reviews. If a thought or a memory survives the death of your physical brain, then it's truly a memory in every sense of the word.

Psyche Is Light

Psyche is the light, intelligence, and consciousness within our spirit body.

The "soul" or "spirit" is typically defined as the combination of both the psyche and the spirit body. Psyche or Intelligence can function separately from the spirit body and actually view your spirit body from a position outside your spirit body.

Psyche Is the Ultimate Causal Agent

I have produced a few books in which I introduce a Psyche Ontology and a Psyche Epistemology wherein Psyche or Non-Local Consciousness is the fundamental unit of reality.

The purpose is to develop a new philosophy of science, and thereby hopefully introduce a new, better, and healthier way of doing science.

I call this new philosophy of science — The Ultimate Model of Reality, or a Psyche Ontology.

Syntropy, Quantum Mechanics, and Science 2.0 Series

Science 2.0: I Upgraded My Science

https://www.amazon.com/dp/B0771K6WTX

Scientific Proof of God's Existence: A Primer

https://www.amazon.com/dp/B071713NNL

https://www.amazon.com/dp/1521325170

Scientific Proof of God's Existence: Finding God Where the Atheists Refuse to Look for Him

https://www.amazon.com/dp/B07B26CRHX

Quantum Mechanics from a Non-Physical Spiritual Perspective.

https://www.amazon.com/dp/B01J023TGU

https://www.amazon.com/dp/1521132380

Quantum Neuroscience: The Answer to Life, the Universe, and Everything

https://www.amazon.com/dp/B079Z6QQQB

God Is in the Light: God is light, and in Him is no darkness at all.

Syntropy in Defense of Quantum Mechanics: The Answer to Life, the Universe, and Everything

The Ultimate Cause Series

Psyche is the ultimate causal agent.

The Ultimate Model of Reality: Psyche Is the Ultimate Cause

Putting Psyche Back into Psychology: Restoring Science to Consciousness

BioPsychoSocial: Including Psyche or Light into our Theoretical Models

NATURE vs. NURTURE vs. NIRVANA: An Introduction to Reality

Author's Website:

I hope you enjoy these books.

Mark My Words

Introduction: Evolution Is Entropy

Now pay attention!

Despite the FACT that I have and will have debunked and falsified the Theory of Evolution thousands of times in hundreds of different ways, don't ever once let me convince you that there is NO such thing as evolution or random mutations.

Evolution is REAL, very REAL. Genetic change and random mutations are very real. Genetic entropy or genetic deterioration is REAL. It is obvious that our gene pool is deteriorating. Mutations are REAL. The vast majority of our mutations take place during the genetic recombination which happens during the production of our gametes – egg and sperm. Natural selection doesn't do anything except wait around for us to die. Natural selection doesn't touch our genes. It can't. Natural selection is effectively Creation by

Death. However, the genetic mutations that take place during genetic recombination are very REAL indeed.

According to our scientists, evolution or random mutation is the same thing as entropy or the second law of thermodynamics. They are both Creation by Chance. According to our scientists, evolution is entropy. They are both the same "creative mechanism" according to our scientists. Our scientists are wrong, but they don't know that. There is no such thing as Creation by Chance, therefore, both the theory of evolution and the second law of thermodynamics are false. However, given the fact that our scientists treat evolution the same way that they treat entropy, as Creation by Chance or Creation by Random Disorder, according to our scientists, evolution is entropy. They treat evolution and entropy as if they were the same thing – the same Designer and Creator.

Although creation by chance, creation ex nihilo, the theory of evolution, abiogenesis, spontaneous generation, macro-evolution, and the second law of thermodynamics are demonstrably false, Entropy is very REAL at the physical level; and, it works as advertised at the physical level. Our scientists erroneously define entropy as "disorder". I have done research and found out what entropy truly is; and, I have written a few books on the topic.

Whenever the Materialists, Naturalists, Darwinists, Nihilists, Behaviorists, and Atheists start talking about evolution, they get most everything wrong, because evolution or entropy cannot design and create. However, these people do indeed get one thing perfectly right. Evolution, or random mutation, or genetic entropy is indeed the CAUSE of ALL of our heritable diseases, developmental diseases, and heritable mental illnesses.

Remember, the Theory of Evolution is FALSE because random mutations or entropy cannot design and create genes, proteins, and life forms. However, evolution or random mutation or genetic entropy is very REAL; and, it can indeed destroy genes, proteins, and life forms. Do you see how that works? It's important to understand.

The Theory of Evolution is a fictional story that they made up out of thin air. There is NO empirical evidence supporting any version of it. In fact, ALL of the empirical evidence and experimental evidence that we have on hand as a race FALSIFIES the different versions of the Theory of Evolution and VERIFIES Quantum Mechanics instead. Quantum Mechanics is Supernatural. Quantum Mechanics is the Priesthood Power of God. Quantum Mechanics and Psyche are Pure Syntropy. There is NO entropy in the spirit realm, the non-local realm, the quantum realm, or the transdimensional realm.

The very existence of something like the Orthodox Interpretation of Quantum Mechanics from Henry P. Stapp FALSIFIES Materialism, Naturalism, and the various versions of the Theory of Evolution. Fictional stories like the Theory of Evolution cannot stand in the light of truth.

The Theory of Evolution is the very pinnacle of fictional ad hoc just-so story telling; and, *ad hoc just-so stories* are logic fallacies. The Theory of Evolution is fictional, because it never happened – none of it happened! The chemical evolution of proteins and genes from atoms is physically impossible. Abiogenesis, spontaneous generation, and the various different forms of macro-evolution are physically impossible. The fictional nature of the story becomes most egregious whenever they try to guesstimate how many millions of years it took for evolution to do something for us. They are making it up as they go along. It's a fictional story, and nothing more.

Since the whole Theory of Evolution is nothing but a fictional story, you can successfully and rightfully make up fictional stories of your own to debunk it. That's the way fiction works!

Comparative Psychology and Evolutionary Psychology are based upon Darwinism and the Theory of Evolution, which means that they too are nothing more than fictional *ad hoc just-so stories* that these people have made up out of thin air.

In fact, in his book *Biopsychology*, John Pinel tells us as much when he tells us that his "evolutionary perspective" is based upon *scientific inferences*. *Scientific inferences* are fictional ad hoc just-so stories. *Scientific inferences* are logic fallacies. The whole of their evolutionary perspective, evolutionary psychology, and comparative psychology is based upon *scientific inferences* or stories that they have manufactured out of thin air. Making up stories is what makes being a scientist fun, according to John Pinel.

From *Biopsychology* page 13, John Pinel writes:

> **Scientific inference is the fundamental method of biopsychology and of most other sciences – it is what makes being a scientist fun. This section provides further insight into the nature of biopsychology by defining, illustrating, and discussing scientific inference.**
>
> **The scientific method is a system for finding things out by careful observation, but many of the processes studied by scientists cannot be observed. For example, scientists use empirical (observational) methods to study ice ages, gravity, evaporation, electricity, and nuclear fission – none of which can be directly observed; their effects can be observed, but the processes themselves cannot. Biopsychology is no different from the other sciences in this respect. One of its main goals is to characterize, through empirical methods, the unobservable processes by which the nervous system controls behavior.**
>
> **The empirical method that biopsychologists and other scientists use to study the unobservable is called scientific inference. Scientists carefully measure key events they can observe and then use these measures as a basis for logically inferring the nature of events that they cannot observe.**
>
> **Like a detective carefully gathering clues from which to recreate an unwitnessed crime, a biopsychologist carefully gathers relevant measures of behavior and neural activity from which to infer the nature of the neural processes that regulate behavior.**
>
> **The fact that the neural mechanisms of behavior cannot be directly observed and must be studied through scientific inference is what makes biopsychological research such a challenge – and as I said before, so much fun. (p. 13.)**

Scientific inference is a logic fallacy; yet, he erroneously calls it an empirical method. The whole Theory of Evolution is based upon *scientific inferences* or *fictional ad hoc stories* that these people have manufactured out of thin air. There's nothing empirical about it! There is NO empirical evidence demonstrating the creative powers of entropy, evolution, or random mutations. Chemical evolution of proteins and genes from atoms is physically impossible thanks to entropy or the second law of thermodynamics, or what is more properly known as the physical limitations of Physical Laws. Chemical evolution or spontaneous generation can't happen, which means that it never happened. Evolution is entropy, or the second law of thermodynamics. It can't design and create. It's physically impossible.

Furthermore, evolution (genetic change), random mutations, and natural selection didn't even exist until AFTER God designed and created the proteins, genes, genomes, brains, eyes, and life forms in the first place. It's physically impossible for something that doesn't even exist yet to design, program, engineer, manufacture, and create proteins and genes out of thin air. Spontaneous generation or abiogenesis is physically impossible. Entropy or the second law of thermodynamics prevents it from happening. Evolution is entropy, which means that it can't design and create anything. Evolution or entropy can only deteriorate and destroy things. It can't design and create. It's physically impossible for evolution or entropy to design and create proteins, genes, genomes, brains, eyes, and life forms. That's just the way it is, because evolution of any kind is entropy, or resistance to acceleration, or resistance to progress.

Our scientists erroneously define entropy as "disorder"; and, it is obvious that disorder can't design, create, and organize things.

It took me years, even decades, to discover that evolution or random mutation is entropy, or the second law of thermodynamics. That discovery also came with a powerful gift. I finally realized that Quantum Mechanics is the exact opposite of Materialism, Naturalism, Darwinism, Classical Physics, and Entropy. Quantum Mechanics and Psyche are Pure Syntropy. Quantum Mechanics is the Power of God, or the Priesthood Power of God. Psyche or Non-Local Consciousness is the only thing that can control Quantum Mechanics at the quantum level or the psyche level.

There is NO aging or entropy in the Quantum Realm, Psyche Realm, Spirit Realm, or Transdimensional Realm. Transdimensional means non-physical and non-local – not located in our physical 3D space-time realm. Everything in the Quantum Realm is Pure Syntropy. It is endless, timeless, eternal, and everlasting because there is NO entropy meaning that nothing ages, gets old, or dies in the Non-Local Realm.

The Gods create physical matter by infusing a particle of spirit matter with space-time and the ability to acquire entropy. The Gods create a particle of physical matter by taking a particle of spirit matter, filling it full of space, slowing it down to sub-light speeds, and making it subject to entropy, the passage of time, and an aging process. According to the theory of relativity, the particles of spirit matter existing at velocities faster than the speed of light experience NO passage of time, meaning that they do not age and are not subject to entropy. Entropy is a function of time or an aging process. Spirit matter and physical matter are the same thing – they are quantum objects. However, spirit matter is pure syntropy; whereas, physical matter has been slowed down by being infused with space-time and made subject to entropy or the passage of time.

I talk about all of this in great detail in my book, *Quantum Neuroscience: The Answer to Life, the Universe, and Everything*. If a comparison between Evolution and Quantum Mechanics interests you, I recommend you take a look at that book:

https://www.amazon.com/dp/B079Z6QQQB

The book, *Quantum Neuroscience: The Answer to Life, the Universe, and Everything*, makes a detailed comparison between Neuroscience and Quantum Neuroscience, which means that it makes a detailed comparison between Classical Physics and Quantum Mechanics.

When properly understood, Quantum Mechanics FALSIFIES Classical Physics, Materialism, Naturalism, Darwinism, Nihilism, Scientism, Behaviorism, Determinism, and even Atheism. Quantum Mechanics is Supernatural. Quantum Mechanics or Syntropy is the exact opposite of Classical Physics, Entropy, Random Mutations, Materialism, Naturalism, Darwinism, and the Theory of Evolution.

Quantum Mechanics is a proven and verified science. Quantum Mechanics is the best-proven and most-used science that we have. In contrast, the Theory of Evolution has NO empirical evidence supporting it. In fact, ALL of the empirical evidence that we have on hand as a race, including Quantum Mechanics, FALSIFIES Materialism, Naturalism, Darwinism, and the Theory of Evolution. Do you see how that works? It's important to understand.

Quantum Mechanics, Spirit Matter, and Quantum Non-Local Consciousness (Psyche or Intelligence) are PURE SYNTROPY. The syntropy has to exist somewhere someplace somehow, because according to the Law of Entropy and the Second Law of Thermodynamics, the physical multi-verse should have burned out and suffered heat death an eternity or two ago; and, there should be NO more physical universes anywhere, but here we are nonetheless. The very existence of this physical universe – it's beginning full of syntropy and its ongoing existence billions of years later – is positive proof that Someone Psyche knows how to do syntropy or Someone Psyche is syntropy. Quantum Mechanics is syntropy or the Priesthood Power of God. God's Psyche knows how to do syntropy or Quantum Mechanics; otherwise, this physical universe would not exist.

This is what Quantum Mechanics or Transdimensional Physics is trying to teach us. Quantum Mechanics, Spirit Matter, and Psyche are pure syntropy. Evolution, Random Mutations, Physical Matter, and Classical Physics are entropy, or resistance to acceleration, or resistance to progress. Entropy or disorder cannot design and create. There is no such thing as Creation by Chance, especially where genomes, brains, and life forms are concerned.

Mark My Words

—

Source

Scientific Proof of God's Existence: Finding God Where the Atheists Refuse to Look for Him

https://www.amazon.com/dp/B07B26CRHX

Reference

Pinel, J. (2014). _Biopsychology_ (9th ed.). New York: Pearson.

Proof of God through a Process of Elimination

A couple of years ago, I developed a Scientific Proof of God that I found extremely convincing. It was the thing that convinced me that God, of necessity, must exist. It was a Proof of God through a Process of Elimination. It was based upon Sherlock Holmes' maxim, which states:

> **How often have I said to you that when you have eliminated the impossible** [and the false], **whatever remains, however improbable, must be the truth? — Sherlock Holmes.**

When it comes to science, this quote became my Bible.

Here's how it works in practice.

If you successfully eliminate _everything_ that is false, then ONLY the truth will remain.

This is Logic 101. It's elementary; and, it works.

It's fascinating to study and observe what remains after you have eliminated everything that is false.

—

The Logical Route

Two years ago, I used logic and common sense to eliminate everything that hasn't been experienced or observed from science.

I originally realized that the different types of Spontaneous Generation have NEVER been experienced nor observed, which means that they should be eliminated and dismissed from science as being unscientific. In fact, Louis Pasteur falsified Spontaneous Generation in 1859, the very same year that Charles Darwin published "On the Origin of Species".

So, what are the different types of Spontaneous Generation that are being taught and promoted right now throughout the world?

The obvious ones are Chemical Evolution, Abiogenesis, Creation Ex Nihilo, Macro-Evolution, Design and Creation by Random Chance or Random Mutations, Materialism, Naturalism, Darwinism, Nihilism, Atheism, Design and Creation by Natural Selection, and the Theory of Evolution. It has been stated that most of the scientists in this world attribute origins and creation to these things. ALL of these things are Creation by Spontaneous Generation, to one extent or another. These different types of Spontaneous Generation are prevented from happening by entropy or the second law of thermodynamics. Proteins, and the matching genes to go along with them, don't just spontaneously generate out of thin air willy-nilly as we go along. Such a thing is physically impossible thanks to entropy or the second law of thermodynamics.

Spontaneous Generation of any type is impossible, so it should be eliminated as a scientific explanation because it has been falsified or proven false. ALL the different types of Spontaneous Generation have NEVER been experienced nor observed, and therefore they should be eliminated from science because they are impossible. The ultimate goal is to eliminate the false, the falsified, and the impossible from science so that ONLY the truth remains.

Since Materialism, Naturalism, Nihilism, Atheism, and the different types of Darwinism cannot design and create anything at all, they must be eliminated from science as being ineffective and unscientific. Since Materialism, Naturalism, and the different types of Evolution are based upon Spontaneous Generation, and since Spontaneous Generation has been falsified, it logically follows that Materialism, Naturalism, Darwinism, and their derivatives should be eliminated from science as being impossible, scientifically inaccurate, unscientific, falsified, speculative, and untrue.

I realized instinctively that if we eliminate these false creators or false gods, then ONLY the truth will remain.

If you successfully eliminate _everything_ that is false, then ONLY the truth will remain. It starts by eliminating everything that hasn't been experienced nor observed, such as Materialism, Naturalism, Atheism, Nihilism, Chemical Evolution, Macro-Evolution, the Theory of Evolution,

Creation Ex Nihilo, Creation by Entropy, Design and Creation by Random Chance, Abiogenesis, and Spontaneous Generation.

It's fascinating to observe what remains once we have eliminated these falsehoods – the truth remains.

Why?

The truth remains because it has been experienced, observed, and verified. The truth remains because it has been caught in the act. The truth remains because it is empirical and obviously true. The truth remains because it has been replicated. Even though it isn't always replicable on demand, the truth has been experienced and observed by many different people throughout the world.

The different versions of the Theory of Evolution or the different types of Evolution are based upon Spontaneous Generation or blind random chance. Spontaneous Generation or Evolution is prevented from happening by entropy or the second law of thermodynamics. It requires some type of Syntropy to design and create, because entropy cannot design and create anything. Evolution is entropy, not Syntropy.

Since Materialism, Naturalism, Darwinism, Entropy, or Spontaneous Generation cannot design and create anything at all, some type of God or Syntropy must of necessity exist in order to have done ALL of the science, design, creation, and production which Materialism, Naturalism, Evolution, and Entropy could NEVER have done.

How often have I said to you that when you have eliminated the impossible [and the falsified]**, whatever remains, however improbable, must be the truth? — Sherlock Holmes.**

It's elementary, and It works.

—

The Scientific Route

A year ago, I developed a Scientific Proof of God that I found extremely convincing. It was based upon the Scientific Method and *negating the consequent*. It works through a process of elimination.

The *negating the consequent* version of the Scientific Method is extremely good at identifying and eliminating anything that is false. We can go that route as well and produce the same results as the Logical Route, because *negating the consequent* is philosophically and logically sound. It works.

If you successfully eliminate underline everything that is false, then ONLY the truth will remain.

The *negating the consequent* version of the Scientific Method can be used to falsify theories or prove them false. Let's run a few of these by you and let you decide for yourself what to make of them. Let's start with Chemical Evolution and use the Scientific Method to falsify it.

Scientific Hypothesis: If Chemical Evolution is true, then we will observe atoms spontaneously generating into DNA, genes, genomes, and proteins.

Scientific Observations: We have NEVER observed atoms spontaneously generating into DNA, genes, genomes, and proteins. It's physically impossible.

Scientific Conclusion: Therefore, Chemical Evolution is false and has been falsified by the Scientific Method. In other words, Chemical Evolution is falsified by a complete lack of observation. Chemical Evolution or spontaneous generation has NEVER been observed.

Now, let's use the *negating the consequent* version of the Scientific Method to FALSIFY Macro-Evolution:

Scientific Hypothesis: If Macro-Evolution is true, then we will observe cats occasionally giving birth to dogs, and we will observe the occasional ape giving birth to humans.

Scientific Observations: We have NEVER observed one species giving birth to a completely different species. It's physically impossible. It's prevented from happening by genetics and random mutations.

Scientific Conclusion: Therefore, Macro-Evolution is false and has been falsified by the Scientific Method. In other words, Macro-Evolution is falsified by a complete lack of observational support.

Thanks to entropy or the second law of thermodynamics, it's physically impossible for random mutations to produce a functionally compatible or genetically compatible Mr. and Mrs. Mutant at the very same time in the very same place. It can't be done, which means that it wasn't done. Thanks to random mutations or entropy, it's physically impossible for chimp-like ancestors to give birth to genetically compatible male and female chimpanzees, or genetically compatible male and female human beings. It can't be done, which means that it wasn't done. That's the REAL science.

The Theory of Evolution is FALSIFIED by a complete lack of observational support. Let's use the Scientific Method to falsify the Theory of Evolution.

Scientific Hypothesis: If the Theory of Evolution is true, then we will observe natural selection and random mutations producing new and unique life-forms in real-time. It should be happening here and now, and not in some distant unseen past.

Scientific Observations: We have NEVER observed one species giving birth to a completely different species. It's physically impossible. It's prevented from happening by genetics and random mutations.

Scientific Conclusion: Therefore, the Theory of Evolution is false and has been falsified by the Scientific Method. In other words, the Theory of Evolution is falsified by the fact that random mutations and natural selection have NEVER been caught in the act of designing and creating new and unique life-forms from scratch.

These are powerful observations, which just happen to be true.

Let's falsify a couple other theories by using the *negating the consequent* version of the Scientific Method:

Scientific Hypothesis: If the claims of Materialism and Naturalism are true, then we will observe atoms spontaneously generating into DNA, genes, genomes, and proteins.

Scientific Observations: We have NEVER observed atoms spontaneously generating into DNA, genes, genomes, and proteins. In fact, spontaneous generation was falsified in 1859 by Louis Pasteur.

Scientific Conclusion: Therefore, Materialism and Naturalism are false and have been falsified by the Scientific Method. In other words, Materialism and Naturalism are falsified by a complete lack of observational evidence supporting their hidden assumptions or major premises which claim that the non-physical, the supernatural, or the quantum does not exist. It's physically impossible to observe that something does not exist.

Materialism, Naturalism, and their derivatives are FALSIFIED by a complete lack of observational evidence supporting their major premises or hidden assumptions which state that the Quantum or the Non-Physical does not exist. The verified and proven existence of Quantum Mechanisms or Supernatural Mechanisms such as Action at a Distance FALSIFIES Materialism, Naturalism, and their derivatives.

Since Materialism, Naturalism, Darwinism, Entropy, or Spontaneous Generation cannot design and create anything at all, some type of God or Syntropy must of necessity exist in order to have done ALL of the science, design, creation, and production which Materialism, Naturalism, Evolution, and Entropy could NEVER have done. Somebody had to do the science, and we KNOW that it wasn't Evolution or Entropy. Evolution and Entropy are NOT Someone Psyche. There's NO intelligence there. Evolution and Entropy have NEVER been caught in the act of doing science, design, and manufacturing; and, they NEVER will, because it's impossible.

How often have I said to you that when you have eliminated the impossible [and the falsified]**, whatever remains, however improbable, must be the truth? — Sherlock Holmes.**

It's elementary, and it works.

God or Syntropy must of necessity exist, or we wouldn't exist.

Mark My Words

—

Source Material

Using THE SCIENTIFIC METHOD to Eliminate the Usual Suspects and to Prove the Truth.

https://www.amazon.com/dp/B01J6STHP0

https://www.amazon.com/dp/1521133581

The Scientific Method Proves That the Theory of Evolution Is False.

https://www.amazon.com/dp/B01IAAIRT2

My Philosophical Proof of God's Existence

I've been working on this Philosophical Proof of God's Existence for three years now, because it wasn't obvious at the beginning how to proceed.

An argument is *valid* if the conclusion necessarily follows from the premises. An argument is *sound* if it is valid and the premises are true.

I have developed a philosophical proof of God's existence – or two – that I find logically compelling, philosophically valid, and scientifically sound. Let's start with the best and most believable conventional philosophical proof of God's existence, the Kalam Cosmological Argument.

Kalam Cosmological Argument

Let's start with William Lane Craig's version of the Kalam Cosmological Argument.

Premises:

Whatever begins to exist has a cause;

The universe began to exist;

Conclusion:

The universe has a cause.

https://en.wikipedia.org/wiki/Kalam_cosmological_argument

https://philosophy-of-science.com/wp-content/uploads/2018/04/Kalam-Cosmological-Argument.pdf

It begs the question, "Which universe began?" And, that's the main problem with the Kalam Cosmological Argument. The term "universe" is too vague and needs to be more carefully defined because we know of some universes that had NO beginning. Nevertheless, the Kalam Cosmological Argument is a good place to start, when it comes to a philosophical proof of God's existence.

This argument or proof is a syllogism. If the premises are true, then the conclusion has to be true. This is a properly constructed syllogism; therefore, it is a *valid* philosophical argument. An argument is *valid* if the conclusion necessarily follows from the premises.

This argument is also believed by many people to be a *sound* argument. Remember, an argument is *sound* if it is valid and the premises are true. This argument is considered TRUE by many people because it is philosophically *valid* and seems to be scientifically or observationally *sound*.

I find this philosophical proof of God's existence logically consistent, philosophically *valid*, and somewhat *sound*. It works in part, because its Premises are based somewhat upon observed truths. However, it can be greatly clarified and improved upon, because I can think of many exceptions to Craig's second premise.

Craig's Kalam Cosmological Argument FAILS to convince me in the end, because I can think of many exceptions to his second premise. There are some universes that had NO beginning; and therefore, they have NO cause. They are uncaused.

Proofs Demand Perfectly Sound Premises

You have got to get your premises perfect – perfectly true – or the proof doesn't work because it isn't *sound*. Craig's Kalam Cosmological Argument is a *valid* or well-constructed argument; but, it isn't *sound* – it isn't true. Craig can't see it nor understand it because he is using a falsehood as one of his hidden premises – namely creation ex nihilo. Falsehoods negate or falsify proofs.

Genesis 1:1-2:

1 In the beginning God created the heaven and the earth.

2 And the earth was without form, and void; and darkness was upon the face of the deep. And the Spirit of God moved upon the face of the waters.

IN THE BEGINNING OF WHAT?

It's extremely important to figure out and KNOW for sure which beginning they are actually talking about. You see, verse one of Genesis is also supposed to be a philosophical proof of God's existence; but, it only works if you know which beginning they are talking about.

Clearly God existed BEFORE this beginning, or He wouldn't have been able to begin it. Verse two of Genesis actually tells us that God was in a spirit form or a spiritual body during the formation of our physical earth.

Clearly, whatever substance God organized our heavens and our earth from had to exist BEFORE the beginning of our heavens and our earth because creation ex nihilo is impossible – meaning that it is scientifically and logically *unsound*. Creation ex nihilo is philosophically and scientifically absurd. It didn't happen because it can't happen. Not even God can create something from nothing. This is what science and logic tells us is true. Something from nothing leaves nothing.

Every aspect of your philosophical proof has to be perfect, perfectly sound or perfectly true, and perfectly understood; or, your philosophical proof isn't going to work and isn't going to be convincing.

SO, IN THE BEGINNING OF WHAT?

In the beginning of our physical universe!

The majority of the scientists are Big Bang proponents, which means that the majority of the scientists really truly believe and KNOW that our physical universe as well as physical matter had a beginning of some kind, which means that our physical universe had a Maker, or Creator, or Organizer, or Ultimate Cause. It's unavoidable. Every beginning has a Cause. If there was a Big Bang, somebody pushed the button! Clearly, our physical universe began, which means that it has a Maker or Creator of some sort.

Once we successfully identify "the beginning" as the Beginning of our Physical Universe, then the first verse of Genesis becomes a successful and believable and convincing Kalam Cosmological Argument because we KNOW that all the premises are true. We scientists KNOW for a fact that our physical universe and physical matter had a beginning of some sort, which means that our physical universe had a Beginner or Maker who made it. It's undeniable, since we KNOW that the premises are true.

Then we also KNOW for a fact that whatever substance God used to make or to construct this physical universe, it wasn't physical matter! It had to be some kind of Dark Matter or Spirit Matter because physical matter didn't exist yet in this universe. We also KNOW that whoever and whatever God is, He pre-dated or pre-existed the beginning of this physical universe. Now, we have a platform of truth and knowledge upon which we can build.

My Adjustments to the Kalam Cosmological Argument

An argument is *valid* if the conclusion necessarily follows from the premises; and, an argument is *sound* if it is valid and the premises are true. Remember, the premises have to be true in order for the argument to be *sound.*

Even though Craig's argument is philosophically valid and scientifically or observationally sound, at least superficially, I found that I had to adjust it a bit so as to make it specifically clear what I'm talking about and what I'm trying to prove. Craig's argument is too vague and universal for my tastes, which allows falsehoods, faulty interpretations, confusion, logic fallacies, and unscientific ideas to creep into it too easily.

So, I chose to adjust it in the following manner and tighten it up a bit.

Premises or Observations:

Whatever begins to exist has a Beginner, Maker, or an Ultimate Cause;

Our physical universe began to exist;

Conclusion:

Therefore, our physical universe has a Beginner, Maker, or an Ultimate Cause.

Now this thing is bullet-proof. The premises are TRUE and universally agreed upon, which means that the conclusion has to be TRUE as well.

These adjustments are important and essential, because I'm trying to bring the premises and conclusion into line with what has been experienced and observed in real life by real people. I'm trying to make my argument scientifically valid and sound, not just philosophically so. A philosophical proof that proves a lie to be true is of no value to anyone. I want this thing solid and sound when I'm done with it so that it actually does prove that some kind of God exists. In other words, I was able to think of exceptions to Craig's second premise that actually falsified and ruined his Kalam Cosmological Argument for me so that it no longer worked as a proof of God's existence. The Premises of a proof have to be empirical, obviously true, and bullet-proof in order for the proof to be efficacious, convincing, real, and true.

Whenever I develop a philosophical proof, I try to switch over to observational evidence, experiential proof, verified reality, scientific evidence, and logical common sense as quickly as I can in order to prevent myself from being tricked and deceived. Self-deception works, and it works every time.

It's the **physical universe** that we are talking about here because the non-physical universe, quantum universe, non-local universe, psyche universe, syntropy universe, transdimensional universe, transcendent universe, or spiritual universe had NO beginning; and therefore, it will have no end. It's the **physical universe** that had a beginning, not the transdimensional universe. This is crucial to get straight and get right.

Craig's argument talks about "the universe" – that's way too vague.

Why?

Well, the Multi-Verse, or the Non-Physical Universe, or the Chaos Construct, or the Quantum Transdimensional Transcendent Universe, or the Spirit Realm had NO beginning, which technically makes Craig's premise #2 false to begin with when he says that "our universe began," because the Primal Universe or Original Universe had NO beginning and therefore had NO cause. However, it has been observed and agreed upon by the scientists in general that our physical universe definitely had a beginning; therefore, our physical universe definitely had to have had an Ultimate Cause or a Person who organized it and brought it into existence. It's our **physical universe** that began, NOT "the universe".

Furthermore, the Beginner or Ultimate Cause has to be a person, or a living entity, or a Psyche, or an intelligent being of some kind because we KNOW that raw matter, or dead matter, or inert matter, or chaotic matter, or entropy cannot spontaneously generate into anything whatsoever. There's no such thing as spontaneous generation or creation ex nihilo where physical matter is concerned.

Psyche is the Ultimate Cause; and, Psyche is the ultimate causal agent. Matter cannot design and create. There is NO such thing as spontaneous generation or creation ex nihilo. Spontaneous generation or creation ex nihilo has been FALSIFIED by the scientific method. Such a concept as Creation Ex Nihilo is philosophically, logically, and scientifically unsound. It doesn't make any logical sense. Something from nothing is illogical. It can't be done, which means that it wasn't done. Not even God can do the impossible. Not even God can do creation ex nihilo!

Our physical universe came from something that already existed, and our physical universe was organized by Someone Psyche or Someone Spiritual who also already existed.

Creation Ex Nihilo and Atheism are kissing cousins. They are of the same kind. You could even say that they are siblings.

Creation Ex Nihilo is the creation or manufacture of something from nothing. Creation Ex Nihilo has NEVER been experienced nor observed; and, it never will be, because it's impossible. Creation Ex Nihilo is illogical and patently absurd. It's magic. It won't happen, because it can't happen. It never happened because it can't happen.

Atheism is creation or manufacture of something by NOTHING. Technically, Atheism has NEVER been experienced nor observed; and, it never will be, because it's impossible.

Atheism is illogical and patently absurd. It won't be verified, because it can never be verified. NOTHING will never be caught in the act of manufacturing something.

Likewise, spontaneous generation, macro-evolution, chemical evolution, and abiogenesis are a type of Atheism – the creation or manufacture of something by NOTHING. It can't be done, which means that it wasn't done. These types of "evolution" are physically impossible and prevented from happening by entropy or the second law of thermodynamics. They can't happen, which means that they didn't happen.

Most people are content to simply provide a *valid* philosophical proof; but when I'm doing proofs, I demand that the Premises be based upon scientific proof, actual observations, and real-life experiences. In other words, I demand that the Premises be scientifically and observationally *sound*. It makes for a better and an infinitely more convincing proof. The *soundness* of a proof is everything, in my humble opinion.

Ultimately, Craig's version of the Kalam Cosmological Argument isn't *sound* because I can think of many exceptions to his second Premise. My second premise doesn't allow for any exceptions because I tightened it up and made it specific.

My version of the Kalam Cosmological Argument does indeed PROVE that some kind of God exists – the God who designed, created, and produced our physical universe.

Why?

My proof works and is convincing because MOST of the scientists in the world agree that our physical universe began in some sort of Big Bang; and, this proof should be convincing to the Christians and Muslims because their God and scriptures tell us that He created the heavens and the earth – namely our physical universe. Physical matter and physical universes have a beginning, which means that they have a Beginner, Maker, or some sort of Ultimate Cause. Big Bangs have a beginning, which means that they have somebody who pushed the button and made them go bang.

God's Psyche didn't have a beginning, because it has always existed. The spirit matter or dark matter from which God organized this physical universe has always existed. But, practically everyone is in complete agreement that this physical universe, physical matter, and our physical earth had a beginning, which means that the Beginner or Ultimate Cause was some type of Psyche, Spirit, or God BEFORE the beginning or organization of physical universes and physical matter.

My adjustment to the Kalam Cosmological Argument is already philosophically *valid* and scientifically *sound*; so, it already works as a philosophical proof of God's existence; but, let's formalize it and expand it just a bit, so that I can comment on it some more.

My Philosophical Proof of God's Existence

Remember, an argument is *valid* if the conclusion necessarily follows from the premises; and, an argument is *sound* if it is valid and the premises are true.

Premise and Observation: Anything that has a beginning has a Beginner or an Ultimate Cause – a Designer, Creator, and Manufacturer who brought it into existence. This is Logic 101. It's inherently logical and sound. It has been experienced and <u>observed</u>.

Observations: It has been <u>observed</u> that our physical universe had a beginning, which means that it had a Beginner or an Ultimate Cause – a Designer and Creator who brought it into existence. That means that physical matter had a beginning, which means that it had a Beginner or Ultimate Cause – a Designer, Creator, and Manufacturer who organized it or brought it into existence.

Conclusion: When it comes to the organization and beginning of our physical universe and physical matter, by definition, in principle or practice, that Beginner, Designer, Manufacturer, Maker, and Creator has to be a God.

This is a logical conclusion since the rest of us here on this physical earth aren't in the habit of making physical matter and physical universes from scratch. As physical beings, we can't touch the sub-atomic or the quantum or the spiritual. Only a transcendent and transdimensional God would have such capabilities. Such a God would have to exist BEFORE the organization of this physical universe from spirit matter, chaotic matter, dark matter, or primal matter. Such a God would have to be transcendent without any physical limitations whatsoever – an omnipotent God who is the master of transdimensional physics or quantum mechanics. Such a God would have to have sufficient knowledge and power for such an endeavor.

The FACT that our physical universe had to have had a Beginner or an Ultimate Cause tells us some important things about the nature and capability the Being, the Psyche, the Spirit, or the God who designed and organized our physical universe. The Person or God who organized our physical universe had to pre-date our physical universe; and, the stuff that He used to form or organize our physical universe also had to pre-date our physical universe. This is Logic 101. First things first!

Technically, according to the rules associated with arguments and proofs, the first premise is allowed to be a Given Truth, Axiom, or a freebie; but, I demand that the first premise be scientifically proven, or observed and experienced by Someone Psyche sometime somewhere. All of the hidden premises and freebies that philosophers use is what gets us into trouble in the first place. I want all of my premises and my conclusion to be proven and verified in one way or another; otherwise, my philosophical proofs and scientific arguments are worthless in my humble opinion – as worthless as the philosophical proofs and scientific arguments being used to support and promote Materialism, Naturalism, Darwinism, Nihilism, and Atheism.

I demand that ALL of the premises be observed, verified, experienced, and proven to be real and true. ONLY then is the conclusion guaranteed to be true. A valid proof is worthless, if it proves a lie to be true. The premises and the conclusion must be *sound*! Consequently, I demand that ALL of the premises and even the conclusion be observed, verified, experienced, and proven to be real and true in Real Life by Real People. ONLY then can you be sure that you have indeed found the truth.

Notice that in this Philosophical Proof of God's Existence, I used OBSERVATIONS as my Premises so that I could be absolutely sure that my conclusion is true. My Premises are not as concise as in the previous examples, but I KNOW that they are true, which is all that really matters to me in the end.

The Materialists, Naturalists, and Theists are simply satisfied to win the argument. But, winning the argument isn't enough for me. I want to have the truth and KNOW the truth when I am done. I'm a scientist. I want to know the truth, not just win the argument.

Try this one! It's got teeth and claws.

Observation: Anything that was obviously made obviously had a Maker or Creator who made it.

Observation: A genome was obviously made.

Logical Conclusion: Therefore, a genome obviously has a Maker or a Creator who designed it, programmed it, engineered it, field-tested it, fine-tuned it, made it, manufactured it, and deployed it.

My philosophical proofs of God's existence really do PROVE that some kind of God exists, because my philosophical proofs of God's existence are in fact Scientific Proofs of God's Existence or OBSERVED Proofs of God's Existence. There's a huge difference there; but, only for those who are actually looking for such a thing. Like I said, MOST philosophers are simply satisfied to provide a *valid* argument and could care less if their Premises are scientifically and observationally *sound*. MOST philosophers and scientists are satisfied with a valid argument and a valid conclusion, and they could care less if their Conclusion has actually been experienced and observed in real life by real people.

I want EVERY PART of my philosophical proofs and scientific arguments to be experienced and observed; or, my philosophical proofs, scientific arguments, and chosen conclusions end up being completely worthless to me. I even want my Conclusions to have been experienced and observed by Someone Psyche, sometime somewhere. Only then can I KNOW that I have finally found the truth.

Remember, an argument is *valid* if the conclusion necessarily follows from the premises; and, an argument is *sound* if it is valid and the premises are true.

While doing philosophical proofs, theological proofs, and scientific proofs, I DEMAND that ALL of the Premises and the Conclusion be *sound*, which means that I demand that all of the Premises and the Conclusion be experienced and observed by Someone Psyche somewhere sometime. That's one of the reasons why Islam FAILS for me, because nobody has ever seen Allah; however, thousands, if not millions, of different people have seen and experienced our resurrected Lord Jesus Christ either in the flesh or during their Near-Death Experiences. Jesus Christ is the being of light and love whom people encounter and experience after they have died, and their brain is clinically dead. Jesus Christ is REAL, and He truly exists, because ALL of the observational evidence is telling us that it is so.

It's the observational evidence that I find convincing, and NOT the philosophical proofs. The philosophical proofs simply give us a way to structure, compound, and then multiply the effects of the observational evidence. It's the observational evidence, knowledge, and truth that I'm after, and NOT victory over my foes. Victory is hollow if what you win ends up being worthless and false.

Commentary on My Philosophical Proof of God

Every beginning has a Person or a Psyche who caused it to happen. Psyche is the Ultimate Cause; and, Psyche is the ultimate causal agent.

See: *The Ultimate Model of Reality: Psyche Is the Ultimate Cause*

https://www.amazon.com/dp/B071NC9JK6

Remember, every beginning has a Person, Intelligence, or a Psyche who caused it to happen or brought it into existence. This is Logic 101.

My Philosophical Proof is a philosophical proof of God's existence that actually makes logical sense to me and works for me. I find it compelling and convincing. It's based upon the Kalam Cosmological Argument, which I also find convincing and believable, as long as it is worded in a manner such that the Premises are based upon observational evidence and lived experiences. It's the Observational Evidence that I find convincing, not necessarily the philosophical proof.

https://en.wikipedia.org/wiki/Kalam_cosmological_argument

https://philosophy-of-science.com/wp-content/uploads/2018/04/Kalam-Cosmological-Argument.pdf

With Observed and Proven Premises in place, I found my version of the Kalam Cosmological Argument convincing; but, just because it works for me is no guarantee that it will work for you.

As with any philosophical proof, if <u>you</u> don't find the premises, arguments, and observations compelling, credible, and convincing, then you won't find the conclusion convincing either. That's the weakness of philosophical proofs. Observation and experience can easily falsify them; and, a lack of personal observational experience can easily make that person a victim of the philosophical proofs that are faulty and false. I seldom find any philosophical proof convincing because they are all subject to legerdemain or trickery. Most of them have NO observational evidence supporting them.

When it comes to philosophical proofs, their conclusion is also subject to *personal interpretation*, which introduces a wide variety of logic fallacies into the mix. Philosophical proofs are worthless if their Premises have NEVER been experienced nor observed. Their conclusions and the interpretations of their conclusions are worthless too if they have NEVER been experienced nor observed. That's why Materialism, Naturalism, Nihilism, and Atheism are worthless, because their hidden assumptions or major premises have NEVER been experienced and observed, nor can they be experienced and observed. Materialism, Naturalism, and their derivatives are FALSIFIED automatically due to a complete lack of observational evidence supporting their hidden assumptions or major premises which claim that the quantum, or the non-local, or the non-physical does not exist.

I was never really convinced by philosophical proofs. Most people aren't. There's too much self-deception going on when it comes to philosophy. Materialism and Naturalism are philosophy or religion, not science. They have to be taken on blind faith as being true. In contrast, whether I'm talking about science or philosophy, the thing that I do find compelling, convincing, and believable is Evidence, Observation, and Experience. Science is observation, or it should be. I've always found a Scientific Proof of God's Existence infinitely more believable and convincing than a philosophical proof of God's existence, because the science is based upon observation and experience.

Furthermore, a complete lack of observational experience or verified proof can easily falsify a premise, a philosophical proof, a hypothesis, or even a theory. That's what happened to Materialism, Naturalism, Darwinism, Nihilism, Behaviorism, Determinism, Scientism, Atheism, Macro-Evolution, and therefore the Theory of Evolution. <u>A complete lack of observational evidence</u> supporting their major premises, primary assumptions, or hidden assumptions completely FALSIFIES them.

It's the observations and experiences that I find convincing, not the philosophical proofs or philosophical sophistry. The wonderful thing about the Kalam Cosmological Proof of God's Existence, especially my modified version, is that it's backed by tons of observational evidence and it's easily transformed into a Scientific Proof of God's Existence simply by finding some observed evidence and experiential evidence to insert as your premises.

For example:

Premises or Observations:

Anything that has been fine-tuned has a Fine-Tuner;

Our physical universe has clearly been fine-tuned;

Conclusion:

Therefore, our physical universe has a Fine-Tuner.

This is a powerful syllogism that works as an excellent philosophical proof of God's Existence because it is based exclusively upon observational evidence and logical common sense. This argument is philosophically and logically SOUND because its premises have been experienced and observed by real-life human beings.

This argument is solidified by the scientific observation that spontaneous generation is physically impossible, thanks to entropy or the second law of thermodynamics. There's no such thing as spontaneous generation, abiogenesis, chemical evolution, creation ex nihilo, or macro-evolution thanks to entropy or the second law of thermodynamics which prevents these types of things from happening in the wild.

This Fine-Tuner Argument is as much a Scientific Proof of God's Existence as it is a philosophical proof of God's existence. The evidence of fine-tuning is all around us and impossible to deny. Fine-tuning is one of the most convincing Scientific Proofs of God's Existence.

Every instance or example of cosmic fine-tuning, mechanical fine-tuning, and biological fine-tuning is a miniature scientific proof of God's existence.

How do we know?

We scientists KNOW because living cells, genomes, genes, proteins, eyes, brains, and life forms have <u>NEVER been observed</u> spontaneously generating from atoms out of thin air. They can't because it's physically impossible for them to do so thanks to entropy, random diffusion, or the second law of thermodynamics.

For me personally, I found all of those precision-tuned Cosmological Constants or Physical Constants to be one of the most convincing Scientific Proofs of God's Existence that I have ever come across, especially since we scientists KNOW for a fact that such precision fine-tuning doesn't spontaneously generate out of thin air, because of entropy or the second law of thermodynamics. Likewise, we scientists KNOW for a fact that the exquisite and precise programming found in our genomes doesn't just spontaneously generate out of thin air either, here in this physical realm – once again thanks to entropy or the second law of thermodynamics. Your genome is God's Signature.

Again, for your convenience.

Observation: Anything that was obviously made obviously had a Maker or Creator who made it.

Observation: A genome was obviously made.

Logical Conclusion: Therefore, a genome obviously has a Maker or a Creator who designed it, programmed it, engineered it, field-tested it, fine-tuned it, made it, manufactured it, and deployed it.

My Genomic Argument is a variation on the Fine-Tuning Argument. They both are excellent Scientific Proofs of God's Existence, which also makes them some of the very best philosophical proofs of God's existence around because they are real and true, having been experienced and observed.

The BEST listing and documentation of God's precision fine-tuning is found on Reasons to Believe, by Hugh Ross, and is freely available to the public. Ross presents a link to this list in the appendix of his book, *Why the Universe Is the Way It Is*, and in the appendix of a couple of his other books. I archived it on a couple of my websites because it kept disappearing or moving around on me, and I could never find it again.

Fine-Tuning Is Scientific Proof of God's Existence

https://philosophy-of-science.com/wp-content/uploads/2018/04/compendium_part1.pdf

https://philosophy-of-science.com/wp-content/uploads/2018/04/compendium_part2.pdf

https://philosophy-of-science.com/wp-content/uploads/2018/04/compendium_Part3_ver2.pdf

https://philosophy-of-science.com/wp-content/uploads/2018/04/compendium_Part4_ver2.pdf

Ross, H. (2008). *Why the Universe Is the Way It Is*. Grand Rapids, MI: Baker Books.

For me personally, the first truly convincing Scientific Proof of God's Existence came to me on the day when I first realized that entropy or the second law of thermodynamics prevents chemical evolution, random mutations, and natural selection from spontaneously generating proteins, genes, genomes, eyes, brains, and life forms out of thin air from scratch. The theory of evolution is spontaneous generation, and spontaneous generation is physically impossible thanks to entropy or the second law of thermodynamics. Functional information-rich genomes, functional proteins, and life forms do not and cannot spontaneously generate from atoms. They just can't. It's physically impossible. It was then that I realized that God must exist in order to have done all of the Science and Fine-Tuning and Programming which chemical evolution, random mutations, and natural selection could NEVER have done. On that day, I simply KNEW that God exists, because I KNEW why He must exist.

Ironically, Louis Pasteur FALSIFIED spontaneous generation (and therefore Materialism, Naturalism, Darwinism, Atheism, Chemical Evolution, Macro-Evolution, Creation Ex Nihilo, and the Theory of Evolution) in 1859 – the very same year that Charles Darwin published "On the Origin of Species". We have KNOWN since the very beginning of the theory of evolution that spontaneous generation, abiogenesis, macro-evolution, chemical evolution, creation ex nihilo, or the theory of evolution is FALSE; but, most scientists have deliberately chosen to ignore that evidence and pursue wishful thinking instead. *Wishful thinking*, or *confirmation bias*, or *blind-faith* is a logic fallacy – one of the logic fallacies upon which Materialism, Naturalism, and Darwinism are based.

Remember, your philosophical arguments and syllogisms are NOT sound if their premises have NEVER been experienced nor observed. They may be logically valid, but they are NOT sound. That's how the Materialists, Naturalists, and Darwinists trick us and deceive us. They give us syllogisms and philosophical arguments that are logically valid, but totally unsound due to a complete lack of observational evidence supporting their chosen premises or hidden assumptions. Consequently, because of faulty premises, these people also *jump to conclusions* that are philosophically, logically, and scientifically impossible and unsound. *Jumping to conclusions* or *begging the question* is also a logic fallacy. Materialism, Naturalism, Darwinism, and Atheism are unscientific because their premises are scientifically unsound, having never been experienced nor observed.

Materialism, Naturalism, Darwinism, and their derivatives are based exclusively upon entropy; and, entropy by definition in principle cannot do Fine-Tuning and Programming. Therefore, God or Syntropy must of necessity exist in order to have done ALL of the Fine-Tuning and Programming that needed to be done. Simple. Logical. Parsimonious. TRUE!

Notice that whenever I present a scientific argument or a philosophical proof, I try to switch over to observed science or proven science for my Premises, as quickly as I can. It's the Observations and Experiences and Science that are convincing, and NOT the philosophical proof or syllogism. Remember, Materialism, Naturalism, Darwinism, Nihilism, Behaviorism, Determinism, Scientism, and Atheism have NO observed and NO proven Premises. Consequently, they FAIL before they even get started. There's no way to support them with science or observation because there is none, where their major premises or hidden assumptions are concerned.

Discussion of fine-tuning and the physical constants.

https://en.wikipedia.org/wiki/Fine-tuned_Universe

https://philosophy-of-science.com/wp-content/uploads/2018/04/Fine-tuned-Universe.pdf

https://en.wikipedia.org/wiki/Dimensionless_physical_constant

https://philosophy-of-science.com/wp-content/uploads/2018/04/Dimensionless-physical-constant.pdf

https://en.wikipedia.org/wiki/Physical_constant

https://philosophy-of-science.com/wp-content/uploads/2018/04/Physical-constant.pdf

https://philosophy-of-science.com/wp-content/uploads/2018/04/Physical-constant.pdf

Remember, the second law of thermodynamics, random mutations, death (natural selection), random diffusion, or entropy cannot do Physical Constants, Precision Fine-Tuning, and Programming. It's physically impossible. It can't be done, which means that it wasn't done.

Now, try this one:

Premises or Observations:

1. We have <u>observed</u> that everything that has a beginning has some kind of Creator or Maker who made it.

2. We have <u>observed</u> that our physical universe had a beginning. Consequently, we have <u>observed</u> that physical matter had a beginning.

Conclusion:

3. Therefore, it is logical to conclude that our physical universe and the physical matter within it had some kind of Creator or Maker who made them, organized them, or brought them into existence.

This argument has teeth because its premises have been experienced and observed. This argument is philosophically and logically SOUND because its premises have been experienced and observed. This Creator Argument or Maker Argument is as much a Scientific Proof of God's Existence as it is a philosophical proof of God's existence.

We have caught intelligent beings in the ACT of design, creation, manufacturing, and production zillions of times in trillions of different ways. Intelligent Design or Intelligent Creation is an OBSERVED, verified, and proven science. We KNOW that it is REAL because it has been EXPERIENCED and OBSERVED. It has been caught in the act. There's NO philosophical speculation, guesswork, or wishful thinking going on here where the observation of intelligent beings (or intelligent psyches) in action is concerned.

I find these Philosophical Proofs of God's Existence equally as compelling as the others because their premises have been experienced and observed in real life by real people. If you want the truth, then go with what has been experienced and observed, because the philosophical speculation, wishful thinking, and blind faith of the Materialists and Naturalist are worthless in the end.

Notice once again that when it comes to My Philosophical Proofs of God's Existence, I try to turn them into a Scientific Proof of God's Existence as quickly as I can. In other words, I try to turn them into an OBSERVED Proof of God's Existence. When it comes to science and proof, observation is where the tires really hit the pavement. Observed Proof of God's Existence is the most convincing and believable proof of God's existence. That's just the way it is.

In contrast, I also found the <u>complete lack of observational evidence</u> supporting their major premises or hidden assumptions equally as convincing when it came time to falsify theories and ideas such as Materialism, Naturalism, Darwinism, Nihilism, Behaviorism, Determinism, and Atheism. These philosophies or hypotheses are FALSIFIED by a complete lack of observational evidence supporting their major premises or hidden assumptions which state that the Non-Local, the Quantum, or the Non-Physical does not exist. The very existence and the verified existence of Quantum Mechanics and Action at a Distance FALSIFIES Materialism, Naturalism, Darwinism, Nihilism, Atheism, Classical Physics, and their derivatives.

Introducing Science 2.0

As a result of these observations, I upgraded my science to Science 2.0.

Science 2.0: I Upgraded My Science

https://www.amazon.com/dp/B0771K6WTX

Science 2.0 allows ALL of the evidence into evidence, and then it pursues a preponderance of that evidence. Evidence or observation is the only thing that has value to us. Philosophical arguments and philosophical proofs are absolutely worthless if the Premises that are employed are false and have been falsified, as has happened in the case of Materialism, Naturalism, Darwinism, and their derivatives.

When it comes to the non-physical sciences, Science 2.0 uses a Burden of Proof Methodology that is based upon a preponderance of the observational evidence. Under Science 2.0, observation and experience of any kind take precedence over philosophical speculation, guesswork, hypothesis, wishful thinking, scientific inferences, and confirmation biases. Science 2.0 is the way that science should have always been done but wasn't.

Science 2.0 is a new and better way of doing science that is based upon observational evidence, eye-witness evidence, and experiential evidence.

Under Science 2.0, the BEST way to find and know the truth is to live it, experience it, and observe it for yourself, or to choose to trust someone who has. The second-best way to find and know the truth is to use the scientific methods to falsify and eliminate everything that is false such as Materialism, Naturalism, Darwinism, Atheism, and their derivatives. If you successfully eliminate and remove everything that is false, then ONLY the truth will remain. It's fascinating to observe what remains after you have eliminated everything that is false and everything that has been falsified.

Remember, a syllogism can be logically valid but totally unsound. If the Premises of your argument, syllogism, or logic proof are NOT backed by observational evidence, then your philosophical argument is unsound. That's yet another serious problem with Materialism, Naturalism, Darwinism, Nihilism, Atheism, Behaviorism, Determinism, Scientism, Atheism, and the Theory of Evolution. There is NO observational evidence supporting their major premises or hidden assumptions, and there can NEVER be any observational evidence supporting their primary assumptions or major premises; therefore, the arguments produced by these philosophies, religions, dogmas, or pseudo-sciences can be forced to be valid, but they will NEVER be philosophically or logically sound. In other words, they will always be false. These falsified philosophies are unscientific and unsound because there can never be ANY observational evidence supporting their hidden assumptions or major premises which claim that the non-physical or the quantum does not exist.

The moral of the story is that when it comes to your philosophical proofs, be sure to use observed evidence, verified scientific evidence, experienced evidence, veridical evidence, and empirical evidence as your Premises if you want to have the greatest chance of your conclusions and subsequent interpretations being true.

My Philosophical Proofs of God's Existence have teeth and claws because their premises are OBSERVED evidence and have actually been experienced in real life. My Philosophical Proofs of God's Existence are both valid and sound because they are in fact Scientific Proofs of God's Existence, meaning that their premises are backed by tons of observational evidence and experience. We KNOW that the experienced and the observed are TRUE because they have been experienced and observed. We KNOW that Materialism, Naturalism, Darwinism, Nihilism, and Atheism are false because their major premises or hidden assumptions have NEVER been experienced nor observed and can't be experienced or observed.

We KNOW that God exists, because the Philosophical Arguments and Scientific Arguments that I have presented here PROVE that He exists. Once I knew that God exists, then I realized that I still have a long way to go before I get to know God. I used to be a Materialist, Naturalist, Nihilist, and Atheist; so, I still have a lot of work to do and a lot to accomplish where knowing God is concerned.

Go Out of Your Way to Get the Right Interpretation

Remember, philosophy has NO value whatsoever if it doesn't match with what has been experienced and observed. Notice how I tried to switch over from logic to what has actually been experienced and observed, before drawing any conclusions. The observations are essential. Science is observation. A philosophical proof has to match with reality, or it's worthless.

It's also important to draw the correct conclusion.

Online, when I provided a scientific proof of God's existence, one of the readers asked me which of all the man-made gods we should believe in.

Therefore, in harmony with his question, I ask, "Which of all the man-made gods is the one who was the Beginner, Designer, Manufacturer, Maker, and Creator of this physical universe and the physical matter in this universe?"

I chose to go with the True and Living God, Jesus Christ, because He has been experienced and observed after He rose from the dead. Thousands have seen Him and touched Him in the flesh; and, thousands have seen Him and embraced Him during their Near-Death Experiences. Should you ever find yourself in hell, remember that Jesus Christ can get you out of there just for the asking.

We KNOW that the Biblical God Jesus Christ exists, because He has been experienced and observed by thousands after He rose from the dead. Jesus Christ claims to be the God who organized the heavens and this earth, as well as all the life forms on this earth.

Notice once again, that when it comes to God's Existence, I try to switch over to observation and experience as quickly as I can, because philosophical speculation and wishful thinking are absolutely worthless in the end.

I define "science" as observation and experience. I have trained myself to switch away from philosophical speculation, wishful thinking, hypothesis, confirmation biases, and sophistry over to observation, experience, and empirical evidence as quickly as I can in order to prevent myself from being deceived. I encourage my readers to do the same.

A Proof or Argument Has to Be Logically Sound

This is most important thing to know and understand about philosophical proofs and scientific arguments – they MUST BE *valid* and *sound*. In other words, their Premises MUST BE true; otherwise, the proof, argument, and conclusion are absolutely worthless and liable to deceive.

An argument is *valid* if the conclusion necessarily follows from the premises. An argument is *sound* if it is valid and the premises are true.

Remember, a philosophy and a philosophical proof are worthless if their Premises are false. Likewise, a scientific argument is worthless if its Premises are false. That's precisely what's wrong with Materialism, Naturalism, Darwinism, Nihilism, Atheism, and their derivatives. Their hidden premises are false and can never be true.

If the premises of your argument don't add up to the conclusion, then your proof is worthless no matter how valid it might seem to be. If your premises have NEVER been

496

experienced nor observed in the wild, then your conclusion is nothing but fiction and is philosophically unsound. Your chosen conclusion is automatically falsified by a complete lack of observational evidence. It's called falsifying a theory!

In order to be logically sound, your Premises have to be experienced, observed, and verified. In order for your Conclusion to be logically sound, your Premises have to be logically sound, and your Conclusion also has to be experienced, observed, and verified in real life by real individuals. Missing evidence cannot be used as evidence. The missing links really are missing, so get used to it.

Remember, a God who has never been experienced nor observed is completely worthless to us. Evolution is a man-made god who has never been experienced, observed, nor caught in the act. A premise or a conclusion has to be experienced and observed in order for it to be philosophically and logically sound. A premise and a conclusion that has never been experienced nor observed is completely worthless to us. If the Biblical God hadn't revealed Himself to us, He would have remained forever unknown. In any philosophical proof, religious argument, or scientific argument, the Premises absolutely MUST BE true, or those arguments and proofs are totally worthless to us. That's just the way things work.

If your Premises completely lack confirming evidence or verified evidence, then your conclusion is automatically false. Design and creation by Chemical Evolution, Spontaneous Generation, Abiogenesis, or Macro-Evolution have NEVER been experienced nor observed, which means that these concepts or theories are automatically false. We KNOW that these things are prevented from happening by entropy or the second law of thermodynamics.

Technically, random mutations and natural selection have NEVER been caught in the act of design and creation, which means that these theories or concepts are automatically false. In other words, random mutations and natural selection cannot design and create. Random Mutations are entropy. Natural Selection ultimately results in death and extinction, which means that natural selection is also a type of entropy. Evolution is entropy. By definition, in principle, Materialism, Naturalism, Darwinism, Nihilism, Atheism, Classical Physics, and the Theory of Evolution are based upon entropy. Entropy cannot design and create anything. So, who did?

The Premises behind Materialism, Naturalism, Darwinism, and their derivatives are false and have been falsified by science and by observation, which means that the Conclusions that these people make are also false and have already been falsified. Materialism, Naturalism, and Darwinism are FALSIFIED by a complete lack of confirmed evidence, verified evidence, or observed evidence supporting their major premises and hidden assumptions, which claim that the quantum or the supernatural does not exist. A claim that something or someone does not exist cannot be verified, which means that it can NEVER be proven true; therefore, such a Premise is assumed to be false and will always be false until someone finds a way to prove it true, which they NEVER will. The Premises behind Materialism, Naturalism, Darwinism, Nihilism, Behaviorism, Determinism, Physical Reductionism, Scientism, Atheism, and their derivatives will always be false because it's impossible to prove them to be true.

In contrast, the verified, observed, and proven existence of Quantum Mechanisms, Action at a Distance, or Supernatural Mechanisms FALSIFIES the major premises of Materialism and Naturalism which claim that the quantum or the supernatural does not exist. The very existence of Quantum Mechanics or Transdimensional Physics FALSIFIES Materialism, Naturalism, and their derivatives such as Nihilism, Darwinism, and Atheism.

I have observed and experienced the FACT that anything that has a beginning has a Beginner or an Ultimate Cause, who is the person who designed, created, manufactured, and produced that thing. This reality has always held true.

A Beginner or Creator or Maker is the person who brings a thing into existence in the first place. This is a FACT that I find fully compelling, believable, and incontrovertible whether we are talking about science, philosophy, logic, religion, reality, or existence in general. I find it convincing, because it has been experienced and observed. We have observed people or intelligent beings bringing things into existence or choosing things into existence zillions of times in trillions of different ways.

Likewise, I find the existence of the Biblical God Jesus Christ compelling and believable because He has been experienced and observed by thousands, if not millions, of different people after He died and rose from the dead. Science is Observation, or it should be. These people have seen Him and touched Him both here on the physical plane as well as on the spiritual, quantum, or transdimensional plane.

Every time that Jesus Christ is seen, experienced, touched, embraced, and observed whether in the flesh, in visions and theophanies, or during our Near-Death Experiences (NDEs), those experiences and observations are Scientific Proof of His Existence. Science is observation, or it should be.

https://www.youtube.com/watch?v=UPj4wci_bcI

https://www.youtube.com/watch?v=Vm647n1360A

https://www.lds.org/scriptures/bofm/3-ne/11?lang=eng

https://www.lds.org/scriptures/pgp/js-h/1?lang=eng

In order for an argument or a proof to be logically sound, the premises have to be true, which means that the premises have to be experienced and observed by Someone Psyche. A philosophical proof that has NO observational evidence supporting it is completely worthless in the end. That's why Materialism, Naturalism, Darwinism, Nihilism, Behaviorism, Determinism, Atheism, and the Theory of Evolution are completely worthless. They don't have and can't have any observational evidence supporting their major premises or hidden assumptions which claim that the quantum or the supernatural does not exist. In fact, the observed and proven existence of Quantum Mechanics and Action at a Distance FALSIFIES Materialism, Naturalism, and their derivatives.

It took me three years to develop a Philosophical Proof of God's Existence that I found completely and totally believable and true.

Why did it take so long?

I realized in hindsight that it took me so long to develop such a proof because I was looking for Premises that I KNOW to be true. I was looking for Scientific Proof of God's Existence, not just philosophical proof of God's existence. I was looking for OBSERVED Proof of God's Existence or EXPERIENCED Proof of God's Existence. It's the observations and experiences that I find convincing, NOT the philosophical proofs. It's the empirical evidence that I find convincing, NOT the philosophical arguments.

I don't trust philosophical proofs, because MOST of them have hidden premises that I KNOW to be false. Materialism, Naturalism, Darwinism, Nihilism, Behaviorism, Determinism, Atheism, and their derivatives ARE philosophical proofs or philosophical arguments. I don't believe them to be true because they have hidden premises that are demonstrably and empirically false. As a race, WE have FALSIFIED Materialism, Naturalism,

Darwinism, Atheism, the Theory of Evolution, and their derivatives trillions of times in thousands of different ways. Falsified theories cannot be used as convincing Premises, especially when you KNOW that they are false and why they are false.

If you successfully eliminate everything that is false, then ONLY the truth will remain. It's fascinating to study and observe what remains after you have successfully falsified and eliminated Materialism, Naturalism, Darwinism, Nihilism, Atheism, and their derivatives. The VERIFIED and the OBSERVED remain. The Truth remains. Suddenly you find yourself looking at TRUE and VERIFIED Premises, as well as philosophically sound, scientifically observed, verified, experienced, and proven Conclusions. Your Conclusions ARE TRUE because they have been experienced and observed, and because their Premises are true and philosophically sound. Your Premises ARE TRUE because they too have been experienced and observed in real life.

Do you see how that works?

It works because it is TRUE.

In the end, all you really want is the truth.

Observational Reality and Common-Sense Logic

Anything that has a beginning has a Beginner or a Creator.

In contrast, God's Psyche and your psyche have always existed and will always exist. Your psyche and God's Psyche have no beginning, which means that they will have no end. This also means that your psyche or intelligence has NO Beginner, Designer, Creator, Manufacturer, Maker, or Ultimate Cause who brought it into existence in the first place. You have always existed, and you will always exist.

Things that have a beginning have a Beginner or a Creator, the person or psyche who brought them into existence. Physical matter, physical universes, physical genomes, and physical life forms have a beginning, which means that they have a Designer, Creator, and Maker who brought them into existence. In contrast, the things that have always existed, such as Psyche and Syntropy and Quantum Mechanics, have NO Creator because they have always existed.

Do you see how that works? This is Logic 101.

The things that have always existed, such as Psyche and Syntropy, have NO Creator. They are without a beginning of days or an end of years. Psyche is Syntropy. Quantum Mechanics or Transdimensional Physics is Syntropy. Syntropy is a type of unity, wholeness, completeness, or perfection. The Atonement of Christ is Syntropy. Syntropy means "without a beginning of days or an end of years". Syntropy means eternal, everlasting, and infinite. The Priesthood Power of God is Syntropy. Quantum Mechanics is the Priesthood Power of God. Syntropy had no beginning, and it will have no end. Syntropy, or Psyche, or Intelligence, or Consciousness has always existed. Syntropy is an organizing force that counteracts the effects of entropy. Syntropy has to exist or entropy wouldn't exist. Our physical universe had to receive an initial infusion of Syntropy, or none of that subsequent entropy would have been possible.

Do you see how that works?

Its explanatory power is through the roof and without limit!

I just provided the answer to life, the universe, and everything. That's the explanatory power of Syntropy, Psyche, and Quantum Mechanics.

Critiquing Craig's Cosmological Argument

William Lane Craig and others have declared him to be the master of the Kalam Cosmological Argument.

https://en.wikipedia.org/wiki/Kalam_cosmological_argument

Notice that William Lane Craig makes an argument that is philosophically valid and semi-sound; but then, he *jumps to a conclusion* at the end that is philosophically illogical and scientifically unsound. Craig is not going to critique his own argument, because he is too close to it and too emotionally invested in it to be able to see what might be wrong with it.

William Lane Craig states the Kalam cosmological argument as a brief syllogism, most commonly rendered as follows:

Whatever begins to exist has a cause;

The universe began to exist;

Therefore:

The universe has a cause.

From the conclusion of the initial syllogism, he appends a further premise and conclusion based upon ontological analysis of the properties of the cause:

The universe has a cause;

If the universe has a cause, then an uncaused, personal Creator of the universe exists who *sans* the universe is beginningless, changeless, immaterial, timeless, spaceless and enormously powerful;

Therefore:

An uncaused, personal Creator of the universe exists, who *sans* the universe is beginningless, changeless, immaterial, timeless, spaceless and enormously powerful.

Referring to the implications of Classical Theism that follow from this argument, Craig writes:

"... transcending the entire universe there exists a cause which brought the universe into being *ex nihilo* ... our whole universe was caused to exist by something beyond it and greater than it. For it is no secret that one of the

500

most important conceptions of what theists mean by 'God' is Creator of heaven and earth."

https://en.wikipedia.org/wiki/Kalam_cosmological_argument

William Lane Craig deliberately left his Kalam cosmological argument vague so that he could sneak a falsehood in at the end.

The only thing Craig gets wrong is that he *jumps to the conclusion* that God brought our universe into being **ex nihilo**. *Jumping to conclusions* or *begging the question* is a logic fallacy. In his conclusion, **ex nihilo** is an *unjustified add-on* or a *special pleading*, which are logic fallacies. In his promotion of Creation Ex Nihilo, Craig is *assuming facts not in evidence*.

There's NO such thing as Creation Ex Nihilo. Technically, Creation Ex Nihilo is Atheism – design and creation from nothing by nothing. It's impossible to create something from nothing. It's also impossible for nothing to design and create something. Even God can't do the impossible. Atheism is also impossible, because it's impossible for nothing to create something from nothing. Creation Ex Nihilo is philosophically and logically unsound because it doesn't make logical sense and because it's the result of *jumping to conclusions*, which is a logic fallacy. Creation Ex Nihilo or spontaneous generation is also scientifically unsound because it has been falsified by the scientific method.

Clearly, God existed BEFORE He organized this physical universe and brought it into existence. The very existence of God, the pre-existence of God, FALSIFIES creation ex nihilo. God did not spontaneously spring into existence from nothing. Creation ex nihilo does not and cannot apply to God, nor can it apply to the dark matter or spirit matter from which God organized this physical universe. God's Psyche has always existed and will always exist. The dark matter or spirit matter has always existed and will always exist. Creation Ex Nihilo or spontaneous generation is false and has been falsified. Creation Ex Nihilo is unnecessary and unjustified, so let's delete it. If you successfully eliminate everything that is false, then only the truth will remain.

Therefore, Craig's concluding statement should read:

"Transcending the entire universe there exists a cause which brought the universe into being. Our whole universe was caused to exist by something beyond it and greater than it. For it is no secret that one of the most important conceptions of what theists mean by 'God' is Creator of heaven and earth."

As long as we get rid of the logic fallacy and scientific falsehood that Craig tacks on at the end of his argument, I find the Kalam Cosmological Argument philosophically, logically, and scientifically valid and sound.

William Lane Craig teaches us by example that a philosophical proof can be valid, the premises can be sound, and the conclusion absolutely true; but, the person can still draw false conclusions and produce faulty and invalid interpretations from that True Conclusion or Proven Conclusion. Interpretation of the evidence or scientific inference is where the Scientific Method always falls down and dies whenever the Scientific Method is used to prove that a lie is true.

Remember, even William Lane Craig falls down and FAILS whenever he switches away from observational evidence over to *wishful thinking* and starts *jumping to conclusions* rather than relying upon observational evidence to make his case. Creation Ex Nihilo has NEVER been experienced nor observed; and, it never will be, because it's impossible. It's the observational evidence that we find convincing, and NOT the philosophical arguments. Philosophical arguments are worthless if their premises are not backed-up by observational evidence, verified evidence, and experiential evidence. Even a valid and sound conclusion from a philosophical argument is worthless, if it has NO observational evidence supporting it.

We find the existence of the Biblical God Jesus Christ compelling, believable, and convincing ONLY because it's backed-up by observational evidence and experiential evidence. Philosophical proof of God's existence is worthless if the premises are NOT backed up by observational evidence, verified evidence, scientific evidence, experiential evidence, eye-witness evidence, and empirical evidence. A conclusion is also worthless if it has NO observational evidence supporting it.

I was able to tighten up my version of the Kalam Cosmological Argument and get even more specific with it because I wasn't trying to insert a falsehood at the end. I was only interested in finding the observed and verified truth. Creation ex nihilo has never been experienced nor observed. It doesn't exist. It's impossible. Even God cannot do the impossible. Since I wasn't trying to support a falsehood with my philosophical argument, I was able to go directly to the Observed and Experienced Truth of the matter.

The theory of evolution is spontaneous generation; and, spontaneous generation is a type of Atheism or Creation Ex Nihilo – creation of something from nothing or by nothing. Something from nothing or something by nothing is impossible. It didn't happen because it can't happen.

If you accept the premises of the Kalam Cosmological Argument as being true, then they PROVE that our physical universe had a Creator, Beginner, or Ultimate Cause; however, they do NOT prove the veracity of creation ex nihilo. In fact, ALL of the observational evidence and experiential evidence that we have on hand as a race FALSIFIES creation ex nihilo, spontaneous generation, or the theory of evolution. Furthermore, it's ONLY our physical universe that had a beginning. The quantum universe, or the supernatural universe, or the spiritual universe, or the non-physical universe, or the non-local universe has always existed and will always exist. It has NO Creator or Ultimate Cause because it has always existed.

The Appearance of Design

Biology is the study of complicated things that have the appearance of having been designed with a purpose. – Richard Dawkins.

They have the appearance of having been designed because they were designed. There's NO other logical explanation for their origin because entropy or physical matter or natural selection cannot design and create anything. It's NEVER been caught in the act of doing so. It can't, which means that it didn't.

Spontaneous generation, creation by entropy, macro-evolution, abiogenesis, creation ex nihilo, chemical evolution, OR the theory of evolution was FALSIFIED in 1859 by Louis Pasteur. Biology appears to have been designed because it was designed.

Now try this philosophical proof of God's existence. It's an extremely powerful syllogism, because it is also a Scientific Proof of God's Existence or an OBSERVED Proof of God's Existence.

First Observation or Premise: Anything that was obviously made obviously had a Maker or Creator who made it. This is Logic 101.

Second Observation or Premise: A genome was obviously made. Such a thing doesn't just spontaneously generate out of thin air. It took some planning, programming, science, and manufacturing to get the job done.

Conclusion: Therefore, a genome obviously has a Maker or a Creator who designed it, programmed it, engineered it, field-tested it, fine-tuned it, made it, manufactured it, and then deployed it.

These observations convinced me that God does in fact exist, because I'm a scientist and I believe in the scientific evidence that has been experienced and observed. In other words, I KNOW that the premises are true; therefore, I KNOW that the conclusion MUST be true as well.

Some type of Syntropy, Psyche, or God MUST exist; or, physical matter, entropy, genomes, proteins, and this physical universe would NOT exist. It's elementary.

Consequently, since WE KNOW for a fact that God does indeed exist, the next task is to figure out who He is. I choose to go with our resurrected Lord Jesus Christ, because He has been EXPERIENCED and OBSERVED both in the flesh and during our Near-Death Experiences (NDEs), after He rose from the dead. I choose to go with the ONE who has been experienced and observed in real life by real people like you and me.

Mark My Words

Conclusions

The Beginner or the Ultimate Cause of this physical universe has to pre-date the beginning of this physical universe, and so does the stuff of which He was made and the stuff by which He made. This is Logic 101.

God organized the heavens and the earth from already pre-existing matter. God transformed some of that pre-existing spirit matter or dark matter into physical matter; and therefore, our physical universe had a physical beginning, even though the spirit matter or dark matter from which it was made has always existed. We KNOW that the dark matter or spirit matter exists by the effect that it has on ordinary physical matter. It has to exist, or our observations and measurements don't make any logical sense.

Do you see how that works?

Everything makes sense to me once we get rid of the logic fallacies and switch over to observations and measurements instead. Then suddenly, everything is philosophically, logically, and scientifically valid and sound.

Verified means that it has been experienced and observed. Falsified means that it has NEVER been experienced nor observed. Materialism, Naturalism, Darwinism, Nihilism, Behaviorism, Determinism, Creation Ex Nihilo, and Atheism have been FALSIFIED by the fact that their major premises or hidden assumptions have never been experienced nor observed and can't be. Materialism, Naturalism, Darwinism, Nihilism, and Atheism are FALSIFIED by the things that have been experienced and observed. They are FALSIFIED by Science.

The lesson in all of this is to switch away from philosophical speculation and sophistry over to observational evidence, experiential evidence, eye-witness evidence, empirical evidence, veridical evidence, scientific evidence, and phenomenological evidence as soon as you possibly can.

The truth has been repeatedly verified, experienced, and observed. The falsified and the false have NEVER been experienced nor observed, which is why they have been falsified. The false is falsified by the truth; and, the truth has been experienced and observed. That's just the way things work. That's the way things should work in science; and, that's definitely the way that things should work in religion, philosophy, and logic as well. Go with the experienced and the observed, and get rid of the wishful thinking, scientific inferences, blind-faith, and confirmation biases as quickly as you can. Go with the best and get rid of all the rest.

If you successfully eliminate everything that is false, then ONLY the truth will remain.

Mark My Words

—

Source

1. *Syntropy in Defense of Quantum Mechanics: The Answer to Life, the Universe, and Everything*

https://www.amazon.com/dp/B07BPT3W8R/

2. Web Page:

https://philosophy-of-science.com/my-philosophical-proof-of-god/

3. The Official Website:

https://philosophy-of-science.com/

Source for the Original One

Some might prefer the tighter version with less commentary; but, it's still pretty long. With this essay, I wanted to explain why I now believe in God's existence. It's the culmination of years of research, study, and thought. I'm not a prophet, so I don't know the Biblical God in person; but, I am a scientist, scholar, philosopher, logician, and theoretician, so I'm able to think things through carefully, logically, and rationally in order to reach a reasonable conclusion on the matter.

1. *The Ultimate Model of Reality: Psyche Is the Ultimate Cause*

https://www.amazon.com/dp/B071NC9JK6

2. *Science 2.0: I Upgraded My Science*

https://www.amazon.com/dp/B0771K6WTX

3. Web Page:

https://philosophy-of-science.com/original-one/

References

1. Slife, B. D. & Williams, R. N. (1995). Science and Human Behavior. In *What's Behind the Research? Discovering Hidden Assumptions in the Behavioral Sciences*, (pp. 167–204). Thousand Oaks, CA: SAGE Publications.

    https://mypsyche.us/wp-content/uploads/2017/04/Science.pdf

    https://philosophy-of-science.com/wp-content/uploads/2018/04/Science.pdf

2. Gantt, E. (2014). Logical Arguments. In *Psychology 353 – LDS Perspectives in Psychology*, (pp. 8-11). Provo, UT: Brigham Young University.

    https://philosophy-of-science.com/wp-content/uploads/2018/04/Logical-Arguments.pdf

3. Gantt, E. (2014). Leveling the Playing Field – Why Science is Not a Trump Card. In *Psychology 353 – LDS Perspectives in Psychology*, (pp. 50-58). Provo, UT: Brigham Young University.

    https://philosophy-of-science.com/wp-content/uploads/2018/04/Verification-vs-Falsification.pdf

4. Rychlak, J. F. (1981a). *A Philosophy of Science for Personality Theory* (2nd ed.). Malabar, FL: Robert E. Krieger Publishing Company.

5. Ross, H. (2008). *Why the Universe Is the Way It Is*. Grand Rapids, MI: Baker Books.

6. **My Amazon Page:**

    https://amazon.com/author/science

7. *Science 2.0: I Upgraded My Science*

    https://www.amazon.com/dp/B0771K6WTX

Obviously Made

A computer, a computer program, a calculator, a battery, a book, a watch, a car, a bridge, a table, a chair, a skyscraper, a house, a phone, a tablet, an airplane, a jet, and even a baby was obviously made. Likewise, a genome was obviously made. They had a beginning which means that they had some kind of Beginner or Maker who brought them into existence.

ALL of these things were obviously made.

NONE of these things has ever been observed spontaneously generating out of thin air ex nihilo, because they can't. Entropy or the second law of thermodynamics prevents spontaneous generation from happening. Furthermore, spontaneous generation was falsified in 1859 by Louis Pasteur. Spontaneous generation is physically impossible. That's the real science behind all of this.

Things that were obviously made obviously had a Maker who made them. This truth is unavoidable, because there is NO such thing as abiogenesis, spontaneous generation, creation ex nihilo, chemical evolution, or macro-evolution. They are physically impossible. They are prevented from happening by entropy or the second law of thermodynamics.

ALL of the things on the list were obviously made.

However, a genome is different than the others in a very obvious and unique way.

Can you tell what that is?

The first things on the list were obviously man-made. However, a genome obviously was NOT man-made. A genome and the proteins which that genome produces obviously pre-date man or human beings, which means that a genome and proteins obviously were NOT man-made. So, who made them? It's obvious that they were made, because such things have NEVER been observed spontaneously generating out of thin air ex nihilo.

Observation or Premise: Anything that was obviously made obviously had a Maker or Creator who made it.

Observation or Premise: A genome was obviously made.

Logical Conclusion: Therefore, a genome obviously has a Maker or a Creator who designed it, programmed it, engineered it, field-tested it, fine-tuned it, made it, manufactured it, and deployed it.

This syllogism is both a Philosophical Proof of God's Existence and a Scientific Proof of God's Existence. If its premises are true, then its conclusion must be true as well.

Genomes were obviously made; so, who made them?

That is the million-dollar question and the 800-pound gorilla, is it not?

There's only one logical answer to this question because genomes cannot spontaneously generate out of thin air as the Atheists and Darwinists claim.

Your genome is God's Signature, and God's Signature is written on every cell in your body. Simple. Logical. Parsimonious. True. Q.E.D.

We shouldn't be embarrassed by the truth. Science or observation makes the truth obvious, or it should.

Natural Selection Is a Blind Watchmaker

Richard Dawkins is the best that the Naturalists, Darwinists, and Atheists have been able to produce. In one of his best-selling books, Dawkins claims that natural selection is a blind watchmaker. How many blind watchmakers have you encountered during your life? They don't exist. Likewise, design and creation by natural selection does not exist. It's science fiction.

All appearances to the contrary, the only watchmaker in nature is the blind forces of physics, albeit deployed in very special way. A true watchmaker has foresight: he designs his cogs springs, and plans their interconnections, with a future purpose in his mind's eye. Natural selection, the blind, unconscious, automatic process which Darwin discovered, and which we now know is the explanation for the existence and apparently purposeful form of all life, has no purpose in mind. It has no mind and no mind's eye. It does not plan for the future. It has no vision, no foresight, no sight at all. If it can be said to play the role of watchmaker in nature, it is the *blind* watchmaker.

Mutation is random; natural selection is the very *opposite* of random.

We have seen that living things are too improbable and too beautifully 'designed' to have come into existence by chance.

The essence of life is statistical improbability on a colossal scale.

— Richard Dawkins

The Blind Watchmaker (1986), 5, 41, 51, 317.

Richard Dawkins claims that a blind watchmaker, called natural selection, made your proteins, genes, genomes, eyes, brain, and physical body. Richard Dawkins is talking about "spontaneous generation by natural selection". These claims are so obviously false that it's simply amazing to find people who actually believe that they are true.

In fact, these quotes are based upon a *category error* logic fallacy, wherein Richard Dawkins places natural selection into the category of "Invisible Gods Who Can Design and Create Anything They Set Their Minds To". Richard Dawkins just elevated natural selection to the status of Godhood by turning natural selection into a Designer and a Creator, but one that is running around blind and half-cocked.

It's obvious that genomes required vision, foresight, and some kind of purpose in mind. Dawkins even says as much in many of his statements. Genomes don't just spontaneously generate out of thin air. And, if you take Dawkins' explanation of natural selection in this quote seriously, you simply KNOW that natural selection could never have designed and created a genome from scratch or from atoms. It's physically impossible. Chemical Evolution and Macro-Evolution of any kind are prevented from happening by entropy or the second law of thermodynamics.

Furthermore, natural selection did not exist and could not exist until after God designed, programmed, engineered, fine-tuned, field-tested, created, manufactured, and deployed the genomes in the first place. This reality is so obviously true that Dawkins' claims don't make any logical sense to the rational mind.

Dawkins portrays natural selection as if it were some kind of magical God. Dawkins describes natural selection is if it were spontaneous generation, creation ex nihilo, or magic. These things don't exist. They have been falsified by science –meaning that they have been falsified by observation and experience.

In fact, natural selection doesn't really exist either. Natural selection is defined into existence by the scientists, but natural selection doesn't exist as a person, entity, psyche, intelligence, field, or force that can actually do something constructive for us. Natural selection is a function of entropy or work; and, entropy cannot design and create. Dawkins portrays natural selection as some type of spontaneous generation. His analogy is flawed, because there is no such thing as spontaneous generation or abiogenesis.

Dawkins is wrong. Natural selection is purely random – purely a function of chance. When you get selected against and die as a result, that's just as much a function of chance as anything else in this universe. The fittest are typically the first to die or the first to get selected against through war, risk-taking, infection, bad luck, and accidents of every kind. The timid, the meek, and the restrained really do inherit the earth.

There's no such thing as "spontaneous generation by natural selection". In fact, there's no such thing as spontaneous generation. Spontaneous generation, abiogenesis, or macro-evolution was falsified in 1859 by Louis Pasteur – the very same year that Charles Darwin published "On the Origin of Species". We have KNOWN from the very beginning of the theory that the Theory of Evolution, Creation by Natural Selection, Abiogenesis, Chemical Evolution, Materialism, and Naturalism are false because they are physically impossible and prevented from happening by entropy or the second law of thermodynamics.

Mankind, human beings, and blind watchmakers obviously did NOT make the genomes, so who did?

Well, WE KNOW that it wasn't natural selection because natural selection cannot design and create. Natural selection doesn't touch our genes. It can't. It's physically impossible. Dawkins doesn't even understand his own theory. The "agent of change" in the theory of evolution is the random mutations which take place during the genetic recombination that is used to produce our gametes – egg and sperm. Natural selection doesn't touch our genes. Natural selection results in entropy, death, and extinction. That's it!

Natural selection doesn't do anything to our genes or for our genes. Natural selection doesn't do anything except wait for us to die. Even sexual selection doesn't do anything for you after you have been conceived. Natural selection is not the panacea or magic potion that Dawkins claims that it is. Richard Dawkins' claims are obviously false to anyone who actually understands the Theory of Evolution, Entropy, Natural Selection, Random Mutations, and Science.

Furthermore, Richard Dawkins' blind watchmaker model is obviously self-refuting. A blind watch maker is an intelligent being; and, intelligent beings obviously qualify as Makers or Creators or Gods. Just because he is blind, it doesn't satisfy the demands of Richard Dawkins' claim that proteins, genomes, and life forms just spontaneously generated out of thin air by natural selection. Furthermore, a blind watchmaker (or natural selection) is definitely NOT going to be making a watch or genome without some help from some other intelligent beings. An intelligent watch maker (blind or otherwise) actually falsifies and refutes Dawkins' claim that our proteins, genes, genomes, eyes, brains, and physical bodies spontaneously generated out of thin air by natural selection.

A watch or genome is NOT going to spontaneously generate out of thin air. Entropy or the second law of thermodynamics prevents it from doing so. Entropy prevents Chemical

Evolution of any kind from happening in the first place. That's the real science behind all of this.

A watch of any kind was obviously made. Likewise, a genome of any kind was obviously made. These things do NOT spontaneously generate out of thin air by natural selection or by any other means. They have NEVER been observed doing so, because they can't. It's physically impossible. Entropy or the second law of thermodynamics prevents watches and genomes from spontaneously generating into existence. Therefore, Richard Dawkins' claims are falsified by science, entropy, random diffusion, and the second law of thermodynamics. His claims are also falsified by logical common sense.

Yes, a blind watchmaker can theoretically make watches because he is an intelligent being. NO, a gene, protein, genome, eye, brain, or physical body cannot spontaneously generate into existence. They are prevented from doing so by entropy. Spontaneous generation, abiogenesis, or macro-evolution of any kind is physically impossible. It can't happen, which means that it didn't happen.

Most people can quickly see where Richard Dawkins goes wrong, because he is not a good philosopher, and he's not much of a scientist either. Most of what he promotes was falsified decades ago.

—

It's Obvious that Genomes Were Made

Even Richard Dawkins seems to agree that genomes were obviously made. So, what's his hang-up? His problem is that he doesn't want the Biblical God to be his Maker. Anything but God!

https://www.youtube.com/watch?v=GlZtEjtlirc

https://evolution-is-entropy.com/wp-content/uploads/2018/05/Richard-Dawkins.zip

In the following quote, Richard Dawkins states in a round-about way that genomes were obviously made. It would take from here to infinity for a genome to self-assemble by sheer higgledy-piggledy luck. I believe it, because it is true.

> **It is grindingly, creakingly, crashingly obvious that, if Darwinism were really a theory of chance, it couldn't work. You don't need to be a mathematician or physicist to calculate that an eye or a hemoglobin molecule would take from here to infinity to self-assemble by sheer higgledy-piggledy luck. Far from being a difficulty peculiar to Darwinism, the astronomic improbability of eyes and knees, enzymes and elbow joints and all the other living wonders is precisely the problem that any theory of life must solve, and that Darwinism uniquely does solve. It solves it by breaking the improbability up into small, manageable parts, smearing out the luck needed, going round the back of Mount Improbable and crawling up the gentle slopes, inch by million-year inch. Only God would essay the mad task of leaping up the precipice in a single bound.**

> **— Richard Dawkins**

> *Climbing Mount Improbable* (1996), 67-8.

Dawkins is trying to make a case for Evolution or Darwinism, but he inevitably ends up refuting it, and debunking it, and falsifying it with truths hidden among the deceptive lies. A genome would take from here to infinity to self-assemble by sheer higgledy-piggledy luck. That is the truth; and, natural selection is nothing but luck – bad luck at that. It's never your lucky day when you get selected against and die.

ALL of Richard Dawkins' books are self-refuting and self-defeating. They contain within them their own seeds of destruction.

Why?

It's because his books were obviously made, which means that they obviously had a Maker, Creator, or God who made them.

In our own limited way, human beings are Gods. We are Creators. Adam was a son of God. That means that both his spirit body and his physical body were descended from the Gods. We human beings are Gods. We are descended from the Gods. Both our spirit body and our physical body are descended from the Gods. We inherited the ability to design and create from the Gods, our ancestors.

In this quote, Dawkins actually explains why Darwinism can't work and doesn't work – because it's a theory of chance or a theory of spontaneous generation. Smeared-out luck is still luck. It is still chance. In fact, it becomes less likely the more smeared-out it is. Infinite luck is even more impossible and improbable than one-time luck. Infinite luck or smeared-out luck has to get lucky each inch by million-year inch, instead of getting lucky only once. That's impossible! It's physically impossible for genetically compatible Mr. and Mrs. Mutants to be born at the same place at the same time year after year for millions of years. Genetics, random mutations, and entropy prevent it from happening! It can't be done, which means that it wasn't done. We don't have chimp-like ancestors because that's physically impossible. Both genetics and random mutations prevented it from happening.

Dawkins is right. ONLY God could have taken the task of designing and creating a genome in a single bound; and, the Gods have admitted that they are scientists and took their time getting our physical world, physical genome, and physical body just right for our existence. Even the Gods didn't do it in a single bound; but, they did do it. The Gods are scientists, not magicians; and, Richard Dawkins repeatedly portrays Darwinism as magic. The contrast couldn't be more obvious.

The Gods watched until they were obeyed.

https://evolution-is-entropy.com/wp-content/uploads/2018/05/The-Gods-Watched-Until-They-Were-Obeyed.pdf

While using God to make his case, Richard Dawkins actually falsifies his case. Dawkins actually admits that ONLY God could do "life" in a single bound; and, we all know instinctively that a genome has to be fully in place and fully functional in the first place, or an organism is going to die and go extinct. Life had to be done in a single bound, because it couldn't have been done over millions of years – one gene at a time. A gene by itself is worthless. It doesn't do anything. It requires all 20,000 of them to make a genome. They all have to be there from the very beginning, or an organism dies and goes extinct.

Likewise, a million apes pounding away on a typewriter could NEVER produce Dawkins' books in a trillion zillion years. First of all, they wouldn't know how to load the paper into the machine. Second of all, that typewriter, that paper, and those apes were obviously made, which means that they each had an intelligent Maker – a fact that falsifies Dawkins' claims that natural selection made them. Natural selection didn't make the typewriter and the paper, and it certainly didn't make the apes.

http://www.blogos.org/thinkabout/infinite-monkey-theorem.html

Dawkins' arguments are self-defeating, because Dawkins' deliberately ignores and excludes the Maker, Creator, or God behind each one of them. Richard Dawkins' arguments and claims are non-starters. Everything that he talks about was obviously made. The blind watchmaker was obviously made. His books and concepts were obviously made. Even natural selection is a made-up concept concocted out of thin air by the scientists or intelligent psyches who thought up the idea.

There's NO getting around it – anything that was obviously made obviously had a Maker, Creator, or God who made it. That's just the way it is. Proteins and the matching genes to go along with them don't spontaneously generate out of thin air. That's physically impossible. It's prevented from happening by entropy or the second law of thermodynamics. Proteins and genes were obviously made.

Richard Dawkins' examples and claims are self-defeating. They are defeated by science, logic, common sense, entropy, and reality itself. They are defeated by observation and experience. Science is observation and experience, not spontaneous generation. Spontaneous generation, chemical evolution, or macro-evolution was falsified by Science, by experience, and by observation.

I buy into the science, and not Dawkins' wishful thinking and confirmation bias.

—

Belief in Darwinism Requires an Infinite Amount of Blind Faith

I have repeatedly observed that belief in Darwinism, Materialism, Naturalism, Nihilism, and Atheism require an infinite amount of ignorance and blind-faith in order to believe in them because their Hidden Assumptions or Major premises cannot be experienced nor observed and have to be taken on blind-faith as being true.

Faith is belief without evidence and reason; coincidentally that's also the definition of delusion.

Do not indoctrinate your children. Teach them how to think for themselves, how to evaluate evidence, and how to disagree with you.

— Richard Dawkins

I love to read Dawkins. He always puts a smile on my face. Practically everything he says is self-defeating in one way or another.

Dawkins debunks himself and falsifies himself as he goes along. Dawkins defeats himself with practically everything he says.

I also love the hypocrisy. He should practice what he preaches, but he's incapable of seeing and understanding the contradictions.

Teach your children how to think for themselves, how to evaluate the evidence, and how to falsify and debunk Richard Dawkins. Teach your children to allow ALL of the evidence into evidence and not just the selective sub-set that Richard Dawkins allows into evidence.

Ironically, Dawkins' beliefs or blind faith are without evidence and run contrary to reason, which according to Dawkins means that they are delusional. That always puts a smile on my face every time I think of it. Everything that Dawkins says is based upon blind-faith and runs contrary to the scientific evidence and observational evidence that we have on hand as a race. Gotta love it! Self-deception works, and it works every time – especially when it comes to PhD scientists and authors. Remember, Dawkins' beliefs or blind faith are contradicted by the scientific evidence or empirical evidence which has been experienced and observed.

We should be thankful to Richard Dawkins for giving us a nearly endless stream of material to falsify and debunk. It's good for us. It forces us to think. It teaches us how to evaluate the evidence and how to disagree with and falsify Richard Dawkins.

Dawkins' books are NOT best-sellers because they are true. They are best-sellers because they are telling their readers what they want to hear – namely that God does not exist. People will pay a lot of money to be told that God does not exist because that's what they want to hear.

In contrast, my books will never be best-sellers, because I'm not telling the Materialists, Naturalists, Darwinists, and Atheists what they want to hear. I'm telling them the truth. These people aren't going to pay money to be told that God exists and why He must exist. And, the Christians and Muslims already know that God exists, so technically they don't need my books and won't be buying my books either. That's just the way it is.

I make and produce these books for my own benefit, so that I have a way of remembering what I have learned. While in a physical package, memory is such a fickle thing – one day you wake up and it is gone. But, it's not going to disappear on me and be gone if I can get it into a book.

—

The Philosophy of Science Falsifies Darwinism

I'm big on the Philosophy of Science. I actually used the Philosophy of Science and *negating the consequent* to falsify Materialism, Naturalism, Darwinism, and Atheism in many of my books.

I'm a generalist. I'm good at everything and master of nothing. The breadth and depth of my education is quite amazing, actually. My problem is that I get bored easily unless I'm learning something new.

I'm a scientist, psychologist, theoretician, logician, philosopher, accountant, mathematician, statistician, and computer scientist.

When I first watched Richard Dawkins debate others, I was surprised when I first realized that I KNOW a lot more science, theology, psychology, theory, logic, history, and philosophy than Richard Dawkins does. I was embarrassed for him. I felt sorry for him. He was making a fool of himself and didn't even know it. His ignorance was simply amazing. He has no breadth. His atheism has stunted his growth. He stopped learning long ago.

I was expecting Dawkins to have something significant or convincing to say, but he didn't. His science is over fifty years old and out-of-date. It expired long ago. And, Dawkins has absolutely no knowledge of philosophy, logic, and the Philosophy of Science. His arguments didn't work, because they were easily falsified as he went along; yet, all the time he actually thought that he was winning the debate. Go figure!

Philosopher of science, Karl Popper, wrote: "Testable theories are scientific, but those that are 'untestable' are not."

512

In *Unended Quest*, Popper declared, "I have come to the conclusion that Darwinism is not a testable scientific theory, but a metaphysical research programme, a possible framework for testable scientific theories."

Karl Popper is right.

I came to the same conclusion myself multiple times in many different ways.

Darwinism is a faulty and falsified religion. Evolution and natural selection are man-made gods.

Darwinism, Materialism, Naturalism, Nihilism, Atheism, and the Theory of Evolution are NOT testable scientific theories. They are philosophy, religion, or metaphysics because they require a ton of blind faith in order to believe in them and because ALL of the observational evidence and experiential evidence falsifies them.

Materialism, Naturalism, and Darwinism are NOT science. They are not testable, and there is NO evidence and can be NO evidence supporting their Hidden Assumptions or Major Premises which claim that the quantum or the supernatural does not exist. ALL of the evidence that we have on hand as a race tells us clearly and conclusively that the quantum or the supernatural does in fact exist in spades. Quantum Mechanics or Supernatural Mechanics or Transdimensional Physics is our best-proven, most-verified, and most-used science that we have. WE KNOW that it exists because it has been experienced and observed. WE KNOW that it exists by the effects that it has on physical matter.

The very existence of Syntropy or Quantum Mechanics or Intelligence falsifies Materialism, Naturalism, Darwinism, Nihilism, and Atheism. The proven, verified, and experienced existence of magnetism, gravity, the strong nuclear force, the weak nuclear force, psyche, dark energy, and dark matter or spirit matter FALSIFIES Materialism, Naturalism, Darwinism, Nihilism, and even Atheism.

The false is falsified by the truth; and, the truth is repeatedly experienced and observed. Science is observation and experience, not philosophical speculation and wishful thinking.

Remember, Darwinism is metaphysics or religion – not science. It requires an infinite amount of blind-faith and ignorance in order to believe in it. The same applies to Materialism, Naturalism, Nihilism, and Atheism. These are metaphysics or religion, and not science. That is what I have experienced and observed. I used to be a Materialist, Naturalist, Nihilist, and Atheist until ALL of the scientific evidence and observational evidence convinced me that I was wrong.

In my psychology courses in college, each author paid homage to Darwinism and the Theory of Evolution in their texts. Most of the things that they said were obviously false. It was so obvious that it was painful at times. It's unfortunate that they chose to call it evolutionary psychology and an evolutionary perspective, because evolution is in dispute. Something that is in dispute shouldn't be a part of science, especially a metaphysical philosophy such as Darwinism.

Evolutionary psychology is actually comparative psychology, because they are comparing animals with humans. They should have called it Comparative Psychology and not evolutionary psychology. Evolution is in dispute. Comparisons are not. Furthermore, they should have called it a Genetic Perspective rather than an evolutionary perspective for the very same reason. Evolution is in dispute. Genes and genomes are not. The genes and genomes have been experienced and observed; whereas, the different types of evolution have not. The different types of evolution – including random mutations and natural selection – have NEVER been caught in the act of designing and creating new unique proteins, genes, genomes, and life-forms from scratch because they can't. It's physically impossible.

Philosophies like Materialism, Naturalism, Darwinism, Nihilism, and Atheism cannot do science. They will never be caught in the act of doing science, design, creation, engineering, field-testing, fine-tuning, manufacturing, distribution, or anything else because they can't. This is so obviously true, that everyone has completely overlooked its significance. Only Psyche, Intelligence, or Syntropy can do design and creation; and, there's no intelligence anywhere within the Theory of Evolution. Richard Dawkins even says as much in the different statements that he has made.

Evolution is entropy. The different types of evolution are based upon entropy. Materialism, Naturalism, Darwinism, Nihilism, Behaviorism, Scientism, Determinism, Classical Physics, Physical Reductionism, and Atheism are based exclusively on entropy. Evolution is prevented from happening by entropy. Entropy or evolution cannot design and create. There is no intelligence in evolution or entropy.

Fascinating, is it not, how the truth rises to the top if you choose to let it do so?

—

Computer Science Falsifies Darwinism

Bill Gates — DNA is like a computer program but far, far more advanced than any software ever created.

A genome is like a computer program. It was obviously made.

Those of us who are Computer Scientists recognize a genome for what it truly is. A genome is both hardware and software. It's a computer program, a radically advanced four-dimensional computer program. A genome is also precision hardware. A genome is both software and a machine.

https://creation.com/four-dimensional-genome

https://evolution-is-entropy.com/wp-content/uploads/2018/05/Four-Dimensional-Genome.pdf

I was a computer programmer for a decade in the 1980's. I KNOW for a fact from experience and observation that elegant code does NOT spontaneously generate out of thin air. Computer programs don't spontaneously generate from nothing or natural selection, as Richard Dawkins claims. Elegant code is designed, planned, and made. It requires a Maker, Creator, or God. Computer programs were obviously made, which means that they obviously have a Maker who made them.

Likewise, a genome is obviously elegant programming code. A genome is also hardware. Programming code and hardware don't spontaneously generate into existence automatically. A million monkeys pounding away on a keyboard won't produce them either, because that keyboard and those monkeys were obviously made by some kind of Maker, Creator, or God – and they still can't produce a genome from scratch. A million monkeys and a keyboard do NOT qualify as an adequate example of spontaneous generation, chemical evolution, or macro-evolution because they were obviously made. Anything that was obviously made obviously had a Maker, Creator, or God who made it. That's logical common sense.

A complex computer program such as a genome was obviously made, which means that it obviously has a Maker who made it.

—

Scientific Proof of God's Existence

I found this philosophical proof of God's existence and Scientific Proof of God's Existence convincing ever since it was first revealed to me. Genomes are what convinced me that God exists. God must of necessity exist in order to have done ALL of the science and made ALL the genomes which chemical evolution, macro-evolution, natural selection, and random mutations NEVER could have done or made. To me personally, this reality is obviously true. Genomes were obviously made. Ever since that revelation or epiphany, I have known that God exists because I have known why He must exist. There's no denying it because I believe the science and KNOW that the premises are true.

Let's take another look at this Proof of God.

First Observation or Premise: Anything that was obviously made obviously had a Maker or Creator who made it. This is Logic 101.

Second Observation or Premise: A genome was obviously made. Such a thing doesn't just spontaneously generate out of thin air. It took some planning, programming, science, and manufacturing to get the job done.

Conclusion: Therefore, a genome obviously has a Maker or a Creator who designed it, programmed it, engineered it, field-tested it, fine-tuned it, made it, manufactured it, and then deployed it.

The first premise is obviously true. If you don't believe that the first premise is true, then you aren't living in the same dimension or the same universe as the rest of us. Anything that was obviously made obviously had a Maker or Creator who made it. This claim is obviously true.

Consequently, only the second premise is in dispute. Obviously, if you don't believe that genomes were made by Someone Intelligent or Someone Psyche, then you are not going to believe the conclusion either. However, most of us are convinced that genomes were obviously made. Even Richard Dawkins said as much, without saying as much.

You don't need to be a mathematician or physicist to calculate that an eye or a hemoglobin molecule [or a genome] **would take from here to infinity to self-assemble by sheer higgledy-piggledy luck. — Richard Dawkins**

Even to Dawkins, it's clear that genomes were obviously made – made by natural selection. The thing he doesn't realize or understand is that natural selection doesn't touch our genes, which means that natural selection cannot design and create genomes. Natural selection doesn't have any teeth. It doesn't touch our genes.

Dawkins is right, it would take from here to infinity for genomes to self-assemble by sheer higgledy-piggledy luck; but, Dawkins is wrong every time he claims or suggests that natural selection did it because natural selection can't touch our genes. Natural selection doesn't do anything for us or against us, before or after the selection has been made. If we are talking about natural selection or survival of the fittest, we can only die once. If we are talking about sexual selection, we can only be conceived once. Either way, selection doesn't touch our genes. Selection doesn't change our genes. Selection of any kind doesn't have anything to do with our genes.

It's the random mutations that affect our genes; and, random mutations are entropy. Every scientist knows, or should know, that entropy cannot design and create anything. Instead, entropy leads us towards disease, disability, mental illness, cancer, death, and extinction. That is what has been experienced and observed. Entropy is death and extinction. Natural selection produces entropy; and, random mutations are entropy. Entropy cannot design and create. It's physically impossible.

Self-assembly of genes, proteins, genomes, eyes, brains, and life forms from atoms is impossible. Spontaneous generation is impossible. Design and creation by random mutations and natural selection is impossible. Chemical evolution is impossible. Macro-evolution of any kind is impossible. Self-assembly of genomes from scratch is impossible. It can't happen, which means that it didn't happen. Genomes were obviously made. There's no denying it. It's obvious.

The mathematicians have indeed run the numbers as Dawkins suggested; and, they have proven beyond a shadow of a doubt that it is physically impossible for proteins and their matching genes to spontaneously generate or self-assemble by sheer luck – even if given an infinite amount of time to do so. Dawkins is right, it would take from here to infinity for genomes to self-assemble by sheer higgledy-piggledy luck. It can't happen, which means that it didn't happen. It's physically impossible. Self-assembly of genomes from atoms is prevented from happening by entropy, random diffusion, or the second law of thermodynamics. Chaos or entropy cannot design and create. Chance or blind luck cannot design and create. Chance or blind luck is also a function of entropy. Natural selection produces death or entropy; and, entropy or natural selection or death cannot design and create as Richard Dawkins says that it does.

The Probability of a Protein Forming by Chance
https://www.youtube.com/watch?v=W1_KEVaCyaA
https://evolution-is-entropy.com/wp-content/uploads/2018/05/Probability-of-a-Protein-Forming-by-Chance.zip

The Probability of Making a Protein
https://www.youtube.com/watch?v=cQoQgTqj3pU
https://evolution-is-entropy.com/wp-content/uploads/2018/05/Probability-of-making-a-protein.zip

Evidence for Creation by Outside Intervention
https://www.youtube.com/watch?v=ci4s75al1Rw
https://evolution-is-entropy.com/wp-content/uploads/2018/05/Evidence-for-Creation-by-Outside-Intervention-01.zip
https://evolution-is-entropy.com/wp-content/uploads/2018/05/Evidence-for-Creation-by-Outside-Intervention-02.zip
https://evolution-is-entropy.com/wp-content/uploads/2018/05/Evidence-for-Creation-by-Outside-Intervention-03.zip

These videos are worth owning, watching, and keeping. I archived them so that they can't disappear on me while I'm alive.

Observation or Premise: Anything that was obviously made obviously had a Maker or Creator who made it.

Observation or Premise: A genome was obviously made.

Logical Conclusion: Therefore, a genome obviously has a Maker or a Creator who designed it, programmed it, engineered it, field-tested it, fine-tuned it, made it, manufactured it, and deployed it.

These observations convinced me that God does in fact exist, because I'm a scientist and I believe in the scientific evidence that has been experienced and observed. In other words, I KNOW that the premises are true; therefore, I KNOW that the conclusion MUST be true as well.

Some type of Syntropy, Psyche, or God MUST exist; or, physical matter, entropy, genomes, proteins, and this physical universe would NOT exist. It's elementary, and it's obvious.

Consequently, since WE KNOW for a fact that God does indeed exist, the next task is to figure out who He is. I choose to go with our resurrected Lord Jesus Christ, because He has been EXPERIENCED and OBSERVED both in the flesh and during our Near-Death Experiences (NDEs), after He rose from the dead. I choose to go with the ONE who has been experienced and observed in real life by real people like you and me. Go with the true and living God – the one who isn't a man-made God, like evolution or natural selection.

Evolution and natural selection are man-made gods, imbued by definition with all of the power and foresight of a God. But, you don't want to go with any of the man-made gods because they are nothing but fiction. A man-made god isn't going to come and get you out of hell. You want to go with the True and Living God, Jesus Christ, because He has been experienced and observed.

https://www.youtube.com/results?search_query=NDE+Jesus

https://www.youtube.com/results?search_query=NDE+Atheist

Thousands of people have seen Jesus during their Near-Death Experiences (NDEs). Check out the videos! Jesus Christ even came and got some of the atheists out of hell during their NDE. Remember, should you ever find yourself in hell, Jesus Christ can get you out of there just for the asking. Jesus Christ could even get Richard Dawkins out of hell if Richard Dawkins were to ask. That's good enough for me. That's the God that I want to go with – the one who can make good on His promises and get me out of a jam.

Go with the best and get rid of all the rest. That's what I have decided to do, and you can too.

—

Genomes Were Obviously Made

Genomes were obviously made, which means that they obviously had a Maker, Creator, or God who made them.

A genome is Scientific Proof of God's Existence. It was the first convincing Scientific Proof of God's Existence that I encountered in science. It's so obvious to me that the genomes were made, that it was obvious to me that some kind of Maker, Creator, or God made them. The truth of it was undeniable. Genomes are God's Signature.

In contrast, it was also obvious to me that spontaneous generation, creation ex nihilo, abiogenesis, natural selection, random mutations, or macro-evolution could never have made or produced a genome. It's physically impossible. It's prevented from happening by entropy. It can't happen, which means that it didn't happen.

517

Chemical Evolution of any kind is prevented from happening by entropy or random diffusion. That's the real science behind all of this. It's obvious that it requires some type of Syntropy, Intelligence, or Psyche in order to be able to design and create anything – including physical matter and genomes.

Furthermore, it was obvious to me that evolution (genetic change), random mutations, and natural selection did not exist until after God designed and created the physical matter, proteins, genes, genomes, eyes, brains, and life forms in the first place. Random mutations and natural selection obviously could never have produced our genomes because they obviously didn't exist until after our genomes were made. Evolution didn't exist until after God made the genomes.

Proteins and the matching genes to go along with them were obviously made, which means that they obviously had a Maker, Creator, or God who made them.

In fact, physical matter, proteins, genes, genomes, eyes, brains, and physical bodies were obviously made, which means that they obviously had a Maker, Creator, or God who made them. This is Logic 101 and positive proof that God does in fact exist.

It's obvious to me that natural selection, random mutations, chemical evolution, and macro-evolution are man-made gods and are typically portrayed as if they were God; but, they can't design and create. Since we KNOW for a fact that God must exist, who is the True and Living God that does exist? Jesus Christ is the only one who qualifies because He has been experienced and observed after He rose from the dead. Jesus Christ has the necessary credentials, qualifications, capabilities, and pedigree. It doesn't get more God-like than rising from the dead or showing up and getting an atheist out of hell.

So, do you want the truth, or do you want convenient fables and science fiction? The choice is yours. That's not a choice that anybody else can make for you. Where your choices and beliefs are concerned, you are their Maker, Creator, and God. God definitely exists. In your own limited way, you are a God; and, you definitely exist. Get used to it! You'd just as well, because that's where we are going as a race if we don't destroy ourselves first.

Mark My Words

—

Source

The Scientific Method Proves That the Theory of Evolution Is False

https://www.amazon.com/dp/B01IAAIRT2

https://www.amazon.com/dp/1521133611

Memory Storage within the Physical Brain

Introduction

When I decided to take on Neuroscience, I kept repeatedly asking myself, "What's the evidence trying to tell me? What's the evidence trying to teach me?"

Over and over again, the same answer kept coming to the foreground, "The evidence is trying to tell me that there is no RAM, there are no wires, and there are no CPUs within a physical brain." That was downright scary. Think of the ramifications! The evidence was taking me into territory where "science" is traditionally forbidden to go. The discomfort or cognitive dissonance was extreme.

Think of it!

I was being called upon to develop a whole new way of doing science.

Science 2.0: I Upgraded My Science

https://www.amazon.com/dp/B0771K6WTX

And, I was being called upon to create and develop a whole new science.

Quantum Neuroscience: The Answer to Life, the Universe, and Everything

https://www.amazon.com/dp/B079Z6QQQB

I was being called upon to do something that had never been done before. I was being called upon to provide the answer to life, the universe, and everything.

Where's the Beef?

While studying Neuroscience, I kept waiting for them to tell me where our memories are being stored within our physical brain and how our memories are being stored within our physical brain. They never did. They can't. They have NO logical explanation for how our memories are actually being stored within our brain.

All the Neuroscientists would say is that we don't know how memories are being stored within the brain, but clearly memories are being stored within the brain because they can't be stored anywhere else. That *begs the question*. These people *jump straight to the conclusion* that our memories are stored within our physical brain, because according to these scientists our memories can't be stored anywhere else. These people define all other alternatives out of existence, a priori, before doing one bit of thought or science on the subject. They actually cheat and use logic fallacies in order to make their case.

Since the Neuroscientists don't know how memories are being stored within the brain and can't explain in any logical fashion how and where our memories are being stored within our brain, that left a maverick like me completely free to explore all the other options that came my way. I wasn't beholden to nor trapped by the Materialists, Naturalists, Darwinists, and Atheists when it came to memory storage within the physical brain. I was free to think outside the box; so, I did.

I eventually observed that, since there are NO wires in our brain connecting the synapses and neurons together into functional BYTES of RAM and programming code, the physical brain is incapable of storing BYTES of memory and programming code. There isn't a single BYTE of memory or programming code anywhere within a physical brain. It's physically impossible. Neurons are stand-alone switches. Each neuron represents a single hardware BIT. A neuron is either ON or OFF. Since there are NO wires in our brain, that means that there are no CPUs, RAM, or hard drives within our brain. It's physically impossible to store bytes of information within a single hardware BIT. It's physically impossible to transfer BYTES of information through a single hardware BIT. It can't be done, which means that it isn't being done. I have been a computer scientist all of my adult life, so I know how these things work. There are NO logic gates within a physical brain. There are NO logic circuits and NO electronics within a physical brain. There are NO computer chips within a physical brain. A neuron is a switch. It is either ON or OFF. It either fires or it doesn't.

I observed that a synaptic-cleft scrambles everything that comes its way. If you were to try to send a message or a memory through a synaptic cleft, it will be scrambled and randomized in the process. Nothing will be left. Therefore, it is physically impossible to store our memories within our synapses in some kind of distributed fashion as some of the Neuroscientists suggest is happening. There's NO physical mechanism in place for distributing memories among synapses, let alone a mechanism for retrieving and using such memories. Synapses can't be used for memory storage, and synapses can't be used for CPU processing either. It's physically impossible thanks to the randomizing effects of a synaptic cleft.

Remember, it's physically impossible to send a memory or a message through a synaptic cleft. It can't be done, which means that it isn't being done.

Furthermore, I observed that ALL of the synaptic input on a single neuron gets reduced to a single Post-Synaptic Potential (PSP), which determines whether a neuron turns ON or stays OFF. ALL of the synaptic input into a single neuron gets reduced to a single BIT of information at the physical level. The physical brain is literally incapable of storing BYTES of memory and programming code at the physical level, which means that ALL of our memory and programming code and processing of information has to be taking place at the quantum level or the psyche level.

ALL of the physical input that comes into a neuron through its synapses gets reduced to a single BIT of information, a Post-Synaptic Potential (PSP) which determines whether a neuron turns ON or stays OFF. Storing BYTES of information or BYTES of memories within a neuron is physically impossible. It can't be done, which means that it isn't being done. A neuron is a single hardware BIT.

Just as a computer chip, RAM, and CPUs need Someone Psyche external to the computer or separate from the computer to organize or MAP all of those different stand-alone switches or transistors (BITS) into functional BYTES of data, memory, and programming code, the very same reality applies to a physical brain. Someone Psyche has to impose order and organization and meaning onto the individual BITS thereby treating them and using them as BYTES; otherwise, memory storage and the processing of programming code would be impossible at the physical level. You can't store and process BYTES of memory and programming code within a stand-alone hardware BIT without some kind of wire or logic gate to connect the BITS into logical BYTES.

https://en.wikipedia.org/wiki/Logic_gate

http://mypsyche.us/wp-content/uploads/2018/03/Logic-gate-Wikipedia.pdf

When it comes to computers and physical brains, this type of order and organization has to be imposed on the system from outside the system. Computers and brains don't spontaneously generate out of thin air, and neither does the software that is made to run on them or through them. Computers, brains, and neurons don't spontaneously organize themselves into functional BYTES of data and programming code. Spontaneous generation of data, programming code, and hardware is physically impossible. Entropy and random diffusion prevent it from happening. You can't get order and organization out of entropy, random diffusion, and spontaneous generation.

There are NO wires within a physical brain connecting the different neurons, switches, synapses, or transistors into functional logic gates and therefore functional BYTES of information, memory, data, and programming code. If there are wires within your brain, they exist at the quantum level or the psyche level because it's obvious that they don't exist at the physical level.

If the stand-alone switches, neurons, or BITS within your physical brain are indeed being organized into functional BYTES of memory and programming code and logic gates, then WE KNOW that that organization and functionality is being implemented and enforced outside of the brain or separate from the brain at the quantum level, because there is NO physical mechanism within the brain that can be used to organize the separate neurons into functional BYTES of programming code and memory at the physical level.

There are NO wires in a physical brain. This means that all of the RAM, CPU functionality, memory storage, and processing of code that are associated with your brain are taking place outside your physical brain in the quantum realm or psyche realm, because Logic Gate functionality is physically impossible at the physical level within the physical brain due to the fact that there are NO wires within your brain.

We arrive at the truth of the situation through a process of elimination.

All you want is the truth, unless of course you are a Materialist, Naturalist, Darwinist, Nihilist, or Atheist; and then, any old lie or deception will do. I KNOW, because I used to be a Materialist, Naturalist, Nihilist, and Atheist. Self-deception works, and it works every time. There's no getting around Materialism and Naturalism, until you decide to allow ALL of the evidence into evidence and then choose to pursue a preponderance of the evidence.

The fact that there are no functional BYTES within a physical brain at the physical level just might be the most significant discovery that I made while studying Neuroscience and trying to figure out how the physical brain really works at the physical level. There are NO wires within a physical brain connecting the stand-alone switches, or neurons, or BITS into functional logic gates and functional BYTES of memory and programming code. It's physically impossible to store bytes of memory and programming code within a single hardware BIT like a neuron.

Remember, there is a syntropy or synergy within the physical brain that isn't visible, nor detectable, nor verifiable at the physical level with our physical instruments. Syntropy originates and exists at the psyche level or the quantum level, because it's physically impossible at the physical level thanks to entropy or the second law of thermodynamics. Action at a Distance is physically impossible according to Materialism, Naturalism, and Classical Physics; but, Action at a Distance, or Syntropy, or Synergy is an integral part and an essential part of Quantum Mechanics or Transdimensional Physics.

The BYTES of memory and programming code within your physical brain exist at the quantum level or the psyche level because it's physically impossible for them to exist at the physical level. The CPUs and RAM within your physical brain exist at the quantum level or the psyche level because it's physically impossible for them to exist at the physical level.

Finally, when it comes to a computer chip, RAM, a CPU, or a physical brain, some pre-existing Intelligence or Psyche external to the system imposed or forced that specific order or structure or schematic onto the system. Logic gates don't just spontaneously generate out of thin air. Someone Psyche forces them into existence and wires them together into that specific manner or functionality, or they don't come into existence at all.

Logic circuits are mapped onto computer chips by Someone Psyche; and, if neurons within the brain are somehow being MAPPED as logic gates within the physical brain and then being assigned meaning or BYTE functionality, most if not all of that functionality is being implemented at the quantum level or the psyche level by Nature's Psyche, because there are NO wires visible at the physical level within a physical brain capable of producing that kind of reliable logic gate functionality at the physical level.

In other words, Someone Psyche is in charge of synaptogenesis and the MAPPING of functionality onto a physical brain because there is NO way to code for the location addressing, timing sequence, and synaptic typing for the quadrillion synapses (or petabytes of data) for a child's brain within our 750-megabyte genome. Someone Psyche has to impose meaning and purpose onto each and every byte within a computer chip and a physical brain, or it's not going to exist in the first place.

Remember, it's physically impossible to store petabytes of data within a 750-megabyte genome. It can't be done, which means it isn't being done, at least NOT at the physical level. There are no limits at the quantum level or the sub-atomic level. At the quantum level or the psyche level, it's theoretically possible to MAP the output from an infinite number of quantum computers onto a single specific neuron or concept cell. Quantum Maps of Physical Functionality might just be the most powerful and versatile concept that I developed while producing Quantum Neuroscience.

I have been a computer scientist for most of my life, so I have some sense for how these things work, or are supposed to work, at the physical level. The physical brain is nothing like a CPU and RAM at the physical level. There are NO visible logic gates and RAM chips within a physical brain at the physical level. There are NO physical wires within a brain connecting everything together into functional BYTES of code and memory. The neurons are nothing but stand-alone switches – either ON or OFF. Neurons turn off automatically, but who or what turns them on? Neurons are physical switches; and, some invisible hand or Someone Psyche is reaching into the physical brain telepathically at a distance and turning on the specific neurons that it wants to have on.

At the quantum level, a physical brain seems to function as if it were a CPU and as if it had RAM or memory storage; but, there's none of that visible at the physical level.

After I abandoned My Naturalism, My Nihilism, and My Atheism, I go after science with an open mind.

Of all my scientific discoveries during the past few years, this one was the hardest one for me to swallow. I'm making the claim that there is NO memory storage RAM, NO CPU cache, NO programming code, NO data, and NO CPUs anywhere within a physical brain. I'm making the claim that the brain is a transceiver. A physical brain registers sensory input from the physical environment, and then it passes that information on to the Human Psyche and Nature's Psyche for processing. Once the Human Psyche has made a choice, and that choice has been confirmed by Nature's Psyche, then Nature's Psyche collapses the necessary wave functions (turns on the necessary neurons) to make that choice a physical reality.

This is in line with the Orthodox Interpretation of Quantum Mechanics by Henry P. Stapp. This is also in line with the Somatosensory Homunculus Map (physical input) and the Somatomotor Homunculus Map (physical output) that the Neuroscientists talk about. These Maps of Physical Functionality are constructed at the quantum level by Nature's Psyche, and then used by Nature's Psyche at the quantum level in order to get things done for us at the physical level – depending, of course, upon the choices that the Human Psyche makes at the quantum level or the psyche level. It's the Human Psyche who makes the choices as to what it wants to do with its physical body and physical brain; and then, it's Nature's Psyche who collapses the necessary wave functions or turns on the necessary neurons to make that choice a physical reality. This is all in line with the Orthodox Interpretation of Quantum Mechanics by Henry P. Stapp.

The physical brain transmits and receives data to and from the Quantum Realm where the psyches or intelligences reside. The physical brain is a transceiver, not a CPU. The brain transmits or transfers physical input to the Human Psyche and Nature's Psyche. Once the Human Psyche makes a choice, then Nature's Psyche transmits the commands to the physical brain (collapses the necessary wave functions), and that transmission is officially received by the physical brain when the specific motor neurons are turned ON by Nature's Psyche and the physical body moves. While the observed "silver cord" is in place connecting the spirit body with the physical body, the autonomic functions of the physical body continue to receive the necessary input from the quantum realm to keep them going. Sever the silver cord, and the physical body turns off and dies.

The Silver Cord of Life

https://www.youtube.com/watch?v=yzVfPQoWJas

https://www.youtube.com/watch?v=m47Z2UX4cLc

https://science-2-0.com/wp-content/uploads/2018/04/The-Silver-Cord.zip

The spirit body is something completely different than the physical body. And, I have encountered a lot of evidence suggesting that the Psyche or the Intelligence can function separately from the spirit body. It's the Human Psyche who has the experiences and makes the memories, NOT the spirit body, and definitely NOT the physical body. The spirit body and physical body have NO experiences and make NO memories while the Psyche is separated from them. Our thoughts and memories are contained within the Human Psyche, and not the spirit body and not the physical brain; therefore, our thoughts, memories, and personalities survive the death of our physical brain.

The brain has NO physical memory storage, and the brain has and does NO CPU processing of data, code, and information at the physical level. The neurons are registers, switches, or gates – not CPUs and RAM. ALL processing and data storage are done outside the brain at the quantum level or the psyche level by the Human Psyche and Nature's Psyche working together to get the job done. Psyche works at the quantum level to get things done for us at the physical level. Psyche is infinitely faster than the physical brain; therefore, Psyche is held back, dampened, limited, and restrained by the physical brain. This reality has been verified by out-of-body travelers and near-death experiencers.

If my theory is right, then that means that ALL of our consolidated memories, as well as our personality and knowledge and intelligence, will survive the death of our physical body and physical brain. This reality has been observed to be so by Near-Death Experiencers (NDErs) and their after-death Life Reviews. My theory has been verified by experience – out-of-body experiences. Our Psyche, or personality, or individuality survives the death of our physical brain. Observation or experience trumps wishful thinking, as it should.

The reason this discovery was so hard for me to accept at first is because it received a lot of resistance from the Neuroscience community and because it is considered Fringe Science rather than being anything mainstream. Furthermore, the brain clearly has tons of memory storage capacity and lots of CPU-like processing functionality. But, ALL of the observed evidence or physical evidence is telling me that there is NO memory storage and NO CPU processing taking place within a physical brain.

Clearly, the brain has tons of memory storage capacity and many quantum-processors' worth of CPU functionality; but, there are NO signs of any of that at the physical level, which means that it must be happening at the quantum level or the psyche level. And thanks to the Human Psyche, this memory storage capacity and CPU-like processing functionality continues to happen or continues to exist long after the physical brain is dead and gone, according to the empirical evidence from NDEs, OBEs, Life Reviews, and Shared-Death Experiences (SDEs).

Do you disagree?

My theory is easily falsifiable.

All the Neuroscientists have to do is find the physical wires connecting the synapses together into functional BYTES of RAM and logic gates for CPU processing; and, my theory will have been falsified. Or, the Neuroscientists have to find the CPU chips, RAM chips, and WiFi adapters within each of the neurons connecting everything together wirelessly into distributed memory and distributed processing at the physical level, in order to falsify my theory. To falsify my theory, they have to find what they so far have not found.

Until they do, though, I'm standing behind my theory that there is NO memory storage, no RAM, NO computer processing, and no CPUs within a physical brain. ALL of that mapped functionality, processing power, coordination, wireless functionality, distributed connectivity, and memory storage is taking place at the quantum level or the psyche level; and, the physical brain is simply a transceiver. Thoughts and memories are quantum waves, which means that they survive the death of our physical brain. Telepathy or Action at a Distance is WiFi at the quantum level. If a thought or memory survives the death of your physical brain, then it's truly a memory in every sense of the word. Is it not?

I discuss these realities and observations in much greater detail within my book, *Quantum Neuroscience: The Answer to Life, the Universe, and Everything*.

https://www.amazon.com/dp/B079Z6QQQB

Materialism, Naturalism, Darwinism, Nihilism, and Atheism deny the existence of the quantum, action at a distance, the non-physical, the non-local, the spiritual, the transdimensional, and the supernatural. However, the proven existence, verified existence, observed existence, and experienced existence of Supernatural Mechanisms, Action at a Distance, or Quantum Mechanics FALSIFIES the claims of Materialism, Naturalism, and their derivatives.

Quantum Mechanics is Supernatural Mechanics. The Quantum or the Supernatural is Syntropy – without beginning of days or an end of years. This means that Psyche, or Psyche Mechanisms, or Quantum Mechanisms are eternal, immortal, and everlasting in nature and origin. In their natural original state, the Quantum or the Supernatural has NO physical limitations, meaning that they are not subject to entropy. The physical body and physical brain severely dampen, limit, restrain, and restrict the capabilities of the spirit body and the Human Psyche.

The truth falsifies the lies. The observations and lived experiences of the human race FALSIFY the philosophical speculation and wishful thinking of the Materialists, Naturalists, Darwinists, Nihilists, and Atheists.

I learn best through comparison and contrast.

This is cool stuff. I'm hoping that someone will find it useful.

Introducing Quantum Neuroscience

How do you convince someone that his or her life-long beliefs are wrong?

You can't, until he or she is actively looking for that kind of evidence or information. Unless a person is willing to look at evidence that contradicts his pre-chosen beliefs, he or she will never be able to falsify or correct any of their theories and ideas.

In the following statement, can you separate the falsehoods from the truths? Most people can't, because they have been trained in school not to be able to do so.

Two perspectives dominate current thinking about human similarities and differences: an evolutionary perspective, emphasizing human kinship, and a cultural perspective, emphasizing human diversity. Nearly everyone agrees that we need both: our genes enable an adaptive human brain – a cerebral hard drive that receives the culture's software.

We're all kin beneath the skin. (*Social Psychology*, p. 158.)

The neurons in our brains are stand-alone switches. The neurons are either ON or OFF. That's it. A neuron represents a single BIT of information – either ON or OFF. That's how neurons function and work. All of the synaptic input coming into a neuron reduces to a single Post-Synaptic Potential (PSP), which determines whether the neuron fires or not.

There are NO wires in our brains connecting the neurons and synapses within our brains into functional BYTES of information, data, and programming code. There are NO hard drives in our brain. There is no RAM in our brain. There are no CPUs in our brain. Our physical brains are unable to process software or programming code because there are NO functional BYTES within a physical brain. Our physical brains are unable to store memories because there are no BYTES anywhere within our brains. Our neurons are switches. ON or OFF is all that our neurons can store and do. You can't store bytes of information within a single hardware BIT. It's physically impossible. You can't store data or memories or programming code within a synaptic cleft either, because a synaptic-cleft scrambles everything that comes its way. The observational evidence tells us that this is true.

So, where is ALL the processing and memory storage being done since it can't be done within a physical brain?

Our memories are not being stored and processed within our physical brain, because it's physically impossible to do so. Through a process of elimination, that reality leaves Someone Psyche at the quantum level or the psyche level using various different quantum mechanisms to do our memory storage, memory processing, and memory retrieval. Since the Human Psyche isn't consciously aware of any of these processes, that reality leaves Nature's Psyche, the Angels in Heaven, or God's Psyche doing the job of our memory storage, memory processing, and memory retrieval for us.

If you eliminate everything that is impossible or false, then ONLY the truth will remain.

Remember, if a memory survives the death of your physical brain, then it's truly a memory in every sense of the word.

See: *Quantum Neuroscience: The Answer to Life, the Universe, and Everything*.

https://www.amazon.com/dp/B079Z6QQQB

Every science discipline is currently based upon a wide variety of misconceptions, deceptions, and lies; and, as a society or as a whole, we don't even know that it is so.

Materialism, Naturalism, Darwinism, Nihilism, Behaviorism, Scientism, Determinism, and Atheism are used to trick us, to deceive us, and to keep us dumb and blind. These people attempt to use classical physics to trump, eliminate, and falsify Quantum Mechanics, Quantum Non-Locality, Action at a Distance, Quantum Non-Local Consciousness, Psyche, Intelligence, Syntropy, and Quantum Mechanisms in general.

However, the observed and proven existence of Quantum Mechanisms FALSIFIES Materialism, Naturalism, and their derivatives. The verified can be used to falsify the things that are false. Quantum Mechanics is our most-verified, best-proven, and most-used science that we have; and, Quantum Mechanics falsifies Materialism, Naturalism, Darwinism, Nihilism, Atheism, and their derivatives.

Mark My Words

—

Source

Science 2.0: I Upgraded My Science

https://www.amazon.com/dp/B0771K6WTX

Source Material

Quantum Neuroscience: The Answer to Life, the Universe, and Everything

https://www.amazon.com/dp/B079Z6QQQB

Bear, M. F., Connors, B. W., & Paradiso, M. A. (2007). Neuroscience: Exploring the Brain (3rd ed.). New York: Lippincott Williams and Wilkins.

Myers, D. G. (2010). *Social Psychology* (10th ed.). New York: McGraw-Hill.

Pinel, J. (2014). Biopsychology (9th ed.). New York: Pearson.

I Stand All Amazed: Love and Healing from Higher Realms

Remember, you really don't have a story to tell unless you have a Psyche somewhere in the mix. A Psyche or Quantum Non-Local Consciousness actually experiences and remembers events, both in the quantum psyche realm and here in the physical realm. Without a psyche, mind, or soul, we would be nothing more than a rock, computer, or robot. We would have no story or no history to share with other psyches if Materialism, Naturalism, Darwinism, Nihilism, Behaviorism, Determinism, and Atheism were true.

—

Durham, E. (1998). *I Stand All Amazed: Love and Healing from Higher Realms*. Orem, UT: Granite Publishing and Distribution.

—

This is my most favorite all-inclusive Near-Death Experience from a single individual, Elane Durham. I learned the most from it, which means that I found it the most helpful and thought-provoking NDE on record.

I love the questions that she chose to ask while in the presence of Jesus Christ and the Angels of Heaven. For example:

An Explosion of Questions

I wanted to know which was the correct Bible or the right church.

The Perfect Church

I was informed that my term the "right" church was not a proper designation. Rather it was the "perfect" church, which was a part of heaven and had been created there. It was designated as perfect because it was composed of a perfect organizational structure and taught as saving doctrines and ordinances the perfect or complete truth as known by God the Father and His Beloved Son.

In its perfected state Christ's church had been brought to earth and given to mortals. However, it didn't take long for us to begin to change the Church and cause it to lose truth.

In that condition the "perfect" church became so weakened that it fragmented into numerous less perfect churches. This had not only happened once, I was told, but on many different occasions and in many different lands. I was told by the angel that all of these churches have a portion of the truth.

He was referring specifically to the various churches that had fragmented away from Christ's perfected church, but who nevertheless continue to seek diligently for greater oneness with God. These have a portion of the truth. But only the Church as it was created in heaven has all of it.

Both our speed [of spiritual progress] and our distance [covered] are determined by the amount of God's truth taught by the church we choose to join.

Only Christ's perfect church — represented by the rocket I had seen — could take us all the way to the highest heavenly realm in the shortest amount of time.

I'm Not Informed Which Is Christ's Perfect Church

Though I was not informed by the angel which church it was, he did tell me that Christ's heavenly church is on the earth again today, organized just like it was organized in days of old. He also told me that if I truly desired to find it, I would.

[As mentioned in her book, Christ and the Angel gave her additional signs during her NDE to help her identify the True Church or Christ's Heavenly Church, when she finally found it fifteen years later.]

Durham, E. (1998). *I Stand All Amazed: Love and Healing from Higher Realms.* Orem, UT: Granite Publishing and Distribution. (pp. 37, 65-68).

(The copyright permits quoting passages as part of articles and critical reviews. I picked one of her questions and followed it to its conclusion, leaving out the unnecessary bits as I went along, creating one seamless whole in the process.)

Her comments disturbed me. I thought it strange at first that God and the Angels don't seem to be all that concerned whether we find and join the True Church or not. It bothered me for a long while, and I let it work on me until I figured out why. I always thought that God and the Angels would be eager for us to find and join the True Church, their Church, since it clearly exists on the earth right now according to the angel that Elane Durham talked with. So, why didn't they tell her which church is the true Church since she asked and wanted to know? That disturbed me.

I had to think about it and pray about it; and, I still continue to do so.

Eventually, some answers started to come to me.

God the Father, Jesus Christ, and the Angels do NOT talk about Christ's Perfect Church, unless you ASK to know which one it is. You will NEVER find it unless you seek it and ask for it. Even when you do ask to know which church is the True Church, God and the Angels will only go as far as to tell you that it exists, but they will NOT tell you which one it is. They will leave it up to you to find it for yourself, or to go without it if that's what you want.

I found God's refusal to tell people, who ask Him during their NDEs which church is the True Church, fascinating, and interesting. Why does God refuse to answer that question?

Apparently, God can't let you join Christ's Heavenly Church if you don't want to be there. You have to want it and ask for it, or you will never receive it.

It ALL begins with desire and agency. God and Christ can help us to overcome everything, except for a lack of desire. If we have NO desire, then there is nothing that God and Christ can do to help us. They have covenanted with us not to interfere with our agency. We literally have to give God and Christ permission to intervene in our lives, or they will simply leave us alone. They have promised us to do so.

Desire is the KEY that unlocks that specific door.

The angel told her that if she truly **desired** to find Christ's Heavenly Church, she would. Desire was the KEY! God refused to violate her agency by telling her which church she has to join.

The angel told her that Christ's Heavenly Church is on the earth again today, organized just like it was organized in days of old; and, that's is all that he was obligated to tell her. In fact, that's all that he was permitted to tell her. The angel couldn't tell her which church it is because that would eliminate her agency and give her a sense of obligation that she has to join that specific church against her will.

God doesn't want us doing anything against our will. It's our desire that God wants, our hearts. God wants Heaven, His Church, His Celestial Kingdom, and His Earthly Kingdom to be comprised of the willing – those who desire and want to be there. It has NO value to us nor to God if we don't want to be there.

While we were in heaven during our pre-mortal existence, when Satan and his followers rebelled against God, they decided that they didn't want to have anything further to do with God, Christ, and their plan of salvation. Satan and his followers no longer had any desire to be a part of God's plans.

What was God's response?

God's response was the only option left open to Him because God had granted each one of us – His children – our agency. Since Satan and his followers, a third of God's children, didn't want to have anything more to do with God and Christ, God invited them to leave; and, they did so of their own free will and choice.

As long as we have a desire to participate in their plan, then God and Christ can still help us, teach us, and save us. When we formally decide that we don't want to have anything further to do with them, then they are obligated to grant our wish. It's our desire and willingness that God and Christ are after. It's pointless for God and Christ to force us into Heaven, and their Kingdom, and their Church against our will, because we will simply choose to leave if they do so. You are free to leave the program or God's Plan, anytime you want to do so. They know it, even if we don't. We can rebel against God any time we want to do so.

When it comes to repentance, salvation, and exaltation, it's not real until we actually own it, integrate it, and deliberately make it a part of ourselves of our own free will and choice.

Since we are here on this earth and now have physical bodies, clearly, we were part of the two-thirds who decided and chose to participate in God and Christ's plans for our salvation and exaltation; and, our agency remains intact. We can turn against God anywhere along the way, which is why God refuses to tell us which church is the True Church until after we actually want to be a part of it and know it. God refuses to violate our agency by telling us which church we have to join in order to be exalted in the Kingdom of Heaven. God allows us to find it for ourselves if we want to find it.

Desire is the KEY that unlocks the additional doors that are ahead of us. If we have NO desire to go through those doors, God and Christ won't make us do so. Christ's Heavenly Church will be comprised of the willing. Those who want to be there, will be there. Those who don't want to be there, won't be there. It's that simple. God is not going to force anyone to be there. It's not real if we are forced to be there. Heaven is hell if you are being forced to be there.

If we don't want to have anything to do with God and Christ, they are contractually obligated to comply with our wish or our desire. If we desire for God and Christ to stay away from us, they will do so. They will just sit back and patiently way for us to ask them for their help. If we never ask them for help, then they will never help us, because they have promised us not to do so. They have promised not to interfere in our lives unless we give them permission to do so.

The true Church has NO meaning and NO value to us unless we find it and embrace it of our own free will and choice. My best friend told me that God will do nothing to eliminate or destroy our agency. If God were to tell us which church is His Perfect True Church, then we would feel compelled to join it and we would no longer be truly free to find it on our own, explore it, and decide for ourselves whether to make it a part of us or not.

Jesus Christ goes out of His way to protect our agency. He wants to leave us free to decide for ourselves whether we desire to be a part of His Kingdom or not. It has NO value to God nor to us if we are forced into it against our will. The Kingdom of God, or the True Church, or the Perfect Heavenly Church has to be wanted and desired, or we are not going to own it and make it an integral part of us; and consequently, it's not going to be efficacious for us. Desiring it and owning it is what makes it REAL. Christ's Heavenly Church is of NO value to us if we don't want to be there. God knows this just as much as we do.

Furthermore, I have also noticed that whenever someone is told which church is the True Church, or which church is Christ's Perfect Heavenly Church, or which church is the Kingdom of God here on the earth right now, our immediate knee-jerk reaction is to reject it automatically, because we have convinced ourselves a priori that it can't possibly be so. Even before we look at it, even before we talk to its missionaries, even before we read the first sentence in the new revelations and the new scriptures, we have already decided that it isn't true and can't possibly be true. That's the immediate "benefits" of being told which church is Christ's Heavenly Church here on the earth again today. We immediately reject it because we have been trained, conditioned, and brainwashed to do so.

Clearly, knowing which church is the True Church has NO value to us, unless we desire to know and truly want to be a part of it. That's the lesson that I got from this particular NDE – a very valuable lesson indeed.

God and the Angels are more interested in our spiritual growth than our speedy delivery into the True Church, or into the Highest Heaven, or into Exaltation and Godhood. God and the Angels know that with the right spiritual preparation and the right spiritual growth will come the Ultimate Destination of Exaltation in the Presence of God in the Highest Heaven, the Celestial Kingdom.

Consequently, God and the Angels are more interested in the growth process than in short-circuiting the growth process by telling each person which Church is the right Church, or which Church to join, or which Church will get us to the Highest Heaven the fastest. God and the Angels want us to become like God, not simply doing things because we are commanded or told to do them. God is more interested in the change of heart or the true conversion than He is in padding His True Church or His Own Church with huge numbers of people simply for the sake of making His Church look big and impressive and important.

If we are only interested in Salvation and have no interest in nor understanding of Exaltation, then God and the Angels are fine with that for now because it's the spiritual growth and true conversion that they are after initially and not self-aggrandizement or making their Church look good in the eyes of the world.

I found it interesting to observe over and over again while studying NDEs and Visions of various kinds that God won't tell us which Church to join unless we ask Him first. Even then, God won't always tell us which church is His Church, even if we do ask – preferring instead for each one of us to go and find the True Church for ourselves on our own. God values our agency that much!

Still, we have to ask, or nothing at all will happen in the first place. Asking God in prayer for His help gives Him formal permission to intervene and help us. The Biblical God

Jesus Christ responds to the heartfelt prayers that are full of sincerity, intentionality, willingness, and desire. If we don't ask for His help, then He'll just let nature take its course. God can't help us if we refuse to ask for His help.

Therefore, we have to ask God for help if we want to receive any answers from God. In the Bible, the Biblical God tells us repeatedly to ask questions and assures us that we will receive answers, of some kind. If we never ask the questions, then we will never receive the answers. So, the next time you find yourself in the presence of God, be sure to ask Him all the questions that have been bugging you all of your life.

Most of us will never find Christ's Perfect Church while here in mortality because most of us refuse to ask Him which one it is – which Church will bring us Exaltation in the Presence of God. We aren't looking for the Perfect Heavenly Church or Christ's Own Church, so we will never find it. Others don't want it even when they do find it. That's just the reality of the situation.

It took Elane Durham over fifteen years after her Near-Death Experience, but she eventually found Christ's Perfect Church or Christ's Own Church as it had been explained to her and shown to her in visions during her Near-Death Experience. She recognized Christ's "perfect" church or heavenly church when she finally found it here on earth. It matched with everything that the angel had told her about it; and, the Spirit confirmed to her soul that it was the real deal.

I have purchased many copies of Elane's book and given them or loaned them to different people, especially the people who are looking for Christ's Perfect Church, or Christ's True Church, or Christ's Own Church here on earth.

Elane Durham asked the right questions and received some interesting answers.

I noticed how the Angel and Christ refused to tell her explicitly which church is Christ's Perfect Church, because they wanted her to find it on her own and didn't want to employ any kind of compulsion or coercion. They gave her visions during her NDE that helped her to recognize it when she finally found it.

It's never real unless it is willfully and deliberately chosen, owned, and made an integral part of us. Christ and the Angels know this to be true, which is why they wanted Elane to find Christ's Perfect Church for herself, and then either join it or reject it of her own free will and choice. God is interested in the change of heart and the spiritual growth, the processes that are real with real changes and real growth taking place. God wants us to become like Him, because He knows that by doing so, we will one day submit to His will and His requirements and arrive where He is as a result.

I, too, have found Christ's Heavenly Church or God's True Church. In fact, I was actually born into it. However, I still had to find it and embrace it, all on my own.

The True Church or Christ's Heavenly Church here on earth IS a paradox.

The truth has no value to us unless we find it and embrace it of our own free will and choice. Salvation, exaltation, eternal life, and the saving ordinances have no value to us, unless we find them, choose them, accept them, incorporate them, and willingly embrace these things of our own free will and choice.

The True Church is a heavy burden when you don't want to be there. It's a heavy burden when you don't understand what it truly is. It's HELL when it's being forced on you. You are highly embarrassed to let anyone know that you are a member of this church, if you don't want to be there, don't understand it, or aren't truly committed or converted to the cause.

However, when you eagerly desire to be a part of it and sincerely want to be a part of it, miraculously it transforms into something desirous and heavenly; it transforms you into something better and more saintly; all your burdens are lifted; peace settles into your soul for the first time in your life; and, it sets you free. It really is all that it claims to be, but only when you desire it and want to be a part of it.

That's why nobody should be forced into God's True Church or Christ's Heavenly Church because it's HELL if you don't want to be there or are being forced to be there. That's why God, Christ, and the Angels refuse to tell us which church is the True Church during our Near-Death Experiences (NDEs). The whole thing is HELL unless you want to be there.

My ongoing observation when it comes to NDEs is that Jesus Christ is NOT pointing people to His Church nor leading them to His Church during their NDEs. If they want to be a part of His Perfect Heavenly Church, that's something that they are going to have to desire on their own, and want on their own, and pursue on their own for themselves. People have to ask for it, seek it, and want it before they are going to find it.

After encountering all the different NDEs on YouTube where the people are meeting Jesus Christ during their NDE, my immediate question was to ask, "Why isn't Jesus telling them which church to join?" The initial impression I got was that God doesn't care which church we join; but, that wasn't it at all. He cares; but, He knows that His Perfect Heavenly Church isn't efficacious and doesn't work right if the person doesn't want it or doesn't want to be there.

When an atheist finds himself in hell during an NDE, Christ comes immediately and gets that person out of hell, the very moment that person asks God to come and save him. However, that same person is NOT told which church go join or even to join a church. When it comes to Christ's Heavenly Church, each person is allowed to find it for himself or herself, and only if he or she wants to find it and is actually asking to find it and desiring to find it. No asking, then no finding.

God and Christ won't force us to join their Perfect Heavenly Church just to pad their numbers, line their coffers, and make themselves look good. It's not real unless it's deliberately wanted and chosen into existence. Choice is what makes it real. Agency is what makes it powerful and life changing. In contrast, compulsion or force destroys its efficacy and makes it undesirable. It has no meaning or value if it's being forced upon us.

Mark My Words

—

For the continuation of this essay, see:

Putting Psyche Back into Psychology: Restoring Science to Consciousness

https://www.amazon.com/dp/B071NC987S

Evolution Is Entropy

Now pay attention!

Despite the FACT that I have and will have debunked and falsified the Theory of Evolution thousands of times in hundreds of different ways, don't ever once let me convince you that there is NO such thing as evolution or random mutations.

Evolution is REAL, very REAL. Evolution or random mutation is entropy, particularly genetic entropy. It's REAL. Genetic entropy is very REAL at the physical level; and, it works as advertised at the physical level. Genetic entropy produces death and extinction.

Evolution, or genetic change, or genetic drift, or genetic entropy is currently producing ongoing deterioration of the human genome. According to the people who study genetic entropy, at the current rate of deterioration, the human race will be sterile and extinct in a few more thousand years. The human genome has a shelf-life of about ten thousand years total, according to the current rate of genetic entropy or the current rate of the deterioration of the human gene pool. We are already 6,000 years into it and only have another 3,000 to 4,000 years to go before the human race is extinct.

Natural selection or death doesn't touch our genes. It is the genetic recombination that takes place during the production of our gametes – eggs and sperm – that's producing the vast majority of our mutations as a race. It's unavoidable. Our genetic entropy or the deterioration of our gene pool is detectable, and its rate of deterioration is measurable. We don't have all that long before evolution, genetic entropy, or random mutations make the human race extinct.

Genetic entropy is a real phenomenon.

While writing the book on Syntropy, I one day realized that Evolution is Entropy, particularly the Creation by Disorder definition for entropy or the Creation by Chance definition for entropy. There are dozens of different mutually exclusive definitions for entropy, which is why nobody seems to know what entropy is. But, genetic entropy or the deterioration of the human genome is a real phenomenon. Our genome was designed with backup genes in place, but as our human genome continues to deteriorate, the human race will become sterile and eventually go extinct. The experts who have been studying genetic entropy and gauging the rate of deterioration have concluded that the human genome has a lifespan or a shelf life of about 9,000 years total, before the collective human genome is so damaged that it can no longer produce viable offspring.

That kind of suggests that something unseen and supernatural has been propping up and sustaining the human genome, if the current rate of measurable deterioration is any indication.

There are literally dozens of different definitions for entropy. The Materialists, Naturalists, Darwinists, Nihilists, and Atheists believe that ONLY physical matter, entropy, death, and chance exist. Consequently, the mechanism behind evolution has to be one of these. Natural selection is creation by death. Therefore, evolution in general ends up being creation by chance or creation by entropy. This essay is based upon the Genetic Entropy definition for entropy, or the constant deterioration of our genomes.

Whenever the Materialists, Naturalists, Darwinists, Nihilists, Behaviorists, and Atheists start talking about evolution, they get most everything wrong, because evolution or entropy cannot design and create. However, these people do indeed get one thing perfectly right. Evolution, or random mutation, or entropy is indeed the CAUSE of ALL of our heritable diseases, developmental diseases, and heritable mental illnesses.

One of my all-time most significant scientific discoveries came to me when I first realized that the Theory of Evolution is based upon entropy.

Obviously, I'm not the first person to discover entropy; but, I seem to be the first person on the planet to realize that Materialism, Naturalism, Darwinism, Nihilism, Atheism, Determinism, Physicalism, Behaviorism, Scientism, Classical Physics, Chemical Evolution, Macro-Evolution, Natural Selection, Random Mutations, and the Theory of Evolution are based exclusively on entropy; and, I seem to be the first person to realize what that truly means for science in general.

You see, these people have chosen to believe that physical matter and entropy are the ONLY thing that exists. They based ALL of their science exclusively on entropy.

Are you starting to see the significance of this reality? Can you see the problem?

Well, ask yourself, "What is entropy?"

Entropy is corruption, disease, disorder, chaos, death, and extinction; and, the Theory of Evolution is based exclusively on entropy. Are you starting to see the problem? It's a BIG ONE! This is HUGE!

Evolution of any type is entropy; and, evolution of any type is based upon entropy. Entropy is death and extinction. When was the last time that you caught **death** in the act of designing, creating, and producing new and unique life forms from scratch? It has never happened, and it will never happen. Entropy or evolution prevents it from happening.

Entropy literally prevents the Theory of Evolution from becoming true. Evolution produces entropy. The different types of evolution produce death and extinction. Entropy prevents the different types of evolution from designing, creating, and producing something new. Entropy cannot design and create, which means that evolution of any kind cannot design and create. Entropy can only destroy, which means that the different types of evolution can only destroy. This reality is so obvious, that I sometimes wonder why nobody has ever thought of it before.

Life, design, creation, organization, order, information, intelligence, proteins, genes, genomes, computer programs, hardware, cars, buildings, computers, phones, televisions, physical matter, and physical life forms REQUIRE an infusion of Syntropy in order to come into existence. We KNOW that Syntropy must exist, or there wouldn't be any entropy. We KNOW that Syntropy must exist, or there wouldn't be any physical matter. We KNOW that Syntropy must exist, or there wouldn't be any life – the whole thing would be death, or chaos, or entropy. We KNOW that Syntropy must exist, or we wouldn't be here to think about it.

The Theory of Evolution can't work as advertised. It's prevented from doing so by entropy, or death and extinction. Evolution is entropy; and, evolution prevents the Theory of Evolution from becoming true.

The Fruits of Evolution or Entropy

Whenever the Materialists, Naturalists, Darwinists, Nihilists, Behaviorists, and Atheists start talking about evolution, they get most everything wrong because evolution or entropy cannot design and create. However, these people do indeed get one thing perfectly right. Evolution, or random mutation, or entropy is indeed the CAUSE of ALL of our heritable diseases, developmental diseases, and heritable mental illnesses.

Remember, the Theory of Evolution is FALSE because random mutations or entropy cannot design and create genes, proteins, and life forms. However, evolution or random mutation or entropy is very REAL; and, it can indeed destroy genes, proteins, and life forms. Do you see how that works? It's important to understand.

The Theory of Evolution is a fictional story that they made up out of thin air. There is NO empirical evidence supporting any version of it. In fact, ALL of the empirical evidence and experimental evidence that we have on hand as a race FALSIFIES the different versions of the Theory of Evolution and VERIFIES Quantum Mechanics instead. Quantum Mechanics is Supernatural. Quantum Mechanics is the Priesthood Power of God. Quantum Mechanics and Psyche are Pure Syntropy. There is NO entropy in the spirit realm, the non-local realm, the quantum realm, or the transdimensional realm. Psyche, Spirit Matter, and Quantum Mechanics ARE Syntropy.

The very existence of something like the Orthodox Interpretation of Quantum Mechanics from Henry P. Stapp FALSIFIES Materialism, Naturalism, and the various versions of the Theory of Evolution. Quantum Mechanics FALSIFIES Materialism, Naturalism, and their derivatives. Fictional stories like the Theory of Evolution cannot stand in the light of truth.

The Theory of Evolution is the very pinnacle of fictional ad hoc just-so story telling; and, *ad hoc just-so stories* are logic fallacies. The Theory of Evolution is fictional because it never happened – none of it happened! The chemical evolution of proteins and genes from atoms is physically impossible. Abiogenesis, spontaneous generation, and the various different forms of macro-evolution are physically impossible. They are prevented from happening by entropy.

The Theory of Evolution is science fiction. The fictional nature of the story becomes most egregious whenever they try to guesstimate how many millions of years it took for evolution to do something for us, because evolution or entropy can't do anything for us – not ever. They are making the Theory of Evolution up as they go along. It's a fictional story, and nothing more.

Since the whole Theory of Evolution is nothing but a fictional story, you can successfully and rightfully make up fictional stories of your own to debunk it. That's the way fiction works!

Scientific Inference

Comparative Psychology and Evolutionary Psychology are based upon Darwinism and the Theory of Evolution, which means that they too are nothing more than *fictional ad hoc just-so stories* that these people have made up out of thin air.

In fact, in his book *Biopsychology*, John Pinel tells us as much when he tells us that his "evolutionary perspective" is based upon *scientific inferences*. *Scientific inferences* are fictional ad hoc just-so stories. *Scientific inferences* are logic fallacies. The whole of their evolutionary perspective, evolutionary psychology, and comparative psychology is based upon *scientific inferences* or stories that they have manufactured out of thin air. Making up stories is what makes being a scientist fun, according to John Pinel.

From *Biopsychology* page 13, John Pinel writes:

Scientific inference is the fundamental method of biopsychology and of most other sciences – it is what makes being a scientist fun. This section provides further insight into the nature of biopsychology by defining, illustrating, and discussing scientific inference.

The scientific method is a system for finding things out by careful observation, but many of the processes studied by scientists cannot be observed. For example, scientists use empirical (observational) methods to study ice ages, gravity, evaporation, electricity, and nuclear fission – none of which can be directly observed; their effects can be observed, but the processes themselves cannot. Biopsychology is no different from the other sciences in this respect. One of its main goals is to characterize, through empirical methods, the unobservable processes by which the nervous system controls behavior.

The empirical method that biopsychologists and other scientists use to study the unobservable is called <u>scientific inference</u>. Scientists carefully measure key events they can observe and then use these measures as a basis for logically inferring the nature of events that they cannot observe.

Like a detective carefully gathering clues from which to recreate an unwitnessed crime, a biopsychologist carefully gathers relevant measures of behavior and neural activity from which to infer the nature of the neural processes that regulate behavior.

The fact that the neural mechanisms of behavior cannot be directly observed and must be studied through scientific inference is what makes biopsychological research such a challenge – and as I said before, so much fun. (*Biopsychology*, p. 13.)

Pinel, J. (2014). *Biopsychology* (9th ed.). New York: Pearson.

John Pinel is trying to lay a scientific foundation for his belief in Evolution, Materialism, Naturalism, and Darwinism; and, he is using *scientific inferences* to do so. He wrote this section of his book as empirical proof that the Theory of Evolution is true. *Scientific inferences* are how the Materialists, Naturalists, and Darwinists prove that the Theory of Evolution is true. So, what are *scientific inferences*?

He defines *scientific inference* as an empirical method or a scientific method. Then he uses *scientific inference* to prove that the Theory of Evolution is true. Because he has never studied the Philosophy of Science, he has NO idea how faulty, fallacious, and weak *scientific inferences* really are. He has NO idea that *scientific inferences* are logic fallacies. Instead, he literally treats *scientific inferences* as empirical evidence, so that he can prove that the Theory of Evolution is true.

Notice carefully how he defines Evolution, Materialism, Naturalism, and Darwinism as events that we cannot observe. These are events that have NEVER been observed. So, we are going to use *scientific inferences* to manufacture "empirical evidence" for these events that cannot be observed and that have never been observed. *Scientific inference* is a logic fallacy; yet, he erroneously calls it an empirical method. They ALL do that in one way or another, without even realizing that they are using logic fallacies to prove that Evolution, Materialism, Naturalism, and Darwinism are true. These people tack on the word "scientific" to make it seem like science, but it is nothing more than "inference" or personal interpretation masquerading as empirical fact. These people are *begging the question* and *jumping to conclusions*.

Inference is the *affirming the consequent* logic fallacy in action. If you ever study the Philosophy of Science and get a professor like Joseph Rychlak who knows what he is talking about, he will teach you that for every single piece of scientific evidence, there are literally an infinite number of interpretations or inferences that can be given to that single piece of scientific evidence or that single event; and, many of them will seem plausible or believable, but only one of them can be true.

Scientific inference is the *affirming the consequent* logic fallacy in action.

Scientific inference is the *jumping to conclusions* logic fallacy in action.

Scientific inference involves *begging the question* or *circular reasoning*, which are also logic fallacies. *Scientific inference* is a logic fallacy; but, John Pinel is choosing to use *scientific inferences* as empirical evidence and as an empirical method or a scientific method. It's a logic fallacy to do so; but, he doesn't know that. He's never been taught that *scientific inference* is a logic fallacy. Scientists typically don't study the Philosophy of Science in college and grad school; and if they do, they are taught the philosophy of science by Materialists, Naturalists, Darwinists, and Atheists who don't know what they are talking about.

When it comes to the origin of life, there are an infinite number of hypotheses or inferences that are being used to explain the origin of life on this planet. In fact, the Intelligent Design Theory is infinitely more plausible and believable than the Theory of Evolution because intelligent design has actually been OBSERVED trillions of trillions of times, whereas the evolution of genes and proteins from atoms has NEVER been observed and NEVER will be because it's physically impossible. Evolution of any kind has NEVER been observed doing spontaneous generation, abiogenesis, macro-evolution, or the creation of new life forms from scratch because it's physically impossible for them to do so.

Intelligent Design Theory is an infinitely more plausible, believable, credible, and parsimonious *scientific inference* than the Theory of Evolution is because the evolution of atoms into genes, proteins, eyes, brains, and life forms is physically impossible. If you are going to use logic fallacies and are determined to use *scientific inferences* as empirical evidence, then you owe it to yourself to go with the BEST explanation or the BEST *scientific inference* – one that has actually been OBSERVED and caught in the ACT – Intelligent Design Theory. Get rid of the ones that have NEVER been observed such as Materialism, Naturalism, Darwinism, Nihilism, Determinism, and Atheism.

The whole Theory of Evolution is based upon *scientific inferences* or *fictional ad hoc stories* that these people have manufactured out of thin air. There's nothing empirical about it! There is NO empirical evidence demonstrating the creative powers of entropy, evolution, or random mutations. Chemical evolution of proteins and genes from atoms is physically impossible thanks to entropy or the second law of thermodynamics. It can't happen, which means that it never happened. Evolution is entropy, or the second law of thermodynamics. It can't design and create. It's physically impossible.

Furthermore, evolution (genetic change), random mutations, and natural selection didn't even exist until AFTER God designed and created the proteins, genes, genomes, brains, eyes, and life forms in the first place. It's physically impossible for something that doesn't even exist yet to design, program, engineer, manufacture, and create proteins and genes out of thin air. Spontaneous generation or abiogenesis is physically impossible. Entropy or the second law of thermodynamics prevents it from happening.

Evolution is entropy, which means that it can't design and create anything. Evolution or entropy can only deteriorate and destroy things. It can't design and create. It's physically impossible for evolution or entropy to design and create proteins, genes, genomes, brains, eyes, and life forms. That's just the way it is, because evolution of any kind is entropy.

It took me years, even decades, to discover that evolution or random mutation is entropy or the second law of thermodynamics. That discovery also came with a powerful gift. I discovered Syntropy! Since evolution or entropy cannot design and create, some type of Syntropy must have done the job.

I finally realized that Quantum Mechanics is the exact opposite of Materialism, Naturalism, Darwinism, Classical Physics, and Entropy. Quantum Mechanics and Psyche are Pure Syntropy. Quantum Mechanics is the Power of God or the Priesthood Power of God. Psyche or Quantum Non-Local Consciousness is the only thing that can control Quantum Mechanics at the quantum level or the psyche level. Physical matter and entropy can't touch nor control the quantum level. Only the smaller can dwell within and control the larger. Psyche is Syntropy. Quantum Mechanics is Syntropy.

There is NO aging or entropy in the Quantum Realm, Psyche Realm, Spirit Realm, or Transdimensional Realm. Transdimensional means non-physical and non-local – not located in our physical 3D space-time realm. Everything in the Quantum Realm is Pure Syntropy. It is endless, timeless, eternal, and everlasting because there is NO entropy which means that nothing ages, gets old, or dies in the Non-Local Realm or the Syntropy Realm.

The Gods create physical matter by infusing a particle of spirit matter with space-time and the ability to acquire entropy. The Gods create a particle of physical matter by taking a particle of spirit matter, filling it full of space, slowing it down to sub-light speeds, and making it subject to entropy or the passage of time. According to the theory of relativity, the particles of spirit matter existing at velocities faster than the speed-of-light experience NO passage of time, meaning that they do not age and are not subject to entropy. Entropy is a function of time or an aging process. Spirit matter and physical matter are the same thing – they are quantum objects. However, spirit matter is pure syntropy; whereas, physical matter has been slowed down by being infused with space-time and made subject to entropy or the passage of time.

What Is the Difference Between a Hypothesis and a Theory?

A hypothesis is an idea. A theory is a hypothesis that has observational evidence supporting it. It really should be called the "hypothesis of evolution" because there is NO observational evidence supporting macro-evolution, chemical evolution, abiogenesis, and spontaneous generation; and, there NEVER will be. Thanks to entropy, random diffusion, or the second law of thermodynamics, it's physically impossible for atoms to spontaneously generate into functional genes, proteins, genomes, eyes, brains, and life forms. It can't be done, which means that it wasn't done.

A hypothesis is a prediction. A hypothesis is a testable proposition.

In their quest for insight, social psychologists propose *theories* that organize their observations and imply testable *hypotheses* and practical predictions.

A theory is an integrated set of principles that explain and predict observed events.

To a scientist, facts and theories are [the same difference]. **Facts are agreed upon statements about what we observe. Theories are ideas that summarize and explain facts.** (*Social Psychology*, p. 17.)

What are the Theory of Evolution's hypotheses or predictions?

According to the science fiction that I have watched on television – *Star Trek*, *Babylon 5*, and *Earth: Final Conflict* – a million years from now human beings are going to evolve into energy beings. There's a serious problem with that prediction. There has been life on this earth for billions of years, and NONE of it has ever evolved into an energy being. It's not going to happen because it's physically impossible. It won't be done because it can't be done.

The theory of evolution makes NO realistic or useful predictions. It just catalogues what happened in the past according to the fossil record, and then it forces a personal interpretation or a scientific inference onto the fossil record after-the-fact. *Scientific inference* is a logic fallacy. *Personal interpretations* introduce a wide variety of logic fallacies into science. The theory of evolution is based upon a wide variety of scientific inferences, logic fallacies, personal interpretations, and wishful thinking.

Macro-evolution, chemical evolution, abiogenesis, and spontaneous generation have NEVER been observed because they are physically impossible thanks to entropy, random diffusion, and the second law of thermodynamics.

Technically, the theory of evolution makes NO testable predictions because ALL of the observational evidence, empirical evidence, experimental evidence, and experiential evidence FALSIFIES the theory of evolution.

The FACT is that the spontaneous generation of atoms into functional genes, proteins, genomes, eyes, brains, and life forms is physically impossible. Since macro-evolution, spontaneous generation, chemical evolution, and abiogenesis are physically impossible, that means they never happened because they couldn't happen.

The FACT is that the major premises or primary hidden assumptions of Materialism, Naturalism, Darwinism, Nihilism, Behaviorism, Determinism, and Atheism – claiming that the quantum or the supernatural does not exist – are FALSIFIED due to a complete lack of observational evidence or a complete lack of supporting evidence. Furthermore, they are FALSIFIED by observational evidence and experimental evidence. The verified and proven existence of Quantum Mechanics and Action at a Distance FALSIFIES Materialism, Naturalism, and their derivatives.

The primary assumptions or hidden premises associated with Naturalism and Darwinism are NOT testable.

There are NO observed events and NO observed facts associated with chemical evolution, design and creation by random mutations, abiogenesis, spontaneous generation, macro-evolution, or design and creation by genes, proteins, RNA, and amino acids. Stand-alone atoms cannot spontaneously generate into functional genes, proteins, genomes, eyes, brains, and physical bodies because spontaneous generation is physically impossible thanks to entropy or random diffusion.

There are NO observed facts associated with creation by evolution, which means that the "theory of evolution" is in fact a falsified hypothesis and NOT a verified theory.

The theory of evolution is based exclusively upon correlation, and NOT observation. There is a huge difference between the two. Correlation does NOT prove causation. The Darwinists have carefully correlated their hypotheses with the fossil record. The Darwinists have deliberately correlated the fossil record with Darwin's Tree of Life. However, NOBODY except for the Biblical God Jesus Christ actually observed the production of the fossil record. Our fossil record could have been produced in many different ways; and with hindsight, all of these different ways or hypotheses can be made to correlate with the fossil record. Only God knows which one of those ways is the actual way by which He produced the fossil record.

Observation trumps correlation, or at least it should. However, when it comes to the theory of evolution, the Materialists, Naturalists, Darwinists, and Atheists make their correlations trump ALL the observational evidence that falsifies their pre-chosen beliefs. They cheat in order to make their case.

ORIGIN: Probability of a Single Protein Forming by Chance

https://www.youtube.com/watch?v=W1_KEVaCyaA

http://www.originthefilm.com/mathematics.php

http://science-2-0.com/wp-content/uploads/2018/03/THE-MATHEMATICS-OF-ORIGIN.pdf

https://www.youtube.com/watch?v=cQoQgTqj3pU

https://www.youtube.com/watch?v=_zQXgJ-dXM4

http://bio-complexity.org/ojs/index.php/main/article/view/BIO-C.2010.1/BIO-C.2010.1

http://science-2-0.com/wp-content/uploads/2018/03/Case-Against-Darwinian-Origin.pdf

http://science-2-0.com/wp-content/uploads/2018/03/Chemical-Evolution-Is-Impossible.zip

http://science-2-0.com/wp-content/uploads/2018/04/Origin-Of-Life.zip

Once you have eliminated everything that is FALSE and everything that is IMPOSSIBLE, then ONLY the Truth remains. It's elementary.

The theory of evolution has been tested and falsified. There's NO practical application for creation by rocks, chemical evolution, or the theory of evolution because spontaneous generation is physically impossible. There are NO observations when it comes to the theory of evolution. The theory of evolution makes NO testable and verifiable predictions because ALL of observed evidence falsifies Materialism, Naturalism, Nihilism, Atheism, and the Theory of Evolution.

Materialism, Naturalism, Darwinism, Nihilism, Behaviorism, Determinism, and Atheism are FALSIFIED due to a complete lack of observational evidence or a complete lack of supporting evidence for their hidden assumptions or major premises which claim that the quantum or the supernatural does not exist. In contrast, it is said that Quantum Mechanics or Action at a Distance is the most-verified, best-proven, and most-used science that we currently have.

Quantum Mechanics and Quantum Neuroscience

On page 25 of *Biopsychology*, John Pinel quotes another worshipper of the theory of evolution:

Evolution is both a beautiful concept and an important one, more crucial nowadays to human welfare, to medical science, and to our understanding of the world than ever before. It's also deeply persuasive – a theory you can take to the bank. The supporting evidence is abundant, various, ever increasing, and easily available in museums, popular books, textbooks, and a mountainous accumulation of scientific studies. No one needs to, and no one should, accept evolution merely as a matter of faith (Quammen, 2004, p. 8).

Ironically, the mountains of evidence supporting the theory of evolution are correlational – designed and manufactured after-the-fact to fit the fossil record. NONE of that evidence is observational. Correlation is not causation. Instead, ALL of the observational evidence and experiential evidence that we have on hand as a race FALSIFIES Materialism, Naturalism, Darwinism, Nihilism, Behaviorism, Scientism, Determinism, and Atheism. Correlation cannot be used to prove causation. All of these falsified philosophies have to be taken on blind faith as being true because there is NO observational evidence supporting them. The theory of evolution is bankrupt. There's no logical reason to believe in the theory of evolution, except for the fact that everybody else has been deceived by it and has chosen to believe in it.

I talk about all of this in great detail in my book, *Quantum Neuroscience: The Answer to Life, the Universe, and Everything*. If a comparison between Evolution and Quantum Mechanics interests you, I recommend you take a look at that book:

https://www.amazon.com/dp/B079Z6QQQB

The book, *Quantum Neuroscience: The Answer to Life, the Universe, and Everything*, makes a detailed comparison between Neuroscience and Quantum Neuroscience, which means that it makes a detailed comparison between Classical Physics and Quantum Mechanics.

When properly understood, Quantum Mechanics FALSIFIES Classical Physics, Materialism, Naturalism, Darwinism, Nihilism, Scientism, Behaviorism, Determinism, and even Atheism. Quantum Mechanics is Supernatural. Quantum Mechanics or Syntropy is the exact opposite of Classical Physics, Entropy, Random Mutations, Materialism, Naturalism, Darwinism, and the Theory of Evolution.

Quantum Mechanics is a proven and verified science. Quantum Mechanics is the best-proven and most-used science that we have. In contrast, the Theory of Evolution has NO empirical evidence supporting it. In fact, ALL of the empirical evidence that we have on hand as a race, including Quantum Mechanics, FALSIFIES Materialism, Naturalism, Darwinism, and the Theory of Evolution. Do you see how that works? It's important to understand.

Quantum Mechanics, Spirit Matter, and Quantum Non-Local Consciousness (Psyche or Intelligence) are PURE SYNTROPY. The Syntropy has to exist somewhere someplace somehow because, according to the Law of Entropy and the Second Law of Thermodynamics, the physical Multiverse should have burned out and suffered heat death an eternity or two ago; and, there should be NO more physical universes anywhere, but here we are, nonetheless. The very existence of this physical universe – it's beginning full of syntropy and its ongoing existence billions of years later – is positive proof that Someone Psyche knows how to do syntropy or Someone Psyche is syntropy. Quantum Mechanics is syntropy or the Priesthood Power of God. God's Psyche knows how to do syntropy or Quantum Mechanics; otherwise, this physical universe would not exist.

This is what Quantum Mechanics or Transdimensional Physics is trying to teach us. Quantum Mechanics, Spirit Matter, and Psyche are pure syntropy. Evolution, Random Mutations, Physical Matter, and Classical Physics are entropy.

Quantum Mechanics is an infinitely better theory than Materialism, Naturalism, Darwinism, Nihilism, Atheism, Classical Physics, and the Theory of Evolution. Old theories are supposed to fall by the wayside whenever a better theory is proposed to account for the findings. Intelligently designed quantum machinery is an infinitely better theory for the origin of life than the theory of evolution because Quantum Mechanics and Intelligence have been experienced, observed, replicated, verified, and proven to be real and true through a preponderance of the evidence. Design and creation by entropy or evolution has NOT. Instead, design and creation by entropy or evolution has been falsified. It's physically impossible.

Evolution Is Entropy

Random mutations are entropy.

It's physically impossible to produce a functional protein by throwing dice. It can't be done, which means that it wasn't done.

Furthermore, designing, engineering, and making a matching gene to go along with a functional protein requires deliberation, planning, engineering, and intelligence. It can't happen through random chance or luck. It's physically impossible. The production of a protein, a matching gene, as well as a functional genome requires a programmer, an engineer, a designer, a fine-tuner, and Syntropy of the highest order. Someone Psyche has to infuse syntropy or order into the equation in order to make it happen because entropy naturally prevents it from happening. Your genome is God's Signature.

Random chance produces disorder, disorganization, chaos, cancer, death, extinction, and entropy.

Random mutations are the mechanism of change when it comes to evolution, genetic change, genetic drift, or genetic entropy – not natural selection. Natural selection doesn't touch our genes and proteins. Natural selection is worthless when it comes time to change our genes and proteins.

Natural selection is also a product of chance or a function of luck. The ultimate product of natural selection is death and extinction. Death and extinction are entropy. The physical mechanism that we call natural selection causes entropy, death, and extinction. Natural selection cannot design and create because natural selection doesn't touch our genes and proteins.

Since the very beginning of the theory of evolution, Darwinists and Evolutionists have erroneously stated that natural selection is the causative agent behind genetic change – the origin of species by means of natural selection. They are wrong. They don't even understand their own theory. Selection or sexual activity determines whether our genes get passed on to the next generation, or not; but, natural selection or selection of any kind does NOT change our genes. Selection or sexual activity does NOT touch our genes!

Natural selection or survival of the fittest doesn't do anything except result in death and extinction. Death and extinction are entropy, not syntropy, design, and creation. Natural selection is NOT a mechanism of change. Natural selection doesn't do anything. Natural selection is death, extinction, and entropy. Natural selection doesn't touch our genes. It can't. It's physically impossible. There's no invisible person called "Natural Selection" who is reaching into our genes and making them come alive.

The majority of our heritable mutations come from the genetic recombination process that takes place during the production of our gametes (egg and sperm). Genetic recombination is the process that shuffles our genes and introduces errors or mutations into the system; and, genetic recombination or the production of gametes takes place BEFORE any type of selection or sexual activity. Remember, natural selection doesn't touch our genes. God is the person who designed and created our proteins and the matching genes to go along with them, which means that God is the one who designed the physical process that we call genetic recombination. Genetic recombination is the primary mechanism of change driving evolution, random mutations, and genetic drift. Random mutations are entropy, an integral part of physical matter.

Combine random mutations with natural selection and we end up with genetic entropy – the devolution of our genome to the eventual point where our species goes extinct. It all ends in death or entropy. Technically, the "theory of evolution" is the origin of species by means of genetic entropy, random mutations, and extinction, which is physically impossible. Natural selection has nothing to do with the origin of species. Death and extinction are the natural result of evolution or genetic change.

[See: Sanford, J. (2014). *Genetic Entropy* (4th ed.). Cornell University: FMS Foundation.]

Evolution or Entropy Puts on the Brakes

Evolution of any kind cannot design and create. It's physically impossible. It's prevented from happening by entropy.

Chemical evolution, abiogenesis, spontaneous generation, or macro-evolution is physically impossible. Entropy prevents these things from happening, thereby making them physically impossible. That's the way it really is in the natural world. The chemical evolution of functional genes and proteins is physically impossible. It can't happen, which means that it didn't happen.

ORIGIN: Probability of a Single Protein Forming by Chance.

https://www.youtube.com/watch?v=W1_KEVaCyaA

I find all of this fascinating. I used to be a Materialist, Naturalist, Nihilist, and Atheist; so, I find it fascinating to observe how wrong I really was in my chosen conclusions at the time. But, I'm not alone. Millions have fallen for the deceptions and the lies. Nevertheless, we have had the truth all the way along, if we were willing to look for it, find it, and accept it.

Spontaneous generation, abiogenesis, chemical evolution, or macro-evolution was FALSIFIED in 1859 by Louis Pasteur – the very same year that Charles Darwin published "On the Origin of Species". We've known that the Theory of Evolution is false from the very beginning; but as a race, we chose to ignore the evidence.

In fact, evolution (genetic change), random mutations, and natural selection didn't even exist until after God designed and created the proteins, the matching genes, the genomes, the eyes, the brains, and the physical bodies in the first place. It's so obvious that I sometimes wonder how I managed to overlook it for the first fifty years of my life; but, I wasn't looking for it, so I wasn't able to see it. Evolution is entropy. Each species has been de-evolving or degenerating ever since God created its genome, thanks to evolution or entropy.

Entropy prevents a genome from spontaneously generating out of thin air. Spontaneous Generation or Chemical Evolution was FALSIFIED in 1859 by Louis Pasteur, which means that Materialism, Naturalism, Darwinism, Abiogenesis, Macro-Evolution, and the Theory of Evolution were FALSIFIED at the same time in 1859. Isn't it fascinating to observe that we KNEW that the theory of evolution is false from the very beginning; but, our scientists chose to go along with it anyway? The truth is that genomes don't spontaneously generate out of thin air. They NEVER have, and they NEVER will. The same can be said of proteins.

Your genome is programming code. Your genome is God's Signature. I was a computer programmer for a decade. I know for a fact that programming code doesn't spontaneously generate out of thin air. That's physically impossible. Entropy or evolution cannot do computer programming. You cannot do computer programming or genome programming by throwing dice, shuffling the deck, random mutations, or random chance. I also know what happens whenever there is a bug in the code. The results are often fatal; and, if the program manages to survive, the results from a bug in the code are less than optimal. Disease, deformity, cancer, pain, suffering, death, and extinction are the natural result of entropy, or random mutations, or evolution.

Finally, take note that entropy or random mutations prevent a genetically compatible Mr. and Mrs. Mutant from coming into existence at the same time in the same place. Thanks to entropy or the randomness of mutations, it's physically impossible for chimp-like ancestors to evolve into chimpanzees and humans, because it's physically impossible to produce the requisite Mr. and Mrs. Mutant each and every time that one is needed, so as to make a sexually reproducing species evolve naturally into some other species. Macro-evolution of this type is prevented from happening by entropy when it comes to a sexually reproducing species.

Remember, the theory of evolution was made to fit the fossil record after-the-fact in hindsight, and it's constantly adjusted every time another fossil is dug up. The theory of evolution is reactive and not predictive. The theory of evolution is correlational, and not observational. Correlation cannot be used to prove causation. The missing links really are missing. There are NO observations supporting any aspect or any version of the theory of evolution. Design and creation by evolution has NEVER been caught in the act. There are infinitely better and more plausible explanations for the origin of life than random chance or blind luck. The only way to make the theory of evolution work as advertised is to get God to intervene and force it to work; and, God has infinitely better and infinitely faster ways of designing and creating life than random mutations and natural selection.

The theory of evolution is wishful thinking in action. The theory of evolution is the product of Confirmation Bias, which is the psychological tendency to search only for information that confirms one's preconceptions or one's pre-chosen conclusions. These people automatically reject and dismiss anything that falsifies the Theory of Evolution, Materialism, and Naturalism. They cheat, in order to make their case. They employ a wide variety of logic fallacies to make their case for the theory of evolution.

Conclusions Regarding Evolution and Entropy

So, what have we learned?

Evolution is entropy. Entropy cannot design and create. These people fail to meet their burden of proof, because the preponderance of the evidence falsifies Materialism, Naturalism, Darwinism, Nihilism, Atheism, and the Theory of Evolution.

A hypothesis is an idea. A theory is a hypothesis that has observational evidence supporting it. It really should be called the "hypothesis of evolution" because there is NO observational evidence supporting macro-evolution, chemical evolution, abiogenesis, and spontaneous generation; and, there NEVER will be. Thanks to entropy, random diffusion, or the second law of thermodynamics, it's physically impossible for atoms to spontaneously generate into functional genes, proteins, genomes, eyes, brains, and life forms. It can't be done, which means it wasn't done.

Isn't it fascinating how we can find and know the truth simply by thinking about things critically, logically, impartially, and rationally?

I don't make my living by preaching and teaching evolution, so I'm free to see through all of its deceptions and lies. I'm not motivated by confirmation bias where the theory of evolution is concerned, so I'm free to allow ALL of the evidence into evidence and free to pursue a preponderance of the evidence.

I have observed that the preponderance of the evidence falsifies Materialism, Naturalism, Nihilism, Atheism, and the Theory of Evolution. I have also observed that there are NO observations wherein any aspect of the theory of evolution has been caught in the act. Evolution of new life-forms from pre-existing life-forms has never been observed, which means that the theory of evolution has never been verified. In contrast, design and creation by Intelligent Beings or Intelligent Psyches has been observed, verified, and experienced trillions of trillions of times. So, which explanation for the origin of life is true – the one that has been falsified or the one that has been verified?

Repeated falsification proves that a theory is false. Constant verification implies that a theory is true.

Entropy FALSIFIES the claims of Materialism, Naturalism, Darwinism, and the Theory of Evolution, because evolution is entropy, and entropy cannot design and create anything. Entropy can only degrade and destroy. Entropy cannot design, program, fine-tune, field-test, manufacture, engineer, and create, which means that evolution of any kind cannot design, program, fine-tune, and create. Evolution is entropy.

So, where did life come from since it obviously didn't come from evolution or entropy?

Life came from Syntropy.

Life is syntropy. Psyche is syntropy. Intelligence is syntropy. Physical matter originated from syntropy or spirit matter. Quantum Mechanics is syntropy. The Priesthood Power of God is Syntropy. Quantum Mechanisms are the Priesthood Power of God.

Programming code or genomes are the result of syntropy. The arrow of progression in the fossil record, or the ever-increasing complexity of life in the fossil record, is the result of syntropy. WE KNOW this is so because entropy or evolution would prevent it from happening. God is Syntropy.

Life came from Syntropy, not entropy or evolution. Evolution or entropy cannot do life. Evolution or entropy can only do disorder, disease, chaos, cancer, death, and extinction. It takes an infinite amount of blind faith to believe that entropy or evolution can design and create, when ALL of the physical evidence we have on hand is telling us that entropy or evolution can only do disorder, disease, deformity, death, and extinction. The Darwinists and Evolutionists say that evolution is a beautiful concept, but there's nothing beautiful about cancer, disease, death, and extinction. Cancer, disease, deformity, death, and extinction are the result of evolution or random mutations.

Ironically, the very existence of entropy or evolution is Scientific Proof of God's Necessity; and therefore, Scientific Proof of God's Existence. Evolution is entropy, which means that evolution of any kind cannot design and create anything. The chemical evolution of proteins, and the matching genes to go along with them, NEVER happened because entropy prevents it from happening. The development of new species through random mutations and natural selection never happened because entropy prevents it from happening. Consequently, we have to look someplace else besides entropy or evolution for the origin of life on this planet. What's the opposite of entropy or evolution? Syntropy is the opposite of entropy. Syntropy is order and organization. Syntropy is intelligence. God is Syntropy.

It's elementary.

God has to exist, because Someone Psyche had to be there in the first place to wind up the clock or wind up this physical universe with Syntropy in the first place. We couldn't have all of that subsequent entropy without in initial infusion of syntropy. The initial syntropy within this physical universe had to come from someplace, and it came from God and the quantum realm. There's no other logical explanation for its origin. The Syntropy had to come from the syntropy realm, or quantum realm, or psyche realm, or spirit realm because it clearly doesn't exist here in this physical realm. The order, organization, or syntropy couldn't have come from evolution because evolution is chaos and entropy. The order, organization, programming, and structure within 3D proteins, genes, eyes, brains, and genomes had to come from Someone Psyche or Someone Syntropy.

This physical realm and classical physics are based upon entropy. Entropy cannot do design and creation. The psyche realm, quantum realm, and Quantum Mechanics are based upon syntropy, order, and intelligence. Syntropy, order, organization, psyche, and intelligence have the innate ability or the inherent capability of doing design and creation.

Remember, evolution of any kind is entropy; and, evolution of any kind is prevented from happening by entropy. The production of functional genes and proteins from atoms is prevented from happening by entropy. Evolution or entropy cannot produce order or syntropy. It's physically impossible. Therefore, we are looking for a non-physical explanation for the origin of life or the origin of syntropy.

Order and Organization are Syntropy. Life is Syntropy. Psyche is Syntropy. Intelligence is Syntropy. Quantum Mechanisms are Syntropy. Functional genomes and functional proteins are the result of Syntropy. Your genome is God's Signature. God is Syntropy.

Quantum Mechanics is Syntropy, which means that quantum mechanisms are supernatural in nature and origin. Quantum Mechanics is the Priesthood Power of God.

Quantum Mechanics is observed, proven, and verified science. It's time that we start using Quantum Mechanics to explain how things really work. The very existence and verified existence of Quantum Mechanics, Syntropy, Psyche, Transdimensional Mechanisms, or Supernatural Mechanisms FALSIFIES Materialism, Naturalism, Darwinism, Nihilism, the Theory of Evolution, and even Atheism.

You are going to need God's help if you want to gain access to transdimensional physics or quantum mechanics. Fallen, mortal, entropic, physical beings such as us typically cannot gain access to nor control over superpowers or transdimensional physics on their own because physical matter and their physical bodies were designed by God to prevent them from accessing and using quantum mechanics, transdimensional physics, or superpowers. Your genes and physical atoms prevent superpowers – they will NEVER give you superpowers. That's one of the fatal flaws of Naturalism and Atheism – these people superstitiously believe that, with the right configuration, genes and physical matter can magically give them superpowers; but, that's physically impossible.

Materialism, Naturalism, Darwinism, Entropy, and Classical Physics completely lack explanatory power when it comes time to explain what the Human Psyche and Nature's Psyche are doing in the quantum realm and how they work in the quantum realm or the syntropy realm. For that explanation, we need quantum mechanisms, supernatural mechanisms, or psyche mechanisms.

Quantum Mechanics, particularly Quantum Field Theory, is our best-proven, most-verified, and most-used science that we currently have. It's time that we start using Quantum Mechanics or Transdimensional Physics to explain what's happening in the quantum realm, syntropy realm, or psyche realm.

If your interpretation of Quantum Mechanics cannot explain what the Human Psyche and Nature's Psyche are doing at the quantum level in order to get things done for us at the physical level, then your interpretation of Quantum Mechanics is worthless because it's based upon Materialism, Naturalism, Nihilism, and Classical Physics. Naturalism and Classical Physics lack explanatory power when it comes to the psyche realm or the quantum realm because Materialism and Naturalism deny the existence of psyche or syntropy.

The ultimate goal of the Materialists, Naturalists, Darwinists, Nihilists, and Atheists is to suppress and destroy any evidence that falsifies their chosen beliefs. My goal is to try to find a scientific explanation for everything that has ever been experienced and observed. Science is observation and experience, or it should be.

You can't start with something that is false and then hope to produce something that is true based upon that falsehood. You can't use something that is incomplete or false to demonstrate and verify the truth. The evidence doesn't bear out reification and deification of Creation by Evolution, natural selection, death, and entropy. Evolution is entropy. The Theory of Evolution is Creation by Death or Creation by Entropy. The proposed eternal nature of death, heat death, and entropy is NOT supported by the observational evidence or scientific evidence that has been obtained by Out-of-Body Experiences (OBEs), Shared-Death Experiences (SDEs), Near-Death Experiences (NDEs), and after-death Life Reviews.

Materialism, Naturalism, Darwinism, Nihilism, Behaviorism, Determinism, Atheism, and Classical Physics cannot be used to verify the truthfulness and usefulness of Quantum Mechanics, Syntropy, Eternal Life, and Psyche. Quantum Mechanics is supernatural in nature and origin, which means that Naturalism and Quantum Mechanics are mutually exclusive. If the one is true, then the other is automatically false.

You are going to have to decide for yourself which one is true, and which one is false; but, I KNOW for myself which one the preponderance of the evidence verifies and which one the preponderance of the evidence falsifies; and, that's good enough for my needs. All I ever really wanted to know is the truth, and now I do.

Mark My Words

—

Source Material

1. *Scientific Proof of God's Existence: Finding God Where the Atheists Refuse to Look for Him*.

https://www.amazon.com/dp/B07B26CRHX

2. Myers, D. G. (2010). *Social Psychology* (10th ed.). New York: McGraw-Hill.

3. Pinel, J. (2014). *Biopsychology* (9th ed.). New York: Pearson.

4. Sanford, J. (2014). *Genetic Entropy* (4th ed.). Cornell University: FMS Foundation.

References

1. *Quantum Neuroscience: The Answer to Life, the Universe, and Everything*.

https://www.amazon.com/dp/B079Z6QQQB

2. *NATURE vs. NURTURE vs. NIRVANA: An Introduction to Reality*

https://www.amazon.com/dp/B01JWRCSVA

https://www.amazon.com/dp/1521132615

3. *BioPsychoSocial: Including Psyche or Light into our Theoretical Models*

https://www.amazon.com/dp/B0713NDHVW

4. *Science 2.0: I Upgraded My Science*.

https://www.amazon.com/dp/B0771K6WTX

The Theory of Relativity Is Limited

We are going to have to rethink this thing if we are going to make it match with what has been experienced and observed by Out-of-Body Travelers.

In the following PBS Space Time videos, they make the case that the force of gravity does not exist.

Curved Spacetime in General Relativity

https://www.youtube.com/playlist?list=PLsPUh22kYmNAmjsHke4pd8S9z6m_hVRur

According to Einstein's General Relativity, the force of gravity does not exist. All orbiting objects are simply following a constant speed straight-line path in curved spacetime. However, while sitting still in your chair, you are actually accelerating in a curved spacetime resulting in an effect that we call "gravity"; but, it's the acceleration that's real and NOT the force of gravity. So, who is making me accelerate in curved spacetime while I sit still in my chair? What's the force behind this if it isn't gravity? Anyway, lots of weirdness and lots of unanswered questions!

My other question, "Since there is NO force of gravity, then for each planet, star, and galaxy that we observe, who set the initial mass and the initial acceleration?" When the Gods made each planet, star, and galaxy, they also had to give it an initial acceleration. They had to set the orbit. But, before the Gods did that, they had to make a curved spacetime at a universal scale in the first place. As I see it, all of this is only complicated by the fact that there is no force of gravity according to General Relativity.

Furthermore, since there is no force of gravity according to Relativity, why are we able to detect gravity waves? What's transmitting that force to our earth when two black holes merge somewhere else? If it isn't gravity transmitting that force, then what is?

By the way, the following is a really cool video of two merging black holes and gravitational lensing:

https://upload.wikimedia.org/wikipedia/commons/a/a4/BBH_gravitational_lensing_of_gw150914.webm

http://origin-science.org/wp-content/uploads/2018/09/Black-Hole-Merger.zip

Anyway, General Relativity is unintuitive; and, there are claims all over the internet from one person to another that they have falsified General Relativity. Even though General Relativity is accepted by the science community in general because they have nothing better to take its place, there have been many different people who have tried to dispute it and refute it. There's something extremely important that's still missing from Relativity demonstrated by the fact that we are still lacking a Unified Theory of Everything.

https://en.wikipedia.org/wiki/Alternatives_to_general_relativity

Furthermore, there are lots of different scientific proofs and observations online where things are actually moving faster than the speed-of-light in violation of Einstein's Special Relativity. It seems obvious and clear to me that Special Relativity ONLY applies to entropic physical matter or space-time, and not necessarily to anything else.

The main idea that I find useful from the Theory of Relativity is the observation that TIME STOPS when an object reaches the speed-of-light. This has been of interest to me because I experienced Quantum Tunneling; and, TIME STOPPED during the event. I was frozen in time, and then instantly I and the car I was driving were someplace else.

Aspects of the Theory of Relativity are in dispute or are obviously incomplete. The NET result from all of this contention is that whenever I find a discrepancy between Relativity (Entropy and Matter and Time) and Quantum Mechanics, I go with Quantum Mechanics or Spiritual Mechanics or Supernatural Mechanics every time. I also choose to go with whatever has been experienced and observed. You see, the Theory of Relativity is the pinnacle of classical physics; and, classical physics has been deprecated and at times even falsified by Quantum Mechanics. If an observation violates Relativity, that's perfectly fine because Quantum Mechanics and Psyche allow it to violate Relativity and Classical Physics. In other words, Quantum Mechanics or Spiritual Mechanics TRUMPS Relativity and Classical Physics every time! By allowing Psyche and Quantum Mechanics in to play, we can literally explain everything that comes our way.

The Theory of Relativity, Entropy, and Gravity are limited exclusively to space-time, locality, and entropic physical matter. There are NO speed limits and there is NO entropy in the Quantum Realm or the Spirit World; and, gravity doesn't seem to apply there either. This is what has been experienced and observed. Maybe there really is NO force of gravity, and gravity is ONLY an emergent property of entropic physical matter, time, or mass. Entropy has NO quantum force and NO quantum field, which means that entropy doesn't really exist as a person, place, or thing; but, it's really hard to visualize gravity as being the same type of thing. Still, the idea that your fatness is accelerating through curved spacetime while sitting in a chair does make gravity sound a lot like entropy after all.

Whatever turns out to be true in regard to the force of gravity, one thing is obvious and has already been experienced and observed – Quantum Mechanics allows God's Psyche to violate, trump, or override Relativity, Spacetime, Entropy, Light-Speed, Time, Distance, Space, Classical Physics, Genetics, Physical Laws, Physical Restrictions, and Gravity at will. If we are ever able to develop a Quantum Theory of Gravity, it's going to be based upon Quantum Mechanics and NOT Classical Physics. I've seen talk online of Quantum Theories of Gravity that are based upon Quantum Entanglement or Quantum Non-Locality. I think those are much more likely to succeed than anything based upon materialism, naturalism, classical physics, relativity, death, and entropy.

Faster-than-Light Falsifies Special Relativity

In 2017, I developed a better way for doing science – I start with what has been experienced and observed, and then I try to find a scientific explanation for it. The theory of relativity is limited, and therefore inadequate and sometimes flawed. It doesn't cover everything that has been experienced and observed.

The professional physicists on PBS Space Time have repeatedly stated that our universe was expanding faster than the speed-of-light during inflation, and that parts of our physical universe are currently expanding faster than the speed-of-light. Their explanation? Space-time isn't physical; therefore, space-time can expand faster than the speed-of-light. So, what else isn't physical and can therefore move faster than the speed-of-light? I can think of a few things. Can you? Let's start with spirit matter and psyche. They've been observed to travel faster than the speed-of-light. In fact, not only can they travel faster than the speed-of-light; but, they can also quantum tunnel or teleport at will, as long as they are not being held back by entropic physical matter and other physical limitations.

The theory of relativity is the very pinnacle of Classical Physics. The theory of relativity and classical physics are limited and flawed because they reject and ignore the non-physical half of the equation – the Syntropy, Psyche, Spirit Matter, Quantum Fields, Dark Energy, Action at a Distance, and Quantum Mechanics, which are some of the things that can exist and function at velocities greater than the speed-of-light. The theory of relativity is limited exclusively to the physical realm; and, it has NO explanatory power when it comes to the quantum realm, psyche realm, or syntropy realm.

The theory of relativity is deliberately limited to velocities less than the speed-of-light, which means that the theory of relativity is deliberately limited to physical matter, space-time, entropy, and classical physics. Because of all the physical limitations associated with the theory of relativity, Relativity completely ignores Syntropy, Psyche, Spirit Matter, Dark Energy, Quantum Fields, Action at a Distance, Non-Locality, and Quantum Mechanics – everything faster than the speed-of-light – everything that is non-physical and has NO physical limitations. Entropy, death, and the theory of relativity are physical limitations.

The theory of relativity correctly states that Syntropy, Quantum Mechanics, Action at a Distance, and Faster-than-Light velocities are physically impossible. The theory of relativity incorrectly states that everything is physical, and therefore everything has physical limitations. Einstein had problems with all of that "spooky" stuff, so he and his followers incorrectly concluded that everything in the universe is limited by the speed-of-light.

The Big Bang Theory is derived by rewinding the current expanding universe backwards in time towards the hypothetical Big Bang Singularity. I have observed that you can stop anywhere along the way and still end up with the universe that we see today. In other words, all you have to do is to rewind to the point where the God's started making physical matter, planets, stars, and galaxies; and, you will still end up with the universe we see today. You don't have to rewind the thing all the way back to the Big Bang Singularity in order to get the observable universe that we have today. The cosmic microwave background radiation was actually produced by the production of physical matter from energy; so, the Big Bang Singularity isn't needed for the CMB either.

Even though the Big Bang Theory is seriously flawed, we keep referring to it because the Materialists and Naturalists have NO other model for origins that they can relate to. We need a common frame of reference so that we can have a discussion in the first place.

So, what do we observe from the Big Bang Theory?

In order to make the Big Bang Theory work, the scientists and astrophysicists have to introduce an inflationary period that lasted at least 380,000 years of cosmic time into the model. What's happening during this inflationary period? Our newly born universe is expanding FASTER than the speed-of-light in complete violation of the theory of relativity.

How's that possible? Can you think of a plausible explanation?

I can. I can think of a few of them. Here's one possibility.

Faster-than-Light Inflation is possible because during the proposed inflationary period, our universe would have still been in a Syntropy and Quantum Mechanics state of existence without any physical matter, space-time, entropy, or physical limitations within it yet. The physical matter, the entropy, the space-time, the sub-light speeds, the physical limitations, and the theory of relativity would have come into play sometime AFTER the 380,000 years of inflation. During inflation, Syntropy and Quantum Mechanics would have ruled supreme making FTL Inflation possible. Logical. Parsimonious. It works.

A scientist friend suggested another possibility.

Our God could have picked a small corner of Chaos, part of the original construct, and shined His Syntropy or His Light or His Glory into it from a chosen center point. That Power of Godliness, or Syntropy, or Quantum Laws would have penetrated from that center point outward spherically in all directions faster than the speed-of-light at first, bringing order or law to chaos. At first, it would move much faster than the speed-of-light, because it is Syntropy, Quantum Mechanics, or the Priesthood Power of God. For the first 380,000 years of "inflation", the Quantum Order or Syntropy would have been spreading into this universe bringing order and organization to this universe. Chaos would have been converted to Quantum Mechanics or Syntropy at velocities greater than the speed-of-light.

I'm trying to bring order, logic, and truth to Science by mentioning and discussing these types of things. In fact, I'm trying to upgrade Science and make Science match with what has actually been experienced and observed. Rather than trying to explain away the evidence, I'm trying to explain the evidence.

Ironically, the very existence of an Inflationary Period falsifies Materialism, Naturalism, Darwinism, Nihilism, and Atheism because the whole thing is moving faster than the speed-of-light which is physically impossible. The proven existence of the physically impossible FALSIFIES Materialism, Naturalism, and their derivatives. Inflation, Faster-than-Light Velocities, Quantum Mechanics, Syntropy, and Psyche are physically impossible which means that they are non-physical in nature and origin. Their very existence and experienced existence FALSIFY Materialism, Naturalism, Darwinism, Nihilism, and Atheism which claim that the non-physical DOES NOT EXIST. That's what the contention is all about, is it not? Observational evidence falsifies Scientific Naturalism.

With that initial Quantum Construct in place and order brought to chaos and that Quantum Construct continuing to spread outward into an already pre-existing Chaos faster than the speed-of-light, then from 380,000 years to about 150 million years during which the central core of our universe was transparent but no large-scale structures had yet been formed, God would have started to transition the inner core of our universe from Spirit Matter and Quantum Mechanics over to Sub-Light Speeds, Entropy, Physical Matter, Classical Physics, and the physical limitations associated with the Theory of Relativity.

However, the outermost portions of our physical universe would continue to expand at velocities faster than the speed-of-light as more and more of Chaos gets ordered or organized into Quantum Mechanics, Syntropy, or the Quantum Construct.

It's a Quantum Wave, and NOT a Big Bang. As that wave expands spherically in all directions, God slows it down towards the core and transitions it from the spiritual into the physical towards the core making planets, stars, and galaxies as He goes along; yet, the outer edges of the Quantum Wave would continue to expand into the already pre-existing Chaos bringing order, laws, light, and syntropy to that Chaos. Since the expansion of order and organization is moving faster than the speed-of-light on the periphery, soon the outer parts expand beyond our observable universe.

I like this Inflation into Chaos Model a lot – which entails Syntropy or Quantum Mechanics or Order spreading spherically from a central core into a small portion of Chaos – because it falsifies creation ex nihilo. Creation ex nihilo is Atheism, creation by nothing from nothing. Creation ex nihilo is magic. Creation ex nihilo is unscientific and illogical. If you believe in creation ex nihilo, you are an atheist, whether you realize it or not. This model of Expansion into Chaos or Inflation into Chaos, bringing order to chaos, also takes into consideration both entropy and syntropy – both physical matter and spirit matter.

Remember, entropy is a function of time or an aging process. Entropy leads to physical death. Spirit matter transitions to entropic physical matter whenever entropy or corruption gets introduced into it. Syntropy transitions to entropy whenever corruption or death gets introduced into the system. Anywhere there is physical matter and sub-light speed physical limitations, entropy has been introduced into the system. There could be entropy and physical matter in the core of our universe while syntropy or dark energy continues to function at the periphery converting Chaos into Quantum Law.

Syntropy means without beginning of days or an end of years. Syntropy means ageless and eternal. Quantum Mechanics, Psyche, the Glory of God, and the Power of God are Syntropy. Classical physics, the theory of relativity, death, and physical limitations are entropy.

Building a Bridge between Spirit Matter and Physical Matter

When I tried to explain to explain to my scientist friend that a lot of my research has been to try to build a bridge between spirit matter and physical matter, he asked me, "Where's the cavern? Where's the gulf between them?"

Having experienced both sides of the equation first-hand in-person, he understood that spirit matter and physical matter are the SAME THING. They are both quantum waves of energy. They are both dust, energy, mass, matter. They simply exist in a different frequency, phase, or dimension; but, they are the same thing. It's ALL matter, and it's ALL energy – all the way down the rabbit hole. Spirit matter and physical matter can exist in the same space at the same time, as long as they are out of phase with each other.

My friend has phase-shifted. A huge elk passed through his truck unscathed. Actually, it was nothing that my friend did. It was something that God did for him in order to save the life of that elk. Two particles of physical matter can occupy the same space at the same time, as long as they are out of phase with each other. Spirit matter is matter that's in a different phase, frequency, or dimension than the physical matter; but, it's ALL matter. Matter can also be converted into energy, or quantum waves, and then back into matter by Nature's Psyche or God's Psyche.

My friend understands all of this, so he was curious to know what kind of gap or gulf I was finding between spirit matter and physical matter. The answer is that the physical matter is capable of taking on corruption, disorder, or entropy; and, the spirit matter is not. There's a huge gap between the physical matter and the spirit matter thanks to entropy or corruption.

In order to raise someone from the dead, all God has to do is to remove the entropy, corruption, or death from the physical matter that used to comprise his or her physical body, and then let Syntropy or Eternal Life take over. If you successfully remove the entropy and other physical restrictions from a dead physical body, it can spontaneously regenerate into the former order or organization that it once had, thanks to syntropy or quantum entanglement. It will be just like running the movie backwards, after you have removed the entropy or physical restrictions. There will be an "at-one-ment" that will take place, an Atonement. Atonement is syntropy – the removal of death, hell, sin, corruption, and entropy. This removal of entropy becomes possible through the Atonement of Christ. It's a quantum mechanism.

It's really simple to understand once a person chooses to do so.

Corruption or physical matter is entropy. Incorruption or spirit matter is syntropy.

It's ALL matter or energy. We have intelligent dust or intelligent energy that can ACT; and, then we have dumb dust or dumb energy that simply REACTS. The intelligent dust or the intelligent energy or the intelligent light would be Psyche or Intelligence. The dumb dust or inert dust or mindless energy would be spirit matter. Spirit matter can be converted temporarily into physical matter by introducing entropy, space-time, and physical limitations into that spirit matter. Spirit matter is converted into physical matter by introducing corruption or entropy into that spirit matter thereby slowing it down to sub-light velocities and converting it into entropic physical matter.

Psyche and spirit matter are eternal, without beginning of days or an end of years. They are syntropy. Physical matter has a beginning, which means that it can have an end. Entropic physical matter is made from spirit matter by introducing entropy or corruption into that spirit matter thereby making that newly created physical matter subject to entropy, corruption, physical limitations, and death. When it comes to physical matter, the process of entropy can be reversed by an infusion of syntropy, energy, information, and/or a removal of the disorder or entropy. The physical matter can be converted back into spirit matter by an infusion of syntropy, order, life, or incorruption.

1 Corinthians 15: 42-44, 50-53:

So also is the resurrection of the dead. It is sown in corruption; it is raised in incorruption: it is sown in dishonor; it is raised in glory: it is sown in weakness; it is raised in power: it is sown a natural body; it is raised a spiritual body. There is a natural body, and there is a spiritual body. Now this I say, brethren, that flesh and blood cannot inherit the kingdom of God; neither doth corruption inherit incorruption. Behold, I show you a mystery; We shall not all sleep, but we shall all be changed, in a moment, in the twinkling of an eye, at the last trump: for the trumpet shall sound, and the dead shall be raised incorruptible, and we shall be changed. For this corruptible must put on incorruption, and this mortal must put on immortality.

Take it literally. Paul knows what he is talking about, because he has seen it in vision. Corruption is entropy. Incorruption is syntropy.

How would a physical body that has died and now lies in the grave get raised from the dead?

It would require an infusion of syntropy – a reversal of entropy or an elimination of entropy. Entropy is death. Remove the entropy and death; and, the body comes alive.

Simple. Parsimonious. Logical. True.

The thing that Paul calls a "spiritual body" is a resurrected physical body that has had all of the entropy removed from it and has received an infusion of syntropy or glory. That body is capable of existing at any phase or level of existence, including both the spirit realm and this physical realm, simply by changing phase at will. That's how the resurrected Lord Jesus Christ was capable of walking through walls and closed doors, levitating or standing in the air, walking on water, and teleporting or quantum tunneling from one location to another within His physical body. All of this becomes possible when you remove the entropy from your physical body and give your physical body an infusion of syntropy or glory. A physical body can even rise from the dead once the entropy has been removed.

God holds all the KEYS to quantum mechanics, syntropy, the removal of entropy, and the resurrection from death. We human beings can't remove the entropy from physical matter; but, God can. It's called Atonement.

"Raised in incorruption" means raised into syntropy or restored to syntropy. It means being given direct access to all the different quantum mechanisms. "Raised in glory and raised in power" means to be supercharged and superpowered – power and glory are syntropy or quantum mechanics.

"It is sown a natural body; it is raised a spiritual body." We are born here in mortality in a physical body and start to take on entropy; and then, that physical body is infused with syntropy during the resurrection from the dead and becomes a spiritual body or a syntropy body. There's NO entropy or corruption in a resurrected body. It's pure syntropy. It's pure Quantum Mechanics. It is pure power and glory.

It has to be transformed from corruption to incorruption, or from entropy to syntropy. We are going to be changed. We are going to inherit incorruption or syntropy. Corruption must put on incorruption. Entropy must put on syntropy, in order to transition from a mortal body to an immortal body. It's right there in plain sight.

The theory of relativity seems to be the pinnacle of classical physics – the ultimate in physical limitations. Resurrection from the dead seems to be the ultimate in Syntropy or Quantum Mechanics, where our physical body and this physical universe are concerned. Resurrection is a reversal of entropy or death. Atonement is a removal of entropy, sin, corruption, or death.

There are many other examples of this phenomenon mentioned in the books that the Biblical God Jesus Christ had a hand in writing and producing.

Corruption Being Raised by Christ into Incorruption:

Your Physical Resurrection from the Dead:

Mosiah 16: 4-13.

2 Nephi 9: 6-14.

Alma 41: 1-8.

Alma 5: 14-16.

Alma 40: 1-26.

As entropy gets introduced to parts of our physical universe, physical matter comes into being. Physical matter is full of corruption or entropy, and it is limited to sub-light speeds and has many other physical limitations. We age and die. That's the nature of entropy or corruption. However, God tells us that through the Atonement of Christ, the physical matter associated with our physical body can receive an infusion of order or syntropy, have ALL of its entropy removed, and be raised as an incorruptible, syntropy-filled, immortal body, spiritual body, or syntropy body.

The same thing can happen to the worlds in our physical universe that went through a physical phase or entropy phase or corruption phase of existence, and then died a heat death. These dead worlds can also be filled full of syntropy, raised from the dead, have the entropy or the heat death removed from them, and be made immortal and incorruptible. Thanks to Syntropy or the Atonement of Christ, the corruption, entropy, and death can be reversed.

Thanks to syntropy, the film really can be run backwards, and the dead person really can come back to order, organization, and life. Thanks to Syntropy or the Atonement of Christ, we are NOT limited to classical physics, entropy, death, and the theory of relativity for all eternity. Thanks to Syntropy, Quantum Mechanics, and Psyche, we really can be raised from the dead and live once again in the flesh.

Ezekiel has also seen it in vision. Resurrection from the dead is Syntropy in action. It's powerful, because it works and makes logical sense. It makes infinitely more sense than creation ex nihilo or atheism. Observe what happens when the entropy is removed from the system. Syntropy or Atonement takes over.

Ezekiel 37: 1-10:

The hand of the Lord was upon me, and carried me out in the spirit of the Lord, and set me down in the midst of the valley which was full of bones, and caused me to pass by them round about: and, behold, there were very many in the open valley; and, lo, they were very dry. And he said unto me, Son of man, can these bones live? And I answered, O Lord God, thou knowest. Again, he said unto me, Prophesy upon these bones, and say unto them, O ye dry bones, hear the word of the Lord. Thus saith the Lord God unto these bones; Behold, I will cause breath to enter into you, and ye shall live; and I will lay sinews upon you, and will bring up flesh upon you, and cover you with skin, and put breath in you, and ye shall live; and ye shall know that I am the Lord. So, I prophesied as I was commanded: and as I prophesied, there was a noise, and behold a shaking, and the bones came together, bone to his bone. And when I beheld, lo, the sinews and the flesh came up upon them, and the skin covered them above: but there was no breath in them. Then said he unto me, Prophesy unto the wind, prophesy, son of man, and say to the wind, Thus saith the Lord God; Come from the four winds, O breath, and breathe upon these slain, that they may live. So, I prophesied as he commanded me, and the breath came into them, and they lived, and stood up upon their feet, an exceeding great army.

The Materialists, Naturalists, and Darwinists laugh at and mock this sort of stuff, because in their complete ignorance, they have limited themselves to physical matter, classical physics, entropy, and the theory of relativity. These people completely reject and ignore the Syntropy, Power of Godliness, Quantum Mechanics, Quantum Non-Local Consciousness, Action at a Distance, Quantum Non-Locality, Spirit Matter, Intelligence, and Psyche. These people deliberately reject the most interesting parts of science, the most-observed parts of science, and the most-verified parts of science.

Quantum Mechanics and Quantum Field Theory are the best-proven, most-verified, and most-used science that we have; and, the very existence of Quantum Mechanisms or Syntropy falsifies Materialism, Naturalism, Darwinism, Nihilism, and Atheism.

The Expansion into Chaos Model

I have learned to like my friend's **Expansion into Chaos Model** as a replacement for the Big Bang Theory, because it's more parsimonious and logical than the Big Bang Theory; yet, it also explains the inflationary period that they typically associate with the Big Bang. The explanatory power of the **Inflation into Chaos Model** is noticeably superior to the Big Bang Model.

The Big Bang Model is based upon entropy, the theory of relativity, spherical shells, weird funnel-shaped shells, and creation ex nihilo – all of which are limited purely to the physical and have NO explanatory power when it comes to the quantum level, the psyche level, or the syntropy level of existence.

Furthermore, there is NO such thing as creation ex nihilo or magic. Creation ex nihilo is Atheism – creation from nothing by nothing. Creation ex nihilo is magic. Nothing can create something from nothing, not even God. The Materialists, Naturalists, Darwinists, Nihilists, and Atheists always portray the Big Bang as creation ex nihilo, spontaneous generation, abiogenesis, a random fluctuation in entropy, a quantum fluctuation, or magic. They'll accept anything but God! However, God, Psyche, or Intelligence is the BEST and most parsimonious explanation.

https://en.wikipedia.org/wiki/Chronology_of_the_universe

https://quantum-neuroscience.com/wp-content/uploads/2018/03/Chronology-of-the-universe-Wikipedia.pdf

To me, it always seemed logical and intuitive to believe that our expanding universe is expanding into something; and it is, according to the **Inflation into Chaos Model** for the origin of this particular physical universe. Under this model for origins, our current universe is expanding into Chaos or expanding into "matter unorganized". The matter or energy is already there. It's just being infused with syntropy, order, quantum fields, and quantum mechanics as the sphere expands. The matter is being transformed from chaotic matter to spirit matter. It's being given order, organization, purpose, and structure. Meanwhile, the Psyches or Intelligences trapped within that chaotic matter are being set FREE as the sphere expands and the chaotic matter is converted into spirit matter.

Within the approximate center of the sphere where we reside, God transitions a small portion of that spirit matter into physical matter by introducing sub-light velocities, physical limitations, and the ability to accumulate entropy into that spirit matter thereby converting it into physical matter. In the periphery as the order and organization expand outwards, the Gods introduce the structure of Quantum Fields into the chaos of unorganized matter and unorganized energy in preparation for bringing physical matter to those parts of the universe. The **Expansion into Chaos Model** is more simple, logical, and parsimonious than the Big Bang Model. It also matches better with the scriptures that the Biblical God Jesus Christ had a hand in producing.

The Biblical God Jesus Christ has repeatedly stated that He organized everything spiritually or at the quantum level BEFORE He organized it and created it at the physical level. The Biblical God Jesus Christ also periodically talks about the "matter unorganized" or the Chaos Realm to which He brought order, organization, syntropy, light, and life. In other words, as our universe continues to expand into Chaos or "unorganized matter", God brings Quantum Mechanics, or Law, or Order, or Life, or Syntropy to that unorganized matter, thereby setting the Psyches or the Intelligences or the Acting Dust within that unorganized matter FREE to act and choose for itself what it wants to be and what it wants to become.

The Gods organize everything spiritually, BEFORE they organize it physically.

Congregating dark matter or spirit matter into clusters is done by God in preparation for turning some of that spirit matter or dark matter into Physical Matter, Galaxies, Suns, Planets, and Moons. Dark matter or spirit matter is holding our galaxies together, keeping them from flying apart, according to the astrophysicists. The **Organization of Chaos Model** or **Expansion into Chaos Model** even explains how the dark matter or spirit matter ends up in a halo surrounding and suffusing each physical galaxy. The Gods organize it spiritually, before they create it physically. The overall order, or organization, or sphere continues to expand into chaos, and it can continue to do so for all eternity.

Within the **Sphere of Organization**, there seems to be an equal infusion of "dark energy", or the expanding force of the universe, or the Light of Christ taking place at every point within the sphere so that each physical galaxy is moving away from every other physical galaxy, and so that the galaxies that are furthest away from us seem to be moving away from us faster than the galaxies that are near us. That's only possible if the dark energy or expanding force is being infused evenly throughout the whole sphere or universe simultaneously, maintaining the expansion of the whole sphere into chaos. The whole sphere is stretching and expanding into chaos, not just the edge of the sphere. The **Inflation into Chaos Model** explains the "horizon problem" better than the Big Bang Theory does, in my humble opinion.

Dark Energy and Quantum Fields proceed forth from the presence of God to fill the immensity of space. Dark Energy is the power by which the planets, stars, and galaxies were made.

<u>Doctrine and Covenants 88: 1-44</u>.

The Gods are currently infusing huge amounts of Dark Energy into our universe from some other dimension causing our physical universe to expand, possibly in preparation for future construction. The Gods can convert some of that Dark Energy into physical matter, planets, stars, and galaxies anytime they choose to do so. Remember, heat death is impossible at the quantum level or the psyche level. Entropy, heat death, or the second law of thermodynamics doesn't exist in the Quantum Realm or Syntropy Realm.

During the **Expansion of Order into Chaos**, God infuses chaos with syntropy or quantum mechanics or order, in preparation for turning a very small part of that Syntropy, or Psyche Realm, or Spirit Realm into a physical realm that is subject to entropy, death, and corruption.

Why would God do such a thing? Why create a corruptible physical realm, when there is already a perfect functional spirit realm or syntropy realm?

It's because a physical realm is the Ultimate Consensus Reality – it's predictable, controllable, reliable, orderly, structured, yet malleable too. Physical matter has memory or entropy – it's contained, constrained, restricted, and limited – thereby making physical matter highly reliable, predictable, dependable, and controllable. In a physical realm, I can depend upon this book being there on my hard drive and my cloud drive tomorrow when I go looking for it. A physical realm also hurts; and, pain is our greatest teacher because it is our greatest motivator. With a physical realm, we actually have skin in the game. It makes it real. The test is a real test.

Now with all of this talk about syntropy, spirit matter, psyche, and God, don't let me convince you that the physical matter isn't real. Physical limitations are REAL. Physical matter really is limited to sub-light speeds, which means that the Theory of Relativity is true at the physical level, even though it's false or incomplete at the quantum level or the psyche level. Physical matter really does fill full of corruption or entropy at the physical level, even though entropy DOES NOT EXIST at the quantum level or syntropy level. Physical matter can and does experience heat death, even though spirit matter is without beginning of days or an end of years.

Matter can be converted to energy or some type of quantum wave – all the way down to Psyche or Intelligence. Thanks to Quantum Superposition, different types of matter can occupy the same space at the same time, as long as they each are out of phase with each other. Psyche or Intelligence can occupy the same space at the same time as its assigned Spirit Body because they are out of phase with each other. Your Spirit Body can occupy the same space at the same time as your Physical Body because they are out of phase with each other or are in a different dimension from each other.

The physical matter is real, and so is chaos, hell, or outer darkness. It's all real; and, it's all comprised of different types of quantum waves in different dimensions or different phases.

Some Psyches or Intelligences or Smart Dust prefer chaos, hell, and outer darkness; and, some of these Psyches or Intelligences prefer the order, organization, syntropy, light, and life. That's just the way it is.

Whenever a God brings order, law, or syntropy into a portion of Chaos, He sets both types of Psyche or Intelligence FREE to act. Some of these Psyches or Intelligences prefer the chaos or entropy or death; and, some of them prefer the order, syntropy, and life. Since these Psyches or Intelligences have been set FREE by God to act, they are free to choose chaos, hell, and death or to choose syntropy and eternal life. We reveal ourselves to ourselves by every choice that we make. We each choose whatever we prefer most; and that's possible, because God set us FREE to do so.

2 Nephi 2: 11-12, 25-29:

For it must needs be, that there is an opposition in all things. If not so, my firstborn in the wilderness, righteousness could not be brought to pass, neither wickedness, neither holiness nor misery, neither good nor bad. Wherefore, all things must needs be a compound in one; wherefore, if it should be one body it must needs remain as dead, having no life neither death, nor corruption nor incorruption, happiness nor misery, neither sense nor insensibility. Wherefore, it must needs have been created for a thing of naught; wherefore there would have been no purpose in the end of its creation. Wherefore, this thing must needs destroy the wisdom of God and his eternal purposes, and also the power, and the mercy, and the justice of God. Adam fell that men might be; and men are, that they might have joy. And the Messiah cometh in the fulness of time, that he may redeem the children of men from the fall. And because that they are redeemed from the fall they have become free forever, knowing good from evil; to act for themselves and not to be acted upon, save it be by the punishment of the law at the great and last day, according to the commandments which God hath given. Wherefore, men are free according to the flesh; and all things are given them which are expedient unto man. And they are free to choose liberty and eternal life, through the great Mediator of all men, or to choose captivity and death, according to the captivity and power of the devil; for he seeketh that all men might be miserable like unto himself. And now, my sons, I would that ye should look to the great Mediator, and hearken unto his great commandments; and be faithful unto his words, and choose eternal life, according to the will of his Holy Spirit; and not choose eternal death, according to the will of the flesh and the evil which is therein, which giveth the spirit of the devil power to captivate, to bring you down to hell, that he may reign over you in his own kingdom.

In the Chaos Realm or Outer Darkness, the Intelligence or Psyche is acted upon or trapped by the unorganized chaotic matter. It is hell.

In this Syntropy Realm or Celestial Realm, God has set the Psyches or Intelligences FREE to act upon and control the matter – both the spirit matter and the physical matter. Technically, spirit matter and physical matter are the same thing – they just exist at a different phase, or a different level, or a different dimension. The physical matter is able to take on entropy or corruption; and, the spirit matter is not.

My friend's **Expansion into Chaos Model** is simple, logical, and parsimonious. It matches better with the observational evidence than the Big Bang Model does. It matches infinitely better with what the Biblical God Jesus Christ has told us about origins and life than the Big Bang Model does. The **Expansion into Chaos Model** has a lot more explanatory power than the Big Bang Model, because the **Inflation into Chaos Model** takes Syntropy, Quantum Mechanics, and Inflationary Faster-than-Light Expansion into consideration, whereas classical physics and the Big Bang Model cannot without breaking classical physics and the theory of relativity.

FTL Expansion or the Inflationary Period cannot be explained in terms of classical physics, entropy, and the theory of relativity. However, FTL Expansion is an integral part of Quantum Mechanics, Syntropy, and Psyche or Intelligence. Quantum Tunneling or Teleportation of physical matter is also an integral part of Syntropy or Quantum Mechanics – we catch it in the act all the time. The only reason the atoms in your physical body don't quantum tunnel away on you is because Syntropy, or God, or God's Command prevents them from doing so. Quantum Mechanics is vastly superior to Classical Physics. Syntropy is vastly superior to entropy. FTL Expansion into Chaos is vastly superior to the theory of relativity and its physical limitations.

By letting Psyche, Syntropy, and Quantum Mechanics in to play, we can literally answer every question that comes our way. That's NOT possible from the perspective of Classical Physics, Materialism, Naturalism, and Entropy. According to the Materialists and Naturalists, Psyche, Syntropy, and Quantum Mechanisms are physically impossible, which means that they exist at the quantum level, or the psyche level, or the syntropy level. Psyche or Intelligence working through Syntropy and Quantum Mechanics can do the physically impossible at will. Yes, the whole thing is physically impossible, but Quantum Mechanics or Syntropy makes the "physically impossible" possible.

Technically, Materialism, Naturalism, Darwinism, Nihilism, Behaviorism, Determinism, Scientism, and Atheism are LIES or deceptions. In contrast, Psyche, Quantum Mechanics, and Syntropy represent the true reality of our existence at every level of existence. I see the Hand of God in Quantum Field Theory.

Remember, Psyche, Eternal Life, Syntropy, the Conservation of Energy, and Quantum Mechanics are vastly superior to chaos, death, entropy, classical physics, the theory of evolution, and the theory of relativity. The one has NO physical limitations, and the other is physical limitations. The one has NO entropy, and the other is entropy.

Simple. Parsimonious. Logical. True.

Mark My Words

—

Source

God Is in the Light: God is light, and in Him is no darkness at all.

https://www.amazon.com/dp/B07168S37N

Science 2.0: I Upgraded My Science

https://www.amazon.com/dp/B0771K6WTX

References

The Ultimate Model of Reality: Psyche Is the Ultimate Cause

https://www.amazon.com/dp/B071NC9JK6

The Uncertainty Principle

Notice carefully how your college professors have been taught to define Quantum Mechanics. They have been taught to define Quantum Mechanics as the uncertainty principle or the indeterminacy principle.

> **In quantum mechanics, the uncertainty principle, also known as Heisenberg's uncertainty principle or Heisenberg's indeterminacy principle, is any of a variety of mathematical inequalities asserting a fundamental limit to the precision with which certain pairs of physical properties of a particle, known as complementary variables, such as position x and momentum p, can be known.**
>
> https://en.wikipedia.org/wiki/Uncertainty_principle

A case in point:

> **Question: Do I think Quantum Mechanics will give a scientific explanation of free agency?**
>
> **Answer: Not unless you want free agency to be random, as I understand Quantum Mechanics. I mean, just to say that it's unpredictable doesn't save agency, it saves unpredictability. You don't get a lot of meaning or moral grounding out of unpredictability. So, I think Quantum Mechanics loosens things up for a more general understanding, but I don't think it is an explanation as it now stands, as I understand it.**
>
> Richard N. Williams on "Science, Religion, and Agency".
>
> https://www.youtube.com/watch?v=wv3v0lro-xo

[He is right. Quantum Mechanics based exclusively on the Uncertainty Principle, the Indeterminacy Principle, Randomness, or Unpredictability is absolutely worthless. It completely lacks explanatory power, which is why the Materialists and Naturalists eagerly embrace it and promote it. The Uncertainty Principle is the ONLY thing the Materialists and Naturalists preach and teach when it comes to Quantum Mechanics, because the Uncertainty Principle is Classical Physics or Random Diffusion in action. The Uncertainty Principle is the materialistic and naturalistic and atheistic version of Quantum Mechanics. It's worthless! Williams admits that he understands Quantum Mechanics as randomness, and he truthfully admits that that explanation for Quantum Mechanics is worthless, because it completely lacks explanatory power.]

These PhDs have been taught to define Quantum Mechanics as randomness, unpredictability, and uncertainty. They have been taught to define Quantum Mechanics as the Uncertainty Principle. That was the way that I originally understood Quantum Mechanics to be – the Uncertainty Principle.

Even the best of our PhD professors have been taught by the Materialists, Naturalists, Darwinists, and Atheists to define Quantum Mechanics as uncertainty, unpredictability, randomness, random mutations, random diffusion, random chance, indeterminism, chaos, entropy, and classical physics. You can't resolve anything with Classical Physics! Materialism and Naturalism destroy everything that they touch, including Quantum Mechanics.

No wonders NONE of these people can figure out how Quantum Mechanics really works, what it is, what it does, and how it does it. Their teachers have deliberately neutered Quantum Mechanics so that it is nothing more than Randomness, Uncertainty, Unpredictability, Materialism, Naturalism, Darwinism, Entropy, and Classical Physics.

Quantum Mechanics IS Syntropy, not entropy. The entropy is limited by God to Classical Physics, Materialism, Physicalism, Naturalism, Determinism, Behaviorism, Scientism, Darwinism, and the Theory of Evolution. These things are based upon entropy. Quantum Mechanics is not.

Quantum mechanisms are syntropy in action. There's NOTHING uncertain about it. The whole thing is deliberately chosen into existence. It all comes down to choice! It all comes down to the Interpretation of Quantum Mechanics that you choose to embrace and promote. Your choice makes ALL the difference in the world!

If your chosen interpretation of Quantum Mechanics is based exclusively on unpredictability and the Uncertainty Principle, then your interpretation of Quantum Mechanics is indeed worthless! If your chosen interpretation of Quantum Mechanics is based upon Materialism, Naturalism, Darwinism, Entropy, Physical Limitations, and Classical Physics, then your interpretation of Quantum Mechanics is absolutely worthless.

If your chosen interpretation of Quantum Mechanics is based exclusively on the Uncertainty Principle and indeterminacy, then your interpretation is worthless. If your interpretation of Quantum Mechanics defines quantum mechanisms as entropy or uncertainty or indeterminism or classical physics, then your definition of Quantum Mechanics is worthless. Quantum Mechanics involves syntropy, psyche, or choice.

If your interpretation of Quantum Mechanics cannot explain what Nature's Psyche and the Human Psyche are doing at the quantum level in order to get things done for us at the physical level, then your interpretation of Quantum Mechanics is worthless. If your interpretation of Quantum Mechanics defines Quantum Mechanics as uncertainty, entropy, random diffusion, random mutations, and classical physics, then your interpretation of Quantum Mechanics is erroneous, fallacious, and worthless because Quantum Mechanics is syntropy.

Remember, if your chosen interpretation of Quantum Mechanics is based exclusively on the Uncertainty Principle or Classical Physics, then it's absolutely worthless when it comes time to explain what's happening at the quantum level or the psyche level. The Uncertainty Principle never made sense from an ontological and epistemological perspective because the Uncertainty Principle doesn't make any logical sense from both the quantum level and the physical level.

At the quantum level, everything is Syntropy! Syntropy is the same thing as infinite possibilities. This means that at the quantum level, everything that has been organized from spirit matter was chosen into existence by Someone Psyche. There's nothing uncertain about it once the choice has been made! Simple. Parsimonious. Logical. It works! It's true.

The math, though interesting to study, is absolutely worthless if it doesn't explain something that has actually been experienced and observed. Science is observation and experience, not philosophical speculation or wishful thinking.

At the physical level, everything is entropy. However, the Uncertainty Principle doesn't make any sense at the physical level, either.

Why?

Well, what do we observe at the physical level?

We observe physical particles that have BOTH a measurable position and a measurable velocity! Technically, the Uncertainty Principle doesn't apply at the physical level, either.

God designed our physical realm to be the Ultimate Consensus Reality – totally predictable, totally reliable, totally dependable, and totally determined. Thanks to this physical reality, I can rely upon this paper being on my hard drive and cloud drive when I go looking for it tomorrow. I can rely upon my house, my wife, and my dog being there when I drive home from work tomorrow. There's NOTHING uncertain or unpredictable about it because a physical reality was designed by God to be the Ultimate Consensus Reality. It's totally predictable.

What did we just learn? We learned that the Uncertainty Principle or the Indeterminacy Principle is absolutely worthless at both the quantum level and the physical level because it really doesn't explain what's happening at either level of existence. The Uncertainty Principle is the WRONG emphasis. That particular interpretation of Quantum Mechanics tries to turn Quantum Mechanics into Classical Physics; and, it FAILS at that as well. The Uncertainty Principle has NO predictive value or predictive capability at the physical level, so it's useless at the physical level; and, everything is chosen into existence by Someone Psyche at the quantum level. So, how do we resolve this classic failure?

At the quantum level or psyche level, the Uncertainty Principle teaches us that everything is unpredictable, or uncertain, or possible, until a CHOICE is made by Someone Psyche, who collapses the wave function thereby making the infinite possibilities collapse into something ACTUAL and REAL at both the quantum level and the physical level. We learned that to resolve this conundrum and contradiction, we have to redefine the Uncertainty Principle as Infinite Possibilities, and then suddenly It works at both levels of existence – the quantum level and the physical level. It's the same difference, but the "uncertainty" definition has NO real explanatory power at either level of existence while the other "infinite possibilities" definition has infinite explanatory power at both levels of existence.

Do you see how that works?

It all comes down to choice. It all comes down to the interpretation that you choose to give to Quantum Mechanics. You can choose an interpretation that has NO explanatory power whatsoever at any level of existence; or, you can choose an interpretation that has infinite explanatory power at every level of existence. The choice is yours.

It ALL depends upon the interpretation that you choose to give to Quantum Mechanics. A materialistic, naturalistic, and atheistic interpretation based exclusively on Classical Physics and the Uncertainty Principle is totally worthless when it comes time to explain what's happening at the quantum level or the psyche level. In contrast, an interpretation of Quantum Mechanics based upon Infinite Possibilities or Syntropy provides us with – well – infinite possibilities.

Can you see the difference?

It's hard if you have been brainwashed into Materialism, Naturalism, Darwinism, Nihilism, and Atheism. I KNOW, because I used to be a Materialist, Naturalist, Nihilist, and Atheist; and, Quantum Mechanics based upon Random Diffusion or the Uncertainty Principle made absolutely no sense to me whatsoever. It was designed to be that way because the Materialists and Naturalists are trying to trick us and deceive us. These people hide behind the obfuscation and confusion that they generate, while pretending the whole time that what they are teaching is pure science when in fact what they are preaching is pure dogmatic religion.

Technically, it's impossible to understand what's happening at either level of existence – the quantum level or the physical level – if you have been brainwashed and conditioned into Materialism, Naturalism, Darwinism, Nihilism, Scientism, Behaviorism, and Atheism.

The "Uncertainty Principle" should be redefined as "Infinite Possibilities". Then it works and explains everything. Infinite possibilities collapse into a single actual reality whenever Someone Psyche makes a CHOICE – this reality and truth applies at both the quantum level and the physical level. It all comes down to the choices we make.

Remember, we need an Interpretation of Quantum Mechanics that explains what's really happening in our brains. I found two. It was hard to do, because the Materialists and Naturalists and Atheists are trying to hide this information from us, and these people are very good at what they do.

The Non-Local Consciousness Interpretation of Quantum Mechanics

Van Lommel, P. (2010). *Consciousness Beyond Life: The Science of the Near-Death Experience*. New York: HarperCollins.

This book taught me what I call the Non-Local Consciousness Interpretation of Quantum Mechanics, which is the science behind Near-Death Experiences (NDEs).

I examine the Quantum Non-Local Consciousness Interpretation of Quantum Mechanics extensively in this book of mine: *Quantum Mechanics from a Non-Physical Spiritual Perspective*. https://www.amazon.com/dp/B01J023TGU

It's worth your time to study NDEs and the Non-Local Consciousness Interpretation of Quantum Mechanics, in my humble opinion; but, you will have to decide for yourself if you agree with it or not. That's not something I can do for you.

The Orthodox Interpretation of Quantum Mechanics

This article introduces the Orthodox Interpretation of Quantum Mechanics.

Schwartz, J. M., Stapp, H. P., & Beauregard, M. (2004). *Quantum Physics in Neuroscience and Psychology: A Neurophysical Model of Mind-Brain Interaction*. Published Online: Phil. Trans. R. Soc. B.

http://www-physics.lbl.gov/~stapp/PTRS.pdf

http://mypsyche.us/wp-content/uploads/2017/10/PTRS.pdf

https://www.researchgate.net/publication/7613549_Quantum_physics_in_neuroscience_and_psychology_A_neurophysical_model_of_mind-brain_interaction

http://escholarship.org/uc/item/4w8665vk

This paper summarizes everything I have been looking for during the past fifty-five years of my life. That's a long time to wander in darkness looking for the truth; but luckily, our generation finally has the truth, if we know where to find it and recognize it as true when we do find it.

This paper, and Henry P. Stapp's other papers and books, taught me what Stapp calls the **Orthodox Interpretation of Quantum Mechanics**, which explains the interplay or the scientific interface between mind and matter.

Examine these two websites, from Henry P. Stapp:

https://sites.google.com/a/lbl.gov/stappfiles/

http://www-physics.lbl.gov/~stapp/

These free articles explain the Orthodox Interpretation of Quantum Mechanics in great detail. You could spend a lifetime studying them, and still have things to learn. Stapp was the one who taught me that "Infinite Possibilities" is an infinitely better definition for the Uncertainty Principle than "uncertainty". I've taken it and run with it ever since.

I promote and APPLY the Orthodox Interpretation of Quantum Mechanics extensively in my Magnum Opus: *Quantum Neuroscience: The Answer to Life, the Universe, and Everything*. https://www.amazon.com/dp/B079Z6QQQB

I rely upon the Orthodox Interpretation of Quantum Mechanics to do science and have it make sense to me when I am done. The Orthodox Interpretation explains what the Human Psyche and Nature's Psyche are doing at the quantum level in order to get things done for us at the physical level. Remember, it should have been called "Infinite Possibilities" rather than the Uncertainty Principle. Tell it as it is rather than trying to hide one's ignorance behind Materialism, Naturalism, Darwinism, Nihilism, and Atheism.

The Best or Most Explanatory Interpretations of Quantum Mechanics

Quantum Mechanics depends upon the interpretation you choose to use. The best, most useful, and most explanatory interpretations of Quantum Mechanics attempt to explain what has already been experienced and observed. Interpretations based upon Classical Physics are worthless because they don't explain anything; and, the Many Worlds interpretation is worthless because it has never been experienced nor observed.

The Non-Local Consciousness Interpretation of Quantum Mechanics.

The Orthodox Interpretation of Quantum Mechanics.

These are the only two interpretations of Quantum Mechanics that are worth your time, in my humble opinion. Everything else ends up being obfuscation, a deception, and even at times a lie. The Naturalists and Atheists are trying to deceive you.

These two interpretations of Quantum Mechanics seem to downplay or ignore the Uncertainty Principle or the Indeterminacy Principle. Actually, what they do is RESOLVE it into one functional whole; but, you really have to study what they are talking about and promoting in order to see how they do so. Instead of emphasizing the randomness or the uncertainty, they emphasize the infinite possibilities that are contained within Quantum Mechanics. It's the same concept, but a completely different emphasis. Randomness and uncertainty focus on Entropy, Materialism, Naturalism, Darwinism, Atheism, and Classical Physics. Infinite Possibilities focus on Psyche, Syntropy, Quantum Mechanisms, and God. The Uncertainty Principle or Indeterminacy Principle focuses on both realities – the randomness of Classical Physics and the infinite possibilities of Quantum Mechanics – but you choose which reality to emphasize and promote.

The Orthodox Interpretation of Quantum Mechanics by Henry P. Stapp explains how Quantum Mechanics and Psyche really work at the quantum level in order to get things done for us at the physical level. The Orthodox Interpretation bridges or interfaces the quantum level with the physical level. It explains what is happening at every level of reality.

Under the Orthodox Interpretation, Quantum Mechanics is defined as infinite possibilities or syntropy! Then what happens is that the Human Psyche makes a decision or a choice, deciding what it wants to do with its physical environment and physical body. After the Human Psyche makes a decision or a choice, Nature's Psyche collapses the necessary wave functions and/or fires the necessary neurons to make that choice physically REAL in the physical world. According to the Orthodox Interpretation, the Human Psyche makes the choice, and Nature's Psyche collapses the wave function; thereby, infinite possibilities become ONE single physical reality each time a choice is made.

There's nothing uncertain or random about it! It is all deliberately chosen into existence. Yes, anything is possible, until a choice or a decision has been made; but, after a choice or a decision has been made by the Human Psyche and collapsed into existence or collapsed into physical reality by Nature's Psyche, then only ONE of the possibilities ends up being physically REAL. It's no longer a possibility, because now it is actual and REAL.

Do you see how that works?

Your physical reality is chosen into existence by the Human Psyche; and, the infinite possibilities are collapsed into a single physical reality by Nature's Psyche based upon what the Human Psyche chooses to do with its physical body and physical surroundings. Simple. Parsimonious. Logical. It works!

The explanatory power of Infinite Possibilities and the Orthodox Interpretation of Quantum Mechanics is through the roof! It simply dwarfs the Uncertainty Principle, Entropy, Classical Physics, Materialism, Naturalism, and Darwinism. The difference is night and day.

These truths FALSIFY Materialism, Naturalism, and their derivatives. In fact, every truth FALSIFIES Materialism, Naturalism, Darwinism, Nihilism, and Atheism.

Infinite Possibilities or One Choice

We are NOT doing anything weird, or unconventional, or new by defining the Quantum Uncertainty Principle as "Infinite Possibilities".

Watching this Space Time video proves that this definition is true.

https://www.youtube.com/watch?v=bG-xu5H6plk

https://quantum-neuroscience.com/wp-content/uploads/2018/08/Infinite-Possibilities.zip

There are literally infinite possibilities in momentum space, until a CHOICE is made, and a position is established. After the CHOICE is made by Someone Psyche – typically Nature's Psyche – then the infinite possibilities in momentum space collapse into a single reality or actuality in position space, or what we call Locality or Space-Time. That's just the way things work at the quantum level and the psyche level. A quantum or a wave packet of energy literally has Infinite Possibilities in momentum space, until Nature's Psyche makes a CHOICE and collapses those Infinite Possibilities into a single physical position in Space-Time. This is what has been experienced and observed.

Said another way, from the other perspective in terms of particle space, a quantum particle is literally omnipresent and exists simultaneously at ALL positions in Space-Time, until Someone Psyche makes a CHOICE and thereby collapses the associated wave function into a single position in Space-Time. When dealing with Quantum Mechanics, we are literally dealing with Infinite Possibilities, or ONE Chosen Position and therefore ONE Chosen Reality. Psyche makes the CHOICES. A particular Psyche or Intelligence is identified by the CHOICES that it makes. This is how things really work at the quantum level in the Quantum Realm, the Psyche Realm, or the Spirit World.

The Naturalists and Atheists constantly describe Quantum Mechanics as "weird" and "unintuitive". It's because they REFUSE to allow Psyche or Choice into the equation.

Someone Psyche or Someone Intelligent is NEEDED to handle the Infinite Possibilities of a "quantum particle" or a "wave packet of energy" moving through Space-Time, as well as to CHOOSE the final resting place of that quantum particle within Space-Time. It's perfectly intuitive and makes perfect logical sense once we allow Psyche or Choice into the science of Quantum Mechanics. In fact, it explains everything that has ever been experienced and observed. Psyche deals in Infinite Possibilities; and, Psyche is constantly choosing among Infinite Possibilities. I'm not going to apologize for figuring this out. This is what scientists are supposed to do. I start with what has been experienced and observed; and then, I try to find a scientific explanation for it. By allowing Psyche and Quantum Mechanics in to play, we can literally explain everything that comes our way.

One of the main goals of this book is to defend Quantum Mechanics with Psyche or Syntropy. The ultimate goal is to find the BEST explanation for what has been experienced and observed. I believe I have done so. Science is observation and experience after all.

Mark My Words

—

Source

God Is in the Light: God is light, and in Him is no darkness at all.

> https://www.amazon.com/dp/B07168S37N

Science 2.0: I Upgraded My Science

> https://www.amazon.com/dp/B0771K6WTX

References

Quantum Neuroscience: The Answer to Life, the Universe, and Everything

> https://www.amazon.com/dp/B079Z6QQQB

Quantum Mechanics from a Non-Physical Spiritual Perspective

> https://www.amazon.com/dp/B01J023TGU

Syntropy versus Entropy

Some type of Syntropy must exist, or all of that subsequent entropy would not have been possible. My contribution to Science is to give it a name. I chose to call it Syntropy; and then, I chose to define it, explain what it is, and explain where it has been hiding all this time.

Physical matter seems to be an accumulation of entropy. Physical was designed to accumulate entropy. Entropy only exists within physical matter. Everything else is Syntropy. Psyche is Syntropy. Quantum Non-Local Consciousness is Syntropy. Spirit Matter is Syntropy. Quantum Mechanics or Supernatural Mechanisms are Syntropy. The different Forces and Fields are Syntropy. Quantum Waves are Syntropy. Gravity and Magnetism are Syntropy. Dark Matter and Dark Energy are Syntropy. Syntropy is eternal and everlasting. Syntropy means "without a beginning of days or an end of years". Syntropy has always existed and will always exist. Action at a Distance is Syntropy. Love is Syntropy. Syntropy is the opposite of entropy, which is disorder, corruption, disease, death, and extinction. Entropy is only found in physical matter. Physical matter had a beginning, and it can theoretically have an end.

You will never learn anything about Syntropy or Psyche from the Materialists, Naturalists, Darwinists, Nihilists, Behaviorists, and Atheists because the verified and proven existence of Syntropy or Quantum Mechanics falsifies Materialism, Naturalism, Darwinism, and their derivatives.

Even though Albert Einstein was an Einstein, he got some things wrong. He fought against Syntropy, Psyche, Quantum Mechanics, and Spooky Action at a Distance until the very end of his life. His theory of relativity is the very pinnacle of classical physics, but its limitations are overcome by Quantum Tunneling or Teleportation. The theory of relativity ONLY applies to physical matter or entropy. Physical limitations are restricted to physical matter. Simple. Logical. Parsimonious. True. In contrast, the Quantum Realm or Spirit World is based upon Syntropy, Psyche, Action at a Distance, and Quantum Mechanisms or Supernatural Mechanisms. These scientific truths have been experienced and observed on both sides of the veil.

One of my greatest scientific discoveries is Syntropy.

Everything is Syntropy except for physical matter!

Physical matter is based upon entropy.

There's NO entropy or death in the Syntropy Realm, Quantum Realm, or Spirit World. Everything there is eternal and everlasting. Syntropy means eternal and everlasting – without a beginning of days or an end of years.

Why is there physical matter, entropy, or death since it is obvious that Quantum Mechanics and Syntropy are vastly superior?

Why would anyone want a body?

Only with a spirit body or a physical body are we psyches able to feel and experience pleasure and pain.

Why would anyone want the pain?

Physical matter or entropy forces us to learn because it hurts. Thanks to entropy or death, there's a time-limit placed upon us for this learning. Entropy is a function of time and results from the passage of time. It has been observed that deadlines motivate people to act now rather than procrastinating everything to another time or era.

Have you ever heard the phrase, "No pain then no gain"? It has been observed that pain is our greatest teacher. Pain, particularly physical pain, motivates us to learn. We Psyches learn most and learn fastest while in a physical realm or an entropy realm. The lessons are real because they hurt. We actually have to work in order to overcome the entropy that surrounds us. While in an entropy realm or physical realm, in order to bring some measure of Syntropy or Love into our lives, we actually have to choose it and work at it.

Near-Death Experiencers (NDErs) while standing in the presence of God or the angels have asked them what the purpose of mortal life is.

The most common answer is that they are meant to learn how to bring Love or Syntropy into their lives and into the lives of others by learning how to love others while at the same time being surrounded by death, pain, disease, and entropy. It's all about Syntropy or Love – learning how to choose the Syntropy and the Love over the entropy, pain, misery, and death.

The second-most common answer that I have encountered from NDErs is that this mortal life or physical life is a school-ground where we can learn by experience to tell the difference between the bitter and the sweet. Pain, entropy, corruption, and death force us to learn by motivating us to learn. It places a deadline on that learning. Physical matter or entropy forces us to learn because it hurts and because it can be fatal.

Doctrine and Covenants 29: 39: **And it must needs be that the devil should tempt the children of men, or they could not be agents unto themselves; for if they never should have bitter they could not know the sweet.**

It's not a comfortable situation, but it is the truth. We learn best and learn fastest by being given choices that actually have serious consequences. Both a spirit body and a physical body provide serious consequences for our choices.

Spirit bodies can feel pain and can be damaged; but, they are easily healed. Not so with a physical body. When it comes to a physical body and entropy, the damage can actually be permanent for the duration of a person's mortal life. That ups the stakes greatly, does it not? While in a physical body, we are forced to pay attention or risk injury and death. The cause and effect nature of a physical realm means that there are real consequences for our actions and our choices. We don't always get a do-over in a physical realm. Many things are permanent.

Doctrine and Covenants 29: 43-45:

And thus did I, the Lord God, appoint unto man the days of his probation — that by his natural death he might be raised in immortality unto eternal life, even as many as would believe; and they that believe not unto eternal damnation; for they cannot be redeemed from their spiritual fall, because they repent not; for they love darkness rather than light, and their deeds are evil, and they receive their wages of whom they list to obey.

It's a test. It's a school-ground. It's a test for us to see what we will choose to do when we are left on our own to make our own choices and decisions; and, there are consequences for our choices.

According to the polls that I have seen, the vast majority of scientists have chosen to believe that only physical matter or entropy exists. These people are Materialists, Naturalists, Darwinists, Nihilists, Behaviorists, Determinists, Physical Reductionists, and Atheists. They have put into place censors or a peer-review system to ENFORCE their beliefs onto others. They place mind-guards into our public-school systems to prevent people from learning anything about Syntropy. Nowhere in Science is the pressure to conform or the groupthink the strongest than when it comes to Materialism, Naturalism, Darwinism, and their derivatives. These people have set up our school systems so that they are now free to force their religion or their beliefs onto the public; and, they do. They put mind-guards or censors into place to prevent people from learning anything that might falsify the Theory of Evolution.

Remember, mind-guarding is a phenomenon that feeds groupthink when some members protect the group from information that would call into question the truthfulness of the Theory of Evolution, Materialism, Naturalism, and Atheism. These religions or philosophies are based upon a refusal to look at evidence – any evidence that falsifies them. The information that I'm giving you here is precisely the information that these people censor, ban, block, and destroy.

If you pay attention and observe carefully, you will notice that the Materialists, Naturalists, Darwinists, Nihilists, and Atheists have to use intimidation, peer-review, threats of unemployment, censorship, and ridicule in order to force their beliefs onto the rest of us because these people can't make their case or meet their burden of proof on the merits alone. There is no evidence and will never be any evidence supporting their claims that Syntropy, Psyche, Non-Locality, Action at a Distance, Quantum Mechanics, and the Non-Physical do NOT exist. ALL of the observational evidence, experiential evidence, empirical evidence, eye-witness evidence, and scientific evidence PROVES that they are wrong. The verified and proven existence of Quantum Mechanics or Supernatural Mechanics falsifies Materialism, Naturalism, and their derivatives.

For these people, Syntropy, Quantum Non-Locality, Eternal Life, Immortality, Love, Free Will, Agency, Resurrection, Atonement, Forgiveness, Psyche, God, and Life are the enemy. These people fight these things religiously with a passion. When it comes to the Materialists, Naturalists, Darwinists, and Atheists, the groupthink or the dark side of the force is strong with these people. The Nazis or Fascists were enforcing Darwinism within their society – survival of the fittest. Militant Atheism is the force behind Communism, and these people will kill you if you refuse to do what they say and refuse to believe what they want you to believe. Communism is enforced atheism. Belief in Materialism, Naturalism, Darwinism, and their derivatives is based exclusively on blind faith, and it is also based upon intimidation, ridicule, and force.

Richard Dawkins wrote something to this effect online:

I suspect that most of our regular readers here would agree that ridicule, of a humorous nature, is likely to be more effective than the sort of snuggling-up and head-patting that Jerry is attacking. I lately started to think that we need to go further: go beyond humorous ridicule, sharpen our barbs to a point where they really hurt.

Michael Shermer, Michael Ruse, Eugenie Scott, and others are probably right that contemptuous ridicule is not an expedient way to change the minds of those who are deeply religious. But I think we should probably abandon the irremediably religious precisely because that is what they are — irremediable. I am more interested in the fence-sitters who haven't really considered the question very long or very carefully. And I think that they are likely to be swayed by a display of naked contempt. Nobody likes to be laughed at. Nobody wants to be the butt of contempt.

You might say that two can play at that game. Suppose the religious start treating us with naked contempt, how would we like it? I think the answer is that there is a real asymmetry here. We have so much more to be contemptuous about! And we are so much better at it. We have scathingly witty spokesmen of the caliber of Christopher Hitchens and Sam Harris. Who have the faith-heads got, by comparison? Ann Coulter is about as good as it gets. We can't lose!

If you can bear to listen to him, take, as an example of a typical faith-head trying to be contemptuous, David Bentley Hart, whose radio interview happened to be posted here at the same time as Jerry's article.

Listen to the stumbling, droning inarticulacy, the abysmal lack of anything approaching wit or intelligence. Imagine this yammering fumblewit coming up against Christopher Hitchens, or Dan Dennett, or PZ Myers — doesn't it make your mouth water?

Maybe I'm wrong. I'm only thinking aloud, among friends. Is it gloves off time? Or should we continue to go along with the appeasers and be all nice and cuddly, like Eugenie and the National Academy?

End of Quotes from Richard Dawkins. [Note: the original is no longer available on the Richard Dawkins Foundation, so this had to be pieced together from different articles online that had quoted Dawkins when he first wrote his comments and proposed his Master Plan of Contempt.]

Notice that it's a plan to intimidate, trick, and deceive the ignorant and the uneducated – the fence-sitters. He wants to go after the vulnerable – those who are most likely to be deceived. It's a program of intimidation and brainwashing that is every bit as nefarious and evil as the one that they accuse the faith-heads of perpetrating. He's definitely chosen the low road to success and is proud of it. Can you sense his bitterness? He hates religion and God.

These people have formally rejected Syntropy or Love. I KNOW because I used to be one of them. Have you even heard of Syntropy – the opposite of entropy? Most likely you haven't because these people have blocked, banned, censored, and suppressed all knowledge regarding Syntropy. In fact, these people will tell you that Psyche or Syntropy does not exist. That's what they truly believe. That's what I used to believe. I spent some time ridiculing and mocking. Unlike these people, though, I considered myself to be an open-minded scientist; and, I was willing to look at opposing evidence. I was willing to allow all of the evidence into evidence; and, I pursued a preponderance of the evidence. It was the scientific evidence – the observational evidence – that convinced me that My Materialism, My Naturalism, My Nihilism, and My Atheism are false. I want evidence – lots and lots of evidence. I cannot live on blind faith alone.

Richard Dawkins also stated:

Mock them! Ridicule them! In public! Don't fall for the convention that we're all too polite to talk about religion. Religion is not off the table. Religion is not off limits. Religion makes specific claims about the universe which need to be substantiated and need to be challenged and, if necessary, need to be ridiculed with contempt.

Why is all of this ridicule and mocking necessary? It's because these people can't make their case on the merits alone. ALL of the evidence that we have on hand as a race falsifies their chosen beliefs. Mocking and ridicule is the only thing they have left in their arsenal because the evidence certainly isn't on their side. These people hold you in contempt and they despise you. They're not interested in Syntropy, Forgiveness, or Love. They have NO desire to uplift and enlighten. They are trying to trick you and deceive you while pretending that they are the ones who have been wronged. They are trying to drag you down to their level. The Christians would say that these people are trying to drag you down to hell.

I learn best through comparison and contrast.

Entropy is based upon force. You have to work and steal and kill in order to overcome it, rise above it, and dominate it. Even then, death or entropy is going to get you in the end. Entropy is the no-win scenario. Its only benefit is that it forces us to learn. Our physical bodies hurt, and the pain motivates us to learn. Pain is our greatest teacher. It forces our psyches to learn and grow. Otherwise, we would be passive.

Syntropy is based upon forgiveness and love. The Atonement of Christ is the pinnacle of Syntropy. There's something very unique about Syntropy or Love – the more you give away, the more you seem to have. That's the exact opposite of entropy. Entropy is death. Syntropy is life. Entropy is. However, Syntropy, Forgiveness, Friendship, and Love have to be chosen into existence by Someone Psyche or they will never exist in the first place.

Jesus Christ is the being of light and love whom people encounter during their Near-Death Experiences (NDEs). How did he get all of that light and love? Syntropy! Jesus Christ received all of that light and love by giving it away to us. Sacrifice and forgiveness are the very pinnacle of Love or Syntropy. Syntropy is something vastly different than entropy. Isn't that fascinating to think about? The more you give it away, the more you have when you are done.

John 10: 10: **The thief cometh not, but for to steal, and to kill, and to destroy: I am come that they might have life, and that they might have it more abundantly.**

Within this simple verse, Jesus Christ summarizes our modern-day scientific discoveries.

The thief is entropy. Entropy steals, kills, and destroys. Jesus Christ came to give us Syntropy. Jesus Christ came to introduce some love, mercy, kindness, grace, glory, forgiveness, redemption, action at a distance, quantum mechanics, supernatural mechanisms, health, and sanity into our lives.

We KNOW that some type of Syntropy must exist, or all of that subsequent entropy would not have been possible. This is logical common-sense. In contrast, the message of Materialism, Naturalism, Darwinism, Nihilism, and Atheism is that Syntropy, Psyche, Spirit Matter, Quantum Non-Locality, and Supernatural Mechanics do not exist.

They both can't be true, because they are mutually exclusive. If one is proven to exist, then the other one has been proven false. The verified and proven existence Quantum Mechanics or Syntropy falsifies Materialism, Naturalism, and their derivatives such as Darwinism, Nihilism, Behaviorism, Determinism, Physical Reductionism, and Atheism.

Syntropy is one of my most significant and useful scientific discoveries. Syntropy puts a dose of reality into our lives. The discovery of Syntropy has powerful ramifications when it comes to Science as well as our favorite science fiction.

I enjoy X-Men and Mutant X as much as the next guy; but, the "science" behind super-powered mutants is fatally flawed. There will NEVER be any mutants with superpowers or quantum abilities because superpowers or quantum abilities do not and cannot derive from our genes. Our genes are based upon entropy. There's nothing within our genes to gift us or endow us with Syntropy, Action at a Distance, Quantum Mechanics, Supernatural Abilities, or Superpowers.

Genes can only be used to produce mechanical abilities or physical abilities, not quantum abilities or superpowers or supernatural abilities. Superpowers are a quantum phenomenon, or spiritual phenomenon, or psyche phenomenon. They don't have anything to do with physical matter and genes.

There's only two known ways for physical beings to get access to Superpowers or Quantum Abilities, and that's God and Satan. God is notoriously stingy when it comes to giving fallen mortal beings Superpowers; and, Satan will betray you and destroy you in the end. We are meant to do without Superpowers for the duration of our mortal lives. They seem to get restored to us, though, when we die and return to the Spirit World or the Quantum Realm.

Physical matter, including our genes, was deliberately and purposefully designed by God to greatly limit and restrict the natural and innate quantum abilities or supernatural abilities which are an inherent part of our spirit body and psyche. Giving us superpowers or quantum abilities would break this physical school-ground which God designed for us to teach us how to become more like Him – humble and teachable.

The purpose of this physical school-ground is to motivate us to learn how to bring Syntropy or Love or God into our lives while at the same time being trapped within and surrounded by entropy, corruption, disease, disorder, death, and extinction.

Mark My Words

Proof of Heaven

I used to be a Materialist, Naturalist, Nihilist, and Atheist. My discovery of the Syntropy Realm or Spirit Realm represents a major paradigm shift for me.

I'm not going to apologize for finding the truth. It's what we scientists are supposed to do, is it not?

One of the most interesting near-death experiences that I have read is "Proof of Heaven: A Neurosurgeon's Journey into the Afterlife" by Eben Alexander. Of this experience, Raymond A. Moody wrote, "Dr. Eben Alexander's near-death experience is the most astounding I have heard in more than four decades of studying this phenomenon. He is living proof of an afterlife."

Eben Alexander wrote this about his near-death experience:

"Om told me that there is not one universe but many – in fact, more than I could conceive – but that love lay at the center of them all. Evil was present in all the other universes as well, but only in the tiniest trace amounts. Evil was necessary because without it free will was impossible, and without free will, there could be no growth – no forward movement, no chance for us to become what God longed for us to be. Horrible and all-powerful as evil sometimes seemed to be in a world like ours, in the larger picture love was overwhelmingly dominant, and it would ultimately be triumphant.

I saw the abundance of life throughout the countless universes, including some whose intelligence was advanced far beyond that of humanity. I saw that there are countless higher dimensions, but that the only way to know these dimensions is to enter and experience them directly. They cannot be known and understood, from lower dimensional space. Cause and effect exist in these higher realms, but outside our earthly conception of them. The world of time and space in which we move in this terrestrial realm is tightly and intricately meshed within these higher worlds. In other words, these worlds aren't totally apart from us because all worlds are part of the same overarching divine Reality. From those higher worlds one could access any time or place in our world.

It will take the rest of my life, and then some, to unpack what I learned up there. The knowledge given me was not 'taught' in the way that a history lesson or math theorem would be. Insights happened directly, rather than needing to be coaxed and absorbed. Knowledge was stored without memorization, instantly and for good. It didn't fade, like ordinary information does, and to this day I still possess all of it, much more clearly than I possess the information that I gained over all my years in school.

That's not to say that I can get to this knowledge just like that. Because now that I'm back here in the earthly realm, I have to process it through my limited physical body and brain. But it's there. I feel it, laid into my very being. For a person like me who had spent his whole life working hard to accumulate knowledge and understanding the old-fashioned way, the discovery of this more advanced level of learning was, alone, enough to give me food for thought for ages to come." (Pages 48 and 49.)

"My awareness was larger now. So large, it seemed to take in the entire universe."

"I was a citizen of a universe staggering in its vastness and complexity and ruled entirely by love." (Page 95.)

"At the heart of the enigma of quantum mechanics lies the falsehood of our notion of locality in space and time. The rest of the universe – that is, the vast majority of it – isn't actually distant from us in space. Yes, the physical space seems real, but it is limited as well. The entire length and height of the physical universe is as nothing to the spiritual realm from which it has risen – the realm of consciousness (which some might refer to as the 'life force').

This other, vastly grander universe isn't 'far away' at all. In fact, it's right here – right here where I am, typing this sentence, and right there where you are, reading it. It's not far away physically, but simply exists on a different frequency. It's right here, right now, but we're unaware of it because we are for the most part closed to those frequencies on which it manifests. We live in the dimensions of familiar space and time, hemmed in by the peculiar limitations of our sensory organs and by our perceptual scaling within the spectrum from subatomic quantum up through the entire universe. Those dimensions, while they have many things going for them, also shut us out from the other dimensions that exist as well.

The ancient Greeks discovered all of this long ago, and I was only discovering for myself what they'd already hit upon: Like understands like. The universe is so constructed that to truly understand any part of its many dimensions and levels, you have to become a part of that dimension. Or, stated a little more accurately, you have to open yourself to an identity with that part of the universe that you already possess, but which you may not have been conscious of.

The universe has no beginning or end, and God is entirely present within every particle of it. Much – in fact most – of what people have had to say about God and the higher spiritual worlds has involved bringing them down to our level, rather than elevating our perceptions up to theirs. We taint, with our insufficient descriptions, their truly awesome nature.

But though it never began and will never end, the universe does have punctuation marks, the purpose of which is to bring beings into existence and allow them to participate in the glory of God. The Big Bang that created our universe was one of these creative 'punctuation marks'. Om's view was from outside, encompassing all of Om's Creation and beyond my even higher-dimensional field of view. Here, to see was to know. There was no distinction between experiencing something and my understanding it.

'I was blind, but now I see,' now took on a new meaning as I understood just how blind to the full nature of the spiritual universe we are on earth – especially people like I had been, who had believed that matter was the core reality, and that all else – thought, consciousness, ideas, emotions, spirit – were simply productions of it.

This revelation inspired me greatly, because it allowed me to see the staggering heights of communion and understanding that lie ahead for us all when each of us leaves the limitations of our physical body and brain behind.

Humor. Irony. Pathos. I had always thought these were qualities we humans developed to cope with this so often painful and unfair world. And they are. But in addition to being consolations, these qualities are recognitions – brief, flashing, but all-important – of the fact that whatever our struggles and suffering in the present world are, they can't truly touch the larger, eternal beings we in truth are. Laughter and irony are at heart reminders that we are not prisoners in this world, but voyagers through it." (Pages 156 and 157.)

Eben Alexander from the book, "Proof of Heaven".

http://ldssoul.com/forum/viewtopic.php?f=65&t=238

This is Science because it is based upon observation and experience. Science is observation and experience, or it should be. To me personally, his words also became scripture.

Eben Alexander's experiences and observations are in complete harmony with everything that I have encountered and learned from other NDErs and OBErs. It's in harmony with the Science associated with Quantum Mechanics, Supernatural Mechanisms, Action at a Distance, Psyche, Quantum Non-Local Consciousness, Spirit Matter, and Syntropy. It's also in complete harmony with the Scriptures that the Biblical God Jesus Christ had a hand in writing and producing – namely, the Bible, the Book of Mormon: Another Testament of Jesus Christ, the Doctrine and Covenants, and the Pearl of Great Price. Truth harmonizes with and is consistent with all other truths.

God is the Father of Lights.

Spirit matter is the same thing as physical matter, except that they exist at different frequencies or in different dimensions. Psyche, Spirit Matter, and Physical Matter can exist in the same space at the same time because they are out-of-phase with each other meaning that they reside at different frequencies in different dimensions of the same space. Eben Alexander observed this reality and truth.

The difference between spirit matter and physical matter is that physical matter is based upon entropy whereas everything else is based upon Syntropy or Love. The Love or the Syntropy is dominant. However, evil, entropy, pain, physical matter, disease, and death are necessary because without them learning and free will are impossible, and without free will, there could be no growth – no forward movement, no chance for us to become what God longs for us to be.

Eben Alexander is talking about the Syntropy Realm, the Quantum Realm, Higher Dimensions, Higher Frequencies, the Primal Construct, the Spiritual Realm, the Realm of Psyche, the Realm of Consciousness, the Realm of the Life Force, and the Realm of Light and Love.

His words are very liberating, compared with the groupthink or dogma that the mind-guards behind Naturalism and Darwinism want to force upon us and make us believe.

Remember, the mind-guards behind Materialism, Naturalism, Darwinism, Nihilism, and Atheism have to use intimidation, mocking, ridicule, contempt, deception, and force in order to make their case and spread their beliefs because there will NEVER be any evidence to support their claim that the Non-Local, the Spiritual, the Quantum, or the Supernatural does not exist. In fact, the verified and proven existence of Quantum Mechanics or Action at a Distance falsifies Materialism, Naturalism, Nihilism, Atheism, and their derivatives such as the Theory of Evolution.

The Materialists, Naturalists, Darwinists, Nihilists, and Atheists will NEVER be able to make their case or meet their burden of proof based upon the merits alone because there is no evidence and will never be any evidence to support their claims that Syntropy, Psyche, Spirit Matter, Quantum Mechanics, and Action at a Distance do not exist. Therefore, these people have to resort to intimidation, mocking, ridicule, contempt, censorship, and force in order to make the rest of us believe that they are right.

Hopefully by learning to tell the difference between Syntropy and Entropy, you will also learn to discern which group is telling us the truth and which group is trying to trick us and deceive us. I wish you well.

Mark My Words

—

Source

NATURE vs. NURTURE vs. NIRVANA: An Introduction to Reality

https://www.amazon.com/dp/B01JWRCSVA

Reference

Myers, D. G. (2010). *Social Psychology* (10th ed.). New York: McGraw-Hill.

Alexander, E. (2012). *Proof of Heaven: A Neurosurgeon's Journey into the Afterlife*. New York: Simon & Schuster.

BioPsychoSocial and Social Relationships

We see Signs of Psyche all throughout Social Relationships. It's all based upon choice! The environment is, and the genes exist; but, social interactions are chosen into existence by the Human Psyche and all those other psyches.

Friendship, mercy, kindness, and love do not exist until Someone Psyche chooses them into existence or believes them into existence. There isn't an atom of friendship, justice, mercy, or love to be found anywhere within our genome or the physical universe. All these things are a function and a product of Psyche or Non-Local Consciousness, and they continue to exist long after our physical brain is dead and gone according to the scientific evidence being obtained from Near-Death Experiences (NDEs), Out-of-Body Experiences (OBEs), Shared-Death Experiences (SDEs), and our after-death Life Reviews. Even our genomes, physical matter, and this physical universe are the product of Someone Psyche or Someone Intelligent.

Entropy, disorder, extinction, and death are restricted to physical matter. Everything else is Psyche or Syntropy. Syntropy is eternal and everlasting, without a beginning of days or an end of years. Psyche is Syntropy. Quantum Mechanics or Transdimensional Physics is Syntropy. All those invisible non-physical forces and fields are Syntropy. Spirit matter or dark matter is Syntropy. Syntropy must exist or all of that subsequent entropy and physical matter would not have been possible in the first place. Physical matter and physical universes have a beginning and can theoretically have an end. Syntropy has no beginning or end.

Social Relationships are based upon Syntropy, NOT entropy and physical matter! We can have and maintain social relationships without a physical body.

Remember, your genes would have to be psychic, social, intelligent, and interactive in order to be able to do all the different things that the Materialists, Naturalists, and Darwinists say that they are doing. According to these people, it's all being run by the genes, and the genes are being run by evolution and chance. There's no such thing as Psyche or Syntropy where these people are concerned. Only physical matter exists.

But, the rocks have never been caught in the act of social interaction, nor have the rocks been caught in the act of designing and creating proteins, genomes, and life forms from atoms or from rocks. Psyche or Syntropy end up being the best and most believable explanation where any type of Science is concerned. The rocks can't do science, but the Human Psyche certainly can.

Mark My Words

The Completed Essays on Social Relationships

I'm working on and completing this essay within my book:

God Is in the Light: God is light, and in Him is no darkness at all.

https://www.amazon.com/dp/B07168S37N

I can't put all of my good stuff into this book alone.

Look under **Social Relationships** within that book for some of the results.

It's huge, which is why it's not included here.

—

Signs of Psyche

While studying Social Psychology, I started seeing Signs of Psyche all throughout my college textbook on the subject.

I started to document what I was discovering in the autobiographical journal entries that I was called upon to present for that particular class in psychology.

The results can be found in my autobiography:

Tripping the Light Fantastic: How Prescription Drugs Almost Killed Me

https://www.amazon.com/dp/B071RJP9T8

Look under the sections:

Social Psychology Autobiographical – Part I

Social Psychology Autobiographical – Part II

—

Brain Wars and the Spiritual Brain

The books, *Brain Wars* and *The Spiritual Brain*, by Mario Beauregard contain scientific evidence that convinced me that Psyche, Mind, or Soul does in fact exist.

I decided to include my notes for these books within my book:

I Am Not a Creationist: So What Am I?

https://www.amazon.com/dp/B071XTM8XY

Mark My Words

—

BioPsychoSocial – The Best Secular Model for Reality and Existence

The BioPsychoSocial model, when untainted by Materialism and Naturalism, includes or integrates the Human Psyche, Syntropy, and Quantum Mechanics into the mix for superior explanatory power. I wrote an essay on the BioPsychoSocial Model for this book exploring my growing understanding of the BioPsychoSocial Model; but, this book doesn't have room for it. That essay can now be found under the above title within my book:

BioPsychoSocial: Including Psyche or Light into our Theoretical Models

https://www.amazon.com/dp/B0713NDHVW

Enjoy!

—

Psyche or Syntropy Is the Best Explanation

Ask yourself how we detect Psyche or Quantum Non-Local Consciousness. How do we catch Psyche in the act? By definition, in principle or practice, Psyche or Intelligence is non-physical which means that it can't be detected nor recorded by our physical instruments. So, how do we know that it exists?

When it comes to the various different invisible and non-physical forces and fields like Psyche, we KNOW that they exist by the effects that they have on physical matter. We KNOW that gravity, wind, magnetism, dark matter or spirit matter, quantum waves, the strong and weak nuclear forces, light, the zero-point field, and dark energy exist by the effects that they have on physical matter. That's just one of the ways that we KNOW that Psyche exists.

As I studied and thought about this subject, I discovered that there are many different ways through which we come to KNOW that the Psyche, Spirit, Self, or Soul does in fact exist.

One of the most obvious Signs of Psyche is CHOICE! It all comes down to choice. Whenever you encounter some kind of choice being made, you have in fact caught some kind of Psyche in the act. Physical matter doesn't have any choice. Physical matter does whatever the different forces and fields make it do.

Mark My Words

—

This was an extensive and interesting essay that I developed for this book, but I had to remove it due to space limitations. You can find the full version in:

BioPsychoSocial: Including Psyche or Light into our Theoretical Models

https://www.amazon.com/dp/B0713NDHVW

NATURE vs. NURTURE vs. NIRVANA: An Introduction to Reality

https://www.amazon.com/dp/B01JWRCSVA

https://www.amazon.com/dp/1521132615

—

Social Psychology Has No Meaning or Value without Psyche

This was another interesting article that I had to remove due to space limitations. The full version can be found in:

The Ultimate Model of Reality: Psyche Is the Ultimate Cause

https://www.amazon.com/dp/B071NC9JK6

—

Science 2.0, Quantum Mechanics, and Syntropy

I have been a scientist all of my life; and after 55 years of life, I finally found what I have been searching for all of my life – an explanation for how everything works at all levels of existence or reality.

I found the answer to life, the universe, and everything. I found what I was looking for in Quantum Mechanics, Action at a Distance, Psyche, and Syntropy. It explains everything! As a result of these discoveries, I upgraded my science.

I have made some fascinating discoveries during my science career. Ironically, most of what I have discovered has already been discovered by someone else before me; but, I chose to do something unique that has never been done before. I chose to give each scientific discovery a quantum mechanical explanation, a psychic explanation, or a spiritual explanation. I chose to take Quantum Mechanics and Psyche seriously, which most scientists have never done before. Consequently, it makes it seem as if I have made a wide variety of different scientific discoveries that have never been discovered before when in fact I have not. I've simply discovered what others have chosen to ignore and reject.

Even Psyche or Intelligence has been experienced and observed, which means that it has already been discovered; but, I chose to do something different that hasn't been done before. I chose to include it in Science; and, I chose to try to define it and explain what it is and how it works.

Psyche is the innate intelligence within all the different forms of energy that have ever been discovered or observed. The psyche within energy gives all the different forms of energy the ability to hear God's commands, understand God's commands, and obey God's commands. Psyche is the "organizing force" or the "law following force" within our universe. Psyche is identified by choice, particularly its choice to obey the quantum laws and the physical laws.

Physical matter, spirit matter, and dark matter are simply different forms of organized energy existing simultaneously in different phases, different frequencies, or different dimensions. They each have a psyche within them that gives them the ability to obey God's Laws. God provides the Laws; and, the psyche within the different forms of energy provides the obedience to those Laws. Matter is organized energy or lawful energy. Someone Psyche within that matter is choosing to obey the Quantum Laws and the Physical Laws; otherwise, it would be nothing but random chaos.

A LAW in order to be a Law requires some kind of intelligent Law Giver and Law Enforcer. Laws are worthless without some type of uniformity, order, and enforcement. The thing that most people don't realize is that Laws also require Someone Psyche with enough intelligence and understanding to actually choose to obey those Laws. Without Someone Psyche or someone intelligent enough to function as the Law Follower or the Law Obeyer, Laws are also completely worthless. Laws are pointless unless there is someone intelligent enough, within the different forms of energy, who is capable of actually choosing to obey those Laws. This is logical common sense. I have a feeling that it has all been discovered before; but, nobody ever takes the time to actually look at everything from the perspective of Psyche.

Psyche is the innate intelligence, within all the different forms of energy, which gives those different forms of energy the ability to obey God's Laws and God's Commands.

When it comes time to explain life, the universe, and everything, there are only two real contenders – Creation by Chance Evolution and Creation by Intelligent Intervention. They are mutually exclusive. If one is verified and proven to be true, then the other one has been falsified in the process. So, ask yourself which one has been verified and which one has been falsified.

This was a thirty-page essay, which was developed for this book, that I had to delete due to space limitations. It can be found in its entirety within the following book:

Science 2.0: I Upgraded My Science

https://www.amazon.com/dp/B0771K6WTX

—

The Biblical God Tried to Teach Us about the Different Types of Light

This is an essay that I had to remove from the book due to space limitations.

Psyche is Energy or Light.

In the revelations that He gave us, the Biblical God Jesus Christ talks a lot about Light. Energy is Light; and, Psyche is the innate intelligence within all the different forms of energy or light. Psyche or Energy cannot be made, and it cannot be destroyed. It has always existed, and it will always exist.

Appropriately, this essay is now found in the following book:

God Is in the Light: God is light, and in Him is no darkness at all.

https://www.amazon.com/dp/B07168S37N

—

Philosophy of Science

An understanding of the Philosophy of Science was essential for developing Science 2.0, Syntropy, and Quantum Neuroscience. The following essays discuss the source for these scientific discoveries:

The Logic or Thinking behind Naturalism and Its Derivatives Is Flawed

The Scientific Method Is Based Upon a Logic Error

The most recent versions of these essays are found in the following book:

The Ultimate Model of Reality: Psyche Is the Ultimate Cause

https://www.amazon.com/dp/B071NC9JK6

—

Source

BioPsychoSocial: Including Psyche or Light into our Theoretical Models

https://www.amazon.com/dp/B0713NDHVW

Mark My Words

Fixing Science with Quantum Mechanics

Quantum Mechanics, or Syntropy, or Psyche is the answer to life, the universe, and everything.

One of the most useful things I ever did was to use Quantum Mechanics to upgrade science from Materialism and Naturalism TO Observation and Experience. I upgraded my science to Science 2.0. Rather than defining science as Materialism, Naturalism, Darwinism, Nihilism, and Atheism as most college professors do, I now define science as Observation and Experience.

Unless it has been experienced and observed by Someone Psyche somewhere sometime, it really isn't science. The major premises or hidden assumptions of Materialism, Naturalism, Darwinism, Nihilism, and Atheism – which claim that the quantum or the supernatural does not exist – have not been and cannot be experienced and observed. Therefore, Materialism and Naturalism are NOT science, because their major premises or hidden assumptions cannot be experienced nor observed but have to be taken on blind faith as being real and true.

The observed, experienced, and verified existence of Quantum Mechanics, Psyche, the Biblical God Jesus Christ, and Action at a Distance FALSIFIES Materialism, Naturalism, Darwinism, Nihilism, Atheism, and their derivatives.

The truth falsifies the false; and, the truth has been repeatedly verified, experienced, and observed.

Mark My Words

—

Reference

Science 2.0: I Upgraded My Science.

https://www.amazon.com/dp/B0771K6WTX

—

The Law of Psyche

Psyche has been experienced and observed by Out-of-Body Explorers and Near-Death Experiencers. The First Law of Thermodynamics tells us that Psyche or Energy is conserved, which means that it is eternal and everlasting. It cannot be made, and it cannot be destroyed. Psyche or Energy is syntropic which means that it is always conserved both at the quantum level and the physical level.

I prefer the scientific approach. I define science as observation and experience. Psyche has been experienced and observed; therefore, we KNOW that it exists. Quod erat demonstrandum!

Law of Psyche: The Law of Psyche states that every massless quantum particle is some type of Psyche or Intelligence. Psyche is the innate intelligence within all the different forms of energy which gives that energy the inherent ability to understand, follow, and obey God's Laws and God's Commands. Every massless elementary particle is some type of Psyche because it is psychic, quantum, telepathic, makes choices, can self-generate or self-propagate, can think, can communicate, can transmute, can move, can phase-shift, can collapse its own wave function, is perceptive, is sentient, is omniscient, is omnipresent, is intelligent, is conscious, is alive, must travel at the speed-of-light from our perspective at the physical level, is prescient, and without any physical limitations imposed upon it can quantum tunnel at will in its native original environment at the quantum level. Remember, every massless quantum particle is some type of Psyche, or Intelligence, or Life Force. Psyche is seen by Out-of-Body Travelers as a pinpoint of light or a Photon. Each Photon is a type of Psyche or Intelligence. A Psyche is a Photon, a massless elementary particle. If it looks like a Photon and acts like a Photon, then it is some type of Psyche. Every massless quantum is a Psyche. Psyche USES mass, energy, or matter to get things done. This is what has been experienced and observed.

Psyche has been hiding in plain sight all the time where nobody can see it nor find it because they aren't looking for it; but, there's no great mystery here. Psyche is seen as a Photon BOTH at the quantum level and the physical level because it is a Photon. Psyche or Intelligence IS what it appears to be, a Photon of Light or a Pinpoint of Light. No surprise there. Psyche is what it has been experienced and observed to be – a massless quantum of energy that looks like and acts like a Photon. Psyche, whenever it has been seen both at the quantum level and the physical level, looks like a Photon or a Pinpoint of Light.

I know that it's hard to believe because we have been trained, brainwashed, and conditioned all of our lives not to believe it; but, it is true, nonetheless. Psyche has been experienced and observed. Every massless quantum or every massless elementary particle looks like a photon and acts like a photon because is a Controlling Psyche who USES energy, mass, or matter to get things done both at the quantum level and the physical level.

Mark My Words

—

Syntropy vs. Entropy

Syntropy is Mind-Over-Matter or the Placebo Effect. If we are paying attention, we will notice that it shows up everywhere in our physical universe. Syntropy or Mind-Over-Matter shows up a lot in Social Psychology. Obviously, it should because Social Psychology is the study of the interaction between the Human Psyche and all those other psyches, on every level of existence.

Syntropy vs. Entropy

In the cyclical nature of the Universe there are two opposing principals, Entropy and Syntropy. To understand syntropy, it is useful to understand its opposing force, entropy.

The Principle of Entropy has effects that are easy to understand and observe. According to this Principle, all organized forms of matter require more energy than those that are less organized. These organized forms will lose their order and initial energy unless they are constantly nourished by energy. Plants need water and sunlight to grow (energy); when that is deprived they begin to breakdown. A new car will fall apart unless energy is applied to maintain it, and a business that doesn't put energy into maintaining itself will close its doors. As life forms age their systems lose energy becoming less efficient allowing for disease and illness to set in. The tendency of nature, systems, and organisms to lose energy and become disorganized over time, essentially the law of death is entropy.

Entropy's opposing force, however, is Syntropy. This Principle is much harder to notice, the best example of which, is life itself. It is the law of order and organization, finality and differentiation, the ability to attract, evolve and bring together ever-increasing complex forms creating something new. A new galaxy forming from the ruins of an older galaxy, building a business, and the bringing together of individual cells to create an organism are all examples of the syntropic nature of the Universe. Consciousness focusing energy to create and maintain a system; the law of life is Syntropy.

In an organism the degree of internal disorganization is entropy and the level of internal organization is syntropy. The sum of these two quantities represents the level of health of an organism in the present and the transformation potential for the evolution of that system or organism.

As far as health is concerned, the more energy we have available to maintain and balance our various systems, the greater our health benefits will be. But if we are overly stressed, in chronic pain, mentally and emotionally drained, most of our energy is being used to sustain these negative states. This means much less energy for our healing and growth processes. Over time an organism's systems begin to break down, dis-ease sets in and entropic forces overtake the syntropic forces until death occurs.

The holistic healing model integrates the whole person in the healing process: body, emotion, mind, and spirit, all of which are different forms of energy. A holistic approach utilizes non-invasive treatments and natural healing techniques to ultimately change the energy associated with disease and dysfunction, bringing balance and a higher degree of homeostasis to the patient.

Aside from practitioner's treatments, a major aspect of holistic healing is the patient's active participation in the process. The outcome is subject to the patient's personal desire, intent, and follow-through in their healing process. Self-healing by means of a balanced diet, exercise, adequate rest, and stress release techniques to name a few are imperative for anyone wanting to heal themselves and become more whole.

Syntropy Energetics is an integrative approach to health and wellness, which utilizes holistic therapeutic modalities and life practices to bring about change in a client. These healing techniques work with different energies of the body at the different layers and levels in which they manifest. *Primal Reflex Release Technique* (PRRT), works with the physical body, utilizing reflexes to quickly release pain, tension, and stress. *Subtle energy healing* accesses our subtle energy fields to create instant change at these primary levels of organization. *Biodynamic Cranial Touch* is a means to access the different levels of consciousness associated with energy, allowing *it* to return the client to a state of wholeness.

http://syntropyenergetics.com/Syntropy_vs._Entropy.html

I'm not promoting syntropy energetics per se; but, I am indeed promoting Syntropy or Mind-Over-Matter. The Placebo Effect is REAL, documented, proven, verified, observed,

584

and powerful. The Placebo Effect or Mind-Over-Matter is an integral part of Science 2.0, Spirit Matter, and Syntropy. Likewise, Action at a Distance or Transpersonal Psychology is an integral part Quantum Mechanics, Transdimensional Physics, Psyche, and Syntropy. It has been experienced and observed, which means that it's REAL and truly exists.

The physicists have observed that it requires a huge infusion of energy to elevate electrons from a lower level to a higher level. That's how spirit matter or dark matter is converted into physical matter – by a huge infusion of energy, or syntropy, or God's glory, power, and light. If that energy is released from the physical matter, then it goes back to being spirit matter. Where does this huge infusion of energy come from? It comes from the Quantum Realm, the Syntropy Realm, or the Zero-Point Field which is the Light of Christ. The explanatory power of Syntropy or Quantum Mechanics is through the roof!

Begging the question and applying *circular reasoning* – which are logic fallacies – the Materialists, Naturalists, Darwinists, Nihilists, and Atheists define "science" as Materialism and Naturalism. Materialism, Naturalism, Darwinism, and their derivatives are based upon a wide variety of different logic fallacies. Materialism and Naturalism are based exclusively on entropy. There's NO Syntropy where Materialism, Naturalism, Darwinism, Nihilism, and Atheism are concerned.

I have upgraded my science to Science 2.0. Science 2.0 defines "science" as observation and experience, both at the quantum level and at the physical level. Unlike Materialism and Naturalism, Science 2.0 works at both levels of existence – both the psyche level and the physical level. Science 2.0 defines "science" as Syntropy. Quantum Mechanics is Syntropy. It works at both the quantum level and the physical level to explain what's happening at both levels of existence. Quantum Mechanics, Transdimensional Physics, or Syntropy has infinitely greater explanatory power than Materialism, Naturalism, Darwinism, Nihilism, Atheism, and Classical Physics. Materialism, Darwinism, and Naturalism are powerless to explain the Placebo Effect or Mind-Over-Matter.

Please let me demonstrate.

To consider the answer, consider how we interpret and label our bodily states.

The principle seemed to be: *A given state of bodily arousal feeds one emotion or another, depending on how the person interprets and labels the arousal.*

A frustrating, hot, or insulting situation heightens arousal. When it does, this arousal, combined with hostile thoughts and feelings, may form a recipe for aggressive behavior. (*Social Psychology*, p. 368).

We = Human Psyches. The person = the Human Psyche.

It's the very same physical input from the environment, the very same physical body, and the very same physical genes making up the physical cells in the physical body; but, we are observing two completely different chosen responses or two different chosen behaviors depending upon how the same Human Psyche decides to interpret and label the incoming physical stimulus.

If the physical harm is interpreted and labeled as deliberate, mean, intentional, and hostile, anger or retaliation tends to ensue.

If the physical harm is interpreted and labeled as accidental, unintentional, non-hostile, and coming from a friend who typically means us well, then compassion and forgiveness tend to be the result.

Everything in these experiments remains the same, except for the way that the Human Psyche chooses to label and interpret the physical harm or the physical stimulus.

Even a dog can tell the difference between a deliberate kick and an accidental trip, even though the physical behavior is precisely the same and has the same physical effect of doing harm to the dog. It's the Psyche that interprets and labels the physical action, and that makes ALL the difference in the world.

This is the smoking gun, where Psychology, Psyche, and Science are concerned. It's the same physical input, but a completely different chosen response or a completely different chosen interpretation, depending upon the way that the Human Psyche or the dog's psyche chooses to interpret and label the physical assault.

Our chosen interpretation of the evidence helps to determine the emotional feelings and the behavioral outcome of every physical event.

Guns prime hostile thoughts and punitive judgments. What's within sight is within mind. This is especially so when a weapon is perceived as an instrument of violence rather than a recreational item. For hunters, seeing a hunting rifle does not prime aggressive thoughts, though it does for non-hunters. (*Social Psychology*, 369.)

This is the smoking gun, where Psychology, Social Psychology, Behaviorism, Biology, and Genetics are concerned. We have the very same physical stimulus, but two completely different chosen interpretations given to that physical stimulus.

How's that possible?

It's NOT according to Determinism, Behaviorism, Darwinism, Genetics, Materialism, Naturalism, Classical Physics, and the Theory of Evolution.

Here we are observing the same people, with the same genetics as before the experiment or observation, with the same physical stimulus, but two completely different chosen interpretations or two completely different chosen responses. It's the Human Psyche who is making the choices and doing the interpretation and the labeling – NOT their genes. Their genes have NO way of telling the difference between a gun that is used for recreational purposes and a gun that is used to kill people; but, the Human Psyche does and can.

The society or environment provides the gun; but, it's the Human Psyche who decides what that gun means to that particular Human Psyche. Some Human Psyches decide that the gun means recreation and fun. Other Human Psyches decide that the gun is a good way to kill other people and commit suicide. It's the very same gun, but a completely different meaning given to that gun by a Human Psyche.

Do you see how that works?

It's called Psyche Determinism or a Psyche Ontology, wherein Psyche is the fundamental unit of reality. Psyche is the Ultimate Cause. Psyche is the ultimate causal agent. It is the Psyche's chosen interpretation or chosen label for the physical stimulus that determines how that gun is used and perceived by that specific Psyche who is doing the using, choosing, and the perceiving. Their chosen interpretation of the evidence helps to determine the behavioral outcome of their chosen actions and chosen beliefs.

The Human Psyche can literally transform a physical object typically used for recreation and fun into a murder instrument if it chooses to do so. The Human Psyche can transform this computer into a murder instrument with enough planning and forethought or anger and rage.

Anger, rage, and hate are chosen into existence by the Human Psyche. So are justice, mercy, kindness, friendship, and love. The very same physical object transforms and morphs into something completely different depending upon whether the Human Psyche chooses to love it or chooses to hate it. This reality and truth FALSIFIES Behaviorism, Materialism, Naturalism, Determinism, and even Classical Physics.

The Human Psyche can trump its nature and nurture at will.

The choice or decision as to how to interpret and label the meaning of physical objects is done by the Human Psyche, and NOT the genes. The Human Psyche is being impacted and influenced by the physical input, and NOT the genes. Decisions and choices change the Human Psyche, NOT the genes.

Arousal tends to spill over: One type of arousal energizes other behaviors.

Other research shows that viewing violence *disinhibits*. In Bandura's experiment, the adult's punching of the Bobo doll seems to make outbursts legitimate and to lower the children's inhibitions. Viewing violence primes the viewer for aggressive behavior by activating violence-related thoughts. Listening to music with sexually violent lyrics seems to have a similar effect.

Media portrayals also evoke *imitation*. (*Social Psychology*, p. 377.)

Thoughts and memories are a product and a function of the Human Psyche.

How do we know?

We KNOW because our thoughts, memories, individuality, and personality survive the death of our physical brain, according to the empirical evidence from Near-Death Experiences (NDEs), Out-of-Body Experiences (OBEs), Shared-Death Experiences (SDEs), and our after-death Life Reviews.

The viewer in these experiments is the Human Psyche, NOT our genes.

Being a cognitive behaviorist, an atheist, a materialist, and a determinist, Bandura denies the existence of the Human Psyche or the human spirit, but the evidence tells us that he is wrong to do so.

We can't explain what's really happening, both here in the physical realm and there in the spirit realm or quantum realm when we remove the Human Psyche from the equation.

Genes and physical matter don't make choices and decisions for us; and, a careful study of the evidence makes this abundantly clear.

Television's Effects on Thinking

We have focused on television's effect on behavior, but researchers have also examined the cognitive effects of viewing violence: Does prolonged viewing *desensitize* us to cruelty? Does it give us mental *scripts* for how to act? Does it distort our *perception* of reality? Does it *prime* aggressive behavior? (*Social Psychology*, p. 378.)

Cognitions or thoughts are a function of the Human Psyche. Mental scripts and aggressive thoughts are a function of the Human Psyche, NOT our genes. Our genes can't make us do anything we don't want to do. Our genes can't make us believe anything we don't want to believe. Our reinforcement history cannot make us do anything we don't want to do. Our beliefs are chosen into existence by the Human Psyche.

Chosen actions or chosen behaviors are a function and product of the Human Psyche. It's the Human Psyche who is getting desensitized, NOT our genes. Mental scripts and anything within the mind have to do with the Human Psyche and NOT our genes. Physical input or physical stimulus comes in from the physical environment and registers on the physical brain, where it is then transferred or transmitted by the physical brain to the Human Psyche. Then the Human Psyche decides how to interpret and label that physical input, and the Human Psyche chooses how to respond to that physical input or chooses to ignore it. Physical stimuli take place at the physical level, and spiritual stimuli take place at the quantum level; but, our **perception of reality** is done by the Human Psyche at every level or dimension of existence.

Do you see how that works?

Our chosen behaviors or chosen actions are chosen into existence by the Human Psyche. The meaning, or label, or interpretation that is given to a physical object or a physical stimulus is chosen into existence by the Human Psyche. Our beliefs are chosen into existence by the Human Psyche. Love or friendship does not exist until it is chosen into existence by the Human Psyche.

You can grind down the whole physical universe into its constituent parts, and you will never find a single molecule or gene of justice, mercy, kindness, friendship, or love. There's no such thing as justice, mercy, kindness, friendship, or love until it is chosen into existence by Someone Psyche. Furthermore, once these things are chosen into existence by the Human Psyche, these things continue to exist long after your physical genes, physical body, physical society, physical world, and physical brain are dead and gone.

That's just the way things work at every level of existence.

Syntropy is Endless and Eternal without a beginning of days or an end of years. Quantum Mechanics and Psyche are Syntropy. Compared with the Quantum Realities or Eternal Realities associated with the Human Psyche, evolution and evolutionary psychology are completely worthless and inadequate, because evolution and evolutionary psychology are restricted and limited exclusively to the physical realm and classical physics. They don't touch the quantum realm.

We can't use the theory of evolution, evolutionary psychology, behaviorism, and determinism to explain what's happening at the psyche level and what's being done at the quantum level by Syntropy or Psyche to get things accomplished for us here at the physical level. However, we can indeed use Syntropy, Quantum Mechanics, and Psyche to explain these things at every level of existence or every dimension of existence, including both the psyche level and the physical level.

When it comes to science in general, Syntropy, Psyche, and Quantum Mechanics or Transdimensional Physics has infinitely more explanatory power than Materialism, Naturalism, Darwinism, Nihilism, Scientism, Behaviorism, Determinism, Evolutionary Psychology, and Atheism. Syntropy, Psyche, and Transdimensional Physics can be used to explain and understand every type of science and every level of existence that comes our way.

Mark My Words

Quantum Mechanics from a Non-Physical Spiritual Perspective

https://www.amazon.com/dp/B01J023TGU

https://www.amazon.com/dp/1521132380

References

Quantum Mechanics from a Non-Physical Spiritual Perspective

https://www.amazon.com/dp/B01J023TGU

https://www.amazon.com/dp/1521132380

Interpretations of Quantum Mechanics

There are literally thousands of different interpretations of Quantum Mechanics, and that is in fact the weakness of Quantum Mechanics. Quantum Mechanics or Transdimensional Physics is a proven and verified science. What hasn't been proven is "what it all means". You cannot prove an interpretation. Interpretation is a function of the Human Psyche, and you cannot use the physical sciences to prove the truthfulness or the falseness of things at the quantum level or the psyche level. The existence of things at the quantum level can only be inferred at the physical level by the effects that these things have on physical matter. That's just the way these things work.

The main problem with Quantum Mechanics is that it has spent decades under the command and control of the Materialists, Naturalists, and Atheists. These people don't have a clue as to how things really work. I KNOW, because I used to be a Materialist, Naturalist, Nihilist, and Atheist. These people neutralize, defoliate, dumb-down, limit, and castrate every theory and idea that comes their way – especially when it comes to science.

The psyche level or the quantum level has REAL effects on physical matter, but Psyche and Quantum Mechanisms cannot be observed directly by our physical instruments, because they are sub-atomic. Remember, when it comes to physics, the smaller can dwell within and control the larger. Transdimensional means non-physical. The sub-atomic or the quantum waves can dwell within and control the quarks, strings, electrons, bosons, gluons, and atoms because the sub-atomic or psyche is smaller than these things. Psyche transmits, receives, and stores quantum waves. Psyche is the fundamental unit of reality – the ultimate elementary particle – because it is the smallest thing that exists.

Materialism, Naturalism, Darwinism, Nihilism, Atheism, Scientism, Behaviorism, Determinism, Atheism, and Classical Physics create unsolvable problems in science.

I was specially prepared to resolve these unsolvable problems in science.

First of all, I successfully FALSIFIED Materialism, Naturalism, Darwinism, and Atheism by using the Scientific Method and *negating the consequent* in order to do so. I was in fact the first person to do so, as far as I know. I learned how to do this by studying the Philosophy of Science, *affirming the consequent*, and *negating the consequent*. Then I used the Scientific Method and negating the consequent in all of my books to FALSIFY Materialism, Naturalism, Darwinism, and Atheism. I observed that ALL of the unsolvable problems in Psychology are immediately resolved by getting rid of Materialism, Naturalism, Darwinism, Nihilism, and Atheism. Materialism and Naturalism caused the binding problem, the mind-brain problem, the nature versus nurture problem, and the free-will versus determinism problem. These problems are automatically solved by eliminating Materialism, Naturalism, Darwinism, and their derivatives. Simple. Fundamental. Logical. Parsimonious.

Second, I was the first person to develop a Psyche Ontology, wherein I demonstrate that Psyche is the fundamental unit of reality. I called it the Ultimate Model of Reality. Look for it in some of my other books. If there is such a thing as a point particle, Psyche or Non-Local Consciousness is it. A point particle would be the smallest unit of reality – the fundamental unit of reality – an infinite singularity. I have hypothesized many times that Psyche, or Quantum Non-Local Consciousness, is an infinite singularity. As such, it is theoretically capable of storing an infinite amount of information in something that has NO physical size and takes up NO physical space whatsoever. The smaller can dwell within and control the larger. Psyche transmits, receives, and stores quantum waves. It's Someone Psyche who chooses the frequency at which each string will vibrate. Thoughts and memories are quantum waves or the vibrations within strings. Simple. Fundamental. Parsimonious. Logical. All of the complex math points to the existence of these types of things.

Third, I was the first person to develop Science 2.0. Science 2.0 is the way that science should have been done but wasn't. Science 2.0 allows ALL of the evidence into evidence and then pursues a preponderance of that evidence. Under Science 2.0, the BEST and FASTEST way to find and know the truth is to observe it, live it, and experience it, or to choose to trust someone who has. Under Science 2.0, the second-best way to find and know the truth is to use the Scientific Methods and *negating the consequent* to eliminate everything that is false. If you successfully eliminate everything that is false, then ONLY the truth will remain. This is logical, parsimonious, simple, and true. It's fascinating to observe and study what remains, after you have successfully falsified and eliminated Materialism, Naturalism, Darwinism, Nihilism, Behaviorism, Scientism, Determinism, and Atheism. I observed when you successfully eliminate everything that is false, then only the VERIFIED or the OBSERVED or the EXPERIENCED remains. Syntropy, Quantum Mechanics, and Quantum Non-Local Consciousness (Psyche) remain after you have successfully eliminated everything that is false, unfruitful, and unproductive. The verified and the observed remains. What you have actually experienced for yourself remains.

Fourth, based upon Science 2.0, I developed a new science discipline called Quantum Neuroscience. When it comes to Quantum Neuroscience, I chose to allow ALL of the evidence into evidence, and then I pursued a preponderance of that evidence. The discoveries were stunning, and quite unbelievable at first. There's nothing like it on the planet that I know of. Quantum Neuroscience taught me how thoroughly brainwashed into Scientism, Naturalism, Nihilism, and Atheism I really was.

Fifth, I observed that if your interpretation of Quantum Mechanics cannot explain what the Human Psyche and Nature's Psyche are doing at the quantum level, then your interpretation of Quantum Mechanics is absolutely worthless. Materialism, Naturalism, Darwinism, Nihilism, Behaviorism, Scientism, Determinism, Atheism, and Classical Physics CANNOT explain what the Human Psyche and Nature's Psyche are doing at the quantum level, and how they work at the quantum level to get things done for us at the physical level. Interpretations of Quantum Mechanics based upon Classical Physics, the Indeterminacy Principle, Random Diffusion, Materialism, Naturalism, and Entropy CANNOT explain what the Human Psyche and Nature's Psyche are doing at the quantum level or the psyche level. Materialism and Naturalism have NO explanatory power at the quantum level or the psyche level, with means that they are worthless when it comes time to explain these things.

I found two interpretations of Quantum Mechanics that actually explain what Nature's Psyche and the Human Psyche are doing at the quantum level and how they work and function at the quantum level in order to get things done for us at the physical level. I discovered the interface or the bridge between the quantum level and the physical level. The BEST interpretation of Quantum Mechanics is the Orthodox Interpretation of Quantum Mechanics by Henry P. Stapp. It's the interpretation that has the most explanatory power at the quantum level or the psyche level.

Henry P. Stapp explained to me that the mathematics for what the Human Psyche is doing at the quantum level are completely different from the mathematics for what Nature's Psyche is doing at the quantum level. The Human Psyche makes choices and requests of Nature's Psyche. It is Nature's Psyche who collapses the wave function thereby producing a single physical reality as a result.

Heisenberg's Uncertainty Principle is worthless and wrong because it's based upon Classical Physics, Random Diffusion, and Entropy. I, Henry Stapp, and some of his followers modified the Uncertainty Principle by redefining it as "Infinite Possibilities". It's the same difference, but it's a huge difference. Out of ALL the infinite possibilities, the Human Psyche makes a choice. Then, depending upon whether it's physically possible and whether God allows it or not, Nature's Psyche reaches out and collapses all of those infinite possibilities into ONE single physical reality. At that point, it's a done deal. At that point, it's physically real.

Rinse and repeat.

Once again, we start with infinite possibilities. The Human Psyche makes a choice deciding what it wants to do with its physical body. Then Nature's Psyche reaches out into those infinite possibilities and collapses them into one single physical reality. By redefining the Uncertainty Principle as "Infinite Possibilities", we convert Quantum Mechanics from classical physics and entropy over to Syntropy. Infinite Possibilities is Syntropy. In contrast, the Uncertainty Principle is classical physics, random diffusion, and entropy. You have to switch over to Syntropy and Quantum Mechanics if you want to explain what the Human Psyche and Nature's Psyche are doing at the quantum level in order to get things done for us at the physical level. The Uncertainty Principle is a dead-end. Entropy is a dead-end. In contrast, Infinite Possibilities are infinite, endless, and eternal – the very definition of Syntropy. Suddenly, you can explain everything, simply by switching over to Syntropy, Quantum Mechanics, and Psyche.

Sixth, my ultimate science discovery is that there are NO BYTES of programming code or memory within a physical brain. I didn't believe it at first. I refused to believe it at first. My science colleagues rejected it at first, because we have been trained, conditioned, and brainwashed to do so. However, there are NO wires in our brain connecting neurons together into functional BYTES of programming code and memory. A neuron is a switch. It is either ON or OFF. ALL 7,000 synapses on a single neuron get compressed into a single Post-Synaptic Potential (PSP), a single BIT of information, which then determines whether a neuron fires or not. NO messages are transmitted through a synaptic cleft. A synaptic-cleft scrambles everything that comes its way in terms of information or memories. Information transmitted through synapses gets reduced to a single BIT of information, which means that the postsynaptic neuron either fires or it doesn't.

I kept waiting for the Neuroscientists to tell me where the programming code is being stored and how it is being executed. They NEVER did. I kept waiting for the Neuroscientists to tell me where the data or the memories are being stored and how this information is being accessed. They NEVER did. I kept waiting for the Neuroscientists to tell me where all the CPUs and RAM are being stored within the physical brain. They NEVER did. The ONLY thing these people would ever say is that we don't know how these things work, but they do. Clearly, they work. Clearly, they exist. But, we don't know how they work or where they exist. These Materialists and Naturalists couldn't answer these questions and explain this science, because there are NO functional BYTES anywhere within a physical brain at the physical level. You can't answer questions when there are no answers – when your chosen paradigm or chosen model prevents you from providing the answers. By design, Materialism, Naturalism, and Classical Physics have NO explanatory power at the quantum level or the psyche level.

If there really are BYTES of programming code and memory within our brain, then it's ALL being stored and processed at the quantum level or the psyche level, because it doesn't exist at the physical level. It can't exist at the physical level. Why? It's physically impossible to construct functional BYTES out of stand-alone switches with NO wires to connect them. Neurons are stand-alone switches. This means that it's physically impossible to construct functional BYTES of programming code and memories out of neurons at the physical level, because there are NO visible physical wires connecting the neurons or switches together into a functional whole. There are no BYTES observable within a physical brain, which means that those BYTES have to exist and must exist at the quantum level or the psyche level, because clearly they do exist judging by the effect that they have on our physical brain and physical body.

You can only explain what's happening at the quantum level by choosing to allow Syntropy, Quantum Mechanics, and Psyche into the mix. Instantly, everything can be explained, once you have allowed Syntropy, Quantum Mechanics, and Psyche to come in to play.

Seventh, I fell in love with Quantum Maps of Physical Functionality due to the massive explanatory power that it provides. Science is only valuable when it has explanatory power. If your science can't explain what Nature's Psyche and the Human Psyche are doing at the quantum level and how they work at the quantum level, then your science is worthless. By choosing to allow Syntropy, Quantum Mechanics, Transdimensional Physics, Quantum Processing, and Psyche in to play, I discovered that I can explain everything in science in a logical and realistic manner.

The output from an infinite number of quantum computers existing at the quantum level can be MAPPED to a single physical neuron or concept cell.

Even with physical supercomputers running the jobs and the image recognition software, it still takes a ton of time for a computer to "recognize" a face or "identify" a person. The Human Psyche does so instantly. On the input side, our neurons are registers of physical information. That information is transferred directly to Nature's Psyche where it is processed at the speed of thought, and the output from those infinite number of quantum computers is sent to a single concept cell or neuron which Nature's Psyche has MAPPED for that physical functionality at the quantum level. That single neuron or concept cell then fires unilaterally "out-of-the-blue" thereby giving us a physical feeling of recognition.

ALL of the processing and memory storage are done by Nature's Psyche completely outside the conscious awareness of the Human Psyche. The only thing the Human Psyche experiences is the physical feeling of recognition. "Ah, I know that person." The Human Psyche-Quantum Processing-Nature's Psyche Symbiosis is vastly superior to anything that our physical computers are able to do. Nature's Psyche MAPS our physical brain at the quantum level, and then uses those MAPS to get things done for us at the physical level. Instantly, the explanatory power of our science goes through the roof! Suddenly, we KNOW how everything is done and how everything works. Suddenly, we have the answer to life, the universe, and everything.

Materialism, Naturalism, Darwinism, Nihilism, Atheism, Scientism, Behaviorism, Determinism, Atheism, and Classical Physics create unsolvable problems in science.

By allowing Quantum Mechanics and Psyche in to play, ALL of the unsolvable problems in science are automatically solved and go away. Through Quantum Mechanics and Psyche, you can answer every scientific question and scientific hypothesis that comes your way. The explanatory power of Syntropy, Quantum Mechanics, Quantum Processing, Quantum Memory Storage, and Psyche are through the roof. There is NO limit to what you can explain. Everything can be solved. Everything is solved.

Mark My Words

—

Source

Quantum Mechanics from a Non-Physical Spiritual Perspective

https://www.amazon.com/dp/B01J023TGU

https://www.amazon.com/dp/1521132380

References

Quantum Mechanics from a Non-Physical Spiritual Perspective

https://www.amazon.com/dp/B01J023TGU

https://www.amazon.com/dp/1521132380

—

Transdimensional Physics

Transdimensional means Non-Local, Non-Physical, or Not-Located in our physical 3D space-time realm. Transdimensional means quantum, or syntropic, or spiritual.

In physics, a quantum (plural: quanta) is the minimum amount of any physical entity (physical property) involved in an interaction. The fundamental notion that a physical property may be "quantized" is referred to as "the hypothesis of quantization". This means that the magnitude of the physical property can take on only discrete values consisting of integer multiples of one quantum. For example, a photon is a single quantum of light (or of any other form of electromagnetic radiation) and can be referred to as a "light quantum", or as a light particle. Similarly, the energy of an electron bound within an atom is quantized and can exist only in certain discrete values. Indeed, atoms and matter in general are stable because electrons can exist only at discrete energy levels within an atom.

Quantization is one of the foundations of the much broader physics of quantum mechanics.

https://en.wikipedia.org/wiki/Quantum

These people get their emphasis or their focus wrong because they have convinced themselves that only physical matter or entropy exists. These people erroneously define Quantum Mechanics is terms of physical matter. That's a faulty and limited interpretation for Quantum Mechanics. Quantum Mechanics should be defined in terms of ENERGY. Energy is eternal and everlasting. Energy is conserved according to the First Law of Thermodynamics, which means that energy is syntropic. Entropic physical matter is just one form of energy. It's a mistake and a catastrophe to define Energy or Quantum Mechanics in terms of only ONE form of energy – physical matter.

A quantum should be defined as ENERGY, not as a physical entity. Physicist David Bohm said that physical matter is frozen light or condensed light. Physical matter is restricted and limited energy. Physical matter is frozen energy or condensed energy. Quanta are packets of energy. A quantum is organized energy. The Gods can form energy into many different types of Quanta. The true and most accurate definition for "quantum" is energy, or a packet of energy. Physical matter is just one type of energy or one type of quantum. Spirit matter, entropy, time, locality, and even space are different types of Quanta or different types of Energy. There are many different forms of energy. Physical matter is just one of them.

Do you see now how Physicalism, Materialism, Naturalism, Darwinism, Nihilism, Behaviorism, Determinism, Physical Reductionism, and Atheism were designed to hide the truth from us and make science incomprehensible? These people destroy science and deliberately limit science exclusively to entropic physical matter just so that they can say that God's Psyche does not exist. The destruction of science is a high price to pay in order to get your way; but, the Naturalists and Atheists are willing to do so because they hate or fear God that much.

I belabor this and put a lot of time and effort into this because these are concepts that you will have never heard about before. Why? It's because the Atheists and Naturalists who control our public schools have chosen to believe that only entropic physical matter exists. You won't learn about Psyche, Spirit Matter, and Non-Physical Physics or Transdimensional Physics from any college professor in our public schools. The purpose of our public schools is to indoctrinate you into Physicalism, Naturalism, Darwinism, Nihilism, and Atheism. You have to educate and train yourself if you want to find the Science behind Psyche, Spirit Matter, the Angels, and the Gods. The Naturalists and Atheists deliberately ban, block, censor, censure, ridicule, mock, and destroy any scientific evidence, observational evidence, or experiential evidence that falsifies Naturalism, Darwinism, and Atheism. You will never learn the truth from these people because they deliberately ban, block, censor, ridicule, and destroy the truth whenever they encounter it.

My purpose in all of this is to set you free.

I had no idea how severely My Nihilism, My Naturalism, My Physicalism, and My Atheism were destroying and limiting my understanding of Science and Quantum Mechanics until I started to allow myself to look at and study the scientific evidence, observational evidence, experiential evidence, revelatory evidence, spiritual evidence, out-of-body evidence, eye-witness evidence, non-physical evidence, quantum evidence, psychic evidence, syntropic evidence, and empirical evidence that falsify Materialism, Naturalism, Darwinism, Nihilism, Atheism, and their derivatives. The false is falsified by the truth; and, the truth is repeatedly experienced and observed.

There's nothing within the hidden assumptions or major premises of Materialism, Naturalism, Nihilism, and Atheism that has ever been experienced or observed; and, there never will be because it's impossible to experience and observe that something does not exist. In fact, observing that the non-physical and the supernatural (quantum) do indeed exist falsifies Materialism, Naturalism, Darwinism, Nihilism, Atheism, and their derivatives. The truth is repeatedly experienced and observed. There's nothing true within Materialism, Atheism, Nihilism, and Naturalism because there's nothing there to experience and observe.

In contrast, our resurrected Lord Jesus Christ has been experienced and observed, both in the flesh and during our near-death experiences, after He rose from the dead. We KNOW that He exists because He has been experienced and observed.

Science is observation and experience. Science is knowledge and truth; and eventually, the truth set me free. I'm no longer brainwashed nor enslaved by My Atheism and My Nihilism like I used to be. I'm now free to study the whole of Science and Reality, not just Materialism and Naturalism. I'm now free to study anything that has been experienced or observed.

Quantum Mechanics is Transdimensional Physics. Thoughts and memories are quantum waves. Quantum waves of energy are an integral part of Transdimensional Physics. Psyche is intelligence or personality. Psyche is the innate intelligence found within all the different forms of energy that have ever been experienced and observed. All these things survive the death of our physical brain, according to the empirical evidence from Near-Death Experiences (NDEs), Out-of-Body Experiences (OBEs), Shared-Death Experiences (SDEs), and our after-death Life Reviews. Psyche has been experienced and observed. Spirit bodies and spirit matter have been experienced and observed. I'm simply trying to find a scientific explanation for it all.

According to Quantum Mechanics, you (your psyche) are everywhere in the universe simultaneously, and your spirit body is never far behind. Quantum Tunneling is faster than light. Quantum Teleportation is instantaneous. Time stops during quantum tunneling or quantum teleportation. According to the Theory of Relativity, time stops at velocities faster than the speed-of-light. Quantum tunneling is an integral part of psyche and spirit matter.

Quantum Tunneling and Quantum Omnipresence also apply to physical matter, but not as much as they apply to spirit matter and psyche because Nature, the Physical Laws, Entropy, and God deliberately and greatly limit the "omnipresence" of physical matter.

The restrictions and limitations placed into physical matter produce some things that the Physicalists and Naturalists truly love, enjoy, trust, worship, and adore. Physical matter is predictable, controllable, reliable, replicable, dependable, and maintains its shape and its presence. You can make money from physical matter. In a physical realm, due to all the different limitations and restrictions that God has placed upon physical matter, I can depend upon this paper being on my hard drive and cloud drive tomorrow or next week when I go looking for it. You can depend upon your house, your car, your dog, and your spouse being there when you go looking for them as well. That's the benefits of physical limitations and physical restrictions. The atoms in your physical body are not going to quantum tunnel away from you at will because God prevents them from doing so by all the different physical restrictions and physical limitations that He has imposed upon entropic physical matter.

Entropic physical matter has been deliberately limited and restricted by the Gods in order to create the ultimate consensus reality, a physical reality. When it comes to science, it's the quantum side that's interesting. All of the superpowers that we covet are found on the quantum side of the equation. The quantum side of reality does not have any physical limitations. There is no entropy in the Syntropic Realm, or the Quantum Realm, or the

Spirit World. Everything in the Transdimensional Realm is eternal and everlasting, without a beginning of days or an end of years. Everything there is a perpetual motion machine. It never wears out and it never runs down. Quantum Tunneling, Telepathy, Superpowers, and Action at a Distance are an integral part of the Transdimensional Realm. The quantum is syntropic which means that it is always conserved.

The following video is one of my most favorite videos because it explains the difference between physical matter and spirit matter at the quantum level. He's deliberately talking about physical matter; but unknowingly, he is also talking about spirit matter because quantum mechanics works on both sides of the veil. Quantum Mechanics works on the spiritual side as well as on the physical side. You just have to train yourself to understand what it all means and how it all works because you won't get the spiritual side or the psychic side from a Materialist, Naturalist, Nihilist, or Atheist. The spiritual side is hidden in plain sight, and all you have to do to find it is to be willing to look for it.

Is Quantum Tunneling Faster than Light?

https://www.youtube.com/watch?v=-IfmgyXs7z8

https://syntropy.site/wp-content/uploads/2018/04/Is-Quantum-Tunneling-Faster-than-Light.zip

In this video, while talking about Quantum Tunneling and your physical body, he states:

Wouldn't it be nice to be everywhere at once? According to Quantum Mechanics, you are. At least a little bit.

Talking about your physical body he says:

Any material object is really a matter wave. You're everywhere in the universe, but not very much.

Remember this!

According to Quantum Mechanics, you are everywhere at once. According to Quantum Mechanics, you are omnipresent. So, ask yourself why you aren't omnipresent right now. What has been done to you, particularly what has been done to your physical body, so that you are NOT omnipresent right now? This IS the answer to life, the universe, and everything. Something has been done to physical matter and your physical body so that you are mostly right where you are right now. Done by whom? And, what precisely has been done?

In this video, he actually answers the question and explains why your physical body isn't everywhere all at once. He also indirectly, though not explicitly, explains why your spirit body and particularly your psyche WERE everywhere at once or omnipresent but aren't anymore.

It has to do with the De Broglie Wavelength of each of these different objects – psyche, spirit body, and physical body. They each have a different De Broglie Wavelength.

This De Broglie Wavelength defines how well determined an object's position is. A large wavelength means a highly uncertain position [omnipresence]**. A small wavelength means a well-defined position** [the locality of physical matter]**. That's true of subatomic particles, and it's sort of true of anything. Right now, I'm mostly right here. But there's also a small chance that I'm here, here, or here. There's an infinitesimal chance that I'm**

on the moon. Observe me, and you'll collapse my wave function and probably find me pretty much exactly where you expect to.

Of course, he is only talking about physical matter because these people don't believe in the existence of spirit matter and psyche; but, this De Broglie Wavelength applies equally as well to spirit matter and psyche as it does to physical matter. Quantum Mechanics works at ALL levels or dimensions of existence. The explanatory power of Quantum Mechanics, Psyche, and Spirit Matter are infinitely greater than anything we can get from being limited exclusively to classical physics, physical matter, entropy, Materialism, and Naturalism.

Out-of-Body Travelers and Near-Death Experiencers have floated outside of their spirit body as an immaterial viewpoint in space and observed their spirit body. They have also interacted with spirit matter in different dimensions of the Spirit Realm; so, they KNOW that spirit matter is real and truly exists. How else are you going to discover spirit matter besides experiencing it and observing it for yourself, or choosing to trust someone who has?

Materialists, Physicalists, Darwinists, Naturalists, and Atheists aren't interested in finding a scientific explanation for spirit matter, quantum matter, or syntropic matter; but, I am because I'm no longer a materialist, naturalist, nihilist, and atheist.

In the De Broglie Wavelength concept, I discovered how spirit matter and physical matter differ from each other, as well as what gives spirit matter its supernatural or quantum mechanical capabilities. We have a good handle on physical matter. I want to understand spirit matter – what it is and how it works. Spirit matter has been experienced and observed. It has specific properties that are different than entropic physical matter.

While making physical matter and entropy, the Gods give physical matter a very short De Broglie Wavelength which greatly limits the distance that a physical atom can quantum tunnel or teleport. In contrast, spirit matter innately and inherently has a very long and possibly an infinite De Broglie Wavelength thereby giving atoms of spirit matter, or dark matter, or tachyons the ability to quantum tunnel to the other side of this universe instantaneously.

All the Gods have to do is give their resurrected physical bodies an infinite De Broglie Wavelength, and then they can quantum tunnel their physical bodies across the universe instantaneously if they so desire. Give your spaceship an infinite De Broglie Wavelength and a psyche to control it, and instantly it could quantum tunnel to a different galaxy if it wanted to.

Psyche is the innate intelligence within all the different forms of energy that have ever been discovered, experienced, and observed. Energy is intelligent, psychic, telepathic, sentient, perceptive, conscious, alive, and aware. Psyche is the Life Force or the Intelligence within energy. Psyche has been experienced and observed. Out-of-Body Travelers have experienced psyche or consciousness as an immaterial viewpoint in space. They have observed psyche as a pinprick of light, a point particle of light, or a spark of light. Psyche is energy. Psyche is also the intelligence within energy. Psyche or Energy is conserved, which means that it is syntropic. Syntropy is conservation of energy. Syntropy is the First Law of Thermodynamics.

By removing the physical limitations or the physical restrictions from an atom of entropic physical matter, instantly that particle of physical matter takes on all of the quantum mechanical capabilities of syntropic spirit matter while at the same time remaining a particle of physical matter. It's still physical matter, but now it is syntropic physical matter with the ability to teleport or quantum tunnel great distances. The innate or inherent teleportation, or quantum tunneling, capabilities of spirit matter have been experienced and observed by Out-of-Body Explorers. The teleportation or quantum tunneling of physical matter has also been experienced and observed. God holds the KEYS to all of this spiritual or quantum mechanical stuff; but, occasionally He chooses to share those KEYS with human beings like us.

Translated Beings. Translation refers to being physically changed by God from a mortal human being to an immortal human being. A person that has been translated is referred to as a translated being. A translated being has had his or her physical body changed from entropic physical matter to syntropic physical matter. They are still physical beings, but they are no longer subject to death or entropy. In other words, God has given these people access to quantum mechanics. Not only are they immortal; but, they are also telepathic, and they can quantum tunnel at will. They can quantum tunnel their physical bodies to different places on this earth, and they can't be killed. God has given them the KEYS to quantum mechanics. God has given them superpowers.

I want to understand Quantum Mechanics or Transdimensional Physics – what it is and how it truly works. I understand physical limitations and physical restrictions. It's the psychic, or the spiritual, or the transdimensional that's interesting because it's foreign to us while we are trapped in this entropic physical format. But, we'll never get any information about the syntropic or the psychic from the Naturalists and the Atheists because these people have convinced themselves that Psyche, Spirit Matter, and Syntropy do not exist. These people erroneously teach that only physical matter or entropy exists.

Subatomic particles ARE spirit matter and/or psyche. Think about it! It's true! It's only when we get to the atoms that we are actually talking about "pure physical matter" if there is such a thing. Physical matter is organized energy. Physical matter is made from energy. What are quarks made of? What are electrons made of? What are physical laws made of? What's entropy made of? What's time made of? What's space made of? What's spirit matter made of? What's gravity made of? What's magnetism made of? What are the physical constants made of? What are vibrating strings made of? What are quantum waves made of? What are quanta made of? It's all made from ENERGY. They are just different forms of Energy. Energy is syntropic which means that energy is conserved. The form is never conserved; but, the energy is always conserved.

According to the Ultimate Law of Thermodynamics, the form is never conserved. Entropic physical matter is comprised of different forms of energy. The form is never conserved. Only the underlying Energy is conserved. The Ultimate Law of Thermodynamics differentiates between what is being conserved and what is not conserved. The form of energy is never conserved.

Physical matter and entropy were made by the Gods from raw energy, which means that physical matter and entropy can be disassembled back into raw energy and then formed into something else instead. The form is never conserved.

Entropy is death. Entropy is an expiration date. The Materialists and Naturalists erroneously teach that entropy or death is conserved. Entropy is a "form" of energy in that entropy is a measure of energy's transition from a useful form to an unavailable form. God's Psyche can make unavailable energy available again because God designed and made the rules which made that energy unavailable in the first place. Entropy is a convention or a construct, which means that the energy that has been made unavailable can be turned back into available usable energy anytime God chooses to do so because the energy is always conserved. There's nothing sacred or sacrosanct about entropy. It's the ENERGY that's being conserved not the entropy. Entropy comes and goes as God sees fit. The amount or the quantity of physical matter and entropy are not conserved. Entropic physical matter can be disassembled into raw energy and then formed into something else instead.

This is the Ultimate Law of Thermodynamics. It explains everything that we have ever experienced and observed. The Ultimate Law of Thermodynamics is the answer to life, the universe, and everything.

The Ultimate Law of Thermodynamics falsifies Materialism, Naturalism, and their derivatives. The Physicalists, Materialists, Naturalists, Darwinists, Nihilists, Behaviorists, and Atheists erroneously teach that only physical matter or entropy exists; consequently, these people erroneously teach that physical matter and entropy are being conserved. One error leads to another, and then another. Entropy is death. These people erroneously teach that death is conserved. Do they not? That's the sum-total of their message. Is it not? These people erroneously teach that entropy, death, and physical matter are conserved. They are wrong. Science is observation and experience. Observation and experience falsify Materialism, Naturalism, and their derivatives.

Physical matter is organized energy. Physical matter has been deliberately limited and restricted. Organized by whom? Restricted and limited by whom? Only Someone Intelligent or Someone Psyche can organize, manipulate, limit, and restrict raw energy.

According to the Orthodox Interpretation of Quantum Mechanics by Henry P. Stapp, the Human Psyche makes ALL of the choices as to what it wants to do with its spirit body and its physical body, while it is Nature's Psyche or God's Psyche who actually collapses the wave function thereby making the Human Psyche's choices actual and real both in the quantum realm and here in the physical realm. Quantum Mechanics works on both sides of the veil; whereas, Classical Physics, Materialism, Naturalism, Darwinism, Nihilism, Behaviorism, Determinism, Physical Reductionism, and Atheism are deliberately restricted to the physical side or the entropic side of the equation. The Materialists and Naturalists erroneously teach that only physical matter or entropy exists. They are wrong. There's the syntropic side, spiritual side, psychic side, transdimensional side, or the quantum mechanical side that they have chosen to reject, dismiss, and ignore.

During the making of this physical universe, God took a small portion of dark matter or spirit matter, slowed it down to sub-light speeds, gave it a short De Broglie Wavelength, and made it subject to entropy or space-time. God imposed limitations or physical laws upon it. God imposed structure and order upon it. Remove those restraints, and it would go back to being dark matter or spirit matter.

It's all defined and explained by the De Broglie Wavelength. The explanatory power of Quantum Mechanics or Syntropic Mechanics is through the roof to infinity and beyond.

A physical body has a short De Broglie Wavelength, which means that it has been localized or made physical and present in this 3D space-time realm. It's mostly here, rather than being everywhere. Physical matter has been made subject to entropy or the passage of time. Entropy is a function of space and time. Entropy doesn't exist in the Quantum Realm or Syntropic Realm. Entropic physical matter has been localized into both space and time. Spirit matter and psyche have not.

In contrast to physical matter, a spirit body has an innately long De Broglie Wavelength, which means that it tends toward being omnipresent all at once just as described in this video and experienced in reality. A psyche or intelligence or consciousness seems to be capable of an infinite De Broglie Wavelength, or TRUE omnipresence.

Physical beings such as us are comprised of three different entities, and each one of them has a different De Broglie Wavelength. The Law of Quantum Superposition allows for all three to exist in the same space at the same time in an additive fashion, because they each are out of phase with the other or exist in a different dimension than the other. These realities explain everything.

This is cool science!

In its native original state, your Psyche, Quantum Non-Local Consciousness, or Intelligence experiences Quantum Omnipresence. It's capable of being simultaneously everywhere in the universe just as this video describes. According to Quantum Mechanics, your Psyche IS everywhere in the universe all at once, in its native original state. A Psyche is a pinprick of light; but, its thought waves or its quantum waves are omnipresent, or its De Broglie Wavelength is infinite. The spirit matter in your spirit body has many of the same capabilities. A large De Broglie Wavelength means a highly uncertain position or Quantum Omnipresence. It all makes sense once we choose to allow all of the observations and experiences of the human race into evidence from both sides of the veil.

A Psyche or Intelligence is restrained by and limited by being assigned to a spirit body; but, out-of-body explorers have observed and experienced the fact that the Psyche or Consciousness can separate from the spirit body and observe the spirit body from an immaterial viewpoint in space. It's the Psyche who has the experiences and forms new memories, and NOT the spirit body, and definitely NOT the physical body.

Being attached to a physical body with a silver cord greatly limits and restrains a spirit body and a psyche even more by making the spirit or the soul subject to a short De Broglie Wavelength or an entropic physical body. God – the Maker of physical matter, physical universes, and physical bodies – deliberately forces physical matter to be entropic, limited to sub-light speeds, and made to have a very short De Broglie Wavelength. We will never be able to propel physical matter faster than the speed-of-light because God deliberately prevents it from happening by the physical limits that He has placed upon entropic physical matter.

However, as the Maker and Enforcer of these physical laws or physical limitations, God doesn't place and hasn't placed these physical limitations onto His own physical body. His physical body is fully capable of Quantum Tunneling, Teleportation, and even Omnipresence. With no physical limitations being imposed upon it, even a physical body is capable of teleportation, quantum tunneling, and omnipresence just like a spirit body and psyche. It's amazing to think about, isn't it?

Most scientists spend little or no time thinking about the spiritual side of the equation, because they have convinced themselves that it does not exist. However, their chosen beliefs don't change the fact that the Psyche and the Spirit Body have been experienced and observed by millions of different people.

Psyche leads, and your spirit body follows; but, both seem capable of experiencing Quantum Omnipresence or a very long De Broglie Wavelength, as well as Quantum Tunneling or Teleportation at will, while they are separated from the physical body. Psyche and a spirit body are capable of an infinite velocity or teleportation because they have little or no mass. The capabilities of Psyche and a Spirit Body seem to be limitless.

However, while contained within or married to a physical body, a spirit body and a psyche are also limited to a short De Broglie Wavelength. They have to temporarily separate from the entropic physical body in order to experience a longer De Broglie Wavelength and omnipresence.

God deliberately and knowingly designed physical matter to significantly limit, inhibit, restrain, and block the effects of Quantum Tunneling, Quantum Teleportation, Quantum Omniscience, and Quantum Omnipresence. Physical matter has been given a short De Broglie Wavelength, which means that it has locality or presence, which means that it has physical limitations and a well-defined position or location in space-time. Those physical limitations or physical laws show up in Classical Physics, the Laws of Thermodynamics, and the Theory of Relativity, which state that it is physically impossible to accelerate physical matter faster than the speed-of-light; but, those physical limitations don't apply to the quantum realm or the spirit world.

Violation of relativity, teleportation, quantum tunneling, quantum omnipresence, omniscience, and faster-than-light travel seem to be an inherent and natural part of the quantum realm or psyche realm. In contrast, physical matter and physical laws greatly reduce quantum uncertainty thereby significantly increasing predictability, dependability, stability, and control. The result of all this is the Ultimate Consensus Reality, a physical reality.

Furthermore, as the designer and creator of these physical laws and physical limitations, God granted himself an exemption allowing Him to continue to teleport or quantum tunnel His physical body instantaneously anywhere in the physical universe that He wants to go. God holds the KEYS to Quantum Tunneling or the teleportation of physical matter; and for obvious reasons, God only gives those abilities or KEYS to those whom He trusts. God also continues to experience Quantum Omnipresence and Quantum Omniscience. God's Psyche and Spirit is capable of being everywhere at once and capable of multitasking. The same reality apparently applies to His resurrected physical body as well.

If our Psyche is natively omnipresent and omniscient, then why in the universe would we ever allow ourselves to be limited by a spirit body or a physical body? We've answered some of this already, but it's still a great deal of fun to think about.

As I see it, the answer lies in Order, Organization, Standardization, Structure, and Consensus. Limitations, rules, and laws open up new possibilities. Spirit body limitations and physical body limitations allow us to experience gender and sexual activity, which really isn't possible if you were to remain as a pinprick of light or a Psyche. A physical reality is the Ultimate Consensus Reality. I can depend on my dog, car, wife, and house being there after I get home from work. In a physical reality, I can depend upon this essay being there on my hard drive and in the cloud when I go looking for it tomorrow. None of that would be possible if I were to have chosen to remain as a Psyche or a pinprick of light. It's hard to interact with physical objects if you have no hands. Limitations, structure, order, commandments, or laws provide new opportunities. A physical body does have its advantages.

God deliberately designed physical matter to greatly limit your quantum mechanical nature and properties. A physical reality is the Ultimate Consensus Reality, meaning that it is highly stable, predictable, controllable, reliable, and dependable. You can actually do science in a physical reality. God deliberately designed physical matter and this physical reality to greatly limit, inhibit, control, restrain, and contain your innate quantum capabilities. It's necessary to provide the order, stability, and predictability of this physical realm. In a physical realm, our actions and choices actually have serious consequences. It's REAL because it hurts. It's all theory, until after it has been experienced and observed. The physical makes it real.

God created this physical universe and your physical body to keep your spirit body and your psyche mostly where you currently are – localized here in this physical 3D space-time realm; but, a physical realm is NOT our original innate state of existence. Transdimensional Physics or Quantum Mechanics is the NORM; and, a physical reality is a highly limited and restricted sub-set of that NORM.

Again, according to Quantum Mechanics, you (your psyche or spirit) exist everywhere in the universe simultaneously. God designed this physical universe and your physical body as a school-ground to keep your spirit or soul caged or penned mostly within your physical body so that you can learn from the experience what it is like to have limitations.

Why is this important?

It's because limitations are the ONLY way to provide us with order, structure, predictability, control, law, stability, organization, creativity, and some type of physical life or physical existence. The Ultimate Consensus Reality, a physical reality, is ONLY possible through limitations, structure, order, organization, laws, and agreed upon restrictions. A consensus reality is the opposite of random chaos and omnipresence. A single physical reality is the opposite of infinite possibilities. We learn best and learn fastest while in a physical body. The goal is to learn self-control and responsibility – to learn how to limit, restrain, direct, respect, and choose our actions and behaviors.

Imagine what your physical life would be like without the restraints being imposed by God and His physical laws. Without restrictions such as entropy, space-time, locality, and the theory of relativity, the atoms within your physical body could quantum tunnel away on you at will, before long leaving you without a physical body. How would you feel if your physical body were to quantum tunnel away on you one atom at a time as you slowly dissolve into thin air? Your physical body has the innate capability to quantum tunnel away on you; but, God prevents it from doing so by the physical laws or short De Broglie Wavelength that He imposes on your physical body.

How would you feel if your hard drive, or your car, or your house, or your spouse were to quantum tunnel away from you one atom at a time? If such a thing were to happen as a regular part of our lives, then this physical reality would no longer be the Ultimate Consensus Reality. God uses the physical laws or the physical limitations of this physical universe to prevent this from happening to you.

Do you see why this is important? Do you see now why the commandments of God, rules, laws, limitations, orders, and restraints are important to you and your physical body?

I find it fascinating to try to figure out how things really work at every level of existence or reality, and why they are made to work that way.

Transdimensional Physics Has NO Physical Limitations

Transdimensional means non-physical, non-local, and not located in our Physical Consensus Reality. Transdimensional means not located in our local physical 3D space-time realm. Transdimensional Physics is Quantum Mechanics – the way that matter ordinarily "acts" in the complete absence of physical laws to restrain it and constrain it.

Remember, physical matter was designed to limit and restrain the effects of Quantum Mechanics. Physical matter was designed by God to localize Action at a Distance or to shorten the De Broglie Wavelength. Physical matter was designed and created to form the Ultimate Consensus Reality – reliable, predictable, dependable, and controllable.

In its original native state, spirit matter or dark matter or quantum matter had full access to all aspects of Transdimensional Physics or Quantum Mechanics. Physical matter retains those same innate capabilities. So, who or what prevents the atoms in your physical body from quantum tunneling away on you? Who or what prevents you from dissolving into thin air?

It's God's physical laws, God's physical restraints, God's physical limitations, or God's control over physical matter that keeps your physical body from quantum tunneling away on you one atom at a time to the nether reaches of the universe. It's that short De Broglie Wavelength which God has placed upon your physical body that keeps it from quantum tunneling away from you at will.

By creating the Ultimate Consensus Reality, a physical reality, God uses the physical laws to dampen, restrain, and limit the Quantum Mechanics or the Transdimensional Physics that are an innate part of all matter. By converting spirit matter into physical matter, God infuses space-time, locality, entropy, physical restraints, physical limitations, a short De Broglie Wavelength, and physical laws into the newly organized physical matter thereby greatly increasing the improbabilities or decreasing the probabilities that the physical matter will teleport away on you one atom at a time.

Quantum Mechanics or Transdimensional Physics is an innate capability of ALL matter. The Physical Laws or Classical Physics greatly dampen, restrain, limit, and minimize the quantum mechanisms that are an innate part of all matter.

So, who or what is preventing your car, your home, your wife, your computer, and your dog from quantum tunneling away on you one atom at a time? It's God, the Physical Laws, and the short De Broglie Wavelength that are dampening, limiting, and restricting the quantum laws or transdimensional physics so as to provide us with the Ultimate Consensus Reality, a physical reality.

A physical reality was designed by God to be reliable, predictable, and controllable. Once again, imagine what would happen to you and your life if the atoms in your hard drive, your physical body, your car, and your home retained the ability to quantum tunnel or teleport at will anywhere in the universe that they wanted to go. The Physical Laws that God put into place prevents them from doing so.

Nevertheless, the physical atoms and physical molecules do indeed retain their innate ability to quantum tunnel or teleport; and, these physical particles can and do quantum tunnel whenever and wherever God permits them to do so.

Fascinating, is it not?

What would it be like if God were to remove the physical limitations or physical restraints from your physical body, yet leave you here on this earth? What would it be like if God were to give you complete psychic control over your own physical body? What would your life be like? What would you be able to do?

Obviously, you would be able to teleport your physical body anywhere you wanted it to go.

What else would you be able to do?

Your life would be the ultimate in mind-over-matter. You would be able to heal your physical body at will.

You would be able to phase-shift or walk through walls.

You would be able to levitate or walk on water.

If you had complete psychic control over every atom in your physical body, you would be indestructible. They wouldn't be able to kill you. If they were to surprise you and blow you up, you could just reassemble yourself. Theoretically, you should even be able to force the atoms in your physical body to stay in place or stay coherent even while you are standing in the middle of a nuclear explosion or in the middle of the sun.

You should be able to turn off all pain signals within your physical body.

You could even let people kill you, and then reassemble yourself and teleport out of there later on if you wanted to do so.

Are there human beings who have demonstrated or experienced any of these quantum capabilities or transdimensional capabilities?

Yes, indeed there are.

My best friend has experienced phase-shifting. An elk passed through his truck unphased. Either God wanted my friend to live, or God wanted that elk to live. His friend also experienced phase-shifting – a whole car and family passed through their car and his family during their vacation. Obviously, God wanted those families to live for whatever reason.

I have experienced quantum tunneling or teleportation. I and my mother's car were teleported to safety. That was an interesting experience, let me tell you. It was nothing that I did. It was something that God did for me in order to save me. Obviously, in the process of saving me from physical harm, God also taught me that it's possible for Him to quantum tunnel physical bodies and physical cars at will.

Both my best friend and I have experienced that "bubble of protection" wherein God makes you an indestructible super-hero for a few seconds.

The Bubble of Protection, Quantum Tunneling, and Teleportation:

https://www.youtube.com/watch?v=DmBnCTuQUOc

This stuff sounds like science fiction; but, thanks to Quantum Mechanics or Transdimensional Physics, it's very real indeed. Quantum Mechanics is the best-proven and most-used science that we have. It's the Materialism, Naturalism, Nihilism, and Atheism that are the science fiction in all of this – whether we realize it or not.

Quantum Weirdness

So, ask yourself, "Do you want the truth, or do you want a convenient fiction? Do you want infinite explanatory power where your science is concerned, or do you prefer to be kept in ignorance and darkness where your science is concerned?" This is the million-dollar question.

When I was a Nihilist and Atheist, I wanted the convenient fiction. I wanted to die and cease to exist. I wanted annihilation. I was a sinner, and I was in a very bad situation. You can't in good conscience end your life when you believe that God exists, and you believe that there is going to be some kind of afterlife or judgment. I desperately needed the convenient fiction at the time; and, that's precisely what I received.

We Materialists, Naturalists, Darwinists, Nihilists, and Atheists call it "quantum weirdness" or "spooky action at a distance". But, there's nothing weird or strange about it once a person chooses to interpret Quantum Mechanics from a spiritual perspective or a non-physical perspective.

Quantum Mechanics from a Non-Physical Spiritual Perspective

https://www.amazon.com/dp/B01J023TGU

https://www.amazon.com/dp/1521132380

Nevertheless, I used to be a Materialist, Naturalist, Nihilist, and Atheist back in 2012. God let me experience that and be that so that I would know what it is and what it is like. At the time, when it came to Transdimensional Physics or Quantum Mechanics, I wasn't seeing it, and I wasn't having it. I didn't believe it. I didn't want it. More than anything else, I didn't want God to exist, and I didn't want there to be an afterlife. I truly wanted to die and cease to exist. I wanted annihilation. That is what I wanted most. And, in a very real sense, I got my wish. That person or personality eventually died and ceased to exist.

Boy, am I seeing things differently now! I can see now that the explanatory power of Transdimensional Physics is through the roof. The explanatory power of Quantum Mechanics is limitless and infinite. It explains everything. There's nothing that you can't explain when you finally choose to bring Quantum Mechanics, Psyche, and Syntropy into the mix.

Eventually, you realize that Quantum Mechanisms or Supernatural Mechanisms have been experienced and observed in real life by real people. They have been caught in the act. They are real; and, they are true. I've even experienced them; but, I didn't know what they were at the time, so I tried to explain them away as a figment of my imagination or thought that maybe I had passed out or some such.

I experienced Quantum Teleportation of my physical body and the car that I was driving. One instant I was traveling 45 mph and about to broadside a car that had pulled out in front of me; and, the next instant I was 30 feet away in the middle of a lawn, with the car completely stopped and the engine turned off. I felt time stop; and, I have NO memory of the intervening travel. It's as if I had passed out and woke up in the middle of the lawn with the car stopped and the engine turned off. I had experienced Quantum Tunneling or Teleportation; but, I didn't know what it was at the time. I explained it away as having passed out, because I didn't have a scientific explanation for it. For all I know, I might have experienced phase-shifting as well and actually passed through that other car unphased. I remember hitting the brakes; but, I have no memory of how I got into the middle of the lawn.

To physical mortal beings like us, Quantum Mechanisms or Supernatural Mechanisms seem like magic or miracles, but only because God uses physical laws or our physical consensus reality to deliberately limit and restrict the normal and natural effects of Quantum Mechanics or Transdimensional Physics while we are here in this physical realm. Physical matter and classical physics put a deliberate damper on Quantum Mechanisms or Supernatural Mechanisms or Spiritual Mechanisms. The physical limitations are real, and they truly have an impact because they were designed to be that way. The physical makes everything seem real, and the physical makes everything dependable and reliable and predictable. When it comes to the physical, cause and effect truly come into play and are obviously real.

However, God doesn't apply the same physical limitations to Himself.

Jesus Christ has been observed walking on water, levitating into the air, teleporting, and walking through walls with His physical body. These are all quantum mechanical phenomena or transdimensional phenomena. Remove the physical limitations, and even physical matter can be made to do these quantum mechanical functions. Quantum

Mechanics and Quantum Tunneling remain an innate part of physical matter, although God typically prevents them from happening as a regular every-day part of our lives.

I'm not going to apologize for finding Quantum Mechanics or Transdimensional Physics interesting. It is science after all, or it should be. It has been experienced and observed. Quantum Mechanics is found all throughout the Bible. Transdimensional Physics has been caught in the act.

Luke 4: 28-30 New International Version (NIV):

All the people in the synagogue were furious when they heard this. They got up, drove him out of the town, and took him to the brow of the hill on which the town was built, in order to throw him off the cliff. But he walked right through the crowd and went on his way.

Jesus phase-shifted and/or teleported to safety. This really happened. It was experienced and observed.

Luke 4: 28-30 King James Version (KJV):

And all they in the synagogue, when they heard these things, were filled with wrath, and rose up, and thrust him out of the city, and led him unto the brow of the hill whereon their city was built, that they might cast him down headlong. But he, passing through the midst of them, went his way.

While here in mortality, Jesus Christ had the ability to phase-shift and teleport. Luke was a physician or a medical doctor, which is one of the reasons why he went out of his way to document the eye-witness accounts of all this quantum mechanical stuff that Jesus could do while here in mortality and after He rose from the dead.

Luke 24 New International Version (NIV)

Jesus Has Risen

24 On the first day of the week, very early in the morning, the women took the spices they had prepared and went to the tomb. 2 They found the stone rolled away from the tomb, 3 but when they entered, they did not find the body of the Lord Jesus. 4 While they were wondering about this, suddenly two men in clothes that gleamed like lightning stood beside them. 5 In their fright the women bowed down with their faces to the ground, but the men said to them, "Why do you look for the living among the dead? 6 He is not here; he has risen! Remember how he told you, while he was still with you in Galilee: 7 'The Son of Man must be delivered over to the hands of sinners, be crucified and on the third day be raised again.'" 8 Then they remembered his words.

9 When they came back from the tomb, they told all these things to the Eleven and to all the others. 10 It was Mary Magdalene, Joanna, Mary the mother of James, and the others with them who told this to the apostles. 11 But they did not believe the women, because their words seemed to them like nonsense. 12 Peter, however, got up and ran to the tomb. Bending over, he saw the strips of linen lying by themselves, and he went away, wondering to himself what had happened.

On the Road to Emmaus

13 Now that same day two of them were going to a village called Emmaus, about seven miles from Jerusalem. 14 They were talking with each other about everything that had happened. 15 As they talked and discussed these things with each other, Jesus himself came up and walked along with them; 16 but they were kept from recognizing him.

17 He asked them, "What are you discussing together as you walk along?"

They stood still, their faces downcast. 18 One of them, named Cleopas, asked him, "Are you the only one visiting Jerusalem who does not know the things that have happened there in these days?"

19 "What things?" he asked.

"About Jesus of Nazareth," they replied. "He was a prophet, powerful in word and deed before God and all the people. 20 The chief priests and our rulers handed him over to be sentenced to death, and they crucified him; 21 but we had hoped that he was the one who was going to redeem Israel. And what is more, it is the third day since all this took place. 22 In addition, some of our women amazed us. They went to the tomb early this morning 23 but didn't find his body. They came and told us that they had seen a vision of angels, who said he was alive. 24 Then some of our companions went to the tomb and found it just as the women had said, but they did not see Jesus."

25 He said to them, "How foolish you are, and how slow to believe all that the prophets have spoken! 26 Did not the Messiah have to suffer these things and then enter his glory?" 27 And beginning with Moses and all the Prophets, he explained to them what was said in all the Scriptures concerning himself.

28 As they approached the village to which they were going, Jesus continued on as if he were going farther. 29 But they urged him strongly, "Stay with us, for it is nearly evening; the day is almost over." So, he went in to stay with them.

30 When he was at the table with them, he took bread, gave thanks, broke it, and began to give it to them. 31 Then their eyes were opened, and they recognized him, and he disappeared from their sight. 32 They asked each other, "Were not our hearts burning within us while he talked with us on the road and opened the Scriptures to us?"

33 They got up and returned at once to Jerusalem. There they found the Eleven and those with them, assembled together 34 and saying, "It is true! The Lord has risen and has appeared to Simon." 35 Then the two told what had happened on the way, and how Jesus was recognized by them when he broke the bread.

Jesus Appears to the Disciples

36 While they were still talking about this, Jesus himself stood among them and said to them, "Peace be with you."

37 They were startled and frightened, thinking they saw a ghost. 38 He said to them, "Why are you troubled, and why do doubts rise in your

minds? 39 Look at my hands and my feet. It is I myself! Touch me and see; a ghost does not have flesh and bones, as you see I have."

40 When he had said this, he showed them his hands and feet. 41 And while they still did not believe it because of joy and amazement, he asked them, "Do you have anything here to eat?" 42 They gave him a piece of broiled fish, 43 and he took it and ate it in their presence.

44 He said to them, "This is what I told you while I was still with you: Everything must be fulfilled that is written about me in the Law of Moses, the Prophets and the Psalms."

45 Then he opened their minds, so they could understand the Scriptures. 46 He told them, "This is what is written: The Messiah will suffer and rise from the dead on the third day, 47 and repentance for the forgiveness of sins will be preached in his name to all nations, beginning at Jerusalem. 48 You are witnesses of these things. 49 I am going to send you what my Father has promised; but stay in the city until you have been clothed with power from on high."

The Ascension of Jesus

50 When he had led them out to the vicinity of Bethany, he lifted up his hands and blessed them. 51 While he was blessing them, he left them and was taken up into heaven. 52 Then they worshiped him and returned to Jerusalem with great joy. 53 And they stayed continually at the temple, praising God.

If you had complete control over every atom in your physical body, you could raise yourself from the dead at will.

John 20:19-20 New Living Translation (NLT)

Jesus Appears to His Disciples

19 That Sunday evening the disciples were meeting behind locked doors because they were afraid of the Jewish leaders. Suddenly, Jesus was standing there among them! "Peace be with you," he said.

20 As he spoke, he showed them the wounds in his hands and his side. They were filled with joy when they saw the Lord!

This required some kind of phase-shifting and teleportation.

Transdimensional Physics in Action

John the Beloved was boiled in oil and didn't die.

This event that I will now recount is not found in the Bible, but a church writer named Tertullian makes mention of it in the 36th chapter of a book that he authored ... But, it appears that Jesus had plans for the John that was placed into a vat of boiling oil, and when God decides that you will not die, then you simply will not die!

If God decides that you are not going to die, then you are not going to die.

John was allegedly banished by the Roman authorities to the Greek island of Patmos, where, according to tradition, he wrote the Book of Revelation. According to Tertullian (in The Prescription of Heretics) John was banished (presumably to Patmos) after being plunged into boiling oil in Rome and suffering nothing from it.

That's physically impossible!

But, through Quantum Mechanics or Transdimensional Physics, the Human Psyche and God's Psyche can do the physically impossible.

John the Revelator and the Three Nephites were given complete psychic control over the physical atoms within their physical bodies.

3 Nephi 28: 1-40:

And it came to pass when Jesus had said these words, he spake unto his disciples, one by one, saying unto them: What is it that ye desire of me, after that I am gone to the Father? And they all spake, save it were three, saying: We desire that after we have lived unto the age of man, that our ministry, wherein thou hast called us, may have an end, that we may speedily come unto thee in thy kingdom. And he said unto them: Blessed are ye because ye desired this thing of me; therefore, after that ye are seventy and two years old ye shall come unto me in my kingdom; and with me ye shall find rest.

And when he had spoken unto them, he turned himself unto the three, and said unto them: What will ye that I should do unto you, when I am gone unto the Father? And they sorrowed in their hearts, for they durst not speak unto him the thing which they desired. And he said unto them: Behold, I know your thoughts, and ye have desired the thing which John, my beloved, who was with me in my ministry, before that I was lifted up by the Jews, desired of me. Therefore, more blessed are ye, for ye shall never taste of death; but ye shall live to behold all the doings of the Father unto the children of men, even until all things shall be fulfilled according to the will of the Father, when I shall come in my glory with the powers of heaven. And ye shall never endure the pains of death; but when I shall come in my glory ye shall be changed in the twinkling of an eye from mortality to immortality; and then shall ye be blessed in the kingdom of my Father. And again, ye shall not have pain while ye shall dwell in the flesh, neither sorrow save it be for the sins of the world; and all this will I do because of the thing which ye have desired of me, for ye have desired that ye might bring the souls of men unto me, while the world shall stand. And for this cause ye shall have fulness of joy; and ye shall sit down in the kingdom of my Father; yea, your joy shall be full, even as the Father hath given me fulness of joy; and ye shall be even as I am, and I am even as the Father; and the Father and I are one; and the Holy Ghost beareth record of the Father and me; and the Father giveth the Holy Ghost unto the children of men, because of me. And it came to pass that when Jesus had spoken these words, he touched

every one of them with his finger save it were the three who were to tarry, and then he departed. And behold, the heavens were opened, and they were caught up into heaven, and saw and heard unspeakable things.

And it was forbidden them that they should utter; neither was it given unto them power that they could utter the things which they saw and heard; and whether they were in the body or out of the body, they could not tell; for it did seem unto them like a transfiguration of them, that they were changed from this body of flesh into an immortal state, that they could behold the things of God. But it came to pass that they did again minister upon the face of the earth; nevertheless, they did not minister of the things which they had heard and seen, because of the commandment which was given them in heaven.

And now, whether they were mortal or immortal, from the day of their transfiguration, I know not; but this much I know, according to the record which hath been given — they did go forth upon the face of the land, and did minister unto all the people, uniting as many to the church as would believe in their preaching; baptizing them, and as many as were baptized did receive the Holy Ghost. And they were cast into prison by them who did not belong to the church. And the prisons could not hold them, for they were rent in twain. And they were cast down into the earth; but they did smite the earth with the word of God, insomuch that by his power they were delivered out of the depths of the earth; and therefore, they could not dig pits sufficient to hold them. And thrice they were cast into a furnace and received no harm. And twice were they cast into a den of wild beasts; and behold they did play with the beasts as a child with a suckling lamb and received no harm. And it came to pass that thus they did go forth among all the people of Nephi and did preach the gospel of Christ unto all people upon the face of the land; and they were converted unto the Lord, and were united unto the church of Christ, and thus the people of that generation were blessed, according to the word of Jesus. And now I, Mormon, make an end of speaking concerning these things for a time.

Behold, I was about to write the names of those who were never to taste of death, but the Lord forbade; therefore, I write them not, for they are hid from the world. But behold, I have seen them, and they have ministered unto me. And behold they will be among the Gentiles, and the Gentiles shall know them not. They will also be among the Jews, and the Jews shall know them not. And it shall come to pass, when the Lord seeth fit in his wisdom that they shall minister unto all the scattered tribes of Israel, and unto all nations, kindreds, tongues and people, and shall bring out of them unto Jesus many souls, that their desire may be fulfilled, and also because of the convincing power of God which is in them. And they are as the angels of God, and if they shall pray unto the Father in the name of Jesus they can show themselves unto whatsoever man it seemeth them good. Therefore, great and marvelous works shall be wrought by them, before the great and coming day when all people must surely stand before the judgment-seat of Christ; yea even among the Gentiles shall there be a great and marvelous work wrought by them, before that judgment day. And if ye had all the scriptures which give an account of all the marvelous works of Christ, ye would, according to the words of Christ, know that these things must surely come. And wo be unto him that will not hearken unto the words of Jesus, and also to them whom he hath chosen and sent among them; for whoso receiveth not the words of Jesus and the words of those whom he hath sent receiveth not him; and therefore, he will not receive them at the last day; and it would be better for them if they had not been born. For do ye suppose that ye can get rid of the justice of an offended God, who hath been trampled under feet of men, that thereby salvation might come?

And now behold, as I spake concerning those whom the Lord hath chosen, yea, even three who were caught up into the heavens, that I knew not whether they were cleansed from mortality to immortality — but behold, since I wrote, I have inquired of the Lord, and he hath made it manifest unto me that there must needs be a change wrought upon their bodies, or else it needs be that they must taste of death; therefore, that they might not taste of death there was a change wrought upon their bodies, that they might not suffer pain nor sorrow save it were for the sins of the world. Now this change was not equal to that which shall take place at the last day; but there was a change wrought upon them, insomuch that Satan could have no power over them, that he could not tempt them; and they were sanctified in the flesh, that they were holy, and that the powers of the earth could not hold them. And in this state, they were to remain until the judgment day of Christ; and at that day they were to receive a greater change, and to be received into the kingdom of the Father to go no more out, but to dwell with God eternally in the heavens.

God is Syntropy. If God decides that you are not going to die, then you are not going to die.

These three were transfigured, or translated, or turned into seraphim. They can't be killed. Many of their capabilities and powers are discussed. Their capabilities are quantum mechanical, supernatural, and transdimensional in nature and origin. They seem to violate the physical laws or trump the physical laws, because God permits them to do so.

As a race, we are never going to master teleportation or quantum tunneling unless God permits us to do so. We are never going to be able to travel faster than the speed-of-light unless God permits us to do so. The physical laws that God put into place prevents us from doing so against His will.

Nevertheless, it has been observed that ALL of these quantum mechanical capabilities return to us when our spirits leave their physical body behind. A spirit or a ghost is a holistic combination of both psyche and spirit body. Psyches and spirit bodies function and exist according to the rules of transdimensional physics or quantum mechanics. In other words, spirit bodies and psyches have NO physical limitations whatsoever.

Out-of-Body Travelers and Near-Death Experiencers have observed that their psyche and their spirit body can teleport or quantum tunnel anywhere at will. At the quantum level or the psyche level, we are dealing with Transdimensional Physics or Quantum Mechanics, and not Physical Limitations or Physical Laws.

Powerful, is it not?

God is Syntropy. If God decides that you are not going to die, then you are not going to die. If God wants to teleport you to safety, He can, and He will. If God wants to translate you or transfigure you, He can, and He will. God can do, and He will do, whatever He needs to do and wants to do in order to accomplish His purposes and His goals. God has NO physical limitations. In their native original state, transdimensional physics, quantum mechanisms, psyche, and spirit matter have NO physical limitations because they are pure syntropy.

God gives certain particles of spirit matter physical limitations thereby converting those particles of spirit matter into physical matter. The physical matter is then used to create or organize the Ultimate Consensus Reality, a physical reality. A physical reality is predictable, controllable, reliable, and dependable. A physical reality is the ultimate in order, law, and organization. A physical reality is powerful because it's limited, ordered, organized, contained, controlled, predictable, dependable, and lawful.

A physical reality, being the Ultimate Consensus Reality, ends up being the opposite of chaos. That's its advantage, and that's also its limitation. Remember, when it comes to a physical reality, its limitations are its advantage. The atoms in your physical body aren't going to quantum tunnel away on you willy-nilly while they are subject to the physical limitations of God's physical laws. The physical limitations or God's Physical Laws make our physical bodies possible, reliable, dependable, controllable, and predictable.

Mark My Words

—

Source

Quantum Mechanics from a Non-Physical Spiritual Perspective

https://www.amazon.com/dp/B01J023TGU

https://www.amazon.com/dp/1521132380

Putting Psyche Back into Psychology: Restoring Science to Consciousness

https://www.amazon.com/dp/B071NC987S

NATURE vs. NURTURE vs. NIRVANA: An Introduction to Reality

https://www.amazon.com/dp/B01JWRCSVA

https://www.amazon.com/dp/1521132615

References

God Is in the Light: God is light, and in Him is no darkness at all.

https://www.amazon.com/dp/B07168S37N

Quantum Mechanics from a Non-Physical Spiritual Perspective.

https://www.amazon.com/dp/B01J023TGU

https://www.amazon.com/dp/1521132380

Quantum Neuroscience: The Answer to Life, the Universe, and Everything.

https://www.amazon.com/dp/B079Z6QQQB

MY FAVORITE SCIENTIFIC DISCOVERIES

Introduction

I have been a scientist all of my life; and after 55 years of life, I finally found what I have been searching for all of my life – an explanation for how everything works at all levels of existence or reality.

I found the answer to life, the universe, and everything.

I found what I was looking for in Quantum Mechanics, Action at a Distance, Psyche, or Syntropy. It explains everything!

As a result of these discoveries, I upgraded my science.

I have made some fascinating discoveries during my science career. Ironically, most of what I have discovered has already been discovered by someone else before me; but, I chose to do something unique that has never been done before. I chose to give each scientific discovery a quantum mechanical explanation, a psychic explanation, or a spiritual explanation. I chose to take Quantum Mechanics and Psyche seriously, which most scientists have never done before. Consequently, it makes it seem as if I have made a wide variety of different scientific discoveries that have never been discovered before when in fact I have not. I've simply discovered what others have chosen to ignore and reject.

Even Psyche or Intelligence has been experienced and observed, which means that it has already been discovered; but, I chose to do something different that hasn't been done before. I chose to include it in Science; and, I chose to try to define it and explain what it is and how it works.

Psyche is the innate intelligence within all the different forms of energy that have ever been discovered or observed. The psyche within energy gives all the different forms of energy the ability to hear God's commands, understand God's commands, and obey God's commands. Psyche is the "organizing force" or the "law following force" within our universe. Psyche is identified by choice, particularly its choice to obey the quantum laws and the physical laws.

Physical matter, spirit matter, and dark matter are simply different forms of organized energy existing simultaneously in different phases, different frequencies, or different dimensions. They each have a psyche within them that gives them the ability to obey God's Laws. God provides the Laws; and, the psyche within the different forms of energy provides the obedience to those Laws. Matter is organized energy or lawful energy. Someone Psyche within that matter is choosing to obey the Quantum Laws and the Physical Laws; otherwise, it would be nothing but random chaos.

A LAW in order to be a Law requires some kind of intelligent Law Giver and Law Enforcer. Laws are worthless without some type of uniformity, order, and enforcement. The thing that most people don't realize is that Laws also require Someone Psyche with enough intelligence and understanding to actually choose to obey those Laws. Without Someone Psyche or someone intelligent enough to function as the Law Follower or the Law Obeyer, Laws are also completely worthless. Laws are pointless unless there is someone intelligent enough, within the different forms of energy, who is capable of actually choosing to obey those Laws. This is logical common sense. I have a feeling that it has all been discovered before; but, nobody ever takes the time to actually look at everything from the perspective of Psyche.

Psyche is the innate intelligence, within all the different forms of energy, which gives those different forms of energy the ability to obey God's Law and God's Commands.

Science 2.0: Redefining Science

Most scientists in this world define "science" as Materialism, Naturalism, Physicalism, Physical Reductionism, Darwinism, Nihilism, Behaviorism, Determinism, Classical Physics, Entropy, and Atheism. Such an act is called *begging the question* or *jumping to conclusions*, which is a logic fallacy.

Materialism, Physicalism, Darwinism, Behaviorism, and Determinism were carefully and purposefully designed to convince people that God and the Supernatural do not exist. They were designed to convince people that Action at a Distance or Quantum Mechanics does not exist. Materialism, Naturalism, and their derivatives were designed to exclude and eliminate any and all evidence demonstrating or proving that Psyche or Syntropy exists.

We'll never be able to use Materialism and Naturalism to prove that God exists or to prove that Action at a Distance exists because these philosophical assumptions start with the pre-chosen conclusion that God, Psyche, Action at a Distance, and Quantum Mechanics do not exist.

I used to be a Materialist, Naturalist, Nihilist, and Atheist; so, I know of what I speak. I have been there and done that.

How we choose to define things makes all the difference in the world. As long as we continue to define "science" as Materialism and Naturalism, then Science's most basic, most fundamental, and most essential questions will remain forever unanswered and unknown to us. It's unavoidable.

Materialism, Naturalism, Behaviorism, and Determinism have NO scientific explanation for something as simple and basic as choice! They can't explain choice, so they say that it does not exist. That's what they do with everything that they can't explain – they say that it doesn't exist, sweep it under the rug, and pretend that it doesn't exist. That's what they tried to do with Quantum Mechanics. They can't explain it, so they just ignore it. As long as we choose to define "science" as Materialism and Naturalism, then Science's most basic concepts will remain unexplained and unexplainable.

However, if we choose to define "science" as Fine-Tuning, then instantly we have thousands (if not millions) of different Scientific Proofs of God's Existence suddenly spring forth before our eyes as if from nowhere. I KNOW because I have done that as well. Science IS Fine-Tuning; and, Fine-Tuning is an infinitely better definition for Science than Materialism, Naturalism, Darwinism, and Atheism. Whenever we do Science, we are in fact doing Fine-Tuning of some sort, are we not?

How we choose to define things makes all the difference in the universe.

When we choose to define "science" as Fine-Tuning, Science's explanatory power literally goes through the roof! The same thing happens when we choose to define "science" as Quantum Mechanics, Action at a Distance, Intelligence, Psyche, and Syntropy. Instantly, we can explain EVERYTHING that comes our way. Suddenly, nothing is beyond our reach.

Mark My Words

—

Source

Science 2.0: I Upgraded My Science

https://www.amazon.com/dp/B0771K6WTX

1. Science Is Observation and Experience

Possibly my most useful and my most powerful scientific discovery is the fact that Science is Observation and Experience, and NOT Materialism, Naturalism, Darwinism, Nihilism, and Atheism.

Most people define science as Materialism and Naturalism. It's a scam! Materialism, Naturalism, and their derivatives are based exclusively on blind-faith and wishful thinking. There is NO evidence supporting the major premises or hidden assumptions of Materialism and Naturalism, which state that the quantum or the supernatural does not exist. Everything that has been experienced and observed falsifies Materialism, Naturalism, Darwinism, Nihilism, and Atheism.

For me personally, this was a major scientific discovery because I used to be a Materialist, Naturalist, Nihilist, and Atheist.

With this discovery or observation, I experienced a paradigm shift and developed a new and better way of doing science, which I call Science 2.0. Science 2.0 is based upon the observed, verified, experienced, and proven aspects of Quantum Mechanics rather than Materialism and Naturalism.

Quantum Mechanics, or Syntropy, or Psyche is the answer to life, the universe, and everything. Syntropy or Quantum Mechanics falsifies Materialism, Naturalism, and their derivatives.

Science is in desperate need of upgrading because Materialism, Naturalism, Darwinism, Nihilism, and Atheism have NO explanatory power when it comes to Quantum Mechanics, Quantum Waves, Near-Death Experiences, Out-of-Body Experiences, Non-Locality, Non-Physicality, Transdimensionality, Consciousness, Thoughts, Dark Matter or Spirit Matter, Dark Energy, Syntropy, Psyche, Mysticism, Quantum Tunneling, the Quantum Zeno Effect, Action at a Distance, and invisible non-physical forces and fields such as Gravity and Magnetism and the Zero-Point Field of Light. Science needs to be upgraded and brought into the modern Quantum Age. I have attempted to do so in the books that I have written.

One of the most useful things I ever did was to use Quantum Mechanics to upgrade science from Materialism and Naturalism TO Observation and Experience. I upgraded my science to Science 2.0. Rather than defining science as Materialism, Naturalism, Darwinism, Nihilism, and Atheism as most college professors do, I now define science as Observation and Experience. Science 2.0 is based upon Quantum Mechanics or Transdimensional Physics, which has been experienced and observed. Science 2.0 is based upon Phenomenology or the Lived Experiences of the human race.

Unless it has been experienced and observed by Someone Psyche somewhere sometime, it really isn't science. The major premises or hidden assumptions of Materialism,

Naturalism, Darwinism, Nihilism, and Atheism – which claim that the quantum or the supernatural does not exist – have not been and cannot be experienced and observed. Therefore, Materialism and Naturalism are NOT science, because their major premises or hidden assumptions cannot be experienced nor observed but have to be taken on blind faith as being real and true.

The observed, experienced, and verified existence of Quantum Mechanics, Psyche, the Biblical God Jesus Christ, and Action at a Distance FALSIFIES Materialism, Naturalism, Darwinism, Nihilism, Atheism, and their derivatives.

The truth falsifies the false; and, the truth has been repeatedly verified, experienced, and observed. This is good stuff!

Science 2.0 allows all of the evidence into evidence and pursues a preponderance of the evidence. I've even written a few books based upon Science 2.0 and what I learned from it. As a scientist, I've found it very useful. I'm no longer trapped by nor restricted by My Materialism, My Naturalism, My Nihilism, and My Atheism. Upgrading my science to Science 2.0 helped me to overcome them.

Mark My Words

—

Reference

Science 2.0: I Upgraded My Science

https://www.amazon.com/dp/B0771K6WTX

Quantum Neuroscience: The Answer to Life, the Universe, and Everything

https://www.amazon.com/dp/B079Z6QQQB

Syntropy in Defense of Quantum Mechanics: The Answer to Life, the Universe, and Everything

https://www.amazon.com/dp/B07BPT3W8R/

2. Philosophy of Science and Affirming the Consequent

When it comes to the Philosophy of Science, Science, Personality Theory, Psychology, and the Scientific Method, I discovered that studying and learning the difference between *affirming the consequent* and *negating the consequent* is the most interesting and most useful concept that one can study and learn about. It's the concept that paid the most dividends in the end.

All of the Psychologists and Personality Theorists turn to Joseph F. Rychlak for this information because he pioneered Humanism, and he was twice a president of the APA's division of Theoretical and Philosophical Psychology.

Most scientists have no idea how important the Philosophy of Science really is. The Materialists, Naturalists, Darwinists, Nihilists, Behaviorists, Determinists, Physical Reductionists, and Atheists are doing science wrong; and, they don't even know it.

One of the most significant and life-changing scientific discoveries that I made during my research came to me while studying the Philosophy of Science when I was introduced to *affirming the consequent* and *negating the consequent*. It totally changed the way that I look at Science and the way that I practice or do Science.

I even had a college professor who was teaching us how to produce syllogisms that use *affirming the consequent* and *negating the consequent* so that we could learn to see the difference between the two.

Knowledge is power; and, this is powerful stuff.

The following articles were KEY when it came to this scientific discovery.

Rychlak, J. F. (1970). The Human Person in Modern Psychological Science. *British Journal of Medical Psychology*, 43(3), 233–240.

https://mypsyche.us/wp-content/uploads/2016/12/Rychlak.pdf

Slife, B. D. & Williams, R. N. (1995). Science and Human Behavior. In *What's Behind the Research? Discovering Hidden Assumptions in the Behavioral Sciences*, (pp. 167–204). Thousand Oaks, CA: SAGE Publications.

http://mypsyche.us/wp-content/uploads/2017/04/Science.pdf

Gantt, E. (2014). Logical Arguments. In *Psychology 353 – LDS Perspectives in Psychology*, (pp. 8-11). Provo, UT: Brigham Young University.

https://philosophy-of-science.com/wp-content/uploads/2018/04/Logical-Arguments.pdf

Gantt, E. (2014). Leveling the Playing Field – Why Science is Not a Trump Card. In *Psychology 353 – LDS Perspectives in Psychology*, (pp. 50-58). Provo, UT: Brigham Young University.

https://philosophy-of-science.com/wp-content/uploads/2018/04/Verification-vs-Falsification.pdf

You see, the way we choose to define science and the way we choose to do science makes all the difference in the world.

Knowledge of *affirming the consequent*, *negating the consequent*, and the Philosophy of Science led me to redefine Science and to remake Science.

Mark My Words

Redefining Science

Knowledge of *affirming the consequent* and the Philosophy of Science led me to redefine Science.

Using the Scientific Method to Eliminate the Usual Suspects and to Prove the Truth

https://www.amazon.com/dp/B01J6STHP0

NATURE vs. NURTURE vs. NIRVANA: An Introduction to Reality

Quantum Mechanics from a Non-Physical Spiritual Perspective

Science 2.0: I Upgraded My Science

Science 2.0 redefines Science as Observation and Experience. Science 2.0 changes Science so that Science allows ALL of the evidence into evidence and pursues a preponderance of that evidence. Science 2.0 is based upon Phenomenology or the Lived Experiences of the human race and NOT Materialism and Naturalism. I basically redefined Science as Action at a Distance, Fine-Tuning, Quantum Mechanics, Syntropy, Psyche, and Intelligence. Science is what the scientists do. Science depends on the people or the psyches who do the Science. The definition for Science and the way we do science has needed upgrading for centuries. I developed and provided that upgrade so that I could develop new Science. Science has been headed this way for decades. I just completed the process and then started using it to make new and interesting scientific discoveries.

Changing Methodologies

Knowledge of *negating the consequent* and the Philosophy of Science led me to change the way that I do Science.

Quantum Neuroscience: The Answer to Life, the Universe, and Everything

Based upon the principles that I gleaned from *negating the consequent*, the Philosophy of Science, and Science 2.0, I updated Neuroscience and brought it into the Quantum Age or the modern age. Neuroscience is based upon Classical Physics, Materialism, Naturalism, Darwinism, Nihilism, Behaviorism, Determinism, Physical Reductionism, Entropy, and Atheism. Quantum Neuroscience is based upon Syntropy, Quantum Mechanics, Action at a Distance, Psyche, and the Orthodox Interpretation of Quantum Mechanics by Henry P. Stapp. There's a bit of a difference between the two. Materialism and Naturalism deny the existence of Quantum Mechanics, Action at a Distance, Syntropy, and Supernatural Mechanisms; and, the verified and proven existence of Quantum Mechanics, Syntropy, Psyche, and Action at a Distance falsifies Materialism, Naturalism, Darwinism, Nihilism, Behaviorism, Determinism, Physical Reductionism, and even Atheism. The one denies God's existence; and, the other verifies God's existence. I asked myself which one provides us with most of the verified and proven Science. Quantum Mechanics subsumes Classical Physics and Naturalism, meaning that Naturalism and

618

Classical Physics are a small sub-set of Quantum Mechanics. Quantum Mechanics or Supernatural Mechanics gives us everything that Classical Physics gives us, and then it gives us infinitely more. Science 2.0 and Quantum Neuroscience are based upon Quantum Mechanics.

Scientific Proof of God's Existence: A Primer"

https://www.amazon.com/dp/B071713NNL

https://www.amazon.com/dp/1521325170

Scientific Proof of God's Existence: Finding God Where the Atheists Refuse to Look for Him

https://www.amazon.com/dp/B07B26CRHX

Syntropy in Defense of Quantum Mechanics: The Answer to Life, the Universe, and Everything

https://www.amazon.com/dp/B07BPT3W8R/

What Is Affirming the Consequent?

When it comes to the Philosophy of Science, Science, Personality Theory, Psychology, and the Scientific Method, I discovered that studying and learning the difference between *affirming the consequent* and *negating the consequent* is the most interesting and most useful concept that one can study and learn about. It's the concept that paid the most dividends in the end. All of the Psychologists and Personality Theorists turn to Joseph Rychlak for this information because he pioneered Humanism, and he was twice a president of the APA's division of Theoretical and Philosophical Psychology.

Scientific Inference, Scientific Verification, Affirming the Consequent, or *Affirming One's Conclusion* involves *jumping to conclusions,* a *blind leap of faith, circular reasoning,* or *begging the question.* It involves using your Consequent or your Conclusion as evidentiary proof that your Conclusion is true.

When we are first presented with the Theory of Evolution, most of us just instinctively KNOW that it is false. I mean, how many human beings have you seen descending from apes and monkeys in our modern age? It doesn't happen because it can't happen. Macro-Evolution of any kind is prevented from happening by random mutations and genetics. Consequently, the Darwinists and Naturalists turn to *affirming the consequent* in order to convince us that the Theory of Evolution is true, and the ruse works. Millions have fallen for the deception and can't see anything wrong with it.

However, thanks to random mutations and genetics, it's physically impossible for chimp-like ancestors to repeatedly and reliably produce genetically compatible Mr. and Mrs. Mutants at the same place and at the same time year after year for millions of years. It can't be done, which means that it wasn't done. Chimp-like ancestors don't give birth to genetically compatible male and female chimpanzees and human beings. It has never been observed, and it never will be because it is physically impossible. Random mutations and genetics prevent it from happening.

I had to understand why most scientists believe that Materialism, Naturalism, Darwinism, Nihilism, Behaviorism, Determinism, Physical Reductionism, Atheism, and the Theory of Evolution are true. There's got to be a logical reason.

Affirming the consequent, begging the question, circular reasoning, special pleading, hindsight bias, and *confirmation bias* have provided me with that reason. After studying the Philosophy of Science, I could finally see it and understand it. I can see what has been done to us and where we went wrong as scientists.

It all makes sense to me now.

Affirming the consequent is the logic fallacy upon which the Scientific Method, Materialism, Naturalism, Darwinism, and Traditional Science are based.

We can't use Science and the Scientific Methods to verify and know the Truth. Verification doesn't work because it's logically flawed. Scientific Methods cannot be used to produce Knowledge and Truth.

Due to the *affirming the consequent* logic fallacy, and the *jumping to conclusions* logic fallacy, and the *category error* logic fallacies which are built into the scientific methods, it is impossible to use Science and the Scientific Methods to know the truth and to prove the truth.

If your ultimate goal in life is to KNOW THE TRUTH and prove the truth, then Science and the Scientific Methods are in fact one of the worst ways for accomplishing that task.

This is how scientific verification or *affirming the consequent* works in practice.

The following logical argument outlines the basic approach that has been taken by traditional scientists throughout the history of science:

Scientific Hypothesis: If Theory X is true, then we will observe Y.

Scientific Observations: We observe Y.

Scientific Conclusion: Therefore, Theory X is true.

This sort of thinking, however, reflects a logical fallacy called *affirming the consequent*. Here's a comparable example to demonstrate:

We hypothesize: If Sally's pet is a cat, it will have a tail.

We observe: Sally's pet has a tail.

We conclude: Therefore, Sally's pet is a cat.

We can easily see that this logic is fallacious. Just because we observe Y (a pet with a tail) that does not mean that our theory X (the pet is a cat) is true. After all, dogs, lizards, birds, and mice have tails too.

Yet, **this IS the scientific method**, and this is exactly how traditional scientists use the scientific methods to demonstrate and prove the truth. The whole enterprise is based upon a logic fallacy or two.

By *affirming the consequent*, you can prove anything to be true – anything – including Materialism, Naturalism, Darwinism, Nihilism, Atheism, and the Theory of Evolution. By *affirming the consequent*, you can prove that Sally's pet is a cat, even though it is in fact a bird. By *affirming the consequent*, you can prove that the theory of evolution is true, even though it is in fact a dog. Do you see how that works?

620

We hypothesize: If the Theory of Evolution is true, it will match with the fossil record and Darwin's tree of life.

We observe that the Theory of Evolution matches with the fossil record and Darwin's tree of life.

We conclude: Therefore, the Theory of Evolution is true.

This is how they prove that the theory of evolution is true. In fact, I see it on almost every page of my college textbooks where the Evolutionists are simply amazed at how miraculously all that evidence supports the truthfulness and predictions of the Theory of Evolution. These people never realize that they are *affirming the consequent*. They never realize that Darwin's tree of life and the theory of evolution were carefully designed and tailor-made after-the-fact to match with any and all evidence – past, present, and future. But, their arguments are fallacious, and they don't even know it.

Why is this argument fallacious?

It's because some version of Darwin's tree of life can also be made to match with Intelligent Design Theory or Christianity; and, the different types of Christianity and Islam can be made to match with the fossil record as well. The whole thing is based upon *circular reasoning, begging the question*, or *affirming the consequent*. It's a logic fallacy.

Affirming the consequent is how the Materialists, Naturalists, Darwinists, Nihilists, Behaviorists, Determinists, and Atheists use the Scientific Method to prove that the Theory of Evolution is true. They cheat; and, most of them don't even realize that they are doing so, because they have never studied the Philosophy of Science.

This is a good point at which to observe that the logic of empirical study is flawed – not fatally, but in a way that limits the certainty with which our explanation of empirically proven facts can be believed in. We never achieve logical necessity [certainty] in the proof garnered by a scientific experiment. This is because we always commit the logical error that Aristotle pointed out long ago, of *affirming the consequent* of an "If, then" line of argument. Another way of saying this is that the empirical findings act as a predicating meaning for our theory, but there are always going to be other theories that can take meaning from this data array as well. (Rychlak, *Artificial Intelligence and Human Reason: A Teleological Critique*, p. 33-34).

Drawing conclusions or interpreting the scientific data is an integral and essential part of the Scientific Method; but, this is also where the logic errors and the flaws are introduced into the process. I have observed that Creation by Rocks is never the best explanation that can be given to scientific evidence; yet, Creation by Rocks or Materialism is the explanation that is most given to scientific evidence. Interesting, is it not?

I have also observed that there are infinitely better and more believable explanations for scientific evidence than Materialism, or Design and Creation by Rocks. Materialism, or Creation by Rocks, always provides a false interpretation or a false explanation to any data array or set of scientific evidence. That has been my observation once I finally started looking at the empirical evidence and the logic associated with the Scientific Method. Materialism, Naturalism, and Atheism are based upon a refusal to look at contradictory evidence and a refusal to look at any other possible explanation for the scientific evidence. Once I started looking at the logic and the evidence, it was easy to see that Materialism and Naturalism are fatally flawed.

There's a lot of money that can be made telling the Atheists and the Materialists exactly what they want to hear; but, it's dishonest. Like Joseph Rychlak's books, my books go largely unnoticed, because I'm not telling the Materialists and the Atheists what they want to hear; but, I sleep well at night with a clear conscience knowing that I have finally found the truth that I have been searching for all of my life. This stuff is really cool, and it has set me free; but, most people will never see it because they don't want to see it. Such is life.

The fallacy of affirming-the-consequent stipulates the fact that it will always be possible for some other explanation to account for any empirically observed fact pattern. This loss of certainty in validation is not fatal for the scientific method, of course. It has not prevented scientists from curing polio or putting people on the moon. It merely alerts us to the fact that some conceptualizer [psyche] **always has to make a decision as to which fundamental grounding** [or interpretation or explanation] **will be used in the sequence of theory formation and testing. The grounds are never 'out there' in the hard data but 'in here' as assumptive frameworks. If we who theorize appreciate that we will never attain certainty in validating our theories, we will be in a better position to see that alternative groundings that explain such empirical evidence can be complementary. To** *complement* **is to fill out or make up for what is lacking in any theoretical understanding of a subject.** (Rychlak, *In Defense of Human Consciousness*, pp. 18-19).

Notice that there must be a conceptualizer, psyche, interpreter, theorizer, observer, decider, chooser, and assumer behind every theoretical hypothesis and the Scientific Method, or the science experiment will never take place. The grounds for doing science take place 'in here' within our psyche. Contrary to the claims of the Darwinists and the Materialists, you can't place the rocks or raw physical matter into the role of conceptualizer, psyche, theorizer, formal cause, final cause, or ultimate cause. The rocks won't go there and can't do that. Once again, logic and the Scientific Method have proven Materialism and Scientific Naturalism inadequate and false. The Scientific Method can definitely be used to prove things false, which has happened in the case of Materialism and Naturalism thousands of different ways. In fact, the falsification of Materialism and Naturalism is complementary, in that Naturalism and Materialism have been falsified thousands of different ways. We have our fill of evidence demonstrating and proving what is lacking or false in Naturalism and Materialism.

Science can only work through a kind of negating procedure of falsifying claims put to nature by the theorist in question. As scientists, says Popper, we never really verify things but continually falsify – or fail to falsify – claims [theories, hypotheses, etc.] **expressed by some investigator** [recognizing, of course, that serendipitous findings occur as well]**. This is why the scientist always restates his hypothesis into the null form. Ultimately the reason we must falsify has to do with the logical fallacy of 'affirming the consequent' of an 'If** [antecedent] **. . . then** [consequent] **. . .' proposition. We like to think our theory has necessarily been verified, but Popper teaches us that it has not. There will always be, in principle, other ways of accounting for the observed data** [the facts] **than our preferred theory.** (Rychlak, *The Psychology of Rigorous Humanism*, pp. 181-182).

There will always be other ways of accounting for the scientific data than our preferred conclusion, Materialism.

Affirming the consequent or verification makes the Scientific Method unreliable; and, *affirming the consequent* is how they prove that Materialism, Naturalism, Darwinism, Nihilism, Behaviorism, Determinism, Physical Reductionism, Atheism, and the Theory of Evolution are true.

By *affirming the consequent*, you can prove anything to be true.

But, Science can ONLY work right by *negating the consequent*, a process that is used to falsify scientific claims and scientific theories such as Materialism, Naturalism, Darwinism, Nihilism, Atheism, and the Theory of Evolution.

Mark My Words

—

Source Material

The Ultimate Model of Reality: Psyche Is the Ultimate Cause

https://www.amazon.com/dp/B071NC9JK6

BioPsychoSocial: Including Psyche or Light into our Theoretical Models

https://www.amazon.com/dp/B0713NDHVW

Rychlak, J. F. (1970). The Human Person in Modern Psychological Science. *British Journal of Medical Psychology*, 43(3), 233–240.

https://mypsyche.us/wp-content/uploads/2016/12/Rychlak.pdf

Slife, B. D. & Williams, R. N. (1995). Science and Human Behavior. In *What's Behind the Research? Discovering Hidden Assumptions in the Behavioral Sciences*, (pp. 167–204). Thousand Oaks, CA: SAGE Publications.

http://mypsyche.us/wp-content/uploads/2017/04/Science.pdf

Gantt, E. (2014). Logical Arguments. In *Psychology 353 – LDS Perspectives in Psychology*, (pp. 8-11). Provo, UT: Brigham Young University.

https://philosophy-of-science.com/wp-content/uploads/2018/04/Logical-Arguments.pdf

Gantt, E. (2014). Leveling the Playing Field – Why Science is Not a Trump Card. In *Psychology 353 – LDS Perspectives in Psychology*, (pp. 50-58). Provo, UT: Brigham Young University.

https://philosophy-of-science.com/wp-content/uploads/2018/04/Verification-vs-Falsification.pdf

Rychlak, J. F. (1979). *Discovering Free Will and Personal Responsibility*. New York: Oxford University Press.

Rychlak, J. F. (1981a). *A Philosophy of Science for Personality Theory* (2nd ed.). Malabar, FL: Robert E. Krieger Publishing Company.

Rychlak, J. F. (1981b). *Introduction to Personality and Psychotherapy: A Theory-Construction Approach* (2nd ed.). Boston, MA: Houghton Mifflin Company.

Rychlak, J. F. (1982). *Personality and Life-Style of Young Male Managers: A Logical Learning Theory Analysis*. New York: Academic Press.

Rychlak, J. F. (1988). *The Psychology of Rigorous Humanism* (2nd ed.). New York: New York University Press.

Rychlak, J. F. (1991). *Artificial Intelligence and Human Reason: A Teleological Critique*. New York: Colombia University Press.

Rychlak, J. F. (1994). *Logical Learning Theory: A Human Teleology and Its Empirical Support*. Lincoln, NE: Nebraska University Press.

Rychlak, J. F. (1997). *In Defense of Human Consciousness*. Washington DC: American Psychological Association.

Rychlak, J. F. (2003). *The Human Image in Postmodern America*. Washington DC: American Psychological Association.

3. Science and the Scientific Method Falsify Naturalism and Darwinism

For me personally, one of the most interesting and significant scientific discoveries that came my way was the result of studying the Philosophy of Science for a year or more.

I discovered that by *negating the consequent*, the Scientific Method falsifies Materialism, Naturalism, Darwinism, and Atheism. Think about it, when done right, Science and the Scientific Method prove that Materialism, Naturalism, Darwinism, Nihilism, and the Theory of Evolution are false. That's huge! It's a game-changer!

I wonder why nobody has ever thought of it before. I wish I would have known how to falsify Materialism, Naturalism, Darwinism, Atheism, and the Theory of Evolution when I was first in college forty years ago. It would have saved me a lot of trouble, time, and grief.

Materialism, Naturalism, Darwinism, and the Theory of Evolution are DEAD; and now, I can use the Scientific Method to prove that it is so. That's a monumental scientific discovery, in my not so humble opinion. I found that one very satisfying when it was first revealed to me.

This amazing development in the Philosophy of Science is documented in one of my most-ignored books, *Using THE SCIENTIFIC METHOD to Eliminate the Usual Suspects and to Prove the Truth*.

I eventually realized that if you successfully use the Scientific Method to eliminate everything that is false – such as Materialism, Naturalism, Darwinism, Nihilism, Atheism, and the Theory of Evolution – then ONLY the truth will remain. It's amazing to study and observe what remains, after these things have been eliminated from science. Their opposite remains. The truth remains. Quantum Mechanics or Supernatural Mechanics remains. Psyche or Quantum Non-Local Consciousness remains. Intelligence remains.

Logic or common sense also FALSIFIES Materialism, Naturalism, Darwinism, Nihilism, and Atheism.

Materialism is the claim that the quantum or the non-physical does not exist. I never found any evidence that verifies or confirms that claim. Nobody has. Instead, the verified and proven existence of Quantum Mechanics and Action at a Distance falsifies that claim. The verified and proven existence of Dark Energy, Dark Matter or Spirit Matter, Magnetism, Gravity, Thoughts, and Light falsifies Materialism or Physicalism.

Naturalism is the claim that the quantum or the supernatural does not exist. I never found any evidence that verifies or confirms that claim. Nobody has. Instead, the verified and proven existence of Quantum Mechanics and Action at a Distance falsifies that claim. Action at a Distance is Supernatural. Quantum Mechanics is Supernatural. The verified and proven existence of Quantum Mechanics falsifies Naturalism.

Nihilism is based upon the claim that an after-life or spirit world does not exist. I've never found any evidence that verifies or confirms that claim. Nobody has. Instead, the veridical experiences and observations from Near-Death Experiences (NDEs), Out-of-Body Experiences (OBEs), Shared-Death Experiences (SDEs), and our after-death Life Reviews falsify that claim. Millions of different people have had an NDE or some type of OBE. The repeatedly verified and proven existence of NDEs falsifies Nihilism. We are going somewhere after all, when we die. Our thoughts and memories and personality continue to exist after our physical brain is dead and gone. That's what all the evidence is telling us. Nihilism is also the philosophical belief that the Psyche, Spirit, or Soul does not exist and that life, therefore, has no meaning or purpose. I never found any evidence that verifies or confirms that claim. Nobody has.

Atheism is the claim that God does not exist. I never found any evidence that verifies or confirms that claim. Nobody has. Instead, the verified and proven existence of the Biblical God Jesus Christ falsifies that claim. Thousands of different people have seen Him in the flesh after He rose from the dead; and, thousands of different people have seen Him, embraced Him, and conversed with Him during their NDEs because He is the being of light and love whom most people encounter during their NDEs. In fact, during their NDEs, Jesus Christ has come and gotten some of the Atheists out of Hell, just for the asking.

Darwinism is a creative combination of Materialism, Naturalism, Nihilism, and Atheism. When those go down, Darwinism or the Theory of Evolution goes down with them as fruit from the poisoned tree. Darwinism is the belief that the rocks or raw physical matter designed and created and produced all of the proteins and their matching genes. Darwinism is a philosophical belief in spontaneous generation. Darwinism is Atheism – the creation of something by "nothing". It can't be done. It's physically impossible.

Creation ex nihilo is also a type of Atheism or Darwinism – the creation of something from "nothing" – magic. Creation ex nihilo is just as absurd as Darwinism or Atheism. Creation ex nihilo has NEVER been experienced nor observed, and it never will be because it is impossible. Not even God can do the impossible.

Back in 2012, I was a Materialist, Naturalist, Nihilist, and Atheist. I knew what an atheist was; but, I didn't know what those other words meant. Getting a proper definition for Materialism, Naturalism, Darwinism, Nihilism, Atheism, and Creation Ex Nihilo was one of my all-time greatest scientific discoveries.

Why?

Well, once you understand the major premises or hidden assumptions upon which Materialism, Naturalism, Darwinism, Nihilism, and Atheism are based, you just sort of automatically KNOW that they are false because you KNOW that their major premises or hidden assumptions have never been experienced nor observed. Science is observation and experience, not wishful thinking, blind-faith, and dogma. Materialism, Naturalism, Darwinism, Nihilism, Atheism, and Creation Ex Nihilo are self-defeating or self-falsifying because their major premises or hidden assumptions have never been experienced nor observed; and, they never will be.

Think about it!

Examine the hidden premises of Materialism, Naturalism, Darwinism, Nihilism, and Atheism as if you were a scientist and a logician. You will quickly observe that these falsified philosophies are self-defeating, once you know what their hidden premises are.

Let me demonstrate.

How are you going to observe that **God does not exist**? How do you go throughout the whole multi-verse, every dimension, every realm, every existence, every time, every atom, and every alternative reality verifying that God does not exist? If you can do that, then you are a God, and God does in fact exist.

How do you observe and verify that the **spirit world does not exist**? In order to do that, you would have to go to the spirit world and observe that it does not exist, and then return to tell about it. That's precisely what people do to verify that the spirit world does indeed exist; and, millions have done so.

How do you observe and verify that **Quantum Mechanics or Supernatural Mechanisms do not exist**? You can't. It's impossible. Instead, the repeated observation and repeated verification of Action at a Distance, the Quantum Zeno Effect, and Quantum Tunneling proves that Supernatural non-physical mechanisms do in fact exist. Even observing the effects of dark matter, dark energy, gravity, magnetism, wind, and light on physical matter proves that non-physical phenomena or supernatural phenomena do in fact exist. You can't use the repeated observation of non-physical phenomena as scientific proof that the **non-physical does not exist**.

How do you prove that the proteins and their matching genes spontaneously generated out of thin air? The phenomenon would have to be experienced and observed; and, it hasn't. In fact, **spontaneous generation**, chemical evolution, macro-evolution, creation ex nihilo, materialism, naturalism, and the theory of evolution were falsified in 1859 by Louis Pasteur – the very same year that Charles Darwin published "On the Origin of Species". We've known that the theory of evolution is false from the very beginning of the theory.

The false is falsified by the truth; and, the truth is repeatedly experienced, verified, and observed by Someone Psyche or Someone Intelligent in any realm of existence.

Your college professors and public schoolteachers are trying to hide this information from you. In fact, they will be fired and lose their jobs if they teach any of this information to you. Materialism, Naturalism, Darwinism, Nihilism, and Atheism are religions; and, religions protect themselves by censoring, banning, blocking, ridiculing, restricting, and firing their opposition. The tables have turned, and the Atheists now control our school-grounds; but, it's just another religion. All we did is replace one religion with a different religion. We've made no social progress where the truth is concerned. In fact, we've gone backwards further into the dark ages as a result. In order to get the truth into our lives, we have to eliminate everything that is false and everything that has been falsified, starting with Materialism, Naturalism, Darwinism, Nihilism, Atheism, Creation Ex Nihilo, and the Theory of Evolution.

Remember, if you successfully eliminate everything that is false, then only the truth will remain. It's fascinating to study and observe what remains after you have falsified and eliminated Materialism, Naturalism, Darwinism, Nihilism, Atheism, Creation Ex Nihilo, and the Theory of Evolution. The truth remains.

Mark My Words

—

Source

Using THE SCIENTIFIC METHOD to Eliminate the Usual Suspects and to Prove the Truth.

https://www.amazon.com/dp/B01J6STHP0

https://www.amazon.com/dp/1521133581

The Scientific Method Proves That the Theory of Evolution Is False

https://www.amazon.com/dp/B01IAAIRT2

https://www.amazon.com/dp/1521133611

4. Your Brain Has Been Mapped

I've been rethinking Science, so that we would end up with something that matches with what has been experienced and observed when I'm done. One of the greatest and most informative scientific discoveries that I ever came across while studying Neuroscience is the observation that the human brain has been mapped.

Why is this important?

Well, we all KNOW for a fact that maps of any kind require an intelligent Mapmaker to make them. Maps don't just spontaneously generate out of thin air from nothing. A map represents something that is real; and, a map is real also. The verified and proven fact that the neurons and synapses within our brain have been mapped and programmed to provide specific physical functions is scientific proof that our brains have a Mapmaker who mapped that specific physical functionality onto them.

Introduction to Brain Functionality

Brain functionality has been MAPPED. Different parts of the physical brain have been MAPPED to provide different types of physical functionality. This reality has been experienced and observed. Two of the MAIN MAPS are what the Neuroscientists call the Sensory Homunculus Map and the Motor Homunculus Map.

https://en.wikipedia.org/wiki/Cortical_homunculus

https://quantum-neuroscience.com/wp-content/uploads/2018/05/Cortical-Homunculus.pdf

It's real. It's true. Brain functionality has been MAPPED. Scientists, Neuroscientists, and Psychologists have caught these different MAPS being used in real life by real people

In some situations, automatic, implicit prejudice can have life or death consequences.

It also appears that different brain regions are involved in automatic and consciously controlled stereotyping. Pictures of outgroups that elicit the most disgust (such as drug addicts and the homeless) elicit brain activity in areas associated with disgust and avoidance.

This suggests that automatic prejudices involve primitive regions of the brain associated with fear, such as the amygdala, whereas controlled processing is more closely associated with the frontal cortex, which enables conscious thinking.

We also use different bits of our frontal lobes when thinking about ourselves or groups we identify with, versus when thinking about people that we perceive as dissimilar to us.

Even the social scientists who study prejudice seem vulnerable to automatic prejudice. (Myers, D. G. (2010). *Social Psychology* (10th ed.), pp. 314-315. New York: McGraw-Hill.)

A physical brain is MAPPED!

What do we know about maps in general?

We know that maps do not spontaneously generate out of thin air. Maps require some kind of Mapmaker.

Brain functionality has been MAPPED. Mapped by whom? Where are these MAPS being stored? How are these MAPS created, and how are they being used? Who is using these MAPS? Maps have no value if they aren't being used by Someone Psyche or Someone Intelligent to get something useful done.

By definition, in principle, physical matter and genes cannot produce maps at the quantum level and then use those maps at the quantum level in order to get things done for us at the physical level. Instead, different parts of the brain have been mapped by Nature's Psyche at the quantum level; and, then Nature's Psyche uses those maps at the quantum level to get things done for us at the physical level. The Human Psyche isn't consciously aware of these quantum maps.

The Map-Maker Argument

The following syllogism summarizes this scientific discovery:

Scientific Observation: Every type of MAP has a Mapmaker who made it. MAPS don't just spontaneously generate out of thin air.

Scientific Observation: It's obvious that each part of a physical brain has been carefully MAPPED to perform a specific physical function.

Scientific Conclusion: Therefore, every physical brain has a Mapmaker, who MAPPED each part of that brain for specific physical functions. This reality has been experienced and observed.

This is a syllogism. If the premises or observations are true, then the conclusion has to be true as well. This syllogism is philosophically valid and observationally sound.

According to the Orthodox Interpretation of Quantum Mechanics by Henry P. Stapp, Nature's Psyche collapses the wave function thereby converting possibilities into physical realities. Genes and physical matter cannot collapse wave functions. According to the Orthodox Interpretation of Quantum Mechanics, Nature's Psyche (or God's Psyche) made your physical brain and then subsequently mapped your brain at the quantum level to perform specific physical functions at the physical level. Nature's Psyche then uses that Quantum Map of Physical Functionality at the quantum level in order to get things done for you at the physical level. This is how your brain really works according to Orthodox Quantum Mechanics.

Your Primitive Brain and Quantum Maps of Physical Functionality

There's nothing primitive about your brain. ALL the functionality had to be there from the very beginning, or it would have had no survival value and the human race would have gone extinct. However, Nature's Psyche (or God's Psyche) has indeed MAPPED different parts of your brain at the quantum level so that your brain can be used to perform different types of physical functionality at the physical level.

The MAPS are created and stored at the quantum level by Nature's Psyche. We KNOW this is so, because there's not enough memory storage capacity within our 750-megabyte genome to store and process the petabytes of MAPPING and INTERFACING that's taking place both at the quantum level and the physical level simultaneously. It's physically impossible to store petabytes of Mapped Functionality within our 750-megabyte genome.

The 2.9 billion base pairs of the haploid human genome correspond to a maximum of about 725 megabytes of data since every base pair can be coded by 2 bits. — Uncle Google.

You can't shove petabytes of QUANTUM MAPPING and/or SYNAPTIC MAPPING into a 725-megabyte to 750-megabyte genome. It's physically impossible. But, there's seems to be NO memory storage limits at the quantum level or the psyche level. The quantum doesn't have any physical limitations. That's the primary message and lesson of Quantum Mechanics – namely, quantum mechanisms have NO physical limitations. It's also interesting to note that neurons have NO physical function for making and then using quantum maps. Neurons and synapses don't really communicate with each other at the physical level beyond the concepts of ON and OFF. All of the telepathic communication between atoms, molecules, and neurons across synapses is taking place at the quantum level.

Some parts of the brain were MAPPED by Nature's Psyche to register physical input or physical sensations from the environment, and other parts of the brain were MAPPED by Nature's Psyche to register and then manifest the Human Psyche's feelings, emotions, desires, choices, and actions onto the physical plane. We KNOW that it was Nature's Psyche who did this MAPPING for us at the quantum level, because the Human Psyche isn't consciously aware of these things. This QUANTUM MAPPING and SYNAPTIC MAPPING is being handled automatically for us by Nature's Psyche, or by God. Some parts of the brain's physical input are consciously perceived by the Human Psyche; and, other parts of the brain are controlled automatically for us by Nature's Psyche outside our conscious awareness. Remember, ALL of the quantum mapping and physical functionality of the brain is handled by Nature's Psyche outside the conscious awareness of the Human Psyche. Nature's Psyche collapses the wave function, NOT the human psyche and NOT our genes.

Nature's Psyche organized the physical brain structure and then MAPPED the different neurons, glial cells, and synapses at the quantum level to have a specific meaning, purpose, or goal at the physical level. Then Nature's Psyche uses those Quantum Maps of Physical Functionality at the quantum level to get things done for us at the physical level by collapsing the necessary wave functions and/or turning on the necessary neurons from the quantum level whenever the Human Psyche makes a choice and wants to get something done at the physical level.

I was very pleased with the explanatory power of Quantum Maps of Physical Functionality when it was first revealed to me. It made sense to me and basically explained everything that we are observing and experiencing when it comes to a physical brain and brain functionality. I think Quantum Maps of Physical Functionality is my favorite scientific discovery, due to its massive explanatory power. I finally feel like I understand how everything really works at every level of existence.

Quantum Maps of Physical Functionality also match perfectly with the Orthodox Interpretation of Quantum Mechanics by Henry P. Stapp, which explains that it is the Human Psyche who makes the choices and it is Nature's Psyche who collapses the necessary wave functions and/or turns on the necessary neurons in order to make that CHOICE actual and real in the physical world.

I eventually concluded that Quantum Maps of Physical Functionality is the BEST explanation for brain functionality that we have discovered to date. It explains what Nature's Psyche and the Human Psyche are doing for us at the quantum level in order to get things done for us at the physical level. Quantum Maps of Physical Functionality provide the interface between the quantum level and the physical level; and, that's all I really needed to know and have.

Quantum Maps of Physical Functionality also match perfectly with and explain the origin of the Sensory Homunculus Map and the Motor Homunculus Map, which the Neuroscientists have discovered at the physical level. Nature's Psyche (or God) creates these MAPS or INTERFACES at both the quantum level and the physical level, and then Nature's Psyche uses these MAPS at the quantum level in order to get things done for us at the physical level. Quantum Maps of Physical Functionality explain everything that I learned about neuroscience, brain functionality, psyche, consciousness, and quantum mechanics. Quantum Maps of Physical Functionality combine everything into one realistic whole, both at the quantum level and the physical level.

Remember, there's nothing primitive about your brain.

If parts of your brain are missing or fail to develop, Nature's Psyche can still MAP physical functionality onto the parts of the brain that remain. It's called Neuroplasticity; and, Neuroplasticity is a proven and verified part of Neuroscience. A properly developed brain really isn't necessary for a fully functional brain. During development while the brain is at maximum plasticity, Nature's Psyche can MAP all functionality to the parts of the brain that are there, while working around those that are not.

Once the MAPS and axonal connections have been established, though, it's basically a done deal. If you cut the axons or destroy the MAP, there are times when the physical functionality does not return. However, depending upon the location and the severity of the damage, there are other times when the physical functionality gets successfully REMAPPED at both the quantum level and the physical level by Nature's Psyche. It all depends upon the situation, as well as the determination of the Human Psyche.

If the Human Psyche is determined to move its finger again, Nature's Psyche can indeed be encouraged to MAP that physical functionality onto a different part of the physical brain. It doesn't always work, but sometimes it does. Sometimes the axons don't regrow and remap properly. Other times, they do. If the axon is still there but the neuron dies during a stroke, that axon can sometimes be REMAPPED to a different neuron or a different part of the brain. It's hit and miss. Sometimes it works, and sometimes it doesn't. This is what has been experienced and observed.

Oligodendrocytes in the brain and spine tend to block, resist, and restrict axonal regrowth and neural remapping. It's as if the central nervous system has been set in plastic. Remapping can be done, but not as easily as it is done in the peripheral nervous system. The Schwann cells in the peripheral nervous system tend to facilitate axonal regrowth and subsequent remapping. The axonal regrowth doesn't always take place and isn't always done right; but, it's more likely to be done in the peripheral nervous system than the central nervous system. The peripheral nervous system is much more fluid and flexible. Nevertheless, Nature's Psyche can compensate for dead neurons easier than it can compensate for severed axons or nerves that have been cut.

In both cases, where there's a WILL (a Human Psyche), there is often a way, even within the physical brain. Sometimes stroke victims can indeed recover and successfully make Nature's Psyche remap their physical brain. Sometimes it doesn't work, or there isn't enough time to make it work before the person dies. Other times, the results seem to be miraculous.

As a teenager, I cut the nerves in the lateral part of my left thumb; and, that side of my thumb was numb and weird feeling for a long time. Forty years later, everything is back to normal. I'm only marginally aware of it if I'm actually thinking about it. Otherwise, the former damage goes unnoticed.

Remember, your brain, brain functionality, and nervous system have been MAPPED. Mapping requires some sort of intelligent Mapmaker in order to get the job done right. Maps don't just spontaneously generate out of thin air. Spontaneous generation or creation ex nihilo is impossible. It can't happen, which means that it didn't happen. The MAPS within your physical brain required some kind of intelligent Mapmaker, or they wouldn't exist, and they wouldn't work right if they did exist.

A Physical Brain Is a Quantum Computer

While studying Neuroscience and developing Quantum Neuroscience, I observed that there are NO physical wires within a physical brain connecting everything together into functional logic gates, transistors, CPUs, and RAM! There are NO wires in our brains.

Our neurons are nothing more than switches – they are either ON or OFF. That's it! A neuron represents a single BIT of information – ON or OFF. Furthermore, all of that synaptic input into a neuron gets reduced to a single post-synaptic potential (PSP) that determines whether a neuron fires or stays OFF.

There are NO functional BYTEs within a physical brain. There are NO logic gates, transistors, CPUs, or RAM within a brain because there are NO physical wires within a physical brain. At the physical level, ON or OFF is the only thing that gets successfully transmitted across a synapse, which is a gap in space. A physical brain is a quantum computer, with most if not all of its functionality, processing, and memory storage taking place at the quantum level or the psyche level. I call it Quantum Neuroscience.

My theory is easily falsified by the Materialists, Naturalists, Darwinists, Nihilists, and Atheists simply by finding the wires within our brains that are tying or binding everything together into functional logic gates, transistors, CPUs, and RAM.

Until they discover the wires within our brains, though, the best explanation for brain functionality is some type of WiFi, Quantum Waves, or Telepathy. Telepathy is WiFi at the quantum level. Until they discover the wires within our brains, along with the logic gates, transistors, CPUs, and RAM, I have concluded that the best explanation for brain functionality, information processing, and memory storage is to be found on the quantum level or the psyche level. With Quantum Maps of Physical Functionality, it's possible to MAP the output from dozens of non-physical quantum computers onto a single concept cell or neuron at the physical level. That potential reality would explain why we (psyches) recognize faces and people instantly rather than taking hours to do facial recognition as our physical supercomputers often have to do. Physical computers have severe physical limitations. The Psyche by definition in principle has NO physical limitations in its native original domain.

Since there are NO wires within a physical brain, the best way to explain the brain's massive memory storage capacity and massively huge information processing capacity is to look to the quantum level or the psyche level for an explanation because Psyche and Quantum Mechanics in their native original format have NO physical limitations. At times, the brain seems to have NO physical limitations which means that most of its functionality has to be taking place within the quantum realm or the transdimensional realm where there are NO physical limitations to hold it back.

When it comes to Quantum Neuroscience, my theory is that Nature's Psyche MAPS functionality onto our brains at the quantum level or the psyche level, and then uses those Quantum MAPS of Physical Functionality at the quantum level in order to get things done for us at the physical level. Nature's Psyche does so by collapsing the necessary wave functions and/or turning ON the necessary neurons. All of this quantum processing and quantum mapping takes place completely outside the conscious awareness of the Human Psyche. The only thing the Human Psyche does is decide what it wants to do with its physical body and physical brain; and then, Nature's Psyche (or God's Psyche) does all the work needed to make the Human Psyches commands actual and real at the physical level.

Since the Neuroscientists have observed that there are NO wires within our brains, some of the Neuroscientists have admitted that they don't really know how our physical brain works. It shouldn't work, but it does. I have observed that in order to explain how our brains really work, we have to bring Nature's Psyche, the Human Psyche, Quantum Mechanics, the Orthodox Interpretation of Quantum Mechanics by Henry P. Stapp, and Quantum Maps of Physical Functionality into play. The brain has been MADE and MAPPED at the quantum level by Nature's Psyche, and then Nature's Psyche uses those MAPS to get things done for us at the physical level. The explanatory power of Quantum Neuroscience is vastly superior to anything that we can get from Materialism, Naturalism, Darwinism, Classical Physics, and ordinary Neuroscience.

Once we allow Psyche, Syntropy, and Quantum Mechanics into play, then instantly we can explain everything that comes our way. In fact, the verified and proven existence of

Action at a Distance, the Quantum Zeno Effect, Quantum Tunneling, and Quantum Mechanisms in general FALSIFIES Materialism, Naturalism, Darwinism, Nihilism, and even Atheism. That is what I have experienced and observed.

Mark My Words

—

Source

Quantum Mechanics from a Non-Physical Spiritual Perspective

https://www.amazon.com/dp/B01J023TGU

Source Material

Quantum Neuroscience: The Answer to Life, the Universe, and Everything

https://www.amazon.com/dp/B079Z6QQQB

Scientific Proof of God's Existence: Finding God Where the Atheists Refuse to Look for Him

https://www.amazon.com/dp/B07B26CRHX

The Ultimate Model of Reality: Psyche Is the Ultimate Cause

https://www.amazon.com/dp/B071NC9JK6

5. Convincing Proof of God's Existence

I used to be a Materialist, Naturalist, Nihilist, and Atheist back in 2012. I was a practitioner of Scientism – the philosophical belief that science and the scientific method is the ONLY way to find and know the truth. I also believed that Natural Selection could do what they say it does. I simply knew that Natural Selection exists and that its effects are real. I had bought into the party line.

At the time, I interacted with Materialists, Naturalists, Darwinists, Nihilists, and Atheists online. I considered them kindred spirits. They were like me. I felt as if I were a part of them. They were my people.

One of my atheistic friends told me that there is NO scientific proof of God's existence and that there will NEVER be any scientific proof of God's existence. I believed him. I'd never found any convincing proof of God's existence. I figured that God's existence had to be taken on blind faith as being real and true. That is what the Atheists teach.

But one has to consider the messenger in order to truly understand the message.

We defined "science" as Materialism, Physicalism, Naturalism, Darwinism, Nihilism, and Atheism. Obviously, Atheism, Naturalism, or "science" will never prove to anyone that God exists. Science defined as Materialism, Naturalism, and Atheism defines God out of existence, a priori, by definition and in principle.

We were RIGHT! Science as we defined "science" would NEVER produce any scientific proof of God's existence.

How you choose to define things makes all the difference in the world! It's a choice! And, our choices have consequences whether we realize it or not. Our beliefs are chosen into existence.

Shifting One's Perspective

Now, consider what happens when we choose to define "science" as Fine-Tuning.

Science is Fine-Tuning.

Wow!

Instantly, we are looking at thousands if not millions of different Scientific Proofs of God's Existence. Each ACT of Fine-Tuning or Science is a miniature scientific proof of God's existence.

Examine the following syllogism. It's a Philosophical Proof of God's Existence. Because it's a syllogism, if the premises are true, then the conclusion has to be true as well.

Scientific Observation or Premise: Anything that was obviously fine-tuned obviously had a Fine-Tuner who fine-tuned it.

Scientific Observations: A genome was obviously fine-tuned. A protein and a gene were obviously fine-tuned with a specific purpose in mind. Our physical universe and the physical constants were obviously fine-tuned. Your computer and your car were obviously fine-tuned. This computer program that I'm using was obviously fine-tuned. Your physical body was clearly fine-tuned for specific purposes. The neurotransmitters within your brain were clearly and obviously fine-tuned. It's undeniable.

Scientific Conclusion: ALL of these things were obviously fine-tuned which means that they obviously have a Fine-Tuner who fine-tuned them.

If you choose to define Science as Fine-Tuning, then instantly this syllogism and philosophical proof of God becomes a convincing, valid, solid, and sound Scientific Proof of God's Existence.

Instantly, we have Scientific Proof of God's Existence; and, it happens automatically by choosing to define Science as Fine-Tuning rather than choosing to define science as Materialism, Naturalism, Darwinism, Nihilism, and Atheism.

Do you see how that works?

It works wonderfully, but only if you choose to allow it to work. It's a choice! Our beliefs are chosen into existence.

However, if you continue to define science as Materialism, Naturalism, Darwinism, Nihilism, and Atheism, then you will NEVER have any scientific proof of God's existence. That's just the way things work.

There's no compelling reason to define science as Materialism, Naturalism, Darwinism, Nihilism, and Atheism. These things are not the best explanation, anyway. I mean think about it! Who did ALL of that Fine-Tuning or Science BEFORE we human beings arrived on the scene?

When we are talking about the physical constants, this physical universe, and physical matter, we are talking about a Fine-Tuner who was doing Science or Fine-Tuning BEFORE this physical universe began. That's the very definition of a God, is it not?

By choosing to define Science as Fine-Tuning, instantly we have compelling and believable and convincing Scientific Proof of God's Existence. That's one of the reasons why I go overboard trying to define everything and compare everything. I learn best and learn most through comparison and contrast.

Mark My Words

—

Source

Scientific Proof of God's Existence: Finding God Where the Atheists Refuse to Look for Him

https://www.amazon.com/dp/B07B26CRHX

6. Syntropy Is Order, Power, Organization, Creation, and Life

One of my most useful scientific discoveries is the fact that Quantum Mechanics or Syntropy is the NORM. Psyche and Spirit Matter are based upon Quantum Mechanics or Syntropy. The primal construct or primal reality is Quantum Mechanics or Syntropy – it's without a beginning of days or an end of years. Quantum Mechanics and Syntropy have NO physical limitations. This reality has been experienced and observed. Observation and experience are Science. Quantum Mechanics, in particular Quantum Field Theory, is one of our best-proven and most-used Scientific Disciplines.

So, why physical matter, or whence physical matter?

Physical matter was not a part of the Primal Construct. Physical matter is not a part of the NORM. God infused a part of his power, energy, psyche, and glory into each particle of physical matter thereby converting it from spirit matter to physical matter.

Why would God do such a thing?

The goal was to create the Ultimate Consensus Reality. A physical reality is the Ultimate Consensus Reality. It is highly reliable, dependable, and predictable. It's ordered and lawful. It keeps the Commandments of God – the physical laws. In a physical reality, I can depend on this paper being here on my computer and cloud drive tomorrow when I go looking for it. I can depend upon my physical body being here tomorrow as well in basically the same shape it was yesterday.

God deliberately infused a part of Himself into each particle of physical matter in order to minimize the uncertainties natively and naturally associated with spirit matter. When it comes to physical matter – all of the physical laws, the sub-light speed limit associated with the Theory of Relativity, and the deliberately short De Broglie Wavelength associated with physical matter greatly limit and reduce the Quantum Uncertainties which are a native, inherent, and original part of spirit matter.

A ton of interesting science, knowledge, and truth can be acquired by making a detailed comparison between Syntropy and Entropy. What we are in fact doing is comparing Quantum Physics with Classical Physics. We are comparing the Primal Construct with the Ultimate Consensus Reality. The one has always existed, and the other one was made by God to reduce the Quantum Uncertainties to a minimum.

Do you see how that works and why it is necessary?

The Ultimate Consensus Reality, a physical reality, wouldn't be possible without the physical laws and classical physics which were deliberately designed and put into place by God to greatly reduce and minimize the Quantum Uncertainties that are an integral and native part of spirit matter to begin with.

God deliberately and purposefully keeps the physical atoms within your physical body from quantum tunneling away from you at will. That's what the physical laws or the Commandments of God do. They greatly reduce and limit the Quantum Uncertainties that are a native and original part of spirit matter.

There are tons of scientific evidence proving that the Physical Constants were precisely fine-tuned by some kind of Fine-Tuner, which means that the Physical Constants were deliberately made by Someone Psyche or Someone Intelligent in order to reduce and minimize the effects of Quantum Uncertainty and thereby greatly increase the reliability, dependability, and predictability of our Physical Consensus Reality. It was done with a purpose in mind. It works because God makes it work.

Scientific Observation or Premise: Anything that was obviously fine-tuned obviously had a Fine-Tuner who fine-tuned it.

Scientific Observations: A genome was obviously fine-tuned. A protein and a gene were obviously fine-tuned with a specific purpose in mind. Our physical universe and the physical constants were obviously fine-tuned. Your computer and your car were obviously fine-tuned. This computer program that I'm using was obviously fine-tuned. Your physical body was clearly fine-tuned for specific purposes. The neurotransmitters within your brain were clearly and obviously fine-tuned. It's undeniable.

Scientific Conclusion: ALL of these things were obviously fine-tuned which means that they obviously have a Fine-Tuner who fine-tuned them.

This is one of the most convincing Scientific Proofs of God's Existence that I have encountered so far. It works because the Premises and the Conclusion have been experienced and observed. It works because it is Science.

Syntropy versus Entropy

Let's compare Syntropy with Entropy and see what we can learn.

Syntropy is Order, Organization, Creation, and Life. Syntropy is one of my greatest scientific discoveries. Quantum Mechanics is Syntropy. Spirit Matter or Dark Matter is Syntropy. Magnetism, Dark Energy, and Gravity are some sort of Syntropy. The Strong and Weak Nuclear Forces are a type of Syntropy. The Zero-Point Field is Syntropy. Quantum Waves are Syntropy. Thoughts and Memories are Syntropy. Psyche or Quantum Non-Local Consciousness is Syntropy. Syntropy is eternal and everlasting, without a beginning of days or an end of years.

Entropy is disorder, randomness, chaos, disease, and death. Materialism, Naturalism, Atheism, Darwinism, Nihilism, Classical Physics, and the Theory of Evolution are based exclusively upon entropy. Entropy cannot design and create. Entropy is death and extinction.

Random mutations are entropy. Natural selection or survival of the fittest results in death or entropy. Entropy cannot design and create. Entropy is death. Death cannot design and create. Chemical Evolution, Macro-Evolution, Spontaneous Generation, Abiogenesis, and Creation Ex Nihilo are prevented from happening by entropy and physical laws. Chemical Evolution or Macro-Evolution is physically impossible thanks to entropy. Evolution of any kind is based upon entropy, and entropy prevents the Theory of Evolution from becoming true.

The very existence of entropy FALSIFIES Materialism, Naturalism, Darwinism, Chemical Evolution, Macro-Evolution, and the Theory of Evolution because entropy proves that these things cannot design and create anything at all. Furthermore, the existence of entropy proves that some sort of Syntropy must exist. All of that subsequent entropy would not have been possible without a massive infusion of Syntropy to begin with. This is logical common sense.

The Fruits of Entropy

One of my greatest scientific observations and scientific discoveries came to me when I finally realized that Materialism, Naturalism, and Darwinism have been falsified trillions of different times in thousands of different ways. That was a major conceptual breakthrough for me because I used to be a Materialist, Naturalist, Nihilist, and Atheist. Materialism, Naturalism, Darwinism, Nihilism, Behaviorism, Determinism, Physical Reductionism, and Atheism are based exclusively upon entropy; and, entropy prevents the Theory of Evolution from becoming true.

Materialism, Naturalism, and Darwinism are BEST defined as "Design and Creation by Physical Matter" or "Creation by Entropy". These people literally believe and teach that the rocks – physical reactions – or entropy designed, programmed, created, engineered, field-tested, manufactured, and deployed ALL of the different genomes and life forms on this planet. That's what these people really believe, and that's what these people teach in all of our public classrooms.

Ironically, Louis Pasteur falsified Materialism, Naturalism, Darwinism, and the Theory of Evolution in 1859 by demonstrating that the rocks – raw physical matter – cannot design and create. Entropy cannot design and create. It's physically impossible! That was the same year that Charles Darwin published, "On the Origin of Species". Louis Pasteur proved the Theory of Evolution or Creation by Rocks false the same year that Charles Darwin introduced his theory to the world. We've known from the very beginning of the theory that the Theory of Evolution is false because entropy cannot design and create. Remember, Science and the scientific methods can be used to prove theories false. It's called falsifying a theory; and, that's what the scientific methods do, or are supposed to do.

Technically, the scientific methods can't be used to prove anything true. Due to a wide variety of different logic fallacies which are built directly into the scientific methods, the scientific methods cannot be used to prove a theory true, as the Materialists, Naturalists, and Darwinists try to do.

Instead, we can use the scientific methods to falsify theories. Using the scientific methods to prove our theories false is philosophically and logically sound. In other words, if you falsify a theory by *negating the consequent*, you have in fact proven that theory false.

Here's how it works in principle:

Scientific Hypothesis: If Theory X is true, then we will observe Y.

Scientific Observations: We don't observe Y.

Scientific Conclusion: Therefore, Theory X is not true.

Here's how it works in practice:

Scientific Hypothesis: If the Theory of Evolution, Materialism, Naturalism, and Darwinism are true, then we will observe the rocks and physical reactions designing, creating, and manufacturing genomes and life forms from scratch.

Scientific Observations: We have NEVER observed the rocks and physical reactions designing and creating genomes and life forms; and, we NEVER will. They can't. It's physically impossible and prevented from happening by entropy.

The Scientific Conclusion: Therefore, the Theory of Evolution, Materialism, Naturalism, and Darwinism are not true. They have been successfully falsified by the Scientific Method.

See: Slife, B. D. & Williams, R. N. (1995). Science and Human Behavior. In *What's Behind the Research? Discovering Hidden Assumptions in the Behavioral Sciences*, (pp. 167–204). Thousand Oaks, CA: SAGE Publications.

http://mypsyche.us/wp-content/uploads/2017/04/Science.pdf

I just successfully falsified Materialism, Naturalism, Darwinism, and the Theory of Evolution; and, I used the Scientific Method to do so! Best of all, my argument is philosophically and logically sound. I have in fact falsified Materialism, Darwinism, Naturalism, and the Theory of Evolution for REAL! I just used the Scientific Method to PROVE these things false. I successfully used the Scientific Methods for what they are good for – falsifying theories that are false. I wish I would have known how to do that forty years ago. It would have saved me a lot of confusion and grief.

In contrast, there is NO way to falsify theories that are true, such as "Design and Creation by Intelligent Beings or by Intelligent Psyches". Instead, true theories are continuously verified over and over again. "Design and Creation by Psyche or by Intelligence" has been observed and verified and experienced trillions of trillions of different times and ways, with an infinite number of more to go. See how that works? That's Science and the Scientific Methods in action for REAL, doing what they are best at!

Entropy and physical matter cannot design and create; but, Syntropy, Psyche, Intelligence, and Quantum Mechanics certainly can. They've been caught in the act of doing so.

This has been and is one of my greatest scientific discoveries and scientific observations of all time, during the whole of my science career.

Quantum Mechanics and Syntropy trump classical physics and entropy. In fact, classical physics and entropy are a very small sub-set of Quantum Mechanics and Syntropy.

638

Quantum Mechanics is supernatural in nature and origin. 95% of our universe is supernatural and non-physical – Dark Energy and Dark Matter. Only 5% of our universe is physical matter. Entropic physical matter is a very small sub-set of the Quantum Mechanics and Syntropy. There are NO physical limitations and NO entropy within spirit matter, psyche, and quantum mechanisms.

I have observed that there are literally thousands of different ways to falsify Creation by Rocks – or Materialism, Naturalism, Darwinism, and the Theory of Evolution; and, they are all philosophically and logically sound. Thanks to the scientific methods and scientific observations, we KNOW that Creation by Rocks, Materialism, Naturalism, and Darwinism are false because we KNOW why they are false – rocks and physical reactions cannot design and create. Entropy cannot design and create. It's elementary my dear friend.

There's NO way that Evolution (Mutation and Selection) could have designed, programmed, engineered, created, and manufactured the first genomes and life forms because Evolution didn't even exist until AFTER God had designed, created, and deployed the first genomes and life forms in the first place.

I have observed that the Scientific Methods or Observation have been used thousands of different ways to falsify Materialism, Naturalism, Darwinism, and even Atheism. Atheism is Creation by Nothing, or Creation by Chance. The Atheists really truly believe that "Nobody" and "Nothing" designed and created everything. Nobody, Nothing, and Chance are the holy trinity of Atheism. These people believe in those false gods or idols with a passion. Materialism, Naturalism, and Atheism are in fact our modern-day form of idolatry; and, these people are idolaters. These people worship the rocks and physical reactions and entropy as if these things were God.

Ironically, God must of necessity exist in order to have done ALL of the science that needed to be done, which the rocks or raw physical matter could never have done. Only Psyche or some type of Syntropy can design and create; and, God's Psyche is the only one we know of who was there at the time and could have done the job.

Finally, as a capstone, the Biblical God Jesus Christ has told us repeatedly in the Bible, Book of Mormon: Another Testament of Jesus Christ, the Doctrine and Covenants, and the Pearl of Great Price that HE designed and organized the heavens, this earth, and all of the life forms on this earth. The Biblical God confessed to doing the job.

https://www.lds.org/scriptures?lang=eng

I'm not going to apologize for finding, learning, and knowing the truth. That's what we scientists are supposed to do, is it not? I wouldn't be doing my job if I kept feeding you the falsehoods, the falsified, and the lies; now would I?

For me personally, falsifying Materialism and Darwinism became my first really convincing Scientific Proof of God's Existence. After falsifying Materialism, Naturalism, and Darwinism, I simply KNEW that God exists. I was finally willing to follow the evidence, wherever it might lead me.

I discovered that entropy or physical matter cannot design and create and manufacture anything at all. That reality is science and truth; and, it is set in stone where our physical universe is concerned.

Mark My Words

—

Source

Mark My Words. (2016). *The Scientific Method: Proves That the Theory of Evolution Is False*. Kindle.

https://www.amazon.com/dp/B01IAAIRT2

Source Material

Mark My Words. (2016). *The Theory of Evolution Proved to Me that God Exists: Why I Am No Longer an Atheist and Why I No Longer Believe in the Theory of Evolution*. Kindle.

https://www.amazon.com/dp/B01HZYBZ7K

7. We Find the Truth by Living It and Experiencing It

Some might argue that my greatest scientific observation and scientific discovery is the realization that Lived Experience IS the BEST way for finding and knowing the truth.

I spent a year studying the Philosophy of Science. In the process, I was introduced to Phenomenology. Phenomenology is the scientific study of events or phenomena, including Out-of-Body Experiences (OBEs), Near-Death Experiences (NDEs), Shared-Death Experiences (SDEs), Visions, Theophanies, and other types of Spiritual Experiences. These are phenomena that have been experienced and observed by millions of different people on this planet. Phenomenology is the scientific study of the Lived Experiences of the human race, including both our spiritual experiences and our physical experiences.

Begging the question, Science is currently defined as Materialism, Physicalism, Naturalism, Darwinism, Nihilism, Classical Physics, and Atheism before any science is actually done. Thanks to Phenomenology and the Lived Experiences of the human race, I realized one day that Science should be redefined as Observation and Experience. In other words, Science should be brought into the modern Quantum Age of Action at a Distance, Phenomenology, Observation, and Experience rather than being left in the restrictive and limited dark ages of Materialism, Naturalism, and Atheism. Materialism, Naturalism, Darwinism, Nihilism, and Atheism have been falsified. Science should be taken out of the falsified and placed into the verified or the observed. I realized that Science should be redefined as Phenomenology.

As a result, I upgraded my science to Science 2.0 which allows all of the evidence (all of the phenomena) into evidence and pursues a preponderance of the evidence.

Science 2.0: I Upgraded My Science

https://www.amazon.com/dp/B0771K6WTX

Lived Experiences or Psyche Experiences are the best way of finding the truth and knowing the truth in EVERY realm of existence, including this physical realm. The BEST and fastest way to find and know the truth is to live it, experience it, and observe it for yourself, or to choose to trust someone who has. Science is Observation and Experience, and NOT

Materialism and Naturalism which are based exclusively upon philosophical speculation and wishful thinking. In other words, there is NO observational evidence supporting the major premises or hidden assumptions of Materialism and Naturalism which claim that the quantum or the transdimensional does not exist.

Due to the fact that the scientific methods cannot be used to find the truth and prove the truth directly, and due to the fact that the scientific methods are typically restricted to the physical realm or the local realm, and due to the fact that there are a wide variety of logic fallacies built into the scientific methods, the scientific methods are in fact a much weaker source of knowledge and truth than Lived Experiences are. Lived Experiences, or Psyche Experiences, or Direct Observations are in fact an infinitely better way for finding and knowing the Truth than science and the scientific methods can ever be. I wish I would have known that forty years ago. It would have saved me a lot of frustration and grief.

Possibly my greatest scientific discovery of all time is that the Lived Experiences of the human race are in fact a much better way of finding and knowing the truth than traditional science and the scientific methods. For me, that was a major epiphany and scientific breakthrough because I used to be a Materialist, Naturalist, Nihilist, and Atheist.

Lived Experience is extremely powerful. Lived Experience is Science, Observation, Knowledge, and Truth in their purest form. Science is supposed to be Observation and Knowledge and the pursuit of the Truth. In other words, Science is supposed to be Lived Experiences or Observations. The Materialists, Naturalists, Behaviorists, and Atheists did the world a great disservice when they hijacked and stole science and limited science exclusively to physical matter. One of my greatest scientific discoveries was to finally realize how wrong these people really are.

I KNOW how it goes, though, because I allowed My Scientism, My Nihilism, My Materialism, and My Atheism to blind me to Lived Experience as a source of evidence and to blind me as to how powerful and useful Lived Experiences really are as a source of knowledge and truth. But, that's what we Materialists and Atheists do. We refuse to allow Lived Experiences into evidence. Naturalism and Atheism of any kind is based upon a refusal to look at evidence.

I was a practitioner of Scientism. I truly believed that Science and the scientific methods were the ONLY way for finding and knowing the truth. I WAS WRONG! In fact, the scientific methods can't be used to prove the truth, which means that Lived Experience is in reality the ONLY way to find and KNOW the truth directly. The truth is KNOWN by living it and experiencing it for yourself, or by choosing to trust someone who has.

I never fully realized until April 2017 that Lived Experience or Direct Observation is a much better way of finding and knowing the truth than the scientific methods and philosophical guesswork will ever be; but, I'm seeing it now and understanding it now.

We KNOW from the Lived Experiences of the human race that the Biblical God Jesus Christ does indeed exist and truly rose from the dead – although it's nice to have Science pointing us to the same truths, in its roundabout way.

For some Lived Experiences of the Divine see: *Bible, Book of Mormon: Another Testament of Jesus Christ, Doctrine and Covenants, and Pearl of Great Price* – available for free at: https://www.lds.org/scriptures/bible?lang=eng

Possibly my greatest scientific discovery came to me when I first realized that Science should be redefined as Phenomenology, Lived Experience, and Observation. Quantum Mechanics is Supernatural Mechanics. Science should be based upon Syntropy, Psyche, and Quantum Mechanics – and NOT Materialism and Naturalism.

Currently, the scientific community at-large define science as Materialism, Naturalism, Darwinism, Nihilism, Behaviorism, Determinism, Physical Reductionism, Classical Physics, Entropy, and Atheism. It's time for us to upgrade science and bring it into the modern age.

It's time for us to redefine Science as Phenomenology, Observation, Experience, Eye-Witness Verification, Quantum Mechanics, Syntropy, Psyche, and the Lived Experiences of the human race. We'll make much faster progress if we do. I KNOW because I have lived it and experienced it on both sides of the fence; and, the grass really is greener on the Syntropy side or Psyche side of the fence.

Mark My Words

Source

Science 2.0: I Upgraded My Science

https://www.amazon.com/dp/B0771K6WTX

8. A Psyche Ontology and a Psyche Epistemology

Science is Observation and Experience.

The BEST and the FASTEST way to find and know the truth is to live it, experience it, verify it, and observe it for yourself, or to choose to trust someone who has. This is Phenomenology – the scientific study of events and phenomena; and, Phenomenology is based upon the Lived Experiences of the human race. These various different phenomena are evidence – they are scientific observations and scientific experiences.

One of my most favorite and most useful scientific discoveries came when I first realized and learned that Psyche, Intelligence, or Quantum Non-Local Consciousness has been experienced and observed. That discovery, revelation, or epiphany changed everything for me. I've never seen things the same way ever since.

[See: Buhlman, W. L. (1996). *Adventures Beyond the Body: How to Experience Out-of-Body Travel*. New York: HarperCollins.]

https://psyche-ontology.com/buhlman/

William Buhlman was the first person to convince me that Psyche exists as a completely different and separate entity than our spirit body and our physical body. It's our Psyche or Intelligence or Non-Local Consciousness who has the experiences and makes the memories, NOT our spirit body, and definitely NOT our physical body and physical brain.

That pinprick of light or infinite singularity or immaterial viewpoint in space, which the psychologists call Psyche, has been seen and it has been experienced first-hand while out-of-body or while separated from the physical body. I started collecting and documenting the different accounts of this phenomenon. Psyche has been experienced and observed!

https://psyche-ontology.com/psyche-observed/

Science is Observation and Experience, NOT Materialism and Naturalism. The major premises or hidden assumptions of Materialism and Naturalism have never been experienced nor observed; and, they never will be because it's impossible. The impossible should be eliminated from science as well as everything that has never been experienced nor observed. The **non-existence of the Psyche** has NEVER been experienced nor observed; and, it NEVER will be. It's impossible to observe the non-existence of something.

In contrast, Psyche has been experienced and observed many times by countless numbers of different people. The existence of Psyche has been verified by direct observation and experience. The observation and experience of Psyche FALSIFIES Materialism, Naturalism, Darwinism, Nihilism, and even Atheism. The Biblical God's presence or psyche has also been experienced and observed by thousands of different Near-Death Experiencers (NDErs). Science is Observation and Experience. Psyche has been experienced and observed.

https://psyche-ontology.com/psyche-observed/

I have started to document this phenomenon online because observational proof and experiential proof that Psyche exists is the foundation stone upon which Science is based. Psyche is the fundamental unit of reality. Psyche is the most elementary of the elementary particles. The smaller can dwell within and control the larger. If there is such a thing as a point particle, Psyche is it. Psyche is smaller than a string, therefore Psyche can dwell within and control a string, thereby explaining the origins of strings and thereby providing a scientific or observational foundation for String Theory.

Science is based upon a Psyche Ontology. Psyche is the fundamental unit of reality and existence. Psyche is the ultimate causal agent or the ultimate causal force. Psyche is the Ultimate Cause. A Psyche Epistemology is the BEST and the FASTEST way of finding and knowing the truth. It's also the most enduring. Any thought or memory that survives the death of your physical brain is truly a memory in every sense of the word, is it not?

As far as I know, I'm the first person on the planet to develop and promote a Psyche Ontology and a Psyche Epistemology. I studied the Philosophy of Science and Personality Theory for over a year in college. When it came to a Psyche Ontology, they danced around it, but none of them went all the way. I'm the first person to declare that Psyche is the Ultimate Cause and that Psyche is the fundamental unit of reality – thereby bringing Philosophy into the quantum age in the process.

This announcement was made in the following book:

The Ultimate Model of Reality: Psyche Is the Ultimate Cause

https://www.amazon.com/dp/B071NC9JK6

I also purchased https://psyche-ontology.com so that I could start promoting the concept directly because it's important that people know about this scientific and philosophical breakthrough.

The contents of these links

https://psyche-ontology.com/psyche-observed/

https://psyche-ontology.com/buhlman/

are found in this book – *Syntropy in Defense of Quantum Mechanics: The Answer to Life, the Universe, and Everything* – under the headings **Psyche or Consciousness**

Without a Spirit Body, Proof of Psyche – William Buhlman, William Buhlman Wrote the Following, and **Experiential and Observational Proof of Psyche**.

I want people to KNOW that Psyche has been experienced and observed.

https://psyche-ontology.com/psyche-observed/

I'm building the master-copy of Psyche Observations online because I want it to become public knowledge. I intend to add more **Psyche Observed** to this online list when they come to my attention because I know that there are more Observations of Psyche than just the ones that I have found so far during my research.

Science is observation and experience. Remember, Psyche has been experienced and observed; whereas, Materialism and Naturalism have not. The contrast couldn't be more glaringly obvious. The one is real and truly exists; and, the other is nothing but wishful thinking and science fiction. This is one of my greatest scientific discoveries. It changed the way that I look at the world and science in general for the better.

Mark My Words

—

Reference Material

https://psyche-ontology.com/psyche-ontology/

https://psyche-ontology.com/psyche-observed/

https://psyche-ontology.com/buhlman/

https://psyche-ontology.com/papers/

Buhlman, W. L. (1996). *Adventures Beyond the Body: How to Experience Out-of-Body Travel*. New York: HarperCollins.

Rychlak, J. F. (1981). *A Philosophy of Science for Personality Theory* (2nd ed.). Malabar, FL: Robert E. Krieger Publishing Company.

Brent D. Slife on the Philosophy of Science:

http://brentslife.com/

http://brentslife.com/books/

William Buhlman, Joseph F. Rychlak, and Brent D. Slife upgraded my Philosophy of Science in a major way, thereby changing the way that I look at science and the world in general. Although none of them directly promoted Psyche as the Ultimate Cause, a Psyche Ontology, or a Psyche Epistemology, they opened the door for me so that I was able to do so.

My Books Focusing on a Psyche Ontology

The Ultimate Model of Reality: Psyche Is the Ultimate Cause

https://www.amazon.com/dp/B071NC9JK6

Putting Psyche Back into Psychology: Restoring Science to Consciousness

https://www.amazon.com/dp/B071NC987S

BioPsychoSocial: Including Psyche or Light into our Theoretical Models

https://www.amazon.com/dp/B0713NDHVW

NATURE vs. NURTURE vs. NIRVANA: An Introduction to Reality

https://www.amazon.com/dp/B01JWRCSVA

https://www.amazon.com/dp/1521132615

9. The Materialists and Naturalists Are Wrong about 95% of this Universe

The Materialists, Naturalists, Physicalists, Darwinists, Nihilists, Behaviorists, Determinists, Physical Reductionists, Classical Physicists, and Atheists teach that only entropic physical matter exists.

I discovered that the Materialists and Naturalists are wrong about 95% of this universe and actually deny the existence of 95% of this universe.

Ouch!

That was a painful scientific discovery. I used to be a Materialist, Naturalist, Nihilist, and Atheist. I never realized how in-the-dark I really was until I started looking at scientific evidence.

Dark Energy and Dark Matter comprise over 95% of our universe; and, ordinary physical matter comprises less than 5% of our universe. That explains a lot, doesn't it? This is one of my most favorite scientific discoveries.

Okay, the discovery of Dark Matter and Dark Energy is NOT my discovery; but, it was an epiphany for me nonetheless when I first learned about them.

Many people have realized that Dark Matter is Spirit Matter – an invisible type of matter that influences physical matter but can't be seen by nor directly detected by our physical instruments. Dark matter keeps the physical galaxies from flying apart. We KNOW that Dark Matter and Dark Energy exist by the effects that they have on physical matter. Dark Energy is the Zero-Point Field of Light, the Vacuum Energy, or what God calls the Light of Christ. The proven and verified existence of Dark Matter and Dark Energy FALSIFIES the claims of Materialism, Naturalism, Nihilism, and their derivatives.

During the Prime Event, or what most scientists call the Big Bang, only 4.6% of this universe was converted into physical matter. According to Astrophysics, the other 95.4% of this universe remained Non-Local or Spiritual as dark matter (spirit matter) and dark energy (the Light of Christ or vacuum energy or the zero-point field of light).

The Materialists, Naturalists, Darwinists, Nihilists, and Atheists are WRONG about 95.4% of this universe! Imagine it! 95.4% of this universe is still spiritual, or non-local, or non-physical. Doesn't it make you wonder what else the Materialists and Atheists might have gotten wrong?

[See: Ross, H. (2008). *Why the Universe Is the Way It Is*. Grand Rapids, MI: Baker Books.]

There's physical matter, and then there is everything else. Only 4.6% of our physical universe is entropic matter, baryonic matter, or physical matter. The rest of our universe is Energy or Syntropy. 95.4% of our universe is syntropic. Physical matter or entropy is the exception, not the NORM! Entropic physical matter only comprises 4.6% of our universe's cosmic density.

Furthermore, entropy and physical matter are not conserved. Entropy and physical matter are not syntropic. Physical matter and entropy were made from raw energy by the Gods which means that physical matter and entropy can be disassembled back into raw energy.

According to the Ultimate Law of Thermodynamics, Energy or Psyche is syntropic which means that Energy or Psyche is conserved. In contrast, physical matter and entropy are not conserved. Entropic physical matter is comprised of many different forms of energy; and, the form is never conserved. The Ultimate Law of Thermodynamics is hidden within the First Law of Thermodynamics which states that energy is conserved.

The first law of thermodynamics is a version of the law of conservation of energy, adapted for thermodynamic systems. The law of conservation of energy states that the total energy of an isolated system is constant; energy can be transformed from one form to another but can be neither created nor destroyed.

https://en.wikipedia.org/wiki/First_law_of_thermodynamics

Energy can neither be created nor destroyed. Psyche can neither be created nor destroyed. Psyche is the intelligence or the life force within energy that gives energy the ability to understand and obey God's Laws and God's Commands. Psyche or Energy is conserved. The First Law of Thermodynamics tells us that the energy is constant, and that the energy is conserved – not the FORM of that energy. Energy can be transformed from one form to another, which means that the FORM is never conserved.

Transformed by whom?

Transformation of any kind requires Someone Psyche or Someone Intelligent to instigate, originate, initiate, and start that transformation. Any type of beginning requires Someone Psyche to cause it to begin; otherwise, random chaos would continue to reign supreme.

The Ultimate Law of Thermodynamics emphasizes that the FORM is never conserved, which means that the different FORMS and emergent properties of energy such as physical matter, entropy, space, locality, and time are never conserved. Whenever something is formed or made or organized, that means that its FORM is not being conserved. Only Psyche or Energy is syntropic or conserved. Energy or Psyche cannot be created, and it cannot be destroyed, which means that it is being conserved. However, energy can be transformed into many different things by the Gods or by Someone Psyche. The FORM is never conserved. Only the underlying energy is conserved.

In contrast, entropic physical matter is created by the Gods from raw energy which means that physical matter and entropy can be destroyed and turned back into raw energy. The Energy, or the Psyche, or the Life Force cannot be made, and it cannot be destroyed, which means that it's always conserved. However, the form of that energy is never conserved. The Gods can transform energy into anything they desire including entropic physical matter. The Ultimate Law of Thermodynamics differentiates between what is being conserved and what is not being conserved, as a result it is good science.

Based upon E = mc² and the First Law of Thermodynamics, the Gods can turn Dark Matter and Dark Energy into entropic physical matter anytime they choose to do so, and the Gods can also turn entropic physical matter into Dark Matter and Dark Energy anytime they choose to do so; consequently, we are not done yet with what the Gods can still choose to do.

The Materialists, Naturalists, Darwinists, and Atheists erroneously teach that physical matter and entropy are being conserved because these people erroneously teach that only entropic physical matter exists. One error or falsehood leads to another. Entropy is death; and, these people erroneously teach that death is conserved. Do they not?

Materialism, Naturalism, Physicalism, Darwinism, Nihilism, Behaviorism, Determinism, Physical Reductionism, and Atheism were designed to prevent us from discovering the Ultimate Law of Thermodynamics; and, they have done just that. Have they not? Have you ever heard of the Ultimate Law of Thermodynamics? Of course not. The Materialists, Naturalists, Darwinists, and Atheists have successfully hidden it from us; and, these are the people who currently control our public schools.

The Ultimate Law of Thermodynamics states that physical matter and entropy are NOT conserved. Anything that was obviously formed or made obviously can be disassembled and formed into something else, which means that the form is not being conserved. Entropic physical matter is comprised of different forms of energy. Physical matter and entropy were obviously formed or made from raw energy, which means that physical matter and entropy can be disassembled, and the resulting raw energy transformed into something else besides entropic physical matter. The First Law of Thermodynamics states that energy or psyche is conserved, not the form of that energy. The form is never conserved according to the Ultimate Law of Thermodynamics. Entropy is death. Entropy is an expiration date. Death is NOT conserved according to the Ultimate Law of Thermodynamics. In contrast, your Psyche or Energy is eternal and everlasting. Your Life Force, Psyche, or Energy is syntropic which means that it's being conserved. Energy or Psyche cannot be made, and it cannot be destroyed which means that it is conserved.

Life Force, Psyche, or Energy is the ultimate perpetual motion machine because it's always conserved. In fact, entropic physical matter, the aging process, the passage of time, and entropy were designed and made by the Gods in order to put a temporary end to perpetual motion machines where physical matter and a fallen physical world are concerned. Do you see how that works? It's the answer to life, the universe, and everything. It explains everything that has ever been experienced or observed. The Ultimate Law of Thermodynamics is one of my greatest scientific discoveries.

Furthermore, the Quantum Law of Thermodynamics states that heat death is impossible at the quantum level because entropy or the second law of thermodynamics doesn't exist and doesn't apply in the Quantum Realm. At the quantum level, Psyche or Energy is always conserved. It cannot be made, and it cannot be destroyed. Technically, this means that heat death is ultimately impossible at the physical level as well. Even at maximum entropy, ALL of the energy is still there waiting to be reused anytime the Gods choose to do so.

The Human Psyche automatically organizes a non-consensus reality whenever it enters into such a system. Psyche trumps entropy. Psyche is Energy after all, and an infusion of energy is all that's needed to eliminate entropy from a system. Psyche is Intelligence, and an infusion of information, knowledge, or intelligence is all that's needed to eliminate entropy or disorder from a system. Psyche is Life Force, and Psyche or Syntropy is all that's needed to eliminate death or entropy from a system. Do you see how that works? It's the answer to life, the universe, and everything. It explains everything that has ever been experienced and observed.

God Reveals Himself in Two Ways – Through Scripture and Through Science

There are indications that a lot of energy has to be poured into spirit matter in order to convert spirit matter into physical matter. In other words, God poured a tiny bit of His Own Psyche, Light, Intelligence, Power, Consciousness, Energy, Glory, or Life into each physical atom that God brought into existence or commanded into existence. ONLY God could have done such a thing. We human beings certainly don't know how to do that kind of Science!

Because of this KNOWLEDGE, I'm now willing to admit Lived Experiences and the Revelations of God into evidence. I'm not afraid of it anymore, like I used to be when I was a Materialist, Naturalist, Nihilist, and Atheist. Give me evidence – lots and lots of evidence!

(See: https://www.lds.org/scriptures/dc-testament/dc/93?lang=eng).

(See: https://www.lds.org/scriptures/dc-testament/dc/88?lang=eng).

(See: https://www.lds.org/scriptures/dc-testament/dc/76?lang=eng).

(See: https://www.lds.org/scriptures/dc-testament/dc/130?lang=eng).

Science is Observation and Experience. Science is meant to be experienced and observed on both sides of the veil.

I find it fascinating to observe that the Real Science – the experienced and verified observations of the human race – repeatedly and steadily FALSIFY the claims of Materialism, Naturalism, Darwinism, Nihilism, and Atheism. EVERYTHING that we have experienced and observed as a race FALSIFIES Materialism, Naturalism, and their derivatives; whereas, there is NO evidence and can be NO evidence supporting the hidden assumptions or major premises of Materialism and Naturalism which claim that dark matter or spirit matter, dark energy or the stretching force of the universe, psyche or energy, choices or quantum waves, and the zero-point field of light or the light of Christ DO NOT EXIST. These people claim that Supernatural Mechanisms or Quantum Mechanisms or Spiritual Mechanisms do not exist. These people claim that Psyche and Spirit Matter do not exist. These people claim that invisible, intangible, non-physical forces and fields do not exist. These people claim that ONLY physical matter or entropy exists. They are wrong.

Every time we observe, verify, and catch Dark Matter and Dark Energy in the act, those scientific observations and experiences FALSIFY Materialism, Naturalism, Nihilism, and their derivatives such as Darwinism and Atheism. Every time we experience and observe the effects of Magnetism, Gravity, Thought, Radio Waves, Microwaves, X-Rays, Quantum Waves, and Light, we FALSIFY the claims of Materialism and Naturalism which state that the non-physical does not exist.

If you successfully eliminate everything that is false, then ONLY the truth will remain. If you successfully eliminate Materialism, Naturalism, Darwinism, Nihilism, Atheism, and their derivatives, then ONLY the truth remains. The false is falsified by the truth; and, the truth is repeatedly experienced and observed.

Mark My Words

—

10. Quantum Mechanics Is Spiritual Mechanics

The goal in all of this is to hook people up and give them a Model of Reality that actually matches with reality. I want to help people find the truth. I put a lot of effort into this so that I would have a Model of Reality that actually matches with the Scientific Evidence and the Observational Evidence that we have on hand as a race.

In contrast, the goal of the Materialists, Naturalists, Darwinists, Nihilists, Behaviorists, Determinists, Physical Reductionists, Atheists, and Classical Physicists is to convince people that certain things DO NOT EXIST. That goal is totally worthless. That's NOT science. That's religion and dogma. It takes an infinite amount of blind faith to believe that something DOES NOT EXIST. Furthermore, it's philosophically and scientifically impossible to prove that something does not exist. It can't be done, which means that they will NEVER be able to prove that Materialism, Naturalism, Darwinism, and Atheism are true.

These people are tilting windmills, and they don't even know it. These people are fighting imaginary enemies which are nothing more than a figment of their imagination. They are also promoting pseudo-sciences that are also nothing more than wishful thinking which they have manufactured within their minds out of thin air. There's nothing tangible within Materialism and Naturalism that we can get our hands on. The whole thing is purely a Grand Illusion. It doesn't match with Reality, which makes it unscientific to begin with.

When it comes to Scientific Naturalism, Darwinism, and Physicalism, these pseudo-sciences are based exclusively on entropy. Entropy is death, which means that these pseudo-sciences are literally a Dead-End to begin with. There's NO life there! Entropy cannot produce life. Entropy cannot produce order. Entropy can only produce death. Entropy IS death. Evolution is entropy. Evolution is death. The theory of evolution is Creation by Death. Please explain to me how that's supposed to work.

One of my greatest and most useful scientific discoveries is that Quantum Mechanics is Spiritual Mechanics or Supernatural Mechanics. The proven and verified existence of Action at a Distance falsifies Materialism, Naturalism, Darwinism, Nihilism, Behaviorism, Atheism, Classical Physics, and their derivatives. All you really want is the truth unless you are trying to deceive yourself.

Interpretations of Quantum Mechanics and Science

I used to be a Physicalist, Naturalist, Nihilist, and Atheist. I wasted a lot of time and effort learning concepts that are demonstrably false.

My Atheism was a very dark time in my life. If I can help to prevent even one person from going through what I went through, then it will have all been worth it!

What was the net effect of My Atheism?

It dumbed me down. It made me ignorant. It made me dumb and blind even though I still had a soul or a mind.

I knew that God does not exist.

Atheism makes you believe that you know more than you really do. Atheism makes you ignorant of all the different things that you need to learn and know about. Atheism made me ignorant of Quantum Mechanics or Supernatural Mechanics. Atheism made me ignorant of Syntropy. I'd never heard of Syntropy. I didn't know what it was. Atheism of any kind is based upon a refusal to look at evidence. Atheism makes you ignorant. Atheism makes you think that you are smart when you are not. You become arrogant and condescending towards others. That's the fruits of Atheism, Materialism, Naturalism, Nihilism, and Darwinism. You become dumber than what you used to be. Some people have called it the "dumbing down of America".

That's why it took me over 55 years to figure out that evolution is entropy. The theory of evolution is based upon entropy. That's why it doesn't work as advertised. It can't! Entropy can't design, create, and manufacture anything! Entropy is death. Entropy can only destroy; and think about it, the theory of evolution is based exclusively on entropy. The theory of evolution is Creation by Death. The theory of evolution is Creation by Entropy. The theory of evolution was false to begin with; but, I couldn't see that nor understand that because I had been educated and trained in Entropy, Materialism, Naturalism, Nihilism, Darwinism, Atheism, and Classical Physics. It made me dumb and blind.

The Materialists and Naturalists deliberately remove the quantum non-local sciences from consideration. These people are actively trying to hide Syntropy, Psyche, Quantum Mechanics, and Supernatural Mechanisms from us. The various different types of macro-evolution, or spontaneous generation, or chemical evolution have NEVER been observed in the wild. They have NEVER been experienced nor observed because they are physically impossible. They are prevented from happening by random diffusion or entropy. The Naturalists and Darwinists are trying to hide these scientific truths from us.

The Physicalists, Naturalists, and Atheists ONLY permit correlational research and experimental research into evidence. They ONLY permit physical evidence into evidence. They only permit the correlational method and the experimental method. They deliberately BAN all other sources of information and knowledge. However, there is also a Burden of Proof Methodology that is used in archeology and the forensic sciences. It is based upon the preponderance of the evidence. We have to use a Burden of Proof Methodology when going after spiritual, supernatural, and quantum realities and forces because we can't use our physical instruments to detect and record these things.

Correlations indicate a relationship, but that relationship is not necessarily one of cause and effect. Correlational research allows us to *predict,* but it cannot tell us whether changing one variable (such as social status) will *cause* changes in another (such as health).

The correlation-causation confusion is behind much muddled thinking in popular psychology.

The great strength of correlational research is that it tends to occur in real-world settings where we can examine factors such as race, gender, and social status (factors that we cannot manipulate in the laboratory). Its great disadvantage lies in the ambiguity of the results. This point is so important that even if it fails to impress people the first 25 times they hear it, it is worth repeating a twenty-sixth time: *Knowing that two variables*

change together (correlate) enables us to predict one when we know the other, but correlation does not specify cause and effect. (*Social Psychology*, pp. 20-21.)

One of my greatest and most useful scientific discoveries came when I first realized that correlation-causation confusion is behind much of the muddled thinking that's associated with Darwinism and the theory of evolution. Materialism, Naturalism, Darwinism, Nihilism, Atheism, Behaviorism, Determinism, Physical Reductionism, and the Theory of Evolution are purely correlational! There's NO evidence backing their claims that that Quantum or the Supernatural does not exist; and, there never will be. These falsified philosophies or falsified religions are purely correlational. They have NO evidence supporting them. They have to be taken on blind faith as being real and true. In fact, the verified and proven existence of Action at a Distance or Quantum Mechanics falsifies Materialism, Naturalism, Darwinism, and their derivatives. ALL the EVIDENCE falsifies Physicalism, Naturalism, and Darwinism. The theory of evolution is based exclusively on correlation-causation confusion. Fascinating, is it not?

Scientific experimentation has its weaknesses too. Sometimes laboratory research and science experiments fail to generalize to real-world settings. Manufacturing amino acids in a beaker in a simulated laboratory setting has absolutely NOTHING to do with getting entropy and physical matter to spontaneously generate into amino acids in the wild.

Experimental Research: Searching for Cause and Effect

The difficulty of discerning cause and effect among naturally correlated events prompts most social psychologists to create laboratory simulations of everyday processes whenever this is feasible and ethical.

We need to be cautious, however, in generalizing from laboratory to life.

Generalizing from Laboratory to Life

As the research on children, television, and violence illustrates, social psychology mixes everyday experience and laboratory analysis. Throughout this book we will do the same by drawing our data mostly from the laboratory and our illustrations mostly from life. Social psychology displays a healthy interplay between laboratory research and everyday life. Hunches gained from everyday experience often inspire laboratory research, which deepens our understanding of our experience.

This interplay appears in the children's television experiment. What people saw in everyday life suggested correlational research, which led to experimental research. Although the laboratory uncovers basic dynamics of human existence, it is still a simplified, controlled reality. It tells us what effect to expect of variable *X*, all other things being equal — which in real life they never are! Moreover, as you will see, the participants in many experiments are college students. Although that may help you identify with them, college students are hardly a random sample of all humanity. Would we get similar results with people of different ages, educational levels, and cultures? That is always an open question.

Nevertheless, we can distinguish between the *content* of people's thinking and acting (their attitudes, for example) and the *process* by which they think and act (for example, *how* attitudes affect actions and vice versa). The content varies more from culture to culture than does the

651

process. People from various cultures may hold different opinions yet form them in similar ways. (*Social Psychology*, pp. 24, 28, 29.)

The ONLY way to distinguish between Psyche's thoughts and the observed physical actions is to accept the fact that both exist and that both are REAL. The Psyche's choices do indeed correlate with a person's physical behavior or physical actions; but if we want a fullness of the truth, we must accept the independent evidence or experiential evidence that verifies and proves the existence of Psyche because we are not going to get any of that from physical matter and our physical instruments. We must treat the observations and experiences associated with Near-Death Experiences and Out-of-Body Experiences as Scientific Evidence and allow them into evidence, if we want to find and know the truth about Psyche Mechanics or Quantum Mechanics; otherwise, we choose to remain in ignorance. That's just the way it is.

The laboratory is NOT a real-world setting most of the time. The variables in our lives aren't typically controlled, constrained, and manipulated in an artificial manner. Furthermore, a physical laboratory is slow to reveal the existence of Psyche or Quantum Mechanics. It can be done if the scientist is smart enough and creative enough; but, direct observation and experience is a better way for finding and knowing the truth about Psyche, Quantum Mechanics, and Syntropy.

Thought is a function of the Human Psyche. Thoughts and memories are quantum waves.

How do we know?

We know because our thoughts and memories survive the death of our physical brain. We know because our thoughts and memories can't be detected nor recorded by our physical instruments.

If a thought or memory survives the death of your physical brain, then it's truly a memory in every sense of the word, is it not?

This is just a small portion of what I learned while studying Social Psychology. Social Psychology is the scientific study of the Human Psyche and its interaction with all those other psyches. Social Psychology is valid on both sides of the veil – both here in this physical reality and afterwards in the spirit world. This is radically advanced science when it is no longer being restrained and limited by Naturalism, Darwinism, and Atheism. Social Psychology has whole chapters on the Self or the Psyche. When it comes time to define and explain the Self or the Mind, Psyche or Quantum Non-Local Consciousness ends up being the BEST explanation. That is what I have experienced and observed. Science is observation and experience after all.

In contrast, Atheism is the making of something by "Nothing". Atheism is patently absurd. Creation ex nihilo or spontaneous generation is the making of something from "Nothing". Creation ex nihilo or spontaneous generation is a type of Atheism, and it's just as absurd as Atheism. Creation ex nihilo assumes facts not in evidence. In fact, it's contradicted by logic and by evidence. The same can be said of Atheism, Physicalism, Darwinism, Nihilism, and Classical Physics. They are contradicted or falsified by logic and by evidence. They are falsified by Quantum Mechanics or Supernatural Mechanisms.

Do you want the truth, or do you want the science fiction? You can't have both. The one is true, and the other has been falsified by the truth. The one is true, and the other has been falsified by observation and experience. The false is falsified by the truth; and, the truth is repeatedly experienced and observed. Science is observation and experience – NOT Physicalism and Naturalism. Technically, the "science" behind Materialism, Naturalism,

Darwinism, Nihilism, Atheism, and their derivatives has NEVER been experienced nor observed.

There are as many different interpretations of Quantum Mechanics as there are scientists, from what I can tell.

https://en.wikipedia.org/wiki/Interpretations_of_quantum_mechanics

https://en.wikipedia.org/wiki/Minority_interpretations_of_quantum_mechanics

Quantum Mechanics or Supernatural Mechanics is our most-used, most-verified, and best-proven Science that we have; but, nobody seems to know what it means.

I have experienced and observed the fact that interpretations of Quantum Mechanics that are based upon Physical Matter, Materialism, Naturalism, Darwinism, Nihilism, Atheism, and Classical Physics completely lack explanatory power and are therefore absolutely worthless.

Quantum Mechanics or Transdimensional Physics ONLY makes sense from a non-local, non-physical, transdimensional, spiritual perspective.

Mark My Words

Source Material

Quantum Mechanics from a Non-Physical Spiritual Perspective

https://www.amazon.com/dp/B01J023TGU

https://www.amazon.com/dp/1521132380

Myers, D. G. (2010). *Social Psychology* (10th ed.). New York: McGraw-Hill.

Entropy or the Passage of Time

According to the theory of relativity, anything traveling faster than the speed-of-light experiences NO passage of time. Time STOPS at the speed-of-light or faster.

Do you fully understand what this means?

It means that everything existing or traveling at frequencies faster than the speed-of-light is Syntropy. It doesn't age because there is NO passage of time. In other words, there is NO entropy. Entropy is a function of time. ONLY physical matter experiences entropy or is subject to entropy. Everything else is Syntropy.

Light is Syntropy. There is NO age limit on the Light that has been traveling for 13.2 billion years to get here from the other galaxies in this universe. There's NO entropy in Light. It doesn't age, get old, and die. From its perspective, it arrives the very moment it launches, even though from our perspective it took 13.2 billion years to get here. In fact, it has been theorized that Light already knows where it is going to land 13.2 billion light years away, or it doesn't launch in the first place.

Likewise, Psyche or Intelligence is Syntropy. It is ageless and timeless. There is NO entropy in Psyche, Intelligence, or Quantum Non-Local Consciousness. It is eternal and everlasting, without a beginning of days or an end of years. It is Syntropy. It doesn't bleed; so, there's no way to kill it or end it.

We will NEVER be able to accelerate or push physical matter faster than the speed-of-light because then it would no longer be physical matter, and it would no longer be subject to entropy or the passage of time. It would be spirit matter or dark matter instead. It would be Syntropy. It would be spiritual matter.

Physical matter has been caught in the act of Quantum Tunneling; so, it still retains that inherent capability. If you stop the passage of time or temporarily remove the entropy from physical matter, then you can teleport it or quantum tunnel it anywhere you want in the universe instantly. During the quantum tunneling process, it experiences NO passage of time. It doesn't age.

Those of us who have experienced Quantum Tunneling or Teleportation KNOW that time STOPS when you Quantum Tunnel. I felt time stop. It was like I had passed out. And then, I was just instantly somewhere else instead, with NO knowledge or memory of how I got there. I and the car I was driving had just jumped, blinked, or teleported. It was nothing that I did, though. It was something that God did for me in order to save me.

The science is REAL, and it's powerful; and, it's totally consistent, coherent, and harmonious with itself. It ALL makes sense. The Truth matches perfectly with all other truths. The Truth has been experienced and observed.

Quantum Tunneling is REAL. It has been experienced and observed. The only drawback or disappointment when it comes to Quantum Tunneling is that ONLY God has access to it, or ONLY God can do it and trigger it at will. ONLY God holds all the KEYS to Quantum Mechanics or Supernatural Mechanics. Mortal fallen beings like us aren't given direct access to these things. God resides within Syntropy. God is Syntropy. Whereas, we mortal fallen beings are stuck within or placed within entropy and physical matter. We age and die. When we die, our psyche or soul is released back into Syntropy.

God imposes entropy, the aging process, or the passage of time onto physical matter in order to keep the physical matter within your physical body from teleporting or quantum tunneling away from you at will. God keeps the physical matter within your body, house, computer, and car from dissolving into thin air by quantum tunneling to the other side of the universe at will. Physical matter still retains that capability; but, God prevents physical matter from quantum tunneling by making it subject to entropy or the passage of time. God deliberately slows the physical matter down to sub-light speeds so that we can live it, experience it, observe it, and remember having done so.

The explanatory power of this Science is through the roof. By choosing to allow Psyche, Syntropy, and Quantum Mechanics in to play, we can literally explain everything that comes our way.

Mark My Words

Quantum Mechanics Is Mind Over Matter

I used to be a Physicalist, Naturalist, Nihilist, and Atheist.

One of my greatest scientific discoveries came when I first realized that Quantum Mechanics is in fact Spiritual Mechanics, or the way that spirit matter really works. When I first realized that Psyche or a Non-Local Conscious Observer and the Word of Command are needed to convert spirit matter into physical matter, then suddenly the whole of Quantum Mechanics became clear to me.

God must of necessity exist in order to have provided the necessary Word of Command or Conscious Observation, which Quantum Mechanics tells us must take place in order to convert spirit matter into physical matter. The Bible actually tells us that Jesus Christ is the WORD or the Word of Command who brought physical matter into existence in the first place. If God's Psyche didn't exist, then there would be NO physical matter.

For me personally, Quantum Mechanics became my first positive Scientific Verification of God's Existence. The scientists keep verifying Quantum Mechanics or Spiritual Mechanics over and over again; so, it must be the truth, because it's impossible to use the scientific methods to falsify the truth. Quantum Mechanics or Spiritual Mechanics has never been falsified, although the materialistic and naturalistic interpretations of Quantum Mechanics have been falsified trillions of times.

After realizing that God's Psyche must of necessity exist in order to convert spirit matter into physical matter, I simply KNEW that God exists.

Here's a list of some of the books and articles that confirmed or verified these scientific observations for me.

See: Van Lommel, P. (2010). *Consciousness Beyond Life: The Science of the Near-Death Experience*. New York: HarperCollins.

http://avalonlibrary.net/ebooks/Pim%20van%20Lommel%20-%20Consciousness%20Beyond%20Life.pdf

https://science-2-0.com/wp-content/uploads/2018/05/Consciousness-Beyond-Life.pdf

https://quantum-neuroscience.com/wp-content/uploads/2018/05/Consciousness-Beyond-Life.pdf

See: Goswami, A. (2008). *God Is Not Dead: What Quantum Physics Tells Us about Our Origins and How We Should Live*. Charlottesville, VA: Hampton Roads.

See also: Mark My Words. (2016). *Quantum Mechanics from a Non-Physical Spiritual Perspective*. Kindle. Retrieve from:

https://www.amazon.com/dp/B01J023TGU

The existence of Light falsifies Materialism and Naturalism, and it ends up pointing us directly to God and God's Psyche.

See: Mark My Words. (2017). *God Is in the Light: God is light, and in Him is no darkness at all*. Retrieve from:

https://www.amazon.com/dp/B07168S37N

Quantum Mechanics is Supernatural Mechanics or Supernatural Mechanisms. Quantum Mechanics or Transdimensional Physics IS Action at a Distance; and, Action at a Distance falsifies the claims of Classical Physics or Naturalism which claim that Action at a Distance is impossible and does not exist.

I find it fascinating to observe that the Real Science – the experienced and verified observations of the human race – repeatedly and steadily FALSIFY the claims of Materialism, Naturalism, Darwinism, Nihilism, and Atheism. EVERYTHING that we have experienced and observed as a race FALSIFIES Materialism, Naturalism, and their derivatives; whereas, there is NO evidence and can be NO evidence supporting the hidden assumptions or major premises of Materialism and Naturalism which claim that the quantum, or the supernatural, or action at a distance DOES NOT EXIST.

Every time we observe, verify, and catch Quantum Mechanisms or Action at a Distance in the act, those scientific observations and experiences FALSIFY Materialism, Naturalism, Nihilism, and their derivatives such as Darwinism and Atheism.

If you successfully eliminate everything that is false, then ONLY the truth will remain. If you successfully eliminate Materialism, Naturalism, Darwinism, Nihilism, Atheism, and their derivatives, then ONLY the truth remains.

Physical matter or entropy cannot explain everything that comes our way. Classical Physics, Physicalism, or Naturalism cannot explain everything that we encounter. Materialism, Naturalism, Darwinism, Nihilism, Behaviorism, Determinism, Physical Reductionism, Atheism, and Classical Physics LACK explanatory power when it comes to Action at a Distance, Psyche, Syntropy, and Quantum Mechanics.

Here are some of the articles and books that convinced me that the Materialists and Naturalists are wrong in their assumptions.

Is Your Brain Really Necessary?

https://quantum-neuroscience.com/wp-content/uploads/2018/05/Is-Your-Brain-Really-Necessary.pdf

A brain really isn't needed for thoughts and memories to take place. Thoughts and memories are quantum waves. We KNOW this is so because it is physically impossible to record our thoughts and memories with our physical devices or physical machines. Thoughts, memories, or quantum waves are NOT a product or epiphenom of physical matter.

Mind Time

https://zodml.org/sites/default/files/%5BBenjamin_Libet%2C_Professor_Stephen_M._Kosslyn%5D_Min.pdf

https://quantum-neuroscience.com/wp-content/uploads/2018/05/Mind-Time.pdf

https://science-2-0.com/wp-content/uploads/2018/05/Mind-Time.pdf

We continue to think, learn, and remember long after our physical brain is dead and gone according to the scientific evidence that has been obtained from Near-Death Experiences (NDEs), Shared-Death Experiences (SDEs), Out-of-Body Experiences (OBEs), and our after-death Life Reviews. Thoughts and memories are a product of the Human Psyche.

The Mind and the Brain: Neuroplasticity and the Power of Mental Force - Jeffrey M. Schwartz, Sharon Begley (2003):

http://publicism.info/psychology/mind/

https://science-2-0.com/wp-content/uploads/2018/05/The-Mind-and-the-Brain.zip

https://quantum-neuroscience.com/wp-content/uploads/2018/05/The-Mind-and-the-Brain.zip

Quantum Mechanics NEVER made any sense to me while I was a Materialist, Naturalist, Nihilist, and Atheist. Eventually I learned that Quantum Mechanics or Action at a Distance is a non-local, non-physical, immaterial, spiritual, transdimensional phenomenon.

When I finally realized that Quantum Mechanics or Supernatural Mechanics is in fact how spirit matter and psyche work at the quantum level or psyche level, then suddenly it became clear to me what Spiritual Mechanics or Quantum Mechanics is and how it works. Quantum Mechanics only makes sense from a Spiritual Perspective. Classical Physics, Materialism, Naturalism, Darwinism, Nihilism, Behaviorism, Determinism, Physical Reductionism, and Atheism CANNOT explain how Quantum Mechanics or Syntropy works, nor can they explain what Quantum Mechanics or Supernatural Mechanics is.

The Spiritual Brain

https://epdf.tips/the-spiritual-brain-a-neuroscientists-case-for-the-existence-of-the-soul4a910f6e71b0354edde057767fea620d82703.html

https://science-2-0.com/wp-content/uploads/2018/05/The-Spiritual-Brain.pdf

https://quantum-neuroscience.com/wp-content/uploads/2018/05/The-Spiritual-Brain.pdf

These are a few books that you must read and understand if you have a desire to move your Science into the Quantum Age.

These are some of the Science Books that the Materialists, Naturalists, Nihilists, and Atheists are trying to convince you do not exist. They don't want you to find this material because it FALSIFIES Materialism, Naturalism, Nihilism, Atheism, and even Darwinism or the Theory of Evolution.

Statistics seem to suggest that the majority of scientists and philosophers have chosen to believe in and promote Scientific Naturalism, Physicalism, Darwinism, Nihilism, Atheism, and Classical Physics. NO amount of evidence will ever convince them that they are wrong until they are actually willing to look at the evidence and accept it as being real and true.

But, we don't need to convince the scientists that they are wrong. We – the public at-large – can skip right past them and figure out for ourselves why they are wrong and how they are wrong. That's what I eventually chose to do. So long as you have found the truth, it doesn't really matter what the scientists choose to believe; now does it? All they really need is classical physics and the physical sciences to do their work of saving world – or destroying the world. God holds ALL the KEYS to Quantum Mechanics in any case; and, God doesn't give us mere mortals direct access to nor direct control over Quantum Mechanics anyway. There's nothing that the scientists can do to get at Quantum Mechanics or Supernatural Mechanics and use that Science against us or against God. Most of them don't need Quantum Mechanics to do their work; so, just let them remain in the dark if they want to.

The beauty of Psyche, Syntropy, and Quantum Mechanics is that you can literally explain everything that happens to you and everything that comes your way once you choose to allow them into play.

Mark My Words

The Self Does Not Die

The Self or the Psyche does not die. It can't die.

Why?

This reality and truth just might be one of the most important scientific discoveries that a person can make during his or her scientific career.

The theory of relativity teaches us that at the speed-of-light or faster than the speed-of-light TIME STOPS. There is NO passage of time at the speed-of-light or faster than the speed-of-light.

Do you know what that means?

Well, being a Materialist, Naturalist, and Classical Physicist, Albert Einstein erroneously concluded that nothing exists faster than the speed-of-light. He was wrong!

Tachyons or spirit matter exist at frequencies faster than the speed-of-light.

In a book entitled *Physics of the Impossible* by Michio Kaku, we find the following description for Tachyons.

> **Tachyons live in a strange world where everything travels faster than light. As tachyons lose energy [mass], they travel faster, which violates common sense. In fact, if they lose all energy [or mass], they travel at infinite velocity. As tachyons gain energy [or mass], however, they slow down until they reach the speed-of-light. (p. 280).**

That's the exact SAME definition for Spirit Matter that I have been giving in all of my books for the past couple of years.

Tachyons ARE Spirit Matter!

Tachyons make perfect sense to me, because I'm no longer a Materialist, Naturalist, Nihilist, and Atheist. At velocities slower than the speed-of-light, objects take on mass or energy, which slows them down. All of that energy is needed to create the space between the nucleus of the atom and its electrons. Some type of force or field is creating that space-time between the nucleus and the electrons. It has been observed that one has to infuse a lot of energy into an atom in order to raise the electrons to a higher state or a higher orbit. The net result of this slow-down process is that these objects also take on entropy and are forced to experience the passage of time or the aging process. They become subject to space-time. In contrast, if objects lose ALL of their mass, then they are capable of infinite velocities and they experience NO passage of time or NO entropy! Entropy is a function of time.

There's the truth of the matter!

There's the answer to life, the universe, and everything. It's been there all along just waiting for someone to see it, discover it, and declare it.

A photon traveling at the speed-of-light experiences NO passage of time. From its perspective, it arrives at its destination instantly the very moment it launches. From our perspective, existing at velocities much slower than the speed-of-light, it took that photon 13.2 billion years to reach us. From its perspective, it took NO time at all.

At the speed-of-light, TIME STOPS. There is NO passage of time. There is NO entropy. At the speed-of-light or faster, everything is SYNTROPY, meaning that everything is eternal and everlasting.

The Psyche, Self, Mind, Soul, Intelligence, or Quantum Non-Local Consciousness exists at frequencies faster than the speed-of-light, which means that it is SYNTROPY. It is eternal and everlasting, without a beginning of days or an end of years.

This just might be one of my most significant and most useful scientific discoveries of all time. It fits and makes sense to me. It's predicted by the theory of relativity.

Entropy is death.

Why is entropy death?

It's because entropy involves the passage of time. Things age and die at velocities slower than the speed-of-light.

In contrast, SYNTROPY is eternal life. The Psyche is eternal and everlasting because it experiences NO passage of time. It doesn't age. It has NO entropy. That pretty much explains life, the universe, and everything – now doesn't it?

What's the answer to life, the universe, and everything?

Well, it's not 42.

The answer to life the universe, and everything is SYNTROPY or PSYCHE or INTELLIGENCE. The question is, "What is eternal and everlasting, without a beginning of days or an end of years?" The answer is Psyche or Syntropy. SYNTROPY is the fundamental basis for life, the universe, and everything!

Quantum Mechanics is a type of Syntropy. It's eternal and everlasting.

These realities and truths have been experienced and observed. Science IS observation and experience. Science or observation is an infinitely better way for getting at the truth than just to say that psyche or syntropy does not exist, as the Materialists and Naturalists do. Saying that psyche or syntropy DOES NOT EXIST is a cop out. It's the way that these people avoid doing the science and the observations that need to be done in order to establish the truth of the matter. It's an evasion or a dodge, NOT science. These people are avoiding the truth or hiding from the truth rather than seeking for the truth. Scientists are supposed to search for and find the truth; but, these people refuse to do so.

The Self Does Not Die: Verified Paranormal Phenomena from Near-Death Experiences

https://www.amazon.com/dp/0997560800

https://www.amazon.com/dp/B01LYMHXCO

Physicalism, Naturalism, Darwinism, Nihilism, and Atheism are based upon a refusal to look at evidence. I KNOW because I used to be a Materialist, Naturalist, Nihilist, and Atheist. We Naturalists and Atheists have to castrate and lobotomize science in order to prevent people from falsifying our theories and our ideas because ALL of the scientific evidence or observational evidence that we have on hand as a race falsifies the claims of Materialism, Naturalism, Darwinism, Nihilism, and Atheism which state that such falsifying evidence does not exist.

Science itself convinced me that Materialism, Naturalism, Darwinism, Nihilism, and Atheism are FALSE. I believe the Science – the observations and the experiences of real people – and not the philosophical wishful thinking of the Materialists, Naturalists, Darwinists, Nihilists, and Atheists.

Mark My Words

Orthodox Interpretation of Quantum Mechanics

The Orthodox Interpretation of Quantum Mechanics by Henry P. Stapp was the FIND of a lifetime. Henry P. Stapp's work is important to know about and understand – everything else pales in comparison. The Orthodox Interpretation of Quantum Mechanics explains what the Human Psyche and Nature's Psyche are doing at the quantum level in order to get things done for us at the physical level. This is GOOD SCIENCE! Its explanatory power is through the roof to infinity and beyond.

This is the information and the scientific evidence that the Materialists, Naturalists, Darwinists, Nihilists, and Atheists are trying to hide from you; and, they are very good at what they do. I KNOW because I used to be a Materialist, Naturalist, Nihilist, and Atheist. These people are actively banning, blocking, and destroying this information about Syntropy and Quantum Mechanics because it falsifies Physicalism, Naturalism, Darwinism, and Nihilism.

Remember, Materialism, Naturalism, Darwinism, Nihilism, Behaviorism, Determinism, Physical Reductionism, and Atheism are based exclusively on Classical Physics and Entropy. Scientific Naturalism or Physicalism IS Classical Physics. Quantum Mechanics falsifies Classical Physics, or Materialism and Naturalism.

Stapp seems to be giving everything away for free on his website.

https://sites.google.com/a/lbl.gov/stappfiles/

http://www-physics.lbl.gov/~stapp/

He seems to be much more interested in getting his message out to the world than he is in maintaining the copyright for his books.

The Orthodox Interpretation explains what the Human Psyche and Nature's Psyche are doing at the quantum level in order to get things done for us at the physical level.

If your interpretation of Quantum Mechanics can't explain what the Human Psyche and Nature's Psyche are doing at the quantum level in order to get things done for you at the physical level, then your interpretation of Quantum Mechanics is absolutely worthless.

Quantum Mechanics makes absolutely NO sense whatsoever whenever it is reduced to Classical Physics, Materialism, Naturalism, Physicalism, and Nihilism. Most of the interpretations of Quantum Mechanics found online and within our college textbooks have indeed been reduced to Materialism, Naturalism, and Classical Physics; and thereby, they have completely lost their explanatory power and their scientific value. Materialism and Naturalism neuter and destroy everything that comes their way, including Quantum Mechanics.

To demonstrate some of these truths, first I will quote from:

NEUROSCIENCE, ATOMIC PHYSICS, AND THE HUMAN PERSON

http://www-physics.lbl.gov/~stapp/2nd.doc

https://quantum-neuroscience.com/wp-content/uploads/2018/05/Neuroscience-and-the-Human-Person.pdf

http://www-physics.lbl.gov/~stapp/phys.txt

The Quantum Approach.

Classical physics is *an approximation* to a more accurate theory - called quantum mechanics - and quantum mechanics makes mind efficacious. Quantum mechanics *explains* the causal effects of mental intentions upon physical systems: it *explains* how your mental effort can produce the brain events that cause your bodily actions. Thus, quantum theory converts science's picture of you from that of a mechanical automaton to that of a mindful human person. Quantum theory also shows, explicitly, how the approximation that reduces quantum theory to classical physics completely eliminates all effects of your conscious thoughts upon your brain and body. Hence, from a physics point of view, trying to understand the mind-brain connection by going to the classical approximation is absurd: it amounts to trying to understand something in an approximation that eliminates the effect you are trying to study.

Quantum mechanics arose during the twentieth century. Scientists discovered, empirically, that the principles of classical physics were not correct. Moreover, they were wrong in ways that no minor tinkering could ever fix. The *basic principles* of classical physics were thus replaced by *new basic principles* that account uniformly both for all the successes of the older classical theory and also for all the newer data that is incompatible with the classical principles.

Physical theory was turned inside out.

The most profound alteration of the fundamental principles was to bring the consciousness of human beings into the basic structure of the physical theory. In fact, the whole *conception of what science is* was turned inside out. The core idea of classical physics was to describe the "world out there," with no reference to "our thoughts in here." But the core idea of quantum mechanics is to describe *our activities as knowledge-seeking and knowledge-using agents*. Thus, quantum theory involves, basically, not just what is "out there," but also what is "in here," namely "our knowledge." Consciousness is thus introduced into contemporary orthodox physical theory, not as something *whose existence needs to be explained,* but as rather as something whose detailed structure and detailed connection to brain activities needs to be *further* explicated.

Science must bridge the psycho-physical divide.

The basic philosophical shift in quantum theory is the *explicit* recognition that science is about *what we can know*. It is fine to have a beautiful and elegant mathematical theory about an imagined *"really existing physical world out there"* that meets a lot of intellectually satisfying criteria. But the essential demand of science is that the theoretical constructs be tied to the experiences of the human scientists who devise ways of testing the theory, and of the human engineers and

661

technicians who both participate in these tests, and eventually put the theory to work. So, the structure of a proper physical theory must involve not only the part describing the behavior of the not-directly-experienced theoretically postulated entities, expressed in some appropriate symbolic language, but also a part describing the human experiences that are involved in these tests and applications, expressed in the language that we actually use to describe such experiences to ourselves and each other. Finally, we need some "bridge laws" that specify the connection between the concepts described in these two different languages.

Classical physics met these requirements in a rather trivial kind of way, with the relevant experiences of the human participants being taken to be direct apprehensions of various gross behaviors of large-scale properties of big objects composed of huge numbers of the tiny atomic-scale parts. And these apprehensions were taken to be *passive*: they had no effect on the behaviors of the systems being studied. But the physicists who were examining the behaviors of systems that depend sensitively upon the behaviors of their tiny atomic-scale components found themselves forced to go to a less trivial theoretical arrangement, in which the human agents were no longer passive observers, but were *active participants* in ways that contradicted, and were impossible to comprehend within, the general framework of classical physics, *even when the only features of the physically described world that the human beings observed were large-scale properties of measuring devices.*

The two-way quantum psycho-physical bridge.

The sensitivity of the behavior of the devices to the behavior of some tiny atomic-scale particles propagates in such a way that the acts of observation by the human observers of *large-scale properties of the devices* could no longer be regarded as passive: these acts were assigned a crucial *selective* action. Thus, the core structure of the basic general physical theory became transformed in a profound way: the connection between physical behavior and human knowledge was changed from a one-way bridge to a mathematically specified two-way interaction that involves *selections* performed by conscious minds.

This profound change in the principles is encapsulated in Niels Bohr dictum that "in the great drama of existence we ourselves are both actors and spectators." (Bohr, 1963: 15 & 1958: 81) The emphasis here is on "actors": in classical physics we, and in particular our minds, were mere spectators.

This revision must be expected to have important ramifications in neuroscience, because the issue of the connection between mind (the psychologically described aspects of a human being) and brain/body (the physically described aspects of that person) has recently become a matter of central concern in neuroscience.

The Copenhagen formulation.

The original formulation of quantum theory was created mainly at an Institute in Copenhagen directed by Niels Bohr and is called "The Copenhagen Interpretation." Due to the profound strangeness of the conception of nature entailed by the new mathematics, the Copenhagen strategy was to refrain from making ordinary ontological claims, but to take, instead, a fundamentally pragmatic stance. Thus, the theory was formulated *basically* as a set of practical rules for how scientists should go about their tasks of acquiring knowledge, and then using this knowledge in practical ways. Speculations about "what the world out there – apart from our knowledge of it - is really like" were regarded as "metaphysics," and hence outside *real* science.

Copenhagen quantum theory is about the relationships between human agents (called *participants* by John Wheeler) and the systems that they act upon. In order to achieve this conceptualization, the Copenhagen formulation separates the physical universe into two parts, which are described in two different languages. One part is the observing human agent and his measuring devices. That part is described in mental terms - in terms of our instructions to colleagues about how to set up the devices, and our reports of what we then learn. The other part of nature is *the system that the agent is acting upon*. That part is described in physical terms - in terms of mathematical properties assigned to tiny space-time regions.

Von Neumann's Process II.

The great mathematician and logician John von Neumann formulated Copenhagen quantum theory in a rigorous way.

Von Neumann identified two very different processes that enter into the quantum theoretical description of the evolution of a physical system. He called them Process I and Process II (Von Neumann, 1955: 418). Process II is the analog in quantum theory of the process in classical physics that takes the state of a system at one time to its state at a later time. This Process II, like its classical analog, is *local* and *deterministic*. However, Process II by itself is not the whole story: it generates physical worlds that do not agree with human experiences. For example, if Process II were the *only* process in nature then the quantum state of the moon would represent a structure smeared out over large part of the sky.

Process I: A dynamical psycho-physical bridge.

To tie the quantum mathematics to human experience in a rationally coherent and mathematically specified way quantum theory introduces *another process*, which Von Neumann calls Process I. It is a *selection* process that is tied to conscious experience, and it is not determined by the micro-local deterministic Process II. It is a selection made by an agent about how he or she will act or attend.

Any physical theory must, in order to be complete, specify how the elements of the theory are connected to human experience. In classical physics this connection is part of a *metaphysical* superstructure: it is not part of the core dynamical description. But in quantum theory this connection of the mathematically described physical state to conscious experiences is part of the essential dynamical structure. And this connecting process is not passive: it does not represent a mere *witnessing* of a physical feature of nature by a passive mind. Rather, the process is active: it injects into the physical state of the system being acted upon properties that depend upon the intention of the observing agent.

Quantum theory is built upon the practical concept of intentional actions by agents. Each such action is expected or intended to produce an experiential response or feedback. For example, a scientist might act to place a Geiger counter near a radioactive source and expect to see the counter either "fire" during a certain time interval or not "fire" during that interval. The experienced response, "Yes" or "No", to the question "Does the counter fire during the specified interval?" specifies one bit of information. Quantum theory is thus an information-based theory built upon the knowledge-acquiring actions of agents, and the knowledge that these agents thereby acquire.

Probing actions of this kind are performed not only by scientists. Every healthy and alert infant is engaged in making willful efforts that produce experiential feedbacks, and he or she soon begins to form expectations about what sorts of feedbacks are likely to follow from some particular kind of effort. Thus, both empirical science and normal human life are based on paired realities of this action-response kind, and our physical and psychological theories are both basically attempts to understand these linked realities within a rational conceptual framework.

The basic building blocks of quantum theory are, then, a set of intentional actions by agents, and for each such action an associated collection of possible "Yes" feedbacks, which are the possible responses that the agent can judge to be in conformity to the criteria associated with that intentional act. For example, the agent is assumed to be able to make the judgment "Yes" the Geiger counter clicked or "No" the Geiger counter did not click. And he must be able to report. "Yes" the counter is in the specified place, or "No" it is not there. Science would be difficult to pursue if scientists could make no such judgments about what they were experiencing.

All known physical theories involve idealizations of one kind or another. In quantum theory the main idealization is not that every object is made up of miniature planet-like objects. It is rather that there are agents that perform intentional acts each of which can result in a feedback that may conform to a certain criterion associated with that act. One bit of information is introduced into the world in which that agent lives, according to whether the feedback conforms or does not conform to that criterion. Thus, knowing whether the counter clicked or not places the agent on one or the other of two alternative possible separate branches of the course of world history.

These remarks reveal the enormous difference between classical physics and quantum physics. In classical physics the elemental ingredients are tiny invisible bits of matter that are idealized miniaturized versions of the planets that we see in the heavens, and that move in ways unaffected by our consciousness, whereas in quantum physics the elemental ingredients are intentional actions by agents, the feedbacks arising from these actions, and the effects of our actions on the physical systems that our actions act upon.

Consideration of the character of these differences makes it plausible that quantum theory may be able to provide the foundation of a scientific theory of the human person that is better able than classical physics to integrate the physical and psychological aspects of his nature. For quantum theory describes the effects of a person's intentional actions upon the physical world, whereas classical physics systematically leaves these effects out.

An intentional action by a human agent is partly an intention, described in psychological terms, and partly a physical action, described in physical terms. The feedback also is partly psychological and partly physical. In quantum theory these diverse aspects are all represented by logically connected elements in the mathematical structure that emerged from the seminal discovery of Heisenberg. That discovery was that in order to get the quantum generalization of a classical theory one must formulate the theory in terms of *actions*. A key difference between *numbers* and *actions* is that if A and B are two actions then AB represents the action obtained by performing the action A upon the action B. If A and B are actions then, generally, AB is different from BA: the order in which actions are performed matters.

The intentional actions of agents are represented mathematically in Heisenberg's space of actions. Here is how it works.

Each intentional action depends, of course, on the intention of the agent, and upon the state of the system upon which this action acts. Each of these two aspects of nature is represented within Heisenberg's space of actions by an action.

The idea that a "state" should be represented by an "action" may sound odd, but Heisenberg's key idea was to replace what classical physics took to be a "being" by a "doing." I shall denote the action that represents the state being acted upon by the symbol S.

An intentional act is an action that is intended to produce a feedback of a certain conceived or imagined kind. Of course, no intentional act is sure-fire: one's intentions may not be fulfilled. Hence the intentional action puts in play a process that will lead either to a confirmatory feedback "Yes," the intention is realized, or to the result "No", the "Yes" response failed to occur.

The effect of this intentional mental act is represented mathematically by an equation that is one of the key equations of quantum theory. This equation represents, within the quantum mathematics, the effect of the Process I mental action upon the quantum state S of the system being acted upon. The equation is:

$$S \rightarrow S' = PSP + (1-P)S(1-P).$$

This formula exhibits the important fact that this Process I action changes the state S of the system being acted upon into a new state S', which is a *sum* of two parts.

The first part, PSP, represents the possibility in which the experiential feedback called "Yes" appears, and the second part, $(1-P)S(1-P)$, represents the alternative possibility "No", this feedback does not appear. Thus, the intention of the action and the associated experiential feedback are tied into the mathematics that describes the dynamics of the physical system being acted upon.

The action P is important. It represents an action upon the system that is being acted upon by the agent, and it depends on *the intention of the agent*. The action represented by the symbol P, acting both on the right and on the left of S, is the action of eliminating from the state S all parts of S except the "Yes" part. That particular retained part is determined by the intentional choice of the agent. The action of $(1-P)$, acting both on the right and on the left of S, is, analogously, to eliminate from S all parts of S except the "No" parts.

The projection operator P is required to satisfy $P = PP$. This implies $P(1-P) = (1-P)P = 0$, which says that the sequence of these two actions, P and $(1-P)$, in either order, leave nothing.

Thus, the action P is an action in the space in which the physical system is represented, and it reduces to zero all components that correspond to the "No" response, but it leaves intact the components corresponding to the "Yes" response to the intentional action. The action of $(1-P)$ is the analogous action with "Yes" and "No" interchanged. The action of P is the representation of an intentional mental action upon a physically described system.

Notice that Process I produces the *sum* of the two alternative possible feedbacks, not just one or the other. Since the feedback must either be "Yes" or "No = Not-Yes," one might think that Process I, which *keeps* both the "Yes" and the "No" parts, would do nothing. But that is not correct! This is a key point. It can be verified by noticing that S can be written as a sum of four parts, only two of which survive the Process I action:

$$S = PSP + (1-P)S(1-P) + PS(1-P) + (1-P)SP.$$

This formula is a strict identity. The dedicated reader can easily confirm it by collecting the contributions of the four occurring terms PSP, PS, SP, and S, and verifying that all terms but S cancel out. This identity shows that the state S can be expressed as a sum of four parts, *two of which are eliminated by Process I.*

But this means that Process I has a *nontrivial effect* upon the state being acted upon: it eliminates the two terms that correspond neither to the appearance of a "Yes" feedback nor to the failure of the "Yes" feedback to appear.

That is the *first key point*: quantum theory has a specific dynamical process, Process I, which specifies the effect upon a physically described system of an *intentional act* by a conscious agent.

Free Choices.

The second key point is this: the agent's choices are "free choices," *in the specific sense specified below.*

Orthodox quantum theory is formulated in a realistic and practical way. It is structured around the activities of human agents, who are considered able to freely elect to probe nature in any one of many possible ways. Bohr emphasized the freedom of the experimenters in passages such as:

"The freedom of experimentation, presupposed in classical physics, is of course retained and corresponds to the free choice of experimental arrangement for which the mathematical structure of the quantum mechanical formalism offers the appropriate latitude." (Bohr, 1958: 73)

This freedom of action stems from the fact that in the original Copenhagen formulation of quantum theory the human experimenter is considered to stand outside the system to which the quantum laws are applied. Those quantum laws are the only precise laws of nature recognized by that theory. Thus, according to the Copenhagen philosophy, *there are no presently known laws that govern the choices* made by the agent/experimenter/observer/participant about how the observed system is to be probed. This choice is, *in this very specific sense*, a "free choice." It is not ruled out that some deeper theory will eventually provide a causal explanation of this "choice."

Probabilities.

The predictions of quantum theory are generally statistical: only the *probabilities* that the agent will experience each of the alternative possible feedbacks are specified. Which of these alternative possible feedbacks will actually occur in response to a Process I action is not determined by quantum theory.

The formula for the probability that the agent will experience the feedback 'Yes' is Tr PSP/Tr S, where the symbol Tr represents the trace operation. This trace operation means that the actions act in a cyclic fashion, so that the rightmost action acts back around upon the leftmost action. Thus, for example, Tr ABC=Tr CAB =Tr BCA. The product ABC represents the result of letting A act upon B, and then letting that product AB act upon C. But what does C act upon? Taking the trace of ABC means specifying that C acts back around on A.

An important property of a trace is that the trace of any of the sequences of actions that we consider must always give a positive number or zero. Thus, this trace operation is what ties the actions, as represented in the mathematics, to measurable numbers.

[The trace operation, and in fact the operation of multiplying together any two operators, is the quantum analog of the classical process of integrating over all of "phase space," giving equal a prior weighting to

equal volumes of phase space. Thus, the trace operation is in effect a statistical sum over all of the "loose ends" that are not fixed in the expression upon which the trace operation acts.]

Von Neumann's psycho-physical theory of the conscious brain.

The Copenhagen approach separates the world into two parts: "The Observer" which includes the mind, brain, and body of the personal observer together with his measuring devices; and "The System" that this observer is acting upon. "The Observer" is described in psychological terms, whereas "The System" is described in physical/mathematical spacetime terms.

This procedure works very well in practice. However, it seems apparent that the body and brain of the human agent, and his devices, are parts of the physical universe. Hence a complete theory ought to be able to include our bodies and brains in the physically described part of the theory. On the other hand, the structure of the theory depends critically also upon the features that are represented in Process I, and that are described in mentalistic language as intentional actions and experiential feedbacks.

Von Neumann showed that it was possible, without significantly disturbing the predictions of the theory, to shift the bodies and brains of the agents, along with their measuring devices, into the physical world, *while retaining. and ascribing to the mind of the agent, those mentalistically described properties of the agents that are essential to the structure of the theory.* The system acted upon by the mind is the brain. Thus, in this von Neumann re-formulation the Process I action is an action of mind upon brain. Hence von Neumann's re-formulation provides us with the core of a science-based dynamical theory of the conscious brain.

It is worthwhile to reflect for a moment on the ontological aspects of Von Neumann quantum theory. Von Neumann himself, being a clear-thinking mathematician, said very little about ontology. But he called the mentalistically described aspect of the agent "his abstract 'ego'." (von Neumann, 1955: 421). This phrasing tends to conjure up the idea of a disembodied entity, standing somehow apart from the body/brain. But another possibility is that consciousness is an *emergent property* of the body-brain. Notice that some of the problems that occur in trying to defend the idea of emergence within the framework of classical physical theory disappear when one accepts the validity of quantum theory. For one thing, one no longer has to defend against the charge that the emergent property, consciousness, has no "genuine" causal efficacy, because anything it does is done already by the physically described process, independently of whether the psychologically described aspect emerges of not. In quantum theory the causal efficacy of our thoughts is no illusion: it's the real thing!

Another difficulty with "emergence" in a classical physics context is in understanding how the motion of a set of miniature planet-like objects, careening through space, can *be a painful experience.* But within the quantum framework the basic physical structure, namely the quantum state, is essentially knowledge or information imbedded in space-time. Hence there is no intrinsic problem with the idea that a sudden increment in a person's knowledge should be represented by a sudden jump in the quantum state of his brain. The identification of conscious actions with physical actions is no longer problematic. This is because the old idea of "matter" has been eradicated and replaced by a mathematical representation of an information-based psycho-physical reality.

In this connection, Heisenberg remarked:

"The conception of the objective reality of the elementary particles has thus evaporated not into the cloud of some obscure new reality concept, but into the transparent clarity of a mathematics that represents no longer the behavior of the particle but rather our knowledge of this behavior." (Heisenberg, 1958).

Conservation of Causality.

The question arises: How can the effect of a psychologically described action be injected into the dynamics of a physically described system without upsetting the causal structure of the latter.

The answer is this: Physicists have discovered an important and unexpected property of nature. It pertains to observable phenomena that depend upon microscopic properties that are *in principle inaccessible to observation.* In such a situation we are *in principle* unable, due to the lack of crucial micro-data, to give a complete causal description of the observable phenomena. However, our principled inability to give a complete causal account of the psychologically described phenomena, due to this inherent gap in the micro-data, can be partially offset by introducing into the theory, *instead of the inaccessible micro-data,* the *psychologically described selection of an action* made upon the system by an agent.

Thus, the loss of causal determination at the microlevel, due to the limitations imposed by Heisenberg's uncertainty principle, allows an alternative (statistical) causal account to be achieved by replacing the inaccessible micro-data by empirically available and controllable data about human selections of actions!

This feature discovered in atomic science should be equally importance in neuroscience. That is because the basic problem in neuroscience is essentially the same as the one in atomic physics. In both cases the problem is to provide a causal account of connections between experiences that depend sensitively upon micro-properties that are in principle inaccessible. But quantum theory shows how the principled loss of information at the microlevel can be partially offset by using, instead, the controllable and reportable variables of the intentional actions of human beings. Nature left open a causal gap for us to occupy.

The Quantum Brain.

The quantum state of a human brain is, of course, a very complex thing. But its main features can be understood by considering first a classical conception of the brain, and then folding in some key features that arise already in the case of the quantum state of a single particle, or object, or degree of freedom.

Source

http://www-physics.lbl.gov/~stapp/2nd.doc

https://quantum-neuroscience.com/wp-content/uploads/2018/05/Neuroscience-and-the-Human-Person.pdf

http://www-physics.lbl.gov/~stapp/phys.txt

https://docslide.com.br/download/link/stapp-mind-matter-and-quantum-mechanics

https://quantum-neuroscience.com/wp-content/uploads/2018/05/Mind-Matter-and-Quantum-Mechanics.pdf

https://science-2-0.com/wp-content/uploads/2018/05/Mind-Matter-and-Quantum-Mechanics.pdf

Stapp, H. P. (2009). *MIND, MATTER, AND QUANTUM MECHANICS* (3rd ed.). Berlin: Springer.

Process 1 consists of the choices that are made by the Human Psyche.

Process 2 happens when Nature's Psyche collapses the necessary wave functions and/or fires the specific neurons needed to make the Human Psyche's choices a physical reality.

Orthodox Quantum Mechanics explains what the Human Psyche and Nature's Psyche are doing at the quantum level in order to get things done for us at the physical level. This is good stuff! It's thousands of years ahead of anything that we have gotten from the Materialists, Naturalists, Darwinists, Nihilists, and Atheists.

If your interpretation of Quantum Mechanics can't explain what Nature's Psyche and the Human Psyche are doing at the quantum level or the psyche level, then your interpretation of Quantum Mechanics is absolutely worthless when it comes time to figure out how Quantum Mechanics or Action at a Distance really works.

Quantum Mechanics explains how your Psyche, Intelligence, Spirit, Soul, Spark, Mind, or Quantum Non-Local Consciousness controls your physical body and your physical brain. That's the purpose of Quantum Mechanics. In other words, explaining Psyche or Non-Local Consciousness scientifically is what Quantum Mechanics is good for. Quantum Mechanics or Supernatural Mechanics completes science, finishes science, and perfects science. Science is incomplete without Psyche or Syntropy.

For me personally, this is one of the most important, most significant, and most powerful scientific discoveries of my science career. It's essential. It's foundational. It changes everything!

If your interpretation of Quantum Mechanics can't explain how the Human Psyche is controlling its physical brain and physical body, then your interpretation of Quantum Mechanics is worthless. That is what I have experienced and observed.

The Human Psyche makes the decisions and the choices. Nature's Psyche collapses the wave functions and/or turns on the specific neurons needed to make the Human Psyche's choices a physical reality. We KNOW that it must be Nature's Psyche (or God's Psyche) who is collapsing the wave functions and firing the neurons because the Human Psyche isn't consciously aware of any of these quantum mechanical processes.

By choosing to allow Psyche, Syntropy, and Quantum Mechanics in to play, we can literally explain everything that comes our way. That is what I have experienced and observed.

I used to be a Materialist, Naturalist, Nihilist, and Atheist. Quantum Mechanics never made any sense to me until after I got rid of My Materialism, My Naturalism, My Nihilism, and My Atheism. Materialism, Naturalism, Darwinism, Nihilism, and Atheism claim that Psyche, Syntropy, and Supernatural Mechanisms do not exist; but, Quantum Mechanics IS Action at a Distance or Supernatural Mechanisms. Materialism, Naturalism, and their derivatives are FALSIFIED by the verified and proven existence of Action at a Distance or Quantum Mechanics.

Materialism, Naturalism, Darwinism, Nihilism, Behaviorism, Determinism, Physical Reductionism, Atheism, Entropy, and Classical Physics CANNOT EXPLAIN how the Human Psyche and Nature's Psyche are interacting with each other at the quantum level in order to get things done for us at the physical level. Therefore, these "sciences" are worthless, incomplete, and completely lack explanatory power when it comes to the Human Psyche, Nature's Psyche, Quantum Mechanics, Action at a Distance, Invisible Non-Physical Forces and Fields, and Supernatural Mechanisms.

Materialism and Naturalism have NO EXPLANATION for invisible non-physical forces and fields such as Psyche, Gravity, Magnetism, Nuclear Forces, Radio Waves, Microwaves, X-Rays, Dark Energy, Quantum Waves, Thoughts, and our after-death Memories which have been experienced, observed, and caught in the act.

It's time for the world at-large to upgrade science to Quantum Mechanics or Supernatural Mechanics. Once you choose to allow Psyche, Syntropy, and Quantum Mechanics in to play, instantly you can explain everything that comes your way.

Mark My Words

The Mindful Universe

The Orthodox Interpretation of Quantum Mechanics by Henry P. Stapp was the FIND of a lifetime. Henry P. Stapp's work is important to know about and understand – everything else pales in comparison. This is the information and the scientific evidence that the Materialists, Naturalists, Darwinists, Nihilists, and Atheists are trying to hide from you; and, they are very good at what they do. These people are actively banning, blocking, and destroying this information because it falsifies Physicalism, Naturalism, Darwinism, and Nihilism.

Stapp seems to be giving everything away for free on his website.

https://sites.google.com/a/lbl.gov/stappfiles/

http://www-physics.lbl.gov/~stapp/

He seems to be much more interested in getting his message out to the world than he is in maintaining the copyright for his books.

Stapp, H. P. (2011). *Mindful Universe: Quantum Mechanics and the Participating Observer*. Heidelberg: Springer.

https://epdf.tips/mindful-universe-quantum-mechanics-and-the-participating-observer-second-edition.html

Snippets from the *Mindful Universe*:

Quote from the *Mindful Universe*:

The World of Actions

Werner Heisenberg was, from a technical point of view, the principal founder of quantum theory. He discovered in 1925 the completely amazing and wholly unprecedented solution to the puzzle: the quantities that classical physical theory was based upon, and which were thought to be numbers, must be treated not as numbers but as actions! Ordinary numbers, such as 2 and 3, have the property that the product of any two of them does not depend on the order of the factors: 2 times 3 is the same as 3 times 2. But Heisenberg discovered that one could get the correct answers out of the old classical laws if one decreed that certain numbers that occur in classical physics as the magnitudes of certain physical properties of a material system are not ordinary numbers. Rather, they must be treated as *actions* having the property that the order in which they act matters!

This 'solution' may sound absurd or insane. But mathematicians had already discovered that logically consistent generalizations of ordinary mathematics exist in which numbers are replaced by 'actions' having the property that the order in which they are applied matters. The ordinary numbers that we use for everyday purposes like buying a loaf of bread or paying taxes are just a very special case from among a broad set of rationally coherent mathematical possibilities. In this simplest case, A times B happens to be the same as B times A. But there is no logical reason why Nature should not exploit one of the more general cases: there is no compelling reason why our physical theories must be based exclusively on ordinary numbers rather than on actions. The theory based on Heisenberg's discovery exploits the more general logical possibility. It is called quantum mechanics, or quantum theory.

The difference between quantum mechanics and classical mechanics is specified by Planck's constant, which is a tiny number on the scale of human actions. Thus, this tweaking of laws of physics might seem to be a bit of mathematical minutia that could scarcely have any great bearing on the fundamental nature of the universe, or of our role within it. But replacing *numbers* by *actions* upsets the whole apple cart. It produced a seismic shift in our ideas about both the nature of reality, and the nature of our relationship to the reality that envelops and sustains us. The aspects of nature represented by the theory are converted from elements of *being* to elements of *doing*. The effect of this change is profound: it replaces the world of *material substances* by a world populated by *actions*, and by *potentialities* for the occurrence of the various possible observed feedbacks from these actions. Thus, this switch from 'being' to 'action' allows – and according to orthodox quantum theory demands – a draconian shift in the very subject matter of physical theory, from an imagined universe consisting of causally self-sufficient mindless matter, to a universe populated by allowed possible physical actions and possible experienced feedbacks from such actions. A purported theory of matter alone is converted into a theory of the relationship between matter and mind.

What is this momentous change introduced by Heisenberg?

In classical physics the center point of each physical object has, at each instant of time, a well-defined location, which can be specified by giving its three coordinates (x, y, z) relative to some coordinate system. For example, the location of (the center point of) a spider dangling in a room can be specified by letting z be its distance from the floor, and letting x and y be its distances from two intersecting walls. Similarly, the velocity of that dangling spider, as she drops to the floor, blown by a gust of wind, can be specified by giving the rates of change of these three coordinates (x, y, z). If each of these three *rates of change*, which together specify the velocity, are multiplied by the *weight* (= mass) of the spider, then one gets three numbers, say (p, q, r), that define the *momentum* of the spider. In classical physics one uses the set of three numbers denoted by (x, y, z) to represent the position of the center point of an object, and the set of three numbers labeled by (p, q, r) to represent the momentum of that object. These six numbers are just ordinary numbers that obey the commutative property of multiplication that we all, hopefully, learned in third grade: $x * p$ equals $p * x$, where $*$ means multiply.

The six-dimensional space of all possible values $(x, y, z; p, q, r)$ is called *phase space*: it is the space of all possible instantaneous 'states' of the particle.

Heisenberg's analysis showed that in order to make the formulas of classical physics work in general, $x * p$ must be different from $p * x$. He found that the difference between these two products must be Planck's constant. (Actually, the difference is Planck's constant divided by 2π and multiplied by the imaginary unit i, which is a number such that i times i is minus one.) Thus, modern quantum theory was born by recognizing, or declaring, that the symbols used in classical physical theory to represent ordinary numbers actually represent actions such that their ordering in a sequence of actions matters. The procedure of creating the mathematical structure of quantum mechanics from that of classical physics, by replacing numbers by corresponding actions, is called 'quantization'.

The idea of replacing the numbers that specify where a particle is, and how fast it is moving, by mathematical quantities that violate the simple laws of arithmetic may strike you – if this is the first you've heard about it – as a giant step in the wrong direction. You might mutter that scientists should try to make things simpler, rather than abandoning one of the things we really know for sure, namely that the order in which one multiplies factors does not matter. But against that intuition one must recognize that this change works beautifully in practice: all of the tested predictions of quantum mechanics are borne out, and these include predictions that are correct to the incredible accuracy of one part in a hundred million. There must be something very, very right about this replacement of numbers by actions.

In classical physical theory each elementary particle is asserted to have at each instant of time a definite location, defined by a set of three numbers (x, y, z), and definite momentum, defined by a set of three numbers (p, q, r). In quantum theory one generally considers systems of many particles, but insofar as one can consider one particle alone the state of that particle at any instant of time would be represented by a *cloud of pairs of numbers*, with one pair of numbers (called a complex number) assigned to each point in three-dimensional (position) space. Someone might choose to perform a phenomenologically (i.e., experimentally/experientially) described probing action on this 'particle'. In quantum mechanics each such possible probing action turns out to have an associated set of distinct *experientially distinguishable* possible outcomes. The cloud of numbers *taken as a whole* determines the probability for the appearance of each of the alternative possible outcomes of that chosen probing action. The theory thus gives specified rules for computing the probabilities for each of the distinct alternative possible empirically described feedbacks from each of the alternative possible experimental probing actions that the human experimenter might chose to perform, but no rules that specify which probing action he or she will choose.

In classical physical theory when one descends from the macroscopic world of visible objects to the microscopic world of their elementary constituents one arrives at a world containing the 'solid, massy, hard, impenetrable moveable particles' that Newton spoke of. But in quantum theory one arrives instead at clouds, or quantum smears, of numbers that *taken as a whole* have empirical meaning in terms of probabilities of alternative possible experiences.

Briefly stated, the orthodox formulation of quantum theory (see Appendix D) asserts that, in order to connect adequately the mathematically described state of a physical system to human experience, there must be an abrupt *intervention* in the otherwise smoothly evolving mathematically described state of that system.

According to the orthodox formulation, these interventions are probing actions *instigated by human agents who are able to 'freely' choose which one, from among various alternative possible probing actions, they will perform*. The physically describable effect of the chosen probing action is to separate (partition) the prior physical state of the system being probed in some particular way into a set of component parts. Each physically described part corresponds to one *perceivable* outcome from the set of distinct alternative possible perceivable outcomes of that particular probing action.

If such a probing action is performed, then *one* of its allowed perceivable feedbacks will appear in the stream of consciousness of the observer, and the mathematically described state of the probed system will then jump abruptly from the form it had prior to the intervention to the partitioned portion of that state that corresponds to the observed feedback. *This means that, according to orthodox contemporary physical theory, the 'free' choices of probing actions made by agents enter importantly into the course of the ensuing psychologically and physically described events. Here the word 'free' means, however, merely that the choice is not determined by the (currently) known laws of physics; not that the choice has no cause at all in the full psychophysical structure of reality. Presumably the choice has some cause or reason – it is unreasonable that it should simply pop out of nothing at all – but the existing theory gives no reason to believe that this cause must be determined exclusively by the physically described aspects of the psychophysically described nature alone.*

If one sets Planck's constant equal to zero in the quantum mechanical equations, then one recovers (the fundamentally incorrect) classical mechanics. Thus, classical physics is *an approximation* to quantum physics. It is the approximation in which Planck's constant, wherever it appears, is replaced by zero. In this approximation the quantum smearing does not occur – each cloud is reduced to a point – and one recovers classical physics, along with the physical determinism (the causal closure of the physical) entailed by classical physics.

In the classical approximation there is no need for, *and indeed no room for*, any effect of any probing action. The *uncertainty* – arising from the non-zero size of the quantum cloud – that in the unapproximated theory needs to be resolved by the intervention of some particular probing action is already reduced to zero by the replacement of Planck's constant by zero. Thus, all effects upon the physically/mathematically described aspects of nature's process that are instigated by the actions 'freely' chosen by agents are eliminated by the classical approximation. *Consequently*, any attempt to understand or explain within the framework of classical physics the physical effects of consciousness is irrational, *because the classical approximation eliminates the effect one is trying to study*.

Intentional Actions and Experienced Feedbacks

The concept of intentional actions by agents is of central importance. Each such action is intended to produce an experiential feedback. For example, a scientist might act to place a Geiger counter near a radioactive source, with the intention to see the counter either 'fire', or 'not fire', during a certain time interval. The experienced response, 'Yes' or 'No', to the query 'Does the counter fire?' specifies one bit of information. The basic move in quantum theory is to shift, *fundamentally*, from the airy plane of high-level abstractions, such as the unseen precise trajectories of invisible elementary material particles, to the nitty-gritty realities of consciously chosen intentional actions and their experienced feedbacks, and to the theoretical specification of the mathematical procedures that allow us successfully to predict relationships among these empirical realities.

Probing actions of this kind are performed not only by scientists. Every healthy and alert infant is engaged in making willful efforts that produce experiential feedbacks, and he or she soon begins to form expectations about what sorts of feedbacks are likely to follow from some particular kind of felt effort. Thus, both empirical science and normal human life are based on paired realities of this action– response kind, and our physical and psychological theories are both basically attempts to understand these linked realities within a rational conceptual framework.

A purposeful action by a human agent has two aspects. One aspect is his conscious intention, which is described in psychological terms. The other aspect is the linked physical action, which is described in physical terms; i.e., in terms of mathematical entities assigned to spacetime points. For successful living the physically described action should be a *functional counterpart* of the conscious intention: after sufficient empirical honing by effective learning processes the physically described aspect of the felt intentional act should have a tendency to produce the intended experiential feedback.

John von Neumann, in his seminal book, *Mathematical Foundations of Quantum Mechanics*, calls by the name 'process 1' the basic probing action that partitions a potential continuum of physically described possibilities into a (countable) set of empirically recognizable alternative possibilities. I shall retain that terminology. Von Neumann calls the orderly mechanically controlled evolution that occurs between interventions by name 'process 2'. This process is the one controlled by the Schrödinger equation. The numbering, 1 and 2, emphasizes the important fact that the conceptual framework of orthodox quantum theory requires *first* an acquisition of knowledge, and *second*, a mathematically described propagation of a representation of this acquired knowledge to some later time at which a further inquiry is made.

There are two other associated processes that need to be recognized. The first of these is the process that *selects the outcome*, 'Yes' or 'No', of the probing action. Dirac calls this intervention a "choice on the part of nature", and it is subject, according to quantum theory, to statistical rules specified by the theory. I call by the name 'process 3' this statistically specified choice of the *outcome* of the action selected by the prior process 1 probing action.

Finally, in connection with each process 1 action, there is, presumably, some process that is not described by contemporary quantum theory, but that determines what the so-called 'free choice' of the experimenter will actually be. This choice *seems to us* to arise, at least in part, from conscious reasons and valuations, and it is certainly strongly influenced by the state of the brain of the experimenter. I have previously called this selection process by the name 'process 4'; but will use here the more apt name 'process zero', because this process must precede von Neumann's process 1. It is the absence from orthodox quantum theory of any description on the workings of process zero that constitutes the causal gap in contemporary orthodox physical theory. It is this 'latitude' offered by the quantum formalism, in connection with the "freedom of experimentation" (Bohr 1958, p. 73), that blocks the causal closure of the physical, and thereby releases human actions from the immediate bondage of the physically described aspects of reality.

Cloudlike Forms

The quantum state of a single elementary particle can be visualized, roughly, as a continuous cloud of (complex) numbers, one assigned to every point in three-dimensional space. This cloud of numbers evolves in time and, taken as a whole, it determines, at each instant, for each allowed process 1 action, an associated set of alternative possible experiential outcomes or feedbacks, and the 'probability of finding (i.e., experiencing)' that particular outcome.

Heisenberg's uncertainty principle specifies that if one squeezes this spatial cloud – the spatial region in which the numbers are nonzero – into a sufficiently small region, it will violently explode outward when the constricting force is removed.

https://epdf.tips/mindful-universe-quantum-mechanics-and-the-participating-observer.html

https://books.google.com/books?id=yVpyM3dJOqYC&pg=PA24#v=onepage&q&f=false

Stapp, H. P. (2011). *Mindful Universe: Quantum Mechanics and the Participating Observer* (2nd ed.). Berlin: Springer-Verlag.

If any of this is true, and I believe that it is, then it completely falsifies Radical Behaviorism and Hard Determinism. It also falsifies Materialism, Physicalism, Naturalism, and Nihilism. The false is falsified by the truth; and, the truth is repeatedly experienced and observed.

The Materialists, Naturalists, and Classical Physicists are right in that there is no such thing as free will or choice at the physical level. Physical matter doesn't make choices. Physical processes such as random mutations, natural selection, evolution, and entropy don't make choices. Your physical brain doesn't make choices because it can't make choices! Physical matter can't do choice. Physical matter has no choice in the matter.

But, these people don't realize why they are right because they don't realize that choice is a function of Psyche and that every choice takes place at the quantum level or the psyche level instead of the physical level. Instead, these people deny the existence of choice and the existence of Psyche because their naturalistic model for reality can't explain psyche nor choice; consequently, these people erroneously choose to state that psyche or choice does not exist, and that's as far as their science can take them.

Materialism and Naturalism completely lack explanatory power when it comes to Psyche, Syntropy, Non-Locality, Action at a Distance, Quantum Mechanics, and Choice.

Mark My Words

On the Nature of Things

"On the Nature of Things: Human Presence in the World of Atoms" by Henry P. Stapp.

http://www-physics.lbl.gov/~stapp/NOT72C.pdf

https://quantum-neuroscience.com/wp-content/uploads/2018/05/On-Nature-of-Things.pdf

Here's a free book from Henry P. Stapp that he is handing out to the public.

The important thing is to get this information out to the public so that they can take advantage of it. Most of the scientists have officially rejected this information, so you won't get any of this information from them. You will have to look someplace else besides our college classrooms and public schools if you want to find this information because it's being banned from our public schools by the Materialists, Naturalists, Darwinists, and Atheists.

Quantum Physics in Neuroscience and Psychology

This article was my first introduction to Henry P. Stapp and the Orthodox Interpretation of Quantum Mechanics. I mention and discuss it elsewhere, so I won't do so here.

Schwartz, J. M., Stapp, H. P., & Beauregard, M. (2004). *Quantum Physics in Neuroscience and Psychology: A Neurophysical Model of Mind-Brain Interaction*. Published Online: Phil. Trans. R. Soc. B.

http://www-physics.lbl.gov/~stapp/PTRS.doc

This paper summarizes everything I have been looking for during the past fifty-five years of my life. That's a long time to wander in darkness looking for the truth; but luckily, our generation finally has the truth, if we know where to find it and recognize it as true when we do find it.

This paper and Henry P. Stapp's papers and books taught me what Stapp calls the **Orthodox Interpretation of Quantum Mechanics**, which explains the interplay and the scientific interface between mind and matter.

Examine these two websites, from Henry P. Stapp:

https://sites.google.com/a/lbl.gov/stappfiles/

http://www-physics.lbl.gov/~stapp/

They explain the Orthodox Interpretation of Quantum Mechanics in great detail. You could spend a lifetime studying them, and still have things to learn.

Mark My Words

The Grand Designer

In the book, *The Grand Designer: Discovering the Quantum Mind Matrix of the Universe*, Graham Smetham gathers together some of the very best from Henry P. Stapp into a very useful and interesting section entitled, "Henry Stapp's Mindful Universe and the Psycho-Physical Bridge".

This section is available on Google Books – just page down and up to populate all the relevant pages.

https://books.google.com/books?id=Z0FOAgAAQBAJ&pg=PA158

I also archived it at this link:

https://quantum-neuroscience.com/wp-content/uploads/2018/05/The-Grand-Designer.zip

I quote from Graham Smetham:

Henry Stapp outlines his version of von Neumann's analysis as the following three stages of the quantum experimental process:

Process 1: The 'free choice' of the experimental setup. Heisenberg called this phase "a choice on the part of the 'observer' constructing the measuring instruments and reading their recording". This choice is "not controlled by any known physical process, statistical or otherwise, but appears to be influenced by understandings and conscious intentions." Whilst this process was originally delineated as a phase within the experimental setting, Stapp indicates that such 'free choice' of 'probing actions' is a part of the general human condition:

> Probing actions of this kind are performed not only by scientists. Every healthy and alert infant is engaged in making willful efforts that produce experiential feedbacks, and he/she soon begins to form expectations about what sorts of feedbacks are likely to follow from some particular kind of effort. Thus, both empirical science and normal human life are based on paired realities of this action-response kind.

The hugely significant point in this 'process 1' 'free choice' is that it poses a question to which 'reality' [or Nature's Psyche] can feedback a 'yes' or a 'no', and the fact that the choice of the question is free means that the 'free choice' actually determines the nature of the possible feedbacks. Thus the 'free choices' determine the nature of the experienced reality:

> The process is active: it injects into the physical state of the system being acted upon the properties that depend upon the intentional chosen action of the observing agent.

Stapp calls this process 1 'a dynamical psychophysical bridge.'

Process 2: The deterministic quantum evolution of the potentialities within the Schrödinger wavefunction, this is the mathematical description of the development of the probabilities associated the potentialities within the quantum realm.

Process 3: This is what Paul Dirac called a 'choice on the part of nature.' It is the yes or no feedback from the experimental setup – yes, reality is this way or no, reality is not this way; Stapp indicates that complex questions can be reduced to yes-no choices. This delineation of the experimental process, which starts in the classical realm, then evolves within the quantum realm, and finally gives its answer within the classical level again, is forced upon us because we are limited to perceptions and language within the classical realm. This, of course, is a fundamental aspect of our embodied existential situation which was often commented upon by the early quantum physicists. In other words, we can only ask questions and receive answers in the experiential classical domain whilst the processes which determine the nature of the answer take place in the 'ultimate" quantum domain.

The Grand Designer: Discovering the Quantum Mind Matrix of the Universe. Graham Smetham.

This three-step process is the BRIDGE between Psyche and physical reality. It's called a psychophysical bridge. The Orthodox Interpretation of Quantum Mechanics explains how the Human Psyche and Nature's Psyche work together at the quantum level in order to get things done for us at the physical level.

This is the answer to life, the universe, and everything.

Process 1 – Choices are made by the Human Psyche. The Human Psyche chooses what it wants to do with is physical body and physical brain.

Process 2 – Nature's Psyche collapses the wave functions and/or turns on the necessary neurons in order to make the Human Psyche's choices a physical reality. We know that it is Nature's Psyche or "Reality" who is controlling all of

this at the quantum level because the Human Psyche isn't consciously aware of any of this quantum mechanical stuff. The firing of neurons and the collapsing of wave functions happen outside our conscious awareness in the quantum realm under the command and control of Nature's Psyche.

Process 3 – Nature's Psyche responds to the Human Psyche's requests by determining what is possible and what is impossible, based upon the physical laws and the quantum mechanical rules which God has set into place. God's Psyche controls and restricts our access to quantum mechanics, and God's Psyche made the physical laws. Nature's Psyche enforces God's will both at the physical level and the quantum level. God and Nature can read our mind.

God holds ALL the KEYS to both the physical laws and the quantum mechanisms. God is Syntropy.

Try as he might, a fallen mortal physical being can't do anything that God, quantum mechanics, and the physical laws won't let him do. God created and enforces the physical laws in order to prevent us from using quantum mechanisms to destroy ourselves and our physical existence. God created the physical laws to prevent the physical matter in our bodies from quantum tunneling away from us at will. The physical laws make physical life possible.

The physical laws force order, structure, restrictions, and limitations into spirit matter or dark matter thereby slowing it down and converting it into physical matter. Entropy or physical law slows down physical matter to sub-light speeds, makes physical matter subject to time or an aging process, localizes physical matter in space, and prevents the physical matter from quantum tunneling away from us at will. The result is the Ultimate Consensus Reality, a physical reality. It is highly reliable, dependable, controllable, predictable, and stable. In a physical reality, I can actually count on this paper being on my hard drive and cloud drive tomorrow when I go looking for it.

Scientific Observation: Anything that began obviously has Someone Psyche or Someone Intelligent who caused it to begin. There's no such thing as spontaneous generation or creation ex nihilo.

Scientific Observation: Physical matter, physical laws, and entropy began.

Scientific Conclusion: Therefore, physical matter, physical laws, and entropy have Someone Psyche or Someone Intelligent who caused them to begin or made them begin.

It is also logical to conclude that physical matter has Someone Psyche or Someone Intelligent who is currently forcing it or making it obey the physical laws, including the second law of thermodynamics.

The physical laws are NOT an inherent part of spirit matter or dark matter – they ONLY apply to physical matter. Someone Psyche or Someone Intelligent is making the physical laws and entropy apply to physical matter. God's Psyche forced the physical laws and entropy (an aging process) into a small portion of the spirit matter or dark matter thereby converting that spirit matter into physical matter. God deliberately slowed down the spirit matter converting it into physical matter and subjecting it to the passage of time (entropy) so that we can live it, experience it, depend on it, control it, learn from it, and remember having done so.

Remember, at velocities or speeds faster than the Speed-of-light, TIME STOPS – there is NO entropy or passage of time. Light doesn't age because it doesn't experience the passage of time, meaning that it has NO entropy. Light is Syntropy. Furthermore, Someone Psyche is forcing the wave functions to collapse into physical realities. This is what Quantum Mechanics is trying to tell us and teach us.

This IS Science at the cutting-edge.

If you consider yourself to be a scientist, then it's extremely important for you to find, read, and understand this material, so that you can bring yourself into the modern age where science is concerned.

My ultimate desire and goal is to set people free so that they can find the truth and know the truth so that they don't have to go through a lifetime of darkness like I did.

Quantum Mechanics or Syntropy is thousands of years ahead of what the Materialists, Naturalists, Darwinists, Atheists, and Classical Physicists have been able to produce.

Mark My Words

Synaptogenesis Is Applied Quantum Mechanics

Synaptogenesis, synaptic mapping, synapse pruning, and synapse elimination are dynamic and alive.

Who or what is overseeing and controlling this process? There's nowhere near enough information in our genome to code for synaptogenesis and synaptic pruning. Our genes code for proteins, NOT synapses. Our synaptic mapping is taking place someplace else besides our genes.

Who or what is mapping the synapses and deciding what each synapse means or what each synapse does? A synapse has no meaning at the physical level. The synapses can't communicate with each other at the physical level. Technically, neurons can't communicate with each other at the physical level.

Who or what is using this synaptic map or synaptic network at the quantum level to get things done for us at the physical level? In other words, who is doing this synaptic mapping, and then using these quantum maps of physical functionality at the quantum level or the psyche level in order to get things done for us at the physical level?

It can't be the Human Psyche because the Human Psyche isn't consciously aware of any of these quantum processes. It can't be our genes because our genomes have nowhere near enough information storage capacity within them to code for and control synaptogenesis, synaptic mapping, and synaptic pruning.

We are in fact looking at Nature's Psyche and/or God's Psyche for an answer to the source and the cause of Synaptogenesis, Synaptic Mapping, Synaptic Meaning, Synaptic Purpose, and Synaptic Pruning.

I'm a scientist. I'm also a generalist. I'm good at everything, master of nothing. Among many other things, I have been a computer scientist all of my life. I understand physical storage capacity.

Our genome is capable of storing 725 megabytes to 750 megabytes of information. In order to successfully MAP or NETWORK one quadrillion synapses – the estimated number of synapses in a newborn child – assign a 3D address to each, and then assign a meaning or purpose to each synapse, as well as regulate its synaptogenesis and any synaptic pruning, do REMAPPING or UPDATING or RE-NETWORKING of what remains, while keeping it all straight in one's mind would require petabytes of information storage capacity to accomplish. It's physically impossible to shove petabytes of information into our puny 750-megabyte genome. It can't be done, which means that it isn't being done.

Furthermore, genes ONLY code for proteins. There is enough memory storage capacity within our genome to code for all the different proteins that are used to make up our physical body; but, where is the information that's needed to organize all those different proteins into 3D machines and 3D cells being stored? Where are all the 3D blueprints being stored? These millions of different 3D blueprints telling the proteins how to assemble themselves into working machines can't be stored within our genome because there's simply not enough memory storage capacity within our physical genome to store all of that information. That, too, would require petabytes of information storage that simply isn't possible within our 750-megabyte genome.

Who, within a living cell, decides that that particular cell needs a new protein? Who sends the transcription enzymes to the right location in the genome for the specific gene needed to code for that protein? How is that information transferred throughout the cell? Once the mRNA molecule has been made and is on its way to a ribosome, who tells the activation enzymes the correct amino acid to join to the correct codon during the production of transfer RNA (tRNA), so that each tRNA molecule arrives at the ribosome in the correct sequence and at the right time for protein synthesis and the translation process? How is all of this information being transmitted from atom to atom, and molecule to molecule? How would you get two physical atoms to communicate with each other and coordinate with each other? How would you get two or more proteins to communicate with each other and coordinate with each other so as to assemble into much larger nanomachines at the right time in the correct 3D location?

https://en.wikipedia.org/wiki/Protein_biosynthesis

These are atoms and molecules that we are talking about here. Who is telling these different atoms and molecules where to go during the construction of amino acids, during the construction of mRNA and tRNA, and during the construction of proteins? Who tells these atoms and molecules where to go? Who tells the different enzymes which molecules to make? Who tells the transcription enzyme which protein is needed and which gene to read? Who is telling the different proteins where to go and what type of nanomachine to become when they get there?

If given the assignment, how would you make the atoms and the molecules communicate with each other and share information with each other at a distance? Remember, there's NO wires connecting the different atoms together at a distance. Communication between the atoms and molecules takes place wirelessly at a distance.

There's only one logical answer to all of this.

Have you seen it yet?

Telepathy or quantum waves are WiFi at the quantum level or the psyche level. Quantum waves are the answer to life, the universe, and everything. Psyche or Intelligence produces, transmits, receives, and stores quantum waves.

The Materialists, Naturalists, Darwinists, Nihilists, and Atheists assure us that physical atoms are the ONLY thing that exists. They are either right, or they are wrong. Which is it?

Proteins are like tinker toys. The genes determine which shape of tinker toy to produce; but, who or what decides how the different tinker toys should be assembled so as to construct different structures or machines? There's NOT enough memory storage capacity within a genome to store all the different 3D blueprints for all the different gadgets, structures, and nanomachines that can be made with those tinker toys. The genome only has enough memory storage capacity to code for the construction of the individual tinker toys. So, where is all of that other information being stored for all those different 3D blueprints? How is all that information being transmitted from atom to atom, molecule to molecule, and protein to protein?

Who is transmitting all of this information from atom to atom? What makes these atoms and molecules move from one location to their target location? Who is telling these atoms and molecules where their target is located? How is all of this information being transmitted wirelessly from atom to atom, and molecule to molecule?

Where is all of this COMMAND and CONTROL information being stored, and how is it being transmitted throughout the cell to the different atoms and molecules within the cell?

It can't be our genes that are storing and transmitting petabytes of this invisible command and control information. The genes are one of the targets of that information but can't be the source of that information. Our physical genes can't store and transmit petabytes of information at the physical level through physical mechanisms. There's no mechanism in place for doing so at the physical level.

So, what are we left with since the source of this information can't be our genes?

The MAPPING and NETWORKING information for Synaptogenesis and synaptic organization, as well as the meaning or purpose of each synapse cannot be stored within our genome. Furthermore, all the different blueprints for all the different 3D nano-machines and 3D cells can't be stored within our genome. Command and control information, target location addressing, and assignments for trillions of different cells and quadrillions of different molecules can't be stored within our genome. So, where is all this information being stored because it clearly must exist and must be getting stored someplace? How is this invisible information being transmitted wirelessly from atom to atom and molecule to molecule since it can't be transmitted by the atoms and molecules themselves at the physical level? It can't be our physical genes doing all of this information storage and information exchange because our genes are just another molecule.

So, once again, what are we left with?

Well, think about it logically. Physical information storage or physical memory storage is limited by the size of an atom. Physical atoms take up space. It has been estimated that a genome is about as densely packed as information can be stored in a physical format. It doesn't get much more optimal than that at the physical level. But, we NEED millions or billions of times more memory storage capacity in the same amount of space that atoms occupy in order to achieve petabytes to exabytes of memory storage capacity within something the size of a physical brain.

The ONLY solution to this conundrum is to go sub-atomic or quantum. You have to go sub-atomic or invisible in order to achieve information storage densities and information storage capacities greater than what's possible with atoms at the physical level. Physical matter greatly limits information storage density and therefore information storage capacity because physical matter takes up space. Those limitations disappear if you go sub-atomic or quantum.

If you were to choose to store information as quantum waves, you could theoretically store exabytes of information in all of that "empty space" between the nucleus of an atom and its orbiting electron shells. I mean, ask yourself what type of "invisible information" is holding those electrons in orbit in the first place. Positive and negative attract, so there should be NO physical space within an atom; yet, there is. Why? Who is forcing the electrons to orbit the nucleus rather than merging with the protons in the nucleus? There has to be some kind of invisible force and intelligent force doing so; otherwise, physical matter wouldn't exist.

Entropy is death. Syntropy is eternal life. Syntropy is eternal and everlasting. There has to be some type of Syntropy or Psyche in existence, or all of that subsequent entropy and physical matter wouldn't have been possible in the first place.

Telepathy or quantum waves are WiFi at the quantum level or the psyche level. Quantum waves or telepathy could be used to transmit information from atom to atom. Quantum waves could also be used to store information invisibly within atoms. Quantum waves or telekinesis could move an atom from one location to a different target location. There is NO answer at the physical level; but, there are an endless number of different answers to be found at the quantum level or the psyche level.

I have hypothesized that Psyche or Quantum Non-Local Consciousness is an infinite singularity. The math points to the existence of such a thing. If I'm right, then that means that Psyche is capable of an infinite amount of memory storage capacity within something that has NO size and takes up NO space whatsoever. Now, that's infinitely dense information storage capacity that we are talking about there within an infinite singularity or Psyche. That's infinitely more information storage capacity than what's possible at the physical level.

Mark My Words

—

Source

Quantum Mechanics from a Non-Physical Spiritual Perspective

> https://www.amazon.com/dp/B01J023TGU

> https://www.amazon.com/dp/1521132380

NATURE vs. NURTURE vs. NIRVANA: An Introduction to Reality

> https://www.amazon.com/dp/B01JWRCSVA

> https://www.amazon.com/dp/1521132615

Reference Material

Quantum Neuroscience: The Answer to Life, the Universe, and Everything

> https://www.amazon.com/dp/B079Z6QQQB

The Ultimate Model of Reality: Psyche Is the Ultimate Cause

> https://www.amazon.com/dp/B071NC9JK6

11. The Truth about Psychiatry

My medical doctors and psychiatrists got me addicted to half a dozen different brain-altering drugs. They had me cycling through four different sleeping pills. They also had me on pain killers. Anytime that I would complain about the side effects, they would give me another drug. On any given day, I would take a dozen different drugs. Half of the drugs had suicidal ideation as one of the side effects. The ONLY time I was ever calm was while I was thinking of different ways to kill myself. That's your brain on drugs. I overdosed on the sleeping pills trying to end it all.

I don't believe in Psychiatry anymore. Psychiatry is the pinnacle of Materialism, Naturalism, Darwinism, and the Physical Sciences. I discovered from observation and first-hand experience that the drugs or medications cannot do Psyche Therapy. In fact, substance-induced psychosis IS a mental illness. Brain altering drugs produce mental illnesses. The Psychiatrists treat your mental illness by giving you half a dozen NEW mental illnesses. The Psychiatrists do harm to your brain chemistry and cause your brain imbalances as a result. It can kill you.

One of my greatest and most useful scientific discoveries is the realization that Psychiatry cannot do Psyche Therapy and that Psychiatry actually causes mental illnesses. When I (my psyche) finally decided to go cold turkey on the drugs, it took six months of withdrawal symptoms, psychosis, hallucinations, delusions, insomnia, and paranoia before my brain normalized and I started thinking rationally once again. Six months is a long time. That was one hell of a withdrawal. One of my most significant scientific discoveries was my discovery that Psychiatry is hell. The Psychiatrists and Pharmacologists can put you into hell faster and more efficiently than any other group on the planet.

The Psychiatrists and Pharmacologists should be forced to take their own drugs for a couple of months each so that they know first-hand the kind of hell that they are putting people through. The drugs can't do Psyche Therapy, so there's really no reason to be taking them.

I got off everything but the Zoloft. They wouldn't let me stop taking the Zoloft. However, my last pair of Psychiatrists were actually conservative.

The one told me that he could give me more drugs, but that his experience and observation had taught him that Therapy did people a lot more good in the long run than the drugs did. So, he encouraged me to limit myself to the Zoloft and to focus on the Psychotherapy instead. He was right. The Therapy does a lot more good for the Psyche or the Soul than the drugs do.

The other Neuropsychiatrist did brain scans and told me that my brain is normal for someone my age. He told me that my medical history proves that I'm overly sensitive to medications. He then said to me not to let any psychiatrist in the future put me on the full dose of any medication. He said, "Pick ONE, and get rid of all the rest." He said that my medical history indicates that I need the Zoloft, but he told me to immediately cut the dose in half, and then a year from now to cut it in half again. He told me to keep cutting the dose in half until I get down to a maintenance dose where I'm getting ALL of the benefits of the drug without any of the side effects.

I did as he instructed, cut the dose in half, and almost immediately I went from being catatonic during the day to being able to function, drive, walk, and think during the day. I took the Zoloft at night. It calmed me, stopped the racing thoughts, helped me to breathe better, and helped me to sleep. When I woke up, I was alert and able to think and function at half the dose. I cut the dose in half a year later and found my maintenance dose – one-fourth the recommended adult dose. I have also experimented with cutting the dose in half again, with mixed results.

If they won't let you get off the drugs, then pick ONE, and then taper that ONE down to a maintenance dose where you are getting all the benefits with none of the side effects. It works. At the much lower maintenance dose, there's actually some evidence that the Zoloft might be doing me more good than harm.

Be warned! The psychiatrists and pharmacologists typically prescribe the maximum dose that they can get away with before that dose turns toxic and starts to kill you. Get as far away from the point of toxicity that you possibly can, and the drugs just might do you some good in the end. But, when they have you on the full dose and have you on twelve different drugs at the same time, that's a prescription for Hell if ever there was one. I KNOW, because I have been there and done that, and it was indeed Hell.

You don't have to die in order to go to Hell. You can do so right here, right now, and the Pharmacologists, Medical Doctors, and Psychiatrists can help you do so if you let them.

Mark My Words

—

Simple Truths about Psychiatry

By Dr. Peter R. Breggin MD

https://www.youtube.com/playlist?list=PLdo601sRKNc70BA5MV51kS-Mr6ypsvd1F

This is one of my all-time most favorite series about Psychiatry. Dr. Peter Breggin documents medically what was done to me personally by my medical doctors and psychiatrists. They just about killed me. Psychiatry and drugs can be extremely dangerous. They can kill you. It's actually refreshing to finally find a Medical Doctor and Psychiatrist who is familiar with that fact and is actually willing to talk about it online. Highly recommended. Dr. Peter Breggin ended up being one of my greatest scientific discoveries.

I archived this series on one of my websites in the hope that it won't ever disappear on me while I'm alive.

01 – Biochemical Imbalance

https://psyche-ontology.com/wp-content/uploads/2018/04/01-Biochemical-Imbalance.zip

https://www.youtube.com/watch?v=ARZ2Wv2BoFs

02 – How Psychiatric Drugs Work

https://psyche-ontology.com/wp-content/uploads/2018/04/02-How-Psychiatric-Drugs-Work.zip https://www.youtube.com/watch?v=W4Xb29geVwE

03 – Medication Spellbinding

https://psyche-ontology.com/wp-content/uploads/2018/04/03-Medication-Spellbinding.zip

https://www.youtube.com/watch?v=cwIDfcc5Z3w

04 – Helping Deeply Disturbed Persons

https://psyche-ontology.com/wp-content/uploads/2018/04/04-Helping-Deeply-Disturbed-Persons-01.zip

https://psyche-ontology.com/wp-content/uploads/2018/04/04-Helping-Deeply-Disturbed-Persons-02.zip

https://www.youtube.com/watch?v=0F_kLbtnWWM

05 – Helping the Suicidally Depressed

https://psyche-ontology.com/wp-content/uploads/2018/04/05-Help-the-Suicidally-Depressed.zip

https://www.youtube.com/watch?v=t7OPWRqjaSQ

06 - Psychiatric Drugs Are More Dangerous than You Ever Imagined

https://psyche-ontology.com/wp-content/uploads/2018/04/06-Psychiatric-Drugs-Are-More-Dangerous-than-You-Ever-Imagined.zip

https://www.youtube.com/watch?v=luKsQaj0hzs&t=5s

07 - Stimulant Drugs Crush Children

https://psyche-ontology.com/wp-content/uploads/2018/04/07-Stimulant-Drugs-Crush-Children.zip

https://www.youtube.com/watch?v=fY7lQDHSzuM

08 – ADHD Kids

https://psyche-ontology.com/wp-content/uploads/2018/04/08-ADHD-Kids.zip

https://www.youtube.com/watch?v=sUXzGQjTyRc

09 - Stop Drugging Children

10 - Electroshock is Brain Trauma

I feel that this series might be the best thing I can "give" to you, especially if the Psychiatrists and Medical Doctors have you on a bunch of different brain-altering drugs. If you are not careful, the junk can kill you.

Mark My Words

—

Source

Tripping the Light Fantastic: How Prescription Drugs Almost Killed Me

https://www.amazon.com/dp/B071RJP9T8

12. Physical Matter or Entropy Began

Scientific Definitions: Entropy is the result of the passage-of-time or an aging process. Entropy is a function of time. Time consists of things that begin and end. Physical matter and classical physics are based upon entropy. Physical matter and entropy seem to be synonymous – they go together, or not at all. Physical matter and entropy began at some point in time. Physical matter typically means entropic physical matter.

To reduce entropy within a system requires an infusion of Psyche, Intelligence, or Energy from something external to the particular system that's currently under observation. Psyche is the innate intelligence within ALL the different forms of energy which gives that energy the inherent ability to understand, follow, and obey God's Laws and God's Commands. Psyche is syntropic which means that it has the inherent ability, or the inherent energy and intelligence, necessary to reverse entropy and produce order out of chaos. Psyche or energy is conserved. It cannot be created nor destroyed. It is eternal and everlasting. Psyche or energy is the ultimate perpetual motion machine. Entropy is a statistical measure or an accounting of unavailable energy. Psyche is perpetually available energy or conserved energy. Entropy was designed, created, and implemented by God's Psyche within fallen matter or entropic physical matter; therefore, entropy can be overridden and even eliminated by Syntropy and Psyche, particularly God's Psyche. Entropy is just a different form of energy. The form can always be changed by God's Psyche.

Scientific Observations: Everything that began had Someone Intelligent who caused it to begin. There is NO such thing as spontaneous generation, creation ex nihilo, abiogenesis, chemical evolution, or macro-evolution. At the

physical level, entropy automatically prevents these things from happening. That is what has been experienced and observed.

Scientific Observation: Physical matter and entropy began. Big Bang proponents, Christians, Muslims, and Jews are united in their belief that physical matter and our physical universe had a beginning.

Logical Scientific Conclusion: Therefore, it is logical and rational to conclude that physical matter, physical universes, and entropy have Someone Intelligent or Someone Psyche who caused them to begin. If the Big Bang truly happened, then Somebody pushed the button that made it go bang. If physical matter and entropy truly began, then they have Someone Intelligent or Someone Psyche who caused them to begin.

This is what has been experienced and observed when it comes to Science.

This is Science 101.

The physical sciences are particularly interested in beginnings and endings – cause and effect. The physical sciences try to figure out what causes things to happen. As a result, spontaneous generation, abiogenesis, creation ex nihilo, chemical evolution, creation by chance, creation by entropy, or macro-evolution was FALSFIED in 1859 by Louis Pasteur. Therefore, we need some other causal explanation for Origins besides spontaneous generation or chance or evolution.

Evolution of any type is based upon entropy. Evolution is entropy; and, we scientists know for a fact that entropy cannot design and create and manufacture anything. It never happened because it can't happen. Entropy or chance cannot design and create.

The idea that physical matter or entropy began has serious and consequential philosophical and metaphysical implications. This scientific observation automatically prompts scientists like me to ask, "Who or what caused it to begin?"

Are Materialism and Naturalism True?

It is absolutely essential to get the Correct Model or the True Paradigm in place before we do science because the True Paradigm is crucial for getting the correct interpretation for Scientific Evidence or drawing the right conclusions from Scientific Evidence.

Before we can properly and accurately interpret Scientific Evidence and get to the truth of any Scientific Hypothesis that comes our way, we must determine whether Materialism and Naturalism are true or false. All of our scientific interpretations depend upon getting this aspect of science right and 100% correct to begin with, or everything else that we derive from "science" is going to be false.

Materialism or Physicalism is the philosophical belief or the scientific paradigm which states that the non-physical does not exist. Naturalism is the philosophical belief or the scientific paradigm which states that the supernatural or the spiritual does not exist. Nihilism is the philosophical belief or the scientific paradigm which states that the afterlife, or the spirit world, or the syntropy realm, or the transdimensional realm does not exist – in other words, life has no purpose or meaning. Behaviorism or Determinism is the philosophical belief and scientific paradigm which states that free will, choice, or agency does not exist. Atheism is the philosophical belief or the scientific paradigm which states that God and God's Psyche do not exist. Materialism, Naturalism, Nihilism, Behaviorism, and Atheism unitedly state that the psyche, spirit, mind, or soul does not exist. Darwinism or the Theory of Evolution is a combination of all these philosophical beliefs or scientific paradigms.

You will come to completely different diametrically opposed interpretations for Scientific Evidence if you choose to believe that Materialism and Naturalism are true than what you will get if you know that Materialism and Naturalism are false.

So, which is it? Are Materialism and Naturalism true, or are they false? Your choice determines the Scientific Interpretations and the Scientific Conclusions that you will make throughout the whole of your scientific career.

In recent years, I used Scientific Observations, Scientific Experiments, Experiential Evidence, Eye-Witness Evidence, the Scientific Methods, cutting-edge Scientific Discoveries, and Logical Common Sense to answer this question.

Do you want to know what I discovered?

I discovered that ALL of the observational evidence, experiential evidence, scientific evidence, empirical evidence, eye-witness evidence, and experimental evidence FALSIFIES the major premises or the primary claims of Materialism, Naturalism, Darwinism, Nihilism, Behaviorism, Physical Reductionism, Determinism, and Atheism. The observed, experienced, verified, and proven existence of Quantum Mechanics, Supernatural Mechanisms, Psyche, Non-Locality, Non-Physicality, Quantum Non-Local Consciousness, Transdimensionality, Gravity, Magnetism, Dark Energy, Dark Matter or Spirit Matter, the Strong and Weak Nuclear Forces, Quantum Waves, Invisible Types of Light, Action at a Distance, the Quantum Zeno Effect, and Syntropy FALSIFIES Materialism, Naturalism, and their derivatives.

I discovered that Science IS observation and experience, not Materialism and Naturalism. Furthermore, I discovered that observation and experience FALSIFY Materialism, Naturalism, Darwinism, Nihilism, Atheism, and their derivatives.

The false is repeatedly falsified by the truth; and, the truth is repeatedly experienced and observed. That is what I discovered and learned from all my scientific research over the years. I learned that Materialism, Naturalism, Darwinism, Nihilism, Behaviorism, Determinism, Physical Reductionism, and Atheism are FALSE.

Now, I finally have the Correct Model or the True Scientific Paradigm that's needed to get the correct, right, and true interpretation for the Scientific Evidence that comes my way. That's something that I didn't have when I was a Materialist, Naturalist, Nihilist, and Atheist. When I was a Naturalist and Atheist, I didn't have a logical explanation nor a scientific explanation for Quantum Mechanics, Supernatural Mechanisms, Psyche, Consciousness, Mind, Action at a Distance, and Syntropy that made sense to me at the time; but, now I do.

I upgraded my science by choosing to allow ALL of the evidence into evidence and by choosing to pursue a preponderance of that evidence.

Mark My Words

One of the Greatest Scientific Discoveries of All Time

One of my greatest and most useful scientific discoveries is that Physical Matter or Entropy had a beginning.

Scientific Observation: Anything that had a beginning had an Instigator who was instrumental in causing it to begin. There's no such thing as spontaneous generation or creation ex nihilo.

Scientific Observation: Physical Matter or Entropy began. Most Big Bang Scientists and most Christians and Muslims are united in their belief that physical matter or entropy had a beginning.

Scientific Conclusion: Therefore, it is logical and rational to conclude that Physical Matter or Entropy had Someone Psyche or Someone Intelligent who caused it to begin or who was instrumental in bringing it into existence.

This is a syllogism.

If the Premises or Scientific Observations are true, then the Scientific Conclusion has to be true as well.

For me personally, this became powerful Scientific Proof of God's Existence.

Only a die-hard Atheist will deny the truthfulness of the first Premise or the first Scientific Observation; but then again, denial should be expected from them because Atheism is totally irrational and illogical. Atheism is creation of something by nothing, which is impossible, illogical, and irrational. Nothing has never been caught in the act of designing, creating, producing, and manufacturing something. Atheism is a blind faith in spontaneous generation, creation ex nihilo, abiogenesis, chemical evolution, or macro-evolution. These people have placed ALL of their faith in entropy, random diffusion, or chance.

Atheism is patently absurd. I KNOW because I used to be a Physicalist, Naturalist, Nihilist, and Atheist. There's no way to provide evidence or proof that "Nothing" designed and created everything. Such a proposition has to be taken on blind faith alone as being real and true. Atheism is a religion that is based exclusively on blind-faith and wishful thinking. There's NO evidence to support it, and there never will be.

Instead of a blind leap of faith into Atheism, what we have done collectively as a race is to use scientific observations to falsify spontaneous generation, or creation ex nihilo, or chemical evolution, or Atheism. Most of us have concluded that physical matter or entropy had a beginning, which means that it has Someone Psyche or Someone Intelligent who caused it to begin.

It is obvious that some type of Syntropy MUST exist; otherwise, all of that subsequent entropy or physical matter wouldn't have been possible in the first place.

The repeated Scientific Observation that physical matter or entropy began has other scientific ramifications. If physical matter or entropy began, then it can theoretically come to an end when it is absorbed back into Syntropy. Physical matter or entropy came from some type of Syntropy; therefore, it is logical to conclude that when that physical matter or entropy ends, it will return to Syntropy.

If you choose to buy into and support the Big Bang Theory, then you simply KNOW that physical matter and entropy had a beginning; and, you also KNOW that Someone Psyche or Someone Intelligent pushed the button and made it go bang. Explosions don't just spontaneously generate out of thin air ex nihilo. That's impossible. Someone Psyche or Someone Intelligent in the Syntropy Realm or the Transdimensional Realm pushed the button and made it go bang. This is logical common sense based upon all of our Scientific Observations to date.

Unlike genetics and genomes, however, the Big Bang Theory is in dispute. The genes have been experienced and observed by contemporary living human beings; whereas, the Big Bang was not. Therefore, the Big Bang is a best-guess or a philosophical supposition. It will always be in dispute to one extent or another. In other words, there are many other methods and means by which physical matter or entropy could have been brought into existence. I explore some of these in my other books.

The Christians, Jews, and Muslims KNOW that Syntropy, Allah, Jehovah, or God has always existed and will always exist. God's Psyche or God's Intelligence is eternal and everlasting, without a beginning of days or an end of years. Syntropy means eternal and everlasting. Psyche is Syntropy. Consciousness, awareness, or intelligence is a type of Syntropy. Spirit matter or dark matter is a type of Syntropy. Gravity, magnetism, and dark energy are a type of Syntropy. Quantum Mechanics is Syntropy. God is Syntropy. They have always existed, and they will always exist.

Yet, Genesis 1: 1: talks about some type of beginning.

So, it is logical to ask, "What began during this beginning?"

Genesis 1: 1: **In the beginning God created the heavens and the earth.**

Whether we realize it or not, Genesis 1: 1: is actually a syllogism or a Scientific Proof of God's Existence.

Scientific Observation: Every beginning has a Beginner, Maker, or Instigator who triggered it, caused it, or began it. This scientific observation is obviously true.

Scientific Observation: The heavens and the earth had a beginning. In other words, physical matter or entropy began. Most of our scientists are in agreement that this scientific observation is true.

Scientific Conclusion: Therefore, it is logical to conclude that the heavens and the earth – physical matter and entropy – had Someone Psyche or Someone Intelligent who made them or brought them into existence from Syntropy or Spirit Matter, which has always existed and will always exist.

Syntropy by definition in principle had NO beginning and will have NO end. Syntropy is eternal and everlasting, without a beginning of days or an end of years. God is Syntropy. Spirit matter or dark matter is Syntropy. Psyche is Syntropy. Quantum Mechanics is Syntropy. Psyche is Intelligence. Psyche is Life.

Genesis 1: 1: talks about some type of beginning; so, it is logical to ask, "What began during this beginning?"

God tells us what began during this beginning. The heavens and the earth began. In other words, physical matter and entropy began. Simple. Logical. Parsimonious. True.

This is a very simple but very powerful Scientific Observation. Its ramifications spill over into everything else.

If entropy truly had a beginning, then it can also have an end when it gets absorbed back into Syntropy. This means that physical matter is based upon entropy; and, everything else is based upon Syntropy. Entropy or physical matter had a beginning; but, everything else did not. Everything else is Syntropy. If this is true, then it explains everything.

Some of the scientists have declared that the protons will never decay. That would indeed be the case if they are based upon Syntropy to begin with and continue to maintain that Syntropy at their very foundation for the duration of their existence. Entropy tells us that all physical matter will eventually end at heat death. But, if that entropy and physical matter were derived originally from Syntropy, then the entropy will cease to exist at the very point that God absorbs that physical matter back into the Syntropy from whence it came. Syntropy is the NORM; and, entropy or physical matter is the exception to that NORM. God made physical matter and entropy to be that exception to the NORM. He did so with a purpose in mind.

When it comes to Science, it's extremely important to get this right from the very beginning because everything else depends on it.

Remember, entropy is a function of time. If there is no passage of time, then there is no entropy and there is no aging process. The theory of relativity tells us that there is NO passage of time, NO entropy, for objects that are traveling faster than the speed-of-light. From a photon's perspective, it experiences NO passage of time while traveling at the speed-of-light. From its perspective, it arrives the very moment it leaves. From our perspective, it took 13 billion years to get here; but, from its timeless and ageless perspective no time has passed at all between its origin and its destination. That's the beauty of Syntropy. It doesn't age. It doesn't get old! It experiences NO entropy. Syntropy or Quantum Mechanics is timeless and ageless, without a beginning of days or an end of years. There is NO entropy in the Syntropy Realm.

The verified and proven existence of Quantum Mechanics or Action at a Distance falsifies the claims of Materialism, Naturalism, and their derivatives. Quantum Mechanics or Syntropy has been experienced and observed.

Furthermore, I know from personal experience that God can teleport or quantum tunnel physical matter anywhere in the universe instantly simply by temporarily removing the entropy or stopping the passage of time for that physical object. If you remove the entropy or stop the passage of time, then that physical object temporarily becomes spiritual or syntropic or supernatural in nature. It becomes subject to Quantum Mechanics or Supernatural Mechanics rather than classical physics and entropy. I felt time stop. Then instantly I and the car I was driving were someplace else when time resumed, and the entropy was restored.

I experienced quantum tunneling or teleportation. It was nothing that I did. It was something that God did for me in order to save me. I was traveling 40 mph when the car pulled out in front of me. I remember hitting the brakes, but there was nowhere to go. I was going to broadside that car. Time stopped. When time resumed, I was in the middle of a lawn forward and to the right thirty feet away with the car stopped, and the car turned off. I was totally calm. There hadn't been enough time for my adrenaline to start pumping. Somehow, I passed through that car unphased; and, I also went from 40 mph to 0 mph instantly without leaving skid marks and a cloud of smoke. Somebody turned my car, stopped my car, and turned it off for me. I was teleported to safety and have no memory or knowledge of how I got there. It was instantaneous. It's by far the most miraculous our supernatural thing that has ever happened to me. I and the car I was driving experienced Quantum Tunneling or Teleportation. I blinked or jumped.

This is cool science, especially in comparison to the watered-down and neutered stuff that we get from the Materialists, Naturalists, and Darwinists. Materialism, Naturalism,

Darwinism, Nihilism, and Atheism have NO explanation for what happened to me. But, Quantum Mechanics, Psyche, and Syntropy can definitely explain it. The explanatory power of Syntropy or Quantum Mechanics is through the roof!

Remember, Quantum Mechanics or Supernatural Mechanics makes the physically impossible possible. Syntropy makes the physically impossible possible. By choosing to allow Quantum Mechanics and Psyche in to play, we can literally explain everything that comes our way. Quantum Mechanics or Syntropy has infinitely more explanatory power than Classical Physics, Materialism, or Naturalism.

It is obvious that some type of Syntropy MUST exist; otherwise, all of that subsequent entropy or physical matter wouldn't have been possible in the first place.

Since we KNOW for a fact that Syntropy must exist, then the next thing to get right when it comes to Science is whether entropy and physical matter began to exist or not. Since most of the scientists and theologians are in complete agreement that entropy, the passage of time, the aging process, and physical matter began, then we can safely, logically, and rationally conclude that they will end when they get absorbed back into Syntropy, or when they get absorbed back into the Timeless and the Ageless from whence they came.

So, why is there all this physical matter or entropy since it is obvious that Syntropy or Quantum Mechanics is vastly superior?

The advantage to physical matter and the physical sciences is that we can control it, experiment with it, use it, and predict what will happen. Physical matter and entropy form the basis for the Ultimate Consensus Reality. It's reliable, dependable, predictable, and controllable. Cause and effect become obvious, controllable, and predictable in a physical reality.

The drawback to Syntropy or Quantum Mechanics is that God holds all the KEYS to these phenomena. We mortal, physical, fallen human beings cannot Quantum Tunnel or Teleport at will. God prevents us from doing so. We can't phase-shift at will either. Our telepathic abilities are severely hampered and restricted by our physical brains and physical bodies. As mortal physical beings, we will NEVER be able to travel faster than the speed-of-light. We can't levitate or walk on water without some kind of help from God. Physical matter and entropy cannot do any of these faster-than-light quantum phenomena on their own without some kind of direct intervention from God. We are going to age and die; and, ONLY God can raise us from the dead. God holds ALL the KEYS to Quantum Mechanics. That kind of takes the fun out of it, doesn't it?

The greatest scientific discovery that fallen, mortal, physical beings could ever make would be to find a way to remove entropy from physical matter thereby turning entropic physical matter into syntropic physical matter resulting in an immortal physical body and perpetual motion machinery.

The fact that we mortal, fallen, physical human beings have NO direct access to Quantum Mechanisms basically ruins and destroys ALL of our science fiction because ONLY God retains direct access to and direct control over Superpowers and Immorality.

There will never be any X-Men or New Mutants with Superpowers, because God kept all the Superpowers or Quantum Mechanisms to Himself when He designed, created, and produced this physical universe, physical matter, and entropy. In other words, the ONLY way to get Superpowers is from God Himself. We will NEVER get them from our genes! Superpowers are a function of Quantum Mechanics and NOT a function of our genes; and,

ONLY God holds ALL the KEYS to Superpowers or Quantum Mechanics. The genes have NO choice in the matter. Fallen, physical, mortal beings like us are restricted and prevented by God from gaining direct access to Quantum Mechanics or Superpowers. We have to go through Him and work with Him if we want to gain access to these things.

This is what has been experienced and observed.

Genetic mutations have been happening for millions of years if not billions of years on this planet; and, they have NEVER given anyone superpowers or super-human capabilities. Quantum mechanical supernatural powers are NOT to be found within our genes. Instead, physical matter and genes were designed to dampen, restrict, and limit the innate quantum mechanical or supernatural capabilities of our Psyche and spirit body. This is what has been experienced and observed. Physical matter and genes were made by God to prevent us from developing Superpowers; and, that's the way it works.

Mark My Words

Energy Is Conserved

One of my greatest, most explanatory, and most useful scientific discoveries came when I first realized that Syntropy is the Conservation of Energy – in other words, Syntropy is the First Law of Thermodynamics. This was the capstone to everything else that I had discovered so far.

Of course, I'm not the first scientist to have discovered the First Law of Thermodynamics or the Conservation of Energy. However, I might be the first scientist on the planet to have discovered Syntropy and its significance in the overall scheme of things. I try to crawl down this rabbit hole and see how far down it goes. I'm a scientist. I want to KNOW what things are and how they work.

OBSERVATION: The Materialists, Naturalists, Darwinists, Nihilists, Atheists, Behaviorists, Determinists, Physical Reductionists, and Classical Physicists teach that ONLY entropic physical matter exists.

Do you fully realize and understand what that means?

It means that these people ERRONEOUSLY teach and believe that physical matter and entropy are conserved!

These people literally equate Conservation of Energy with the conservation of physical matter and entropy; and, that's how these people destroy Science. These people make NO distinction whatsoever between physical matter, entropy, energy, conservation, and the First Law of Thermodynamics. These people literally teach that entropy is conserved – that entropy is eternal and everlasting.

The Physicalists and Scientific Naturalists literally teach that physical matter and entropy are eternal and everlasting. Do they not? These people literally teach that entropy is irreversible and everlasting. Do they not? These people literally teach that ONLY physical matter or entropy exists. Do they not? These people literally teach that physical matter and entropy are conserved.

And, there's where they fall down never to recover ever again. These people teach that entropy or death is eternal and everlasting. They teach that entropy or death is conserved. They teach that entropy or death is the First Law of Thermodynamics. Do they not? Yes, they do. That's exactly what they teach. They teach that entropy or death is conserved.

WOW! When I saw that and understood that, instantly everything became clear to me! These people ERRONEOUSLY teach that physical matter, entropy, and death are conserved. These people ERRONEOUSLY teach that physical matter, entropy, and death are eternal and everlasting.

They are wrong!

Physical matter and entropy are NOT conserved! Locality and entropy come and go as physical matter comes and goes. Locality is associated with space, and entropy is associated with time. Space-time comes and goes as physical matter comes and goes. Physical matter, space-time, locality, and entropy are NOT conserved! Their quantity changes over time according to the physical laws or the physical restrictions that are being imposed upon Raw Energy by God.

Energy is Syntropy. It's the Energy or the Syntropy that's actually being conserved according to the First Law of Thermodynamics. It's the Energy or the Syntropy that's eternal and everlasting – NOT the physical matter and NOT the entropy. Energy is an invisible, immaterial, intangible, non-physical force, or field. Energy is Syntropy. Energy, Psyche, Syntropy, or Life is being conserved, while physical matter and entropy are not. Remember, Energy or Syntropy is being conserved, NOT physical matter and entropy; and, there you have the answer to life, the universe, and everything.

Physical matter and entropy are NEVER conserved. This should be engraven in stone as one of the Laws of Thermodynamics! Physical matter and entropy are NEVER conserved.

Physical matter is organized energy, entropic energy, limited energy, localized energy, restricted energy, frozen energy, physicalized energy, or energy that has been limited to space-time. Physical matter, or entropy, or locality, or space-time comes and goes as God sees fit. They are NOT conserved. We humans make new physical matter in our particle accelerators or atom smashers; and, we humans destroy physical matter and the associated locality and entropy in our atomic bombs. Physical matter and entropy are NOT conserved. They are NOT eternal and everlasting! Entropy and death are NOT conserved – they are not eternal and everlasting. The conservation of entropy is the illusion, and not the reality.

Entropic physical matter is are NOT the only thing that exists! Remove the physical limitations from physical matter, and it goes back to being Raw Energy or Syntropy. Energy is Syntropy. Syntropy or Energy is conserved, NOT the physical matter or entropy.

Energy or Syntropy can take on many different FORMS, including psyche, invisible forces and fields, entropy, gravity, magnetism, electricity, quantum waves, dark energy, the zero-point field, the strong and weak nuclear forces, visible and invisible light, spirit matter or dark matter, and physical matter. Physical matter or entropy is simply one FORM of Energy. Physical matter is entropic energy or localized energy. Remember, physical matter or entropy is simply one FORM of Energy. The FORM is never conserved! The FORM changes over time. It's the underlying Energy or Syntropy that is being conserved, and NOT the FORM.

Energy or Syntropy is conserved, NOT physical matter and entropy. Entropy or death is a state of existence; and, states are never conserved. They can change anytime that God decides to change them. Entropy is NOT conserved. Its amount and its existence changes over time. This is very good news! Physical matter is a FORM of Energy. Physical matter is NOT conserved. Its amount and its existence changes over time. Entropic physical matter is a FORM of Energy, which means that entropic physical matter is NEVER conserved! The FORM is never conserved.

Entropy is a "form" of energy in that entropy is a measure of energy's transition from a useful FORM to an unavailable FORM. The Materialists and Naturalists teach and believe that the Unavailable Forms of energy are always conserved. They are wrong. The FORM is never conserved. God's Psyche can make unavailable energy available again because God designed and made the rules which made that energy unavailable in the first place. The FORM of the Energy is never conserved.

Energy or Syntropy can take on many different FORMS. Psyche, Intelligence, or Life is a FORM of Energy or Syntropy. So is spirit matter and physical matter. In fact, Psyche or Intelligence IS Energy or Syntropy; and, Psyche or Syntropy FORMS the basis for everything else. Entropy is death. The Ultimate Law of Thermodynamics states that entropy or death is NEVER conserved, while the underlying Psyche or Syntropy is always conserved. This Reality explains everything that has ever been experienced or observed, does it not?

Entropy is like a clock that measures the passage of time. The clock comes and goes as God sees fit. Entropy and Locality are simply a FORM of Energy. Space-time is simply a FORM of Energy. Physical matter is simply a FORM of Energy. Physical matter and entropy can be changed into a different FORM of Energy whenever God desires to do so. Entropy is NOT conserved. Entropy is NOT eternal and everlasting! The amount of entropy or death comes and goes as God sees fit, while the underlying Energy, Syntropy, or Psyche is always conserved. The FORM of Energy is constantly changing while the underlying Energy, Psyche, or Syntropy is always conserved. This is the answer to life, the universe, and everything. Is it not?

When I finally realized that it's the Energy, the Psyche, the Intelligence, or the Syntropy that's being conserved, and NOT the physical matter or entropy, instantly everything became clear to me. Finally, I had the answer to life, the universe, and everything. This just might be the GREATEST scientific discovery of ALL TIME. It certainly was for me.

Mark My Words

A Genome is Software and Hardware

I used to be a Materialist, Naturalist, Nihilist, and Atheist; but, I'm no longer hiding from the evidence like I used to do. I'm thinking much more rationally and logically now. I have chosen to allow ALL of the evidence into evidence. I'm a lot more open-minded than I used to be. I no longer define science as Materialism, Naturalism, and Scientism as I used to do. I upgraded my science to Science 2.0; and now, I define science as observation and experience.

The majority of scientists still define "science" as Materialism, Naturalism, Darwinism, Nihilism, Atheism, and Classical Physics. These people refuse to come into the age of Quantum Consciousness, Quantum Mechanics, and Syntropy. Unlike them, I upgraded my science to Science 2.0; and now, I define science as observation and experience. Supernatural non-physical invisible forces and fields have been experienced and observed.

There are many wonderful gifts that derive from upgrading the definition of Science to include observation and experience, rather than defining "science" as Materialism and Naturalism. For example, it soon became clear to me that it should have been called a Genetic Perspective rather than an evolutionary perspective. The genes have been experienced and observed; whereas, the different types of evolution have not. The different types of macro-evolution or spontaneous generation have NEVER been observed and have never been caught in the act because they are physically impossible and prevented from happening by genetics and entropy.

The scientists should have called it the Genetic Perspective rather than the evolutionary perspective. The theory of evolution is in constant dispute; but, the genes are not. The genes have been experienced and observed; whereas, macro-evolution or chemical evolution or spontaneous generation has NOT.

Here's another Scientific Proof of God that has some teeth and claws because it is founded on observations and experience. It's based upon the idea that anything that has a beginning has a Maker, Creator, or Manufacturer who brought it into existence. It's based upon logical common sense.

Scientific Observation: Everything that began obviously had Someone Psyche or Someone Intelligent who made it or brought it into existence.

Scientific Observations: Physical matter, proteins, their matching genes, and genomes obviously had a beginning.

Scientific Conclusion: Therefore, it is logical and rational to conclude that physical matter, proteins, genes, eyes, brains, and genomes had Someone Psyche or Someone Intelligent who made them or brought them into existence.

Some of the scientists have been smart enough and open-minded enough to observe that a genome is a radically advanced software program, that DNA is in fact like a radically advanced hard drive or memory storage device, and that a living cell is a radically advanced piece of computer hardware or nanotechnology. We, who have made this realization, also realize that ONLY Psyche or Intelligence is capable of designing, creating, engineering, and manufacturing hard drives and hardware; and, ONLY Intelligent Beings or Psyche Beings can do computer programming or the creation of software and genomes.

God must of necessity exist in order to have designed, engineered, manufactured, and created all of the DNA molecules, proteins, genes, genomes, and living cells on this planet – the hardware. These things don't just put themselves together from scratch. Computers don't put themselves together either. They have never been observed doing so! In fact, Louis Pasteur used Science in 1859 to prove that these things can't put themselves together, by falsifying Spontaneous Generation or Materialism, Naturalism, and Creation by Rocks.

Furthermore, God must of necessity exist in order to have written or programmed each and every genome that is encoded as software in the DNA of each life form on this planet. ONLY Psyche or intelligent beings could have done such a thing. We have NEVER caught the rocks in the act of designing, programming, engineering, and manufacturing genomes and life forms from scratch; and, we never will. They can't! The rocks or elements are prevented from doing so by entropy or the second law of thermodynamics. Entropy or random diffusion prevents genomes from spontaneously generating out of thin air from atoms or from scratch.

A Genome IS God's Signature!

What's most interesting about all of this is that God wrote His signature into each one of your cells. Your genome IS God's signature. Each genome on this planet is God's signature. God has written Himself upon you and within every life form on this planet. The rocks or raw physical matter can't write genomes or computer software; but, Psyche certainly can. God's Psyche must of necessity exist, or you wouldn't have a genome. God's Psyche or God's Intelligence must exist, or you wouldn't have a physical body. That's what Science is trying to teach us!

Mark My Words

Young-Earth versus Old-Earth Creationists

I'm not a creationist. I don't believe in creation ex nihilo or creation from nothing. And, I don't believe in creation by nothing either – spontaneous generation, abiogenesis, chemical evolution, macro-evolution, or the theory of evolution. Creation from nothing and creation by nothing are impossible. I'm not a creationist in any traditional sense or secular sense of the word.

That being said, I have sampled all the different views on the subject of origins while studying both science and theology; and, during my research, I have gotten a feel for which one's are true and which ones are science fiction.

It really doesn't matter whether God organized our universe and earth slowly or quickly. The only reason it matters is because if one of them is true then the other one is false. They are mutually exclusive.

I'm good at research and withholding judgment until the facts are in. I'm also becoming very good at finding and explaining the truth. Some people are interested in my take on the war between the Young-Universe Creationists (YUCs) and the Old-Universe Creationists (OUCs).

I have learned to love the Young-Earth Creationists (YECs). Unlike the Materialists, Naturalists, Darwinists, Nihilists, and Atheists, the YECs study and teach ALL the science rather than a small sub-set of the science. Nothing is off-limits for the YECs, or the OECs either. I have learned more science and more truth from a handful of YECs than I have learned from ALL of the Materialists, Naturalists, Darwinists, and Atheists combined.

As an amateur astronomer and telescope maker, I love the science in these videos entitled, "What You Aren't Being Told about Astronomy", even if they are presented by a Young-Earth Creationist:

https://www.youtube.com/watch?v=CzyQbOQ0dv0

https://www.youtube.com/watch?v=E66409i-yn4

All of that being said, in the war between the YUCs and the OUCs, one of them is right and one of them is wrong; and, I KNOW which group is right, and which group is deceiving themselves. Remember, self-deception works, and it works every time. More importantly, I can explain why the one is true and the other one is false.

Observation and Experience

While upgrading my science to Science 2.0, I redefined science as observation and experience, rather than defining science as Materialism and Naturalism as I used to do. Under Science 2.0, observation and experience take precedence over philosophical speculation, sophistry, personal interpretation, and wishful thinking.

Science 2.0: I Upgraded My Science

https://www.amazon.com/dp/B0771K6WTX

Now, let's resolve this controversy through observation and experience.

My best friend is a seer. While in the spirit, he has seen and connected with many of the countless number of inhabited worlds that God the Father and Jesus Christ have organized. There are no speed-limits in the Syntropy Realm, Spirit Realm, or Quantum Realm. There are NO physical limitations in the Syntropy Realm. There is NO entropy in the Syntropy Realm. Instantaneous transpersonal psyche-to-psyche or telepathy is the NORM in the Quantum Realm, Spirit World, or Psyche Realm.

There are human beings like us on those other planets. My friend connected with them, communed with them, and was one with them for a while. There's a universal family of human beings out there in the universe right now. Their spirit bodies and their physical bodies are descended from God the Father just like us. Our Adam was a son of God – both his physical body and his spirit body. We are the children of God just like the Apostle Paul taught. We are descended from the Gods. Our surname is God.

Most of these worlds, that my friend saw, are far more advanced spiritually than we are. In a Syntropy Realm, there is no entropy which means that ALL of the energy is available all of the time. The Celestial Worlds or the worlds where the resurrected Gods and the Angels live don't need technology because they have direct access to Syntropy or Quantum Mechanics; yet, these people also have physical bodies as well.

The ordinary matter or physical matter is ONLY 4% of our universe. The rest of it is Syntropy. Syntropy is eternal and everlasting, without a beginning of days or an end of years. Syntropy, spirit matter, or dark matter is the NORM; and, the physical matter or entropy is the exception to the NORM.

Another couple of seers saw all of this in vision in 1832.

Doctrine and Covenants 76: 19-24

And while we meditated upon these things, the Lord touched the eyes of our understandings and they were opened, and the glory of the Lord shone round about. And we beheld the glory of the Son, on the right hand of the Father, and received of his fulness; And saw the holy angels, and them who are sanctified before his throne, worshiping God, and the Lamb, who worship him forever and ever. And now, after the many testimonies which have been given of him, this is the testimony, last of all, which we give of him: That he lives! For we saw him, even on the right hand of God; and we heard the voice bearing record that he is the Only Begotten of the Father — that by him, and through him, and of him, the worlds are and were created, and the inhabitants thereof are begotten sons and daughters unto God.

There's the truth of the matter. It has been experienced and observed by many different seers throughout human history. Under Science 2.0, experience and observation take precedence over sophistry, philosophical speculation, and wishful thinking.

Your spirit body and your physical body are descended from the Gods. Your surname is God. You are one of the Gods.

These same truths have been confirmed or verified by multiple different seers – my best friend is one of them. What my best friend has experienced and observed – what he has seen – is in complete harmony with what these other seers have seen. That's good enough for me.

Many other worlds are inhabited by human beings like us – descendants of the Gods; but, each world is in a different stage of development or advancement, with most of the world's being much further advanced than we are when it comes to spirituality, goodness, and righteousness. There are other fallen worlds like ours, too.

Our Adam and Eve were NOT the first and were not the ONLY Adam and Eve to have fallen from a higher state to a lower physical state like ours. There are Adams many and many different fallen earths or telestial worlds in existence right now. There are also many different higher worlds – Terrestrial Worlds and Celestial Worlds – each in different stages of advancement.

The higher worlds or Celestial Worlds exist out-of-phase with our physical universe. These Celestial people, with their physical bodies, are angels and Gods. They reside in a higher dimension at a higher frequency. The Higher can experience, see, communicate with, administer to, and sense the lower. The Higher Realms are Syntropy Realms or Quantum Realms, meaning that these people have NO physical limitations. These people and their physical bodies are not limited to sub-light speeds. They have already proven themselves worthy of these blessings.

So, what did I learn from this?

I learned that physical universes, physical matter, and entropy have a beginning, which means that they can also have an end when they get absorbed back into the Syntropy from whence they came. The Syntropy – the eternal and the everlasting – is the NORM, and the physical matter and physical universe is an exception to the NORM.

ONLY 4% of our universe is physical matter or entropy. The rest of our universe is Spirit Matter (Dark Matter) and Dark Energy (the Light of Christ). Only 4% of our universe had a beginning. Only 4% of our universe is entropic. The rest of our universe is syntropic, without a beginning of days or an end of years. The rest of our universe is eternal and everlasting. The rest of our universe is Syntropy.

The YUCs and YECs teach that our earth and our universe began 6,000 years ago, when our Adam and Eve began. They teach that our earth is the only inhabited earth in our universe. They teach that there was only one Adam and Eve.

My seer friend tells me that the Young-Earth Creationists and Young-Universe Creationists have gotten the wrong interpretation of the Bible. The light really has been traveling for 13.2 billion of our years to get here to this earth. The light really has been traveling for 2.5 million years to get here from the Andromeda Galaxy. The scientists really have dug up ice cores in Antarctica and Greenland that are 800,000 years old or more at the bottom of them. The earth and the universe really are much older than 6,000 years old. In fact, at the fundamental core or the syntropic core of our universe, our universe is in fact eternal and everlasting. It is Syntropy.

ONLY the physical matter or the entropy had a beginning, and it will someday have an end.

Entropy is marked by the passage of time. Entropy is a function of time. Only the physical matter or the physical universe experiences entropy or the passage of time. Everything else is Syntropy. Everything else is eternal and everlasting. That means that ONLY physical matter or entropy had a beginning at some point in time in the distant past. Everything else has always existed and will always exist.

Entropy is associated with the speed-of-light limitations that have been built into physical matter. With entropy and the speed-of-light, we can experience and measure the passage of time. Through the speed-of-light, we can also measure distance.

Mark My Words

Messing with the Speed-of-Light

The Young-Earth Creationists (YECs) and Young-Universe Creationists (YUCs) work their magic by messing with the speed-of-light. In the process, the YUCs and the YECs turn God into a liar, trickster, and deceiver. These people literally destroy the speed-of-light as a scientific measurement tool just so that they can say that our universe and earth are only 6,000 years old.

It's called the **anisotropic synchrony convention**, and it was developed by Jason Lisle to give us a 6,000-year-old universe as well as a 6,000-year-old earth. They make the speed-of-light variable depending upon which way it is traveling. Then they say that light travels at an infinite velocity whenever it is moving towards our earth from any direction in space. Yuck! It literally destroys the speed-of-light as a measurement tool for time and distance by destroying the theory of relativity, physics, physical constants, and science! The **anisotropic synchrony convention** has become synonymous with an infinite speed-of-light.

I just can't go there. He deliberately breaks Science, Physics, the Physical Constants, and the Theory of Relativity just so that he can say that our universe is only 6,000 years old. He makes the whole of Science completely meaningless and worthless in the process.

The **anisotropic synchrony convention** turns God into a liar. God Himself told us in scripture that we can depend upon the lights in the heavens as signs for both the measurement of time and the measurement of distance. The **anisotropic synchrony convention** literally destroys the speed-of-light as a time or season measurement tool and as a sign or a measurement of distance.

Jason Lisle, the creator of the **anisotropic synchrony convention**, states that it can't be falsified, which gave me all the incentive I needed to falsify it. I've falsified it a dozen different ways, and so have others. The "purpose" of the **anisotropic synchrony convention** is falsified every time we present something that is clearly over 6,000 years old. An infinite speed-of-light is falsified by Science, Physics, and the Theory of Relativity. Falsifying an infinite speed-of-light falsifies the **anisotropic synchrony convention** and destroys the purpose for which it was made.

The following YouTube videos provide some background on this topic.

Distant Starlight Problem:

https://www.youtube.com/watch?v=5JEFy-ZtEzg

https://www.youtube.com/watch?v=5vUel1AIM8w

https://www.youtube.com/watch?v=0BGAaLm52kg

https://scientific-proof-of-god.com/wp-content/uploads/2018/06/Distant-Starlight.zip

These YUCs and YECs have all of the light travel at an infinite velocity towards the earth and at some other adjusted slower speed while moving away from the earth; and, they have the whole thing average out to be the speed-of-light. They force this paradigm or convention onto the speed-of-light so that they can have a 6,000-year-old universe and an earth-centric model when they are done. I don't buy it.

If light really worked that way – infinite speed coming from all directions towards earth – it would turn our earth into a universe-sized black hole or gravity well, which it is not. There's nothing special about our earth where the speed-of-light is concerned. The speed-of-light is constant no matter which direction it travels. Our universe would be chaos otherwise. Think about it! The **anisotropic synchrony convention** is counter-intuitive and self-defeating once we start to take it seriously and actually apply some science and logic to it.

First, let's apply some logic. The **anisotropic synchrony convention** destroys the speed-of-light as a distance measurement tool, as a speed limit for physical matter or entropy, and as a measurement of the passage of time. When you destroy the speed-of-light as a scientific measurement tool, then you can literally have any age for the universe that you want. There's NO standard anymore. The physical universe could just as easily be six trillion years old as 6,000 years old. Take your pick. It's all the same after you have destroyed the speed-of-light as a measurement tool, as the YUCs and the YECs do with their **anisotropic synchrony convention** tool.

Once you use the **anisotropic synchrony convention** to destroy the speed-of-light as a measurement tool by making all incoming light infinite in velocity, you can just as easily and correctly state that the galaxies at the edge of the known universe are in fact 13.2 quadrillion years away from us and that our universe is in fact quadrillions of years old. Or you could just as easily say that the Andromeda Galaxy is in fact 2.5 trillion light years away from us instead of 2.5 million light years away from us. You could say that our Milky Way Galaxy is billions of light years in diameter rather than 100,000 light years in diameter. You can literally have anything you want for the age of this universe, galaxy, sun, earth, and stars once you have used the **anisotropic synchrony convention** to destroy the speed-of-light as a measurement tool for distance and the passage of time.

The **anisotropic synchrony convention** was designed by the YUCs and YECs so that they can have a 6,000-year-old universe and a 6,000-year-old earth when they are done. It's bad science.

I don't like being lied to. I despise the deception, trickery, and lies whether it's coming from the Creationists or the Naturalists and Atheists. That's why I turned against the Atheists, Naturalists, and Darwinists in a BIG way because I have reams of evidence and books of evidence proving that these people are lying to us and deliberately trying to deceive us. I don't like the **anisotropic synchrony convention** for the very same reasons. It's deceptive and demonstrably false.

The "purpose" of the **anisotropic synchrony convention** is falsified by 800,000-year-old ice cores from Antarctica and Greenland.

https://www.youtube.com/watch?v=8VU1rFobj7o

http://www.sciencemag.org/news/2017/08/record-shattering-27-million-year-old-ice-core-reveals-start-ice-ages

https://www.youtube.com/watch?v=3v-uvh_KWkg

https://scientific-proof-of-god.com/wp-content/uploads/2018/06/Scientific-Evidence-for-Old-Earth.zip

https://www.youtube.com/watch?v=0BGAaLm52kg

https://scientific-proof-of-god.com/wp-content/uploads/2018/06/Distant-Starlight.zip

The age of the ice cores is in constant dispute, especially when the ice is compressed under gravity at the deeper levels; but, the evidence for 50,000-year-old ice from cross-matched samples of different ice cores from different continents is very convincing. The debate is often mean-spirited on both sides, and it's hard to sort the truth from the posturing and fiction; but, I think there is solid convincing proof that the ice cores are at least 50,000 years old at the bottom of them before they start getting compressed and unreliable further down. The 100,000-year cycle that Hugh Ross has noticed in the ice cores is interesting and convincing as well.

http://www.reasons.org/explore/blogs/todays-new-reason-to-believe/read/tnrtb/2005/06/30/deep-core-tests-for-the-age-of-the-earth

http://www.reasons.org/explore/blogs/todays-new-reason-to-believe/read/tnrtb/2007/03/25/antarctic-and-greenland-ice-core-dates-match

The "purpose" of the **anisotropic synchrony convention** is falsified by the ice cores. YUC and YEC are falsified by the ice cores.

Next, I have observed that the "purpose" of the **anisotropic synchrony convention** is falsified by Radiometric Dating and Carbon 14 Dating.

Now, granted, the evolutionists apply some very loose assumptions to radiometric dating thereby forcing the age of the fossils and rocks to be much older than they really are. I own and have read "Thousands . . . Not Billions" by Don DeYoung. They do an excellent job demonstrating that the evolutionists are cheating and deliberately inflating the age of different fossils and rocks in order to support the theory of evolution. Using Carbon 14 dating, they convincingly demonstrated that rocks and fossils that the evolutionists date as being millions or billions of years old by other methods end up being less than 100,000 years old using Carbon 14 dating. Their Carbon 14 dating, of coal and diamonds that are supposed to be hundreds of millions of years old or even a billion years old, clearly demonstrates that the coal and diamonds are between 44,000 and 57,000 years old.

Then what do these Young-Earth Creationists do once they have successfully proven that the diamonds and the coal are only 50,000 years old according to Carbon 14 dating? They cheat. They grossly tighten the assumptions and break the half-life of Carbon 14 which is 5,730 years, so that they can then say that the coal and diamonds are between 4,400 and 5,700 years old, so that they can then say that our earth is less than 6,000 years old. Sorry, but I hate cheating and lies whether it is for or against the theory of evolution. Carbon 14 actually dates those "billion-year-old rocks" as being between 44,000 and 57,000 years old, which means that our earth is at least 57,000 years old according to Carbon 14 dating, thereby falsifying Young-Universe Creationism (YUC) and Young-Earth Creationism (YEC).

With Radiometric Dating, you can literally have any age you want for an object simply by adjusting your assumptions; and, these people do indeed play fast and loose with their assumptions. When it comes to Evolutionists and Darwinists, I have observed that you typically have to **divide** their Radiometric Ages for rocks and fossils by 10,000 in order to match with the ages that Carbon 14 is producing for the same objects. In contrast, when it comes to the YUCs and the YECs, you typically have to **multiply** their age guestimates by 10 in order to make them match with the ages that Carbon 14 is producing. The Darwinists and Evolutionists are clearly cheating more than the YUCs and YECs are, but they BOTH are cheating! They BOTH cheat in order to produce the age for objects that they want to have.

By insisting that our earth is only 6,000 years old, the YUCs and YECs make the half-life of Carbon 14 completely meaningless, along with ALL of the other half-lives that we have discovered. In other words, the YUCs and YECs completely destroy Science, Physics, Radiometric Dating, Half-Lives, the Physical Constants, the Theory of Relativity, Astronomy, Distance Measurements, Age Measurements, and Time Measurements in order to produce an age for the earth that is less than 6,000 years.

Remember, the half-life of Carbon 14 (5,730 years), if taken seriously, falsifies a 6,000-year-old earth. ALL other half-lives, if taken seriously, also falsify a 6,000-year-old earth. Consequently, the half-life of Carbon 14 and half-lives in general falsify Young-Earth Creationism and Young-Universe Creationism. Remember, half-lives have been experienced and observed, and so has a 13.4 billion-year-old galaxy and universe.

https://en.wikipedia.org/wiki/List_of_radioactive_isotopes_by_half-life

https://en.wikipedia.org/wiki/Chronology_of_the_universe

https://en.wikipedia.org/wiki/Observable_universe

https://en.wikipedia.org/wiki/List_of_the_most_distant_astronomical_objects

https://en.wikipedia.org/wiki/List_of_largest_cosmic_structures

Whenever YUC and YEC are assumed to be true, it completely destroys Science, Physics, Radiometric Dating, Half-Lives, the Physical Constants, the Theory of Relativity, Astronomy, Distance Measurements, Age Measurements, and Time Measurements. NOTHING can be depended upon as being true once YUC and YEC are assumed to be true. You can't even trust God to be telling you the truth because God Himself reveals himself to us through Scripture and Science; and, YUC and YEC completely destroy Science at all levels and all types of Science.

The purpose of the **anisotropic synchrony convention** and an infinite speed-of-light is to "prove" that our universe and earth are only 6,000 years old. Carbon 14 dating and all of the other Radiometric Dating methods falsify the "purpose" of the **anisotropic synchrony convention** by demonstrating clearly and convincingly that our earth is much older than 6,000 years.

The **anisotropic synchrony convention** is falsified by our GPS satellites and moon missions. Thanks to our GPS satellites and our moon missions, it would have been noticed if the speed-of-light changed depending upon whether it was outgoing or incoming.

The **anisotropic synchrony convention** is falsified by the variable star Mira. The variable star Mira has a visible tail that is 13 light-years long. At the speed at which Mira is moving, it would have taken over 30,000 years to produce that tail. Mira is moving counter to galactic rotation as if it had been shot out of a gun 30,000 years ago. There is evidentiary proof that the variable star Mira is at least 30,000 years old. There's NO proof that Mira is billions of years old; but, there is definite proof that Mira is at least 30,000 years old. That's the end of a 6,000-year-old-universe for me! The Andromeda Galaxy really IS 2.5 million light years away; and, the distant galaxies that we see in the Hubble Deep Space Field really were over 13 billion light years away when their light launched towards us. Mira is amazing, and it falsifies the **anisotropic synchrony convention**.

The Strangest Star in the Universe:

https://www.youtube.com/watch?v=XyuXBYWZegY

https://scientific-proof-of-god.com/wp-content/uploads/2018/06/Mira.zip

The **anisotropic synchrony convention** is falsified by the Observable Universe.

The radius of the observable universe is therefore estimated to be about **46.5 billion light-years** and its diameter about 28.5 gigaparsecs (**93 billion light-years**, 8.8×10^{23} kilometers or 5.5×10^{23} miles).

https://en.wikipedia.org/wiki/Observable_universe

https://science-2-0.com/wp-content/uploads/2018/06/Observable-universe.pdf

An Observable Universe would not be possible and would not be measurable if the speed-of-light is infinite, as the **anisotropic synchrony convention** claims that it is.

The **anisotropic synchrony convention** is falsified by the theory of relativity. The theory of relativity proves that NOTHING can travel faster than the speed-of-light in the physical realm. An object has to be massless or completely spiritual in order to travel faster than the speed-of-light. The **anisotropic synchrony convention** has ALL incoming light traveling at an infinite velocity towards our earth. That's physically impossible according to the theory of relativity.

http://www.reasons.org/explore/publications/tnrtb/read/tnrtb/2013/07/02/stronger-and-more-comprehensive-tests-affirm-the-universe-s-unchanging-physics

You literally have to break physics, astronomy, the theory of relativity, the physical constants, and science in order to make YUC and YEC true and produce a 6,000-year-old universe as these people try to do.

Let's give them the **anisotropic synchrony convention**. Let's conclude for the sake of argument that it is true. Let's conclude that ALL light travels to our earth at an infinite velocity from all directions in space. Let's assume that c, the speed-of-light, is infinite whenever light is traveling towards our earth from any direction in space.

$E = mc^2$. Energy equals mass times the speed-of-light squared. What is infinity squared? How much Energy does infinity produce when it is squared?

Now ask yourself, "How much Energy should be traveling towards our earth right now from every direction in space since the **anisotropic synchrony convention** is 100% true and the speed-of-light is infinite while traveling towards our earth?" If c is infinite while traveling towards our earth, then there should be an infinite amount of Energy traveling towards our earth right now from every direction in space. That's what we should observe if the **anisotropic synchrony convention** is true.

What would our earth be like with an infinite amount of Energy coming at it from every direction in space all at once? Our earth would be infinitely hot, would it not? Our earth would be a Big Bang singularity, would it not? How long could that go on before our earth would explode and we would have yet another Big Bang to have to deal with?

$E = mc^2$. If c is infinite, then E is infinite. That's what the theory of relativity is telling us must be true.

The observable fact that there is NOT an infinite amount of Energy converging on our earth from all directions in space is scientific proof that the **anisotropic synchrony convention** is FALSE and that the speed-of-light headed towards our earth is NOT infinite. I just used $E = mc^2$ and the theory of relativity to falsify the **anisotropic synchrony convention**.

With NO infinite speed-of-light while traveling towards our earth, the purpose of the **anisotropic synchrony convention** is broken and the whole idea is falsified. The purpose of the **anisotropic synchrony convention** is to give us a 6,000-year-old universe; but, the only way that's possible is if the speed-of-light approaches infinity while traveling towards our earth. The **anisotropic synchrony convention** and an infinite speed-of-light is self-defeating once it is taken seriously and applied directly to $E = mc^2$. An infinite amount of Energy headed towards us from all directions in space would not be good for us; and, I think it would have been noticed by now.

Jason Lisle, the creator of the **anisotropic synchrony convention**, states that it can't be falsified. He is wrong. It has already been falsified by $E = mc^2$, the theory of relativity, and the impossibility of an infinite speed-of-light in a physical realm. It is falsified by taking it seriously and assigning an infinite speed-of-light to $E = mc^2$ and then measuring to see if there is currently an infinite amount of Energy coming towards our earth from all directions in space. There is not an infinite amount of Energy coming towards our earth from all directions in space, so the **anisotropic synchrony convention** and an infinite speed-of-light have been falsified by scientific observations. The Science falsifies it or proves that it is false. That means that our physical universe is at least 13.2 billion years old just as the speed-of-light suggests. That means that Young-Universe Creationism has been successfully falsified. Our physical universe is indeed much older than 6,000 years. That also means that the physical matter that was used in the organization of this earth could in fact be 13.2 billion years old or older. It's NOT a guarantee, but it is a possibility.

Once you have successfully eliminated everything that is false, then ONLY the truth remains.

I'm a scientist. I've always been a scientist. The **anisotropic synchrony convention** felt wrong and anti-science from the very moment that I encountered it. It deliberately breaks the speed-of-light as a physical constant, as a distance-measurement tool, and as a measurement of time passage just so that they can get a 6,000-year-old universe out of the deal. It deliberately breaks the theory of relativity – the pinnacle of classical physics. It also breaks and violates the First Law of Thermodynamics. Its ultimate goal is to replace the truth with a bunch of different lies. I don't like it. The **anisotropic synchrony convention** turns God into a liar, trickster, and deceiver. If it were true, you wouldn't be able to trust anything that you observe with your physical eyes. That's not good. In fact, that's the very definition of evil. It breaks trust and faith, and at the same time the **anisotropic synchrony convention** demands an infinite amount of blind faith in order to believe that it is true. A variable speed-of-light or an infinite speed-of-light would literally turn order into chaos. Finally, the **anisotropic synchrony convention** has been falsified by $E = mc^2$ and the fact that there is NOT an infinite amount of Energy converging on our earth from all directions in space.

I have chosen to fight against and falsify the **anisotropic synchrony convention** and an infinite speed-of-light because it is falsified by observational evidence, science, and common-sense logic. Like I said, I hate deceptions and lies whether they are coming from the Creationists, Naturalists, or Darwinists. I have chosen to fight against them whenever I encounter them. I particularly despise the **anisotropic synchrony convention** because it deliberately destroys the speed-of-light as a time-measurement tool and a distance-measurement tool in order to promote and support a deceptive lie that is unnecessary in the first place. The word "day" in Genesis 1 was interpreted as an "undetermined amount of time", or "years", or an "epoch", or a "season", a "thousand years" during Biblical times; so, get used to it.

<u>Genesis 2: 16-17</u>:

> **And the Lord God commanded the man, saying: Of every tree of the garden thou mayest freely eat: but of the tree of the knowledge of good and evil, thou shalt not eat of it: for in the day that thou eatest thereof thou shalt surely die.**

So, when did Adam die? He died nearly a thousand years later, which means that the word "day" is translated as a "thousand years" in this particular verse of scripture.

<u>2 Peter 3: 8</u>: **But, beloved, be not ignorant of this one thing, that one day is with the Lord as a thousand years, and a thousand years as one day.**

In this particular scripture from the Bible, a "day" is once again interpreted as being "a thousand of our years".

A "day" is translated as a "season" and an "indeterminate amount of time" elsewhere in the Bible too.

I dislike the **anisotropic synchrony convention** because it deliberately breaks Physics, the Physical Constants, the Speed-of-Light, and Science in order to give these people a 6,000-year-old universe. Some of the YUCs and YECs are particularly unpleasant and obstinately closed-minded too, which doesn't help either.

I'm a scientist. Science itself convinced me that God exists. The Science convinced me that creation ex nihilo or spontaneous generation is physically impossible and therefore false. The Science also convinced me that our earth is millions of years old and our universe is billions of years old. I believe the Scientific Evidence, Experiential Evidence, and Observational Evidence – not the wishful thinking of the YUCs, YECs, Materialists, Naturalists, Darwinists, and Atheists.

There are other ways to falsify the purpose or the intention of the **anisotropic synchrony convention**. Let's mention a couple of them that I found convincing.

I think a case can be made that the 25 mass extinction events found in the fossil record can be used to falsify the 6,000-year-old earth theory; and, they definitely falsify the theory that the Genesis flood is the ONLY catastrophe and extinction event on record.

<u>https://en.wikipedia.org/wiki/Extinction_event</u>

Anything that is discovered and verified to be over 6,000 years old FALSIFIES the goals and intentions of the **anisotropic synchrony convention** which was designed by the YUCs and the YECs to give them a 6,000-year-old universe and a 6,000-year-old earth.

<u>https://en.wikipedia.org/wiki/List_of_longest-living_organisms</u>

<u>https://en.wikipedia.org/wiki/Pando_(tree)</u>

I feel sorry for the PhD scientists in general. Most of what they were taught in college has been falsified. I have observed that for most PhDs, their PhD actually does them a great disservice when it comes to science and prevents them from finding and knowing the truth. This is because they got their PhDs and education from Materialists, Naturalists, and Atheists who were deliberately trying to hide the truth from them. I, too, got my education from Physicalists, Naturalists, and Darwinists; so, I was forced to supplement my education by falsifying what they were trying to get me to believe. I've gotten pretty good at it.

A person can never seem to see what's wrong with his own theories and chosen beliefs; but, you can get very good at seeing what's wrong with other people's theories and beliefs. Despite their PhDs, the YUCs and YECs are often just as wrong as the Naturalists, Darwinists, and Atheists – just at the other extreme. Extremism combined with self-deception can turn any PhD into a deceiver, and they often don't even know it.

The YUCs and the YECs claim that the one-way speed-of-light cannot be measured. They are wrong. If you were to set up a computer and laser device here on earth and one on the moon, then send a time stamp in the laser beam to the moon. When that time stamp gets to the moon, it is recorded and compared with the clock's time on the moon. Meanwhile, the moon's computer and laser device on the moon could send a time stamp to earth. When the laser beam reaches earth, the time stamp in the beam as well as the time stamp in the receiving computer could be compared. By comparing the time stamps of both the transmission and the reception, both on the moon and on the earth, the clocks could eventually be synchronized, and the one-way speed-of-light determined.

Furthermore, we have already run this experiment with radio waves and radio broadcasts – transmitting to and from the moon, as well as with all the different GPS satellites we have in orbit. The delay is the same either way we choose to broadcast. A difference would have been noticed if there were one.

https://syntropy.site/wp-content/uploads/2018/05/Measuring-the-Speed-of-Light.pdf

There are NO speed limits in the Quantum Realm or the Syntropy Realm. That reality has been experienced and observed by Seers, Near-Death Experiencers, and Out-of-Body Explorers. However, God deliberately placed entropy or the passage of time into place where physical matter is concerned, and the speed-of-light and many other physical laws are held constant all directions and at all times by God in order to prevent physical matter from quantum tunneling away from us at will.

God made it physically impossible to travel faster than the speed-of-light in a vacuum any direction we choose to go. That speed limit holds all directions and at all times. We couldn't use the lights in the sky as a measurement tool for times, seasons, and distance signposts if the speed-of-light is variable. These people claim that their variable speed-of-light can't be falsified; but, it has already been falsified by the synchronization being done with all of our GPS satellites all around the earth and our moon missions. A variable speed-of-light would have been noticed if transmissions from the moon were instantaneous; whereas, transmission to the moon took seconds or hours or some such. A variable speed-of-light has never been experienced nor observed; and, direction has no effect on the speed-of-light.

Yes, when it comes to the **anisotropic synchrony convention**, the science is interesting – a lot more science and thought than we typically get from the Materialists, Naturalists, and Atheists; but, I don't think the YUCs and YECs should be messing with and destroying the speed-of-light, just so they can get a 6,000-year-old universe and earth out of the deal. Here's why:

Genesis 1: 14: And God said, "Let there be lights in the firmament of the heaven to divide the day from the night; and let them be for signs, and for seasons, and for days, and years."

This verse of scripture is telling me that God put the stars and the galaxies into the heavens to be used to measure time – the amount of time that it took for the light from these objects to reach our earth. They are meant to be used as signs or as distance measurements. They are meant to be used to measure the passage of time! This scripture is telling me that some of those galaxies really were 13.2 billion light years away when they sent their light towards our earth 13.2 billion years ago. This scripture is telling us that we really can rely upon the speed-of-light as a sign for the number of days and years it took for that light to reach us.

Obviously, the YUCs and YECs have chosen to interpret this scripture differently than the way I chose to interpret it; and, that's the whole issue in a nutshell – which interpretation is true and which interpretation is false?

I KNOW, based upon the revelations from God and the observations of the Seers who have seen the universe and God's creation, that the interpretations which the YUCs and YECs have made are wrong. My best friend is a Seer. He and other seers have actually seen and interacted with the Children of God, other humans, who live on other planets throughout this physical universe. They are more advanced than we are. They no longer need physical tools because God has given them direct access to and control over Quantum Mechanics, Spiritual Mechanics, or Supernatural Mechanics. They are translated beings and resurrected beings with translated and resurrected physical bodies – not fallen mortal beings like us. These people have what Paul calls a "spiritual" physical body, which means that their physical body is based upon Quantum Mechanics or Spiritual Mechanics rather than classical physics. Our physical bodies are based upon classical physics because we are fallen mortal beings. We are entropic beings. When we are resurrected from the dead, our physical body will be a Spiritual Body or a Quantum Mechanical Body and will be based upon syntropy rather than entropy. We will be syntropic beings.

From our perspective, a translated physical body, a resurrected physical body, and a translated or resurrected galaxy would look NO different than a fallen mortal one. Translated and resurrected stars, galaxies, and bodies are based upon Syntropy. Fallen mortal bodies like ours are based upon entropy. That's the only real difference between them. Physical bodies are limited by classical physics; whereas, resurrected and translated bodies have NO physical limitations and are based upon Syntropy or Quantum Mechanics instead. We have no idea from our perspective which galaxies or stars have been resurrected and which ones are still fallen, mortal, and entropic. They both would look the same to us.

The YUCs and YECs claim that our earth is the ONLY inhabited earth in this physical universe. These people also claim that our whole universe fell when our particular Adam and Eve fell. They are wrong. God the Father, His world, His planet, His star, and His people did NOT fall when our Adam and Eve fell. All of the other inhabited worlds in this physical universe where the translated and resurrected beings (angels) live did NOT fall when our Adam and Eve fell. The tree of life and Eden did NOT fall when our Adam and Eve fell. Only our earth seems to be a fallen earth at this point in time. From what the different Seers have seen, the rest of the earths in this universe have been translated and/or resurrected. They have already become New Earths like the one that our earth will someday become when it is translated and then later resurrected.

Scripture, the Revelations from God, and the Seers are telling me that YUC and YEC are false.

The ice core science is also telling me that the YUCs and YECs are wrong. Most everything the YECs say about evolution and God is good stuff until they start trying to smash everything down into a 6,000-year-old universe and earth. Then they lose me because I'm a scientist and I know better. I really do believe that some of the ice cores that the scientists have gotten from Antarctica and Greenland ARE in fact over 800,000 years old at the bottom of them, which means that the rocks underneath those ice cores are even older. Our earth doesn't have to be billions of years old as the atheists and secularists claim; but, our earth does have to be millions of years old as the scientific evidence demonstrates it to be.

I like to think that the Apostle Peter saw the YUCs and the YECs in vision and warned us in advance against them. He was a Seer.

2 Peter 3: 8: **But, beloved, be not ignorant of this one thing, that one day is with the Lord as a thousand years, and a thousand years as one day.**

Peter warns us not to be ignorant in this one thing. Peter warns us not to be messing with the speed-of-light as a measurement tool. One Biblical Day of creation is at least a thousand of our years. Furthermore, back in those days, the word "thousand" was a code-word for a really huge amount of time or people.

Facsimile 2

Fig. 1. Kolob, signifying the first creation, nearest to the celestial, or the residence of God. First in government, the last pertaining to the measurement of time. The measurement according to celestial time, which celestial time signifies one day to a cubit. One day in Kolob is equal to a **thousand** years according to the measurement of this earth, which is called by the Egyptians Jah-oh-eh.

The fatal flaws that the YUCs and YECs make are to assume that our earth is the center of the universe, our earth is the ONLY inhabited planet in the universe, that the speed-of-light can and should be messed with thereby turning God into a liar and a deceiver, that our Adam and Eve were the only ones, that our whole universe fell when our particular Adam and Eve fell, that our earth and our universe is only 6,000 years old, and that the scientific evidence proving otherwise is false. Young-Earth Creationism demands that we change the physical constants; that we mess with the signs; that we turn God into a trickster and deceiver; that we destroy Science and Physics and the Theory of Relativity; and that we deny and dismiss the scientific evidence that proves that our earth and our universe are much older than 6,000 years. They want us to believe that there was only one extinction event on this planet instead of the 25 that have been documented. They want us to believe that the 800,000-year-old ice cores are only 6,000 years old – or 4,000 years old if all of that ice was laid down during the flood. They want us to believe that when our Adam and Eve fell, it was a universal fall affecting everything in this universe including our earth.

Think about it logically, though.

God the Father, Jesus Christ, the angels in heaven, and their Celestial World or Heavenly Kingdom did NOT fall when our Adam and Eve fell. It was our Adam and Eve who fell, and nothing else in this universe or the other universes. Even the Garden of Eden and the Tree of Life did NOT fall when our Adam and Eve fell. The world that Adam and Eve fell into was already a lone and dreary world – a fallen world, or a telestial world. The Garden of Eden did not fall with them! That's why God had to put cherubim and a flaming sword to prevent them from entering Eden, partaking of the fruit of the Tree of Life, and living forever in their sins. The Garden of Eden and the Tree of Life did NOT fall when our Adam and Even fell!

Our earth did not fall when our Adam and Eve fell – our earth was already a fallen world before they fell. There is only one mass-extinction event mentioned in the Bible; but, there are over 25 different mass-extinction events recorded in the fossil record, which means that those other mass-extinction events took place BEFORE our Adam and Eve were placed into the Garden of Eden. If it took God a thousand years to set up each mass-extinction event in the fossil record, then our earth is at least 30,000 years old.

Unlike what the YUCs and the YECs claim and believe, our earth is NOT the only inhabited planet in this universe; and, our Adam and Eve were NOT the only Adam and Eve to have fallen over the millennia. When our Adam and Eve fell, those many other inhabited worlds did NOT fall with them! The world where God the Father and Jesus Christ live did NOT fall with them! There are innumerable inhabited worlds throughout this universe in different stages of advancement and/or different phases of existence. They did NOT fall when our Adam and Eve fell. A fall is a localized event, not a universal event. Contrary to what the YUCs and the YECs seem to claim, God the Father didn't fall when our Adam and Eve fell. The Garden of Eden and the Tree of Life did not fall when our Adam and Eve fell. Other earths did not fall when our Adam and Eve fell. Our earth did not fall when our Adam and Eve fell.

WE KNOW that the Biblical God Jesus Christ exists because He has revealed himself to thousands of us, if not millions of us, in the flesh, during our Near-Death Experiences, and through visions and revelations.

When it comes to Origin Science, the Biblical God Jesus Christ and God the Father said it best:

Moses 1: 30-35:

And it came to pass that Moses called upon God, saying: Tell me, I pray thee, why these things are so, and by what thou madest them? And behold, the glory of the Lord was upon Moses, so that Moses stood in the presence of God, and talked with him face to face. And the Lord God said unto Moses: For mine own purpose have I made these things. Here is wisdom and it remaineth in me. And by the word of my power, have I created them, which is mine Only Begotten Son, who is full of grace and truth. And worlds without number have I created; and I also created them for mine own purpose; and by the Son I created them, which is mine Only Begotten. And the first man of all men have I called Adam, which is many. But only an account of this earth, and the inhabitants thereof, give I unto you. For behold, there are many worlds that have passed away by the word of my power. And there are many that now stand, and innumerable are they unto man; but all things are numbered unto me, for they are mine and I know them.

Many other galaxies and worlds have come and gone, long before our earth was formed. That's what this scripture is telling us. Each inhabited earth is a separate planting at a separate point in time.

All you really want is the truth; and, God revealed the truth to Moses thousands of years ago. Good enough! There's no greater authority than God. I believe that God told us the truth here; and, I believe that the seers are telling us the truth as well. I choose to go with God and the seers on this one and not the YUCs and the YECs. God knows how it really happened, anyway.

There have been worlds without number; and, many of them have passed away or died, only to be resurrected from the dead and/or phased into a higher realm of existence. There are many worlds that now stand.

On each one of these different worlds, the first man is called Adam; and, there have been many different Adams. There have been many different falls on many different worlds. When they fell, our world didn't fall with them – our Adam and Eve didn't fall with them. According to God, someday our physical mortal earth will die; and, then it will be resurrected from the dead and become a Celestial World and the home of the future Gods and angels that came from this world or lived on this world during their mortal probation. That same process has already happened to many different worlds throughout our physical universe. We are NOT the first, and we will NOT be the last!

Nothing else fell when our Adam and Eve fell – not even the Garden of Eden and the Tree of Life – not even our earth! Only our Adam and Eve fell when they fell. They were cast out of Eden onto our earth or into our earth which was already in a fallen or telestial state of existence – a mortal physical-matter entropic state of existence.

The Seers throughout history have falsified YUC and YEC.

I have redefined Science as observation and experience. The science is clear. The speed-of-light is clear. The ice cores make it clear that our earth is much older than 6,000 years. The observations and the experiences of the seers are clear. Now that I'm in, I'm in all the way. We should go with the observations and the experiences – the science – rather than the wishful thinking, personal interpretations, faulty interpretations, and philosophical speculation of the Materialists, Naturalists, Darwinists, Nihilists, Atheists, YUCs, and YECs. That is what I have concluded after studying all of the scientific evidence and the observational evidence that has come my way in recent years. I have chosen to go with what has been experienced and observed.

I discuss Origin Science in much greater detail in the following book:

Scientific Proof of God's Existence: Finding God Where the Atheists Refuse to Look for Him

https://www.amazon.com/dp/B07B26CRHX

I hope you enjoy it.

I used to be a Materialist, Naturalist, Nihilist, and Atheist. I've done that as well. I can tell you from experience that Scientific Naturalism is not a good platform upon which to build a scientific discipline or a belief system. It's flawed. It's self-defeating. YUC and YEC are also flawed and self-defeating.

Extremism of any kind destroys itself over time. Extremism has limits; and, these limits eventually falsify it and destroy it. An infinite speed-of-light is physically impossible because it would result in an infinite amount of Energy traveling within every beam of light. An infinite speed-of-light would destroy this earth. If the **anisotropic synchrony convention** were true, it would destroy this earth. Yet, here we are. I reject extremism of any kind.

I guess you could say that I believe in a balanced approach to science or a realistic approach to science – one that has actually been experienced and observed.

In the summer of 2017, I upgraded my science to Science 2.0. I redefined Science!

The Materialists, Naturalists, Darwinists, Nihilists, and Atheists define "science" as Materialism, Naturalism, Darwinism, Behaviorism, Determinism, Physical Reductionism, Nihilism, and Atheism.

Science 2.0 defines Science as observation and experience.

The Physicalists, Scientific Naturalists, Darwinists, and Atheists deliberately destroy Science in order to convince people that the Theory of Evolution is true. In other words, these people deliberately destroy and dismiss Observations and Experiences in order to convince people that the Theory of Evolution is true.

The YUCs and the YECs do the same exact thing but at the other extreme. The YUCs and the YECs deliberately destroy Science – they deliberately destroy and dismiss Observations and Experiences – in order to convince people that our physical universe and our earth are only 6,000 years old.

The Biblical God Jesus Christ has been experienced and observed, after He rose from the dead. The Naturalists and Atheists have to dismiss and destroy these observations and experiences in order to convince people that the theory of evolution is true.

Half-lives and radioactive decay have been experienced and observed:

https://en.wikipedia.org/wiki/List_of_radioactive_isotopes_by_half-life

The YUCs and the YECs literally have to destroy and dismiss the Science behind Carbon 14 and half-lives in order to convince people that our earth is only 6,000 years old because Carbon 14 Dating and Radiometric Dating falsify YUC and YEC. The YUCs and the YECs also have to destroy and dismiss the Science behind the ice cores in order to convince people that our earth is only 6,000 years old. The YUCs and the YECs have to destroy the Theory of Relativity and make the speed-of-light infinite in order to convince people that our universe is only 6,000 years old. YUC and YEC literally destroy astronomy and astrophysics in order to convince people that our universe is only 6,000 years old. YUC and YEC do just as much destruction as Atheism and Naturalism – just at the opposite extreme.

Atheism, Naturalism, Darwinism, Nihilism, and Atheism destroy and dismiss everything at the sub-atomic level, the psyche level, or the quantum level. YUC and YEC destroy everything at the astronomical level and the geological level. They BOTH end up destroying Science, or observation and experience.

WE KNOW that Physicalism, Naturalism, Darwinism, Nihilism, Atheism, Determinism, Behaviorism, Physical Reductionism, Young-Earth Creationism, and Young-Universe Creationism are FALSE because they are Extremism – they are based upon a dogmatic refusal to look at and accept ANY evidence that falsifies them or contradicts them. These people are extremists.

Materialism and Scientific Naturalism are based upon a selective refusal to accept an endless number of observations and experiences of a sub-atomic, quantum, intangible, supernatural, non-physical nature that prove beyond a shadow of a doubt that God exists. YUC and YEC are based upon a selective refusal to accept observations and experiences that prove beyond a shadow of a doubt that our earth and our physical universe are older than 6,000 years. All of these falsified philosophies are based upon a refusal to look at ANY evidence that falsifies them. That's NOT science. That's Dogmatic Extremism instead.

The New Atheists are extremists; the YUCs and the YECs are extremists – just at different extremes. They BOTH refuse to look at and accept the observations and experiences of the human race whenever those observations and experiences falsify their pre-chosen conclusions and pre-chosen beliefs. Meanwhile, the truth is found somewhere between the two extremes. Isn't that the way it always works?

The extremists MUST REFUSE to look at any and all evidence that falsifies their point of view because there's a ton of Observational Evidence, Experiential Evidence, Eye-Witness Evidence, Empirical Evidence, and Scientific Evidence that FALSIFIES their pre-chosen conclusions and beliefs.

In contrast, Science 2.0 allows ALL of the evidence into evidence and attempts to pursue a preponderance of that evidence. Science 2.0 defines Science as observation and experience. Science 2.0 allows into evidence anything that has been experienced or observed by any member of the human race. It's a better way to do Science!

Only you can decide for yourself if I'm telling you the truth or if the Materialists, Naturalists, Atheists, YUCs, and YECs are telling you the truth. We both can't be telling you the truth because what they are teaching and preaching contradicts and falsifies what I'm teaching and preaching. We both can't be right. Something has to give.

I KNOW for myself, though, that I have finally found the truth because what I'm teaching and preaching nowadays has actually been experienced and observed. In fact, I now define Science as observation and experience rather than Materialism and Naturalism. It's a better way to go when it comes to Science. Observations and experiences have infinitely more explanatory power than the philosophical speculation, guesswork, and wishful thinking of the Materialists, Naturalists, Darwinists, Nihilists, Atheists, YUCs, and YECs.

Mark My Words

Entropy Is Death

Entropy is death. Syntropy is life.

The Materialists, Naturalists, Darwinists, Nihilists, Behaviorists, Determinists, Physical Reductionists, Atheists, and Classical Physicists teach and preach that ONLY physical matter or entropy exists.

These people have literally enshrined entropy or death and made it the foundational principle for their science, their philosophy, their religion, their dogma, and their worldview.

Imagine it! These people have made entropy or death the cornerstone of their science and their belief system. They worship, promote, and protect entropy or death with a passion and a religious zeal that rivals anything you will ever find from a theist. Consequently, these people have NO scientific explanation for where all of the Life or Syntropy came from in the first place. Entropy or death cannot be used to explain the origin of life. By definition, in principle, entropy or death cannot produce life.

The theory of evolution is based upon entropy or death. We KNOW for a fact that that is NEVER going to work. Entropy or death cannot design and create. It's a proven fact!

This is possibly the most powerful, most truthful, and most useful scientific discovery that a person can make. The opposite of entropy is Syntropy. We KNOW for a fact that some type of Syntropy MUST exist, or all of that subsequent entropy wouldn't have been possible.

This is extremely powerful and useful Science to know about and understand.

So, where is all that Syntropy since we KNOW that it MUST exist?

Well, that's easy to answer.

Entropy is death.

So, what's the opposite of death?

Syntropy or Life is the opposite of death.

Psyche, Consciousness, Intelligence, or Life IS Syntropy. Quantum Mechanics or Supernatural Mechanics is a type of Syntropy. Spirit matter or dark matter is Syntropy. Syntropy is eternal and everlasting, without a beginning of days or an end of years. Syntropy is Eternal Life. That means that resurrection from the dead is a type of Syntropy. Furthermore, the Atonement of Christ is Syntropy. The Atonement of Christ can be used to restore us back to Eternal Life or Syntropy, and thereby make us immortal once again.

Entropy is a function of time – or a product of the passage of time or the aging process. Physical matter is the only thing that contains entropy and is subject to the passage of time or the aging process. Everything else is Syntropy!

Mark My Words

Action at a Distance

In this essay, we have established that Syntropy is Conservation of Energy or the First Law of Thermodynamics. Energy is Syntropy. Conservation of Energy is Syntropy. Psyche, Intelligence, or Quantum Non-Local Consciousness is the fundamental unit of Syntropy, Existence, and Reality. According to the First Law of Thermodynamics, Energy, Psyche, or Syntropy is conserved; whereas, the different FORMS of Energy are NOT conserved. Space-time is a FORM of Energy. Physical matter and spirit matter are different FORMS of Energy. Entropy and locality are different FORMS of Energy. Physical matter, space-time, entropy, and locality are NOT conserved. The FORM is never conserved. ONLY the underlying Psyche, Syntropy, or Energy is conserved. This is the Ultimate Law of Thermodynamics. It explains how everything really works.

When it comes to Science, the explanatory power of Psyche, Syntropy, Quantum Mechanics, and Supernaturalism is vastly superior to anything that we could ever get from Materialism, Naturalism, Darwinism, Nihilism, Behaviorism, Determinism, Physical Reductionism, Atheism, and Classical Physics because these things deny the existence of Psyche, Syntropy, Action at a Distance, and Supernatural Mechanisms.

However, knowledge of Psyche, Syntropy, Quantum Mechanics, Action at a Distance, and Supernatural Mechanisms is vastly inferior to Classical Physics in one major noticeable way.

Do you know what that is?

We fallen mortal physical beings do NOT have direct access to nor direct control over Quantum Mechanics, Psychic Mechanisms, and Syntropy. For us, knowledge of Quantum Mechanics has no real practical application or value at the physical level in our everyday lives.

Quantum Mechanics is a teaser of what may someday be; but, Quantum Mechanics has very few practical applications for fallen mortal physical beings like us.

Why?

It's because God deliberately designed physical matter, physical laws, physical restrictions, and classical physics to greatly dampen, limit, and restrict the innate quantum mechanical capabilities of our psyche and spirit body.

The purpose of physical matter, physical restrictions, physical limitations, entropy, and locality is to prevent us from gaining direct access to and control over Quantum Mechanics and Syntropy. Physical matter was designed by God to prevent us from turning into X-Men with superpowers.

It works.

In fact, it works so well that it successfully convinces the Materialists, Naturalists, Nihilists, Darwinists, and Atheists that Action at a Distance, Psyche, Spirit Matter, and Syntropy DO NOT EXIST.

One of our purposes for being here on this earth as fallen mortal beings is to convince God that we can be trusted with direct access to and direct control over Quantum Mechanics or Syntropy at all levels of existence. Imagine what would have happened if God had given control over the elements to someone like Stalin, Pol Pot, Hitler, or any of the other hundreds of Militant Atheists that have come along throughout history. These people will kill you unless you believe what they want you to believe and do what they want you to do. A Militant Atheist is the most dangerous and deadly creature on the planet – sometimes more dangerous than all of the religious fanatics combined.

Thankfully, it requires Faith in God and obedience to His Commandments in order for a fallen mortal being to gain access to and control over Quantum Mechanics, Psychic Mechanisms, and Syntropy. God holds all the KEYS to Quantum Mechanics and Syntropy; and, God only gives those KEYS to those whom He has learned to trust.

The Materialists, Naturalists, Darwinists, Nihilists, Behaviorists, Determinists, Physical Reductionists, and Atheists will NEVER gain access to nor control over Quantum Mechanics, Psychic Mechanisms, Action at a Distance, God's Power, and Syntropy because they have chosen to believe that these things do not exist.

Thereby, God can hide His presence and His Ultimate Power in plain sight where the Materialists, Naturalists, Darwinists, Nihilists, and Atheists will NEVER see it nor find it.

Thankfully, the Militant Atheists will never develop nor gain direct control over the physical atoms around us. The physical atoms won't obey them. These people are effectively limited to the physical level and have no real access to the Quantum Level.

People were greatly afraid of nanobots, nanomachines, and nanotechnology when they first became aware of the possibility.

Thankfully, we never have to worry about the Militant Atheists gaining access to such capabilities. Should they ever successfully develop a Universal Solvent that turns everything into a Gray Goo, they and their lab will be the first thing to go. They would have no way to control and contain these nanobots at the psyche level or the quantum level.

You see, in order to construct a nanomachine or nanobot that can actually seek out and destroy cancer cells successfully, it would require developing the quantum mechanical ability to interact with the atoms within those nanobots telepathically at a distance telling those nanobots where to go and what to do. The Militant Atheists, Materialists, Naturalists, and Nihilists will NEVER develop such a thing because they have chosen to believe that telepathy and other psychic mechanisms do not exist.

Likewise, in order to successfully construct a nanobot that can actually seek out and destroy cancer cells, it would require some kind of quantum mechanism in place that would grant the atoms within that nanobot the quantum ability to identify cancer cells at a distance, move towards those cancer cells, attach to those cancer cells, and then destroy those cancer cells. In other words, it would require some kind of telepathic communication as well as invisible and non-physical forces and fields at the psyche level or the quantum level in order to successfully command and control nanobots at the quantum level or the sub-atomic level. The Materialists, Naturalists, Darwinists, Nihilists, and Atheists will NEVER develop such a thing because they have chosen to believe that Psychic Mechanism, Quantum Mechanisms, Supernatural Mechanics, Psyche, Telepathy, and Syntropy DO NOT EXIST.

At the quantum level or the psyche level, valences are invisible, intangible, immaterial, non-physical forces and fields that act as tractor beams or repulsor beams between atoms and molecules. Valences are how atoms and molecules are drawn together, or pulled apart, at the quantum level or the syntropic level. The proven and verified existence of valences FALSIFY Materialism, Naturalism, Darwinism, Nihilism, and Atheism which state that the non-physical or the intangible DOES NOT EXIT.

Remember, targeting of any type requires Intelligence or Psyche; and, the Militant Atheists and Behaviorists have chosen to believe that Psyche or Syntropy does not exist. Communicating with atoms directly and controlling atoms directly requires some kind of Psyche and Telepathic Action at a Distance (Telekinesis); and, the Militant Atheists have chosen to believe that such a thing DOES NOT EXIST.

Their chosen beliefs and lack of faith greatly limit what the Militant Atheists are able to do to us; and, God deliberately designed it to be that way in the first place in order to protect us from them. It's bad enough what they can do at the physical level. Imagine what they could do if they had access to and control over the quantum level or the sub-atomic level with their minds. Some of the Old Testament prophets are a case in point. They were bad from the perspective of the Atheists and Unbelievers; but at least the Old Testament prophets limited their use of God's Power or their use of Quantum Mechanics to defending themselves and the House of Israel from attack thereby preventing the complete extinction of God's chosen people.

The bad news in all of this is that we NEED God's help and God's permission in order to be able to use Quantum Mechanics or God's Priesthood Power at the physical level. We are NEVER going to gain superpowers through the manipulation of and mutation of our genes. It's physically impossible. God put those genes and that physical matter into place in order to prevent us from gaining access to and control over superpowers. There will NEVER be any X-Men. Superpowers do not and cannot come from our genes. Superpowers are a function of Psyche, Quantum Mechanics, and Syntropy.

The good news in all of this is that God designed physical matter, physical limitations, entropy, and locality to prevent the Militant Atheists and other religious fanatics from gaining direct access to and control over Quantum Mechanics, Psychic Mechanisms, Superpowers, Energy, and Syntropy. That's a very good thing indeed.

Mark My Words

Quantum Law of Thermodynamics

The Materialists, Naturalists, Darwinists, Nihilists, Behaviorists, Determinists, Physical Reductions, Classical Physicists, and Atheists preach and teach that "Heat Death Is Coming". These people also teach and believe that entropic physical matter is the only thing that exists, and therefore that entropy or death is always conserved.

They are wrong.

One of my greatest scientific discoveries is the Quantum Law of Thermodynamics or Syntropy, which states that Heat Death is impossible at the quantum level or the psyche level. Out-of-Body Travelers and Near-Death Experiencers have observed that there is NO heat death, NO second law of thermodynamics, and NO thermodynamics at the quantum level. ONLY the First Law of Thermodynamics – the Conservation of Psyche or the Conservation of Energy – applies at the quantum level. Even though disorder does exist within the non-consensus realities in the Spirit Realm or the Quantum Realm, it has been observed that the Human Psyche automatically organizes any non-consensus reality that it enters into. The Human Psyche and God's Psyche eliminate entropy or disorder at will. This is what has been experienced and observed. Science is observation and experience after all.

Psyche is Energy. In fact, Psyche is the innate intelligence within all the different forms of energy which gives that energy the inherent ability to understand, follow, and obey the Human Psyche's commands and God's Commands. Energy is all that's needed to reverse entropy and eliminate entropy, and the Quantum Realm or Spirit World is Pure Energy, Pure Exergy, and Pure Syntropy. Heat Death is impossible in the Syntropy Realm or the Quantum Realm. Your Psyche or Intelligence will never die! This is what has been experienced and observed. This is the Quantum Law of Thermodynamics.

Psyche is Energy. All you need is energy to remove entropy from any system. Psyche is Intelligence or Knowledge. All you need is knowledge or information to remove entropy or disorder from any system. Psyche is Eternal Life or Life Force. All you need is Eternal Life or Life Force to remove entropy or death from any system. Psyche is Syntropy. All you need is Syntropy to remove entropy from any system.

The Quantum Law of Thermodynamics states that heat death is impossible at the quantum level and the psyche level in the Quantum Realm or Spirit World. Throughout all of my study, research, thought, and prayer, one of the greatest revelations that I received is that heat death is impossible at the quantum level or the psyche level. This reality has to be true, or we wouldn't exist. The Syntropy or Conservation of Energy has to exist somewhere for real; otherwise, we wouldn't exist. This is Logic 101. We observe entropy at the physical level, which means that the Syntropy has to be happening at the quantum level because there is nowhere else for it to happen. The Syntropy has to exist; otherwise, NONE of that subsequent entropy would have been possible in the first place. This reality and truth is obvious.

Exergy is energy that is available to be used. At the physical level, after the system and surroundings reach thermal equilibrium, the exergy is zero. Determining exergy was the first goal of thermodynamics. Energy is neither created nor destroyed during a process. Energy changes from one form to another. In contrast, exergy is always destroyed when a process is irreversible, for example loss of heat to the environment. This destruction is proportional to the entropy increase of the system together with its surroundings. The destroyed exergy has been called anergy. Anergy is unavailable energy. For an isothermal process, exergy and energy are interchangeable terms, and there is no anergy in an isothermal system. In an isothermal system, ALL of the energy is available for use all the time.

There is no anergy and no entropy in an isothermal system. Technically, there is NO thermodynamics in an isothermal system. The second law of thermodynamics doesn't exist in an isothermal system. It has to be this way; otherwise, we wouldn't exist. The Quantum Realm, Psyche Realm, Spirit World, Transdimensional Realm, or Syntropy Realm is an Isothermal System. There is NO heat death in the Quantum Realm; and, Psyche, Intelligence, or Life Force cannot die. This is the Quantum Law of Thermodynamics. It explains everything that has ever been experienced and observed.

The Big Bang Singularity, if it existed, was an Isothermal System comprised of Pure Syntropy or Pure Energy and made by the Gods. Only a God could make such a thing. By definition, in principle, there was NO entropy or NO anergy within the Big Bang Singularity. It was Pure Syntropy or Pure Exergy. ALL of the energy was available for use. The Quantum Law of Thermodynamics makes the Big Bang Singularity a realistic possibility because there is NO heat death and NO entropy at the quantum level or the psyche level. At the quantum level, ALL of the energy is available for use all of the time. This is what has been experienced and observed.

Syntropy is ZERO entropy, ZERO anergy, and ZERO death. Syntropy is also Eternal Life or INFINITE LIFE. The math behind the Quantum Law of Thermodynamics or Syntropy has already been discovered. It's just been sitting there waiting for someone to discover it and identify it for what it truly is.

The Quantum Law of Thermodynamics or Syntropy is the answer to life, the universe, and everything. It explains everything that has been experienced and observed.

Mark My Words

—

Source

Using the Scientific Method: To Eliminate the Usual Suspects and to Prove the Truth

> https://www.amazon.com/dp/B01J6STHP0

I Am Not a Creationist: So What Am I?

> https://www.amazon.com/dp/B071XTM8XY

Science 2.0: I Upgraded My Science

> https://www.amazon.com/dp/B0771K6WTX

13. Entropy Falsifies Evolution

One of the most fascinating and useful of my scientific discoveries is that Evolution of any kind is based upon entropy. Evolution is Entropy. Random mutations are entropy; and, natural selection leads to death and extinction which is entropy.

Of course, it's obvious that Materialism, Naturalism, Darwinism, Nihilism, Atheism, Behaviorism, Scientism, Determinism, Physical Reductionism, and Classical Physics are based exclusively on entropy. What isn't immediately obvious is the true ramifications of such a scientific observation.

Entropy cannot do the Quantum Mechanical or the Supernatural. In other words, entropy cannot do design and creation. Entropy cannot do order and organization. Entropy cannot do Psyche or Intelligence. Entropy cannot do Syntropy. Entropy can only do death, which means that entropy cannot do life.

One of my greatest and most useful scientific discoveries came when I first realized that entropy, natural selection, and creation by evolution DON'T REALLY EXIST as a person, place, or thing. Entropy, natural selection, and creation by evolution ARE NOT quantum forces NOR quantum fields. Entropy, natural selection, and creation by evolution are NOT fundamental forces in nature; and, they are NOT psychic or intelligent Psyches either. Entropy, natural selection, and creation by evolution are philosophical concepts or figments of the imagination rather than something that's ACTUAL and REAL. Entropy, natural selection, and creation by evolution DO NOT EXIST in any useful form. This is the truth of the situation whether we like it or not.

Remember, entropy is death. Entropy, death, natural selection, and the different types of evolution CANNOT FUNCTION as a Designer, Creation, Maker, and Manufacturer; but, the Human Psyche or Human Intelligence certainly can. This is what has been experienced and observed.

Chemical Evolution is the spontaneous generation of genes and proteins from atoms. In order for something like Chemical Evolution to have happened for real, it would have required some kind of Quantum Mechanical or Supernatural intervention because entropy prevents Chemical Evolution from taking place naturally in the wild. Chemical Evolution would have required some type of Syntropy, or Creation, or Engineering, or Intelligent Intervention in order to make it happen for real. The fact that Chemical Evolution or Spontaneous Generation is prevented from happening by entropy or random diffusion is scientific proof that the Theory of Evolution is false. The fact that Spontaneous Generation has been falsified by Science itself is scientific proof that the Theory of Evolution is false.

Macro-Evolution is one species giving birth to a completely different genetically incompatible species – two cats giving birth to a dog, or two chimp-like ancestors giving birth to chimpanzees and human beings. Macro-Evolution is prevented from happening both by genetics and by random mutations or entropy. In other words, Macro-Evolution is physically impossible. There's no way in the universe that Macro-Evolution, or Random Mutations, or Entropy would be able to produce a genetically compatible Mr. and Mrs. Mutant at the same time in the same place over and over and over again year after year for millions of years. It's physically impossible, which means that it didn't happen because it couldn't happen. Genetics or standardization prevents Macro-Evolution; and, Random Mutations also prevent Macro-Evolution. Macro-Evolution never happened, because it's physically impossible and prevented from happening by both genetics and entropy. The fact that Macro-Evolution of any kind is prevented from happening by entropy or random mutations is scientific proof that the Theory of Evolution is false.

This is one of my most significant and life-changing scientific discoveries, especially since I used to be a Materialist, Naturalist, Nihilist, and Atheist. These philosophies are based exclusively on entropy; and, entropy cannot design and create anything at all. Materialism, Naturalism, and their derivatives such as Darwinism or the Theory of Evolution are self-defeating and self-falsifying because they are based upon entropy and entropy prevents them from happening in the first place. Proteins and the matching genes to go along with those proteins don't just spontaneously generate out of thin air from scratch as the Naturalists and Darwinists claim, because entropy prevents them from doing so. Entropy prevents spontaneous generation or macro-evolution from happening in the first place. When it comes to Evolution, entropy provides the seeds for its destruction or the seeds for its falsification.

Fascinating, is it not?

Well, I think it is because I used to be a Materialist, Naturalist, Nihilist, and Atheist; and at times, I was even on the fence when it came to Darwinism and the Theory of Evolution.

Evolution is entropy. Entropy or evolution cannot do Syntropy or Quantum Mechanics. Entropy or evolution cannot do Syntropy or Psyche and Intelligence. Evolution or entropy cannot do Syntropy or Organization and Life. Entropy or evolution cannot do Science, Manufacturing, Engineering, Fine-Tuning, Construction, Creation, and Life. Entropy produces death and extinction, NOT life. Random mutations introduce entropy into our genomes; and, natural selection facilitates death or entropy rather than order, organization, life, and creation. Furthermore, natural selection doesn't touch our genes. Under the stress and pressures of random mutations and natural selection, entropy leads every physical life form towards death and extinction.

Entropy cannot produce Syntropy; but, Quantum Mechanics and Psyche certainly can. Entropy cannot design and create; but, Psyche or Intelligence certainly can. Psyche or Intelligence has been caught in the act of design and creation; and, entropy has not. This means that entropy or evolution could NEVER have created or produced our genome; but, some type of Intelligent Being or God-Like Psyche certainly could have.

We KNOW that some type of Syntropy must exist because entropy exists; and, we also KNOW that entropy falsifies Evolution or prevents the different types of Evolution from becoming actual or real. Therefore, some type of Syntropy or Organizing Force had to be used in order to produce our proteins, genes, and genomes in the first place.

Here we have a prime example of how Science itself FALSIFIES Materialism, Naturalism, Darwinism, Nihilism, Atheism, and the Theory of Evolution. This is one of my most significant scientific discoveries. It was hiding there in plain sight where nobody seemed to be able to see it or understand it. Evolution is entropy, and entropy prevents the Theory of Evolution from becoming true.

Mark My Words

—

Source

The Scientific Method Proves That the Theory of Evolution Is False.

https://www.amazon.com/dp/B01IAAIRT2

https://www.amazon.com/dp/1521133611

14. Entropy Is Death and Syntropy Is Eternal Life

Syntropy is by far my greatest, most useful, most powerful, and most explanatory scientific discovery that I have made so far. Logical common sense tells us that some type of Syntropy MUST exist, or all of that subsequent entropy wouldn't have been possible. NONE of our physical reality would have been possible without some type of Syntropy.

Entropy is death. Syntropy is Eternal Life. Syntropy is eternal and everlasting, without a beginning of days or an end of years.

Chances are good that you have never heard of Syntropy.

Why haven't you ever heard of Syntropy? It's the foundation of ALL science, after all.

The reason is that the Materialists, Naturalists, Darwinists, Nihilists, Behaviorists, Determinists, Physical Reductionists, and Atheists have BANNED Syntropy or Eternal Life from science. These people state axiomatically a priori that Syntropy or Psyche does not exist. These people define science as Materialism and Naturalism.

Are they right?

They can't be right. Some type of Syntropy MUST exist, or all of that subsequent entropy wouldn't have been possible. Think about it logically and rationally, rather than emotionally; and, you will KNOW that this is true.

Entropy is death. Physical matter or entropy results in death. This reality has been experienced and observed.

Physical Matter, Materialism, Naturalism, Darwinism, Nihilism, Behaviorism, Determinism, Reductionism, Atheism, and Classical Physics are based exclusively on entropy. Entropy is death. These "scientific disciplines" are literally based upon Creation by Death. Now, please explain to me how Creation by Death or Creation by Entropy could ever be made to work!

It can't! It's impossible! Not even God can do the impossible!

This is an extremely powerful Scientific Observation. Its explanatory power is through the roof!

The Theory of Evolution IS Creation by Death, or Creation by Entropy. It's NEVER going to work! I just resolved millennia of scientific controversy and debate with that simple observation.

Materialism, Naturalism, Darwinism, Nihilism, Atheism, Classical Physics, and their derivatives ARE Creation by Death or Creation by Entropy. That's NEVER going to work! It's impossible. Creation by Death or Creation by Entropy is synonymous with magic, or creation ex nihilo, or creation by chance. That's just never going to work while we are stuck here in the physical realm at the physical level of our existence. Not even God can do creation ex nihilo, or creation by death, or creation by entropy.

Some type of Syntropy or Eternal Life or Psyche MUST exist in order to MAKE the physical laws and classical physics WORK. Syntropy MUST exist in order to make entropy and physical laws possible in the first place. I just solved ALL of the mysteries, confusion, problems, and controversies that stand at the foundation of modern-day materialistic, naturalistic, and atheistic sciences. It was solved by the simple observation that entropy is death, and Syntropy is some type of intelligence or life.

Since we KNOW for a fact that entropy exists, we also KNOW for a fact that its opposite Syntropy MUST also exist. All that entropy wouldn't have been possible without a massive initial infusion of Syntropy or Life. That's just the way it is. The truth of it is undeniable; yet, the Materialists, Naturalists, Darwinists, Nihilists, and Atheists choose to deny the existence of Syntropy or Psyche anyway. These people are constantly telling us that Psyche, Consciousness, and Life do not exist – only physical matter or entropy exists.

These people are demonstrably wrong.

Quantum Mechanics, Spiritual Mechanics, Transdimensional Mechanics, or Supernatural Mechanics is a type of Syntropy – eternal and everlasting, without a beginning of days or an end of years. The verified and proven existence of Quantum Mechanics, Magnetism, Gravity, Dark Matter or Spirit Matter, Dark Energy, Invisible Light, the Strong and Weak Nuclear Forces, the Zero-Point Field, Invisible Forces and Fields, Radio Waves, Microwaves, X-Rays, Gamma Rays, Thoughts or Quantum Waves, Psyche or Intelligence, Space or the Vacuum, Time, the Quantum Zeno Effect, Action at a Distance, and Syntropy FALSIFIES the major premises of Materialism, Naturalism, Darwinism, Nihilism, Atheism, and their derivatives which state that these various different Quantum Mechanisms and Non-Physical Realities DO NOT EXIST.

Well of course they exist! Their existence has been proven and verified by scientific observations or scientific experiments. It's the Materialism, Naturalism, and Atheism that are FALSE and WRONG, not the scientific observations! Go with the science and NOT the wishful thinking of the Materialists, Naturalists, and Atheists. Science is observation and experience, NOT Materialism and Naturalism. The proven and verified existence of all these different non-physical and invisible forces and fields FALSIFIES Materialism, Naturalism, Darwinism, and their derivatives. You want the truth, and not the fiction and the wishful thinking.

Entropy is restricted to or limited to physical matter and space-time. Everything else is Syntropy. Everything else is eternal and everlasting. This reality and truth has been experienced and observed during Out-of-Body Experiences and Near-Death Experiences. Only physical matter or entropy is subject to the passage of time or the aging process. Everything else is eternal and everlasting. Everything else is Syntropy.

I just solved ALL of the problems that Materialism, Naturalism, and Atheism created for us in the first place. Entropy is death. Syntropy is Eternal Life.

Within the books that the Biblical God Jesus Christ had a hand in writing and producing, we observe Him talking about Eternal Life or Syntropy ALL the time. It has to exist because He has observed it, lived it, and experienced it for Himself. Science is observation and experience. Jesus Christ wouldn't be talking about Eternal Life or Syntropy if it didn't exist. Resurrection from the dead is a type of Syntropy. Christ rose from the dead. The Atonement of Christ is Syntropy – or restoring everything back to its original natural Eternal Order. Entropy is death; and, death was introduced into our physical bodies when Adam and Eve fell. Death was introduced into our physical bodies and our lives when we chose to leave God's Presences or chose to leave Syntropy.

With Syntropy or Psyche or Quantum Mechanisms, we can literally explain everything that comes our way. This IS powerful science! Yet, the Materialists, Naturalists, and Atheists assure us that it doesn't exist. They are wrong. Of course, it exists. It has been experienced and observed! Syntropy or Life has been experienced and observed on both sides of the veil. It's real and truly exists.

Quantum Mechanics is God's Priesthood Power. Quantum Mechanics is Syntropy. God's Psyche or God's Intelligence used Quantum Mechanisms to create

physical matter from spirit matter and to do Cosmic Fine-Tuning. The physical laws as we know them are the result of Syntropy or Fine-Tuning.

Fine-Tuning is a type of Syntropy. Physical matter or entropy cannot DO Fine-Tuning. Entropy or physical matter can only do death.

Entropy is disorder, randomness, chaos, and death. Materialism, Naturalism, Atheism, Darwinism, Nihilism, Classical Physics, and the Theory of Evolution are based exclusively upon entropy or death. Evolution (genetic change) or random mutation is entropy. Natural selection eventually results in death or entropy. Random Mutations and Natural Selection cannot produce life because they are in fact entropy or death. Chemical evolution, abiogenesis, spontaneous generation, and macro-evolution are prevented from happening by entropy. Evolution of any kind is entropy. Entropy or evolution cannot do fine-tuning, teleology, purpose, planning, design, programming, engineering, manufacturing, production, and creation. Entropy is death! Entropy or death cannot produce functional proteins nor the matching genes to go along with them! This is logical common sense.

In many of his books, Hugh Ross uses Cosmic Fine-Tuning as Scientific Proof of God's Existence. The whole thing made logical sense to me because ONLY Psyche can convert or transmute spirit matter into physical matter. ONLY Psyche can organize physical matter into useful and productive forms such as planets, stars, galaxies, genomes, and life forms. ONLY Psyche can do Cosmic Fine-Tuning. God's Psyche must of necessity exist, in order to have done all of the cosmic fine-tuning and science that needed to be done; otherwise, you and I would not be here right now in this physical realm. Entropy or death cannot do creation and fine-tuning! Science has made it obvious and clear to me that this must be so. It's obvious that the rocks or raw physical matter cannot do Cosmic Fine-Tuning. That kind of process requires an active, living, agentic, conscious, intelligent Being or Psyche, whom many of us tend to call God or the Ultimate Scientist.

For the best list of the Fine-Tuning that has been done by God see:

http://www.reasons.org/explore/blogs/todays-new-reason-to-believe/read/tnrtb/2010/11/16/rtb-design-compendium-2009

I archived it because it kept disappearing on me.

Archival Copies:

https://syntropy.site/wp-content/uploads/2018/04/compendium_part1.pdf

https://syntropy.site/wp-content/uploads/2018/04/compendium_part2.pdf

https://syntropy.site/wp-content/uploads/2018/04/compendium_Part3_ver2.pdf

https://syntropy.site/wp-content/uploads/2018/04/compendium_Part4_ver2.pdf

Only some type of Syntropy, Psyche, or Intelligence can do Fine-Tuning!

Remember, entropy is death, and Syntropy is Eternal Life. If you are able to remember that simple truth, then you will be able to explain every scientific observation that comes your way.

I just fixed science and upgraded science with this simple yet powerful scientific observation. It explains everything that has ever been experienced or observed.

Mark My Words

—

Source

Science 2.0: I Upgraded My Science

https://www.amazon.com/dp/B0771K6WTX

15. Genetic Entropy

Genetic Entropy is one of my most favorite scientific discoveries. This scientific discovery doesn't belong to me, though. It originates with John Sanford. After reading his book, *Genetic Entropy*, I simply KNEW that the theory of evolution is false because I now KNOW why it is false.

One of my greatest scientific discoveries came to me when I first realized that the different types of evolution cannot design and create anything whatsoever. Evolution of any type is based exclusively on entropy. Evolution is entropy; entropy is death; and, entropy or death can't design and create anything. Creation by entropy, death, natural selection, or evolution is physically impossible. The Theory of Evolution is Creation by Entropy or Creation by Death; and, this is physically impossible. This is what has been experienced and observed. Is it not?

The theory of evolution is typically defined as Creation by Mutation/Selection – particularly, 'the origin of species by means of natural selection'. The first part of the title of Darwin's book is, "On the Origin of Species by Means of Natural Selection".

Creation by Natural Selection IS science fiction. Natural selection doesn't touch our genes! It can't. It's physically impossible for natural selection to get at our genes and change them. Natural selection cannot design and create anything, let alone a genome.

In truth, natural selection is NOT the mechanism of change behind the theory of evolution. It's the random mutations that produce genetic change, NOT natural selection! Natural selection or survival of the fittest doesn't do anything. It just waits for you to die. Natural selection is entropy or death. They built a whole "science" on a fictional, immaterial, invisible process that doesn't even touch our genes – natural selection! And, they literally give natural selection ALL the credit for designing, programming, creating, and producing our genomes and our physical bodies. For these people, natural selection is their god. They worship it with a passion. Natural selection is a man-made god, an idol.

Natural Selection: The evolutionary process by which heritable traits that best enable organisms to survive and reproduce in particular environments are passed to ensuing generations.

Everyone who has taken introductory psychology has learned that nature and nurture together form who we are. As the area of a rectangle is determined by both its length and its width, so do biology and experience together create us.

As *evolutionary psychologists* remind us, our inherited human nature predisposes us to behave in ways that helped our ancestors survive and reproduce. We carry the genes of those whose traits enabled them and their children to survive and reproduce. Thus, evolutionary psychologists ask how natural selection might predispose our actions and reactions when dating and mating, hating and hurting, caring and sharing. Nature also endows us with an enormous capacity to learn and to adapt to varied environments. We are sensitive and responsive to our social context.

To explain the traits of our species, and all species, the British naturalist Charles Darwin (1859) proposed an evolutionary process. Follow the genes, he advised. Darwin's idea, to which philosopher Daniel Dennett (2005) would give "the gold medal for the best idea anybody ever had," was that natural selection enables evolution.

Natural selection implies that certain genes — those that predisposed traits that increased the odds of surviving long enough to reproduce and nurture descendants — became more abundant.

Natural selection, long an organizing principle of biology, has recently become an important principle for psychology as well. *Evolutionary psychology* studies how natural selection predisposes not just physical traits suited to particular contexts — polar bears' coats, bats' sonar, humans' color vision — but also psychological traits and social behaviors that enhance the preservation and spread of one's genes. We humans are the way we are, say evolutionary psychologists, because nature selected those who had our traits — those who, for example, preferred the sweet taste of nutritious, energy-providing foods and who disliked the bitter or sour flavors of foods that are toxic. Those lacking such preferences were less likely to survive to contribute their genes to posterity.

As mobile gene machines, we carry not only the physical legacy but also the psychological legacy of our ancestors' adaptive preferences. We long for whatever helped them survive, reproduce, and nurture their offspring to survive and reproduce.

"The purpose of the heart is to pump blood," notes evolutionary psychologist David Barash. "The brain's purpose," he adds, is to direct our organs and our behavior "in a way that maximizes our evolutionary success. That's it." (*Social Psychology*, p. 8, 159.)

Everything they wrote here is false or incomplete.

It's NOT a rectangle, it's a triangle! It's not just nature and nurture that form us. There's an essential third component!

Do you know what it is?

NATURE vs. NURTURE vs. NIRVANA: An Introduction to Reality

The third component is deliberately eliminated from science by the Materialists, Naturalists, Darwinists, Nihilists, and Atheists. These people state that it does not exist.

The BioPsychoSocial Model tells us that it does exist. Somebody is right, and somebody is wrong. They both can't be right.

These people have been teaching for over 150 years that Natural Selection made you, that evolution made you; but, that's physically impossible. Natural selection can't make anything. Natural selection and the theory of evolution are based exclusively on entropy. Natural selection results in entropy or death. Evolution is entropy. Entropy is death. Entropy cannot make anything at all. Entropy or death can only destroy. The different types of evolution can only destroy. The different types of evolution or entropy can only produce death and extinction.

Natural selection doesn't predispose anything! Natural selection doesn't organize anything! It can't. Natural selection doesn't touch our genes! Natural selection doesn't endow us with anything! Natural selection has NO ability to learn anything. Natural selection doesn't enable anything. Natural selection doesn't do anything. Natural selection is supposed to be dumb and blind without a soul or a mind, according to the Darwinists. Creation by Natural Selection wins the rotten tomato for the most stupid, illogical, irrational, and ineffective idea ever created.

The theory of evolution is correlational, NOT observational. NO type of evolution has ever been caught in the act of design and creation. It's physically impossible for natural selection and random mutations to design and create something. Entropy prevents them from doing so. Chemical evolution or macro-evolution is prevented from happening by random diffusion or entropy. Macro-evolution is also prevented from happening by genetics. The genes are there to prevent macro-evolution from happening. Evolution of any type is entropy and death. Death cannot create life!

In fact, evolution (genetic change), random mutations, and natural selection didn't even exist until AFTER God designed, programmed, engineered, field-tested, fine-tuned, manufactured, created, and produced the proteins and their matching genes in the first place.

The theory of evolution is a fictional story that they made up out of thin air after-the-fact to fit the facts. NO part of it can actually design and create. It's a fictional story, not science.

We are sensitive and responsive to our social context, NOT our genes!

Who is this **WE** that they keep talking about in our Social Psychology textbooks? **WE** can't be our society, environment, or social context that **WE** are sensitive to and responsive to! And, it's definitely NOT our genes. Our genes aren't sensitive and responsive to anything according to the Evolutionists! The genes, natural selection, and random mutations are supposed to be dumb and blind without a soul or a mind. Our genes can't be sensitive nor responsive to anything.

Personal pronouns imply a person or a psyche – NOT our genes (nature) and NOT our environment (nurture).

Natural selection and your genes DO NOT and CANNOT pass your Psyche Legacy (psychological legacy) from one generation to the next! Natural selection doesn't touch your genes, and it definitely doesn't touch nor change your Psyche either. It's science fiction to imply that it does. The theory of evolution is science fiction. "Design and creation by natural selection" is science fiction. The idea that your genes carry your "longings" or "desires" from one generation to the next is science fiction. There's no such thing as genetic memory, at least not at the physical level. All of our thoughts and memories are carried as quantum waves from one generation to the next through our Psyche or Quantum Non-Local Consciousness.

Your genes don't care whether you live or die. Only YOU care whether you live or die. Your Psyche cares whether you live or die; but, your genes do not. In order for your genes to care, they would have to have some sort of Psyche, or Intelligence, or Consciousness, or Awareness. But, if your genes have a Psyche, then the very existence of that Psyche falsifies Materialism, Naturalism, and Darwinism which claim that Psyche does not exist. Physicalism, Naturalism, and the Theory of Evolution are self-defeating. They don't work as advertised because they can't work as advertised.

Natural selection results in entropy and death, NOT X-Men and new unique life forms. Natural selection cannot design and create and program genomes. Natural selection doesn't touch our genes.

Random mutations are also entropy; but, at least random mutations by definition in principle actually change our genes. However, random mutations cannot design and create anything either.

Entropy is death. Death cannot design and create new unique genomes and life forms. That's physically impossible! Mutation and Selection can only produce entropy or death. They are based exclusively on entropy or death. Death cannot design and create life. Death can only end life.

Remember, the theory of evolution is Creation by Entropy or Creation by Death. That's NEVER going to work because it's physically impossible!

Isn't it refreshing to finally have access to the truth, rather than all the science fiction that the Evolutionists have been feeding us throughout our lives?

Well, I think it is.

I used to be a Materialist, Naturalist, Nihilist, and Atheist until I finally started to study the evidence. The evidence and the truth set me free! The Science and the Scientific Evidence convinced me that God must exist in order to have done all the Science and Fine-Tuning which natural selection and evolution could NEVER have done.

Mark my Words

Can Natural Selection Create?

The Physicalists, Naturalists, Darwinists, Nihilists, and Atheists teach that natural selection can create anything that it sets its mind to. These people teach that natural selection made you.

Are they right?

They are not!

They are deceiving themselves and trying to trick us and deceive us as well.

Creation by Natural Selection is demonstrably false, which means that it has been falsified by Scientific Evidence. Natural selection doesn't do anything. Natural selection doesn't touch our genes, nor does it pre-determine our future. Natural selection doesn't have a mind. There is NO intelligence or psyche within natural selection. There may (or may not) be some type of intelligence or psyche within our genes; but, there is NOTHING there when it comes to Natural Selection or Evolution. Natural selection is a fictional concept that they made up out of thin air. It doesn't really exist as a person or an entity. The same can be said of evolution. I'm not the only scientist to have figured this out by now. Natural selection is worthless as a creative agent and can't function as a creative agent.

[Editorial Note: I have written permission from John C. Sanford to use all of the quotes from John C. Sanford which I use in my books, so long as I cite the sources which I have done.]

START OF THE QUOTE FROM "GENETIC ENTROPY" BY JOHN SANFORD — USED BY PERMISSION FROM THE AUTHOR JOHN SANFORD.

Chapter 9: Can Natural Selection Create?

Newsflash — Mutation/Selection cannot even create a single gene.

We have been examining the problem of genomic degeneration and have found that deleterious mutations occur at a very high rate. Natural selection can only eliminate the worst of these, while all the rest accumulate — like rust on a car. Might beneficial mutations at other sites in the genome compensate for this continuous and systematic erosion of genetic information? The answer is that beneficial mutations are much too rare, and are much too subtle, to keep up with such relentless and systematic erosion of information. This is carefully documented by Sanford et al. (2013), and Montañez et al. (2013). It is very easy to systematically destroy information, but apart from the operation of intelligence it is very hard (arguably impossible) to create information.

This problem overrides all hope for the forward evolution of the whole genome. However, some limited traits might still be improved via Mutation/Selection. Just how limited is such progressive ("creative") Mutation/Selection? By now it should be clear that random spelling errors in an instruction manual could never give rise to an airplane component (say a molded aluminum part), which then resulted in a significantly improved overall performance of a jet plane. Not even with an unlimited number of flight trials/crashes and an unlimited budget. So, it is certainly reasonable to ask the parallel biological question, "Could Mutation/Selection create a single functional gene from scratch?"

A gene is like a book, book chapter, or an executable program — and minimally consists of a text string with 1,000 characters. Mutation/Selection could not create a single gene because of the enormous preponderance of deleterious mutations, even within the context of a single gene. The net information must always still be declining, even within a single gene or linkage block. Even if a gene was 50% established, deleterious mutations would degrade the completed half of the gene much faster than beneficials could create the missing half of the gene. However, to better understand the limits of forward selection, let us for the moment discount all deleterious mutations and only consider beneficial mutations. Could Mutation/Selection then create a new and functional gene?

1. Defining our first desirable mutation. The first problem we encounter in trying to create a new gene via Mutation/Selection is defining our first beneficial mutation. By itself, no particular nucleotide (A, T, C or G) has more value than any other, just as no letter in the alphabet has any particular meaning outside of the context of other letters. So, selection for any single nucleotide can never occur except in the context of the surrounding nucleotides (and in fact, within the context of the whole genome). A change of a single letter within a word or chapter can only

729

be evaluated in the context of the surrounding block of text. This brings us to an excellent example of the principle of "irreducible complexity" within the genetic realm. In fact, it is irreducible complexity at its most fundamental level. We immediately find we have a paradox. To create a new function, we will need to select for our first beneficial mutation, but we can only define that new nucleotide's value in relation to its neighbors — and we are going to have to be changing most of those neighbors also. We create a circular path for ourselves. We will keep destroying the "context" we are trying to build upon. This problem of the fundamental inter-relationship of nucleotides is called epistasis. True epistasis is almost infinitely complex, and virtually impossible to analyze, which is why geneticists have always conveniently ignored it. Such bewildering complexity is exactly why language and information (including genetic language and genetic information) can never be the product of chance, but always requires intelligent design. The genome is literally a book, written literally in a language, and short sequences are literally sentences. Having random letters fall into place to make a single meaningful sentence, by accident, would require more tries (more time), than earth history can provide (i.e., "methinks it is like a weasel" would take 27 ^ 28 tries — that is 10 followed by 40 zeros). The same is true for any functional string of nucleotides. If there are more than a dozen nucleotides in a functional string, we know that realistically they will never just "fall into place". This has been mathematically demonstrated repeatedly. But as we will soon see, neither can such a sequence arise by selecting one nucleotide at a time. A pre-existing "concept" is required as a framework upon which a sentence or a functional sequence must be built. Such a concept can only pre-exist within the mind of the author. Starting from the very first mutation, we have a fundamental problem even in trying to define what our first desired beneficial mutation should be.

2. Waiting for the first mutation. Let's assume we can know the first desired mutation. How long do we have to wait for it to happen? Human evolution is generally assumed to have occurred in a small population of about 10,000 individuals. The mutation rate for any given nucleotide, per person per generation is exceedingly small (very roughly about one mutation per 30 million individuals, for a given nucleotide site). Within a population of 10,000, one would have to wait 3,000 generations (at least 60,000 years) to expect a specific nucleotide to mutate. But two out of three times, it will mutate into the "wrong" nucleotide. So, to get a specific desired mutation at a specific site just in one individual will take three times as long, or at least 180,000 years. Once the mutation arises in one individual, it has to become "fixed" (such that each individual in the population will eventually have a double dose of that mutation). Because a newly arisen mutation arrives in a population as just a single copy, it arrives on the brink of extinction. The vast majority of new mutations soon drift back out of the population, even the ones that are beneficial. So, any specific desired mutation must arise many times before it "catches hold" in the population. Only if the mutation is dominant and has a very distinct benefit does selection have any reasonable chance to rescue it from random elimination via drift. According to population geneticists, apart from effective selection, in a population of 10,000, our given new mutant has only one chance in 20,000 (the total number of non-mutant nucleotides present in the population) of NOT being lost via drift. Even with some modest level of selection operating, there is a very high probability of random loss, especially if the mutant is recessive or is weakly expressed (we actually know that most mutations will be both recessive and nearly neutral). Therefore, even a beneficial mutation will be randomly lost due to genetic drift most of the time. Our numerical simulations suggest a weakly beneficial mutant will be lost about 99 out of 100 times. So, a typical mildly-beneficial

mutation must happen about 100 times before it is likely to "catch hold" within the population. So, on average, in a population of 10,000 we would have to wait 180,000 × 100 = 18 million years to stabilize our first desired beneficial mutation, to begin building our hypothetical new gene. So, in the time since we supposedly evolved from chimp-like creatures (6 million years), there would not be enough time to realistically expect our first desired mutation to go to fixation in the genomic location where our required gene is hopefully going to arise. A vast amount of mutations would arise during 18 million years, but only once would that specific nucleotide mutate to that specific new nucleotide — such that it's not lost due to genetic drift and is fixed.

3. Waiting for the other mutations. After our first desired mutation has been found and fixed, we need to repeat this process for all the other nucleotides encoding our hoped-for gene. A gene is minimally 1,000 nucleotides long. More realistically, a human gene is on average about 50,000 nucleotides long, when regulatory elements and introns are included. To be extremely generous we will only consider a gene of 1,000 nucleotides (and we assume each nucleotide is by itself selectable). If this process was a straight, linear, and sequential process, it would require about 18 million years × 1,000 = 18 billion years to create the smallest possible gene. This is more than the time since the reputed Big Bang! So, it is a gross understatement to say that the rarity of desired mutations limits the rate of evolution. Furthermore, single nucleotides do not carry any information by themselves, and cannot be selectively favored. Specified information requires many characters (minimally, a sentence or similar text string is needed). Like any message, a genetic message which specifies some life function requires many nucleotides to reach its "functional threshold". Functional threshold is the minimal number of characters (or nucleotides) needed to convey a meaningful message. Below the functional threshold, individual letters or nucleotides have no benefit and cannot be favored by selection. This means that realistically, waiting time will be much, much longer — because no selection can happen until the minimum string of nucleotides falls into place by chance. If the functional threshold for selection is 12 (no selection until all 12 letters are in place), the waiting time in our hypothetical human population becomes trillions of years.

Sanford, John (2015-02-23). Genetic Entropy (Kindle Locations 1684-1755). FMS Publications. Kindle Edition. USED BY PERMISSION.

END QUOTE.

—

Trillions of years!

Well, that's the END of the Theory of Evolution, isn't it?

The Darwinists NEVER use their God-given brains to stop and think about these kinds of things. At the best possible average pace, with God making sure that there are NO deleterious mutations and NO devolution taking place, it would take on average 18 million years to fixate and stabilize a SINGLE beneficial mutation through "Natural Means" into a population of 10,000 apes which God has already designed and created in the first place and kept alive and functional during those 18 million years, just so that population of 10,000 God-created apes can achieve their first beneficial mutation through "Natural" Hands-off Mutation and Selection. 18 million years on average per beneficial mutation! Think about it!

If those apes need 1,000 such beneficial mutations in order to become men, then you are looking at 18 billion years on average to produce those targeted 1,000 beneficial mutations; and, that's with a population of 10,000 apes that God has already designed and created in the first place and that God is making sure receive ONLY beneficial mutations and NO devolution or deleterious mutations. And, that's also with God keeping that population of 10,000 apes alive during those 18 billion years so that they can indeed "evolve" their necessary 1,000 beneficial mutations and become men all on their own through "Natural Means".

Furthermore, it has been estimated that it would in fact take at least 20 million such beneficial mutations to convert chimpanzees into humans through "Natural Means". With that targeted goal in mind and assuming NO deleterious mutations or extinctions along the way, how long would it take on average to convert 10,000 chimpanzees into 10,000 humans using Natural Selection and Random Mutations to do the job? So, what do you get if you multiply 20 million beneficial mutations with 18 million years per beneficial mutation? At the BEST possible pace, with God keeping those 10,000 chimpanzees alive all along the way, and with God making sure that there is NO devolution, NO extinction, and NO deleterious mutations taking place, the quickest on average that Mutation and Selection could convert a chimpanzee into a human through "Natural Means" is 360 trillion years.

John Sanford isn't exaggerating whenever he says that it could take trillions of years for Mutation/ Selection to design and create something useful "naturally". And, it really isn't Natural Evolution if God has to design and create the 10,000 chimpanzees in the first place, and then keep them alive for 360 trillion years by blocking ALL deleterious mutations and preventing ALL extinctions that might take place during that period of time, just so He can convert 10,000 chimpanzees into 10,000 humans "naturally" or through "Natural Means".

Think about it! At the BEST possible average pace, it would take at least 360 trillion years for Mutation and Selection to convert a population of 10,000 chimpanzees into 10,000 humans through "Natural Means", with God keeping those 10,000 mutants alive and preventing deleterious mutations during the whole time. And, that's with 10,000 chimpanzees that God has already designed and created in the first place! How old did they say our universe is? How long would it take to convert a bacterium into a human through "Natural Means" when billions of beneficial mutations are needed? Wouldn't it be easier and faster to just let God design and create those 10,000 humans in the first place?

YES, it would be!

The Darwinists and Materialists NEVER stop and use their God-given brains to think about and calculate these kinds of things. You will NEVER get these kinds of calculations and truths from the Darwinists because they don't DO this kind of science. It's too difficult and painful for them. I'm willing to wrap my mind around these kinds of things. Your typical Darwinist isn't. Your typical Darwinist is afraid of it because they don't want to be proven wrong. For the Materialists and Darwinists, ignorance is bliss! But, ignorance is the reason why the Darwinists and Materialists truly believe that Mutation/Selection can design and create anything that it sets its mind to. The rest of us KNOW BETTER!

If you think about it, this is radically advanced science — the best that humans are able to come up with! Can you see and understand now why the 9th Chapter of "Genetic Entropy" put an END to the Theory of Evolution for me? It's because I understood what John Sanford was talking about and chose to believe that it is true. Now the onus is on you.

What do the Darwinists typically do when presented with these kinds of Statistical Models of the Mutation/Selection Process?

Assuming that they don't go head-in-the-sand and actually study them instead, the Darwinists try to shave the figures in half or by one-tenth, which is exactly what they do when designing Mutation/Selection Models of their own. They cheat. They choose parameters that are scientifically inaccurate and don't match with Reality in order to shave those estimates down to something that they might be willing to accept. They keep shaving and cheating until they get the numbers that they want, and then they call the results "Science".

If they really take John Sanford's Models seriously, the Darwinists will demand that God artificially accelerate the Mutation/Selection Process so as to make it possible for the Theory of Evolution to be true. But, even if God were to speed up the process a thousand-fold, it's still going to take 360 billion years for chimpanzees to evolve into humans through "Natural Means"; and, the Darwinists are still going to complain, even though we are starting with Chimpanzees that God has designed and created in the first place.

It can be fascinating and entertaining to watch a Darwinist try to shave trillions of years off a Statistical Estimate that he doesn't like, all in an attempt to increase the possibility that the Theory of Evolution might be true.

And, that's just the beginning of the problems for the Theory of Evolution. It only gets worse from there on forward, because John Sanford actually has TEN points in chapter 9 of "Genetic Entropy", each of which decreases the likelihood of Mutation/Selection creating anything at all, even if it has an infinite number of years to do so. Evolution by random mutations and evolution by natural selection CANNOT design, create, and deploy anything! It has been conclusively and finally demonstrated that it is so. It has been empirically and logically observed to be so. The Theory of Evolution doesn't work and can't create new unique genomes from scratch, so we have no choice but to declare the whole thing to be FALSE.

Many different scientists taught me that Random Mutations and Natural Selection (Evolution) cannot design and create genomes and life forms. They met their burden of proof and demonstrated to me that it must be so. (See the partial list of Reference Materials below for a selection of some of the best scientific evidence that falsifies the Theory of Evolution.)

I chose to believe the scientific evidence rather than the claims of the Materialists, Naturalists, and Darwinists who teach that Psyche, Intelligence, and Syntropy do not exist.

Evolution Is Creation by Death

I did make a scientific discovery in recent days (May 2018) that I think has merit and value to the Scientific Community.

I do seem to be the first person on the planet to realize that NO type of evolution can do selection. Selection of any kind involves choice; and, choice is exclusively the product of a Psyche or a Mind. Evolution of any type by definition, in principle, is dumb and blind without a soul or a mind. The very definition of Evolution eliminates its ability to do selection, a priori. Evolution can't do selection because evolution can't do Psyche or Choice. Evolution doesn't exist as a Psyche, Person, Intelligence, Mind, or Soul capable of doing choice or selection.

Furthermore, any attempt to imbue evolution with a soul, psyche, or mind automatically FALSIFIES Materialism, Naturalism, Darwinism, Nihilism, and the Theory of

Evolution which claim that Psyche or Syntropy does not exist. The theory of evolution is a non-starter because evolution of any kind cannot do selection or choice.

Always remember, creation by natural selection is science fiction and wishful thinking because evolution of any kind cannot do selection or choice.

Materialism, Naturalism, Darwinism, Nihilism, Behaviorism, Determinism, Physical Reductionism, Atheism, Classical Physics, and the Theory of Evolution are based exclusively on entropy. Entropy is death. Death cannot design, create, and produce life. Such a thing has never been experienced nor observed. Darwinism or the theory of evolution is Creation by Death, or Creation by Entropy. Creation by Death or the theory of evolution is impossible. It can't happen, which means that it didn't happen.

Therefore, the theory of evolution cannot be used to explain the origin of life. The most that the theory of evolution can explain is death, extinction, devolution, and genetic entropy. Evolution of any kind is a function of death or entropy. Evolution is entropy. Evolution is death. Evolution or death, of any kind, can't be used to explain the origin of life. The truthfulness of this reality becomes obvious, once a person realizes and accepts the fact that the Theory of Evolution, Physicalism, Naturalism, Atheism, Nihilism, Classical Physics, and Darwinism are based exclusively on entropy. Entropy or death cannot produce life. The different types of evolution or entropy cannot produce life. It's impossible for entropy or death to produce life.

At times I've wondered why nobody else has been able to see and understand these obvious truths; but then, I used to be a Materialist, Naturalist, Nihilist, and Atheism, and there I find my answer. At the time, I wasn't able to see nor understand these obvious scientific truths because I didn't want to see them, understand them, nor accept them. No seeking, then no finding. I wasn't looking for any of this, so I never found it. I only found it after I started looking for it. I had convinced myself that this type of information doesn't exist. Self-deception works, and it works every time, especially when it comes to scientists like me.

The axiom stating that evolution is dumb and blind without a soul or a mind, if taken as being true, prevents evolution of any type from being able to do selection. In other words, the very definition of evolution as being dumb and blind FALSIFIES the Theory of Evolution by preventing evolution of any kind from being able to do selection or choice. Evolution is entropy; and, entropy is death. Death cannot do selection or choice. Death cannot do life. Entropy or death can only destroy. That is what has been experienced and observed.

Meanwhile, the Materialists, Naturalists, Darwinists, Nihilists, and Atheists DEMAND that you accept on blind faith that their claims – that death, entropy, or evolution can produce life – are true.

We scientists have FALSIFIED the Theory of Evolution trillions of times in thousands of different ways, but we choose to ignore the evidence because it isn't telling us what we want to hear. That's the way we do science in this world – by ignoring the evidence, discounting the evidence, banning the evidence, and destroying the evidence. We do our science this way so that we can prove to ourselves and to others that the Theory of Evolution is true.

There is another way to do science, though – a better way of doing science. I call it Science 2.0; and, it involves allowing ALL of the evidence into evidence. Once we choose to do so, then ALL of the evidence that we have on hand as a race FALSIFIES the claims of Materialism, Naturalism, Darwinism, Nihilism, Behaviorism, Determinism, Physical Reductionism, and Atheism which claim that this evidence does not exist.

Once we have eliminated all of the falsehoods such as Materialism, Naturalism, Atheism, Nihilism, and Darwinism, then we are left staring at THE TRUTH, which is that ONLY Psyche can design, program, engineer, field-test, fine-tune, manufacture, create, and do science.

By eliminating all of the falsehoods or pseudo-sciences, it becomes obvious that God's Psyche must of necessity exist in order to have DONE all of the Science that needed to be done, which evolution and the rocks could NEVER have done. Remember, evolution and the rocks can't do science because they can't do selection or choice.

The observation and realization, that evolution of any type cannot do selection, just might be one of my greatest scientific discoveries even if I don't end up being the first person on the planet to have made this discovery.

Obviously, everyone across the world is now starting to make these kinds of scientific discoveries right and left because we have finally started to take our blinders off. More and more of us are willing to see, which makes us able to see. Nowadays, it's obvious that the Theory of Evolution is false; whereas, we couldn't see it before because we didn't want to see it, and we didn't know where to look.

All you want is the truth. Everything else is worthless in the end.

Mark My Words

—

Source

Science 2.0: I Upgraded My Science

https://www.amazon.com/dp/B0771K6WTX

The Scientific Method Proves That the Theory of Evolution Is False

https://www.amazon.com/dp/B01IAAIRT2

https://www.amazon.com/dp/1521133611

NATURE vs. NURTURE vs. NIRVANA: An Introduction to Reality

https://www.amazon.com/dp/B01JWRCSVA

https://www.amazon.com/dp/1521132615

Myers, D. G. (2010). *Social Psychology* (10th ed.). New York: McGraw-Hill.

Reference Materials

Wells, J. (2000). *Icons of Evolution: Science or Myth? Why Much of What We Teach About Evolution Is Wrong*. Washington, DC. Regnary.

Sanford, J. (2014). *Genetic Entropy* (4th ed.). Cornell University: FMS Foundation.

Sanford, J. C., Marks, R. J., Behe, M. J., Dembski, W. A., & Gordon, B. L. (Eds.). (2013). *Biological Information: New Perspectives*. Hackensack, NJ: World Scientific.

Meyer, S. C. (2010). *Signature in the Cell: DNA and the Evidence for Intelligent Design*. New York: HarperCollins.

Meyer, S. C. (2013). *Darwin's Doubt: The Explosive Origin of Animal Life and the Case for Intelligent Design*. New York: HarperCollins.

Mark My Words. (2016). *The Scientific Method: Proves That the Theory of Evolution Is False*. Kindle. Retrieve from: https://www.amazon.com/dp/B01IAAIRT2

Mark My Words. (2016). *The Theory of Evolution Proved to Me that God Exists: Why I Am No Longer an Atheist and Why I No Longer Believe in the Theory of Evolution*. Kindle. Retrieve from: https://www.amazon.com/dp/B01HZYBZ7K

Kin Selection

Natural Selection is science fiction. By definition, in principle, Natural Selection can't do CHOICE or selection. Natural selection can't do what they say it does. By definition, in principle, natural selection or survival of the fittest is supposed to be dumb and blind without a soul, psyche, or mind. Therefore, natural selection can't do choice or selection. Instead, natural selection results in entropy or death. Natural selection, or entropy and death, cannot design and create and produce anything. "Design and creation by Natural Selection" is prevented from happening by entropy, random diffusion, or the second law of thermodynamics. Random mutations are also entropy. Entropy is death. The theory of evolution is Creation by Entropy, or Creation by Death. That's not going to work because it's physically impossible. Death or entropy cannot design and create life. This reality is obviously true.

Do you want the truth, or do you want the science fiction?

Can you handle the truth? Most of our scientists can't.

Kin Selection is another smoking gun where Psyche is concerned.

Kin Selection: The idea that evolution has selected altruism toward one's close relatives to enhance the survival of mutually shared genes.

Our genes dispose us to care for relatives. Thus, one form of self-sacrifice that *would* increase gene survival is devotion to one's children. Compared with neglectful parents, parents who put their children's welfare ahead of their own are more likely to pass their genes on. As evolutionary psychologist David Barash wrote, "Genes help themselves by being nice to themselves, even if they are enclosed in different bodies." Genetic egoism (at the biological level) fosters parental altruism (at the psychological level). Although evolution favors self-sacrifice for one's children, children have less at stake in the survival of their parents' genes. Thus, according to the theory, parents will generally be more devoted to their children than their children are to them. (*Social Psychology*, p. 452.)

By definition, in principle, evolution is dumb and blind without a soul or a mind. This means that evolution of any type cannot DO choice or selection. It's a deceptive lie to say that evolution can do choice or selection. It's science fiction.

Without realizing it, these people have painted themselves into a corner with this one. The idea of genes helping each other even if they are enclosed in different bodies is physically impossible. There's NO physical mechanism in place whereby the genes can communicate with each other, especially the genes in different bodies! Such an idea is ludicrous. If the genes are communicating with each other and recognizing each other, then they are doing so at the quantum level or the psyche level because they can't do so at the physical level.

The ONLY way to make Kin Selection true is if we all axiomatically agree in advance that the genes (and evolution) are psychic and are therefore capable of determining telepathically which genes are related to them and which genes are not. In order for the genes to be nice to each other, they have to know each other, perceive each other, recognize each other, and show favoritism to each other. They have to be psychic and have some kind of psyche. The genes can't be dumb and blind if we want Kin Selection to work as advertised. The genes have to communicate with each other, know each other, and recognize each other in order for them to be able to help themselves and be nice to themselves. The genes also have to be subliminally communicating their desires to the Human Psyche in order to make the Human Psyche be nice to the genes too. The only way to make Kin Selection work as advertised is if we agree in advance that the genes have some sort of Psyche or Mind and are telepathically connected with each other.

However, if we agree in advance axiomatically that genes have a psyche, that genes perceive each other telepathically, that genes know each other and recognized each other, and that genes are therefore psychic, then we have in fact FALSIFIED Materialism, Naturalism, Darwinism, Nihilism, Atheism, even the Theory of Evolution in the process.

These people personify the genes – imbue them with Psyche and Intelligence – and in the process falsify the major premises of Materialism, Naturalism, Darwinism, Nihilism, and Atheism which state that Psyche or Syntropy does not exist. Thereby, these falsified philosophies or falsified religions end up being self-defeating. The truth cannot be built upon falsehoods

When it comes to Science, we observe that Psyche, Quantum Mechanics, or Syntropy always ends up being the best possible explanation that can be given to ALL of the evidence that we are observing and experiencing, including the physical evidence. Remember, entropy and physical matter would not exist without a massive initial infusion of Syntropy, Intelligence, Power, Quantum Mechanics, or Psychic Intervention somewhere sometime along the way.

Remember, perception is a function and a product of Psyche. At the physical level, the genes have no way of knowing or perceiving which genes are related to them and which genes are not; and at the physical level, the genes have NO way to pass that information on to the physical body or physical brain even if the genes were to know who is related to them and who is not. At the physical level, the genes have no way to perceive each other, thereby falsifying the claims of Kin Selection. By restricting and limiting everything to the physical level, the Materialists and Naturalists automatically falsify anything and everything that needs Psyche or Intelligence or Perception in order to become true – things such as Kin Selection and Creation by Mutation/Selection.

Remember, evolution can't do selection! By definition, in principle, evolution cannot do CHOICE! Selection requires some type of choice, and choice requires some type of Psyche or Mind! By definition, in principle, evolution is dumb and blind without a soul or a mind. The theory of evolution is self-defeating because evolution of any kind can't do selection or choice. Evolution doesn't exist as some type of Psyche or Person who is capable of making choices or doing selection. The false is falsified by the truth; and, the truth is repeatedly experienced and observed. The theory of evolution is obviously false because evolution of any type by definition in principle cannot do selection, psyche, or choice.

Evolution of any type cannot do science, but the Human Psyche and God's Psyche certainly can. They've been caught in the act of doing so.

With just a bit of scientific observation, it's easy to see that Kin Selection is a *fictional ad hoc just-so story* that they made up out of thin air after-the-fact to match with what we have experienced and observed from Intelligent Beings or the Human Psyche. They took Kin Selection, a philosophical idea or religious idea, and they personified it, humanized it, anthropomorphized it, and deified it. Kin Selection and the Theory of Evolution are man-made idols or man-made gods. The Theory of Evolution is our modern-day form of idolatry, which is a belief in false gods that are incapable of delivering the goods.

The theory of evolution is self-defeating because it can't do what they say it does. The genes can't be nice to each other without a psyche or a soul; and, if the genes have a psyche or a soul, then the very existence of Psyche or Syntropy or Soul falsifies Materialism, Naturalism, Darwinism, Nihilism, Behaviorism, Determinism, Physical Reductionism, Atheism, and the Theory of Evolution. The false is falsified by the truth; and, the truth is repeatedly experienced and observed.

Mark My Words

—

Source

> ***God Is in the Light: God is light, and in Him is no darkness at all.***
>
> https://www.amazon.com/dp/B07168S37N

Reference

Myers, D. G. (2010). *Social Psychology* (10th ed.). New York: McGraw-Hill.

16. Scientific Proof of Psyche

One of my greatest, most useful, and most explanatory scientific discoveries came to me when I personally first discovered Psyche or Quantum Non-Local Consciousness. The ramifications cascaded all throughout my scientific endeavors.

Psyche is the innate intelligence within all the different forms of energy which gives that energy the inherent ability to understand, follow, and obey God's Laws and God's Commands.

Science is all about observation and experience. Go with the experienced and the observed; and then, get rid of all the rest – the philosophical speculation and wishful thinking.

Remember, Psyche or Quantum Non-Local Consciousness has been experienced and observed. Psyche or Intelligence is a massless quantum or a massless elementary particle; and, Psyche looks like and acts like a Photon

whenever it is seen or observed, both at the quantum spiritual level and at the macro physical level. At every level of existence, Psyche or Intelligence uses mass, energy, or matter in order to get things done. Photons and other massless quantum particles have been experienced and observed which means that different types of Psyches or Intelligences have also been experienced and observed. The massless elementary particles within fermions, leptons, quarks, electrons, and neutrinos are Controlling Psyches or Controlling Particles who command and control the mass, energy, forces, and fields that surround them. This is what has been experienced and observed.

This is one of my greatest, most useful, and most explanatory scientific discoveries of all time. Every massless elementary particle is some type of Psyche or Intelligence. It commands and controls everything else that surrounds it.

The Gods can form energy into anything, including physical matter and entropy. Entropy is a Physical Law. It was made. Time is a construct. Entropy is a function of time. Entropy is a construct. Physical matter is Organized Energy or Restricted Energy. Organized and restricted by whom? Anything that was obviously organized obviously has an Organizer to organize it. God's Psyche is the Organizer.

Psyche is syntropic meaning that Psyche is conserved. The First Law of Thermodynamics is the Conservation of Psyche. Syntropy is the First Law of Thermodynamics, which means that Psyche or Energy is conserved. Syntropy is Eternal Life. Your Psyche or Intelligence or Life Force was not made, and it cannot be destroyed. It is eternal and everlasting. It is conserved.

Entropy is death. I eventually discovered that the Materialists, Naturalists, Darwinists, Nihilists, Behaviorists, Determinists, and Atheists use and define the Second Law of Thermodynamics as the Conservation of Entropy or the Conservation of Death. These people literally teach, preach, and believe that entropy or death is conserved. These people use the Second Law of Thermodynamics, the Conservation of Entropy, and the Conservation of Death to FALSIFY the First Law of Thermodynamics or the Conservation of Psyche and the Conservation of Eternal Life. That's precisely what they do.

He makes this point in the following video:

https://www.youtube.com/watch?v=UwYSWAlAewc

https://syntropy.website/wp-content/uploads/2018/07/Anti-gravity-and-the-True-Nature-of-Dark-Energy.zip

In this video, he states:

It takes work to expand the volume of the universe, and the new energy turns into dark energy. But, who's doing that work? Where does the energy come from? The answer is maybe nowhere. This is where intuition goes completely out the window. See, the law of conservation of energy no longer applies in an expanding universe. This law is a property of a Newtonian universe, in which space and time are fixed static dimensions. In a universe governed by general relativity, this is no longer true. Energy can be forever lost or gained from nothing within an expanding curved spacetime.

Wow! He used Creation Ex Nihilo and his Atheism to FALSIFY the First Law of Thermodynamics, the Conservation of Psyche, God's Psyche, Eternal Life, Syntropy, and the Conservation of Energy.

Creation Ex Nihilo is NOT a scientific explanation. It's not an explanation at all. Creation Ex Nihilo is a blind leap of faith into nothing. Why did "nothing" decide to fill our universe full of dark energy instead of filling our universe full of concrete? Creation by nothing from nothing would produce nothing. Why would nothing produce something rather than nothing? Creation Ex Nihilo is not an explanation for what things are, why they are, and how they work. Creation Ex Nihilo lacks explanatory power. It fails to satisfy. Yet, these people truly believe that nothing can produce anything and in just the right amounts. It's magic. It's science fiction.

The realization, that the Naturalists and Atheists define the Second Law of Thermodynamics as the Conservation of Death or the Conservation of Entropy and then use the Second Law of Thermodynamics to FALSIFY the First Law of Thermodynamics or Eternal Life, ended being one of my most interesting and most useful scientific discoveries.

Consequently, Scientific Proof of Psyche ended up becoming one of my primary endeavors once I discovered the existence of Psyche, Syntropy, Life Force, or Eternal Life.

—

Back in the 1970's using brain stimulation, Neuroscientists demonstrated that Psyche or Mind is a completely different entity than the physical brain.

What did the Scientific Community do with this amazing scientific discovery? They swept it under the rug and hid it from the people because it wasn't telling them what they wanted to hear. That's the way the Materialists and Naturalists do science – by hiding and destroying any evidence that falsifies Materialism, Naturalism, and Darwinism. It works. It prevents us from finding the truth.

Any Science which points us to Psyche, Mind, Eternal Life, Syntropy, and Quantum Nonlocality is going to end up being vastly superior to the watered-down stuff that we get from the Materialists, Naturalists, and Atheists which is precisely why these people try to hide the First Law of Thermodynamics or the Conservation of Psyche from us.

Scientific Proof of Psyche:

(See: Penfield, W. (1978). *The Mystery of the Mind: A Critical Study of Consciousness and the Human Brain*. Princeton, NJ: Princeton University Press.)

(See: Eccles, J. C., & Popper, K. R. (1977). *The Self and Its Brain: An Argument for Interactionism*. New York: Routledge.)

(See: Eccles, J., & Robinson, D. N. (1984). *The Wonder of Being Human: Our Brain and Our Mind*. New York: The Free Press.)

(See: Eccles, J. (1985). *Mind and Brain: The Many-Faceted Problems*. New York: Paragon House Publishers.)

These scientists solved the mind-brain problem over forty years ago; but, the Materialists and Atheists refuse to read the memo. Materialism and Atheism of any kind are based upon a refusal to look at evidence. I KNOW, because I used to be a Materialist and an Atheist, until I started looking at the evidence.

Once you KNOW that Psyche or Mind exists, then you are suddenly free to go looking for the Mind of God.

Mark My Words

—

Wilder Penfield

Penfield, W. (1975). *The Mystery of the Mind*. Princeton, New Jersey: Princeton University Press.

https://muse.jhu.edu/book/39368

http://www.chabad.org/library/article_cdo/aid/113106/jewish/Appendix-5-Neurology-Medicine-and-the-Soul.htm

http://mypsyche.us/Neurology-Medicine-and-the-Soul

This book resolves the Mind-Brain Problem, using Scientific Experiments and Scientific Observations to do so. Wilder Penfield has the distinction of getting there first.

By using scientific experimentation, scientific methods, brain stimulation, and observation of the results, the Mind-Brain Problem was officially solved in 1975 by Wilder Penfield. Others have replicated and confirmed Wilder Penfield's findings; but, he got there first, because he was brave enough to report what he was finding.

The Materialists and Naturalists spend all of their time suppressing and ridiculing this type of observational evidence in an attempt to explain it away or convince us that it does not exist.

Remember, other neuroscientists like John Eccles, Mario Beauregard, and Eben Alexander have come to the very same conclusions, but Wilder Penfield has the distinction of getting there first. These men tend to quote Wilder Penfield and this book, *The Mystery of the Mind*, in many of their books and presentations.

The following is some of what they quote:

Is the mind merely a function of the brain? Or is it a separate but closely related element?

Throughout my own scientific career, I, like other scientists, have struggled to prove that the brain accounts for the mind.

(*The Mystery of the Mind*, Preface.)

This is the million-dollar question and the ultimate quest.

Is there any evidence of the existence of neuronal activity within the brain that would account for what the mind does?

Before venturing to answer, it may be of interest to refer again to action that the mind seems to carry out independently, and then to reconsider briefly our experience with stimulation of the cortex of conscious patients and our experience of what effects are produced by epileptic discharge in various parts of the brain. This should give some clue if there is a mechanism that explains the mind.

"(a) What the Mind Does

It is what we have learned to call the mind that seems to focus attention. The mind is aware of what is going on. The mind reasons and makes new decisions. It understands. It acts as though endowed with an energy of its own. It can make decisions and put them into effect by calling upon various brain mechanisms. It does this by activating neuronal

mechanisms. This, it seems, could only be brought about by expenditure of energy.

(b) What the Patient Thinks

When I have caused a conscious patient to move his hand by applying an electrode to the motor cortex of one hemisphere, I have often asked him about it. Invariably his response was: "I didn't do that. You did." When I caused him to vocalize, he said: "I didn't make that sound. You pulled it out of me." When I caused the record of the stream of consciousness to run again and so presented to him the record of his past experience, he marveled that he should be conscious of the past as well as of the present. He was astonished that it should come back to him so completely, with more detail than he could possibly recall voluntarily. He assumed at once that, somehow, the surgeon was responsible for the phenomenon, but he recognized the details as those of his own past experience. When one analyzes such a "flashback" it is evident, as I have said above, that only those things to which he paid attention were preserved in this permanently facilitated record.

(c) What the Electrode Can Do

I have been alert to the importance of studying the results of electrode stimulation of the brain of a conscious man, and I have recorded the results as accurately and completely as I could. The electrode can present to the patient crude sensations. It can cause him to turn head and eyes, or to move the limbs, or to vocalize and swallow. It may recall vivid re-experience of the past, or present to him an illusion that present experience is familiar, or that the things he sees are growing large and coming near. But he remains aloof. He passes judgment on it all. He says "things seem familiar," not "I have been through this before." He says, "things are growing larger," but he does not move for fear of being run over. If the electrode moves his right hand, he does not say, "I wanted to move it." He may, however, reach over with the left hand and oppose his action.

There is no place in the cerebral cortex where electrical stimulation will cause a patient to believe or to decide...

I am forced to conclude that there is no valid evidence that either epileptic discharge or electrical stimulation can activate the mind...

The mind seems to act independently of the brain in the same sense that a programmer acts independently of his computer...

For my own part, after years of striving to explain the mind on the basis of brain-action alone, I have come to the conclusion that it is simpler (and far easier to be logical) if one adopts the hypothesis that our being does consist of two fundamental elements. If that is true, it could still be true that energy required comes to the mind during waking hours through the highest brain-mechanism.

Because it seems to me certain that it will always be quite impossible to explain the mind on the basis of neuronal action within the brain, and because it seems to me that the mind develops and matures independently throughout an individual's life as though it were a continuing element, and because a computer (which the brain is) must be programmed and operated by an agency capable of independent understanding, I am forced to choose the proposition that our being is to be explained on the basis of two fundamental elements. This, to my mind, offers the greatest likelihood of leading us to the final understanding toward which so many stalwart scientists strive...

The nature of the mind presents the fundamental problem, perhaps the most difficult and most important of all problems. For myself, after a professional lifetime spent in trying to discover how the brain accounts for the mind, it comes as a surprise now to discover, during this final examination of the evidence, that the dualist hypothesis seems the more reasonable of the two possible explanations.

Since every man must adopt for himself, without the help of science, his way of life and his personal religion, I have long had my own private beliefs. What a thrill it is, then, to discover that the scientist, too, can legitimately believe in the existence of the spirit!

(From Wilder Penfield, *The Mystery of the Mind*.)

Source:

http://www.chabad.org/library/article_cdo/aid/113106/jewish/Appendix-5-Neurology-Medicine-and-the-Soul.htm

Wilder Penfield set out to prove Materialism and Naturalism true; but, he ended up FALSIFYING Materialism and Naturalism instead.

This is the smoking gun – catching psyche in action completely separate from the physical brain. This effectively solves the Mind-Brain Problem.

However, if science continues to be defined as Materialism, Naturalism, Darwinism, Nihilism and Atheism, then science will NEVER be able to solve the Mind-Brain Problem, because Materialism and Naturalism cause the Mind-Brain Problem, not solve it.

But if instead, we choose to upgrade our science to Science 2.0 and choose to define science as Observation, Experience, Knowledge, Truth, Fact, and the Lived Experiences of the human race, then we quickly realize that the Mind-Brain Problem has already been solved by the collective Lived Experiences of the human race as a whole, including those presented to us by Wilder Penfield.

Good enough.

Since I have upgraded my science to Science 2.0, in all things I choose to go with the observational evidence rather than the philosophical guesswork of the Materialists and Naturalists. I simply find and present the observational evidence, and then leave it up to you my reader to decide for yourself what to make of it. Wilder Penfield made me a believer in the non-local or the non-physical, but your mileage may vary because you are a completely different person than I am.

Of course, these observed and proven realities show us the desperate need that we have for something like Quantum Neuroscience, so that we can figure out what's really going on.

There are others out there who are smarter and much more experienced than I am; and, they could probably do a much better job of introducing and explaining Quantum

Neuroscience than I can – if they actually wanted to do so, which apparently they don't. Therefore, it's left up to someone like me to introduce it to the world, since nobody else is going to.

Mark My Words

The Quantum Law of Thermodynamics

The Quantum Law of Thermodynamics states that heat death is impossible at the quantum level and the psyche level. Entropy or the second law of thermodynamics doesn't exist in the Quantum Realm or the Spirit World. Your Psyche, Mind, or Intelligence cannot die.

Psyche is the innate intelligence within all the different forms of energy which gives that energy the inherent ability to understand, follow, and obey God's Laws and God's Commands. A Spirit or a Ghost is a spirit body and psyche combination. The Psyche or Intelligence within a spirit body is co-eternal with God's Psyche. Psyche, Intelligence, Mind, Life Force, or Energy cannot be made; and, it cannot be destroyed.

Syntropy is the Conservation of Psyche or the Conservation of Energy. The Quantum Law of Thermodynamics or Syntropy has to be real and has to be true, or we wouldn't exist. Our physical universe had to receive a massive infusion of Syntropy, Exergy, or Energy 13.8 billion years ago, or NONE of that subsequent entropy would have been possible. This is Logic 101. We see entropy at the physical level, which means that the Syntropy has to exist at the quantum level.

God revealed the Quantum Law of Thermodynamics, the Impossibility of Death at the Quantum Level, and the Conservation of Psyche to Joseph Smith over 170 years ago. Joseph Smith talks about what God taught him in the King Follett Discourse. Here is one of the most pertinent parts:

The Immortal Intelligence

I have another subject to dwell upon, which is calculated to exalt man; but it is impossible for me to say much on this subject. I shall therefore just touch upon it, for time will not permit me to say all. It is associated with the subject of the resurrection of the dead — namely, the soul — the mind of man — the immortal spirit. Where did it come from? All learned men and doctors of divinity say that God created it in the beginning; but it is not so: the very idea lessens man in my estimation. I do not believe the doctrine; I know better. Hear it, all ye ends of the world; for God has told me so; and if you don't believe me, it will not make the truth without effect. I will make a man appear a fool before I get through; if he does not believe it. I am going to tell of things more noble.

We say that God Himself is a self-existing being. Who told you so? It is correct enough; but how did it get into your heads? Who told you that man did not exist in like manner upon the same principles? Man does exist upon the same principles. God made a tabernacle and put a spirit into it, and it became a living soul. How does it read in the Hebrew? It does not say in the Hebrew that God created the spirit of man. It says, "God made man out of the earth and put into him Adam's spirit, and so became a living body."

744

The mind or the intelligence which man possesses is co-equal [co-eternal] with God himself. I know that my testimony is true; hence, when I talk to these mourners, what have they lost? Their relatives and friends are only separated from their bodies for a short season: their spirits which existed with God have left the tabernacle of clay only for a little moment, as it were; and they now exist in a place where they converse together the same as we do on the earth.

I am dwelling on the immortality of the spirit of man. Is it logical to say that the intelligence of spirits is immortal, and yet that it has a beginning? The intelligence of spirits had no beginning, neither will it have an end. That is good logic. That which has a beginning may have an end. There never was a time when there were not spirits [intelligences]; for they are co-equal [co-eternal] with our Father in heaven.

I want to reason more on the spirit of man; for I am dwelling on the body and spirit of man — on the subject of the dead. I take my ring from my finger and liken it unto the mind of man — the immortal part, because it had no beginning. Suppose you cut it in two; then it has a beginning and an end; but join it again, and it continues one eternal round. So, with the spirit [intelligence] of man. As the Lord liveth, if it had a beginning, it will have an end. All the fools and learned and wise men from the beginning of creation, who say that the spirit of man had a beginning, prove that it must have an end; and if that doctrine is true, then the doctrine of annihilation would be true. But if I am right, I might with boldness proclaim from the housetops that God never had the power to create the spirit [intelligence] of man at all. God himself could not create himself.

Intelligence is eternal and exists upon a self-existent principle. It is a spirit from age to age and there is no creation about it. All the minds and spirits that God ever sent into the world are susceptible of enlargement.

The first principles of man are self-existent with God. God himself, finding he was in the midst of spirits and glory, because he was more intelligent, saw proper to institute laws whereby the rest could have a privilege to advance like himself. The relationship we have with God places us in a situation to advance in knowledge. He has power to institute laws to instruct the weaker intelligences, that they may be exalted with Himself, so that they might have one glory upon another, and all that knowledge, power, glory, and intelligence, which is requisite in order to save them in the world of spirits.

This is good doctrine. It tastes good. I can taste the principles of eternal life, and so can you. They are given to me by the revelations of Jesus Christ; and I know that when I tell you these words of eternal life as they are given to me, you taste them, and I know that you believe them. You say honey is sweet, and so do I. I can also taste the spirit of eternal life. I know that it is good; and when I tell you of these things which were given me by inspiration of the Holy Spirit, you are bound to receive them as sweet, and rejoice more and more.

Source: https://www.lds.org/ensign/1971/05/the-king-follett-sermon?lang=eng

The mind of man, the intelligence, or the psyche is the immortal part of us because it had NO beginning; and, it cannot die nor be destroyed. The spirit body comprised of spirit matter was made or born; but, NOT the intelligence, mind, or psyche that dwells within it. Nevertheless, spirit matter is made from energy; and, the energy had NO beginning, and the energy cannot be destroyed.

Psyche or Mind is the innate intelligence within all the different forms of energy which gives that energy the inherent ability to understand, follow, and obey God's Laws and God's Commands. Psyche or Energy cannot be made; and, it cannot be destroyed.

This is also what the Quantum Law of Thermodynamics teaches us. Psyche cannot die. Psyche, Mind, or Intelligence is conserved. Heat death is impossible at the quantum level or the psyche level. This is what has been experienced and observed.

The Quantum Law of Thermodynamics is Syntropy; and, Syntropy is the Conservation of Psyche or the Conservation of Energy. This means that Psyche or Energy cannot be made, and it cannot be destroyed because it is always conserved. Some of that energy can be made temporarily unavailable to us at the physical level; but, the energy is still there because the energy is always conserved. At the quantum level, though, ALL of the energy is available for use, all of the time. This is what the Quantum Law of Thermodynamics or Syntropy is trying to teach us.

In contrast, the purpose of Physicalism, Materialism, Naturalism, Darwinism, Nihilism, Behaviorism, Determinism, Physical Reductionism, Atheism, and their derivatives is to PREVENT us from discovering the Quantum Law of Thermodynamics, Psyche, Syntropy, Mind, the Conservation of Psyche, Spirit Bodies, Spirit Matter, Action at a Distance, Immortality, Eternal Life, and God. These people erroneously teach and believe that ONLY entropic physical matter exists, and that entropy or death is conserved. They are wrong.

Remember, the BEST and FASTEST way to find and know the truth is to live it, experience it, observe it, and witness it for yourself, or to choose to trust someone who has. Science is observation and experience after all. When it comes to Materialism, Naturalism, Darwinism, Nihilism, Atheism, and their derivatives, there is NOTHING to experience and NOTHING to observe.

Mark My Words

—

Source

The Ultimate Model of Reality: Psyche Is the Ultimate Cause

https://www.amazon.com/dp/B071NC9JK6

Other Books that Provide Scientific Proof of Psyche

Kelly, E. F., Kelly, E. W., Crabtree, A., Grosso, M., & Gauld, A. (2007). *Irreducible Mind: Toward a Psychology for the 21st Century*. Plymouth, United Kingdom: Rowman and Littlefield.

Buhlman, W. L. (1996). *Adventures Beyond the Body: How to Experience Out-of-Body Travel*. New York: HarperCollins Publishing.

Marshall, P. D., Kelly, E. F., & Crabtree A. (Eds.). (2015). *Beyond Physicalism: Toward Reconciliation of Science and Spirituality*. London, United Kingdom: Rowman and Littlefield.

17. Mind-Over-Matter

Mind-over-Matter is Action at a Distance; therefore, Mind-over-Matter is Quantum Mechanics.

Any Science that demonstrates Mind over Matter is going to point us directly to Psyche and establish the fact that Psyche is something completely different than matter, whether we are talking about spirit matter or physical matter.

The Placebo Effect is scientific proof of Psyche's existence.

(See: Dispenza, J. (2014). *You Are the Placebo: Making Your Mind Matter*. USA: Hay House Inc.)

(See also: McTaggart, L. (2002). *The Field: The Quest for the Secret Force of the Universe*. New York: HarperCollins.)

A complete understanding of Quantum Mechanics explains how God's Psyche was able to convert or transmute spirit matter into physical matter at the locations in this universe where He wanted there to be physical matter. It explains how the galaxies came into existence all at once!

If the Big Bang Theory were 100% true, then every particle in this universe should be moving away from every other particle in this universe; and, there would be NO planets, stars, and galaxies! This universe should be nothing more than one huge homogeneous ball of gas. It required God's Psyche in order to transmute spirit matter into physical matter at the current LOCATION of the planets, stars, and galaxies. God's Psyche is the ONLY way to explain the existence of planets, stars, and galaxies, because they shouldn't exist at all if the Big Bang Theory as typically presented to us were 100% true.

The Big Bang Theory "proves" that this physical universe had a beginning. That's the part of the Big Bang Theory that I truly believe in. However, the Big Bang Theory FAILS to explain how the stars, planets, and galaxies came together; and, for an explanation of that reality we have to turn to God's Psyche and Mind-over-Matter.

A complete understanding of Mind over Matter ends up explaining how God's Psyche was able to organize planets, stars, galaxies, genomes, and life forms from raw unorganized physical matter. These are all things which the rocks or raw physical matter could NEVER have done!

For me personally, John Pratt's sacred calendars provided Scientific Proof of God's Existence and scientific proof of mind over matter at a cosmic scale:

http://www.johnpratt.com/items/docs/lds/dates.html

http://www.johnpratt.com/items/docs/article_nums.html

That scientist certainly had a vision, which would ONLY have been possible by choosing to believe in God's existence in the first place.

The LDS Scriptures convinced me that the Gods are Scientists; and, the Gods are capable of mind-over-matter functionality or quantum mechanics. Quantum Mechanics IS mind-over-matter, whether we realize it or not.

(See: https://www.lds.org/scriptures/pgp/abr?lang=eng).

You particularly want to look at Abraham 4: 10, 12, 18, 21, and 25; where the Gods watched and waited to make sure that the physical matter obeyed them, during the organization and terraforming of this earth.

The Gods are Scientists, not magicians! There is NO such thing as Creation Ex Nihilo or magic.

(See also: Mark My Words. (2017). *I Am Not a Creationist: So What Am I?*

https://www.amazon.com/dp/B071XTM8XY).

There are a lot of books about Mind-Over-Matter. ALL of my books touch upon the subject to one extent or another. It's one of the most interesting aspects of Science in my humble opinion. Mind-Over-Matter is Quantum Mechanics; and, supernatural mechanisms or quantum mechanisms are mind-over-matter mechanisms. ALL of my books touch upon Quantum Mechanics or Mind-Over-Matter to one extent or another.

https://stealth-ascendancy.com/monism-versus-dualism/

My Author Page on Amazon:

http://amazon.com/author/science

Enjoy!

Mark My Words

18. Out-of-Body Travel

The book, *Adventures Beyond the Body: How to Experience Out-of-Body Travel*, by William Buhlman was a game-changer for me. There was my first encounter with someone who can go out-of-body basically "at will". He turned the whole experience into a Science. I'm talking about a REAL SCIENCE of direct observation, experimentation, and lived experience – NONE of that garbage where they deliberately restrict their "science" to entropic physical matter, philosophical speculation, and wishful thinking.

This book about Out-of-Body Experiences (OBEs) provided me with some of my very first evidentiary PROOF that Materialism and Naturalism are false. Buhlman ran science experiments and tried things out while out-of-body on the astral plane or in the spirit realm. He did REAL SCIENCE, doing direct observation of the spirit realm and direct experimentation with the spirit realm, while out-of-body. I thought this whole thing was really cool and fascinating.

With Lived Experiences, you go directly to KNOWLEDGE, TRUTH, and PROOF. Lived Experiences are vastly superior to the scientific methods for getting at truth and proof and knowledge. Technically, you can't use the scientific methods to prove anything due to the wide variety of logic fallacies and biases which are built into the scientific methods. But, you can definitely use Lived Experiences or Direct Observations as knowledge and proof of the truth.

This was the first book to help me see what Science should really be all about – Observation and Lived Experience, which document KNOWLEDGE and PROVE the TRUTH.

Buhlman helped to set me up, so that I was ready to take the NDErs and their out-of-body experiences seriously.

There are at least 13 million documented cases of Near-Death Experiences (NDEs) in the record of Lived Experiences for the human race. The Materialists and Naturalists have chosen to deny and reject every single one of them. Denial of evidence and rejection of evidence is NOT science! But, this is exactly the way that the Materialists and Atheists do science, by denying and rejecting evidence or Lived Experiences that falsify their pre-chosen point of view.

The discovery of all those different NDEs and OBEs is in fact one of my greatest scientific discoveries!

The best way to KNOW God is to live Him and experience Him directly, or to choose to trust someone who has. Science and Scientific Methods can point us to God as the BEST EXPLANATION for the scientific evidence. However, through Lived Experience we can KNOW the truth and KNOW God directly if we want to.

That's very powerful Science there!

Mark My Words

—

Here are a few of my favorite books about Near-Death Experiences and Out-of-Body Exploration:

Buhlman, W. L. (1996). *Adventures Beyond the Body: How to Experience Out-of-Body Travel*. New York: HarperCollins Publishing.

Durham, E. (1998). *I Stand All Amazed: Love and Healing from Higher Realms*. Orem, UT: Granite Publishing and Distribution.

See also: Storm, H. (2000, 2005). *My Descent into Death: A Second Chance at Life*. USA: Random House.

These books convinced me that Out-of-Body Travel is real. In fact, William Buhlman treats out-of-body exploration as a Science, an experiential and observational Science. He actually experiments with the phenomenon as if it were a science. He was the first person to convince me with Scientific Evidence, Experiential Evidence, and Observational Evidence that My Materialism, My Naturalism, and My Nihilism are FALSE.

After reading his book, there was NO going back to My Materialism, My Naturalism, and My Nihilism. I found the Scientific Evidence or Observational Evidence too convincing to dismiss and ignore.

Mark My Words

—

My best friend is a Seer, has had Near-Death and Out-of-Body Experiences, and has seen and conversed with Jesus Christ. I gave him a copy of William Buhlman's book. When my friend realized that I was interested in this subject, he introduced me to Near-Death Experiences on YouTube which I NEVER knew existed because I didn't believe that such things existed and therefore wasn't looking for them.

Particularly impressive for me personally were the hundreds of different accounts of Near-Death Experiences where people encountered and talked with Jesus Christ, the Being of Light and Love. https://www.youtube.com/results?search_query=NDE+Jesus

My best friend watched some of these with me and told me which ones matched with his experiences in the presence of Jesus Christ as well as which ones were fake. I had NO idea that any of this stuff existed because I had been led to believe that such things do not exist because I used to be a Materialist, Naturalist, Nihilist, and Atheist. I had a lot of falsehoods that I had been taught that I had to abandon and overcome. It took a few years, once I finally decided to do so.

I still have TONS of that Materialism, Naturalism, Nihilism, and Atheism within me that I get blocked by and held back by from time to time. It has taken some work to break fifty years of conditioning, brainwashing, and groupthink; but, it's possible if one chooses to do so. I now treat Near-Death Experiences, Out-of-Body Experiences, Shared-Death Experiences, and after-death Life Reviews as observational evidence, experiential evidence, and scientific evidence. I now allow ALL of the evidence into evidence and have chosen to pursue a preponderance of that evidence. It led me to Psyche, Quantum Mechanics, and Quantum Fields which led me to the Biblical God Jesus Christ. Science itself convinced me that God MUST exist. The observations and experiences of the human race convinced me that Jesus Christ really exists and truly rose from the dead.

Mark My Words

19. Quantum Neuroscience – My Scariest Scientific Discovery

My scariest, most radical, and most controversial scientific discovery came to me when I first realized and accepted the fact that there are NO wires within a physical brain connecting everything together into functional logic gates, transistors, CPUs, and RAM. A human brain has NOWHERE to store long-term memories or consolidated memories. There is NO flash memory and NO RAM chips within a physical brain.

A physical brain DOES NOT HAVE memory engrams, which means that a physical brain does not store memories.

Think about it!

That's frightening!

I used to be a Materialist, Naturalist, Nihilist, and Atheist.

That scientific discovery or scientific observation is frightening!

I was taught that our brain is wired together into functional circuits, CPUs, and RAM. I was taught that our brain is a complex computer.

It is not! I was taught wrong.

There are NO physical wires in our brains. There are no CPUs and RAM within our brain. Our after-death Life Review memories are NOT stored within our brain. Those types of memories survive the death of our physical brain; so, they can't be stored within our brain.

Neurons are nothing more than stand-alone switches. They are either ON or OFF. And, the neurons aren't even physically connected to anything. The neurons aren't wired together into any kind of CPU or RAM functionality. The neurons are switches; and, there is NO physical indication as to who or what is turning them ON. Their default state is OFF, so that takes care of itself; but, who or what turns on that first neuron in a cascade of action potentials? There are NO wires to turn on that first neuron and get things going.

That's scary to think about!

A neuron is a stand-alone hardware BIT. It is either ON or OFF. It's a switch. It is either OPEN or CLOSED. That's it!

It's physically impossible to store even a BYTE of programming code or memory within a single hardware BIT, let alone the petabytes of information that are supposedly being stored within our brain. This is insane! Our physical brain doesn't even have a single BYTE of physical memory storage capacity because our physical brain is comprised of nothing but stand-alone BITS.

Furthermore, a synaptic-cleft or synapse scrambles and randomizes everything that comes its way. It's physically impossible to store memories and information within a synapse! If are synapses are indeed organized to produce some type of Distributed Memory System as the neuroscientists claim, then that functionality is taking place at the quantum level because it's impossible at the physical level.

Why?

ALL of those thousands of synaptic "connections" that feed into a neuron get reduced to a single Post-Synaptic Potential (PSP) which determines whether the neuron fires or stays OFF. All of that synaptic input into a neuron gets reduced to a single BIT of information – ON or OFF. That's it. ON or OFF is the sum-total of memory storage and CPU processing that a physical brain is capable of performing! Your brain can do ON or OFF. That's it! So, where's all of that CPU processing and memory storage taking place since it isn't taking place within your brain?

Scary, isn't it?

God went out of His way to make it obvious and clear that our physical brains can't do CPU processing and RAM-type memory storage. There are NO wires in our brains. Our brains can only do ON or OFF. That's it. There's nothing more there within our brains. You can't get complex CPU processing and memory storage out of ON or OFF.

A physical brain DOES NOT HAVE memory engrams, which means that a physical brain does not store memories. A physical brain DOES NOT HAVE logic gates, transistors, CPUs, and RAM, which means that a physical brain can't do any type of computer processing or the running of programming code. A physical brain can only do ON or OFF. Neurons can only do ON or OFF. That's it!

So, how does this thing really work?

ALL of our brain functionality is MAPPED at the quantum level or the psyche level by Nature's Psyche. I learned to call it Quantum Maps of Physical Functionality. Nature's Psyche has MAPPED functionality onto our brains at the quantum level or the psyche level, and then Nature's Psyche uses those quantum MAPS of physical functionality at the quantum level to get things done for us at the physical level. This is the ONLY thing that ever made any sense to me, after I finally realized and accepted the fact that there are NO wires, NO CPUs, and NO RAM within a physical brain. Clearly, our brain functions as if it were a computer with RAM; but, since that type of functionality is physically impossible within a physical brain at the physical level, ALL of that computer-like memory storage and processing power has to be taking place at the quantum level or the psyche level under the command and control of Nature's Psyche. That's just the way it is.

Nature's Psyche (or God's Psyche) MAPPED functionality or the layout of our neurons and synapses at the quantum level; and, only Nature's Psyche knows what each neuron does and what each synapse means because Nature's Psyche is the one who MAPPED the functionality or the meaning onto our brain's neurons and synapses in the first place. Nature's Psyche is the only one who knows which neuron it has MAPPED to be used or fired in order to raise your finger into the air. When you want to raise your finger, Nature's Psyche fires the appropriate neuron that's needed to get the job done. Nature's Psyche MAPPED that physical functionality onto that specific neuron in the first place, so it knows which neuron it needs to fire in order to raise your finger off the table. Nature's Psyche simply waits for you, the Human Psyche, to decide that it wants to raise its finger; and, then Nature's Psyche springs into action collapsing the necessary wave functions in order to get the job done for you. Your finger rises off the table into the air.

On the input side, the neurons function as registers of physical input, which the Human Psyche and Nature's Psyche pick up on and process at the quantum level. Nature's Psyche does any processing that needs to be done by collapsing the necessary wave functions; and, the Human Psyche chooses what it wants to do with all of that physical input.

Once the Human Psyche makes a decision or a choice – decides to raise its finger – then the Human Psyche issues the command and Nature's Psyche collapses the necessary wave functions needed to make the specific finger-raising neuron FIRE or turn ON. Your finger rises off the table.

There's the sum-total of physical brain functionality in a couple of paragraphs. The brain registers physical sensations on the input side of the equation which the Human Psyche and Nature's Psyche experience and use for information gathering purposes. Then when the Human Psyche makes a choice, Nature's Psyche turns on the specific neurons needed to accomplish that physical task by collapsing the necessary wave functions. This explains how a physical brain really works.

Nature's Psyche designed, mapped, and made the physical brain to facilitate the sensation of physical events, which the Human Psyche and Nature's Psyche process and choose to remember. The Human Psyche chooses what it wants to remember. If the part of the physical brain that Nature's Psyche assigned to facilitate the processing of physical events is damaged, then there are no sensations, events, or long-term memories for the Human Psyche to remember. In contrast, it's Nature's Psyche and the Angels in Heaven who produce our after-death Life Review because the Human Psyche doesn't seem to have direct access to its after-death Life Review. The Human Psyche only has access to the physical events that it chooses to remember; and, if the physical machinery gets damaged, the Human Psyche may not have access to that either.

A brain is a transceiver. The brain registers and transmits physical input to the Human Psyche and Nature's Psyche; and, the brain receives commands from the Human Psyche through Nature's Psyche which collapses specific wave functions thereby turning specific neurons ON.

All of our brain functionality is being done at the quantum level or the psyche level by Nature's Psyche (or God's Psyche) completely outside the conscious awareness of the Human Psyche. It's the Human Psyche who makes the choices as to what it wants to do with all of that physical input and its physical body. Nature's Psyche executes the commands.

The brain is MAPPED by Nature's Psyche. Finding and stimulating the specific neuron or concept cell, that Nature's Psyche has MAPPED to a specific group of memories or concepts, can indeed trigger the restoration or re-experiencing of certain memories. Our physical brain was MAPPED by Nature's Psyche to facilitate the retrieval of long-term memories from the quantum level, just as other parts of our physical brain were MAPPED by Nature's Psyche to facilitate the consolidation and transfer of certain memories into long-term storage at the quantum level. If those parts of the brain are damaged, then the long-term memories can no longer be retrieved or can no longer be stored depending upon what part of the brain has been damaged.

A physical brain is needed for the experiencing of physical events, which means that a functional physical brain is also needed for the storage of the memories of those physical events at the quantum level. While the Human Psyche is married to its physical body and physical brain, the Human Psyche is dependent upon the physical brain for the physical input that's being used in memory storage and memory retrieval. If the physical machinery is no longer there, then memory storage and memory retrieval will no long work properly while the Human Psyche is married to and dependent upon its physical body for the experiencing and remembering of physical events. This is logical common sense. It certainly makes a lot more sense than the Naturalists' ongoing search for hidden engrams or silent engram within the physical matter. There's NOTHING in the physical matter at the physical level that can store long-term memories. That's just the way it is!

Furthermore, the discovery of memory engrams or RAM within a physical brain at the physical level won't explain how a person's memories, personality, individuality, identity, perceptions, intelligence, knowledge, wisdom, experiences, ability to choose, and after-death life review manage to survive the death of their physical brain. Materialism, Naturalism, Darwinism, Nihilism, Behaviorism, Determinism, Physical Reductionism, Atheism, and Classical Physics have NO EXPLANATORY POWER when it comes time to explain how a person's memories, personality, intelligence, knowledge, ability to think, ability to perceive, and after-death life review SURVIVE the death of their physical brain.

Remember, the memories that survive the death of your physical brain are NOT being stored within your physical brain. They are being stored within your Human Psyche and Nature's Psyche instead.

All of this is completely in line with the Orthodox Interpretation of Quantum Mechanics by Henry P. Stapp. According to the Orthodox Interpretation, the Human Psyche makes the decisions or the choices; and, Nature's Psyche executes the commands by collapsing the necessary wave functions or turning on the necessary neurons to get the job done.

The explanatory power of Quantum Maps of Physical Functionality is through the roof. It explains everything in a logical scientific fashion. Nature's Psyche MAPS our brains at the quantum level for specific physical functionality at the physical level; and then, Nature's Psyche uses those MAPS at the quantum level to get things done for us at the physical level.

Ironically, Quantum Maps of Physical Functionality provides the answer to life, the universe, and everything; and, the concept was discovered when I finally realized and accepted the fact that there are NO physical wires within a physical brain. When it comes to our physical brain, all of our CPU processing and RAM memory storage are taking place at the quantum level because it's physically impossible to do these things at the physical level within a physical brain. A synapse is a gap-in-space, and it's physically impossible to map meaning, purpose, and functionality onto a gap-in-space at the physical level, which means that this functionality and mapping has to be taking place at the quantum level instead.

Scary, is it not?

Well, I thought it was scary when I first realized the truth of it. There are no physical wires within our brains connecting everything together into functional logic gates, transistors, CPUs, and RAM which means that all of that functionality is taking place and being handled by Nature's Psyche at the psyche level or the quantum level completely outside the conscious awareness of the Human Psyche.

The only thing the Human Psyche has to do is decide what it wants to do with its physical body and all of that physical input or those physical sensations coming its way. Nature's Psyche handles the rest.

Mark My Words

—

My Magnum Opus

Quantum Neuroscience: The Answer to Life, the Universe, and Everything.

https://www.amazon.com/dp/B079Z6QQQB

20. Psyche and the Ultimate Law of Thermodynamics

One of my greatest scientific discoveries is Psyche and its relationship to the Ultimate Law of Thermodynamics. Psyche is the Ultimate Cause which means that Psyche is the ultimate causal agent or the ultimate causal force. Psyche is the innate intelligence within all the different forms of energy which gives that energy the ability to obey God's Laws and God's Commands. Psyche is seen or observed as some type of photon or particle of light.

Introduction

I have been a scientist all of my life; and after 55 years of life, I finally found what I have been searching for all of my life – an explanation for how everything works at all levels of existence or reality.

I found the answer to life, the universe, and everything.

I found what I was looking for in Quantum Mechanics, Action at a Distance, Psyche, or Syntropy. It explains everything!

As a result of these discoveries, I upgraded my science.

I have made some fascinating discoveries during my science career. Ironically, most of what I have discovered has already been discovered by someone else before me; but, I chose to do something unique that has never been done before. I chose to give each scientific discovery a quantum mechanical explanation, a psychic explanation, or a spiritual explanation. I chose to take Quantum Mechanics and Psyche seriously, which most scientists have never done before. Consequently, it makes it seem as if I have made a wide variety of different scientific discoveries that have never been discovered before when in fact I have not. I've simply discovered what others have chosen to ignore and reject.

Even Psyche or Intelligence has been experienced and observed, which means that it has already been discovered; but, I chose to do something different that hasn't been done before. I chose to include it in Science; and, I chose to try to define it and explain what it is and how it works.

Psyche is the innate intelligence within all the different forms of energy that have ever been discovered or observed. The psyche within energy gives all the different forms of energy the ability to hear God's commands, understand God's commands, and obey God's commands. Psyche is the "organizing force" or the "law following force" within our universe. Psyche is identified by choice, particularly its choice to obey the quantum laws and the physical laws.

Physical matter, spirit matter, and dark matter are simply different forms of organized energy existing simultaneously in different phases, different frequencies, or different dimensions. They each have a psyche within them that gives them the ability to obey God's Laws. God provides the Laws; and, the psyche within the different forms of energy provides the obedience to those Laws. Matter is organized energy or lawful energy. Someone Psyche within that matter is choosing to obey the Quantum Laws and the Physical Laws; otherwise, it would be nothing but random chaos.

A LAW in order to be a Law requires some kind of intelligent Law Giver and Law Enforcer. Laws are worthless without some type of uniformity, order, and enforcement. The thing that most people don't realize is that Laws also require Someone Psyche with enough intelligence and understanding to actually choose to obey those Laws. Without Someone Psyche or someone intelligent enough to function as the Law Follower or the Law Obeyer, Laws are also completely worthless. Laws are pointless unless there is someone intelligent enough, within the different forms of energy, who is capable of actually choosing to obey those Laws. This is logical common sense. I have a feeling that it has all been discovered before; but, nobody ever takes the time to actually look at everything from the perspective of Psyche.

Defining Psyche

Psyche is the innate intelligence, within all the different forms of energy, which gives those different forms of energy the ability to obey God's Laws and God's Commands.

There's also another way to look at Psyche that works equally as well.

Psyche has been observed by Out-of-Body Travelers as a pinpoint of light, a pinprick of light, a point particle of light, or a spark of light. That's precisely the way that the physicists describe the fermions – quarks, leptons, neutrinos, and electrons. These "particles" of light OBEY the Pauli exclusion principle. Obedience to Laws requires intelligence! The components of the quarks and the electrons are individual pinpricks of Psyche or Intelligence; and, the psyches within fermions USE the bosons, gluons, photons, quantum waves, and light to communicate with each other, interact with each other, and bind themselves to each other. The elementary parts of Fermions are Psyche or Intelligence. That would make Psyche or Intelligence the fundamental unit of reality and existence. Photons, gluons, and W and Z bosons are the four force-carrying gauge bosons of the Standard Model. There's NO mass in the elementary components of the fermions, the psyches, or the pinpricks of light. Over 99% of the mass within an atom is

associated with the gluons. Psyche or Intelligence is driving the mass or driving the gluons. It's all made from Energy which is always conserved.

From a different angle, we observe that the elementary components within quarks, fermions, electrons, neutrinos, leptons, gluons, and bosons are massless, timeless, ageless, syntropic "quantum particles," pinpoints of light, quantum waves, or Photons that we call Psyche or Intelligence or Quantum Non-Local Consciousness. Thoughts are quantum waves. Psyche or Intelligence is there in the models and the math just waiting to be identified as such.

Photons are Psyche or Intelligence, which means that the massless, timeless, ageless, syntropic elementary components or "elementary particles" within quarks, fermions, electrons, neutrinos, leptons, gluons, and bosons are in fact Photons, or Psyche, or Intelligence, or Quantum Non-Local Consciousness. Psyche or Intelligence is Light and Truth. Psyche is a Photon; and, Photons are psychic. Psyche is observed by Out-of-Body Travelers as a pinpoint of light, a point particle of light, or a Photon. Furthermore, the existence of intelligence is obvious. It all fits together perfectly and makes logical sense.

Psyche is the innate intelligence within all the different forms of energy which gives that energy the inherent ability to understand, follow, and OBEY God's Laws and God's Commands. Obedience to Law requires some type of intelligence. This is what has been experienced and observed.

Remember, any massless quantum particle is a Psyche or an Intelligence. Any massless elementary particle is a Psyche. A Photon is a Psyche. The massless components of quarks and electrons are Psyches. Whenever they are seen or observed both at the spiritual level and the physical level, Psyches look like and act like Photons or Particles of Light. Psyches USE energy or mass to get things done. This is what has been experienced and observed.

This is one of my greatest and most interesting scientific discoveries of all time. Its explanatory power is through the roof to infinity and beyond.

Mark My Words

The Ultimate Law of Thermodynamics

Time and entropy were made by God, which means that God can change them, manipulate them, and even destroy them. Entropy is like a clock. It keeps track of the passage of time within physical matter or entropic matter. A clock is made, and a clock can be destroyed or come to an end. Entropy was made, which means that entropy can be disassembled and turned back into raw energy. Physical matter was made, which means that physical matter can be disassembled and turned back into raw energy. Entropy and physical matter are NOT conserved.

Physical matter, space-time, space or locality, and time or entropy were made by God, which means that God can change them, manipulate them, and even eliminate them.

It's the underlying Energy or Syntropy or Psyche that are being conserved, not the physical matter nor the entropy. Physical matter, spirit matter, entropy, space, and time are simply different forms of energy. The form is never conserved. Physical matter and entropy are never conserved. They are temporary. The form is constantly being changed by God as God sees fit. This is the Ultimate Law of Thermodynamics. The Ultimate Law of

Thermodynamics differentiates between what is being conserved and what is not conserved. Psyche, Syntropy, or Energy is conserved. Entropic physical matter is NOT conserved.

It's the Energy, or the Syntropy, or the Psyche that cannot be made nor destroyed. God cannot make Energy, Syntropy, or Psyche, and God cannot destroy Energy, Syntropy, or Psyche. It's the Energy, or the Syntropy, or the Intelligence that's eternal and everlasting without a beginning of days or an end of years. It's the Energy, or the Syntropy, or the Psyche who is being conserved. However, God can transform raw energy or unorganized energy into anything He wants it to be, including physical matter, physical bodies, genomes, proteins, planets, stars, galaxies, spirit bodies, forces, fields, space, time, locality, and entropy. The form is never conserved. The form can change as God sees fit. It's the underlying Energy, Syntropy, or Psyche that's being conserved.

This is called the Ultimate Law of Thermodynamics. It explains everything that we human beings have ever encountered. It's one eternal round.

Time or entropy, space or locality, physical constants, physical laws, physical matter, physical genomes, physical proteins, physical planets, physical stars, physical galaxies, and physical universes are made and disassembled by God as He sees fit. Only the underlying Psyche, Syntropy, or Energy is being conserved. The energy cannot be made, nor can it be destroyed. The energy or psyche has always existed and will always exist. It is syntropic, without a beginning of days or an end of years. It is eternal and everlasting. God's Psyche has always existed and will always exist. Your psyche or your intelligence has always existed and will always exist. Energy or psyche is syntropic which means that it is always being conserved; whereas, the form of that energy is never conserved. The form is constantly being changed by God as God sees fit.

This is called the Ultimate Law of Thermodynamics. It is the answer to life, the universe, and everything.

The Ultimate Law of Thermodynamics is hidden within the First Law of Thermodynamics which states that energy is conserved.

The first law of thermodynamics is a version of the law of conservation of energy, adapted for thermodynamic systems. The law of conservation of energy states that the total energy of an isolated system is constant; energy can be transformed from one form to another but can be neither created nor destroyed.

https://en.wikipedia.org/wiki/First_law_of_thermodynamics

Energy can neither be created nor destroyed. Psyche can neither be created nor destroyed. Psyche is the intelligence or the life force within energy that gives energy the ability to obey God's Laws and God's Commands. Psyche or Energy is conserved. The First Law of Thermodynamics tells us that the energy is constant and that the energy is conserved – not the FORM of that energy. Energy can be transformed from one form to another, which means that the FORM is never conserved.

Transformed by whom? Transformation of any kind requires Someone Psyche or Someone Intelligent to instigate, originate, initiate, and start that transformation. Any type of beginning requires Someone Psyche to cause it to begin; otherwise, random chaos would continue to reign supreme.

The Ultimate Law of Thermodynamics emphasizes that the FORM is never conserved, which means that the different FORMS of energy such as physical matter, entropy, space, locality, and time are never conserved. Whenever something is formed or made or organized, that means that its FORM is not being conserved. Only Psyche or Energy is syntropic or conserved. Energy or Psyche cannot be created, and it cannot be destroyed. However, energy can be transformed into many different things by the Gods or by Someone Psyche. The FORM is never conserved. Only the underlying energy is conserved.

Remember, the Gods can form energy into anything. The form is never conserved because the Gods can always transform that energy into something else whenever they choose to do so. The Gods have complete control over everything at the quantum level or the psyche level, except for your choices. Your psyche or your energy is being conserved, which means that your choices and your ability to choose is being conserved. It's eternal and everlasting.

These realities and truths explain everything that we human beings have ever experienced or observed.

Conclusion

All of this was a long time in coming, fifty-five years, but I finally got there in the end. Still, I wish I would have known about all of this stuff forty years ago. It would have saved me a lot of time and grief.

ALL of my greatest scientific discoveries and scientific observations have pointed me directly to God because ONLY God's Psyche could have done ALL of the science which needed to be done at the time that these various things came into existence. This is obvious and clear to me now; but, it wasn't when I was a Materialist, Nihilist, and Atheist. Go with the BEST and get rid of all the rest. That's what I finally decided to do; and, you can too. All of this came open to my view, once I got rid of My Materialism, My Scientism, My Nihilism, and My Atheism.

If you really want to do Science, you start by getting rid of the pseudo-sciences, such as Darwinism, Materialism, and Naturalism. Then you switch over to Observation or Lived Experience for your scientific evidence and your Science. Through Lived Experiences, including spiritual experiences, you can go directly to KNOWING the TRUTH which is something that you can't do with the scientific methods and philosophical speculation. I never realized how powerful and revelatory Lived Experience can be, until after I got rid of My Materialism, My Atheism, My Nihilism, and My Scientism.

I never realized how weak and insubstantial the scientific methods and science really are until AFTER Joseph Rychlak, Brent Slife, and Edwin Gantt pointed out all the different logic fallacies that are built into the scientific methods. These are concepts which you will never get from your materialistic and atheistic college professors because these people want you believing that science and the scientific methods are infallible.

There was a time in my life when I truly believed that Science and the scientific methods are the ONLY way of finding and knowing the truth. Boy, was I wrong! Over fifty years of being wrong! I'd been duped! Due to all the different logic fallacies that are built into the scientific methods, we can't use the scientific methods to prove the truth and to know the truth. If we want to KNOW the TRUTH, then we have to switch over to Lived Experience or Direct Observation, or we have to choose to trust someone who has had the kinds of Lived Experiences that we wish we would have had. I KNOW of what I speak, because I have lived it and experienced it, on both sides of the fence.

Mark My Words

—

Source

The Ultimate Model of Reality: Psyche Is the Ultimate Cause

https://www.amazon.com/dp/B071NC9JK6

A Philosophy of Science for Personality Theory or Psyche Theory

INDIVIDUAL QUANTUM OBJECTS

y = tan(x)   -π/2   0   π/2   2π

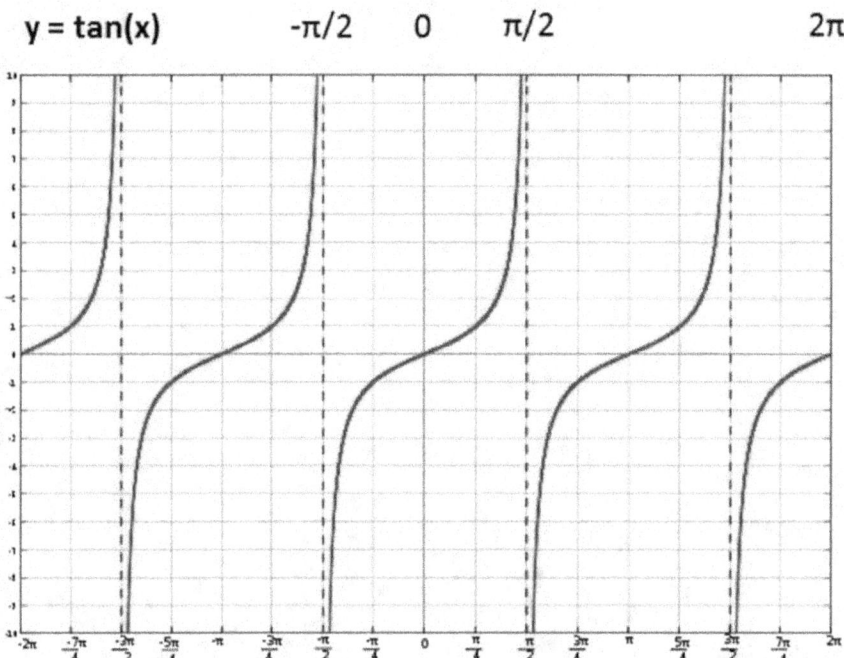

     Psyche has been observed by Out-of-Body Travelers as a pinpoint of light, a pinprick of light, a point particle of light, or a spark of light. That's precisely the way that the physicists describe the fermions – quarks, leptons, neutrinos, and electrons. These "particles" of light OBEY the Pauli exclusion principle. Obedience to Laws requires intelligence! The components of the quarks and the electrons are individual pinpricks of Psyche or Intelligence; and, the psyches within fermions USE the bosons, gluons, photons, quantum waves, and light to communicate with each other, interact with each other, and bind themselves to each other. The elementary parts of Fermions are Psyche or Intelligence. That would make Psyche or Intelligence the fundamental unit of reality and existence. Photons, gluons, and W and Z bosons are the four force-carrying gauge bosons of the Standard Model. There's NO mass in the elementary components of the fermions, the psyches, or the pinpricks of light. Over 99% of the mass within an atom is associated with the gluons. Psyche or Intelligence is driving the mass or driving the gluons. It's all made from Energy which is always conserved.

Technically, the elementary particles or psyches that make up a gluon are massless like photons; but, the force mediated or controlled by the gluon has or produces a lot mass or resistance to acceleration. Likewise, the elementary particles or psyches that make up the quarks and the electrons are massless like photons; but, the Higgs Field that interacts with the quarks and the electrons gives them their small amount of mass. Massless particles are psyches.

From a different angle, we observe that the elementary components within quarks, fermions, electrons, neutrinos, leptons, gluons, and bosons are massless, timeless, ageless, syntropic "quantum particles," pinpoints of light, quantum waves, or Photons that we call Psyche or Intelligence or Quantum Non-Local Consciousness. Thoughts are quantum waves. Psyche or Intelligence is there in the models and the math just waiting to be identified as such.

Photons are Psyche or Intelligence, which means that the massless, timeless, ageless, syntropic elementary components or "elementary particles" within quarks, fermions, electrons, neutrinos, leptons, gluons, and bosons are in fact Photons, or Psyche, or Intelligence, or Quantum Non-Local Consciousness. Psyche or Intelligence is Light and Truth. Psyche is a Photon; and, Photons are psychic. Psyche is observed by Out-of-Body Travelers as a pinpoint of light, a point particle of light, or a Photon. It all fits together perfectly and makes logical sense.

Psyche is the innate intelligence within all the different forms of energy which gives that energy the inherent ability to understand, follow, and OBEY God's Laws and God's Commands. Obedience to Law requires some type of intelligence. This is what has been experienced and observed.

Remember, any massless quantum particle is a Psyche. A Photon is a Psyche. The massless components of quarks and electrons are Psyches.

A Psyche Ontology is based upon the idea that Psyche, or Quantum Non-Local Consciousness, is the fundamental unit of reality. If there really is such a thing as a point particle or an infinite singularity, Psyche is it. Psyche or consciousness is the "elementary particle" upon which everything else is based. Psyche determines or chooses everything else.

I use $y = \tan(x)$ to mathematically model my Psyche Ontology or my Ultimate Cause Model of Reality, quantum objects, spirit matter, physical matter, spirit bodies, physical bodies, universes, infinite singularities, psyche, and non-local consciousness. Each wave on the graph is a different Quantum Object, or a different particle of matter, or a different universe.

This model extends all the way from the infinitely small to the infinitely large; and, I was extremely satisfied with its explanatory power when it was revealed to me.

A single Universe, or Quantum Object, or Particle of Matter, and/or its lifeline exists in the domain between the -π/2 and π/2 asymptotes. Matter or a Quantum Object has some limitations. Matter in any format, whether spiritual or physical, reacts. There's a time lag associated with Matter or a Quantum Object, because it waits for and then reacts to Psyche's Word of Command.

When a psyche or consciousness enters a non-consensus reality, that spiritual reality conforms itself to that psyche's demands and commands. In contrast, a physical reality is the ultimate consensus reality. It doesn't reshape itself automatically to the demands of our psyche. Instead, we have to work it in order to reshape it.

Again, a single Universe, or Quantum Object, or Particle of Matter (and its size or the amount of space it takes up) exists in the domain between the -π/2 and π/2 asymptotes. That Quantum Object, or matter, or particle, or wave, or material is spiritual and takes up little space when it is below (0, 0) on the graph; and, that same Quantum Object or matter is physical and can theoretically take up infinite space and become infinite in size when it is above (0, 0) on the graph above – although it's prevented from reaching infinite size and infinite velocity by the restraints of physical law, on the physical side of the equation.

Psyche resides ON the -π/2 and π/2 asymptotes. Think about it. That means that Psyche or Non-Local Consciousness has NO size and takes up NO space, yet is simultaneously infinite in size, range, scope, and presence. It's a paradox, but it's Reality. It's one eternal round! Psyche goes down the rabbit hole and climbs the stairway to heaven simultaneously, because Psyche exists at the infinite and IS infinite.

Psyche or Intelligence is something completely different from a spirit body; but, they can be combined together into a functional whole. The spirit body and psyche combination are typically called a Spirit, Ghost, or Soul.

At the -π/2 asymptote, Psyche or Quantum Non-Local Consciousness is an infinite singularity or point particle which has NO physical size and takes up NO physical space whatsoever; yet, simultaneously at the π/2 asymptote, Psyche is capable of being omnipresent and omniscient. Psyche resides in the realm of the infinite. Psyche is infinite and instantaneous. In its native format, there are no speed limits placed upon Psyche. Psyche is capable of being instantaneously and simultaneously everywhere. Psyche is infinitely small; but, it is also simultaneously capable of infinite omnipresence and infinite omniscience.

Psyche is viewed or seen as a spark or a pin-point of light. Psyche is experienced as an immaterial viewpoint in space. Psyche or Intelligence is a whole other animal besides matter or dust; yet, both are comprised of different types and/or different frequencies of quantum waves.

Quantum waves are also thoughts and memories. They survive the death of our physical brain according to the observational evidence obtained from Near-Death Experiences (NDEs), Out-of-Body Experiences (OBEs), Shared-Death Experiences (SDEs), and our after-death Life Reviews. A Psyche is capable of transmitting, receiving, and storing quantum waves, which are thoughts and memories.

In contrast, Quantum Objects, particles, or matter have two states of existence according to the Quantum Law of Complementarity – a non-local spiritual state and a localized physical state. According to the Quantum Law of Complementarity, a Quantum Object can be either spirit matter or physical matter, but it can't be in both states simultaneously. A choice has to be made by Someone Psyche to determine which of the two states it will be in. Psyche ACTS. Quantum Objects REACT to psyche's demands and intervention.

According to the Quantum Law of Superposition, a Quantum Object or Particle of Matter can be phase-shifted, dimension-shifted, or frequency-shifted along the domain between the -π/2 and π/2 asymptotes. Matter is finite and reacts to psyche's demands, particularly to God's Psyche who provided matter with Laws, Order, Structure, Restrictions, and Purpose.

Thanks to the Quantum Law of Superposition, a psyche, a spirit body, and a physical body can occupy the same space at the same time because they are out of phase with each other, exist at different frequencies, and therefore exist in different dimensions, but in the very same space at the exact same time.

A Quantum Object or particle of matter can be transitioned, or phase-shifted and moved, all along the x-axis between the -π/2 and π/2 asymptotes. Out of body travelers have observed that on the spiritual side, their spirit body can phase-shift or go into different dimensions at will. Being in a different phase or at a different frequency, their spirit body can walk through physical walls and doors; and, they can levitate and rise into air through physical ceilings and roofs.

However, on the physical side, the physical laws that God put into place prevents us from phase-shifting our physical bodies at will and then walking through walls and doors. God retains the capability of phase-shifting physical matter for himself and apparently doesn't share that capability with mortals. The same thing applies to the teleportation or the quantum tunneling of physical matter. God put the physical laws into place in order to restrain it and contain it so that your physical body doesn't quantum tunnel away on you one atom at a time. God put the sub-light speed limitations and entropy into place on physical matter in order to create the Ultimate Consensus Reality for us to live and experience.

In a physical reality, I can depend on my house, my car, this paper, and my dog being there when I go looking for them tomorrow. A physical reality is the ultimate consensus reality because it is dependable, reliable, and predictable. It provides an excellent school-ground upon which to learn how to interact with other psyches in a social manner.

Physical matter is still capable of phase-shifting and teleportation – changing frequencies – but the whole thing is under God's control. God teleported me and my car to safety one time. It wasn't anything I did. It was something that God did for me to save me. My best friend has experienced phase-shifting. An elk passed through his truck un-phased. Again, it wasn't anything that my friend did. It was something that God did for him in order to save him, or the elk.

Quantum Objects – spirit matter and physical matter – reside in the realm of the finite. Psyche, Intelligence, Consciousness, or Life is the Master of the Infinite!

This IS the Ultimate Model of Reality and the Grand Unified Theory of Everything; and, it's really quite simple to visualize and understand once a person chooses to do so.

Remember, any massless quantum particle is a Psyche. A Photon is a Psyche. The massless components of quarks and electrons are Psyches. This is the answer to life, the universe, and everything. It explains everything that has ever been experienced and observed.

Mark My Words

—

Source

The Ultimate Model of Reality: Psyche Is the Ultimate Cause

https://www.amazon.com/dp/B071NC9JK6

A Psyche Ontology – The Grand Unified Theory of Everything

The universe is a conscious thought. Thought, psyche, or consciousness is the fundamental unit of reality. Energy is Psyche; and, Psyche is Energy. Energy or Psyche is the fundamental unit of reality. Psyche is intelligence, light, energy, and truth. Psyche is the innate intelligence within all the different forms of energy which gives that energy the inherent ability to understand, follow, and obey God's Laws and God's Commands. Psyche, Energy, Light, or Intelligence is syntropic, meaning that it cannot be made nor destroyed.

Doctrine and Covenants 93: 29, 36:

Man was also in the beginning with God. Intelligence, or the light of truth, was not created or made, neither indeed can be. The glory of God is intelligence, or, in other words, light and truth.

Rychlak, J. F. (1981a). *A Philosophy of Science for Personality Theory* (2nd ed.). Malabar, FL: Robert E. Krieger Publishing Company.

I used this book from Joseph Rychlak to point me to a Philosophy of Psyche or a Philosophy of Personality; and, this book was instrumental in helping me to develop and present my Psyche Ontology or Ultimate Model of Reality. Every truth points to every other truth. By observing what was missing from all the different models of reality or personality theories, this book from Joseph Rychlak and its discussion of Aristotle's four physical causes pointed me directly to Psyche as the Ultimate Cause, because physical causes CAN'T DO psyche, non-local consciousness, spirituality, intelligence, science, philosophy, teleology, and life.

Ever since a Psyche Ontology was revealed to me, I have wondered why nobody has ever thought of it before. Why didn't God reveal it to someone thousands of years ago? Why do I see no mention of a Psyche Ontology in any of the literature?

The only answer I have is that people weren't wanting a Psyche Ontology; and therefore, people weren't looking for a Psyche Ontology. God gives us what we want most; and, most people don't want an ontology that points them to God's Psyche or that reduces to God's Psyche. Depending upon how much you are resisting all of this, you might be one of the persons who doesn't want a Psyche Ontology.

Nevertheless, this Psyche Ontology or Ultimate Model of Reality or Ultimate Paradigm ends up being the Grand Unified Theory of Everything because it explains everything and subsumes the whole of existence, reality, knowledge, lived experience, and truth.

The scientists and mathematicians are NEVER going to be able to unite the Theory of Relativity (Gravity) with Quantum Mechanics directly, because Classical Physics and the Theory of Relativity explain the LIMITATIONS of physical matter whereas Quantum Mechanics or Spiritual Mechanics or Non-Local Mechanics explains to us that the quantum or the spiritual or the non-local has NO physical limitations. The scientists will NEVER be able to explain Quantum Mechanics or Non-Local Mechanics in materialistic or physical terms because Quantum Mechanics is a non-local, non-physical, transdimensional, spiritual phenomenon. Quantum Mechanics explains how spirit matter works, and how spirit matter becomes physical matter by taking upon itself some additional limitations.

To unify Classical Physics with Quantum Mechanics we have to provide a mathematical model, like $y = \tan(x)$, to model how Matter or Quantum Objects REALLY work. We have to explain how spirit matter becomes physical matter! When we do so, this

Psyche Ontology or Ultimate Model of Reality becomes the Grand Unified Theory of Reality, which everyone has been looking for.

When Matter or a Quantum Object is below (0, 0) and approaching infinite velocities, infinite frequencies, and zero size, that particle of Matter or that Quantum Object can function instantaneously and simultaneously with NO distance and speed limitations whatsoever, because the stuff is spirit matter, quantum waves, and NOT physical matter. The only real limitation which spirit matter has is that it reacts and therefore it lags just slightly behind Psyche's Word of Command, where the timing is concerned.

When Matter or a Quantum Object is above (0, 0) and approaching infinite mass and infinite inertia and zero velocity, that particle of physical matter is in fact approaching LIMITATIONS which it can't get beyond. The ONLY way that that particle of physical matter or that Quantum Object can get beyond those limitations is to transform back into spirit matter where it will have no such limitations.

It is my observation that Seraphim, Translated Beings, Resurrected Beings, God the Father, and Jesus Christ can convert or transmute their physical bodies into spirit matter at will, and then convert that spirit matter back into physical matter when they reach their chosen destination. That's how these people do what they do. While their bodies are spirit matter, these people can travel across the universe instantaneously at the speed of thought. When these people reach their chosen destination, they then convert their bodies back into physical matter. It's a Quantum Jump Drive or a Blink Drive, and it is infinitely faster than a warp drive which tries to keep everything in a physical matter format.

Warp drive will NEVER be possible because it's impossible to push physical matter faster than the speed-of-light. However, when we temporarily convert physical matter back into spirit matter, then we are dealing with Quantum Leaps and a Blink Drive or a Quantum Jump Drive, which has NO physical limitations! It's teleportation and it works. It has also been EXPERIENCED and OBSERVED. It solves everything that we have been looking for.

The ONLY problem is that God or God's Psyche is in control of it all; and at the physical level, God gifts this "teleportation ability" only to His righteous followers whom He has selected for the gift or the experience. Teleportation or Blink Drives are something that we mortal beings will never gain control of because God is in control of this gift or this "technology"; and, it is God who decides whom He is going to teleport or blink from one location to another. ONLY God's Psyche controls this spiritual gift. This is an ability or a gift that is ONLY God's to give.

I and my car have teleported. I blinked, or I jumped; BUT, it was nothing that I consciously did or consciously chose. God did it for me. God is the ONLY one who can do these kinds of things for us. If we are going to blink or jump out of harm's way or walk through fire unscathed, this is something that God is going to have to choose to do for us, because it is something that we mortal fallen beings can't do for ourselves. God is KNOWN by what He chooses to do for us; and, God does for us the things that we can't do for ourselves.

This is a Psyche Ontology, wherein Psyche is the fundamental unit of reality and God's Psyche is the Ultimate Psyche or the Ultimate Cause. This IS the Ultimate Model of Reality and the Grand Unified Theory of Everything. A Psyche Ontology subsumes ALL of the other ontologies because ONLY Psyche can do ontology, reality, truth, knowledge, life, transmutation of matter, teleportation, lived experience, memories, and existence. This really IS the Ultimate Model of Reality, which is why nobody has ever thought of it before.

Mark My Words

The Ultimate Model of Reality: Y = TAN(X)

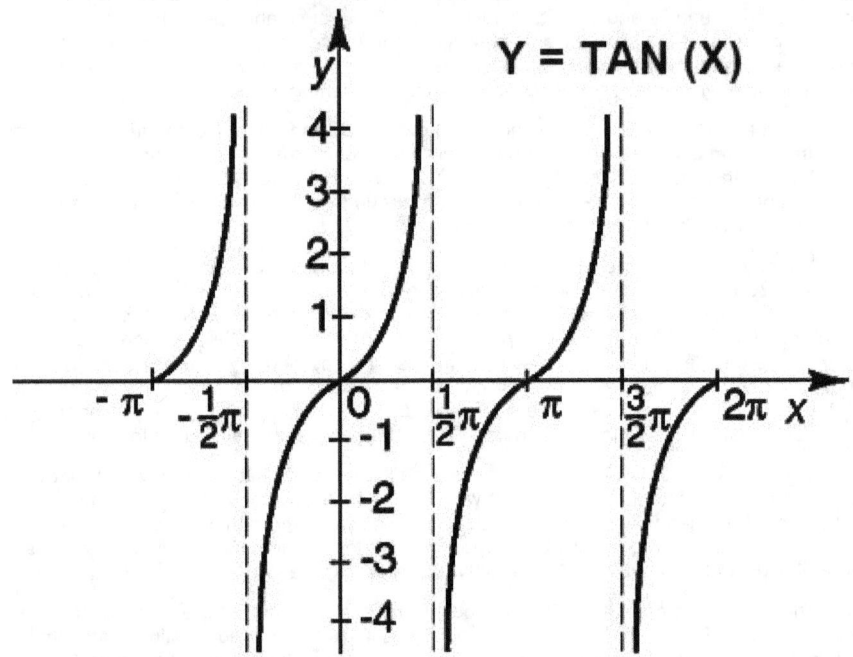

Y = TAN (X)

Psyche has been observed by Out-of-Body Travelers as a pinpoint of light, a pinprick of light, a point particle of light, or a spark of light. That's precisely the way that the physicists describe the fermions – quarks, leptons, neutrinos, and electrons. These "particles" of light OBEY the Pauli exclusion principle. Obedience to Laws requires intelligence! The components of the quarks and the electrons are individual pinpricks of Psyche or Intelligence; and, the psyches within fermions USE the bosons, gluons, photons, quantum waves, and light to communicate with each other, interact with each other, and bind themselves to each other. The elementary parts of Fermions are Psyche or Intelligence. That would make Psyche or Intelligence the fundamental unit of reality and existence. Photons, gluons, and W and Z bosons are the four force-carrying gauge bosons of the Standard Model. There's NO mass in the elementary components of the fermions, the psyches, or the pinpricks of light. Over 99% of the mass within an atom is associated with the gluons. Psyche or Intelligence is driving the mass or driving the gluons. It's all made from Energy which is always conserved.

Technically, the elementary particles or psyches that make up a gluon are massless like photons; but, the force mediated or controlled by the gluon has or produces a lot mass or resistance to acceleration. Likewise, the elementary particles or psyches that make up the quarks and the electrons are massless like

766

photons; but, the Higgs Field that interacts with the quarks and the electrons gives them their small amount of mass. Massless particles are psyches.

From a different angle, we observe that the elementary components within quarks, fermions, electrons, neutrinos, leptons, gluons, and bosons are massless, timeless, ageless, syntropic "quantum particles," pinpoints of light, quantum waves, or Photons that we call Psyche or Intelligence or Quantum Non-Local Consciousness. Thoughts are quantum waves. Psyche or Intelligence is there in the models and the math just waiting to be identified as such.

Photons are Psyche or Intelligence, which means that the massless, timeless, ageless, syntropic elementary components or "elementary particles" within quarks, fermions, electrons, neutrinos, leptons, gluons, and bosons are in fact Photons, or Psyche, or Intelligence, or Quantum Non-Local Consciousness. Psyche or Intelligence is Light and Truth. Psyche is a Photon; and, Photons are psychic. Psyche is observed by Out-of-Body Travelers as a pinpoint of light, a point particle of light, or a Photon. It all fits together perfectly and makes logical sense.

Psyche is the innate intelligence within all the different forms of energy which gives that energy the inherent ability to understand, follow, and OBEY God's Laws and God's Commands. Obedience to Law requires some type of intelligence. This is what has been experienced and observed.

Remember, any massless quantum particle is a Psyche. A Photon is a Psyche. The massless components of quarks and electrons are Psyches.

This is a Psyche Ontology wherein Psyche is the Fundamental Unit of Reality and Existence. Psyche is the Ultimate Cause and the ultimate causal agent. Psyche or Intelligence is the fundamental unit of the Ultimate Model of Reality.

The Human Psyche is the guiding intelligence or controlling intelligence or head intelligence within its spirit body, physical body, and physical brain. It has been experienced and observed that in the non-consensus realities in the spirit world, the Human Psyche is God. Whenever, the Human Psyche enters into a non-consensus reality in the spirit world, that region of space organizes itself to the demands and expectations of the Human Psyche. In contrast, the fermions, quarks, leptons, neutrinos, and electrons are components of Nature's Psyche or the Universe's Psyche. They choose to OBEY God's Laws and God's Commands.

I use $y = \tan(x)$ to mathematically model this Ultimate Cause Model of Reality, quantum objects or particles of matter, infinite singularities, spirit bodies, universes, psyche, and non-local consciousness. This model is a lot more powerful than its simplicity implies. The nuances multiply, the more that one thinks about it.

The domain between $-\pi/2$ and $\pi/2$ is the measure of a single quantum object or a single particle of matter. A quantum object is a particle of matter. It is matter all the way down, with one important exception.

Below (0, 0), the quantum object or particle of matter is in a spiritual state of existence; in other words, it is spirit matter. Astral Travelers have observed that spirit matter exists at different frequencies, wave-lengths, phases, or levels. There are theoretically an infinite number of different frequencies at which spirit matter can exist or reside.

Above (0, 0), the quantum object or particle of matter is in a physical state of existence; in other words, it is physical matter. Sticking with the phase or level or frequency analogy, this implies that physical matter can exist at different phases or can phase-shift. In other words, there could theoretically be an infinite number of physical worlds like ours existing in the exact same space where you are right now, and we wouldn't even know it because these different physical worlds would be at a different phase or at a different level or frequency. This is science fiction at its best.

In fact, the spirit world exists in the exact same space in which you currently exist as a physical being, but it's simply in a different phase of existence. That's how your psyche and your spirit body can reside within your physical body in the exact same space! Your spirit body exists at a different phase than your physical body. Your Psyche exists at a different phase than your spirit body and your physical body. They can all exist simultaneously in the same space.

In the domain between $-\pi/2$ and $\pi/2$, this single quantum object or particle of matter can theoretically exist at any frequency or phase or level or dimension. The domain between $-\pi/2$ and $\pi/2$ thus becomes a measure of this particle's physicality or materiality or density. It has been theorized and observed that as more energy is poured into a particle or an atom, the levels or the orbits of its electrons rise or increase. Likewise, if one were to find a way to pull energy or light out of a hydrogen atom, the orbit of its electron would decrease, and that hydrogen atom would become a hydrino, which is dark matter or spirit matter.

A Quantum Object is a particle of matter. MATTER IS LIGHT that exists at different frequencies or different wavelengths. Matter IS light that exists in different phases. Some have suggested that matter is vibrating strings of light. Physicist David Bohm said that physical matter is frozen light or condensed light. That means that spirit matter is high frequency and high velocity unfrozen light. The frequency and the velocity of the matter decreases from nearly infinite at the $-\pi/2$ asymptote to nearly zero and nearly frozen at the $\pi/2$ asymptote. When measuring the velocity of this quantum object or particle of matter, the speed-of-light is at (0, 0).

That means that physical matter exists at velocities slower than the speed-of-light; and, it means that spirit matter exists at velocities faster than the speed-of-light. This also means that spirit matter exists and travels faster than the speed-of-light, which means that spirit matter can theoretically travel over nearly infinite distances instantaneously or travel at nearly infinite velocities. This also means that physical matter cannot travel faster than the speed-of-light unless it is first transmuted back into spirit matter. Physical matter has to have the mass and inertia and extra energy pulled out of it before it can become spirit matter and travel faster than the speed-of-light. According to Quantum Mechanics, spirit matter becomes physical matter through the Word of Command or conscious observation. God pours a bit of Himself, His energy or light, into that particle of spirit matter transforming it into a physical atom.

Physical matter is frozen or condensed light. Spirit matter is light that exists at velocities and frequencies faster than the speed-of-light. Spirit matter exists in a more refined phase or state than physical matter. Physical matter has massive limitations placed upon it and is placed under God's control so that it can be used to create the Ultimate Consensus Reality – a reality which can be relied upon to function according to LAW. The beauty of a physical consensus reality is that I can rely upon the fact that this book will be there on my flash drive and hard drive tomorrow morning when I turn on my computer and go looking for it. A physical reality is reliable, predictable, and controllable.

Warp drive is impossible, because it is impossible to push physical matter faster than the speed-of-light. However, a Quantum Drive, Blink Drive, Jump Drive, Phase-Shifting, or Teleportation is not only possible but has actually been experienced. I have blinked, or jumped, or teleported, or tunneled, or some such. Actually, it was nothing that I consciously did. God did it for me, in order to save me. God did it to prevent me from becoming one with the car that had pulled out in front of me. I teleported. Time stopped; and when time resumed, I and my car were in the middle of a lawn, twenty or thirty feet away with my car completely stopped and the engine completely off. I have NO memory of the intervening travel. I just blinked or teleported. The whole thing was calm and peaceful. I didn't even have time to get my adrenaline to start pumping. Teleportation is something that I have experienced and therefore I KNOW that it is real and can truly happen for real.

I had another friend caught in a fire, and it didn't burn him. It burned the clothes off of him but didn't burn him and his underwear. Again, it was nothing that he did. God did it for him. I have a couple of friends who raised their children from the dead using the priesthood of God. Again, it was nothing that they did. God did it for them. The child had drowned and was dead, but there was no water in her lungs when God brought her back to life.

My best friend has phase shifted. Again, it was nothing that he did consciously. God did it for him, in order to save him from becoming one with the elk that he ran into at 70 mph. The elk passed right through him and his truck unscathed. He and his brother actually saw the elk pass through them. That's phase shifting.

LIVED EXPERIENCE is the best kind of scientific evidence. Through Lived Experience, we can go directly to knowing the truth. The truth is KNOWN by living it and experiencing it for yourself, or by choosing to trust someone who has.

Under God's control or through God's intervention, quantum objects or matter can phase-shift and teleport. That's how resurrected beings travel from one end of this universe to the other in the blink of an eye. It's infinitely more efficient than a warp drive or a worm-hole will ever be.

Now for the special exception.

At the $-\pi/2$ asymptote and the $\pi/2$ asymptote, LIGHT transmutes into Psyche or some type of fermion. It becomes a pinprick of light, a point particle of light, or a spark of light. It becomes infinitely small while at the same time becoming infinitely large in range and scope. It becomes an infinite singularity.

Should the LIGHT reach the $-\pi/2$ asymptote, it becomes infinitely immaterial and infinitely non-physical, which means that it transmutes, transforms, and becomes something completely different. It becomes a unity and a singularity and takes on NEW PROPERTIES. It comes alive, conscious, and aware. Existing at an infinite frequency and an infinite velocity, it comes alive. It's completely immaterial and totally conscious and alive. It has moved completely to the Limit, and it has become Integral. It becomes living light, a spark of life, and some type of Life Force. It becomes Psyche, Intelligence, or Non-Local Consciousness. Psyche has zero mass, zero size, zero materiality, and takes up zero space. It's an infinite singularity. At the asymptote, the LIGHT or ENERGY or PSYCHE is no longer matter. It's completely immaterial. It is something else. It transmutes and takes on new and different properties and abilities.

The same thing happens to it should the LIGHT reach the $\pi/2$ asymptote. It becomes infinitely solid or infinitely condensed and infinitely real. It's a paradox which helps to bring the thing alive. In this state, Psyche has infinite scope, infinite range, infinite influence, infinite contact, and can permeate the whole of space and time. It becomes an infinite singularity, with infinite potential. Once again, it comes alive. It becomes Psyche, Intelligence, or Non-Local Consciousness. It gains the Word of Command and the ability to ACT on a universal scale. In other words, if you understand the math, the -π/2 and the π/2 asymptotes are one and the same. At the asymptote, we are looking at Psyche, Intelligence, Fermions, Pinpoints of Light, or Point Particles of Light just as has been experienced and observed by Out-of-Body Travelers and Near-Death Experiencers.

ALL of our math moves towards infinity or moves towards zero (which is another type of infinity or unity). Y = TAN(X) is one eternal round. The infinite and the asymptote points to Psyche. This means that ALL of our math points to the existence of Psyche, Intelligence, or Non-Local Consciousness. The math isn't lying! The Materialists freak out whenever their mathematical equations start moving towards infinity or towards zero; but, where else are they going to go? The mathematical equations will go to Psyche, Intelligence, and the Truth every time!

Remember, Psyche is the innate intelligence within all the different forms of energy which gives that energy the inherent ability to understand, follow, and obey God's Laws and God's Commands. Psyche or Intelligence is required for the ability to UNDERSTAND and OBEY Laws.

Most scientists are trapped by Materialism. When I told my scientist friend that Psyche is completely immaterial, he complained that that's impossible – "it has to be something," he responded. Oh, it's SOME THING! It's infinite LIGHT. It's infinite TRUTH. It's infinitely INTELLIGENT. It's infinitely ALIVE, CONSCIOUS, and AWARE. It is infinite memory storage capacity taking up NO SPACE whatsoever. It's the Ultimate Reality! It's an infinite singularity. It's the ether. It's existence! It's big enough to fill the universe, but small enough to dwell in your heart. It's a paradox, which is what makes it so powerful and what makes it come alive. It's LIVING LIGHT. It's the glory of God. It's the Light of Christ. It's everything rolled into one. It's the Spark of Life! ALL of our math keeps pointing to it!

It has been suggested and observed that there is life, consciousness, and awareness in ALL types of light – in all phases of matter. Light is alive, so technically matter is alive to one extent or another. However, THE LIGHT existing on the asymptotes IS a whole other level of life, intelligence, knowledge, awareness, unity, infinity, agency, potential, and truth. It has reached the limit. It has integrated. It has arrived! That kind of infinite LIGHT is the master and the living agent. That kind of LIGHT is Psyche and it has the innate ability to Command and to Act, rather than just reacting as matter does.

Psyche is an Infinite Singularity. At the -π/2 asymptote, the emphasis is on Singularity or Unity. At the π/2 asymptote, the emphasis is on the Infinite or Omnipresence. At the asymptotes, the LIGHT is no longer a quantum object. It's no longer matter. It transforms and transmutes. It's a living entity! It's a living Agent. It's Psyche, Intelligence, or Non-Local Consciousness. Psyche acts. Matter reacts. The ability to COMMAND and the ability to ACT is what happens to LIGHT when it reaches the asymptote or becomes the asymptote. It goes down the rabbit hole and climbs the stairway to heaven all at once! Cool, huh?

Psyche, or Intelligence, or Agency, or Non-Local Consciousness, or Quantum Consciousness is the most interesting Science to study through the scientific evidence which is provided to us by Lived Experiences or Direct Observations.

Lived Experience or Direct Conscious Observation IS the BEST and most effective type of Scientific Evidence. Through Lived Experience, or Psyche Experience, or Lived Science, we human beings go directly to KNOWING the TRUTH by living and experiencing the truth for ourselves, or by choosing to trust someone who has. Human beings or human psyches are unique, in that we can write down our Lived Experiences or Scientific Observations and then share them with other human psyches. This is the best way for doing science! It's much better and more reliable than the Scientific Method.

The Law of Psyche

Psyche has been experienced and observed by Out-of-Body Explorers and Near-Death Experiencers. The First Law of Thermodynamics tells us that Psyche or Energy is conserved, which means that it is eternal and everlasting. It cannot be made, and it cannot be destroyed. Psyche or Energy is syntropic which means that it is always conserved both at the quantum level and the physical level.

I prefer the scientific approach. I define science as observation and experience. Psyche has been experienced and observed; therefore, we KNOW that it exists. Quod erat demonstrandum!

Law of Psyche: The Law of Psyche states that every massless quantum particle is some type of Psyche or Intelligence. Psyche is the innate intelligence within all the different forms of energy which gives that energy the inherent ability to understand, follow, and obey God's Laws and God's Commands. Every massless elementary particle is some type of Psyche because it is psychic, quantum, telepathic, makes choices, can self-generate or self-propagate, can think, can communicate, can transmute, can move, can phase-shift, can collapse its own wave function, is perceptive, is sentient, is omniscient, is omnipresent, is intelligent, is conscious, is alive, must travel at the speed-of-light from our perspective at the physical level, is prescient, and without any physical limitations imposed upon it can quantum tunnel at will in its native original environment at the quantum level. Remember, every massless quantum particle is some type of Psyche, or Intelligence, or Life Force. Psyche is seen by Out-of-Body Travelers as a pinpoint of light or a Photon. Each Photon is a type of Psyche or Intelligence. A Psyche is a Photon, a massless elementary particle. If it looks like a Photon and acts like a Photon, then it is some type of Psyche. Every massless quantum is a Psyche. Psyche USES mass, energy, or matter to get things done. This is what has been experienced and observed.

Psyche has been hiding in plain sight all the time where nobody can see it nor find it because they aren't looking for it; but, there's no great mystery here. Psyche is seen as a Photon BOTH at the quantum level and the physical level because it is a Photon. Psyche or Intelligence IS what it appears to be, a Photon of Light or a Pinpoint of Light. No surprise there. Psyche is what it has been experienced and observed to be – a massless quantum of energy that looks like and acts like a Photon. Psyche, whenever it has been seen both at the quantum level and the physical level, looks like a Photon or a Pinpoint of Light.

I know that it's hard to believe because we have been trained, brainwashed, and conditioned all of our lives not to believe it; but, it is true, nonetheless. Psyche has been experienced and observed. Every massless quantum or every massless elementary particle looks like a photon and acts like a photon because is a Controlling Psyche who USES energy, mass, or matter to get things done both at the quantum level and the physical level.

Mark My Words

Lived Experiences

Lived Experiences ARE the pinnacle of knowledge and truth. There's NO better way for finding and knowing the TRUTH than Lived Experience or Direct Observation.

Science is supposed to be Observation, or Lived Experience, although for most people Science is treated as if it is nothing more than philosophy or sophistry, and they limit Science exclusively to our physical reality. In other words, these people use Science to try to deceive us and to deceive themselves as well; and, they are extremely successful at doing so. Self-deception works, and it works every time. That type of limiting, materialistic, and exclusive "science" really isn't Science – it is sophistry, metaphysics, religion, and philosophy. Science based upon falsehoods such as Materialism and Naturalism are the very definition of BAD SCIENCE or pseudo-science.

In contrast, the TRUTH is KNOWN by living it, witnessing it, and experiencing it for yourself, or by choosing to trust someone who has done so for himself.

By Lived Experiences, we mean ALL of our psyche experiences, thoughts, dreams, choices, decisions, actions, spiritual experiences, conscious experiences, memories, out-of-body experiences (OBEs), mystical experiences, psychic experiences, physical experiences, science experiments, unexplainable "miraculous" events, healing, near-death experiences (NDEs), shared-death experiences (SDEs), theophanies, revelations of God, and revelations from God. Through Lived Experience, we can KNOW the TRUTH directly by experiencing it directly for ourselves.

We KNOW from the Lived Experiences of the human race that Psyche exists and that our Psyche or Intelligence is something completely different than our Spirit Body. We also KNOW from the Lived Experiences of the human race that our Human Psyche or Non-Local Consciousness has been assigned to a Spirit Body or united holistically with a Spirit Body, which looks like our Physical Body. What people call our "Spirit", or our "Ghost", is in fact our Psyche or Personality which has been united with a Spirit Body as a single functioning unit or whole. We KNOW from the Lived Experiences of the human race that Psyche can temporarily separate from its Spirit Body and look at its Spirit Body from an immaterial third-person viewpoint in space.

This is a Psyche Ontology. This is a holistic ontology or a holistic model of reality which takes into consideration ALL of our Lived Experiences as a race. The Human Psyche is unique in that it can record and then share its Lived Experiences with other human beings. The animals can't. This Psyche Ontology, wherein Psyche or Intelligence or Non-Local Consciousness is the fundamental unit of reality, is the Ultimate Ontology and the Ultimate Model of Reality. ONLY Psyche can do ontology, reality, choice, and existence. Without Psyche or Non-Local Consciousness, there would be no existence.

Due to space limitations the rest of this essay was removed. The complete version is found in my book, ***Tripping the Light Fantastic: How Prescription Drugs Almost Killed Me***, where it serves as an introduction to a whole section on the topic of Lived Experiences: https://www.amazon.com/dp/B071RJP9T8

Benefit from the information in that book; and, have a great day and a great life!

Mark My Words

One-Time Events

One-Time Events are often described as "Acts of God"; whereas, ordinary physical events are common and even at times predictable. Events have causes. Events are caused.

The Materialists, Naturalists, Darwinists, Nihilists, Atheists, and especially the Behaviorists are BIG on prediction and control. For these people, it's all about prediction and control.

A physical reality makes replication, prediction, and control possible, because a physical reality is the Ultimate Consensus Reality. A physical reality was designed by God to be dependable, replicable, predictable, and controllable. There are REAL consequences in a physical reality. Science experiments are actually replicable, dependable, and possible in a physical reality.

In contrast, there's NO way to predict and control One-Time Events. There's NO way to predict and control spiritual events, supernatural events, or quantum events. There's NO way to predict and control God, Jesus Christ, and the Holy Ghost. There's NO way to predict and control the Human Psyche.

There was NO way to predict that God the Father and Jesus Christ would appear to Joseph Smith in-person and answer his questions in-person. Likewise, since it was a One-Time Event and a Supernatural Event, there was NO way to control it or prevent it either.

When dealing with Quantum Mechanics or supernatural events, prediction and control go out the window and cease to exist. It's as if these various different quantum mechanisms have a mind of their own. It's as if the various different quantum objects and spirit matter have a mind of their own.

It has been observed that huge chunks of the Quantum Realm, Syntropy Realm, Supernatural Realm, Spirit Realm, or Transdimensional Realm are non-consensus realities where prediction is impossible. Out-of-Body Travelers have observed that whenever they enter into a non-consensus reality, it reforms itself to fit the demands of their Psyche. In a non-consensus reality, the Human Psyche controls and determines what's going to happen and what's going to manifest or become real; and, there is no way for an outside observer or a second person to guess or predict what that might be. In a non-consensus reality in the Spirit Realm, the Human Psyche is God. There's NO way to predict and control the Human Psyche. Every entry into a non-consensus reality by a Human Psyche ends up being a One-Time Event. There's no way to predict, and control, and maintain One-Time Events.

Can you see now why God designed and created a physical reality for us? He did it in order to infuse a measure of order, organization, prediction, reliability, stability, and control into our lives. Consensus, Unity, Standards, Commandments, and Laws provide us with an ordered, organized, replicable, dependable, semi-predictable, and semi-controllable existence for us to experience and learn to enjoy. A physical reality is the Ultimate Consensus Reality. As long as you are alive, and your psyche is tied to or bound to your physical body, you can depend upon your physical body being there tomorrow when you go looking for it. That's the advantage of a Consensus Reality. It provides us with a bit of prediction and control.

Mark My Words

The Altruistic Gene

Do we have altruistic and selfish genes?

> **DNA neither knows nor cares. DNA just is. And we dance to its music.** — Richard Dawkins

According to this quote, there's no such thing as a selfish gene or an altruistic gene. Genes don't care one way or the other whether we live or die.

The same Richard Dawkins wrote a best-seller entitled "The Selfish Gene". Within the title of his book, he personifies the gene giving it personhood and personality. He turns the gene into some sort of psyche or non-local consciousness. He imbues it with a desire to survive and to be selfish.

> **Let us try to teach generosity and altruism, because we are born selfish. Let us understand what our own selfish genes are up to, because we may then at least have the chance to upset their designs, something that no other species has ever aspired to do.**

> **We are survival machines – robot vehicles blindly programmed to preserve the selfish molecules known as genes. This is a truth which still fills me with astonishment.** — Richard Dawkins, *The Selfish Gene*

Can a gene be selfish? Does a gene have a motivational imperative to survive and replicate into the future? Are our selfish genes up to something as he claims? Do our selfish genes have designs and plans? Do our genes make music? Can molecules be selfish? Since according to Richard Dawkins, we have selfish genes, does that mean that we also have helping genes or altruistic genes? Everything has its opposite after all.

Does the title of Dawkins' book have any foundation in reality, or is it nothing but science fiction? The theory of evolution and the evolutionary perspective permeate the whole of science; but, do they have any basis in reality or are they simply a fictional story?

Do you sense the inconsistency within these quotes?

In order for genes to be selfish, they have to know and care! His work is sloppy, incoherent, inconsistent, and inconstant.

So, which is it? Are the genes selfish, or do the genes neither know nor care what's going on? It can't be both. If the one is true, then the other is false.

It's my contention that the theory of evolution is a fictional story that they made up out of thin air. My observation or experience is that everything Richard Dawkins writes contains within it the seeds of its own destruction or the seeds of its own falsification. The same reality applies to these particular quotes.

If we are simply dancing to our DNA as Dawkins claims and our DNA is selfish, then why should we bother to try to teach generosity and altruism to our children? Aren't we already pre-programmed to be selfish and have no other choice in the matter? Aren't we just robots and survival machines for our genes as he claims?

You see, he is inconsistent and contradictory here! There's no coherence to his arguments. Either we are robots, or we are individual psyches capable of teaching our children new tricks that have never been learned before. Which is it?

"Let us understand", "let us try to upset their designs", and "let us aspire to do something that no other species has done before". Who is this "us" that he keeps talking about? It isn't our genes! According to Dawkins, it's our genes that we are trying to overcome. So, who is this "us" that's trying to overcome its selfish genes?

"We are born selfish". "We are survival machines". Who's this "we"?

This "us" and "we" are Human Psyches. Human Psyches are the things that are trying to overcome their selfish genes – mind over matter. We also KNOW that it is Someone Psyche or Someone Intelligent who designed and created the proteins and their matching genes in the first place, because functional genes and proteins have never been observed spontaneously generating out of thin air.

By carefully analyzing the theory of evolution, it quickly becomes obvious and apparent that it is nothing but a fictional story that they made up out of thin air as they went along. The internal contradictions prove that it is so. In one book, he is talking about DNA that neither knows nor cares; and, in another book he is talking about selfish genes that have to know and care or they wouldn't be selfish to begin with. In one book, the genes are dead and inert; and, in the other book the genes are dynamic and alive. So, which is it? Which one is the truth, and which one is the falsehood? They can't both be true because they are mutually exclusive. The theory of evolution is riddled with such inconsistencies all throughout the whole of it. Can you sense the significance of these observations?

The theory of evolution has some serious holes in it. Notice that the theory of evolution is in dispute, but the existence of the genes is not. The scientific community should abandon the theory of evolution and go with the genes instead, because the genes have been experienced and observed. Science is supposed to be about observation and experience, not wishful thinking and philosophical speculation.

Dawkins' "truths" are contradictory, inconsistent, and self-defeating. Do we even have selfish genes; or, is that a fictional story too? What do you think? It doesn't matter what I think. What truly matters to you is what you choose to think and believe! That's the thing that has an impact on your life and destiny.

Now, I'll tell you what I think.

—

The Evolutionary Perspective

Myers, D. G. (2010). *Social Psychology* (10th ed.). New York: McGraw-Hill.

In his chapter about Helping Others, Myers has the obligatory section dedicated to the evolutionary perspective and evolutionary psychology. The implication is that we have helping genes or altruistic genes that are motivated to preserve and protect our kin. Do we have altruistic genes interested in preserving our kin, or is that claim nothing but science fiction? We either do or don't, so which is it?

After having falsified and debunked the Theory of Evolution thousands of times, I no longer feel generous, kind, and altruistic towards such a section within my science books because I know why it's false and I know that they are trying to deceive us. My attack is swift and decisive. I don't have an altruistic gene in my body where the theory of evolution is concerned.

Materialism, Naturalism, Darwinism, Nihilism, Classical Physics, Behaviorism, Determinism, Physical Reductionism, Scientism, Atheism, and the Theory of Evolution are based exclusively on entropy. They all reduce to entropy. Evolution of any type is based upon entropy. Evolution IS entropy. Entropy is death and extinction.

There's no way in the universe that entropy or death is going to lend a helping hand or show an altruistic streak. We don't have an altruistic gene any more than we have a selfish gene.

I'm not generous and kind to the theory of evolution because it's a fictional story, and not a very good one at that. The whole thing is stupid and illogical.

Let's think about it logically for once because nobody else does.

A genome is both hardware and software. A genome is a computer. Does your computer have a selfish gene? Does your computer have an altruistic gene? Does your computer do kinship selection? Do we have selfish computers and altruistic computers? Of course not. Your computer is based upon physical matter or entropy; and, so is your genome. Entropy is death. Entropy doesn't have emotions. Entropy doesn't have selfish genes nor altruistic genes. Entropy has no motivational imperative to survive. Entropy doesn't show favoritism. Entropy doesn't care which one is the fittest. Entropy doesn't do selection in any active sense. The consequences of Natural Selection are death and extinction. When you get selected against, you die. Natural selection doesn't touch our genes.

Your computer and your genome don't care about you one way or the other. They have no motivations, desires, or feelings. They have no psychic imperative or personal motivation to survive. Instead, entropy eventually does them in, and they have no feelings about their death or extinction one way or the other. This is a genome or a computer that we are talking about, and not a human being or a human psyche. Your genome doesn't have any selfish genes, nor does it have any altruistic genes either. Selfishness and altruism are a function and a product of the Human Psyche, and not our genes. Genes, computers, and physical matter are based upon entropy or physical matter. Entropy or evolution doesn't care whether you live or die.

Did your computer spontaneously generate out of thin air, or was your computer designed, created, engineered, and manufactured by Someone Psyche? Obviously, Someone Intelligent made your computer. Well, the same exact truth and reality applies to your genome. Your genome is God's Signature. Genomes don't just spontaneously generate out of thin air. Genomes are designed, programmed, engineered, field-tested, fine-tuned, created, manufactured, and deployed. The same reality applies to your computer.

Your genome is both hardware and software – a computer. Your genome and your computer don't have feelings, motivations, nor desires. They don't have selfish genes nor altruistic genes. Your genome and your computer don't do kinship selection or kinship favoritism. They are based upon entropy or physical matter; and, entropy cannot design and create and effectively do anything. Entropy is death and extinction. Natural selection or survival of the fittest eventually leads to death and extinction. Random mutations are entropy. Evolution is death and extinction. Evolution is entropy.

Saying that the theory of evolution is true doesn't make it true. Saying that most scientists believe in evolution doesn't make the theory of evolution true, either. Most scientists are wrong. Evolution or entropy cannot design and create anything. The theory of evolution is Creation by Evolution or Creation by Entropy. Saying that natural selection or random mutations have been observed doesn't make Creation by Entropy or the Theory of Evolution true. Random mutations are entropy, and entropy cannot design and create anything whatsoever.

Here among all of their favorite fictions, we finally have the truth. Evolution or entropy cannot design and create; and, evolution or entropy doesn't care whether you live or die. There it is in black and white.

One of the most useful and truthful scientific discoveries that I encountered is a concept called Genetic Entropy by John Sanford.

Sanford, J. (2014). *Genetic Entropy* (4th ed.). Cornell University: FMS Foundation.

This book uses science and common-sense logic to falsify the theory of evolution. This book teaches that entropy cannot design and create; and, physical matter including our genome is based upon entropy. Genomes don't create themselves because they can't. They are prevented from doing so by entropy.

It should have been called The Genetic Perspective and not the evolutionary perspective. The genes are real and true – the theory of evolution is not. The theory of evolution is a *fictional ad hoc just-so story* that they made up out of thin air after-the-fact. The theory of evolution doesn't exist as any kind of physical entity, physical object, or person; but, our genes have been experienced and observed. The one truly exists; and, the other one has been falsified. The theory of evolution is in dispute. The genes are not. It should have been called the Genetic Perspective and not the evolutionary perspective.

—

Evolutionary Psychology

Let's critique the following quote about evolutionary psychology. I learn best through comparison and contrast.

Kin Selection: The idea that evolution has selected altruism toward one's close relatives to enhance the survival of mutually shared genes.

Evolutionary Psychology

Another explanation of helping comes from evolutionary theory. Evolutionary psychology contends that life's essence is gene survival. Our genes drive us in adaptive ways that have maximized their chance of survival. When our ancestors died, their genes lived on, predisposing us to behave in ways that will spread them into the future.

As suggested by the title of Richard Dawkins's (1976) popular book *The Selfish Gene*, evolutionary psychology offers a humbling human image — one that psychologist Donald Campbell called a biological reaffirmation of a deep, self-serving "original sin." Genes that predispose individuals to self-sacrifice in the interests of strangers' welfare would not survive in the evolutionary competition.

Genetic selfishness should, however, predispose us toward two specific types of selfless or even self-sacrificial helping: kin protection and reciprocity.

KIN PROTECTION

Our genes dispose us to care for relatives. Thus, one form of self-sacrifice that *would* increase gene survival is devotion to one's children. Compared with neglectful parents, parents who put their children's welfare ahead of their own are more likely to pass their genes on. As evolutionary psychologist David Barash wrote, "Genes help themselves by being nice to themselves, even if they are enclosed in different bodies." Genetic egoism (at the biological level) fosters parental altruism (at the psychological level). Although evolution favors self-sacrifice for one's children, children have less at stake in the survival of their parents' genes.

Thus, according to the theory, parents will generally be more devoted to their children than their children are to them. Other relatives share genes in proportion to their biological closeness. You share one-half your genes with your brothers and sisters, one-eighth with your cousins.

Kin selection — favoritism toward those who share our genes — led the evolutionary biologist J. B. S. Haldane to jest that although he would not give up his life for his brother, he would sacrifice himself for *three* brothers — or for nine cousins. Haldane would not have been surprised that genetic relatedness predicts helping and that genetically identical twins are noticeably more mutually supportive than fraternal twins. In one laboratory game experiment, identical twins were half again as likely as fraternal twins to cooperate with their twin for a shared gain when playing for money.

The point is not that we calculate genetic relatedness before helping but that nature (as well as culture) programs us to care about close relatives.

Some evolutionary psychologists note that kin selection predisposes ethnic in-group favoritism — the root of countless historical and contemporary conflicts. E. O. Wilson noted that kin selection is "the enemy of civilization. If human beings are to a large extent guided . . . to favor their own relatives and tribe, only a limited amount of global harmony is possible". (*Social Psychology*, pp. 452-453.)

Is any of this true, or is it all science fiction?

I don't see any truth within it. It all seems to be correlational. Correlation is NOT causation!

Remember, Kin Selection Theory is correlational. There's NO kin selection, or helping hand, or favoritism taking place between monozygotic twins who were separated at birth and never see each other again. Altruism or helping others is better explained by proximity, the reciprocity norm, and the social responsibility norm than it is by kin selection. These social norms are a function of the Human Psyche and have nothing to do with our genes. Our genes can't make us help other people, and our genes have NO way of knowing which genes are related to them and which genes are not. In order for the genes to know and care, there would have to be a telepathic connection between all the genes and some sort of Psyche or Intelligence within those genes broadcasting and receiving those preferences. If such a psychic system exists for real, then its very existence falsifies

Materialism, Naturalism, and the Theory of Evolution. Kin Selection Theory is nothing but a fictional ad hoc story that they made up out of thin air in hindsight after-the-fact in an attempt to explain the observational evidence. That's the way the whole Theory of Evolution works. It's a fictional story and nothing more.

Selection is some type of choice. How does evolution make choices? It can't. Evolution isn't alive. There's no psyche or intelligence there within evolution. Evolution or entropy can't make choices. Evolution or entropy can't do selection!

How does a gene know that those other genes are its close relatives? Is there some kind of telepathic connection between the genes letting them know that they are related to each other? Is there some kind of psyche within a gene making it altruistic towards its close relatives and desirous for their survival? If there is, then we have just falsified Materialism, Naturalism, and the Theory of Evolution. The very existence of selfish genes and altruistic genes would falsify the Major Premises of Materialism, Naturalism, Darwinism, Nihilism, and Atheism which claim that psychic imperatives or motivational imperatives do not exist. The theory of evolution is self-defeating because the genes don't care one way or the other whether we survive, or they survive. The genes have no way of knowing which genes are related to them and which genes are not.

Our genes don't drive us! Driving implies some sort of intelligence or psyche. Our genes don't predispose us to do anything. Predisposition implies some sort of psyche or intelligence with foresight and purpose and goals in mind. Disposition implies some sort of preference or choice. Genes or physical matter can't make choices. If your genes do in fact have intelligence or psyche, then we have once again falsified Materialism, Naturalism, Darwinism, Nihilism, and Atheism. Materialism, Naturalism, and Darwinism are self-defeating. They have to give these things psyche or intelligence; or, they can't do anything. Yet, if they give Evolution and Naturalism psyche, intelligence, teleology, motivational imperatives, and purpose, then they automatically falsify Materialism, Naturalism, Darwinism, Nihilism, and Atheism.

The theory of evolution and natural selection do not work as advertised. Heroism and self-sacrifice are chosen into existence by the Human Psyche and NOT the genes. We aren't driven by our genes. Our physical body is driven by the Human Psyche.

Our genes do NOT predispose us to care for relatives. Our genes have no way of knowing who our relatives are. What in reality predisposes us to care for our relatives? It's the very fact that our parents protected us, and we were raised with our siblings, which predisposes us to care for them. It's a cultural thing, and not a genetic thing. Proximity is necessary for caring. We Human Psyches don't care about the things that we don't know about. Caring is a function of the Human Psyche and not our genes. Self-sacrifice produces love. We love the people most for whom we sacrifice the most. We care about our friends and the people that we work with because we spend time with them. It has nothing to do with our genes!

"Genes help themselves by being nice to themselves." They personify the genes – turn them into little people. It's a story. It's fiction! He's telling us that the genes are little Who's living in Whoville. If the genes are in fact little people or little psyches, then we have in fact falsified the Major Premises of Materialism, Naturalism, Darwinism, Nihilism, and Atheism which claim that psyche or non-local consciousness does not exist. If there really are selfish genes and altruistic genes, then their very existence falsifies Materialism, Naturalism, and the Theory of Evolution which claim that our genes don't have intelligence or psyche and don't care whether we live or die. Materialism, Naturalism, Darwinism, and their derivatives are self-defeating or self-falsifying. These falsified philosophies are internally incoherent. They are nothing but fictional stories, and bad ones at that.

It's all correlational! We are mutually supportive of the people that we live with and work with due to proximity. It has nothing to do with our genes! Identical twins separated from birth have done NO mutual support because they never made contact with each other. Caring (and hating) is due to proximity and NOT our genes. The genes have no way of knowing who their relatives are; but, Human Psyches certainly do.

Our genes can't guide us. They aren't Prophets and Gods! Our genes can't do selection. Selection of any kind involves choice; and, the genes or physical matter can't do choice. Choice or selection has nothing to do with our genes. Our genes don't do our thinking and choosing for us! Evolution is entropy. Evolution and genes don't make choices for Human Psyches. Evolution or entropy cannot do choice. Physical matter or entropy cannot do choice. Choices are chosen into existence by the Human Psyche.

Nature and culture influence us, for sure; but, they do not and cannot program our choices into us. Choice is a function of the Human Psyche, and our genes and society do not choose our beliefs and desires for us. Some people choose to hate their close relatives the most. It has nothing to do with our genes and everything to do with choice and proximity. If kin selection were true, then Cain would never have killed Abel. If the theory of evolution were true, then there would be no homosexuality on this planet. Survival of the fittest would have seen to their extinction.

Materialism, Naturalism, Darwinism, and Survival of the Fittest are the root of countless historical and contemporary conflicts. Fascism and Nazism were based upon Darwinism or Survival of the Fittest. Communism is based upon Militaristic Atheism or Enforced Atheism. In *V for Vendetta*, one of the characters asked why the Fascists hate the homosexuals so much that they are willing to exterminate them. It's because the homosexuals falsify Darwinism, Naturalism, Survival of the Fittest, and the Theory of Evolution. Their very existence falsifies the very foundation of Fascism or Darwinism.

Again, empathy or altruism comes down to proximity and the Human Psyche. It has nothing to do with our genes! Our genes don't care one way or the other if we survive or die. It's the Human Psyche who is worrying about whether it is going to survive or die – not our genes. Empathy and altruism are best explained by Psyche, Syntropy, and Quantum Non-Local Consciousness because physical matter and genes cannot do choice, empathy, altruism, or love.

Mark My Words

—

Source

The Scientific Method Proves That the Theory of Evolution Is False

> https://www.amazon.com/dp/B01IAAIRT2

Source Material

Dawkins, R. (1976). *The Selfish Gene*.

Myers, D. G. (2010). *Social Psychology* (10th ed.). New York: McGraw-Hill.

The Ultimate Model of Reality

Technically, the Naturalists, Physicalists, Atheists, Behaviorists and Determinists are right whenever they emphatically state that physical matter and entropy cannot do CHOICE. These people are right; but, they don't know why they are right because they don't understand science.

Materialism, Naturalism, Darwinism, Nihilism, Classical Physics, Behaviorism, Determinism, and Atheism cannot explain choice. These people have NO scientific explanation for choice. Instead, they simply claim that choice does not exist; and, they leave it at that. They sweep the problem under the rug and choose to ignore it because they can't explain it. Their problem comes from the fact that physical matter or entropy cannot do choice; and, these people deliberately and knowingly limit themselves and their science exclusively to physical matter or entropy.

Physical matter and entropy cannot do choice. They simply do what they are told to do.

Told by whom?

Told by some kind of Psyche, Syntropy, or God! Remember, the physical laws – including entropy – were chosen into existence by God's Psyche or God's Intelligence BEFORE our physical universe began and BEFORE our physical universe was brought into existence by God. Physical matter just does what it is told to do by Someone Psyche. Physical matter and entropy have no choice. They just do what they are told.

Physical matter and entropy cannot do choice; but, Psyche or Syntropy certainly can. Quantum Mechanics or Supernatural Mechanics is based upon choice. Choice is Syntropy. Some kind of Psyche or Consciousness is needed within Quantum Mechanics to make the choice and collapse the wave function. It's some kind of Psychic Choice that collapses the wave function both here In the physical realm and there in the quantum spiritual realm. It all comes down to choice! Even the physical laws and entropy and physical matter were chosen into existence by Someone Psyche. Some type of Syntropy has to exist, or entropy and physical matter would not exist. This is Science 101.

By choosing to include Psyche, Syntropy, and Quantum Mechanics into science, instantly we can use science to explain CHOICE! If we choose to define "science" as Psyche, Syntropy, and Quantum Mechanics then instantly we can use science to explain everything that comes our way on both sides of the veil, including choice.

Powerful, is it not?

This just might be one of the greatest scientific discoveries of all time.

Our interpretations or our definitions really matter.

If we choose to define "science" as Materialism, Naturalism, Darwinism, Nihilism, Behaviorism, Entropy, Determinism, Physical Matter, Classical Physics, and Atheism, then we will NEVER have a scientific explanation for choice.

If we choose to define Science as Psyche, Syntropy, Intelligence, and Quantum Mechanics, then instantly we have a scientific explanation for choice. It's fascinating how that works!

Mark My Words

—

What is the greatest thing that we can learn from science or observation?

We are, in truth, more than half what we are by imitation. The great point is to choose good models and to study them with care.

LORD CHESTERFIELD, LETTERS, JANUARY 18, 1750

The Great Point or the Great Truth is that the Human Psyche chooses things into existence on both sides of the veil.

We are talking about a Psyche Ontology here, wherein Psyche is the fundamental unit of reality and existence. Psyche is the Ultimate Cause, meaning that Psyche is the ultimate causal agent. This is the Ultimate Model of Reality.

It's the Human Psyche who chooses to do imitation or chooses to rebel. It's the Human Psyche who chooses what models to follow and studies them with care. Only the Human Psyche can choose to imitate someone. Imitation can only be explained by Psyche or Syntropy.

When it comes to imitation, your genes aren't doing any of this for you! Your society isn't doing this for you either. You (your Psyche) are choosing what you want to believe and whom you want to model, study, follow, believe, trust, and imitate.

Once again, when it comes to imitation, Psyche provides the best scientific explanation for what's really going on.

Wallenberg became Swedish ambassador to Hungary, where he saved tens of thousands of Hungarian Jews from extermination at Auschwitz. One of those given protective identity papers was 6-year-old Ervin Staub, now a University of Massachusetts social psychologist whose experience set him on a lifelong mission to understand why some people perpetrate evil, some stand by, and some help. (*Social Psychology*, p. 478.)

It's all a function of choice – a function of the Human Psyche! Society can't make us perpetrate evil, stand by, or help; and, choice certainly has nothing to do with our genes because our genes can't make any choices. Physical matter and molecules don't make any choices. They simply react to the physical laws when there's no Psyche there making the molecules do something else besides what they would otherwise be doing when left to themselves.

Personality Traits

We also need to consider the helpers' dispositions, including, for example, their personality traits and religious values. Surely some traits must distinguish the Mother Teresa types from others. Faced with identical situations, some people will respond helpfully, while others won't bother. (*Social Psychology*, p. 469.)

Notice that he said that the situation is the same!

The physical situation is the same, but the chosen response is different!

Where are these choices coming from?

Our genes can't make choices! Physical matter and entropy cannot do choice. Our society or environment can't make our choices for us either. Our society can force us to do things against our will, but it can't force us to believe that those things are good for us. Our beliefs are chosen into existence by the Human Psyche. They don't exist until they are chosen into existence.

Once again, we are looking at the Human Psyche for an explanation for what we are observing – when faced with the same identical situation, some people choose to respond helpfully, and others choose not to.

We are looking at the Human Psyche who can make choices and does make choices. Once again, Psyche or Syntropy ends up providing us with the best explanation for what's really going on because physical matter or entropy cannot do choice.

According to the Physicalists, Materialists, Naturalists, Darwinists, Nihilists, Behaviorists and Determinists, there is no such thing as choice because physical matter can't do choice; and, physical matter is the only thing that exists according to these people. Materialism, Naturalism, Classical Physics, and Behaviorism have absolutely NO scientific explanation for choice. Psyche, Syntropy, and Quantum Mechanics certainly do!

Quantum Mechanics is based upon choice. Choice is needed to collapse the wave function; and, Someone Psyche provides the choices that Quantum Mechanics needs.

Fascinating, is it not, what we can achieve when we finally choose to allow ALL of the evidence into evidence?

Well, I think it is because I used to be a Materialist, Naturalist, Nihilist, and Atheist until I actually started looking at and studying the scientific evidence.

The Human Psyche can and does trump social pressure at will. Most of our scientists want us to be Materialists, Naturalists, Darwinists, Nihilists, and Atheists. I was until the evidence convinced me that I was wrong.

One day I decided that I didn't like where My Atheism was taking me, so I chose to turn around and go the other way. My genes didn't make that choice for me. They can't make choices. They are computer programs and simply do what they are told. My society and atheistic friends online didn't make that choice for me either. I made that choice. My Psyche made that choice.

CHOICE is Proof of Psyche no matter where we might happen to encounter it – both here in the physical realm and there in the quantum spiritual realm as well. Now, that's powerful science that can be used to explain everything that comes our way. The choices that Psyche makes are what collapse the wave functions, when it comes to Quantum Mechanics or Supernatural Mechanics.

The Human Psyche makes choices; and, the Human Psyche can trump or override its nature and nurture at will.

Mark My Words

—

Source Material

The Ultimate Model of Reality: Psyche Is the Ultimate Cause

https://www.amazon.com/dp/B071NC9JK6

Myers, D. G. (2010). *Social Psychology* (10th ed.). New York: McGraw-Hill.

A Visual Representation of Syntropy

Whether we realize it or not, we actually have a visual representation of Syntropy within Light. Everything is comprised of Light or Quantum Waves.

1 John 1: 5: "This then is the message which we have heard of him, and declare unto you, that God is light, and in him is no darkness at all." **God is Psyche, Syntropy, or Light.**

Spirit is Light; and, Light is Spirit. Spirit matter is a type of light. Dark matter or spirit matter has been proven to exist. Dark matter or spirit matter has been there all the time just waiting for someone to discover it. The proven or verified existence of dark matter and dark energy falsifies Materialism, Naturalism, Darwinism, Nihilism, and Atheism because dark matter and dark energy are non-physical and are invisible to our physical instruments.

God is called the Father of Lights. A spirit body is comprised of spirit matter or light. All matter is a type of light.

Gravity, magnetism, the strong and weak nuclear forces, dark energy, the zero-point field of light, the quantum zeno effect, and Psyche are non-physical invisible Forces and Fields. They are comprised of different types of quantum waves or light. The proven and verified existence of these invisible and non-physical Forces and Fields falsifies the assertions of Materialism, Naturalism, Darwinism, Nihilism, Behaviorism, Physical Reductionism, and Atheism which claim that these invisible and non-physical Forces and Fields do not exist. The verified and proven existence of Light or Quantum Waves falsifies Materialism, Naturalism, Darwinism, Nihilism, and even Atheism.

ALL of this science already exists, meaning that it has already been experienced and observed. Even Psyche has been experienced and observed by Near-Death Experiences and Out-of-Body Exploration. The science is real. It truly exists. It's just been sitting there waiting for someone to come along and unite it all into one functional whole that makes logical sense when we are done with it.

So, how does physical matter fit into all of this?

Physicist David Bohm has stated that physical matter is condensed light or frozen light. Its frequency has been changed by God; and, it has been slowed down to sub-light speeds and made subject to entropy or space-time. Entropy is a function of time. Physical matter is entropic light. The theory of relativity teaches us that light or matter existing faster than the speed-of-light experiences NO passage of time, meaning that it experiences NO entropy. The science is already in place to explain all of this – just waiting for someone to come along and see it as it really is.

God creates physical matter from spirit matter, by slowing down the spirit matter, changing its frequency or phase, and making it subject to space-time or entropy. God deliberately forces the physical matter to reside at sub-light speeds. Physical matter is entropic light, or entropic spirit matter. Physical matter has been made subject to entropy or the passage of time. Entropy is a function of time. If you stop the passage of time – stop the aging process – then there is NO entropy. Then everything becomes Syntropy. Physical matter is based upon entropy. Everything else is based upon Syntropy.

We human beings will NEVER be able to push physical matter faster than the speed-of-light because God designed physical matter to reside within space-time, to be subject to entropy, and to be restricted to sub-light velocities. The only way to get physical matter, or any type of matter, to move faster than the speed-of-light is through Quantum Tunneling or Teleportation.

I have experienced Quantum Tunneling or Teleportation. It's real. It truly exists. God teleported me and the car I was driving to safety. Time stopped. I felt the passage of time stop; and, then I was someplace else instantly. It was nothing that I did. It was something that God did for me in order to save me. Quantum Mechanics is a type of Syntropy. God can Quantum Tunnel or Teleport physical matter at will. Since God created the physical laws and the physical matter in the first place, God still retains the ability to Quantum Tunnel that matter anywhere He wants it to be.

Thankfully, God deliberately places limitations or restrictions on the quantum tunneling of physical matter where your physical body is concerned, thereby preventing the atoms within your physical body from quantum tunneling away from you at will. Physical matter is given a very short De Broglie Wavelength thereby locking it or localizing it basically in-place for the duration of its existence. Atoms and electrons can quantum tunnel through small barriers as any scientist knows; but, God prevents them from quantum tunneling huge distances at will. Nevertheless, God can quantum tunnel the atoms within your physical body across the universe instantly at will because God retains the KEYS to that ability or power. God is the one who created the physical restrictions in the first place.

As Out-of-Body Travelers and Near-Death Experiencers have observed, under the command and control of the Human Psyche, spirit matter or a spirit body can Quantum Tunnel anywhere it wants to go at the speed of thought. Quantum Mechanics or Syntropy is the NORM; whereas, entropic physical matter is the exception to the NORM. Do you see how that works? That's the sum-total of Science in a nutshell.

Psyche or Quantum Non-Local Consciousness is a type of Living Light. Consciousness or Life is a type of Syntropy or Light.

Everything is Light or Syntropy from bottom to top.

Syntropy means eternal and everlasting, without a beginning of days or and end of years. Syntropy is the opposite of entropy. Entropy is disorder, aging, and death. Entropy is a function of time. If you stop the passage of time – stop the aging process – then there is NO entropy. Syntropy means NO aging and NO passage of time, which means NO entropy.

It all fits together perfectly once we have the proper definition for everything – once we have gotten our science straight.

We KNOW that some type of Syntropy must exist, or all of that subsequent entropy wouldn't have been possible. Since we KNOW that Syntropy must exist, our task as scientists is to learn how to identify it whenever we come across it. Quantum Mechanics or Supernatural Mechanics is a type of Syntropy. Quantum Waves are Syntropy. Syntropy is Psyche, Spirit, and Light. It all comes down to Light or Syntropy.

Other Examples of Light or Syntropy

I find it interesting how the truth is always consistent with all the other truths that we discover and embrace. It all fits together perfectly into one united whole.

Tachyons

For a very long time, I thought I was the ONLY one on the planet who realized that Spirit Matter exists at velocities faster than the speed-of-light and is therefore undetectable by physical instruments directly.

Last night 29JUL2017, I was reading in my most favorite introduction to Popular Physics or Speculative Physics – a book entitled *Physics of the Impossible* by Michio Kaku – the following description for Tachyons.

> **Tachyons live in a strange world where everything travels faster than light. As tachyons lose energy [mass], they travel faster, which violates common sense. In fact, if they lose all energy [or mass], they travel at infinite velocity. As tachyons gain energy [or mass], however, they slow down until they reach the speed-of-light. (p. 280).**

That's the exact SAME definition for Spirit Matter that I have been giving in all of my books for the past year or so.

Tachyons ARE Spirit Matter!

Of course, Michio Kaku doesn't realize that Tachyons are Spirit Matter, because as a Materialist and Physicalist he isn't looking for Spirit Matter. But, I instantly recognized the definition for Spirit Matter that I have been providing for the past year in my books within the definition that Michio Kaku provided for Tachyons.

Cool, huh?

Well, at least I thought it was cool.

Tachyons live in a strange world that we have been calling the Non-Local Realm or the Spirit Realm. As tachyons or spirit matter lose energy, their mass or resistance to acceleration decrease, which makes perfect sense to me. In fact, if the tachyons or spirit matter were to lose ALL of their mass, they would become immaterial Psyche and exist at an infinite velocity being instantaneously and simultaneously everywhere in their native original format.

Remember, ZERO MASS = Infinite Velocity = Quantum Tunneling = Teleportation. MASS or Inertia is resistance to acceleration. No resistance to acceleration results in an infinite velocity. It's always made perfect sense to me, ever since I learned about it.

Furthermore, according to the theory of relativity, whenever a quantum object is moving faster than the speed-of-light or has a frequency faster than the speed-of-light, then it experiences NO passage of time meaning that it experiences NO entropy. Everything existing faster than the speed-of-light is Pure Syntropy. Quantum Mechanics is Syntropy. Psyche is Syntropy. Syntropy is the opposite of entropy. We KNOW that Syntropy must exist, otherwise all of that subsequent entropy wouldn't have been possible.

Light and Truth Are Synonymous

The Light has been experienced and observed by Someone Psyche. Psyche, Intelligence, the Life Force, or Quantum Non-Local Consciousness is a type of Syntropy or Light. Seeing the light is the same thing as finding and knowing the truth.

> **Urim and Thummim: A Hebrew term that means "Lights and Perfections." An instrument prepared of God to assist man in obtaining revelation from the Lord and in translating languages.**

An urim and thummim or a seer stone is a Quantum Computer. Most of its functionality resides within the quantum realm or the psyche realm. Likewise, a physical brain is a Quantum Computer, and most of its functionality takes place at the quantum level or the psyche level. For further information see, "Quantum Neuroscience: The Answer to Life, the Universe, and Everything" located at:

An urim and thummim is Light and Truth. The truth is always perfect, without falsehoods or flaws. Is it not?

It is fascinating how all of this fits together perfectly once we get our science or our definitions straight.

Whenever we are dealing with science or religion, it all comes down to light and truth.

The Biblical God Jesus Christ revealed all of this to us again, a century or two before our scientists started to discover it. All of the following has been confirmed and verified by our modern-day scientists and our research experiments in Quantum Mechanics.

Doctrine and Covenants 93: 7-10, 28-30, 36-37, 39-40, 42:

And he bore record, saying: I saw his glory, that he was in the beginning, before the world was; therefore, in the beginning the Word was, for he was the Word, even the messenger of salvation — the light and the Redeemer of the world; the Spirit of truth, who came into the world, because the world was made by him, and in him was the life of men and the light of men. The worlds were made by him; men were made by him; all things were made by him, and through him, and of him.

And now, verily I say unto you, I was in the beginning with the Father, and am the Firstborn; and all those who are begotten through me are partakers of the glory of the same and are the church of the Firstborn. Ye were also in the beginning with the Father; that which is Spirit, even the Spirit of truth; and truth is knowledge of things as they are, and as they were, and as they are to come; and whatsoever is more or less than this is the spirit of that wicked one who was a liar from the beginning.

The Spirit of truth is of God. I am the Spirit of truth, and John bore record of me, saying: He received a fulness of truth, yea, even of all truth; and no man receiveth a fulness unless he keepeth his commandments. He that keepeth his commandments receiveth truth and light, until he is glorified in truth and knoweth all things.

Man was also in the beginning with God. Intelligence, or the light of truth, was not created or made, neither indeed can be. All truth is independent in that sphere in which God has placed it, to act for itself, as all intelligence also; otherwise there is no existence. Behold, here is the agency of man, and here is the condemnation of man; because that which was from the beginning is plainly manifest unto them, and they receive not the light. And every man whose spirit receiveth not the light is under condemnation. For man is spirit. The elements are eternal, and spirit and element, inseparably connected, receive a fulness of joy; and when separated, man cannot receive a fulness of joy. The elements are the

tabernacle of God; yea, man is the tabernacle of God, even temples; and whatsoever temple is defiled, God shall destroy that temple.

The glory of God is intelligence, or, in other words, light and truth. Light and truth forsake that evil one. Every spirit of man was innocent in the beginning; and God having redeemed man from the fall, men became again, in their infant state, innocent before God. And that wicked one cometh and taketh away light and truth, through disobedience, from the children of men, and because of the tradition of their fathers. But I have commanded you to bring up your children in light and truth. You have not taught your children light and truth, according to the commandments; and that wicked one hath power, as yet, over you, and this is the cause of your affliction.

The sum-total of our modern-day science – all of our observations and experiences as a race – were revealed to us in advance in 1833 by the Biblical God Jesus Christ.

Doctrine and Covenants 88: 5-13:

Which glory is that of the church of the Firstborn, even of God, the holiest of all, through Jesus Christ his Son — he that ascended up on high, as also he descended below all things, in that he comprehended all things, that he might be in all and through all things, the light of truth; which truth shineth. This is the light of Christ. As also he is in the sun, and the light of the sun, and the power thereof by which it was made. As also he is in the moon, and is the light of the moon, and the power thereof by which it was made; as also the light of the stars, and the power thereof by which they were made; and the earth also, and the power thereof, even the earth upon which you stand. And the light which shineth, which giveth you light, is through him who enlighteneth your eyes, which is the same light that quickeneth your understandings; which light proceedeth forth from the presence of God to fill the immensity of space — the light which is in all things, which giveth life to all things, which is the law by which all things are governed, even the power of God who sitteth upon his throne, who is in the bosom of eternity, who is in the midst of all things.

The information is there – the science is already there – for anyone who is willing to look for it and see it. We already had the truth until the Materialists, Naturalists, Darwinists, Nihilists, and Atheists started to hide the truth from us in the name of science.

Doctrine and Covenants 131: 6-8

It is impossible for a man to be saved in ignorance. There is no such thing as immaterial matter. All spirit is matter, but it is more fine or pure, and can only be discerned by purer eyes; we cannot see it; but when our bodies are purified we shall see that it is all matter.

It's ALL matter, meaning that it's ALL comprised of light, whether we are talking about spirit matter, psychic dust, living light, psyche, or physical matter.

David Bohm is right. Physical matter is condensed or frozen light. It's all light. It's all matter! Physical matter is entropic light. Physical matter is spirit matter that has been localized into our 3D space-time realm, phased into our physical universe, and has been made subject to the passage of time (entropy).

"Mass is a phenomenon of connecting light rays which go back and forth, sort of freezing them into a pattern. So, matter, as it were, is condensed or frozen light." — David Bohm, Ph.D., Theoretical Physicist, *Dialogues with Scientists and Sages,* 1986.

David Bohm discovered what God revealed to us over 150 years ago.

God Is Light

St. Ambrose wrote:

CHAPTER XIV

Each Person of the Trinity is said in the sacred writings to be Light. The Spirit is designated Fire by Isaiah, a figure of which Fire was seen in the bush by Moses, in the tongues of fire, and in Gideon's pitchers. And the Godhead of the same Spirit cannot be denied, since His operation is the same as that of the Father and of the Son, and He is also called the light and fire of the Lord's countenance.

160. But why should I argue that as the Father is light, so, too, the Son is light, and the Holy Spirit is light? Which certainly pertains to the power of God. For God is Light, as John said: "For God is Light, and in Him is no darkness."

161. But the Son, too, is Light, because "the Life was the Light of men." And the Evangelist, that he might show that he was speaking of the Son of God, says of John the Baptist: "He was not light, but [was sent] to be a witness of the Light. That was the true Light, which lighteth every man that cometh into this world." So, then, since God is Light, and the Son of God the true Light, without doubt the Son of God is true God.

162. And you find elsewhere that the Son of God is Light: "The people that sat in darkness and in the shadow of death have seen a great light." But, which is still more clear, it is said: "For with Thee is the fount of Life, and in Thy light, we shall see light," which means that with Thee, O God the Father Almighty, Who art the Fount of Life, in Thy Son Who is the Light, we shall see the light of the Holy Spirit. As the Lord Himself shows, saying: "Receive ye the Holy Spirit," and elsewhere: "Virtue went out from Him."

163. But who can doubt that the Father is Light, when we read of His Son that He is the Brightness of eternal Light? For of Whom but of the Father is the Son the Brightness, Who both is always with the Father, and always shines, not with unlike but with the same radiance.

164. And Isaiah shows that the Holy Spirit is not only Light but also Fire, saying: "And the light of Israel shall be for a fire." So, the prophets called Him a burning Fire, because in those three points we see more intensely the majesty of the Godhead; since to sanctify is of the Godhead, to illuminate is the property of fire and light, and the Godhead is wont to be pointed out or seen in the appearance of fire: "For our God is a consuming Fire," as Moses said.

165. For he himself saw the fire in the bush and had heard God when the voice from the flame of fire came to him saying: "I am the God of Abraham, and the God of Isaac, and the God of Jacob." The voice came from the fire, and the voice was in the bush, and the fire did no harm. For the bush was burning but was not consumed, because in that mystery the Lord was showing that He would come to illuminate the thorns of our body, and not to consume those who were in misery, but to alleviate their misery; Who would baptize with the Holy Spirit and with fire, that

He might give grace and destroy sin. So, in the symbol of fire God keeps His intention.

166. In the Acts of the Apostles, also, when the Holy Spirit had descended upon the faithful, the appearance of fire was seen, for you read thus: "And suddenly there was a sound from heaven, as though the Spirit were borne with great vehemence, and it filled all the house where they were sitting, and there appeared unto them cloven tongues like as of fire."

Saint Ambrose. *On the Holy Spirit* (Illustrated) (pp. 35-36). Aeterna Press. Kindle Edition.

https://books.google.com/books?id=rzwIDAAAQBAJ&pg=PT39&lpg#v=onepage&q&f=false

Fire is the visual representation of the Holy Ghost; and, fire is a type of light. God the Father, Jesus Christ the Son, and the Holy Ghost are a type of fire, glory, power, or light that doesn't burn up bushes and trees.

The truth testifies of the truth, corresponds with the truth, and is consistent with the truth. The truth has been experienced and observed. Do you want the truth, or do you want the fiction? I'm a scientist. I want the truth. I want the Light.

The false is falsified by the truth. The false is falsified by observation and experience. The observation and experience of Psyche, Syntropy, or Light falsifies the claims of Materialism, Naturalism, Darwinism, Nihilism, Behaviorism, Determinism, Physical Reductionism, and Atheism which state that Psyche or Syntropy does not exist. According to these people, only physical matter or entropy exists. Obviously, they are wrong. There's a lot more to existence than just physical matter or entropy. This reality becomes obvious once we have seen the light.

Physical matter and entropy would not exist without some type of Syntropy or Psyche to bring them into existence. This is what the science is really trying to teach us. It's plain as day for anyone who is willing to look and see.

Mark My Words

—

Source

Quantum Mechanics from a Non-Physical Spiritual Perspective

https://www.amazon.com/dp/B01J023TGU

God Is in the Light: God is light, and in Him is no darkness at all.

https://www.amazon.com/dp/B07168S37N

Source Material

Quantum Neuroscience: The Answer to Life, the Universe, and Everything.

https://www.amazon.com/dp/B079Z6QQQB

References

Kaku, M. (2008). *Physics of the Impossible: A Scientific Exploration into the World of Phasers, Force Fields, Teleportation, and Time Travel*. New York: Anchor Books.

Saint Ambrose. *On the Holy Spirit*. Aeterna Press. Kindle Edition.

Doctrine and Covenants: https://www.lds.org/scriptures/dc-testament?lang=eng

No Explanatory Power

Observe carefully and you will see that Materialism, Naturalism, Darwinism, Nihilism, Atheism, and their derivatives have NO explanatory power when it comes to Psyche or Syntropy. Psyche is Syntropy. It is without a beginning of days or an end of years. It is eternal and everlasting.

Psyche has been experienced and observed by Near-Death Experiencers and Out-of-Body Explorers. Psyche is observed as being a pin-prick of light or a spark. Psyche or Intelligence is experienced first-hand or first-person as an immaterial viewpoint in space. A spirit body (and later a physical body) expands the range and the scope of a Spark or a Psyche exponentially. Doesn't it?

Quantum Mechanics is Syntropy – it's eternal and everlasting, without a beginning of days or an end of years. Physical matter and entropy had a beginning which means that they can theoretically have an end, when they get absorbed back into Syntropy.

Classical Physics or Physicalism has NO explanatory power when it comes to Quantum Mechanics, Psyche, and Syntropy. The Materialists and Naturalists can't explain Psyche, Quantum Mechanics, Consciousness, nor Syntropy in any way that makes logical sense to them or to us. Instead, these people assure us that these things do not exist. That's how they explain them, by stating that they do not exist. But, does that really explain them? Is that really science?

Materialism, Naturalism, Physicalism, Darwinism, Nihilism, Behaviorism, Determinism, Physical Reductionism, and Atheism make for Bad Science because they are severely lacking when it comes to explanatory power.

Science is observation and experience. In contrast, the major premises or hidden assumptions of Materialism, Naturalism, Darwinism, Nihilism, and Atheism have NEVER been experienced nor observed. They can't be because they are a figment of the imagination. They don't exist in any real physical sense.

Technically, natural selection doesn't even exist – not in any real tangible physical form. It's an idea – survival of the fittest – and nothing more. Natural selection doesn't exist as a person, or a force, or a physical mechanism that can produce genetic change or genetic programming. Natural selection doesn't do anything. Natural selection doesn't touch our genes! The only time it comes into play is when you get selected against and die. Natural selection results in death or extinction – it's entropy. There's NO physical mechanism for growth, or change, or progress, or programming, or genetic design associated natural selection. Natural selection is nothing but luck or chance; and, every scientist knows that blind luck cannot design and create anything at all.

In principle, natural selection is a philosophical assumption – a figment of the imagination. It's something that they made up out of thin air to explain their philosophical ideas or their religion. They *explained it by naming it*. There's NO physical mechanism associated with natural selection that can produce genetic change or evolution. Scary, is it not? They built a whole "science discipline" on something that doesn't touch our genes and can't touch our genes.

The idea that aggression is an instinct collapsed as the list of supposed human instincts grew to include nearly every conceivable human behavior. Nearly 6,000 supposed instincts were enumerated in one 1924 survey of social science books. The social scientists had tried to *explain* social behavior by naming it. It's tempting to play this *explaining-by-naming* game: "Why do sheep stay together?" "Because of their herd instinct." "How do you know they have a herd instinct?" "Just look at them: They're always together!"

Instinct theory also fails to account for the variations in aggressiveness from person to person and culture to culture. How would a shared human instinct for aggression explain the difference between the peaceful Iroquois before White invaders came and the hostile Iroquois after the invasion? Although aggression is biologically influenced, the human propensity to aggress does not qualify as instinctive behavior. (*Social Psychology*, p. 356.)

Natural selection, the theory of evolution, and survival of the fittest are also based upon *explaining-by-naming* or a carefully crafted *tautology*. It's a house of straw with no foundational support.

"Why did those chimp-like ancestors survive when everything else went extinct?" "Because of natural selection or survival of the fittest." "How do you know that it was natural selection that preserved them and kept them alive?" "Well just look at them: They survived!"

There's the sum-total of the theory of evolution right there in a nutshell. It's obvious that the theory of evolution and natural selection are true – just look at all the critters that survived! It's obvious and clear that natural selection and random mutations can design and create anything they set their minds to, because here you are.

They *explained it by naming it*! That's a logic fallacy!

The whole theory of evolution is based upon a bunch of different logic fallacies such as these. Logic fallacies or falsehoods can't be used to build the truth; but, that doesn't stop the Materialists, Naturalists, Darwinists, and Atheists from trying to do so. Nevertheless, the whole of their "science" is nothing more than a carefully crafted deception and lie. It's a *fictional just-so story* that they made up out of nothing. They *explained it by naming it*.

Natural selection doesn't do anything. Natural selection doesn't touch our genes. The physical process that does the change within our genes is the random mutations that take place during genetic recombination while our gametes (egg and sperm) are being made. Genetic recombination is the mechanism of change where evolution or genetic drift is concerned.

Yes, the genes are real. Yes, random mutations do indeed happen. But, random mutations are entropy; and, entropy cannot design and create. Entropy can only do disorder, disease, cancer, death, and extinction. There's NOTHING there within the theory of evolution that can actually design and create. There's nothing within the theory of evolution that can do programming, or the production of brand new genes and proteins from atoms. There's NO physical mechanism within the theory of evolution that can produce proteins and their matching genes from scratch or from atoms. There's NO science! The theory of evolution was explained into existence by being named into existence. In other words, it's a fictional story that they made up out of thin air.

By the way, when it comes to psychology, it should have been called a Genetic Perspective and not the evolutionary perspective. The theory of evolution is in constant dispute; whereas, the genes are not. The genes are science because they have been experienced and observed; whereas, creation by Mutation/Selection has not. Random mutations and natural selection cannot design and create anything. It's physically impossible. They are prevented from doing so by entropy or the second law of thermodynamics. Materialism, Naturalism, Darwinism, Nihilism, Atheism, and Classical Physics are based upon entropy. Entropy cannot design and create. It requires some type of Syntropy in order to be able to design and create proteins and their matching genes from scratch.

ONLY Syntropy, Intelligence, or Someone Psyche can design and create. By choosing to bring Psyche, Quantum Mechanics, and Syntropy into play, we can literally explain everything that comes our way. Try it. You might like it. I know I did. I'm a scientist; and, I have observed that Psyche or Syntropy has an infinite amount of explanatory power. In comparison with that, Materialism, Naturalism, and Darwinism are absolutely worthless because they have NO explanatory power when it comes to the science that matters most.

Ask yourself, "How do the Materialists and Naturalists explain Psyche, Syntropy, and Action at a Distance?" What's the answer?

Remember, Action at a Distance as well as the Quantum Zeno Effect have been experienced and observed in science experiments. They are real. They truly exist.

The Materialists, Naturalists, Darwinists, Nihilists, and Atheists explain Psyche, Syntropy, and Action at a Distance by labeling them or categorizing them as NON-EXISTENT. Once again, these people *explain-it-by-naming-it* or explain it by labeling it as non-existent; but, is that really a scientific explanation, or is it a dodge?

So, ask yourself, "When it comes to science, who are the Real Scientists and who are the fakes?" Well, what do we observe?

We observe the New Atheists calling themselves the Brights. The Atheists in general label themselves as the wealthiest, the smartest, the most educated, the most logical, and the most rational. Do they not? The Materialists, Physicalists, Naturalists, Darwinists, Nihilists, Behaviorists, Determinists, Physical Reductionists, and Atheists label themselves as the Real Scientists and call everyone else the fakes or the creationists. These people define "science" as Materialism, Naturalism, Darwinism, Nihilism, and Atheism.

Do you see a pattern here?

These people are *explaining-it-by-naming-it*. These people are defining their opposition out of existence by categorically denying their existence. These people teach and preach that Psyche, Non-Locality, Non-Physicality, Quantum Non-Local Consciousness, Syntropy, Action at a Distance, the Quantum Zeno Effect, Quantum Entanglement, and various other Quantum Mechanisms or Supernatural Mechanisms DO NOT EXIST. They *explain it by naming it*! They explain it by calling it non-existent. That's a logic fallacy because it's NO explanation at all.

Science is all about explaining the phenomena that have been experienced and observed. Science IS observation and experience, or it should be. Gravity, magnetism, the strong and weak nuclear forces, light waves, dark matter, dark energy, the zero-point field of light, the quantum zeno effect, quantum waves, action at a distance, and Psyche are invisible non-physical Forces and Fields that have been experienced and observed. Materialism and Naturalism have NO explanation for them except to say that they do not exist; yet, their verified and proven existence falsifies Materialism, Naturalism, Darwinism, and their derivatives.

The false is falsified by the truth; and, the truth is repeatedly experienced and observed. It's fascinating how that works!

Observation and experience ARE a better explanation for Science than Materialism and Naturalism. Materialism and Naturalism provide NO explanation at all when it comes to the non-physical and the immaterial Sciences that have been experienced and observed.

Mark My Words

—

Reference

Myers, D. G. (2010). *Social Psychology* (10th ed.). New York: McGraw-Hill.

Source

God Is in the Light: God is light, and in Him is no darkness at all.

    https://www.amazon.com/dp/B07168S37N

The Atonement of Christ Is Syntropy

The Atonement of Christ is Syntropy. The Atonement of Christ reverses and eliminates the effects of entropy and death. Syntropy is Eternal Life or Infinite Life.

The Complete Essay

The completed essay will be located in the following book:

The Second Comforter: Supping with Our Resurrected Lord Jesus Christ

https://www.amazon.com/dp/B01IAKHTY6

I just included the following part, which documents where I was first introduced to the term "Syntropy". I wanted to show where I got the word and the idea from. Syntropy is Eternal Lives. The Biblical God Jesus Christ has been talking about Eternal Life all the way along throughout the whole of human history. It's just that most of us aren't listening. We don't know anything about it because we don't want to know about it.

The Atonement Is Syntropy

Who are we?

Doctrine and Covenants 76: 22-24:

And now, after the many testimonies which have been given of him, this is the testimony, last of all, which we give of him: That he lives! For we saw him, even on the right hand of God; and we heard the voice bearing record that he is the Only Begotten of the Father — that by him, and through him, and of him, the worlds are and were created, and the inhabitants thereof are begotten sons and daughters unto God.

We are Gods. Our surname is God. We are the children of the Gods. The Human Psyche is literally a begotten son or daughter unto God. Both our physical body and our spirit body are descended from the Gods. We are Gods. Whenever a Human Psyche or Human Intelligence enters into a non-consensus reality in the Spirit World, that region of space automatically organizes itself according to the demands and expectations of that Human Psyche. In a non-consensus reality, a Human Psyche or a Human Intelligence is God. We are Gods. Each massless particle is some type of psyche. There's intelligence or psyche within every type of energy; and, physical matter is made from many different forms of energy. The Human Psyche is the controlling psyche or the God within its own spirit body and physical body.

The Biblical God Jesus Christ has been experienced and observed after He rose from the dead, both in the flesh and during our Near-Death Experiences. Jesus Christ has achieved Godhood or Godhead. He is one of the Head Gods.

Remember, science is observation. Jesus Christ has been experienced and observed.

God is Light and Truth. Syntropy is Light and Truth.

God is Syntropy. Syntropy is Eternal Lives.

The Atonement of Christ is much easier to understand once we realize that it is Syntropy – restoring everything to its proper order and function, including our psyche, spirit body, and physical body.

The prophets of the Biblical God Jesus Christ understand Syntropy or the Atonement of Christ better than anyone else, because God Himself teaches them about it and how it works in practice.

The Atonement of Christ is Syntropy. Syntropy or Divine Light has the power to enlighten and heal our Psyche, our Spirit Body, and our Physical Body. Syntropy is enlightenment. Let's discuss these three aspects of enlightenment.

First, it becomes clear that by keeping the Commandments of God to the best of our ability, the Atonement of Christ, repentance, or turning one's life around ends up being the ultimate form of Psyche-Therapy. Nothing brings calm, peace, and joy faster than the Atonement of Christ. Whenever it has been seen, the Psyche has been described as a pinprick of light. The Atonement of Christ can heal or enlighten our Psyche or our Soul.

Through repentance and the Atonement of Christ, our spirit bodies can be purged of darkness, corruption, entropy, death, karma, or sin. Our spirit bodies can shine a whole lot brighter than before, thanks to the Atonement of Christ. A purified spirit is Divine Light.

Unenlightened energies show some type of defilement or corruption of the Divine Light. They are divine energies that have somehow been misused and are now void of the divine essence they were originally imbued with.

Every one of us, at one point or another, has created dark light. It's part of the human experience to rise above these conditions. The thing to remember about unenlightened energies is that they are not part of God's divine emanations. They're artificial and not part of your true, divine nature. You do not need to claim these negative energies as your own. You just need to clear them out of your aura and avoid creating any new destructive vibrations. The best part of all is that these negative energies are powerless before the Divine Light. They have no real influence unless you invite them to operate in you. The more good you do, the more of the high colors you attract and accumulate, and the more you displace these lower vibrations.

Gray energy is connected with fear, gloom, and depression. Gray is the color of illness. When someone expresses gray energy, that person can be a worrier. Gray appears in the auras of people who can't see their way out of whatever dark cloud they've fallen into.

This square-shaped aura reflects a soul that has consistently misused its spiritual powers. As a result, many of the energies have been corrupted. Such an aura has become square-shaped because, by its persistent refusal to connect with its higher nature, it has retarded its evolution and reverted back to a more primitive, instinctual state of consciousness. The dark energies have restricted and choked the spiritual flow to the point where little Divine Light can get through.

The things he is doing takes him further away from his spiritual source rather than toward it. We are not born with an aura like this. As the auric power is degraded over time, we devolve to this spiritual level. This soul has a long climb back into the light, not only because of all the work needed to redeem it, but also because living in the dark propels it on to ever darker thoughts and deeds. The

person with such a soul could be a tyrant or mass murderer but doesn't have to be. I have seen this aura around what you would call "normal" people who hold no conspicuous positions in life.

The devolved aura seen in the illustration is very muddled; the lower half is especially disorganized. The black compartment holds sustained hatred, which he has tried to control. It may also indicate that he has been involved in murder. The dark green and brown compartments on the lower left indicate a conniving, cheating nature, compounded by cruelty. His emotions are all mixed up. There's a lot of anger, shown by the blotchy energy to his left at the emotional division. The health lines are turning gray and beginning to droop at the bottom, which means that his misdeeds are leading to physical illness. The head area has a brownish yellow energy, showing pettiness and a lethargic quality to his thinking. The dark green shooting from his forehead shows he is planning some sort of deception. The black thought forms around his aura show hatred toward someone that has recently been expressed. The dark, dirtied blue compartment across his head area shows he's constantly moody and brooding over his life. The sparkles of his color division are much smaller than in the color division of the mixed aura, and many of the points of light are dulled and dirtied. Notice, too, that the energies radiating from his centers are filled with darker colors and do not reach as far as with the mixed aura. This shows he's in an active state of depression. His spiritual division remains unaffected, but he has no access to those powers until he turns his life around.

Fortunately, a soul in such a devolved state is something of a rarity. I have seen maybe a hundred devolved auras in my life, but that's a hundred too many. A soul like this is certainly not past redemption, but it has a lot of work ahead to turn things around.

Martin, B. Y. (2016). *Change Your Aura, Change Your Life*. New York: Penguin Random House.

https://books.google.com/books?id=JSooCgAAQBAJ&pg=PT48&lpg=PT48&dq#v=on epage&q&f=false

Most people define the "soul" or the "spirit" as the combination of both the Psyche and the Spirit Body. Without Syntropy or without input from the Divine Light, the light within the soul can turn dark and even go black. That's not good. An infusion of Syntropy or an infusion from the Atonement of Christ can help to turn that around. It starts by turning to God and asking for His help and forgiveness. We have to give God permission to help us; otherwise, He has covenanted with us to leave us alone and let us do our own thing.

Your aura is a visual representation of your Syntropy. Some people like Barbara Martin can see another person's aura. Your Syntropy or Aura is really there, radiating from your spirit or your soul. We carry our darkness or our light within us.

My best friend has seen both Satan and Jesus Christ. He describes Satan as dark light. In contrast, the being of light and love whom people encounter during their Near-Death Experiences is Divine Light or Divine Love; and, when that being identifies himself by name, He is typically Jesus Christ, or one of the Archangels, or our Mother in Heaven.

Finally, through the Atonement of Christ and the resurrection from the dead, our physical bodies will someday be raised from the dead and become what the Apostle Paul calls a "spiritual body".

<u>1 Corinthians 15: 40-58</u>:

So also is the resurrection of the dead. It is sown in corruption; it is raised in incorruption: it is sown in dishonor; it is raised in glory: it is sown in weakness; it is raised in power: it is sown a natural body; it is raised a spiritual body. There is a natural body, and there is a spiritual body.

<u>1 Corinthians 15: 40</u>:

There are three degrees of glory in the Resurrection.

40 Also celestial bodies, and bodies terrestrial, *and bodies telestial;* but the glory of the celestial, one; and the terrestrial, another; *and the telestial, another.*

Our spirit body is our pre-mortal body. It is a son or daughter of Heavenly Father and Heavenly Mother. Our Psyche or Intelligence is eternal and everlasting – it cannot be made, and it cannot be destroyed. Our Psyche or Intelligence was clothed upon by a spirit body.

Our physical body is what Paul and Jesus Christ call our "natural body".

So, what's a spiritual body?

A spiritual body is a resurrected physical body. A spiritual body is a physical body that has had its entropy removed thereby making it immortal. Our physical body is made from entropic physical matter. A resurrected body is made from one of the many different types of syntropic physical matter or immortal physical matter.

A physical body that is a spiritual body is something completely different than a spirit body which is limited to functioning at the quantum level in the spirit realm. A resurrected physical body or a resurrected spiritual body can exist and function at every phase, level, or dimension of existence including this physical level.

In other words, a resurrected physical body can phase shift into every dimension, level, frequency, or realm that we can possibly imagine. That means that it no longer has entropy, which means that it is immortal and no longer has any physical limitations. A resurrected physical body can travel faster than the speed-of-light at the speed of thought. A resurrected spiritual body or resurrected physical body can quantum tunnel or teleport instantaneously to any destination in the universe, just like a spirit body and psyche can. A spiritual body is a quantum body. It has NO physical limitations. A spiritual body is comprised of syntropic physical matter – NOT entropic physical matter. There's a difference.

A resurrected physical body is indistinguishable from a spirit body, except that it has presence and can be handled while it is here in this physical phase of existence at the physical level or while it's in the physical dimension.

A spiritual body is Syntropy. A spiritual body or syntropic physical body has direct access to Quantum Mechanics or Transdimensional Physics. Telepathy, Teleportation, and Action at a Distance are an integral part of Quantum Mechanics or Syntropy. Some of us retain or regain a small part of our original ESP, Psychic Abilities, or Telepathic abilities while here in physical mortality; but for most of us, our physical bodies and physical brains greatly limit and restrict our Quantum Mechanical abilities or our Spiritual abilities. Quantum Mechanics is Spiritual Mechanics.

798

According to Quantum Mechanics, YOU (your psyche and your physical body) are everywhere in the universe simultaneously, and your spirit body is never far behind. Quantum Tunneling is faster than light. Quantum Teleportation is instantaneous.

In the following video he states: **Wouldn't it be nice to be everywhere at once? According to Quantum Mechanics, you are.**

https://www.youtube.com/watch?v=-IfmgyXs7z8

https://syntropy.site/wp-content/uploads/2018/04/Is-Quantum-Tunneling-Faster-than-Light.zip

This reality applies directly to Psyche in its original natural state of existence; but, he's actually talking about your physical body, particularly the atoms within your physical body. According to Quantum Mechanics, your physical body is everywhere in the universe at once – at least a little bit.

However, God designed physical matter to significantly limit, inhibit, restrain, and block the effects of Quantum Tunneling or Quantum Teleportation. Your physical body and physical brain are the VEIL; and, they also greatly limit and restrict our Telepathic, ESP, Psychic, or Spiritual abilities as well. Our Psyche's quantum mechanical abilities are greatly reduced and restrained by our physical body and physical brain. Nevertheless, in its original native state of being, your psyche and God's Psyche have the innate capability of being omnipresent everywhere in this universe simultaneously. God's Psyche retains that capability, because He has a spiritual body or a resurrected body, rather than a physical body.

According to this video: **Any material object is really a matter wave. You're everywhere in the universe, but not very much.**

https://www.youtube.com/watch?v=-IfmgyXs7z8

https://syntropy.site/wp-content/uploads/2018/04/Is-Quantum-Tunneling-Faster-than-Light.zip

Again, he's talking about physical matter. They never talk about spirit matter. I like to talk about spirit matter, in order to complete quantum theory. The spirit matter is always there at the other extreme. Physical matter exists at the low velocity extreme; and, spirit matter as well as psyche exist towards the high velocity or instantaneous velocity extreme, respectively.

In this video, he talks about the De Broglie Wavelength, and then he gives a very effective visual demonstration as to why physical matter is localized or present or tied-down, whereas spirit matter and psyche are not. The De Broglie Wavelength is the science that describes the difference between spirit matter and physical matter. He doesn't realize that, of course, but I do.

In its native unorganized state, spirit matter has an extremely large De Broglie Wavelength and therefore an undefined omnipresent position; whereas, physical matter has a very small De Broglie Wavelength and a well-defined position. I like to visualize Psyche as having an infinite De Broglie Wavelength, meaning that in its native original state Psyche is instantaneous and omnipresent. Spirit matter and a spirit body tends to be non-local or transdimensional, although it can phase-shift into different dimensions; whereas, physical matter is highly localized thanks to its very small De Broglie Wavelength.

Quantum Physicists tend to define Non-Locality as Action at a Distance or Entanglement. I like to define Non-Locality as not-located here in our physical 3D space-time realm. Either definition works just fine. Either way, we know that God has deliberately localized physical matter here in this 3D space-time realm in order to create the Ultimate Consensus Reality, a physical reality.

The physical is obvious to most of us – it's the quantum that requires some thought. In this video, he calls Quantum Mechanics "weird". It's NOT weird, it's non-physical. The non-physical simply works differently than the physical. The physical has limitations and restrictive laws placed upon it by God in order to create the Ultimate Consensus Reality, a physical reality. That's the trade-off. The "weirdness," or the non-physical, or the quantum had to be temporarily removed or restrained in order to create or organize this physical reality.

We Quantum Physicists call matter of any kind a "quantum object"; but, in the case of physical matter, it's greatly restrained and contained, because it has low velocity. Since physical matter has been slowed down to sub-light velocities by God, physical matter has extremely low probabilities of being anywhere else besides where it currently is thanks to its small De Broglie Wavelength. Physical matter has been localized, in contrast to spirit matter, which is inherently non-local, quantum, instantaneous, and omnipresent. Instantaneous suggests a velocity faster than the speed-of-light; and, quantum tunneling, or teleportation, is instantaneous. In their native original state of existence, both Psyche and Spirit Matter have velocities faster than the speed-of-light, which makes them inherently non-local, instantaneous, and omnipresent.

The speed-of-light limitation or theory of relativity ONLY applies to entropic physical matter. In contrast, Psyche has an infinite or instantaneous velocity, and spirit matter lags only slightly behind. Spirit matter conforms itself to the demands of Psyche. Meanwhile, physical matter conforms itself to classical physics or the theory of relativity, which reduces physical matter to sub-light velocities and very short De Broglie Wavelengths and a well-defined position. Spirit matter is non-local, whereas physical matter has been localized. Finally, from observation and experience, we KNOW that a translated syntropic physical body and a resurrected syntropic physical body have no physical limitations such as entropy and relativity. They, too, can quantum tunnel or teleport.

God slowed down or restrained physical matter in order to create or organize the Ultimate Consensus Reality – a physical reality. A physical reality is highly controllable, dependable, reliable, replicable, and stable unlike a psyche reality or a quantum reality which can be everywhere in the universe simultaneously while in its original native format or state of existence. Physical matter brings a sort of Order or Symmetry or Stability to reality, especially in contrast to the spiritual or quantum side of the equation which originated directly from Chaos. Nevertheless, Syntropy or the Atonement of Christ brought "order to chaos" on BOTH the quantum side and the physical side of the equation. Syntropy works on both the psyche side and the physical side, which is why Syntropy or Quantum Mechanics is the truth. Syntropy or Quantum Mechanics is complete; whereas, Classical Physics or Naturalism is incomplete.

When God created physical matter and this physical universe, He infused a great deal of Syntropy into it in order to do so, which made the subsequent entropy possible. As far as I am concerned, Syntropy and Quantum Mechanics are the pinnacle of science. They are infinitely superior and a whole lot more explanatory than anything that can be produced by Materialism, Naturalism, Darwinism, Nihilism, Atheism, and their derivatives. Materialism and Naturalism provide NO explanation for any of these things, which makes them rather worthless, doesn't it? Materialism, Naturalism, Darwinism, Nihilism, Atheism, Scientism, Behaviorism, Determinism, and the Theory of Evolution have NO explanatory power from what I can tell. They are Bad Science.

Remember, the Atonement of Christ and the Resurrection can restore our physical bodies to their original state of Syntropy, wherein the matter within our bodies has NO entropy and is no longer restricted by any physical limitations such as speed-limits, distance limitations, or what we call the speed-of-light. In other words, Entropy and the Theory of Relativity do NOT apply to Syntropy, Quantum Mechanics, Psyche, Spirit Matter, and Spiritualized or Resurrected Matter. Entropy and the Theory of Relativity do not apply to a Spiritual Body, a Syntropic Physical Body, or a Resurrected Body. The Theory of Relativity and the Laws of Thermodynamics only apply to ordinary physical matter and our mortal physical bodies.

Though technical at times, this is really easy to understand once a person chooses to do so. It's a lot more simple, parsimonious, logical, and explanatory than anything the Materialists, Naturalists, Darwinists, Nihilists, Atheists, Behaviorists, Determinists, and Classical Physicists have been able produce. Quantum Mechanics provides a scientific explanation for how spirit matter and psyche work at the quantum level or the psyche level. Quantum Mechanics can even explain the difference between a physical body and a spiritual body, even though both are capable of functioning here in this physical realm.

In contrast, Materialism and Naturalism state that psyche and spirit matter do not exist; but, that claim is NOT science. It can't be experienced nor observed. It has NO explanatory power. That claim is pure philosophy, sophistry, religion, dogmatism, wishful thinking, and blind-faith. Materialism, Naturalism, and Atheism are NOT science – they are philosophy, religion, dogmatism, and blind-faith rolled into one. Materialism and Naturalism claim that Syntropy, Quantum Mechanics, Psyche, and God do not exist. There's NO way to verify or confirm those claims. They have to be taken on blind faith as being true. That's not science. That's blind faith and religion.

Syntropy is Quantum Mechanics. Syntropy is the Atonement of Christ. Syntropy is the Divine Light. Syntropy means "without a beginning of days or an end of years". Syntropy is eternal and everlasting. Syntropy has explanatory power. Materialism and Naturalism do not.

Once we understand how things really work, we no longer have to apologize for all of the wonderful things that the Atonement of Christ can do for one's psyche, body, and soul. The Atonement of Christ is Syntropy, which means that the Atonement of Christ can restore everything to its original perfect, pristine, and uncorrupted state of existence. The Atonement of Christ works on the Human Psyche, your Spirit Body, and your Physical Body. It's all-inclusive, because it is Syntropy. Syntropy works. Entropy does not.

Doctrine and Covenants 88: 1-47:

Verily, thus saith the Lord unto you who have assembled yourselves together to receive his will concerning you: Behold, this is pleasing unto your Lord, and the angels rejoice over you; the alms of your prayers have come up into the ears of the Lord of Sabaoth and are recorded in the book of the names of the sanctified, even them of the celestial world. Wherefore, I now send upon you another Comforter, even upon you my friends, that it may abide in your hearts, even the Holy Spirit of promise; which other Comforter is the same that I promised unto my disciples, as is recorded in the testimony of John. This Comforter is the promise which I give unto you of eternal life, even the glory of the celestial kingdom; which glory is that of the church of the Firstborn, even of God, the holiest of all, through Jesus Christ his Son — he that ascended up on high, as also he descended below all things, in that he comprehended all things, that he might be in all and through all things, the light of truth; which truth shineth. This is the light of Christ. As also he is in the sun, and the light of the sun, and the power

thereof by which it was made. As also he is in the moon, and is the light of the moon, and the power thereof by which it was made; as also the light of the stars, and the power thereof by which they were made; and the earth also, and the power thereof, even the earth upon which you stand.

And the light which shineth, which giveth you light, is through him who enlighteneth your eyes, which is the same light that quickeneth your understandings; which light proceedeth forth from the presence of God to fill the immensity of space — the light which is in all things, which giveth life to all things, which is the law by which all things are governed, even the power of God who sitteth upon his throne, who is in the bosom of eternity, who is in the midst of all things. Now, verily I say unto you, that through the redemption which is made for you is brought to pass the resurrection from the dead. And the spirit and the body are the soul of man. And the resurrection from the dead is the redemption of the soul. And the redemption of the soul is through him that quickeneth all things, in whose bosom it is decreed that the poor and the meek of the earth shall inherit it.

And again, verily I say unto you, the earth abideth the law of a celestial kingdom, for it filleth the measure of its creation, and transgresseth not the law — wherefore, it shall be sanctified; yea, notwithstanding it shall die, it shall be quickened again, and shall abide the power by which it is quickened, and the righteous shall inherit it. For notwithstanding they die, they also shall rise again, a spiritual body. They who are of a celestial spirit shall receive the same body which was a natural body; even ye shall receive your bodies, and your glory shall be that glory by which your bodies are quickened. Ye who are quickened by a portion of the celestial glory shall then receive of the same, even a fulness. And they who are quickened by a portion of the terrestrial glory shall then receive of the same, even a fulness. And also, they who are quickened by a portion of the telestial glory shall then receive of the same, even a fulness. And they who remain shall also be quickened [or reborn]; nevertheless, they shall return again to their own place, to enjoy that which they are willing to receive, because they were not willing to enjoy that which they might have received.

This scripture from the Biblical God Jesus Christ gives us a bit more detail and instruction about the redemption and resurrection that will take place for each of us thanks to the Atonement of Christ or Syntropy. God calls Syntropy the Light of Christ. This Light or Syntropy "proceedeth forth from the presence of God to fill the immensity of space". It exists. It's REAL.

While studying science, I came to realize that Dark Energy or the Zero Point Field of Light is the Light of Christ or Syntropy. It's a necessary and essential organizing force that operates unseen here in this physical universe providing structure and order to this physical universe. Dark Energy is the expanding force within our universe, and Dark Energy is the counteracting force to Gravity. Both Dark Energy and Gravity are invisible non-physical quantum mechanical functions, forces, or fields that involve instantaneous, simultaneous, omnipresent, universal interaction between physical particles or physical atoms. The Quantum Fields that fill the immensity of space are also the Light of Christ. The Quantum Fields are NEEDED to successfully make and sustain physical matter of any kind. The Quantum Fields or Light of Christ is the power by which physical matter, the planets, stars, and galaxies were made.

This scripture from the Biblical God Jesus Christ actually tells us that Gravity is an integral part of the Light of Christ, which means that Dark Energy is also an integral part of the Light of Christ. Quantum Fields are the Light of Christ. These things provide order, structure, and syntropy to physical matter; otherwise, this physical universe would be nothing but disorder, randomness, chaos, and entropy. The Light of Christ is Syntropy.

Both Gravity and Dark Energy are different types of Syntropy, as are ALL of the different forces and fields that we know about. Think about magnetism, for example. It's a different type of Syntropy. These forces and fields are instantaneous and simultaneous, meaning that they are supernatural or quantum mechanical in nature and origin. They are Action at a Distance. Magnetism, Gravity, and Dark Energy are Action at a Distance. They provide structure, order, or syntropy to physical matter and this physical universe.

Action at a Distance is Syntropy. It proceeds from the presence of God to fill the immensity of space. The Light of Christ or the Light from Christ is Syntropy. Jesus Christ is the being of love and light whom people encounter during their Near-Death Experiences (NDEs). Light is Love. Love is Syntropy. The deed for this whole earth has been given to Jesus Christ because He paid the price for its redemption.

We KNOW that these different forces and fields exist by the effect or influence that they have on physical matter. We can't see these forces or fields with our eyes nor our physical instruments because they are non-physical in nature and origin; but, we can definitely see the effects that these forces and fields have on the physical matter that surrounds us. Technically, the wind is a type of Syntropy. The very existence of Gravity, Dark Energy, Magnetism, and other invisible supernatural non-physical forces and fields FALSIFIES the claims of Materialism, Naturalism, Darwinism, Nihilism, Determinism, Behaviorism, Scientism, Atheism, and Classical Physics which claim that the non-physical DOES NOT EXIST. Remember, all of the invisible forces and fields that we know about are simply different types of Syntropy or different types of Quantum Mechanics. They tend to control physical matter and move physical matter around from the quantum level or the spiritual level.

The Syntropy MUST EXIST, or we wouldn't exist. Take magnetism again for an example. Magnetism is a type of Syntropy. It is Action at a Distance. Without the magnetic fields protecting this earth, our earth would have been fried to a crisp, and we wouldn't exist. It is the very height of arrogance and ignorance to assume that Gravity and Magnetism are the only types of Syntropy in this universe, just because they are the most obvious to us. Soon, we discovered the strong and weak nuclear forces, which are a type of Syntropy, Order, or Organizing Force within this universe. We discovered Quantum Fields which are essential for making and sustaining physical matter. Later, we discovered that there must be Dark Energy, or an expanding force in this universe. Dark Energy or the Zero Point Field of Light is a type of Syntropy.

Syntropy helps to bring order out of Chaos. Well, the most ubiquitous type of Syntropy is the Psyche, or Life Force, or our Quantum Non-Local Consciousness, which tends to reside in the third-eye position just behind our forehead. That's where you are, is it not? When you close your eyes, isn't that where YOU reside, in that third-eye position just behind your forehead?

Notice that those of us who choose not to get off the wheel and choose instead to pursue a better resurrection will remain here in the physical universe rather than being resurrected into the Celestial Kingdom, or Terrestrial Kingdom, or the Telestial Kingdom.

And they who remain shall also be quickened [reborn]**; nevertheless, they shall return again to their own place, to enjoy that which they are willing to receive, because they were not willing to enjoy that which they might have received.**

There are those of us who will **return again** to a physical life without going to a Kingdom of Glory. They choose to **remain** here on the physical plane. Those who choose to remain will also be quickened, or reincarnated, or reborn. These reincarnated or reborn people are in a different category than those who choose to be quickened or resurrected into a Telestial Glory, or a Terrestrial Glory, or a Celestial Glory.

This is important to understand, because the Biblical God Jesus Christ hints at the possibility of Reincarnation or Rebirth all throughout the Bible, Book of Mormon, and Doctrine and Covenants. Some of us choose NOT to undergo another mortal probation but choose to accept resurrection and a Kingdom of Glory instead. But others choose to pursue a better resurrection by choosing to submit to yet another mortal probation here on earth.

This scripture from the Biblical God Jesus Christ tells us that those who choose to **remain** here in this physical realm in pursuit of a better resurrection will also be quickened, reincarnated, or reborn; and, they shall **return again** to their own place either on this earth or on some other physical earth in the distant future. They are NOT willing to end the process of trying to better themselves or improve themselves. They stay on the wheel and keep going until they achieve the type of resurrection that they desire to achieve. To be quickened means to be made physically alive again. Quickening takes place either through rebirth or resurrection. Both are valid and useful principles to know about and understand.

We learn best and learn fastest while we are in a mortal physical body, because the pain hurts. Pain is our greatest teacher. It motivates us to learn. The Doctrine of Eternal Lives, Eternal Progression, and Multiple Mortal Probations is hidden here in plain sight. Those who choose to remain in this physical state shall also be quickened or reborn; and, they will return again to this physical realm in a physical body on this earth or some future physical earth. The death of the physical body is NOT the end. Entropy is not the end. Entropy is overcome by Syntropy or the Atonement of Christ.

Hebrews 11: 35: **Women received their dead raised to life again: and others were tortured, not accepting deliverance; that they might obtain a better resurrection.**

According to Paul, some chose to be resurrected; and, others chose to be reborn, and chose to go through the tortures of yet another mortal physical life in the hope **that they might obtain a better resurrection**. Some chose to get off the rack and accept a resurrection and a degree of glory, and others chose to give the rack or a physical mortal life another go. We are NOT reincarnated as a flea, frog, or fish. We are reborn as a son or a daughter of God. We are reborn as a human being. That's why rebirth is a better description of the process than reincarnation.

Remember, the Spirit Body, Spirit Matter, and the Psyche are based upon Syntropy or Quantum Mechanics. Quantum Mechanics or Syntropy is the opposite of entropy.

Syntropy is one of the most interesting subjects to study, especially once you know and understand what entropy is. Materialism, Naturalism, Darwinism, Nihilism, Atheism, the Theory of Evolution, and Classical Physics are based exclusively upon entropy; and, these people formally reject the existence of Syntropy or Psyche. However, ALL of the scientific evidence, observational evidence, experiential evidence, eye-witness evidence, and empirical evidence tells us clearly and conclusively that Syntropy is REAL and that Materialism and Naturalism are false. Syntropy or Quantum Mechanics is the opposite of Materialism and Naturalism. They are mutually exclusive. The proven existence of one falsifies the other. Syntropy, Psyche, Quantum Fields, and Quantum Mechanics FALSIFY Materialism, Naturalism, Darwinism, and their derivatives.

Borrowing a phrase that many will recognize – Syntropy is more than meets the eye. Syntropy is the Spark of Life or the Human Psyche. It's an invisible power or force that has a direct impact on physical matter – commands and controls physical matter. Psyche is a type of Syntropy. It's an organizing force, or a source of order, light, and life. Psyche is the direct opposite of Entropy, Classical Physics, Materialism, Naturalism, and Atheism. If the one truly exists, then the other one has been falsified.

It's NOT that Materialism and Naturalism are falsifiable that makes them interesting to science, but the fact that science or observation has already FALSIFIED Materialism and Naturalism is the thing that we should be paying attention to as scientists. The different types of Syntropy or Quantum Mechanics, as well as the different types of forces and fields that we have detected, FALSIFY Materialism, Naturalism, Darwinism, Nihilism, Atheism, and their derivatives.

I'm not the first person on this planet to figure this out. There have been others. The following is from Hugh Nibley. I wrote this book on Syntropy before I found the following from Hugh Nibley; but, Nibley seems to confirm or verify my thoughts and ideas on the subject of Syntropy.

Dialogue: Do you think it's the atonement itself that causes syntropy?

Nibley: Well, "atonement" means syntropy — bringing back to its former state, restoring to its former state. You see, when something breaks down, it becomes disorganized and fractured. "At one" means unified again — returned to its unity, returned to its former integrity and structure. It sounds like a Latin word or Greek, but it isn't — it's pure old English, nothing else. Atonement is not one of the technical terms, not even like resurrection. It is an old English word. It's like "bless"— it comes from no other language but English — "woman" comes from no other language but English. "Lord," the word "Lord" — no trace of it in any other language. Isn't that funny?

https://dialoguejournal.com/wp-content/uploads/sbi/articles/Dialogue_V12N04_12.pdf

https://syntropy.site/wp-content/uploads/2018/04/Dialogue_V12N04_12.pdf

Atonement means Syntropy. Entropy means disorganized, fractured, and corrupted. Atonement or Syntropy means to be unified again, to be restored to its former integrity and structure. That's how a physical universe is created, ordered, or organized out of spirit matter or chaotic matter. It gets an infusion of Syntropy to get it going. The entropy or disorder is removed from it. All you want is to know how things really work, and in this dialogue or discussion, Hugh Nibley is trying to explain how things really work.

John Rector also explains that the Atonement of Christ is Syntropy.

The Atonement of Christ: Vehicle of Paradox

The ideas presented thus far help explain both sin's necessity and its transcendable nature. However, the degeneracy and estrangement caused by sin would remain a permanent, iron-clad reality without the one true source of syntropy in the universe: the atonement of Jesus Christ. It is only by and through Christ's atonement that sin can ever be turned on its head — that good can be brought out of evil, light out of darkness, fullness out of emptiness, health out of sickness, and perfection out of imperfection. It is only through the atonement of Christ that evil does not remain a permanent fixture on our eternal landscape, shutting us out from God's presence forever. Christ's great and last sacrifice is the

renewing life-source, the cleansing power extended to humans by which the pain, suffering, and damage done by sin can be eradicated and ultimately transcended. — John Rector.

https://www.byui.edu/documents/instructional_development/Perspective/V1n3PDF/v1n3rector.pdf

https://syntropy.site/wp-content/uploads/2018/04/v1n3rector.pdf

When it comes to origins, this Atonement or Syntropy has infinitely more explanatory power and is a whole lot more sensible, credible, and logical than Materialism, Naturalism, Atheism, Darwinism, Nihilism, Classical Physics, Physical Matter, Creation Ex Nihilo, and Entropy. I still remember the terror that I felt forty years ago when I was first taught about entropy and first understood what it is and what it really means to us. The scientists knew about entropy, but they didn't know anything about Syntropy. The vast majority of them had rejected the Atonement of Christ and had turned to Materialism, Naturalism, and Atheism instead. Sin produces entropy – chaos and disorder. The Atonement of Christ or Syntropy counteracts the sin or the entropy. There must be balance in all things. There must be an opposition in all things. I wish I would have known and understood that forty years ago, when I first learned about entropy. Entropy is death.

Whenever you are dealing with science, explanatory power is good; and, philosophical wishful thinking such as Materialism, Naturalism, and Darwinism is bad. All you want is the truth. Everything else is worthless. Syntropy has value to me because of its explanatory power. I can actually visualize it within my mind; whereas, I cannot visualize how to demonstrate or prove the non-existence of God, or the non-existence of Psyche, or the non-existence of the Spirit Realm. Syntropy is Eternal Life.

Hugh Nibley discusses Syntropy, or Atonement, or the Power of God in other articles as well. Syntropy is the opposite of entropy. Syntropy counter-balances and counter-acts entropy. The Syntropy has to exist, or there couldn't be entropy. This physical universe had to receive a huge infusion of Syntropy or Order or Organization, or all of that subsequent entropy would not have been possible. When it comes to Syntropy, it all makes good sense to me from a scientific standpoint; and, I'm not alone.

In the following article, Hugh Nibley discusses entropy and syntropy.

Matthews continues:

The relevance of this to our problem is that one may think of a proton at rest as a very highly ordered condition of a certain amount of energy — the rest energy of the proton — which can exist in just one state (strictly two if we allow for two possible orientations of the proton spin). If the proton can decay by any mechanism into two or more lighter particles, these serve to define an alternative condition of the system which is relatively highly *disordered*, since it can exist with all conceivable orientations. The number of allowed states depends on the relative momentum of the decay products much as the number of points on the circumference of a circle depends on its radius. *The decay interaction is the shuffling agent* . . . If it exists and operates on a time scale comparable with the age of the universe, then by relentless operation of the Second Law, essentially every proton would by now have decayed into lighter particles . . . . Clearly the opposite is the case, and there must be some very exact law which is preventing this from happening.

Had all the protons decayed, there would be no stable atoms, no elements, no compounds, no earth, no life. When the biologist said that life was wildly improbable, a rare unreasonable event, who would have guessed how improbable it really was? "A human being," writes Matthews, "is at very best, an assembly of chemicals constructed and maintained in a state of fantastically complicated organization of quite unimaginable improbability." So improbable that you can't even imagine it. So "wildly improbable" that even to mention it is ridiculous. So, we have no business being here. That is not the natural order of things. In fact, he says that "the sorting process — the creation of order out of chaos — against the natural flow of physical events is something which is essential to life." So, the physical scientists and the naturalists agree that if nature has anything to say about it, we wouldn't be here. This is the paradox of which Professor Wald of Harvard says, "The spontaneous generation of a living organism is impossible . . . In this colloquial, practical sense I concede the spontaneous origin of life to be 'impossible.'" The chances of our being here are not even to be thought of, yet here we are.

So, as I say, in my school days it was fashionable to brush aside Paley's watch argument with a snort of impatience. If you're walking on the beach and find a beautifully made Swiss watch, you should not with Archdeacon Paley conclude that some intelligent mind has produced the watch. It proves nothing of the sort. Finding the watch only proves, quite seriously, that mere chance at work, if given enough time, can indeed produce a fine Swiss watch or anything else. Indeed, when you come right down to it, the fact that Swiss watches exist in a world created and governed entirely by chance proves that blind chance can produce watches. There is no escaping this circular argument, and some people use it. Today Professor Matthews states the same problem more simply:

If, after seeing a room in chaos, it is subsequently found in good order, the sensible inference is not that time is running backwards, but that some intelligent person has been in to tidy it up. If you find the letters of the alphabet ordered on a piece of paper to form a beautiful sonnet, you do not deduce that teams of monkeys have been kept for millions of years strumming on typewriters, but rather that Shakespeare has passed this way.

But to Professor Huxley or Professor Simpson this is sheer heresy or folly. It was the evolutionist who seriously put forth the claim that an ape strumming on a typewriter for a long enough time could produce, by mere blind chance, all the books in the British Museum, but did any religionist ever express such boundless faith? I don't know any religious person who ever had greater faith than that. Yet serious minds actually believed such an impossibility. They say it is impossible, but then it happens.

Remember, "the decay interaction is the shuffling agent [and] . . . by the relentless operation of the Second Law, essentially every proton would by now have decayed into lighter particles . . . Clearly the opposite is the case." Now "there must be some very exact law which is preventing this from happening."

Kammerees new law of seriality is in direct opposition to the second law: there is "a force that tends toward symmetry and coherence by bringing like and like together." That is a very interesting point. We say that light cleaves unto light, etc. What is that force? Nobody knows. They say it is there because you see it working. Buckminster Fuller calls it *syntropy*. The greatest Soviet astrophysicist today, the Soviets' foremost man in that field, Nikolai Kozyrev, has been working for years on this question. He claims that the second law of thermodynamics is all right, but it doesn't work. Something works against it, something stronger. He says,

Some processes unobserved by mechanics and preventing the death of the world are at work everywhere, maintaining the variety of life. These processes must be similar to biological processes maintaining organic life. Therefore, they may be called vital processes and the life of cosmic bodies or other physical systems can be referred to as vital processes in this sense.

We noted that both the physicist and the biologist were aware of an ordering and organizing agent that opposes the second law. Matthews pays tribute to the Pythagoreans: "Why is it then that when we come to examine the inanimate world we find it controlled by laws which can only be put in mathematical terms?" For that matter, what do I know about it? Yet all inanimate nature conducts itself according to mathematical principles conceived of as pure theory by the human mind. Somebody must be working things out. And so, we begin with the creation story.

There is matter. That is the first law [of thermodynamics]: matter was always there. There is unorganized matter. Or as Lyall Watson says, "The normal state of matter is chaos." It always is, and it always will be. The normal state of matter is to be unorganized. There is unorganized matter; let us go down and organize it into a world. That mysterious somebody is at work, bringing order from chaos. It would be easy to say we were making up a story if we didn't have a world to prove it. Somebody went down and organized it. Matter was always there, always in its normal state of chaos; and long ago the protons should have all broken down, yet here is the world. Matter is unorganized. The temple represents that organizing principle in the universe which brings all things together. It is the school where we learn about these things.

https://publications.mi.byu.edu/pdf-control.php/publications/transcripts/I00058-The_Meaning_of_the_Temple.html

https://syntropy.site/wp-content/uploads/2018/04/The_Meaning_of_the_Temple.pdf

Something or someone has to be opposing the chaos, entropy, and death; otherwise, we wouldn't exist. The Temple represents Syntropy and is there to teach us about Syntropy or the Atonement of Christ. Atonement means to bring all things together into one. The Atonement restores or redeems that which was lost to entropy and death. The Atonement reverses entropy and death. The Atonement of Christ is Syntropy.

It is all based upon Syntropy, or Quantum Mechanics, or the Priesthood Power of God. Quantum Mechanics is Syntropy in action. The Atonement of Christ is really easy to understand once we understand that it is Syntropy – returning to or being at-one with our former state of existence or original state of existence – where we lived in the presence of God and Christ in the Celestial Kingdom, during our pre-mortal life.

In contrast, matter's natural original state is chaos. Our physical universe has some serious limitations – both in size and duration; but, at least it has had an infusion or Order or Syntropy to get it going in the first place. Our observable universe must have received an infusion of Syntropy in the beginning, or there could have been no subsequent entropy.

So, what exists outside of our physical universe or separate from our physical universe besides the other physical universes?

Chaos or matter-unorganized seems to be what exists between the Organized Universes or Syntropic Universes. An Organized Universe or a Creation has a beginning; and in the case of physical universes, it can also have an end. However, Syntropy or Light or Life is something completely different than chaos and entropy.

Jesus, standing at the altar, began the prayer by facing the four directions and crying in an unknown tongue, "Iao, Iao, Iao!"

The Pistis Sophia interprets the three letters of this word as signifying

(1) *Iota*, because the universe took form at the Creation;

(2) *Alpha*, because in the normal course of things it will revert to its original state, alpha representing a cycle;

(3) *Omega*, because the story is not going to end there, since all things are tending towards a higher perfection, "the perfection of the perfection of everything is going to happen" — that is "syntropy" (*Pistis Sophia*, 358).

The eternal process is thus not a static one but requires endless expansion of the universe, since each dispensation is outgoing, tending to separation and emanation, that is, fissure, so that "an endless process in the Uncontainable fills the Boundless".

This is the Egyptian paradox of expanding circles of life that go on to fill the physical universe and then go on without end. Such a thing is possible because of a force that is primal and self-existent, having no dependence on other matter or its qualities. This is that "light stream" that no power is able to hold down and no matter is able to control in any way (*Pistis Sophia*, 227). On the contrary, it is this light that imposes form and order on all else; it is the spark by which Melchizedek organizes new worlds; it is the light that purifies contaminated substances, and the light that enables dead matter to live.

Reduced to its simplest form, creation is the action of light upon matter (*hyle*); matter of itself has no power, being burnt-out energy, but light reactivates it; matter is incapable of changing itself — it has no desire to, and so light forces it into the recycling process where it can again work upon it — for light is the organizing principle.

If Melchizedek is in charge of organizing worlds, it is Michael and Gabriel who direct the outpouring of light to those parts of chaos where it is needed. As light emanates out into space in all directions, it does not weaken but mysteriously increases more and more, not stopping as long as there is a space to fill.

In each world is a gathering of light ("synergy"), and as each is the product of a drive toward expansion, each becomes a source of new expansion, "having its part in the expansion of the universe".

https://rsc.byu.edu/sites/default/files/pubs/pdf/chaps/3.pdf

https://archive.bookofmormoncentral.org/sites/default/files/archive-files/pdf/nibley/2016-11-07/hugh_w._nibley_nibley_on_the_timely_and_the_timeless_2004.pdf

https://syntropy.site/wp-content/uploads/2018/04/Nibley-on-the-Timely-and-the-Timeless.pdf

Symmetry, Order, Organization, and Synergy all have ties to Syntropy. Syntropy is order and organization. Syntropy is life; whereas, entropy is chaos and death. Materialism, Naturalism, Darwinism, Nihilism, Atheism, Classical Physics, and the Theory of Evolution are based exclusively upon entropy, chaos, randomness, and death. Naturalistic Evolution or

Darwinism is entropy. Random mutations are entropy. Natural selection results in death or entropy.

Syntropy is the exact opposite of entropy. Syntropy works in direct opposition to entropy, death, corruption, chaos, and extinction. Syntropy is Light, or Quantum Mechanics, or Transdimensional Physics. Syntropy is what gave this physical universe structure, order, and form. When we die, when our physical body dies, we revert to Syntropy, or our original state of existence. But, the story doesn't end there. We, our psyche and spirit body, continue to progress towards perfection or completion, which is Syntropy. This is really cool science, but only when you are ready and willing to understand it and accept it. If you are afraid of it, then you are going to see it differently than you would if you were to choose to accept it and embrace it.

I find this science or these observations fascinating. Of course, your mileage may vary depending upon what you are trying to prove. Nevertheless, Syntropy, or Quantum Mechanics, or Action at a Distance is real and truly exists whether you believe in it or not.

God knew about Syntropy and was teaching us about Syntropy or the Atonement of Christ thousands of years ago. The following is from around 600 B.C.

2 Nephi 2: 6-13:

Wherefore, redemption cometh in and through the Holy Messiah; for he is full of grace and truth. Behold, he offereth himself a sacrifice for sin, to answer the ends of the law, unto all those who have a broken heart and a contrite spirit; and unto none else can the ends of the law be answered. Wherefore, how great the importance to make these things known unto the inhabitants of the earth, that they may know that there is no flesh that can dwell in the presence of God, save it be through the merits, and mercy, and grace of the Holy Messiah, who layeth down his life according to the flesh, and taketh it again by the power of the Spirit, that he may bring to pass the resurrection of the dead, being the first that should rise. Wherefore, he is the firstfruits unto God, inasmuch as he shall make intercession for all the children of men; and they that believe in him shall be saved. And because of the intercession for all, all men come unto God; wherefore, they stand in the presence of him, to be judged of him according to the truth and holiness which is in him. Wherefore, the ends of the law which the Holy One hath given, unto the inflicting of the punishment which is affixed, which punishment that is affixed is in opposition to that of the happiness which is affixed, to answer the ends of the atonement — for it must needs be, that there is an opposition in all things. If not so, my firstborn in the wilderness, righteousness could not be brought to pass, neither wickedness, neither holiness nor misery, neither good nor bad. Wherefore, all things must needs be a compound in one; wherefore, if it should be one body it must needs remain as dead, having no life neither death, nor corruption nor incorruption, happiness nor misery, neither sense nor insensibility. Wherefore, it must needs have been created for a thing of naught; wherefore there would have been no purpose in the end of its creation. Wherefore, this thing must needs destroy the wisdom of God and his eternal purposes, and also the power, and the mercy, and the justice of God. And if ye shall say there is no law, ye shall also say there is no sin. If ye shall say there is no sin, ye shall also say there is no righteousness. And if there be no righteousness there be no happiness. And if there be no righteousness nor happiness there be no punishment nor misery. And if these things are not there is no God. And if there is no God we are not, neither the earth; for there could have been no creation of things, neither to act nor to be acted upon; wherefore, all things must have vanished away.

I find it fascinating that the prophets of God knew about Syntropy, or the Redemption and Resurrection, over 2,500 years ago. They were millennia ahead of our modern-day Materialists, Naturalists, and Atheists. They KNEW about entropy and syntropy. A "compound in one" is chaos, disorder, death, or entropy – the primal state of everything before God's Light or Syntropy entered into our corner of the universe. "Corruption" is entropy and chaos. "Incorruption" is syntropy, order, organization, and life. If there were no Syntropy, then we wouldn't exist, because God or Syntropy wouldn't exist. Without Syntropy, or the Power of God, or Quantum Mechanics, everything would vanish away into chaos, disorder, destruction, and death. It's fascinating that these people KNEW about Syntropy or the Atonement of Christ over 2,500 years ago. They knew about it because the Biblical God Jesus Christ told them about it.

The following also comes to us from 2,500 years ago.

2 Nephi 9: 4-14:

For I know that ye have searched much, many of you, to know of things to come; wherefore I know that ye know that our flesh must waste away and die; nevertheless, in our bodies we shall see God. Yea, I know that ye know that in the body he shall show himself unto those at Jerusalem, from whence we came; for it is expedient that it should be among them; for it behooveth the great Creator that he suffereth himself to become subject unto man in the flesh, and die for all men, that all men might become subject unto him. For as death hath passed upon all men, to fulfil the merciful plan of the great Creator, there must needs be a power of resurrection, and the resurrection must needs come unto man by reason of the fall; and the fall came by reason of transgression; and because man became fallen they were cut off from the presence of the Lord. Wherefore, it must needs be an infinite atonement — save it should be an infinite atonement this corruption could not put on incorruption. Wherefore, the first judgment which came upon man must needs have remained to an endless duration. And if so, this flesh must have laid down to rot and to crumble to its mother earth, to rise no more. O the wisdom of God, his mercy and grace! For behold, if the flesh should rise no more our spirits must become subject to that angel who fell from before the presence of the Eternal God, and became the devil, to rise no more. And our spirits must have become like unto him, and we become devils, angels to a devil, to be shut out from the presence of our God, and to remain with the father of lies, in misery, like unto himself; yea, to that being who beguiled our first parents, who transformeth himself nigh unto an angel of light, and stirreth up the children of men unto secret combinations of murder and all manner of secret works of darkness. Oh how great the goodness of our God, who prepareth a way for our escape from the grasp of this awful monster; yea, that monster, death and hell, which I call the death of the body, and also the death of the spirit. And because of the way of deliverance of our God, the Holy One of Israel, this death, of which I have spoken, which is the temporal, shall deliver up its dead; which death is the grave. And this death of which I have spoken, which is the spiritual death, shall deliver up its dead; which spiritual death is hell; wherefore, death and hell must deliver up their dead, and hell must deliver up its captive spirits, and the grave must deliver up its captive bodies, and the bodies and the spirits of men will be restored one to the other; and it is by the power of the resurrection of the Holy One of Israel. Oh how great the plan of our God! For on the other hand, the paradise of God must deliver up the spirits of the righteous, and the grave deliver up the body of the righteous; and the spirit and the body is restored to itself again, and all men become incorruptible, and immortal, and they are living souls, having a perfect knowledge like unto us in the flesh, save it be that our knowledge shall be perfect. Wherefore,

we shall have a perfect knowledge of all our guilt, and our uncleanness, and our nakedness; and the righteous shall have a perfect knowledge of their enjoyment, and their righteousness, being clothed with purity, yea, even with the robe of righteousness.

This individual also knew about entropy (corruption) and syntropy (incorruption). He also knew about the resurrection from the dead. He knew about the Redemption of mankind or the Atonement of Christ.

What more do we really need to know?

God has in place a plan to counteract the effects of death or entropy. That's why they call it the Good News or the Gospel of Christ. That's why they call it the Plan of Salvation. Christ can save and redeem all of the rest of us who are lost, as long as we are willing to accept His gift. However, if we reject Him and His gift, there's nothing that He can do about it. He can't force us to accept His gift, His mercy, and His forgiveness. If we don't want to have anything further to do with God, He is contractually obligated to obey our wish.

Many people are deathly afraid of this information – this knowledge. I KNOW, because I used to be one of them. I used to be a Materialist, Naturalist, Nihilist, and Atheist. But, I'm not afraid of it anymore. I've taken my science to a whole new level – something that I call Science 2.0, which allows ALL of the evidence into evidence, and then pursues a preponderance of that evidence. Give me evidence, lots and lots of evidence. Science is observation, or it's supposed to be.

Science 2.0: I Upgraded My Science: https://www.amazon.com/dp/B0771K6WTX

The Atonement of Christ and the Resurrection can restore our physical bodies to Syntropy, Completion, or a Fulness of Power and Glory. Returning to God's presence, from whence we came, is the ultimate fulfillment of Syntropy. Such individuals are finished, or completed, or made perfect. ALL corruption, or entropy, or sin is removed from their physical bodies and their spirits. I'm not afraid of this anymore, like I used to be. I KNOW and sense that I have an advocate with the Father in Jesus Christ, my Savior and Redeemer.

When it comes to our own personal progression, God the Father and Jesus Christ will take each one of us as far as we want to go – all the way to the throne of God if that's where we really want to go. The following comes from to us from nearly 2,000 years ago.

Revelation 3: 19-22:

As many as I love, I rebuke and chasten: be zealous therefore, and repent. Behold, I stand at the door, and knock: if any man hear my voice, and open the door, I will come in to him, and will sup with him, and he with me. To him that overcometh will I grant to sit with me in my throne, even as I also overcame, and am set down with my Father in his throne. He that hath an ear, let him hear what the Spirit saith unto the churches.

This isn't an empty promise. To those of us who overcome the flesh or overcome the temptations and trials of this physical world, Jesus Christ will grant him or her to sit down with Him on his Father's throne. They will be Gods.

Isn't it ironic that it's always the people who have seen God, touched God, embraced God, and talked with God in person who seem to know the most about God and His ways? In comparison with that, what the Materialists, Naturalists, and Atheists have to tell us and teach us is absolutely worthless.

God will continue to work with us for as long as we are willing to work with Him. God will take us as far as we want to go.

Lorenzo Snow wrote:

Was taken sick with the fever (25th of May.) I never had such a severe fit of sickness before since my recollection. My friends and family had given up most all hopes of my recovery. Father Huntington, the President of the Place, called on his Congregation to pray for me. He also with Gen. Rich and some others clothed themselves in the garments of the Priesthood and prayed for my recovery. I believe it was through the continued applications of my family and friends to the throne of Heaven that my life was spared.

In my sickness I went through in my mind the most singular scenes that any man ever did. My family generally believed that I was not in my right mind. But the scenes through which my spirit traveled are yet fresh in my memory as though they occurred but yesterday. And when my people supposed me in the greatest pain and danger I am conscious of having a great many spiritual exercises sometimes partaking of the most acute suffering that heart can conceive and others the most rapturous enjoyment that heart ever felt, or imagination ever conceived.

I suppose at first, I must have been left in the hands of an evil spirit, in fact I was administered to upon this supposition. I was led into the full and perfect conviction that I was entirely a hopeless case in reference to salvation, that eternities, upon eternities must pass and still I saw my case would remain the same. I saw the whole world rejoicing in all the powers and glories of salvation without the slightest beam of hope on my part but doomed to a separation from my friends and family all I loved most dear to eternity upon eternity. I shudder even now at the remembrance of the torments and agony of my feelings. No tongue can describe them, or imagination conceive. Those who were attending me at that time describe me as being in a condition of body. I remained several hours refusing to speak. My body was cool, and my eyes and countenance denoted extreme suffering.

After this scene ended I entered another of an opposite character. My spirit seems to have left the world and introduced into that of Kolob. I heard a voice calling me by name saying, "he is worthy, he is worthy, take away his filthy garments." My clothes were then taken off piece by piece and a voice said, "let him be clothed, let him be clothed." Immediately I found a celestial body gradually growing upon me until at length I found myself crowned with all its glory and power. The ecstasy of joy I now experienced no man can tell, pen cannot describe it. I conversed familiarly with Joseph, Father Smith, and others, and mingled in the society of the Holy One.

I saw my family all saved and observed the dispensations of God with mankind until at last a perfect redemption was effected – though great was the sufferings of the wicked, especially those that had persecuted the saints. My spirit must have remained I should judge for days enjoying the scenes of eternal happiness.

Lorenzo Snow

The Iowa Journal of Lorenzo Snow

Edited and with an Introduction by Maureen Ursenbach Beecher, BYU Studies, vol. 24 (1984), Number 2 - Spring 1984.

https://rjnorman.blogspot.com/2009/09/power-of-resurrection.html

https://scholarsarchive.byu.edu/cgi/viewcontent.cgi?article=2296&context=byusq

https://syntropy.site/wp-content/uploads/2018/04/The-Doctrine-of-Eternal-Lives.pdf

https://syntropy.site/wp-content/uploads/2018/04/Teachings-of-the-Doctrine-of-Eternal-Lives.pdf

https://syntropy.website/wp-content/uploads/2018/04/the-doctrine-of-eternal-lives.pdf

The book, "Teachings of the Doctrine of Eternal Lives", provides the BEST, most comprehensive, most useful, and most instructive Model of Reality that I have found so far in a book. The anonymous author has given an older copy of the book away for free, so be sure to get yourself a copy and read it, if the subject interests you. Don't be afraid of it. There's nothing to be afraid of. Dig in and learn about it. It's enlightening.

https://syntropy.site/wp-content/uploads/2018/04/Teachings-of-the-Doctrine-of-Eternal-Lives.pdf

A "perfect redemption" is Syntropy. We return to our original state of existence, the Celestial Kingdom, hopefully infinitely better than we were before. Each one of us is going to have to go through hell in order to get to heaven. That's just the way things work. This physical life is a living hell for most of us; but, that's how we learn. There are rewards for overcoming it. These physical experiences enlarge or enhance or improve our spirit.

In the end, after all the physical dispensations have come and gone, God is going to save and exalt every one of us who wants to be saved and exalted. God and Christ are winners. They are going to save every one of us in the end; but, that reality applies to the willing. God will force no man and no spirit into heaven. If you don't want to be there, you are invited to leave. You can leave anytime you want to. Satan did.

God will save every one of us, as long as we stick with the plan. As long as we choose to stick with the program, and choose to stick with God, Christ, and their Plan of Salvation, they can take us all the way to the throne of God if that's where we want to go. The Bible tells us as much – for Jesus will save all except the sons of perdition.

The only ones God can't save are the Sons of Perdition – the ones who choose to no longer have anything to do with God, Christ, and the Atonement. The Atonement of Christ can overcome everything, except for our lack of desire. If we turn against God and have no further desire to participate in His Plan of Salvation, then we are done, and Outer Darkness awaits us. Those who are in hell or outer darkness chose to go there. They choose to make their surroundings a living hell. They are there because they wanted to go there. They refused to let the Atonement of Christ purify their spirits. They don't want to have anything further to do with God and His Christ. They chose to take themselves out from under the umbrella or the protection of the Atonement of Christ. We get what we choose. God gives each one of us what he or she wants most. Heaven is a wonderful place, but only for those who want to be there. Even heaven is hell for those who don't want to be there. It all depends upon your chosen frame of mind.

The problem is choice. It all comes down to choice. If you choose to muck things up, you certainly can. If you choose to turn your spirit, soul, or aura dark and grey, you certainly can. If you choose to learn how to enlighten and love, you can do that as well.

Enlightenment is Syntropy. The Atonement of Christ is Syntropy. The Divine Light is Syntropy.

All of this is foolishness to those who don't want to have anything to do with it; and, there are those who don't want to have anything to do with it. God grants their wish. If they don't want to have anything to do with it, then they will never have anything to do with it. Simple. Parsimonious. Logical. True.

The same reality applies to those who go to the Celestial Kingdom and the Throne of God. They are there because they want to be there and chose to be there. They chose to submit to the requirements of the place. They chose to become that type of people. They chose to let the Atonement of Christ and His forgiveness work upon them until they became sanctified and pure in the inner vessel or the spirit body.

What are we going to be doing for the rest of Eternity? God tells us that as well.

Doctrine and Covenants 76: 81, 86-88:

And again, we saw the glory of the telestial, which glory is that of the lesser, even as the glory of the stars differs from that of the glory of the moon in the firmament. These are they who receive not of his fulness in the eternal world, but of the Holy Spirit through the ministration of the terrestrial; and the terrestrial through the ministration of the celestial. And also the telestial receive it of the administering of angels who are appointed to minister for them, or who are appointed to be ministering spirits for them; for they shall be heirs of salvation.

So, what do we see?

We see from this revelation from Jesus Christ that the higher will minister to the lower. Those in the Celestial Kingdom will minister unto those in the Terrestrial Kingdom; and, those in the Terrestrial Kingdom will minister unto those in the Telestial Kingdom. We are going to spend the rest of Eternity ministering to those who are lower than us in an attempt to help them climb the ladder to Heaven or the Celestial Kingdom.

How is this to be done? What is the logistics for this?

Well, in order for those in the Celestial Kingdom to successfully minister to those in the Terrestrial Kingdom, the Celestial people will actually have to go to, and live with, and spend time with the people in the Terrestrial Kingdom. Meanwhile, those in the Terrestrial Kingdom who want to progress to a higher level will do so by going into the Telestial Kingdom, living with the people, and trying to help raise them to a higher level.

By ministering to those who are lower than us, we slowly raise ourselves to a higher level.

Again, what is the logistics for ministering to those who are lower than us?

Well, the Terrestrial Translated Beings can be sent as ministering angels from the Terrestrial Kingdom to minister to those who are living in the Telestial Kingdom. Our earth is a Telestial Kingdom. It is a kingdom of glory.

This revelation teaches us that the resurrected angels in the Celestial Kingdom can also be sent as ministering angels to minister to those in the lower kingdoms, including those in a Telestial Kingdom. But, there are other ways for us to minister to those who live in a lower kingdom than ours. We can be born or reborn as one of them. We can condescend. Such a thing has already been experienced and observed.

Adam was Michael in the pre-existence. Adam or Michael was a member of the Godhead. Michael was a God in the Celestial Kingdom before he condescended and came down to our level in the Telestial Kingdom to be the father of our race in the hope of raising some of us up to his former level in the Celestial Kingdom. Eve was a Goddess in the Celestial Kingdom before she condescended and came down to our level to become the mother of our race. These Gods descended to our level to be with us – to be the parents of our physical bodies.

The Gods can condescend. They can descend to our level to be with us and live with us as one of us. The Gods can be sent as missionaries to lower kingdoms.

1 Nephi 11:16, 26:

16 And the angel said unto me: Knowest thou the condescension of God?

26 And the angel said unto me again: Look and behold the condescension of God!

Do you know and understand the condescension of God?

One of the Gods can choose to come down to our level, be reborn as a babe in a manger, in the hope of raising many of us up to His former level. The Gods can set aside their exalted resurrected physical body and choose to be born again as a babe in a manger on fallen mortal telestial world such as ours in the hope of saving us and exalting us to their former level. The Gods can condescend. The Gods can descend to our level and be reborn as a babe in a manger on a fallen physical mortal earth such as ours. It's plain as day. The angels from heaven have even asked some of us if we understand the condescensions of God. Most of us don't get it because we don't want to understand it and accept it. It's true nonetheless.

What do the Gods get out of condescension, or descending to our level to be with us?

Well, they get to save us and raise us to a higher level; and if successful, they also get a better resurrection in their next life. We raise ourselves to a higher level by ministering to and loving those who are at a lower level. This same reality applies to the Gods as well. These truths have already been revealed to us; and, Jesus Christ has already set the example or the pattern for us. The Gods can condescend and be reborn again as a babe in a manger. It has already been done before; and, it can be done some more.

Remember, the Gods can set aside their exalted station, their exalted resurrected physical body, and their exalted life style; and then, the Gods can choose to condescend or to come down to our level, to be born as a babe in a manger on a telestial fallen world, to live among us as one of us in the hope of being a Savior for us and in the hope of raising some of us up to their exalted level. Now we begin to understand the condescension of God. The Gods can descend to our level, be reborn at our level, and live with us as one of us. Amazing! Is it not?

The Gods can condescend.

The Gods can also rise again. If you find some way to kill a God, or more accurately, if a God lets you kill Him, it's not going to hold. If He rose from the dead once before, He can always do it some more. Once again, Jesus Christ is an accurate case study on this quantum phenomenon associated with resurrection. If you kill a God, He just comes back better than before. That's why He is a God.

This is how things really work. We will spend the rest of Eternity ministering to those who are lower than us in the hope of raising them up to our level; and, by ministering to others and raising them up to our level, our own level or status increases and improves as a result. We increase our own status or level by ministering to those who are further behind than us and need our help climbing the ladder to Heaven, or the Celestial Kingdom, or the Kingdom of God. It's one eternal round. It is Eternal Lives!

This is the way things are done; and, Jesus Christ has already set the example for us. It's Syntropy or Eternal Life. We must do what Christ has done if we want to go where Christ has gone.

D&C 76: 40-44:

And this is the gospel, the glad tidings, which the voice out of the heavens bore record unto us — that he came into the world, even Jesus, to be crucified for the world, and to bear the sins of the world, and to sanctify the world, and to cleanse it from all unrighteousness; that through him all might be saved whom the Father had put into his power and made by him; who glorifies the Father, and saves all the works of His hands, except those sons of perdition who deny the Son after the Father has revealed him. Wherefore, he saves all except them — they shall go away into everlasting punishment, which is endless punishment, which is eternal punishment, to reign with the devil and his angels in eternity.

All of us will eventually be saved in one of the Degrees of Glory, except for the sons of perdition. They can't be saved because they rejected the Atonement of Christ while standing in the presence of God the Father and Jesus Christ. If they don't want to have anything to do with God, then God grants their wish. When Satan and his followers, the devils or evils spirits, rebelled against God and rejected the Atonement of Christ, God let them leave. God couldn't force them to stay, and God couldn't force them to accept the Atonement of Christ and the Plan of Salvation. Heaven is of no value to anyone if they don't want to be there. The Atonement of Christ is cool stuff, but only to those who desire to be a part of it. Those who have formally rejected it receive no value from it. Instead, they choose to go and live with the others who have rejected the Atonement of Christ.

When it comes to the rest of us, God will take us as far and as fast as we want to go. It's not over for us, until we tell God that we are done. We'll just keep coming back, one physical life after another, until we finally achieve the resurrection that we want to achieve for ourselves. This physical life is a school. We stay in school for as long as we want to be in school. God is going to save and exalt every one of us who wants to be saved and exalted. The Atonement of Christ or Syntropy works. It's powerful and everlasting, without a beginning of days or an end of years.

It's all there in plain sight for anyone who is willing to look and see.

Mark My Words

—

Source

God Is in the Light: God is light, and in Him is no darkness at all.

https://www.amazon.com/dp/B07168S37N

The Second Comforter: Supping with Our Resurrected Lord Jesus Christ

https://www.amazon.com/dp/B01IAKHTY6

Original Version: 25MAR2018
Current Version: 07DEC2019